Lecture Notes in Computer Science 11956

More information about this series at http://www.springer.com/series/7409

Danijela Milošević · Yong Tang ·
Qiaohong Zu (Eds.)

Human Centered Computing

5th International Conference, HCC 2019
Čačak, Serbia, August 5–7, 2019
Revised Selected Papers

Editors
Danijela Milošević
University of Kragujevac
Čačak, Serbia

Yong Tang
South China Normal University
Guangzhou, China

Qiaohong Zu
Wuhan University of Technology
Wuhan, Hubei, China

ISSN 0302-9743   ISSN 1611-3349 (electronic)
Lecture Notes in Computer Science
ISBN 978-3-030-37428-0   ISBN 978-3-030-37429-7 (eBook)
https://doi.org/10.1007/978-3-030-37429-7

LNCS Sublibrary: SL3 – Information Systems and Applications, incl. Internet/Web, and HCI

This Springer imprint is published by the registered company Springer Nature Switzerland AG
The registered company address is: Gewerbestrasse 11, 6330 Cham, Switzerland

Preface

We are happy to present HCC 2019, a successful event of the Human Centered Computing conference series. HCC 2019 was held during August 5–7, 2019, in Čačak, Serbia, and enjoyed wholehearted support from the local research communities as well as researchers from across the Western Balkan region.

As in previous events, HCC 2019 drew a wide range of thoughts, research, and practical work, aiming not only to support humans in our everyday activities but also to enhance our capacity through automation enabled by advanced learning algorithms, smart infrastructure and transportation, as well as deepened our understanding of sociotechnical problems and impacts by studies on profound issues due to the penetration of web communities into offline societies.

HCC 2019 attracted submissions from a number of countries and regions. Each submission was reviewed by at least two Program Committee members and invited scholars. The committee decided to accept 29 long papers and 42 short papers and posters, including invited papers to the doctoral symposium.

Among the submissions, we were pleased to find an increased number of papers on deep learning and its applications on a variety of real-life problems, ranging from image/video analysis, to human-computer interaction, and to logistics and supply chain management. We also saw new frontiers of machine learning algorithms, broadened by in memory computing and availability of "super" computers. In particular, we are certain that the application of machine learning algorithms into manufacturing and other "conventional" industrial sectors will inspire more innovations and foster fruitful long-term collaborations after the conference.

We are grateful to all members of the Technical Program Committee and conference Organization Committee. The most significant task fell upon their experienced shoulders. During the months leading to the conference, all the committee members worked hard to identify a set of high-quality submissions reflecting the trends and interests of the related research fields. Our special thanks also go to the external reviewers, student volunteers, and local support team, who played a key role in making HCC 2019 a successful event and Čačak a welcoming conference venue. We are also grateful to Springer's editorial staff for their hard work in publishing these post-conference proceedings in the LNCS series.

August 2019

Danijela Milošević
Yong Tang
Qiaohong Zu
Hu Bo
Vladimir Mladenović

Organization

Conference Co-chairs

Danijela Milosevic — University of Kragujevac, Serbia
Yong Tang — South China Normal University, China

Conference Organizing Committee

Vladimir Mladenovic — University of Kragujevac, Serbia
Matjaž Debevc — University of Maribor, Slovenia
Qiaohong Zu — Wuhan University of Technology, China
Bo Hu — Fujitsu Labs of Europe, UK

Conference Program Committee

Gaoyun An — Beijing Jiaotong University School of Computer and Information Technology, China
Gonzalo Arce — University of Delaware, USA
Tie Bao — Jilin University, China
Yigang Cen — Beijing Jiaotong University School of Computer and Information Technology, China
Dingfang Chen — Wuhan University of Technology, China
Yuan Cheng — Wuhan University, China
WeiHui Dai — Fudan University, China
Vladan Devedzic — University of Belgrade, Serbia
Karen Egiazarian — Tampere University, Finland
Nenad Filipovic — University of Kragujevac, Serbia
Qian Gao — Huazhong University of Science and Technology, China
Fazhi He — Wuhan University, China
Chunping Hou — Tianjin University, China
Wenjun Hou — Beijing University of Posts and Telecommunications, China
Weidong Huang — Nanjing University of Posts and Telecommunications, China
Mirjana Ivanovic — University of Novi Sad, Serbia
Dragan Jankovic — University of Nis, Serbia
Zongpu Jia — Henan Polytechnic University, China
Hai Jin — Huazhong University of Science and Technology, China
Dragan Kukolj — University of Novi Sad, Serbia

Publication Chair

Partners

University of Kragujevac, Serbia
South China Normal University, China
Wuhan University of Technology, China
University of Niš, Serbia
University of Maribor, Slovenia

Reviewers

James Anderson
Jose Albornoz
Natasha Aleccina
Angeliki Antonio
Juan Carlos Augusto
Yunfei Bai
Roberto Barcino
Paolo Bellavista
Adams Belloum
Marija Blagojević
Natasa Bojkovic
Zoran Bojkovic
Luis Carriço
Jingjing Cao
Qinghua Cao
Guohua Chen
Tianzhou Chen
Yiqiang Chen
Lizhen Cui
Aba-Sah Dadzie
Marco De Sá
Matjaž Debevc
Luhong Diao
Monica Divitini
David Dupplow
Haihong E.
Talbi El-Ghazali
James Enright
Henrik Eriksson
Chengzhou Fu
Yan Fu
Shu Gao
José G. Rodríguez García
Mauro Gaspari

Bin Gong
Horacio González-Vélez
Chaozhen Guo
José María Gutiérrez
Chaobo He
Fazhi He
Hong He
Andreas Holzinger
Bin Hu
Cheng Hu
Changqin Huang
Zongpu Jia
Mei Jie
Hai Jin
Lucy Knight
Hiromichi Kobayashi
Ines Kožuh
Roman Laborde
Hanjiang Lai
Thomas Lancaster
Victor Landassuri-Moreno
Liantao Lan
Bo Lang
Agelos Lazaris
Chunying Li
Jianguo Li
Shaozi Li
Wenfeng Li
Xiaowei Li
Zongmin Li
Xiaofei Liao
Hong Liu
Lianru Liu
Lizhen Liu

Yongjin Liu
Alejandro Llaves
Yanling Lu
Hui Ma
Haoyu Ma
Yasir Gilani
Mohamed Menaa
Marek Meyer
Danijela Milošević
Dragorad Milovanovic
Vladimir Mladenović
Maurice Mulvenna
Mario Muñoz
Aisha Naseer
Tobias Nelkner
Sabri Pllana
Xiaoqi Qin
Klaus Rechert
Uwe Riss
Andreas Schraber
Stefan Schulte
Susan Scott
Beat Signer
Matthew Simpson
Mei Song
Xianfang Sun
Yuqing Sun

Wenan Tan
Menglun Tao
Shaohua Teng
Yinglei Teng
Boris Villazon Terrazas
Maria Vargas-Vera
Coral Walker
Jizheng Wan
Qianping Wang
Yun Wang
Yifei Wei
Ting Wu
Xiaoxue Wu
Zhengyang Wu
Toshihiro Yamauchi
Bo Yang
Yanfang Yang
Linda Yang
Zhimin Yang
Xianchuan Yu
Guanghui Yue
Yong Zhang
Gansen Zhao
Shikun Zhou
Shuhua Zhu
Tingshao Zhu
Gang Zou

Abstracts

HOLOgrams for Personalised Virtual Coaching and Motivation in an Ageing Population with BALANCE Disorders

Nenad Filipovic

Faculty of Engineering, University of Kragujevac, Serbia
fica@kg.ac.rs

Abstract. We have developed of hologram coaches, as augmented reality models in combination with three-dimensional biomechanical models of a balance physiotherapist for virtual coaching. Holograms are offering new forms of accessible user interaction. First demo represent virtual coach with mocked edge server presenting the sitting exercise. When the session begins, the hologram physio sits in front of the patient, provides the instructions and shows the exercise. Then the hologram sits close to the patient, at his left, and the patient can see it when he looks at the direction where the hologram physio is. At the end, and without interruptions during the exercise, the hologram verbally asks the patient about his symptoms and provides a reward.

First prototype architecture demonstrate capability of proposed approach to establish communication between virtual coach and edge computer responsible to estimate patient condition based on which virtual coach will make interruptions and promote accessible interaction. It was applied to mobile phone with headset equipment. With this augmented reality technology we expect to help patient at their home for physiotherapy session.

Acknowledgment. This study was funded by the grant from the EC HORIZON2020 769574 HOLOBALANCE project.

Short bio

Nenad D. Filipovic is full Professor at Faculty of Engineering and Head of Center for Bioengineering at University of Kragujevac, Serbia. He was Research Associate at Harvard School of Public Health in Boston, US. His research interests are in the area of biomedical engineering, vertigo disease, cardiovascular disease, fluid-structure interaction, biomechanics, multi-scale modeling, data mining, software engineering, parallel computing, computational chemistry and bioprocess modeling. He is author and co-author 11 textbooks and 5 monographies, over 250 publications in peer review journals and over 10 software for modeling with finite element method and discrete methods from fluid mechanics and multiphysics. He also leads a number of national and international projects in EU and US in area of bioengineering and software development.

He is currently Rector of University of Kragujevac and leads joint research projects with Harvard University and University of Texas in area of bio-nano-medicine computer simulation. He also leads a number of national and international projects in area of bioengineering and bioinformatics. He is a Managing Editor for Journal of Serbian Society for Computational Mechanics and member of European Society of Biomechanics (ESB) and European Society for Artificial Organs (ESAO).

INPRESEC - Paradigm Shift in Information Security and Privacy with AI and ML

Dragan Pleskonjić

INPRESEC
dragan@conwex.org

Abstract. The team, led by INPRESEC intiator and founder Dragan Pleskonjic (Personal Website, LinkedIn), works on research and development of the solution that predicts, prevents and detects security threats and attacks before they actually affect the live system, with demonstrable accuracy of approximately 99%.

The solution utilizes Artificial Intelligence (AI), Machine Learning (ML), predictive analytics, and threat intelligence with a specific approach developed by the team with decades of combined experience in scientific research, academia and professional experience in enterprise security, AI and ML (patent applications in progress). Novel approach to cyber security has been developed to predict the most likely cyber-attacks and to plan optimal proactive cyber-security defensive measures.

The solution improves security posture of the client system by minimizing risks and impacts of security threats and attacks. It significantly reduces the work of security teams, while improving accuracy, response time and performances of the security system.

Short bio

Dragan Pleskonjić is Senior Director Application Security at international company IGT (previously GTECH). In his current role, he directs, coordinates and oversees application security efforts on global organization level. Dragan is a well-known expert and influential strategic thinker in the area of information security, privacy, machine learning (ML) and artificial intelligence (AI). He is an experienced leader and has held top positions at international companies, working with clients and partners from various sectors worldwide, including: finance and banking, technology, telecommunications, services, lotteries, gaming, education, government and others. He possesses rich experience in creating and managing start-ups, new businesses development, proven leadership and talent for creation, management and organization of successful teams. Initiated and has held leading positions in a number of industry projects, as well as in research and development projects. Dragan is an adjunct professor for various cyber-security and computer science courses. He is author of ten books so far, including well-known university textbooks on topics such as cybersecurity, operating systems, and software. Dragan is inventor with set of patents granted by USPTO and also CIPO, EPO and WIPO patent offices. He published more than seventy scientific and technical papers at conferences and journals. His current research and development focus is intelligent predictive security (INPRESEC), exploring the paradigm shift in information security and privacy with artificial intelligence (AI) and machine learning (ML). Dragan is initiator and founder of INPRESEC project and solution. For more information, please visit his personal website at https://www.dragan-pleskonjic.com/.

Contents

Reasoning Based Virtual Machine Mapping Toward Physical Machine

Adeel Aslam, Hanhua Chen$^{(\boxtimes)}$, Jiang Xiao, Song Wu, and Hai Jin

National Engineering Research Center for Big Data Technology and System,
Cluster and Grid Computing Lab, Services Computing Technology and System Lab,
School of Computer Science and Technology,
Huazhong University of Science and Technology, Wuhan 430074, China
{adeelaslam,chen,jiangxiao,wusong,hjin}@hust.edu.cn

Abstract. Cloud computing is an arising paradigm to run and hosts a number of applications and services. These computing services are accommodated by a set of virtual machines. These virtual machines are an abstraction of real servers or physical machines. A physical machine can hosts a number of virtual machines, depending on its capacity. Virtual machine placement has a direct effect on the quality of the services for both end-users and cloud service providers. In this study, we address the problem of virtual machine placement and migration for minimization of resources and power consumption. We formulate this problem as a multi-objective optimization and propose a resource-aware reasoning based scheme with state-of-the-art solutions. The simulations results with real-world traces show that the integrated schemes use a fewer number of physical servers to accommodate more virtual machines.

Keywords: Case Base Reasoning · Physical machine · Resource consumption · Virtual machine placement

1 Introduction

Cloud computing motivates users to move computational data and get services from local to Internet data centers [1]. It provides several benefits such as on-demand response, flexibility, platform independence, cost savings for end users to buy new hardware. Meanwhile, enterprises such as Google, Amazon, and Microsoft are providing cloud services to customers. These are termed as *Cloud Service Providers* (CSP). Cloud computing services include *Infrastructure as a Service* (IaaS), *Software as a Service* (SaaS), *Platform as a Service* (PaaS). However, IaaS provides virtualized computing resources over the web. Google Compute Engine, Microsoft Azure, and Amazon EC2 are well-known IaaS service providers. In IaaS, the users request for cloud services to CSP. Each service depends on its needs and budget. CSP allocate resources against their requests in the form of *Virtual Machines* (VMs). These VMs are isolated from each other. Every new machine has its own operating system, CPU, and memory. It behaves

© Springer Nature Switzerland AG 2019
D. Milošević et al. (Eds.): HCC 2019, LNCS 11956, pp. 1–12, 2019.
https://doi.org/10.1007/978-3-030-37429-7_1

like a physical server which accommodates a number of requested applications [2,3].

A physical machine hosts a number of VMs requests depends on its capacity. A hypervisor or *Virtual Machine Manager* (VMM) inside a host machine allows to support multiple guest VMs. These VMs can migrate among servers and resized dynamically depending on applications request and resource usage. Several placement algorithms have been designed for the allocation and migration of VMs to PMs [4]. The objective of these placement algorithms is to assign multiple VMs against a limited number of hosts without performance degradation. It also helps to switch-off idle servers to save energy, cost, and promote green computing [5].

However, resource requirements of VMs are dynamic and uncertain. Similarly, the energy utilization of the cloud data centers directly relates to a number of active servers. Optimal placement and migration of the VMs from least overloaded PMs to other hosts, enhance the efficiency of data centers. It also assists in load balancing among PMs. The dynamic behavior of resource consumption of VMs, their placement, and migration among PMs are active areas of research [5,6].

From the above discussion, it is clear that service providers have more concerned about the placement of VMs. Numerous research works discuss the importance of placing VMs [7,8]. In recent studies, it has been classified into two types such as static *Virtual Machine Placement* (VMP) and dynamic VMP. The static allocation considers their initial placement whereas, dynamic VMP includes migration of VMs from least utilized hosts to other PMs [7]. Static VMP is more frequent while dynamic VMP depends on pre-set conditions. It includes completion of the applications running on the existing VMs. Similarly, VMs from underutilized servers can be migrated to some others which have the capacities of holding new machines [7–9].

Satpathy *et al.* [12], studies the problem of VMs mapping to PMs. First, they propose a queue structure for managing large number of VMs and then use a crow search based VMP. The objectives of this study include the minimization of resource wastage and power consumption. Similarly, another study by Shaw *et al.* [3], states that a key problem for VMP algorithm is the forecasting of actual resource consumption by VMs due to its dynamic behavior. This study also provides a comparative analysis of existing predictive models. CPU and bandwidth are used as input parameters for the prediction of resource consumption. The results provide 18% power efficiency than other predictive algorithms. A study by Wu *et al.* [13], uses a genetic algorithm for VMP in cloud data centers. This study claims that *First Fit Decreasing* (FFD) is one of popular and the effective heuristic scheme for bin packing problem. The FFD is used as a benchmark for comparison with newly proposed schemes for VMP [12,13].

In this study, we propose a predictive solution for VMP. Heuristic schemes for bin packing problems such as *First Fit* (FF), *Best Fit* (BF), and FFD are integrated with *Case Base Reasoning* (CBR) model [14]. A CBR is a reasoning based predictive model which continuously learns about resource consumption of different VMs. Allocation decisions of resources are based on a prediction of the CBR model. It utilizes previous knowledge and user requests as a case base. A CBR model consists of four phases such as retrieve, reuse, revise, and retain [14]. It first learns and then predicts resource consumption for applications. Besides this, CBR also assists in minimizing SLA violations, fewer resource consumption, power savings. The results of the integrated prediction model along with baseline schemes shows improvement in better utilization of resources and power consumption.

This paper proceeds in the following way. Section 2 includes state-of-the-art work about VMP problem. In Sect. 3, we provide a system model and problem formulation. Similarly, in Sect. 4 we discuss baseline algorithms and integrated CBR approach for VMP. The rest of the sections present results and discussion while the final section is about the conclusion.

2   Related Work

In this section, we briefly describe some of the multi-objective heuristics and other schemes used for solving the VMP problem.

Zhou et al. [10] provides a fault tolerant VMP algorithm. They first formulate VMP as an integer linear problem and propose a differential evaluation algorithm for solving the problem. Similarly, another study by Roh et al. [11] proposes a threshold based wireless link-aware VMP algorithm. It can be termed as traffic-aware load balancing in IaaS cloud infrastructure. Guo et al. [17] studies a shadow routing based scheme for adaptive routing VMP in large data centers. This scheme is simple and continuously adapts changes depending on the needs for VMs. Similarly, a study by Hao et al. [4] provides a scheme for VMs allocation relying on their future arrival request.

In the past few years, many optimization schemes for the VMP problem have been proposed. These include crow search based VMP [12], ant colony optimization [5,7], anti-correlated VMP [3], genetic algorithm [13], knee point-driven evolutionary algorithm [6], space-aware best fit decreasing [8], particle swarm optimization [18]. Similarly, Masdari et.al [9] provides a comprehensive overview of VMP strategies. They develop a taxonomy and classify their placement algorithm with respect to their objectives. Another survey by Filho et al. [19] targets different optimization techniques for placement and migration for VMP. However, their demand for resources are dynamic. In our proposed solution, we are continuously learning from the past historical VMs resource consumption and predict their future demands.

3   System Modeling and Problem Formulation

In this section, we provide discussion about system model and then formulate objectives of VMP.

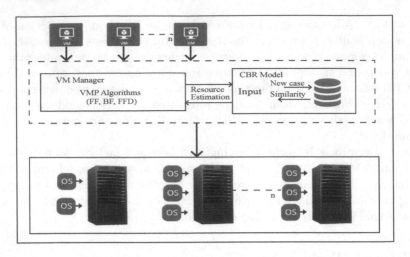

Fig. 1. CBR-based virtual machine placement model

3.1   System Model

In this subsection, we provide a brief description of the system workflow. Figure 1 shows the system architecture. VMs requests depend on end users or cloud services. It can be an application or workload provided by CSP. New requests are arranged in ascending order. The job of the VMP manager is the arrangement of the VMs to appropriate PMs, which have the capacities of accommodating the newly arrived VMs. A CBR module is integrated with the VMP manager. It forecasts the actual resource consumption of requested VMs. Normally, the resource consumption by VMs are dynamic. Generally, end users reserve more resources than their actual need for the application. However, their requirements may be less than the requested resources. Similarly, many applications need more computation requirements and reserved resources are not enough to accommodate their dynamic requests. It may cause performance degradation or SLA violations.

CBR module predicts the actual resource consumption of VMs by learning from the previous history of VMs executions. It assigns more resources to resource constraints applications. It helps to avoid performance degradation, switch-off idle servers, and migrate VMs from underutilized machines to others PMs, which have the capacities of withholding new requests.

3.2   Problem Formulation

In this subsection, we mathematically formulate VMP and its migration. Our objectives include minimization of power consumption which is directly proportional to a number of hosts and minimization of resource wastage. We customize the formulation and used notations from [20,21]. Table 1 presents notions used in problem formulation.

Resource utilization of particular physical machine at any instant is calculated by using the following formulas, including Eqs. (1), (2), and (3).

$$CPU_i^U = CPU_c^P - \sum_{i=1}^{n} CPU_d^V \tag{1}$$

$$RAM_i^U = RAM_c^P - \sum_{i=1}^{n} RAM_d^V \tag{2}$$

From Eqs. (1) and (2), we calculate the resource utilization of a single machine by executing a number of VMs. In case of an integrated approach, CPU_d^V and RAM_d^V depend on return of Algorithm 1. Finally, demanded resources are swap with predicted CPU_a^V and RAM_a^V. Underutilized resources of single PMs running in the data center is calculated using Eq.(3) [20].

$$RW = \frac{CPU_i^U - RAM_i^U}{\sum\limits_{i=1}^{n} CPU_d^V + \sum\limits_{i=1}^{n} RAM_d^V} \tag{3}$$

Table 1. Notations used during problem formulation

Notions	Description
PMs^t	Collection of Physical Machine in data center
VMs^t	Collection of Virtual Machine
PM^i	Single Physical Machine
VM_p^i	Single Virtual Machine arriving at single machine
VM_p^m	Virtual machine migration
CPU_c^P	Total capacity CPU of PM^i
RAM_d^P	Total capacity RAM of PM^i
CPU_d^V	Virtual machine demand for CPU
RAM_d^V	Virtual machine demand RAM
CPU_a^V	Virtual machine actual for CPU
RAM_a^V	Virtual machine actual RAM
CPU_i^U	CPU utilization of single physical machine while running VMs
RAM_i^U	RAM utilization of single physical machine while running VMs
RW	Resource Wastage during VMP and execution
p^T	Total Power consumption
P_i^I	Power consumption of single physical machine in idle state
P_i^C	Power consumption after executing virtual machine
P_I^R	Power consumption of physical machine from its idle state to running
P_i^M	Power consumption while migrating virtual machine

Power consumption of PMs when they are idle and ready to start is calculated using Eq.(4).

$$P_i^I = \sum_{i=1}^{n} P_i^I \tag{4}$$

Similarly, when machines transit from idle to switch-on state, they consume constant C, amount of power as calculated using Eq.(5).

$$P_i^C = \sum_{i=1}^{n} P_I^R + C \tag{5}$$

Energy is consumed when VMs are migrated. They also consume while communication. This energy is calculated using Eq.(6)

$$P_i^M = \sum_{i=1}^{n} VM_p^m \times UnitCost \tag{6}$$

From the above-mentioned power consumption, the total power consumption is calculated by adding all of its units using Eq.(7).

$$P^T = P_i^I + P_i^C + P_i^M \tag{7}$$

Objective functions of VMP is to minimize resource consumption or resource wastage and power consumption.

$$\begin{aligned} &\min_{i} & &P^T{}_i \\ &\min_{p} & &RW^T{}_p \\ &\text{subject to} & &\forall PM^i \in PM^t \\ & & &\forall VM^i \in VM^t \\ & & &CPU_d^V \leq CPU_p^C \\ & & &RAM_d^V \leq RAM_p^C \\ & & &PM_r^i < PM_r^i/2 \Rightarrow VM_i^m \\ & & &CPU_i^C \geq VM_i^m \end{aligned} \tag{8}$$

We have defined a number of constraints in Eq. (8). First two constraints describe that all machines must belong to their collection. Third and fourth constraints define the capacities of PMs. The VMs allocated to a single host must be less than its capacity. Whereas, the last two constraints are about VMs migrations. Migrations are initiated when resource consumption of the PMs are less than a certain threshold. In our study, we consider less than half utilization of host servers. The machines which adopt the VMs during migration must have the capacities of holding new requests.

4   Baseline and Integrated CBR-Based Approach

In this section, we provide a brief description of baseline schemes used for VMP such as FF, BF, and FFD. These techniques are normally used for solving the bin packing problem. Most of the existing proposed schemes for VMP have been compared with these benchmark algorithms. Finally, we provide discussion about CBR model and then integration of model with benchmark schemes.

4.1 First Fit

FF algorithm considers all PMs are in sequential order. A new request for the machine is placed in an ascending order onto the PMs and depends on their capacity. It provides faster results, however, the solution may not be optimal at a large scale. VMs are allocated against a limited number of servers. Each machine has some resource capacities, whereas, a new request which is initiated by some users or by service providers may exceeds their resources demand dynamically.

4.2 First Fit Decreasing

FF algorithm provides better results by sorting PMs from higher capacity to the least capacity. Once they are arranged then FF process takes place. VMs demand includes both CPU and RAM requirements. However, their placement depends on the capacities of PMs. First, we take L1-Norm using Eq.(9) and place PMs according to the results of L1-Norm in decreasing order. FF algorithm is applied to the sorted results. It helps to reduce the number of active servers. However, this algorithm does not guarantee the optimal solution due to the change in resource requirements.

$$L1Norm = \sum_{i=1}^{n}(CPU_d^V + RAM_d^V) \qquad (9)$$

4.3 Best Fit

In the BF algorithm, VMs find their best placement. Initially, this algorithm works like FF algorithm. Latterly, it finds out the best resource containing server which can accommodate the request of VMs to achieve load balancing among active servers. It is calculated by executing the VMs and finding physical server having more resources to accommodate the new request. The main advantage of this scheme is the minimization of SLA violations.

4.4 Case Base Reasoning

CBR is a prediction model which utilizes the historical data for forecasting. In this study, we integrate the module of CBR with above-mentioned algorithms. Resources demanded by VMs and their actual consumption have a difference. The applications may suffer performance degradation. CBR continuously learns about the resource consumption of VMs from historical traces of both input demand and their actual resources consumption.

In this study, we use Euclidean distance for finding the similarity between the new case and the existing case. The similarity is calculated using Eq.(10).

Algorithm 1. CBR-based virtual machine placement

    Input   : PMs^t, VMs^t, $PM_i^C = CPU_c^P \cup RAM_c^P$, $VM_i^d = CPU_i^d \cup RAM_i^d$, C^t
    Output: VMs^t to the PMs^t
1 $vm_i^d \leftarrow VM_i^d \in VM's^t$;
2 **for** $C\ c_i : C^t$ **do**
3      $ed[\ ] = \sqrt{(vm_i - c_i)^2}$;
4      $num = d[\,0\,]$;
5      $position = 0$;
6   **for** $each\ ed, ed_i \in ed$ **do**
7   **if** $num > ed[\,i\,]$ **then**
8             $num = ed[\,i\,]$;
9             $position = i$;
10   **else**
11   null;
12   **end**
13   **end**
14 **end**
15 **if** $c_i^p \geq threshold$ **then**
16   Base Algorithms(VM_i^a);
17 **else**
18   Base Algorithms(VM_i^d) update C^t ;
19 **end**

CBR works in four phases (4R) which are retrieve, revise, retain, and reuse. In retrieve, the similarity is checked between new and existing cases reside in the case base. If the new case has a similarity >90% as a threshold, then it reuses prediction results [14]. Otherwise, the new case is retained into the case base for future inputs. Case base is the central part of the CBR model. It consists of historical traces of resource usage by VMs. Algorithm 1, presents the workflow of the integrated CBR-based scheme.

$$ED = \sqrt{(vm_i - c_i)^2} \tag{10}$$

5 Experimental Setup and Results

In this section, we discuss the dataset and the experimental setup. Furthermore, we provide results and discussion.

5.1 Dataset

We use Google cluster data for the evaluation of proposed schemes [15]. It includes traces of 12,000 PMs, which consists of actual resource consumption against workloads. The dataset contains information about a number of jobs,

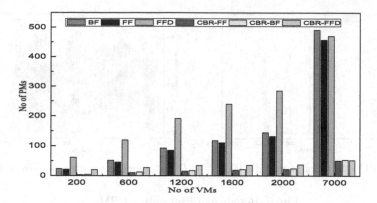

Fig. 2. Virtual machines placement

whereas, each job has many associated tasks running independently on heterogeneous servers. Moreover, according to the information provided by trace document, there is no direct relationship or mapping of VMs to PMs. However, we utilize the assumption provided by [16], where different tasks are considered as VMs. Each task has resource information about its execution. For the complete simulation setup, we use three files such as machine information file, task file, and resource file. Machine information files provide detail about capacities of the PMs including both CPU and RAM. Task files include the exact resource consumption of each task (VMs).

5.2 Experimental Setup

For designing and construction of whole schemes, we use JAVA as a programming language. Eclipse 5.0 is used as an integrated development environment having JVM 8.8. Object-oriented techniques are utilized for modeling simulation. Initially, baseline algorithms such as FF, BF, and FFD are implemented in the model with the same dataset. Finally, we integrate a CBR module with baseline algorithms and then start the simulations. We use Microsoft Window 10 with 64 bit OS, JDK-64 bit having 4 GB RAM, and i5 Toshiba machine having processor 2.6 GHz.

5.3 Results and Discussion

Figure 2 presents results about physical servers used by the different schemes separately. FF, FFD, and BF are baseline algorithms. Results of FF and FFD are almost similar. However, in case of BF, it is not performing as well as FF and FFD. They are fast and provide a solution more quickly as compared to alternative schemes. FF utilized 457 PMs for the entertaining requests of more than 7,000 VMs. Similarly, FFD used 470 and BF utilized 491 PMs for VMs execution. New PMs instances depend on VMs requests and existing capabilities of PMs.

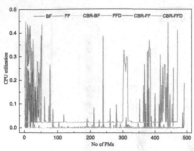
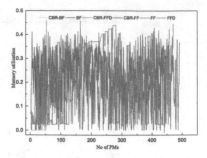

(a) CPU utilization per single PM   (b) Memory utilization per single PM

Fig. 3. Resource utilization w.r.t PMs

In FF and FFD schemes, VMs are allocated to any host machine which has enough resources. However, BF initially works as similar to FF, later on, it searches all active PMs and their capabilities. Newly VMs requests are allocated to the machines having more resource capacities. All VMs who completed their jobs must release resources of PMs. Finally, these machines are also added to the set of pre-active servers. BF has more time complexity and it takes much power due to its large execution for finding an optimal solution.

In our study, VMs are migrated dynamically from one server to another depending upon their server resource utilization. If VMs are consuming half resources of the active machines then these machines are migrated from one server to another for power and resource saving. FF algorithm is used for finding suitable machine during migration.

Similarly, Fig. 2 also shows the results for integrated schemes which are used for handling the same number of VMs. CBR-FF, CBR-FFD, and CBR-BF are those schemes which do not depend upon the requests provided by the users. They maintain their own case base and learn continuously during the new arrival of VMs. Case base contains information about VMs resource consumption. Results of the proposed schemes outperform against the baseline approaches. Integrated schemes utilize less number of PMs against the same requests of VMs. The proposed approach learns dynamically about the new case for further prediction.

Figure 3(a) presents results about CPU utilization of different schemes. Active servers have very less resource wastage. However, simulations results demonstrate that when the number of VMs requests increases than the proposed integrated approaches have less CPU wastage as compared to the other schemes. Especially CBR-BF is outperforming against all these discussed schemes. CBR-BF is outperforming due to its finding and searching for the most appropriate machine with respect to the resources demand. Similarly, results of CBR-FF and CBR-FFD are also good as compared to other baseline algorithms.

Figure 3(b) shows different memory utilization results. Simulations results present that the CBR-BF performed better in term of memory utilization due to learning and search space effectiveness. However, other integrated schemes such as CBR-FF and CBR-FFD also performed better as compared to other baseline algorithms.

6 Conclusion

In this paper, we propose a reasoning based solution for VMP among PMs. Initially, we formulate VMP problem as a multi-objective optimization problem. Next, we integrate the CBR-based model with baseline bin packing algorithms such as FF, FFD, and BF for VMP. It includes data about resource demanded by the VMs and their actual resource consumption. We utilize Google traces for the construction of the case-base. Additionally, we also implement and compare baseline algorithms with integrated schemes. Results demonstrate that CBR along with its learning property provides the effectiveness of the integrated approach.

Acknowledgements. This research is supported in part by the National Key Research and Development Program of China under grant No. 2016QY02D0302.

References

1. Ponraj, A.: Optimistic virtual machine placement in cloud data centers using queuing approach. Future Gener. Comput. Syst. **93**(1), 338–344 (2019)
2. Alharbi, F., Tian, Y.C., Tang, M., Zhang, W.Z., Peng, C., Fei, M.: An ant colony system for energy-efficient dynamic virtual machine placement in data centers. Expert Syst. Appl. **120**(1), 228–238 (2019)
3. Shaw, R., Howley, E., Barrett, E.: An energy efficient anti-correlated virtual machine placement algorithm using resource usage predictions. Simul. Model. Pract. Theory (2019). https://doi.org/10.1016/j.simpat.2018.09.019
4. Hao, F., Kodialam, M., Lakshman, T.V., Mukherjee, S.: Online allocation of virtual machines in a distributed cloud. IEEE/ACM Trans. Netw. **25**(1), 238–249 (2017)
5. Liu, X.F., Zhan, Z.H., Deng, J.D., Li, Y., Gu, T., Zhang, J.: An energy efficient ant colony system for virtual machine placement in cloud computing. IEEE Trans. Evol. Comput. **22**(1), 113–128 (2018)
6. Ye, X., Yin, Y., Lan, L.: Energy-efficient many-objective virtual machine placement optimization in a cloud computing environment. IEEE Access **5**(1), 16006–160020 (2017)
7. Zhao, H., Wang, J., Liu, F., Wang, Q., Zhang, W., Zheng, Q.: Power-aware and performance-guaranteed virtual machine placement in the cloud. IEEE Trans. Parallel Distrib. Syst. **29**(6), 1385–1400 (2018)
8. Wang, H., Tianfield, H.: Energy-aware dynamic virtual machine consolidation for cloud datacenters. IEEE Access **6**(1), 15259–15273 (2018)
9. Masdari, M., Nabavi, S.S., Ahmadi, V.: An overview of virtual machine placement schemes in cloud computing. J. Netw. Comput. Appl. **66**(1), 106–127 (2016)
10. Zhou, A., Wang, S., Hsu, C.H., Kim, M.H., Wong, K.S.: Virtual machine placement with (m, n)-fault tolerance in cloud data center. Cluster Comput. **22**(4), 1–13 (2017). https://doi.org/10.1007/s10586-017-1426-y
11. Roh, H., Jung, C., Kim, K., Pack, S., Lee, W.: Joint flow and virtual machine placement in hybrid cloud data centers. J. Netw. Comput. Appl. **85**(1), 4–13 (2017)

12. Satpathy, A., Addya, S.K., Turuk, A.K., Majhi, B., Sahoo, G.: Crow search based virtual machine placement strategy in cloud data centers with live migration. Comput. Electr. Eng. **69**(1), 334–350 (2018)
13. Wu, G., Tang, M., Tian, Y.-C., Li, W.: Energy-efficient virtual machine placement in data centers by genetic algorithm. In: Huang, T., Zeng, Z., Li, C., Leung, C.S. (eds.) ICONIP 2012. LNCS, vol. 7665, pp. 315–323. Springer, Heidelberg (2012). https://doi.org/10.1007/978-3-642-34487-9_39
14. Raza, B., Kumar, Y.J., Malik, A.K., Anjum, A., Faheem, M.: Performance prediction and adaptation for database management system workload using case-based reasoning approach. Inf. Syst. **76**(1), 46–58 (2018)
15. Google cluster data. https://github.com/google/cluster-data. Accessed 20 Dec 2018
16. Mann, Z.A., Szabó, M.: Which is the best algorithm for virtual machine placement optimization? Concurrency Comput.: Pract. Exp. **29**(10), e4083 (2017)
17. Guo, Y., Stolyar, A.L., Walid, A.: Shadow-routing based dynamic algorithms for virtual machine placement in a network cloud. IEEE Trans. Cloud Comput. **6**(1), 209–220 (2018)
18. Luo, J., Song, W., Yin, L.: Reliable virtual machine placement based on multi-objective optimization with traffic-aware algorithm in industrial cloud. IEEE Access **6**(1), 23043–23052 (2018)
19. Silva Filho, M.C., Monteiro, C.C., Inácio, P.R., Freire, M.M.: Approaches for optimizing virtual machine placement and migration in cloud environments: a survey. J. Parallel Distrib. Comput. **111**(1), 222–250 (2018)
20. Gupta, M.K., Amgoth, T.: Resource-aware virtual machine placement algorithm for IaaS cloud. J. Supercomput. **74**(1), 122–140 (2017)
21. Duong-Ba, T.H., Nguyen, T., Bose, B., Tran, T.T.: A dynamic virtual machine placement and migration scheme for data centers. IEEE Trans. Serv. Comput. (2018). https://doi.org/10.1109/TSC.2018.2817208

Mobile Offloading in Vehicular Networks: A Reward-Based Incentive Caching Scheme

Xuyan Bao[1]($\boxtimes$), Di Han[1], Chen Wang[1], Tai Liu[1], Wenxin Li[1], Mei Song[2], and Jun Ma[1]

[1] Department of Product and Infrastructure, China Telecommunication Technology Labs, China Academy of Information and Communications Technology, Beijing, China
baoxuyan@caict.ac.cn
[2] Beijing University of Posts and Telecommunications, Beijing, China

Abstract. The massive growth of mobile vehicles arises immense mobile traffic demand on the current cellular communication system. To cope with this challenge, a new transmission pattern that vehicle nodes cache some data contents in advance and provide data access for future requests via direct contacts has become an appealing solution for data offloading. However due to the limited resources, the fact that vehicle nodes might not be willing to assist data caching and transfer is plausible. Therefore in this paper, we build a reward-based caching-offloading framework, which efficiently offload cellular traffic by exploiting the storage capability and contact opportunity among vehicle nodes. The whole framework is divided into request phase and caching phase, an auction mechanism which comprises the bidding strategy, the response-for-bids strategy, and the prediction method is designed in the former phase and the optimal caching decision can be derived by solving a 0–1 knapsack problem in the latter phase. We conduct simulations on time-evolving topology traces, which validates the feasibility and efficiency of the proposed offloading paradigm.

Keywords: Mobile offloading · Vehicular network · Incentive cache · Auction mechanism

1 Introduction

The massive growth of mobile vehicles arises immense mobile traffic demand on the current cellular communication system. Specifically in vehicle-to-vehicle (V2V) network, there exists a variety of data applications such as real-time transportation information, live news, location-based advertisements, video clips of local attractions, etc. These data services have imposed a heavy burden on cellular backbone, since they are generally accessed using cellular infrastructure (3G/4G) or roadside infrastructure based on Dedicated Short Range Communication (DSRC) links [1]. Such solution for data access faces many challenges:

© Springer Nature Switzerland AG 2019
D. Milošević et al. (Eds.): HCC 2019, LNCS 11956, pp. 13–25, 2019.
https://doi.org/10.1007/978-3-030-37429-7_2

limited per user data rate, the cost of data downloads under cellular pricing policy, intermittent deployment of infrastructure, etc. [2]. These challenges motivate us to seek for new transmission paradigm, where data transfer can be achieved via vehicle-to-vehicle (V2V) contact when two vehicles move into reciprocal communication range of each other [3]. A 'store-carry-forward' transmission pattern then can be exploited to construct a caching-offloading system, in which the vehicle nodes cache some data contents in advance and provide data access for future data requests via direct contacts [4]. This paradigm is an appealing alternative for cellular traffic offloading, however due to the limited energy and storage capability, the fact that vehicle nodes might not be willing to assist data caching and delivery is plausible. Therefore, it is necessary to build a scheme incentivizing vehicle nodes to cache and forward data such that the offloading performance can be significantly improved.

In this paper, our proposed caching-offloading paradigm for data access can be described by an applicable scenario, as show in Fig. 1. Generally, the base station (BS) is able to provide access to all data items and be regarded as a 'content factory' that periodically disseminates data items. This provides odds for all vehicle nodes to select a number of data items to cache such that data access by opportunistic contacts becomes available. As shown in left part of Fig. 1, the BS publishes data items indexed by 1, 2, 3,..., due to limited storage, each vehicle nodes are only able to cache limited number of data items, in this case, nodes v_1, v_2, v_3, v_4 select data items $(1,3,5), (1,2,3), (3,4), (3,5,6)$ to cache respectively. This naturally arises an issue: *what is the criteria and motive for vehicle nodes to select some data items to cache?* To answer this question, we turn to the right part of Fig. 1, which represents the request part. As a node broadcasts its data request (access data item 3) to nodes nearby, those nodes who have the requested data claim their own prices for providing data access, in this case, node v_3 wins this competition by offering the lowest bid price. This arises another issue: *how to design a bidding strategy for caching nodes to decide which one provides the data access and wins the reward from requester?* These two issues indeed connected by the core element in our model, the reward from data requester, this reward could be a certain virtual credit or an amount of bit-coins. In offloading context, the payment to get data access from caching nodes should be much cheaper than directly access from BS. Under this model, on one hand, the cellular administrator can not only make profit by selling virtual money to requesters, but also alleviate cellular traffic burden (the requester is more willing to access data from caching node with lower cost). On the other hand, to earn rewards from potential requesters, each vehicle node is willing to cache some date items for future data access. Therefore, this caching-offloading paradigm yields a win-win situation between two parties.

1.1 Literature Review

There are a plethora of research about incentive design in vehicular networks. Data access by node contacts in our offloading paradigm follows a 'store-carry-forward' pattern, specifically, some data items are pre-stored in some caching

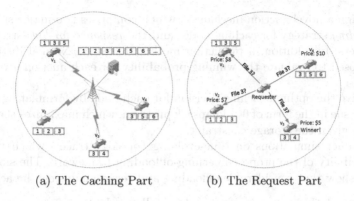

(a) The Caching Part   (b) The Request Part

Fig. 1. An illustration for caching-offloading paradigm

nodes, and these nodes move, carry, and wait for data request to serve by forwarding the data. This is the core concept in delay-tolerant networks, in which the design of incentive scheme is well developed. Under two-hop relaying scheme, a non-cooperative game is proposed in [5], in which the source offers to the relays a positive reward if they successfully deliver a given file to the destination and the relays can choose to accept or reject the offer according to the reward and the expected energy cost. Furthermore, Yu et al. [6] design both the reward and punishment methods to propose a Node-dependence-based dynamic gaming incentive algorithm, which motivates relay nodes forwarding data packets. The above work mainly focus on how to incentivize intermediate nodes participating transmission such that the data packet is guaranteed to be delivered. However in our proposed caching-offloading paradigm, any node is only able to pre-select limited number of data items to cache, the new perspective for incentive design is to maximize its caching revenue by serving future data requests from other nodes. In [7], Zhao et al. propose a similar incentive framework to our proposal, still there exits some drawbacks: (1) the powerful A-type node mentioned in [7] who can freely download a large amount of data might hardly exist due to limited storage space and privacy; (2) In [7], two nodes exchange their stored data objects when they come into contact to maximize their revenue, then frequent contacts might incur extra transmission cost and expenditure for maintaining network information. An incentive mechanism based on reverse-auction Win-Coupon is proposed in [8] in which the auction parties are 3G provider (BS) and the users respectively, while in our proposal the auction is used to model the incentive procedure between users.

1.2 Contribution

Motivated by the above challenges, in this paper, we build a reward-based caching-offloading framework, which efficiently offloads cellular traffic by exploiting the storage capability and contact opportunity of vehicle nodes. Our main contributions are summarized as follows:

- We design a novel auction mechanism which comprises two major strategies: the *bidding* strategy for caching nodes and the *response-for-bids* strategy for requesters. In addition, a prediction method based on Normal distribution is proposed to evaluate the winning probability for each auction given a bid price.
- We derive the optimal caching decision for each node by formulating an allocation issue in the form of 0–1 knapsack problem, which maximizes the reward revenue given the storage constraint.
- We conduct simulations on time-evolving topology traces, which validates the feasibility of the proposed caching-offloading framework. The simulation results show that considerable offloading and reward gains can be achieved.

The rest of this paper is organized as follows. In Sect. 2, we propose the system model for the whole caching-offloading paradigm. Specifically, the optimal caching problem is formulated in Sect. 2.2, the bidding strategy, the response-for-bids strategy and the prediction method are represented in Sect. 2.1. The efficiency of the proposed caching-offloading framework is validated in Sect. 3. Section 4 concludes the paper.

2   System Model

We assume there are $\mathcal{N}$ mobile vehicle nodes, each of which could be both data requester and caching node. Let $\mathcal{M} = \{1, 2, \ldots, M\}$ denote the set of M contents (data items) in the network, without loss of generality, each data item is assumed to have identical data size. Each requester randomly generates a data request for item $d \in \mathcal{M}$ with probability p_d. It is assumed that the popularity distribution follows Zipf distribution [9]. Hence, the request probability of the d-ranked data item is defined as $p_d = d^{-s} / \sum_{d=1}^{M} d^{-s}$, $\forall d \in \mathcal{M}$, where s is the exponent skewness parameter. Each vehicle node $i \in \mathcal{N}$ has different limited storage capacity with the size of K_i data items. For any requester, who carries a data request associated with time constraint t_d, it is prioritized to seek data access from caching nodes nearby, once the data request expires, the BS will immediately push the requested data to the requester. This data access design aims at improving the offloading performance. Now, we focus on solving the aforementioned issues by designing the auction mechanism and caching policy.

2.1   The Design of Auction Mechanism

The whole mechanism comprises three major parts: the *bidding* strategy for caching nodes, the *response-for-bids* strategy for data requesters, and the prediction method to determine the auction-winning probability. When any requester initiates a data request for data item d, it will broadcast this request along with the bid-request (BR) message to all neighbor nodes. The BR message indicates the following information: (1) the maximum reward (B_d) that the requester is willing to pay for a successful data access of d; (2) the auction timeout (AT) before which every node who response to the BR submits its bid. All the neighbor nodes who have cached d and overhear the request are obligated to bid their price lower or equal than the announced B_d to provide data access. All the bids are open and can be detected by any node within communication range of the node who sent the bid. After receiving all the bids within AT, the requester will choose a winner based on its own preferential strategy. If the requested data item can be retrieved before the request expires, the corresponding reward based on winner's bid price will be issued. Otherwise, the requester will immediately download the data from the BS and the agreed reward will not be issued to the winner node. Based on this auction design, two questions naturally arise: (a) for a caching node who overhears the BR message, how the value of the bid price should be set to win the auction with higher chance; (b) For the requester who broadcasts the request, how to determine who wins the auction towards all received bids.

The Bidding Strategy. When a requester denoted by r announces its BR message, we denote all the nodes who overhear this BR by N_r, which contains the set of potential competitors involved in this auction. Furthermore, within N_r, we denote the sub-set of caching nodes who have the requested data item by N_r^*, where $N_r^* \subseteq N_r$. Every node belongs to N_r^* is able to provide data access for r. However, the status of wireless channel differs between each caching node and the requester, which results in different Quality of Service (QoS) for data access. Such distinction should be reflected on the value of bid price. Moreover, given N_r^*, we attempt to estimate how competitive we are in terms of reception probability compared to other caching nodes. It is reasonable to expect that the reception probability plays a key role in any auction decision.

Based on empirical measurements on received signal strength in vehicular environment [11], we assume that the wireless channel undergoes Rayleigh fading, the path loss exponent is α and all nodes have the same transmission power η. For $\forall i \in N_r^*$, let γ_{ir} be the received power for the transmission from i who has inter-vehicle spacing l_{ir} away from r, then γ_{ir} is an exponentially distributed random variable with mean $\eta(l_{ir})^{-\alpha}$ with the following probability density function (pdf) [3]:

$$f(\gamma_{ir}) = \frac{1}{\eta(l_{ir})^{-\alpha}} exp\left(-\frac{\gamma_{ir}}{\eta(l_{ir})^{-\alpha}}\right), \forall \gamma_{ir} \geq 0 \qquad (1)$$

We assume there is no distributed medium contention prior to transmission (the analysis of MAC efficiency is not relevant to this paper), and the requester r can successfully decode the requested data item if and only if its received signal-to-noise-ratio (SNR) exceeds a decoding threshold, i.e.,

$$P_{suc} = \Pr(\gamma_{ir}/n_0 \geq \varepsilon_{th}), \varepsilon_{th} > 1 \qquad (2)$$

where n_0 is the noise power, and ε_{th} is the threshold whose value is determined by the channel coding and modulation, then the reception probability that requester r could successfully retrieve the data item from caching node i, say Q_{ir}, can be simply written as:

$$Q_{ir} = P_{suc} = \exp\left(-\varepsilon_{th} n_0 \eta^{-1} l_{ir}{}^\alpha\right) \qquad (3)$$

Based on Eq. (3), for any $i \in N_r^*$, we can figure out how competitive we are with respect to (w.r.t) the reception probability over N_r^*, specifically, it is reasonable to compute the relative reception probability (RQ_{ir}) as:

$$RQ_{ir} = \frac{Q_{ir}}{\overline{Q}|_{N_r^*}} = \frac{Q_{ir}}{\sum_{i \in N_r^*} Q_{ir} / |N_r^*|} \tag{4}$$

Observe that, if $RQ_{ir} < 1$, our competitors are better positioned than us on average, they are likely to provide a higher bid price, since they can deliver the requested data item with higher probability. In a word, to win this auction, we need to set a lower bid price to increase competitive power. If $RQ_{ir} \approx 1$, which means we have similar channel condition compared to others, it is possible to win the auction with a lower price bid as well. However, if $RQ_{ir} > 1$, we are better positioned than other competitors, we could win the auction even with a higher bid price (close to B_d) since the requester might have willing to pay for a better QoS.

Given the value of B_d, we can design a non-decreasing smooth curve to indicate the relation between the bid price and the relative reception probability, any explicit mathematical function having a characteristic 'S'-shape curve can be scaled to describe this relation such as logistic function, arctangent function, error function, etc. Without loss of generality, in this paper, the bid price p_{ir} will be given by an error function in the form of:

$$p_{ir}(RQ_{ir}) = \frac{B_d}{2} + \frac{B_d}{2} \cdot \mathrm{erf}(a_i(RQ_{ir} - 1)) \tag{5}$$

where $0 < p_{ir}(RQ_{ir}) < B_d$, $\mathrm{erf}(x) = (2/\sqrt{\pi}) \int_0^x e^{-t^2} dt$, the curve is centered on $RQ_{ir} = 1$, and the scale is controlled by a_i. It is noteworthy that if there is no competition, we straightly set our bid price as B_d, and if $RQ_{ir} \approx 0$, it means we can hardly deliver the data item before expiration, we simply set the bid price to 0 (free data access service). Now each caching node who overhears the BR message is able to declare its own bid price according to its own reception probability, in the following, we attempt to design the response function for the data requester to determine the winner in the auction.

The Response-for-Bids Strategy. To determine who wins the auction, the requester r needs to consider both the bid price p_{ir} and the relative reception probability RQ_{ir}. It is preferable to value a high relative reception probability rather than a good bid price, since it is more important to guarantee the delivery of the requested data item. Accordingly, we first identify the following special cases in which the preference of r can be pre-set:

- $p_{ir} = 0, RQ_{ir} = \max_{i \in N_r^*} RQ_{ir}$. which means node i has the largest relative reception probability and willing to provide 'free' data access, then i should be given the highest preference value denoted by k_a;
- $p_{ir} = B_d, RQ_{ir} = \max_{i \in N_r^*} RQ_{ir}$. which means node i has the largest relative reception probability and provides data access if the highest reward can be issued, we denote the preference value under this case by k_b;

- $p_{ir} = 0, RQ_{ir} \approx 0$. which means node i has the smallest relative reception probability but willing to provide 'free' data access, we denote the preference value under this case by k_c;
- $p_{ir} = B_d, RQ_{ir} \approx 0$. which means node i has the smallest relative reception probability but still wants the highest reward, then i should be given the lowest preference value (zero) denoted by $k_d = 0$.

The above cases indeed represent the boundary points of the *response function* of r which express the preference for all received bids. Case (a) and (d) determine the range of all preference value, in addition, to reflect the tendency to favor the relative reception probability against bid price, the relative values of preference under case (b) and (c) are set as $k_b > k_c$, in general, we have $k_a > k_b > k_c > k_d = 0$. Considering a fact that the requester is always inclined to choose a caching node who offers a lower bid price with higher relative reception probability, we define a plane that intersects all these boundary points to represent the *response function* of r:

$$f_r\left(p_{ir}, RQ_{ir}\right) = k_c - (k_c/B_d) \cdot p_{ir} + \left(k_b / \max_{i \in N_r^*} RQ_{ir}\right) \cdot RQ_{ir} \qquad (6)$$

Where k_b and k_c can be arbitrarily set under the condition $k_b > k_c$, and $k_a = k_b + k_c$ in Eq. (6). It can be seen that although the *bidding* and *response-for-bids* strategies are well-defined, a closed-form expression for auction-winning probability is still hard to derive, in the following, we introduce a prediction method exploiting the historical information of past auctions in order to build a probability function capturing the belief that a given bid price may win a given auction.

The Prediction Method for Winning Probability. The auctions are randomly generated from all data requesters, which means the reward budget B_d might be different even for identical data item. Besides, for a given auction, the bid price of a caching node is also affected by other competitors' strategy. Thus we adopt a probabilistic approach to keep track of bidding history of past auctions, then extrapolate the winning probability of a given auction. As time increases, the sample space of past auction records becomes large enough, we consequently approximate the winning price for the auction w.r.t data item d follow a Normal distribution with mean μ_d and standard deviation σ_d, that is to say, the winning price is modelled as a random variable denoted by f_p^d whose probability distribution is $N\left(\mu_d, \sigma_d\right)$. Considering our proposed auction mechanism follow a reverse auction pattern, the winning odds with a bid price p in an auction for data item d is then given by the value at p of the corresponding complementary cumulative normal distribution (CCDF):

$$P_d\left(p\right) = 1 - P\left(f_p^d \leq p\right) = 1 - \frac{1}{\sqrt{2\pi}\sigma_d} \int_{-\infty}^{\frac{p-\mu_d}{\sigma_d}} e^{-x^2/2} dx \qquad (7)$$

The advantage of this statistical method, is that we can scale up a large set of past auctions. If the auction history covers a long period of time, it is practical to use time-weighted averages and standard deviations instead of plain ones. However, there exists a drawback of this method: the winning probability could only be derived until the bidding history of past auctions become relatively large. For illustration, we set a threshold $\alpha_{min} = 50$, which means the winning price for a data item can be modelled as normal distribution only if the number of past auction records reaches α_{min}. However there is no data item initially cached in the network, which implies no auction will happen. *A contradiction arises that the node has to make caching decisions based on the winning probability as known parameter, but there is no auction happened at that time.* To solve this issue, we specifically define $P_d(p)$ under the case $\alpha_A^d < \alpha_{min}$, where α_A^d represents the number of past auction records for data item d. A viable method is to calculate the winning probability under assumption that all neighbor nodes have the same competitiveness at the very beginning. Assume the bid price announced by i for the auction initiated by r is p_{ir}, then the winning probability, say $P_d(p_{ir})$, simply is:

$$P_d(p_{ir}) = \begin{cases} \frac{1}{|N_r|}, & \alpha_A^d < \alpha_{min}, \\ 1 - \frac{1}{\sqrt{2\pi}\sigma_d} \int_{-\infty}^{\frac{p_{ir}-\mu_d}{\sigma_d}} e^{-\frac{x^2}{2}} dx, & \alpha_A^d \geq \alpha_{min}, \end{cases} \quad \forall r \in N_i \qquad (8)$$

Since the auction mechanism for the request process has been elaborately designed, next we are able to propose the optimal caching problem which is tightly connected with the auction process.

2.2　Maximizing the Reward Revenue

An optimal choice for any node is to cache those data items that might yield the maximum reward revenue obtained from potential data requesters. For a caching node $i \in \mathcal{N}$, we denote the set of neighbor nodes who have contact with i by N_i. Then if any neighbor node becomes a requester and queries for data item in the future, i has chance to win the auction and obtain the reward. Accordingly, which data items does i select from $\mathcal{M}$ to fill up its available storage space so as to maximize its reward revenue? This problem can be formulated as 0–1 knapsack problem:

$$\max \sum_{d \in \mathcal{M}} x_d \sum_{r \in N_i} p_d \cdot P_d(p_{ir}) \cdot p_{ir} \qquad (9)$$

$$s.t. \sum_{d \in \mathcal{M}} x_d \leq K_i, x_d \in \{0, 1\}, d \in \mathcal{M}. \qquad (10)$$

Where x_d denotes if i caches data item $d(x_d = 1)$ or not $(x_d = 0)$, and the second summation term $\sum_{r \in N_i} p_d \cdot P_d(p_{ir}) \cdot p_{ir}$ represents the expected reward revenue that node i could obtain by caching data item d, hence the objective goal is to maximize the overall reward revenue by putting a set of data items into the storage space with capacity K_i. Here p_d is the request probability for d, p_{ir}

is the bid price when neighbor node r initiates request and auction process, and $P_d(p_{ir})$ is the probability of winning the auction, this expected reward revenue is influenced by the auction design as shown in Sect. 2.1. The constraints include three parts: K_i indicates the maximum number of data items i could cache, x_d is a binary variable, and the cached data item could only be selected from finite data item set $\mathcal{M}$. This 0–1 knapsack problem is NP-hard [10]. A viable method is to adopt a heuristic greedy algorithm that we calculate the expected reward revenues for all data items, rank them in decreasing order, and fill in the storage space of node i one by one until no more data item can be allocated.

3 Simulation Results

In this section, we evaluate the performance of our proposed caching-offloading scheme using MATLAB simulation platform on synthetic trace with time-evolving topology. We consider a number of vehicles distributed within a single cell, where the radius of coverage of the BS is fixed. To emulate changing topology, we continuously re-generate the position of all vehicle nodes periodically in which the coordinates follow 2-dimensional Random distribution. In addition, we assume the topology remains stable to complete one auction round and data delivery. Two vehicle nodes are considered as neighbor nodes when the distance between them is less than a threshold distance. All vehicles use the same transmission power, the wireless channel undergoes Rayleigh fading and the bandwidth is the same between any pair of vehicles. The maximum auction reward for each item is initially set the same, each node has the same form of bid price function with different scaling parameter (a_i), and a virtual 'wallet'. We generate a request process over the whole simulation period randomly distribute requests among all nodes. Specifically, all system parameters are shown in Table 1.

Table 1. System parameter initialization.

Parameter	Value	Parameter	Value
Topology period	$Period_t = 20$	Radius of BS	$r_{bs} = 500\,\text{m}$
Threshold distance	$d_{th} = 300\,\text{m}$	Number of data items	$M = 20$
Size of data item	$s_d = 50\,\text{Mb}$	Number of vehicles	$N = 50$
Simulation length	$T = 1000$	Transmission power	$\eta = 10^{-5}\,\text{W}$
Noise power	$n_0 = 4 * 10^{-14}\,\text{W}$	Path loss exponent	$\alpha = 3$
Decoding threshold	$\varepsilon_{th} = 2$	Arrival rate	$\lambda_r = 0.5$
Zipf parameter	$s = 1$	Storage capacity	$K_i \in [5, 15]$
Time constraint	$t_d = 15$	Maximum auction reward	$B_d = 50$
Scaling parameter	$a_i \in [1, 20]$	Preference values	$k_b = 10, k_c = 5$
Past auction records	$\alpha_{min} = 30$	—	—

At each time-slot, two important tasks should be executed: (1) each caching node re-calculates its 0–1 Knapsack problem to determine which data item to

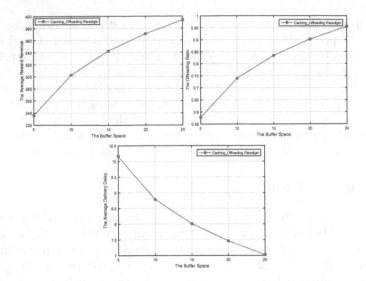

Fig. 2. Performance of caching-offloading scheme with different buffer space

be cached according to current network status; (2) The requester who carries
the data request that generated at current time-slot need to initiate an auction
process. The following metrics are used for performance evaluation: (1) *offloading ratio*: the ratio of requests being served by the helper nodes over all requests
being served; (2) *delivery delay*: the average delay for obtaining the requested
data item either by helper nodes or BS; (3) *average reward revenue*: the average
reward of caching nodes, which is earned from data requesters. Each simulation is repeated multiple times with randomly generated data and requests for
statistical convergence.

3.1   The Impact of Buffer Space

We investigate the system performance of our proposed caching offloading
paradigm under varying buffer conditions. The simulation results are shown in
Fig. 2. It can be seen that as the buffer space increases, each node is allowed to
accommodate more data items. To earn potential rewards as much as possible,
each caching node would make the best of its buffer space to cache more data.
On one hand, when any requester generates a data request, it is more likely to
be served by other caching nodes instead of retrieving from base station, which
improves the level of average reward revenue and the offloading ratio simultaneously. On the other hand, more buffer space brings more data replicas, when
an auction request is initiated along with a data request, if the buffer space is
limited, there will be inadequate caching nodes joining this auction, otherwise
there will be more competitors, which increases the odds that the node who has
higher reception probability will win this auction and complete data delivery
with short delivery delay. Specifically, when the buffer space ranges from 5 to

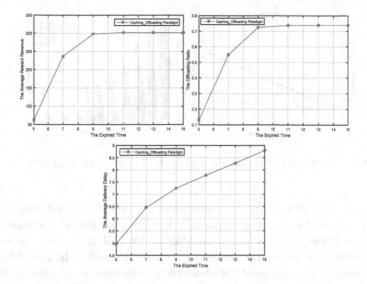

Fig. 3. Performance of caching-offloading scheme with different expired time

25, the average reward revenue and the offloading ratio increases by 66.88% and 65.53% respectively, and the delivery delay decreases by 30.91%.

3.2 The Impact of Expired Time

We investigate the system performance of our proposed caching offloading paradigm under varying expired time. The simulation results are shown in Fig. 3. In this experiment, the buffer space for all nodes is set as 10, which means each of them can only choose 1/3 number of data items to cache. It can be seen that the figures of average reward revenue and offloading ratio appears to be positive-correlated. They both approach their upper limitations when the expired time is beyond 9. It is obvious that the requester has more chance to retrieve data from other caching nodes as expired time increases, which results in higher reward revenue and offloading ratio. More data requests are served by vehicle-to-vehicle contacts instead of immediately downloading from base station, thus the average delivery delay increases. However, the reason why the curves of reward revenue and offloading ratio remain stable as the expired time reaches a certain value (9) is indeed restricted by the caching decisions. Each node is only able to cache 10 data items, they are inclined to choose popular data items to cache such that maximum potential reward can be obtained. Then the data requests of less popular data item can only be served by base station, which means this part of cellular traffic can never be offloaded.

3.3 The Impact of Maximum Auction Reward W.r.t Caching Decisions

This simulation case is designed to investigate the trade-off mentioned in Sect. 1 that how will the reward settings affect the caching decision of each node.

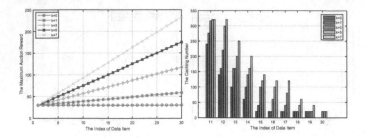

Fig. 4. The impact of maximum auction reward w.r.t caching decisions

To achieve this, we first design five maximum auction reward settings as shown in the left part of Fig. 4, we continuously increase the auction reward for less popular data items to see whether caching nodes will favor them to cache or not. Specifically, the auction reward settings are represented by linear functions with increasing slope, and we statistically record the number of data items being cached. In the right part of Fig. 4, we present the caching number of data items indexed from 11 to 20, it can be seen that as more auction reward are assigned to these data items, more caching nodes choose to cache them so as to increase their potential earning reward. We omit the part of representing the caching number of data items from 1 to 10, since they are still attractive to many nodes due to their high popularity and their caching number remains nearly stable with a subtle reduction. To the least popular data items from 21 to 30, since the request probability is so low that no caching node is willing to take risk caching them to earn reward from occasional data request.

4   Conclusions

In this paper, we propose an incentive caching scheme, which is reward-based and applied to offloading scenario in vehicular networks. The basic idea is to pre-cache some data items at vehicle nodes by exploiting storage capability, then serve future data requests via vehicle-to-vehicle contact instead of accessing data from BS. To further implement this caching-offloading paradigm, we first provide a novel auction mechanism including the *bidding* strategy for caching nodes and the *response-for-bids* strategy for data requesters, then derive the optimal caching decision by formulating an allocation issue in the form of 0–1 knapsack problem. Last, the simulation results validate the feasibility of the proposed scheme and show that considerable offloading and reward gains can be achieved.

Acknowledgements. This work was supported by Department of Product & Infrastructure, China Telecommunication Technology Labs, China Academy of Information and Communications Technology.

References

1. Jiang, D., Delgrossi, L.: IEEE 802.11p: towards an international standard for wireless access in vehicular environments. In: 2008 Vehicular Technology Conference, VTC Spring, pp. 2036–2040. IEEE (2008)
2. Ye, F., Roy, S., Wang, H.: Efficient data dissemination in vehicular ad hoc networks. IEEE J. Sel. Areas Commun. **30**(4), 769–779 (2012)
3. Yao, H., Zeng, D., Huang, H., Guo, S., Barnawi, A., Stojmenovic, I.: Opportunistic offloading of deadline-constrained bulk cellular traffic in vehicular DTNs. IEEE Trans. Comput. **64**(12), 1–1 (2015)
4. Mao, G., Anderson, B.D.O.: Cooperative content dissemination and offloading in heterogeneous mobile networks. IEEE Trans. Veh. Technol. **65**(8), 6573–6587 (2016)
5. Ezzahidi, S.A., Sabir, E., Kamili, M.E., Bouyakhf, E.H.: A non-cooperative file caching for delay tolerant networks: a reward-based incentive mechanism. In: IEEE Wireless Communications and Networking Conference. IEEE (2016)
6. Yu, R., Wang, P. :. Node-dependence-based dynamic incentive algorithm in opportunistic networks. In: International Conference on Computer Communication & Networks, pp. 1–8. IEEE (2014)
7. Zhao, G., Chen, M., Wei, X.: Ris: a reciprocal incentive scheme in selfish opportunistic networks. Wirel. Pers. Commun. **70**(4), 1711–1734 (2013)
8. Zhuo, X., Gao, W., Cao, G., Dai, Y.: Win-coupon: an incentive framework for 3G traffic offloading. In: Proceedings of the 19th annual IEEE International Conference on Network Protocols, ICNP 2011. In: IEEE (2011)
9. Breslau, L., Cao, P., Fan, L., et al.: Web caching and Zipf-like distributions: evidence and implications. In: Eighteenth Joint Conference of the IEEE Computer and Communications Societies, INFOCOM 1999, pp. 126–134, Proceedings. IEEE (2002)
10. Connolly, D.: Knapsack problems: algorithms and computer implementations. J. Oper. Res. Soc. **42**(6), 513–513 (1991)
11. Yin, J., Holland, G., Elbatt, T., Bai, F., Krishnan, H.: DSRC channel fading analysis from empirical measurement. In: International Conference on Communications and Networking in China. IEEE Xplore (2007)

Early Prediction of Student Success Based on Data Mining and Artificial Neural Network

Marko Bursać[1]([⊠]) , Marija Blagojević[2]([⊠]) ,
and Danijela Milošević[2]

[1] High Railway School of Vocational Studies, Belgrade, Serbia
marko.bursac@vzs.edu.rs
[2] University of Kragujevac, Faculty of Technical Sciences Čačak, Čačak, Serbia
{marija.blagojevic,danijela.milosevic}@ftn.kg.ac.rs

Abstract. This paper presents an overview of the research related to the prediction of the success of the participants in the Technical Drawing course. In order to determine the student's success, a data mining model was created supported by artificial intelligence. The proposed model gives an overview of the input data on the basis of which it is possible to determine the success of the student's using artificial neural networks. The results of the prediction give a presentation of the performance of students at the beginning of the course, which gives professors enough time to influence the students and encourage them.

Keywords: Student performance · Neural network · Data mining

1 Introduction

The aim of the research is the early prediction of the students' success in the observed course based on the results achieved in previous years of education. Motivation for this work is reflected in the need of a professor or teacher to have achieved number of points at the end of the course. The results obtained at the very beginning of the semester (two weeks) give the professors enough time (thirteen weeks) to adapt their work. Based on the obtained results the professor can influence at the work of the students, motivate it, organize additional classes and improve students knowledge etc. The analyses conducted by way of artificial neural networks point to the possibility of predicting selected parameters with considerable accuracy [1]. According to [2], these analyses assume the existence of knowledge of various areas, such as information technology, agriculture and data mining.

2 Literature Review

Results in the field of research on students' success can be achieved using artificial neural networks. In paper [3], authors define factors (high school, assessment of placements, number of credits won etc.) that have an impact on performance assessment of students' success. Based on these factors, the use of artificial neural networks

© Springer Nature Switzerland AG 2019
D. Milošević et al. (Eds.): HCC 2019, LNCS 11956, pp. 26–31, 2019.
https://doi.org/10.1007/978-3-030-37429-7_3

provides accuracy of 80%. In the research [4], the authors used as input data as many as 1407 student profiles, to determine the success of the study for 307 students. The obtained results are within the limits of 68–77% of the accuracy of the data. According to [5], the artificial neural network modeling used 10 parameters for input data. The predictive precision rate is 74%. The authors in the paper [6] from the knowledge base draw out the features of students that influence the passing of the final exam. A model with 17 input data was created, 14 neurons in the hidden layer and one output. In the paper [7], the model of artificial neural networks foresees the future profession of graduates for the purposes of enrollment policy, the graduates completed a question-naire to determine the factors that influence the enrollment. The paper [8] use different types of learning algorithms of artificial neural networks in order to predict the most appropriate model for predicting the success of students. Authors as key factors identify success from high school, as well as subjects that are outstanding in the first year of study.

3 Methodology

This study aims to estimate total number of points that the students can achieved during teaching process using artificial neural network defined in three layers. Between the input layer (receive input) and output layer (send data to the user program) are hidden layers. Hidden layer neurons are only connected to other neurons and never directly interact with the user program and it has the opportunity to affect processing. Processing can occur at any layer in the neural network [10]. The output for the single neuron is described in by,

$$O_n = \varphi\left(b_k + \sum_{i=1}^{m} w_{ki} * x_i\right) \tag{1}$$

where, O_n is the output, φ is the activation function, b_k is the threshold bias, w_k are the synaptic weights, x_i are the inputs from the previous layer of neurons, and i is the number of inputs. Each neuron is capable of editing weights supplied to it based upon the accuracy of the entire network. Setting the weights in accordance to the chosen learning algorithm, the learning capability of neuron is obtained [9]. This enables the neural network to learn the behavior of data provided [11]. The neural network is defined through two phases, the learning/training and testing phase. Prior to learning, it is necessary to define the input and output variables and to collect data to which the backpropagation algorithm will be applied [12]. The backpropagation algorithm uses supervised learning, which means that we provide the network with examples of inputs and outputs [13, 14]. The input variables (Table 1) identified are those which can simply be obtained from student file and registrar system.

Table 1. Input and output variables

Variables	Description	Values
Input	Gender	1 - female; 2 - male
	Points from high school	Minimum value - 20 points; Maximum value - 40 points;
	Study program	1 - Railway transport; 2 - Railway engineering; 3 - Electrical engineering in traffic; 4 - Railway construction; 5 - Public urban and industrial traffic; 6 - Environmental Engineering
	Activity in class (Computer and Informatics course)	Minimum value – 0%; Maximum value – 100%;
	Total number of points achieved (Computer and Informatics course)	Minimum value – 0%; Maximum value – 100%;
	Activity in class (Technical drawing - only first two weeks)	Minimum value – 0%; Maximum value – 100%;
	Points achieved at the entrance test	Minimum value - 0 points; Maximum value - 20 points;
Output	The total number of points the student achieves at the end	Minimum value - 0 points;

In this research it is used a single-layer neural network (Fig. 1) with one hidden layer of neurons and with feed-forward backpropagation (developed using MATLAB). The type of algorithm that was selected is Levenberg-Markuardt.

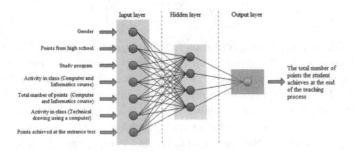

Fig. 1. Design of the neural networks

Predictive precision measures used in this paper is Medium Square Error (MSE), according to the equation [6]:

$$MSE = \sum_{i=1}^{N} \frac{(P_s - P_p)^2}{N} \tag{2}$$

where N represents the number of samples, P_s real results and P_p results obtained by predicting a neural network. During the design and testing of the model, the best results are obtained by working with four neurons in the hidden layer, which is in accordance with [15], the number of neurons in the hidden layer needs to be twice as low as the input data.

4 Results and Discussion

As a result of training, acceptable correlation coefficient R values were obtained, for training R = 0.74619; for validation R = 0.78097; and for testing R = 0.7243 (Fig. 2).

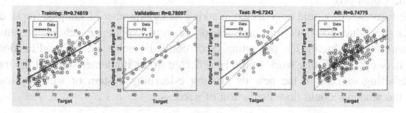

Fig. 2. Results of training

According to the obtained data it is possible to determine the mean square error (MSE) in the following way: for student 1 - $P_s = 79$, $P_p = 84.01$; for student 2 - $P_s = 57$, $P_p = 60.61$ and so on for 238 observing students - the MSE is 23.2.

Of the thirty papers that dealt with the prediction of students' success, the root mean square error (RMSE) was presented in [16] and is 0.46. In the paper [17], an overview of the RMSE for three different learning algorithms of the neural network is provided for Naive Bayes - 0.4204, Multilayer Perceptron - 0.4969 and 0.4431 for decision tree C4.5. In the paper [4], the authors cite the RMSE as 0.22. In other papers dealing with forecasting and related to the prediction of students' success, the MSE can be found in values from 0.585 to 38.235.

Based on the created model, the total number of points achieved during the teaching process is predicted (red color - Fig. 3) for 25 students.

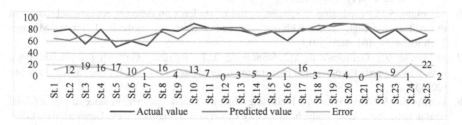

Fig. 3. Results of prediction (Color figure online)

Students eventually achieve results similar to those envisaged (blue color).

5 Conclusions

This paper introduced a model, an artificial neural network approach for identification student success in early stage of performing teaching process (in second week). On the basis of the obtained research results, it is possible to determine whether some of the students need assistance in learning and adopting learning materials in order to achieve a positive grade at the end of the educational process. Using models in the second week of the teaching process gives the ability to timely prepare and assist students in acquiring knowledge. The model can be used in the planning of teaching and knowledge testing as well as subsequently used after re-examining knowledge during the educational process.

The future work would be reflected in taking a large number of samples (input data) for example: education data from high school, years of experience, achieved points for all exams etc. Reducing the predicting error could be achieved by modification of neural network. The future work would be reflected in the implementation of the model for predicting success in the new or existing e-learning system or web application where the students could do a prediction of success.

References

1. Blagojević, M., Blagojević, M., Ličina, V.: Web-based intelligent system for predicting apricot yields using artificial neural networks. Sci. Hortic. **213**, 125–131 (2016)
2. Marinković, B., Crnobarac, J., Brdar, S., Jaćimović, G., Crnojević, V.: Data mining approach for predictive modeling of agricultural yield data. In: First International Workshop on Sensing Technologies in Agriculture, Forestry and Environment (BioSense09), Novi Sad (2009)
3. Naser, S.A., Zaqout, I., Ghosh, M.A., Atallah, R., Alajrami, E.: Predicting student performance using artificial neural network: in the faculty of engineering and information technology. Int. J. Hybrid Inf. Technol. **8**(2), 221–228 (2015)
4. Karamouzis, S., Vrettos, A.: An artificial neural network for predicting student graduation outcomes. In: World Congress on Engineering and Computer Science, San Francisco (2008)
5. Oladokun, V.O., Adebanjo, A.T., Charles-Owaba, O.E.: Predicting students' academic performance using artificial neural network: a case study of an engineering course. Pac. J. Sci. Technol. **9**(1), 72–79 (2008)
6. Mazinani, S.M.: Prediction of success or fail of students on different educational majors at the end of the high school with artificial neural networks methods. J. Innov. Manage. Technol. **4**(5), 461–465 (2013)
7. Miljković, Z., Gerasimović, M., Stanojević, L.J., Bugarić, U.: Using artificial neural networks to predict professional movements of graduates. Croatian J. Educ. **3**(13), 117–141 (2011)
8. Isljamovic, S., Suknovic, M.: Predicting students' academic performance using artificial neural network: a case study from faculty of organizational sciences. Eurasia Proc. Educ. Soc. Sci. **1**, 68–72 (2014)
9. Abraham, A.: Artificial neural networks. In: Handbook of Measuring System Design. London. Wiley Ltd. (2005}
10. Hagan, M., Demuth, H., Beale M.H.: Neural Network Design (2014)

11. Lu, C.: Artificial neural network for behavior learning from meso-scale simulations, application to multi-scale multimaterial flows. MS (Master of Science) thesis, University of Iowa (2010)
12. Roberts, C., Okine, N.A.: A comparative analysis of two artificial neural networks using pavement performance prediction. Comput. Aided Civ. Infrastruct. Eng. **13**(5), 339–348 (1998)
13. Gershenson, C.: Artificial Neural Networks for Beginners, Neural and Evolutionary Computing, pp. 1-8 (2003). https://datajobs.com/data-science-repo/Neural-Net-[Carlos-Gershenson].pdf
14. Wani, S.M.A.: Comparative study of back propagation learning algorithms for neural networks. Int. J. Adv. Res. Comput. Sci. Softw. Eng. **3**(12), 1151–1156 (2013)
15. Karsoliya, S.: Approximating number of hidden layer neurons in multiple hidden layer BPNN architecture. J. Eng. Trends Technol. **3**(5), 714–717 (2012)
16. Simeunović, V., Preradović, L.J.: Using data mining to predict success in studying. Croatian J. Educ. **16**(2), 491–523 (2013)
17. Osmanbegović, E., Suljić, M.: Data mining approach for predicting student performance. Econ. Rev. J. Econ. Bus. **10**(1), 3–12 (2012)

A Method of Vehicle Fault Diagnosis Supporting Multi-value-chain Collaboration

Xiaofeng Cai[1,2], Huansheng Ning[2], Tingyu Liu[3], Changmao Wu[1], and Changyou Zhang[1(✉)]

[1] Laboratory of Parallel Software and Computational Science, Institute of Software, Chinese Academy of Sciences, Beijing, People's Republic of China
changyou@iscas.ac.cn
[2] School of Computer and Communication Engineering, University of Science and Technology Beijing, Beijing, People's Republic of China
[3] Computer School, Beijing Information Science and Technology University, Beijing, People's Republic of China

Abstract. To improve the collaborative efficiency of multi-value-chain in vehicle maintenance, we designed a case-based vehicle fault diagnosis method by analyzing the traditional diagnosis methods of vehicle fault and collecting user's actual requirements. In this paper, we summarized corpus of vehicle fault descriptions, analyzed user's language features and extracted proper nouns in vehicle field as an extended dictionary to improve the accuracy of text segmentation. We also built a database of vehicle maintenance cases and a vehicle structure tree to support this diagnosis method. Then using vectorization method to process fault description corpus and trained the semantic vectorization model with both statistics and topics. After that, we used vector distance algorithm to compute semantic similarity and return the optimal case in database. Finally, we located the exact position of current fault in vehicle structure tree. The experimental result shows that the accuracy of vehicle fault diagnosis achieved 86.7% by this method. The accuracy of vehicle fault locating is also achieved 81.8%. This case-based vehicle fault diagnosis system can be used for online fault consultation. It can also expand the collaborative business mode of vehicle service value chain and improve the quality and efficiency of vehicle maintenance service.

Keywords: Multi-value-chain · Vehicle fault diagnosis · Semantic vectorization model · Semantic similarity · Structure tree · Vehicle fault locating

1 Introduction

With the continuous development of vehicle functions and technologies, vehicle's internal structure become more complex. The fault types of vehicle are also more diversified, which brings great challenges to traditional vehicle maintenance mode. Traditional mode is lack of efficiency and waste too much time on fault diagnosis. Vehicle fault diagnosing includes receiving information on a vehicle from a customer regarding an actual vehicle fault and accessing data regarding the actual vehicle fault

© Springer Nature Switzerland AG 2019
D. Milošević et al. (Eds.): HCC 2019, LNCS 11956, pp. 32–43, 2019.
https://doi.org/10.1007/978-3-030-37429-7_4

from the vehicle [1]. Therefore, in this paper we design and build a semantic relation-based vehicle fault diagnosis system, focusing on customer's description. The main works and researches include the construction of vehicle fault case database, the establishment of vehicle proper noun dictionary, the representation of vehicle structure tree and the design of fault diagnosis and location algorithm. This vehicle fault diagnosis system is designed for both consumer and vehicle service provider. The research in this paper effectively expands the collaborative business model of vehicle multi-value-chain, improves the quality and efficiency of vehicle maintenance service and constructs an efficient information platform for vehicle fault diagnosis between consumers and service providers.

2 Related Works

In the field of vehicle fault diagnosis, there are five main fault diagnosis methods which are empirical diagnosis, simple instrument diagnosis, specialized instrument diagnosis, component-specific diagnosis and intelligent fault diagnosis [2]. The first four fault diagnosis methods are being widely used, but the intelligent fault diagnosis combines still in developing. Combines with machine learning and other edge technologies in computer field, intelligent fault diagnosis will change the situation of vehicle fault diagnosis industry.

Chinese word segmentation can be divided into three categories which is string matching-based segmentation method, statistics-based segmentation method and understanding-based segmentation method. In realistic tasks, statistics-based segmentation method always combines with dictionary to take advantages of both. "Jieba" Chinese text segmentation is a mature tool which support three segmentation mode and allows user to modify the dictionary [3].

Text vectorization is the process of transforming text into a series of vectors that can express text semantics [4, 5]. BOW (Bag of Word) model is a traditional text vectorization method which consider word as basic unit [6].

The conception of Value Chain is first proposed by Michael E. Porter. Company's value behaviors have mutual contacts in marketing activities and constitute the chain of company's marketing behaviors, it called value chain. In value chain, a behavior's implementing manner will influence the cost and effect of other behaviors. Multi-value-chain collaboration in vehicle market refers to the data and service interaction among multiple vehicle value chains which can bring more profit [7–9].

3 The Architecture of Vehicle Fault Diagnosing System

The overall architecture of vehicle fault diagnosis system is shown in Fig. 1. As preparation, we collect proper nouns as customized dictionary to improve the effect of word segmentation. We summarized fault case from Internet and set a standard data format to build fault case database. Following the numbering standard of vehicle parts, we summarized a vehicle structure tree to make a visualized expression of vehicle's system structure.

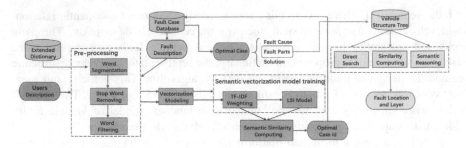

Fig. 1. The architecture of vehicle fault diagnosis system.

The function of pre-processing part is to transform user's description and fault descriptions in database into the basic unit of NLP. Then using vectorization model to map the words into vector space. For fault descriptions in case database, using TF-IDF and LSI model to train semantic vectorization model. In semantic similarity computing step, we choose cosine distance to measure the similarity between two semantic vectors and find the optimal case. According to the optimal case id, searching the corresponding case from database and return relative information to user. The fault parts information is useful to the fault location. We designed three methods which are direct searching, similarity computing and semantic reasoning to determine fault location and layer. Last, the system will return the result of fault location to help maintenance personnel locate fault parts.

4   Implementation of Vehicle Fault Diagnosis Method

4.1   Construction of Vehicle Fault Case Database

Vehicle fault case database is the fundamental of whole fault diagnosis system. The source of fault cases is mainly from the vehicle maintenance case on Internet. The vehicle fault case database is summarized in four dimensions which are fault description, fault cause, fault part and solution.

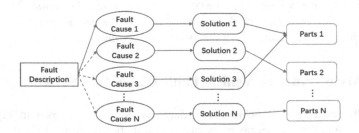

Fig. 2. Logical relationship of fault cases.

The logical relationship of the vehicle fault case database is shown in Fig. 2. Each case corresponds to a unique fault description which is the main basis of semantic reasoning. A fault description may have different fault causes, each possible fault cause corresponds to a unique solution and fault part.

The vehicle fault case database contains four tables, which are fault descriptions table, fault causes table, fault parts table, and solution table. For now, the vehicle fault case database included 439 cases and expand continuously. Against to less data quantity and data acquisition difficulties, we designed a learning ability to the system. After the system is put into use, it can record each diagnosis information of users and summarized as typical fault case to expand database. On the one hand, it can expand the quantity of database, on the other hand, a more complete database will benefit the accuracy of diagnosis result. It's a positive cycle for both customers and service provides.

4.2   The Pre-processing of Fault Description

The vehicle fault descriptions are generally conversational expression, including one or more sentences. We select jieba as the tool of word segmentation, it supports many extra functions includes three segmentation modes, customized dictionary and POS (Part of Speech) tagging [3]. By testing and comparing the effort of different segmentation modes in jieba, we selected the precise mode. Search engine mode and full mode intend to list more possible vocabularies to promote recall rate, which is not helpful for our works but influence the result of the semantic information extracting and similarity computing.

This research contains lots of professional knowledge of vehicle field, so the word segmentation results may be inaccurate or wrong when facing those proper nouns if we only use default dictionary. To solve this problem, we need to build a customized dictionary of vehicle proper nouns. The vehicle proper nouns dictionary comes from following sources: (1) related vehicle dictionaries of Sogou; (2) parts name from vehicle structure tree; (3) vehicle proper nouns on Internet. The vehicle proper nouns dictionary we built contains 2654 nouns from Sogou, 1092 nouns from vehicle structure tree and 1207 nouns from Internet, 4953 nouns in total.

User's fault description is a conversational expression and remains a lot of redundant words in the result of word segmentation. Stop words are those words have higher frequency but meaningless like auxiliary words, adverbs, prepositions and conjunctions. The filtering of stop words mainly relies on stop word list, we remove words from segmentation results which in stop word list. The stop word list we built contains 1599 items, covered most of Chinese stop words. After we removed stop words, we still need to have a selective POS filtering. In vehicle fault diagnosis field, nouns play a more important role of sentence semantic extraction. So, we keep nouns in the result of word segmentations to improve the accuracy of subsequent works.

Verification of Vehicle Proper Nouns Dictionary. To verify the effect of customized dictionary, we designed a group of comparative experiments to compare the word segmentation results. We select 121 fault descriptions from database as test set. The word segmentation results of this comparative experiment are listed in Table 1.

Table 1. Word segmentation results of comparative experiment.

Experiment condition	Daily vocabulary false	Proper nouns false	Accuracy
Default dictionary only	1	35	70.2%
Loaded customized dictionary	1	7	93.4%

After loaded vehicle proper nouns dictionary, the effect of word segmentation on proper nouns false made an obvious progress. The proper nouns false decreased from 35 to 7 and have no influence of daily vocabulary segmentation. For the test set, after loaded vehicle proper nouns dictionary the accuracy has up to 93.4% from 70.2%. This comparative experiment verified that the vehicle proper nouns dictionary is effective for fault descriptions segmentation and will be helpful to the following works.

4.3 The Semantic Vectorization Model of Fault Description

The Vectorization of Fault Description. Vectorized modeling is an important method of text representation in natural language processing. The descriptions can be represented as a series of vectors that express the semantics of the text. In this paper, we choose BOW model to vectorize vehicle fault descriptions. The principle of the BOW model is to count the frequency of each word in fault case database, assign a different word id to each word and create a dictionary with the format of "word id, word frequency". After that, word segmentation results are expressed as a sparse vector set according to the dictionary. Each different word is represented as a dimension in the vector space, each fault description corresponds to a unique vector. The vectorized representation of fault description segmentation results is a necessary step for TF-IDF weighting, latent semantic analysis and semantic similarity computing.

Weighting Model with TF-IDF. TF-IDF algorithm is a calculation method based on word statistics information, always being used to estimate word's importance in a collection of documents [10, 11]. TF algorithm counts the frequency of every word which belongs to current document, the more times a word appears in this document, the more important it is. IDF algorithm counts how many documents a word appeared in, if a word appears fewer in the collection of documents, its ability of distinguish between documents is stronger. Depend on TF-IDF algorithm, we can calculate the statistical weight of each word in the vehicle case database. Using the TF-IDF results to weight each description vector, giving the vectors corresponding statistical weight. Different from traditional TF-IDF weighting method, we considered the influence of POS during the process of weighting. In this research, we consider nouns are more representative to the fault description. We endowed nouns a higher weight by POS tagging to make the selection of key words more reasonable.

The Extraction of Semantic Topics. In the fault description corpus, the semantic information will not always appear explicitly. Only relying on the statistical information to select keywords can not guarantee the selected keywords can accurately reflect the semantics of fault description. Therefore, we use the LSI model to discover the latent semantic information in the text which related to the topics. The mapping relationship of topic model is shown in Fig. 3. From the mapping relationship of topic model, we can see clearly that each description corresponds to one or more topics, and each topic also has a corresponding word distribution.

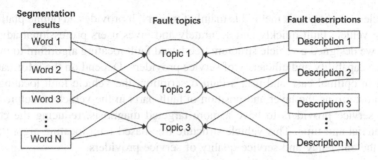

Fig. 3. Mapping relationships of topic model.

After the training steps of LSI model, each word and fault description can be represented as a correspond point in the vector space constitute with k topics [11, 12]. Then, by calculating the similarity between each word and the fault description, we can obtain N words with the highest similarity as the keywords of fault description. Compared with the traditional space vector model, LSI maps words and descriptions into a low-dimensional semantic space through singular value decomposition [13–15]. LSI model can discover the latent semantic information of fault descriptions, further have an essential expression of vehicle fault descriptions.

4.4 Similarity Computing Based on Fault Case Database

After training the fault descriptions in vehicle fault case database by TF-IDF and LSI algorithm, we obtained a semantic vectorization model which includes both statistical distribution information and implicit semantic information. The vehicle fault diagnosis system will use user's fault description as input to compute the semantic similarity. For user's fault description, it also first through the pre-processing and vectorized into vector space. Then weight user's description by TF-IDF to obtain a description vector with statistic information. Last, put user's description vector into the semantic vectorization model we trained to compute the semantic similarity with each fault case's description. The semantic similarity is calculated by comparing the cosine distance between user's description vector and the fault description vector in each fault case. When the angle between two vectors is smaller, the corresponding cosine value is smaller, which means their similarity in vector space is higher [16, 17]. The semantic vector space's dimensions are determined by the quantity of dictionary after pre-processing. The computing of semantic similarity is to calculate the cosine distance of user's description vector and all case fault description vectors. After the semantic similarity computing, fault diagnosis system will return the result in the format "case id, similarity score" according to the similarity score. According to the searching workflow of database, the vehicle diagnosis system will return corresponding fault causes, fault parts and solutions depend on the case id.

5 Vehicle Structure Tree-Based Fault Locating

The vehicle fault diagnosis method is mainly for users, it provides an online platform to determine vehicle fault quickly and accurately and gives users professional advice. In addition, we designed a vehicle structure tree-based fault locating algorithm to improve the service capability and efficiency of service providers. Depend on the information of fault part in optimal case and user's fault description, the vehicle fault locating algorithm can determine the layer and position of fault parts in the vehicle structure tree. It can help service providers to have a more targeted diagnosis, reducing the costs of labor, material and time. This vehicle structure tree-based fault locating algorithm can improve the efficiency and service quality of service providers.

5.1 The Construction of Vehicle Structure Tree

The theoretical basis of the vehicle structure tree in this paper is mainly based on the JB-SJGF-0003-2011 standard auto parts numbering rules issued by the Fuzhou Automobile Group in 2011. The Fig. 4 is a hierarchical representation of the vehicle structural tree we built.

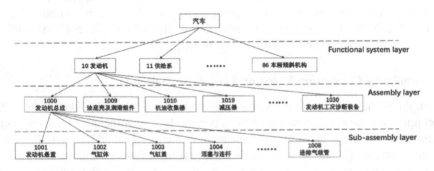

Fig. 4. Hierarchical representation of vehicle structural tree.

According to this standard, the part group number indicates the classification of each functional system of the car. It contains all the functional systems which requires to constitute an integrated vehicle. The third layer is the assembly, which is a combination of several parts and sub-assemblies, or the combination of a certain functional form which has the assembly decomposition characteristics. The last layer is the sub-assembly, which is generally processed by two or more parts through assembly processes or combinations. We consider that a four-layer structure tree is enough to determine the fault location to an acceptable range and ensure a better accuracy and universality. We numbered all functional systems in second layer, using a two-digit group number considered to the number of actual functional systems. The group number is also used as the prefix number of the assembly and sub-assembly of each functional system. The classification standard of Fuzhou Automobile Group divides the vehicle into 86 functional systems. The third layer is the assembly which contained in each functional system, and each assembly is numbered 0 and 9 as the last digit. The reason of using 0 and 9 as the last digit is to distinguish the left and right assemblies, 0 refers to the left assembly, 9 refers to the right assembly. The fourth layer is the subdivision of each assembly, including the sub-assembly and parts, the last digit is generally represented by 1–8. When the assembly distinguishes the left and right, parts with the odd last digit belongs to the left and the even last digit belongs to the right. This numbering rule makes it easier to distinguish between different functional systems and unlikely to lead confusion. After locating to the child nodes in structure tree, the upper layer information can be quickly found according to the relationship between the hierarchical layers of vehicle structure tree.

5.2   Vehicle Structure Tree-Based Fault Locating Algorithm

The vehicle structure tree-based fault locating algorithm contains three methods which are direct searching, similarity computing and semantic reasoning. According to the condition of input, the algorithm will adopt proper methods to ensure the accuracy and success rate of locating result. The logical process of fault locating algorithm is shown in Fig. 5.

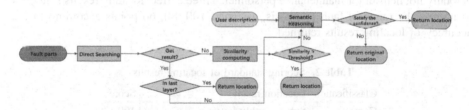

Fig. 5. The working process of fault locating algorithm.

The direct searching method is the pre-step and the most direct and quick fault locating method. The direct searching according to the fault parts information in optimal case to search the corresponding location in vehicle structure tree. If the fault position is at the last layer of structure tree, it can be considered as a successful locating. All the information belongs to this branch will be returned according to the vehicle structure tree. When fault position is at the functional system layer or the assembly layer, it is necessary to use the user description corpus to continue fault locating. The similarity computing method uses the semantic vectorization model to compute the semantic similarity between fault parts information and each element in vehicle structure tree to find the parts with similar semantics. Combined with the semantic vectorization model, compute the cosine distance of fault parts and elements in structure tree. Set a threshold to keep the result with high similarity and return the corresponding branches to the maintenance personnel of service providers. The reason for not further screening is the results computed by semantic similarity may deviate from the actual results. The actual fault location is not necessarily the result with the highest similarity. It still relies on maintenance personnel to have an examine of all possible similarity results, but the range of fault diagnosis is effectively being narrowed.

As a supplement to two methods above, the semantic reasoning method makes a further reasoning relies on user's fault description in addition. User's fault description contains potential fault location, fault phenomenon and many useful information. The semantic reasoning method classified and extracted the useful information, and jointly determine the location of the fault with fault parts information. The extraction of potential location information relies on the vehicle proper nouns dictionary to filter the useful information in user's fault description. It also needs to set a confidence interval to ensure the reliability of the semantic reasoning results. If the results are out of confidence interval it means that the semantic reasoning is not established. In this case, the original results are maintained, waiting maintenance personnel to diagnosis the exact fault location.

5.3   The Evaluation Standard of Locating Result

The vehicle fault locating algorithm is designed to help maintenance personnel narrow the range of fault diagnosis, improve efficiency and service quality of service providers. It requires a standard to evaluate the positioning results to measure the accuracy of fault locating algorithm. The score of each condition are shown in Table 2. According to the accuracy of the locating results, two wrong locating results are both 0 points, because it is totally not helpful for maintenance personnel. Three correct locating results is helpful to narrow the range of fault diagnosis, assigned 100, 80, 60 points according to the accuracy of locating results returned.

Table 2.   Scoring standard of locating results.

Classification	Location result	Score
Correct	Sub-assembly layer	100
	Assembly layer	80
	Functional system layer	60
Wrong	Other functional system branches	0
	Not in structure tree	0

6   The Verification Experiments

6.1   The Verification Experiment of Vehicle Fault Diagnosis

To ensure the test set is more representative and comprehensive, we choose different types of fault as many as possible in test set. We collected 30 actual fault cases on the net-work for comparison with the results of the fault diagnosis system. The result of the comparison experiment is shown in Table 3. For this test set, the correct rate of vehicle fault diagnosis system is 76.7% when only using the default dictionary. After we loaded the vehicle proper nouns dictionary, the accuracy of diagnosis system has up to

86.7%. The result of this experiment proves that the vehicle proper nouns dictionary is helpful for the vehicle fault diagnosis system.

Table 3. Fault diagnosing results.

Experiment condition	Correct result	Wrong result	Accuracy
Default dictionary only	23	7	76.7%
Loaded customized dictionary	26	4	86.7%

6.2 The Verification Experiment of Vehicle Fault Locating

According to the workflow of vehicle fault diagnosis system, when the fault diagnosis is executed, the system will continue fault locating depend on the case returned by diagnosis algorithm. Based on the fault diagnosis experiment, we continue to verify the results of vehicle fault locating. As shown in Table 4, we calculated the score according to the scoring standard in Sect. 5. When only using the default dictionary, the average score of all locating results is 70.9, and the correct rate is 78.1%. After we loaded the vehicle proper nouns dictionary, the average score of all locating results is up to 73.8, and the correct rate also up to 81.8%.

Table 4. Fault locating results.

Experiment condition	100	80	60	0	Average	Accuracy
Default dictionary only	30	6	7	12	70.9	78.1%
Loaded customized dictionary	31	6	8	10	73.8	81.8%

Through the result of contrast experiments, the vehicle proper nouns dictionary improves the accuracy of fault locating. By analyzing the wrong results of fault locating, we found that the reason is mainly because the fault parts are more detailed than the sub-assembly layer in structure tree. It requires maintenance personnel to diagnosis and locate according to the actual fault circumstance. Through the average points and correct rate of test set, the fault locating algorithm can help maintenance personnel to narrow the range of fault and improve the vehicle service ability and efficiency of service providers.

7 Supporting Multi-value-chain Collaboration

Traditional diagnosis methods are applicable to different conditions, but all need a technician during the fault diagnosis process. It leads to a result that regardless of the existence or severity of fault it always consumes the time of maintenance personnel. The vehicle fault diagnosis system we designed in this paper can be used as a pre-diagnosing step before the traditional manual fault diagnosis. It can return a preliminary diagnosis result to maintenance personnel, according to that technicians can have a manual diagnosis and filter out those have no need to repair. For the faults need to be repaired, the fault locating method can narrow the range of fault, helps technicians to have a targeted vehicle maintenance. The utilization of vehicle fault diagnosis system can effectively reduce the time technician spent in fault diagnosis and save the waiting time of the consumer in maintenance shop. This vehicle fault diagnosis method based on fault cases and machine learning algorithm can provide more useful information to technicians, also promote the service quality and efficiency of vehicle maintenance. In our research, the vehicle fault diagnosis system not only support the vehicle service chain but also build a collaboration with vehicle parts chain.

After the diagnosis process, the system records the information of corresponding case, including the fault parts information. The system can record the cumulative number of each fault parts and connect with the database of service provider to query the quantity of each parts. When the requirement of a certain part rising, the system will examine the stock and provide a warning of the risk of insufficient inventory of this certain part. This system can also provide an automation stock management for service providers, effectively optimize the allocation of inventory. For parts suppliers on the vehicle parts chain, it is also helpful to adjust the production according to the requirements in service chain, reducing the production of redundant parts.

8 Conclusion and Prospect

In this paper, we designed and implemented a semantics-based vehicle fault diagnosis method, the system has achieved a relatively high-accuracy in both fault diagnosing and locating. It has the potential and valuable for practical application in vehicle maintenance market. For consumers, it provides a convenient online diagnosis platform, which can return the corresponding fault causes and maintenance suggestions back. For vehicle service providers, this system will save the time cost on fault diagnosis, narrow the scope of fault diagnosis and improve their service capability and quality. For the vehicle maintenance industry, this fault diagnosing system establishes a positive collaboration mechanism between the service chain and parts chain, supports the collaboration of multi-value-chain in the field of vehicle maintenance.

The vehicle fault diagnosing system still has some limitations and deficiencies. First, the scale of the current vehicle fault case database is relatively small. Second, it still has space for further improvement on the choice of corpus vectorization method. In the following works, we will continue to optimize the accuracy and efficiency of vehicle fault diagnosing method and apply it to the actual vehicle maintenance business to verify its reliability and stability.

Acknowledgment. The authors would like to thank the anonymous referees for their valuable comments and helpful suggestions. This paper is supported by The National Key Research and Development Program of China (2017YFB1400902).

References

1. Shumate, D.M.: Vehicle diagnosis system and method, uS Patent 7,590,476, 15 Sept 2009
2. Narasimhan, S., Biswas, G.: Model-based diagnosis of hybrid systems. IEEE Trans. Syst. Man Cybern.-Part A: Syst. Hum. **37**(3), 348–361 (2007)
3. Sun, J.: Jieba'Chinese word segmentation tool (2012)
4. Mikolov, T., Chen, K., Corrado, G., Dean, J.: Efficient estimation of word representations in vector space, arXiv preprint arXiv:1301.3781 (2013)
5. Maas, A.L., Daly, R.E., Pham, P.T., Huang, D., Ng, A.Y., Potts, C.: Learning word vectors for sentiment analysis. In: Proceedings of the 49th Annual Meeting of the Association for Computational Linguistics: Human Language Technologies-volume 1, pp. 142–150. Association for Computational Linguistics (2011)
6. Erk, K.: Vector space models of word meaning and phrase meaning: a survey. Lang. Linguist. Compass **6**(10), 635–653 (2012)
7. Roy, R., Whelan, R.: Successful recycling through value-chain collaboration. Long Range Plan. **25**(4), 62–71 (1992)
8. Normann, R., Ramirez, R.: From value chain to value constellation: designing interactive strategy. Harvard Bus. Rev. **71**(4), 65–77 (1993)
9. Dania, W.A.P., Xing, K., Amer, Y.: Collaboration behavioural factors for sustainable agri-food supply chains: a systematic review. J. Clean. Prod. **186**, 851–864 (2018)
10. Ramos, J., et al.: Using tf-idf to determine word relevance in document queries. In: Proceedings of the First Instructional Conference on Machine Learning, vol. 242, pp. 133–142, Piscataway (2003)
11. Zhang, W., Yoshida, T., Tang, X.: A comparative study of tf* idf, lsi and multiwords for text classification. Expert Syst. Appl. **38**(3), 2758–2765 (2011)
12. Dumais, S.T., et al.: Latent semantic indexing (LSI) and TREC-2, p. 105. NIST Special Publication Sp (1994)
13. Hofmann, T.: Probabilistic latent semantic indexing. In: ACM SIGIR Forum, vol. 51, no. 2, pp. 211–218. ACM (2017)
14. Landauer, T.K., Foltz, P.W., Laham, D.: An introduction to latent semantic analysis. Discourse Process. **25**(2–3), 259–284 (1998)
15. Hofmann, T.: Unsupervised learning by probabilistic latent semantic analysis. Mach. Learn. **42**(1–2), 177–196 (2001)
16. Tata, S., Patel, J.M.: Estimating the selectivity of tf-idf based cosine similarity predicates. ACM SIGMOD Rec. **36**(2), 7–12 (2007)
17. Dehak, N., Dehak, R., Glass, J.R., Reynolds, D.A., Kenny, P.: Cosine similarity scoring without score normalization techniques. In: Odyssey, p. 15 (2010)

Reinforcement Learning Based Signal Quality Aware Handover Scheme for LEO Satellite Communication Networks

Menting Chen[1(✉)], Yong Zhang[1], YingLei Teng[1], Baoling Liu[1], and Lili Zhang[2]

[1] School of Electronic Engineering,
Beijing University of Posts and Telecommunications,
Beijing, People's Republic of China
{mengting.chen,yongzhang,lilytengtt,blliu}@bupt.edu.cn
[2] Shuangchuang Center (Laboratory),
State Grid Shenwang Location Based Service Co., Ltd.,
Beijing, People's Republic of China
lily3286000@126.com

Abstract. With the increasing of Low Earth Orbit (LEO) satellites emission, utilizing existing LEO satellite network systems have lower CAPEX/OPEX than deploying fixed terrestrial network systems in the remote area. Due to the high mobility of LEO satellites, mobility management mechanisms such as handover schemes are the key issue should be settled on leveraging the merits of LEO satellites telecommunications (e.g. lower propagation delay than GEO satellites). In traditional handover schemes, choosing which one is the next-hop satellite for one user is only determined by evaluating some specific criteria in the current state, not guaranteeing long-term and global optimization. To solve this problem, we use the cumulative signal quality that involves the remaining service time and signal quality and propose a Q-Learning based handover scheme. The simulation results show that the proposed scheme improves the overall signal quality and reduce the average handover number of users compared with other handover schemes.

Keywords: LEO satellite network · Handover · Q-Learning · Signal quality

1 Introduction

A satellite network system is a viable option to cover remote areas where there are not adequate available terrestrial communication networks [13], an integral part of the communication system to achieve truly ubiquitous coverage. The LEO satellite as a typical small satellite is becoming fashionable and changing the

This work is supported by National Key R&D Program of China No. 2018YFB1201500, National Natural Science Foundation of China under Grant No. 61771072, and Beijing Natural Science Foundation under granted No. L171011.

D. Milošević et al. (Eds.): HCC 2019, LNCS 11956, pp. 44–55, 2019.
https://doi.org/10.1007/978-3-030-37429-7_5

economics of space [18]. LEO satellites introduce lower end-to-end delay, require lower power and exert more efficient spectrum allocation than the geostationary orbit satellites, making them suitable for future personal telecommunication [16]. Even for fixed users, handover in non-geostationary satellite communication system will be continuous.

To improve the QoS of users, diverse handover schemes have been extensively developed in link-layer and network-layer. Satellite handover occurs in the link layer when the existing connection of one end user with the satellite transfers to the other satellite [6,19]. Due to the high mobility of the LEO satellite system, the handover happens frequently in many circumstances, causing call interruption, thus influencing directly the quality of experience of the users.

For satellite handover schemes, most of the works only considered a specific handover criterion (e.g. remaining service time [8], number of free channels [7], elevation angle [10]) or highly relied on geometric information, lacking an overall solution [20]. It true that there are some schemes combined with two criteria. For instance, Zhao et al. take the user position and the signal strength into consideration to reduce the call termination probability [24]. Besides, Wu et al. firstly presented a graph-based handover framework for LEO satellite network [20], integrating all kinds of satellite networks into one topology graph, where the process for the end-user to switch between different serving satellite during its call period can be considered as to find a way among these consecutive covering satellites. However, the handover strategy of the proposed scheme is the same as that of the traditional largest service time scheme. Among these schemes, single handover criteria leave the user shortsighted and make the network is unable to meet the complex QoS requirements of users.

Reinforcement Learning (RL) method has been used in many problems of network communication [22]. All of the problems (e.g. base station on-off switching strategies [11], routing [1], adaptive tracking control [12] and power allocation [23]) can be formulated as Markov Decision Processes (MDPs). Specifically, Wang et al. first formulated the traffic variations as an MDP, then used actor-critic algorithm (a typical RL algorithm) to improve energy efficiency in radio access networks [11]. In [1], Al-Rawi et al. introduce that RL can address routing challenges by gathering information, learning, and making routing decision efficiently. Besides, Liu et al. propose an online RL method to achieve adaptive tracking control for nonlinear discrete-time MIMO systems [12], and Zhang et al. maximize the overall capacity of a multi-cell network by deep RL method [23].

In horizontal handover, where users switch in the same network layer such as satellite handover in one LEO satellite network system, the signal quality is a key criterion. Moreover, the remaining service time of satellites contributes a lot to satellite handover. The optimal trade-off between the two criteria needs to be achieved. For example, if a user always chooses the satellite of best signal quality but he is about to leave the satellite's coverage area, the user has to switch to another satellite soon. In this way, the ping pong switch occurs, which is intolerable in satellite networks due to a long propagation delay. Therefore, how to lowest the ping pong switching rate by balance the signal quality and remaining service time is our most concerned.

To best of our knowledge, the satellite handover problem for optimal cumulative signal quality is untouched. Our contributions are summarized as follows: (1) we model the signal quality by Ornstein-Uhlenbeck process, then first propose a criterion that combines signal quality and remaining serving time; (2) we propose a handover scheme based on reinforcement learning method to maximize the overall signal quality and minimum the ping pong switching rate in the long term.

2   System Model

2.1   LEO Satellite Network Model

Consider an LEO satellite constellation with a specific topology structure, which operated in slotted time $t \in \{0, 1, 2, ..., T\}$, containing N satellites and M users. The speed of the mobile terminals, which is much lower than the speed of the LEO satellite (about 25000 km/h relative to the earth rotation), can be ignored.

2.2   Satellite Handover Process

A satellite handover is divided into three steps: handover information collection, handover decision-making, and handover execution.

Handover Information Collection. Assume the user connected with LEO satellite can easily obtain its exact position by using the existing Global Position System (GPS) infrastructure (or using the other ways), and also the covering satellites in the future period by predicting the motion of them. The prediction can be made by the predicting method [2,3], or by a centralized controller in Software-Defined Network (SDN) and Network Function Visualization (NFV) architecture. Under such conditions, users can obtain the Received Signal Strength (RSS), remaining service time and user elevation of the connected satellites.

When the RSS of a user's connection with a satellite that may be switched is less than a certain hysteresis threshold RSS_{min} for a period of time, the satellite is placed into the user's candidate satellite sequence and the switching decision is triggered. This threshold needs to be set properly: if the hysteresis threshold is too large, it will lead to frequent switching, causing large switching delay; if it is too small, the satellite will be blocked due to stranded users' service.

Handover Decision-Making. In view of uneven user distribution restricted by the imbalance of regional development and the vast ocean area, certain restrictions should be made on the fairness of handover, i.e. the channel usage $q(t)$ at any time should be less than a threshold:

$$q_n(t) \leq \xi q_{max} \tag{1}$$

q_{max} is the channel number of the satellite, and ξ is the maximum satellite channel utilization rate.

The satellite for next hop from the candidate satellite sequence should be determined by the collected network information. Due to the limited satellite coverage area, the remaining service time directly affects the number of users' handover during a service period. For example, if the satellite time remaining for each hop is very short during a service period, users may be in the process of continuous handover and ping-pong handover is more likely to occur. Therefore, we should consider the compromise between channel quality and remaining time.

In the actual scenario, channel changes are more dynamically. We consider a mean-reverting Ornstein-Uhlenbeck process [4] as the instantaneous dynamics of non-stationary channel model, where a time-varying additive Gaussian channel model is given by $|h_{i,t}|^2$. Thus, the dynamics of the channel are given by:

$$d|h_{i,t}| = \theta(\mu_h - h_{i,t})dt + \sigma_h d\mathcal{B}_{i,t} \tag{2}$$

where $\mu_h > 0$, $\sigma_h > 0$, and $\mathcal{B}_{i,t}$ is a standard Brownian motion. μ_h refers to the geocentric angle α between users and satellites, which related to the elevation angle of users. Since the larger the elevation angle is, the smaller the geocentric angle is, μ_h can be defined as:

$$\mu_h = h_0 \cos\alpha - \cos\alpha_0 \tag{3}$$

where h_0 represents the channel gain, α_0 is the maximum geocentric angle in the coverage area of one satellite. Equation 3 means that the smaller the geocentric angle is, the better the channel quality is.

Handover Execution. Once the next satellite to be handed over is determined, the user will start the soft handover process, and many signaling exchanges will take place during the handover process. Each signaling exchange may fail. The handover failure rate of each user should be less than a threshold to avoid frequent handover:

$$1 - (1 - p)^{n_m} \leq \mu \tag{4}$$

where p is the failure probability of one handover, and n_m is the total number of user m's switches in his communication period.

This possibility of failure is also related to the channel quality, which can be defined by SNR as shown in Eq. 5.

$$\gamma_{i,t} = \frac{p_0|h_{i,t}|^2}{n_0} \tag{5}$$

Then, integrating SIR representing cumulative signal quality can be obtained:

$$\int \gamma_{i,t}dt = \frac{p_0}{n_0} \int |h_{i,t}|^2 dt \tag{6}$$

Therefore, the utility function of one handover (from satellite i to satellite j) can be defined as follows:

$$u_m(t) = \int_t^{t_j} \gamma_{j,t} \mathrm{d}t \tag{7}$$

where t_j are the last coverage times of satellites j.

2.3 Problem Formulation

To sum up, considering the whole process of handover, the overall goal is determined as follows:

$$\begin{aligned}
&\min_{\Omega} \liminf_{T \to \infty} \tfrac{1}{T}\mathbb{E}[\sum_{m=1}^{M} u_m(t)] \\
&\text{s.t.}\quad q_n(t) \le \xi q_{max}, \forall n, n = 1, ..., N \\
&\qquad 1 - (1-p)^{n_m} \le \mu, \forall m, m = 1, ..., M
\end{aligned} \tag{8}$$

This problem is a stochastic programming problem that is control dependent state evolution where the control action (handover) will influence the state process. Also, in the dynamic scenario of LEO satellite handover, the transition probability is unknown and the state space is infinite. In this way, the problem is NP-Hard, difficult to find the best solution, but we can try to make a policy to minimize the effective transmission rate of all users. Also, the handovers of users are independent as a Markov decision process. Thus, a model-free learning technique such as QL is needed to find a policy Ω to better adapt to the dynamic scenario and gain long-term control.

3 Reinforcement Learning Handover Scheme

3.1 Decision-Making Process in Q-Learning

QL is a technique of reinforcement learning where the agent learns to take actions to maximum the cumulative reward (Eq. 9) by trial-and-error interactions with its environment:

$$R_t = \sum_{k=0}^{\infty} \gamma^k r_{t+k+1} \tag{9}$$

where $0 \le \gamma \le 1$ indicates the weight of the experience value and r is the numerical reward obtained at each optimization epoch which comprised by one iteration. Basically, QL has two steps: policy evaluation and policy improvement. It evaluates the policy by calculating value function (the expected value of the return at state S accumulated by action S in a limited period of time):

$$Q_\pi = \mathbb{E}_\pi \left[R_t | s_t = s, s_t = a \right], \tag{10}$$

it then improve the policy by updating the value in the action's corresponding index in the Q matrix. In this model-free approach, the premise of fully evaluating the policy value function is that each state can be accessed, so an exploration strategy called $\epsilon - greedy$ is adopted to update Q value in step of policy improvement. For an iteration for the user m, we have:

$$Q_{t+1}(s, a) = (1 - \alpha)Q_t(s, a) + \alpha Q_t(s, a') \tag{11}$$

where the higher the learning rate $\alpha(0 \leq \alpha \leq 1)$ is, the less the previous training results will be retained.

3.2 Maximum Cumulative Signal Quality Handover Scheme (MCSQ)

Using the QL framework discussed above, the agent, state $\mathcal{S}$, action $\mathcal{A}(s)$ and reward $r : \mathcal{S} \times \mathcal{A}(s) \times \mathcal{S} \rightarrow \mathbb{R}$ can be designed as follows:

- **Agent**: the handover controller.
- **States**: $s = (s^{(1)}, s^{(2)}, ..., s^{(4)}) \in \mathcal{S} := \hat{\mathcal{U}} \times \mathcal{I} \times \mathcal{J} \times \mathcal{H}$. $\hat{\mathcal{U}}$ denotes the discretized space of the user position in slot t, also following the initial consideration of QL, where the space of states is a set of discretized value. $\hat{\mathcal{U}}$ is defined as the area discretized by a grid g_1, g_2 on longitude and latitude correspondingly, which is represented by

$$\hat{\mathcal{U}} := \{(\lfloor u_1/g_1 \rfloor, \lfloor u_2/g_2 \rfloor) | (u_1, u_2) \in \mathcal{U}\} \tag{12}$$

  where $\lfloor \cdot \rfloor : \mathbb{R} \rightarrow \mathbb{N}$ is the floor function. $\mathcal{I} = \{i_0, ..., i_m\}$ represents the connecting satellites set, where i_m represents the number of user m's connecting satellite. $\mathcal{J} = \{J_0, ..., J_m\}$ indicates the set of adjacent satellites that can exert handover in next hop, where J_m represents the set of user m's adjacent satellites. $\mathcal{H} := \{0, 1\}$ indicates whether in each decision epoch the handover is being processed or not.
- **Actions**: $a \in \mathcal{A}(s)$ is the number of satellite selected in state $\mathcal{S}$. The set of actions is defined as

$$\mathcal{A}(s) = \begin{cases} \mathcal{I}, \text{if } \mathcal{H} = 0 \\ \mathcal{J}', \text{Otherwise} \end{cases} \tag{13}$$

  which means that the user only can connect with one of the satellites when the handover is not processed, and otherwise the user can choose an adjacent satellite from the set $\{s^{(3)}\}$ in the time slot of handover. $\mathcal{J}'$ represents the selected satellites.
- **Reward**: $r(s', a, s)$ is defined as the cumulative signal quality of one link defined before, which is influenced by the RSS, the remaining service time:

$$r(s', a, s) = \begin{cases} u_m(t), \text{if } \mathcal{H} = 1 \\ 0, \text{Otherwise} \end{cases} \tag{14}$$

where $r : \mathbb{R} \rightarrow \mathbb{R}$ is for controller to choose the optimal handover satellites of maximum cumulative quality of users received signal.

After many iterations, the Q matrix tends to converge and be stable, since almost every state is accessed at least once.

4 Simulation Results

4.1 Simulation Setup

The simulation referring to the Globalstar system runs using python to conduct the comparison. Satellites and users are randomly sprinkled in the Region $\mathcal{U}$ by a Poisson Point Process (PPP). The maximum geocentric degree α_0 of satellite coverage is determined by the height of satellite orbit and the user-visible elevation angle α_e:

$$\alpha_0 = \sin(\arccos(\frac{R_e}{R_e + h})) \cos \alpha_e - \alpha_e \tag{15}$$

Consider users moving following a straight certain route, and assume the relative velocity between users and satellites is 60 km/s for better simulation effect (about 6 km/s in reality).

Table 1 summarizes the simulation parameters of the network model.

Table 1. Simulation parameters.

Parameter	Value
Height of the satellite orbit (h)	1414 km
User visible elevation angle (α_e)	10°
Radius of satellite coverage (R_0)	2670 km
Radius of the Earth (R_e)	6378 km
Relative velocity	60 km/s
Duration of user service period	120 s
Region $\mathcal{U}$	6000 km×6000 km
Number of satellites	25
Satellite load threshold (ξp_{max})	5
Failure probability of one handover (p)	99%
RSS threshold (RSS_{min})	0.5×10^{-8} w
Hysteresis threshold	3 s
Transmit power of satellite (p_0)	50 w
Channel gain (h_0)	−1.85 dB
θ in the dynamics of channel	1
σ_h in the dynamics of channel	2.5×10^{-5}
Learning rate (α)	0.4
Greedy policy (ϵ)	0.6
Experience rate (γ)	0.9
Granularity of the location (λ)	15
Number of iterations	120

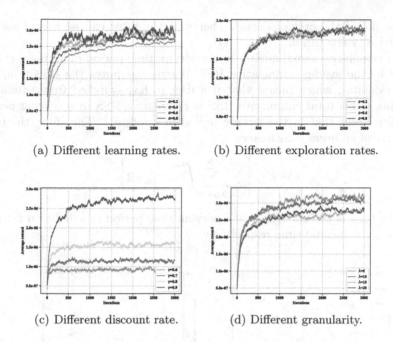

(a) Different learning rates.   (b) Different exploration rates.

(c) Different discount rate.   (d) Different granularity.

Fig. 1. Performance comparison of parameters in the MCSQ scheme.

4.2   Simulation Results

Figure 1 illustrates the convergence property and the parameter evaluation of the MCSQ algorithm. According to the impact of the learning rate can be shown in Fig. 1(a), we choose $\alpha = 0.4$ as our learning rate of all QL methods because the average reward performs more stable. Similarly, we choose $\epsilon = 0.6$ as the exploration rate because of the performance comparison in Fig. 1(b) and $\gamma = 0.9$ as the discount rate according to Fig. 1(c). The granularity of the location λ in the sate $\mathcal{U}$ is determined by grid g_1, g_2. Suppose $g_1 = g_2 = 2000/\lambda$, so the impact of λ is shown in Fig. 1(d). It can be seen that when $\lambda = 5$, average reward remains at a higher level because of more user in one grid region and cumulated Q in the same position of Q table. In consideration of the area of the grid region, we choose $\lambda = 15$ as the granularity of location.

It is worth noting that, in all simulations, the signal quality will be generated randomly according to Eq. 2 in every iteration, and the initial location of satellites and users and relative movement relationship will remain the same in one epoch.

Although the Q table will be stable after about over thousands of iterations, the complexity of the proposed algorithm is low because this paper is dedicated to giving an optimal handover strategy for an initial status of satellites and users. Once we obtain the optimal policy Ω, it acts as a look-up table. Besides, the dynamic of Earth orbit is periodic, several time periods of satellite running can construct a cycle, wherein the location of satellites will be the same as that in the same slot of the previous cycle. Therefore, the table can be reused and be calculated very few times.

Here, cumulative signal quality in the service time period of all users is represented by the average cumulative RSS. It only computes the RSS from the serving satellites, which means that if a user m has switched from satellite i to satellite j, his total cumulative RSS is cumulative RSS in a serving period of satellite i plus that in the serving period of satellite j. Therefore, the total cumulative of all users is as follows.

$$c = \mathbb{E}_M \left[\sum_{m=0}^{M} \sum_{n=0}^{N} \int_{t_{nms}}^{t_{nme}} \gamma_{n,t} \mathrm{d}t \right] \tag{16}$$

where the t_{nms} is the start time of a serving time period of satellite n for user m, and t_{nme} is the end time respectively.

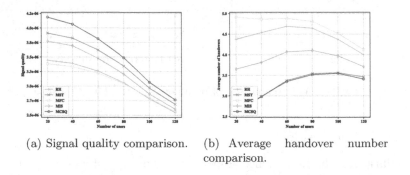

(a) Signal quality comparison.   (b) Average   handover   number comparison.

Fig. 2. Performance comparison of handover schemes.

Figure 2 shows the performance comparison of the proposed algorithm and all baseline algorithms. In this comparison, 100 independent tests are conducted, where the topology of satellites and users is renewed in every test. The baseline algorithms are several traditional schemes:

- **Maximum number of free channels scheme (MFC)** [8]: In this scheme, users will choose the satellite that reserves most free channels. In this way, it tends to achieve the highest fairness index between satellites and uniform distribution of the telecommunication traffic in the LEO network [14]. Therefore, it will avoid overload satellites, but for users, they should find the satellite with more free channels regardless of location or signal quality, leading to more number of handovers.
- **Maximum service time scheme (MST)** [7]: According to this criterion, the user will be served by the satellite that offers the maximum remaining service time [14]. It will get the lowest number of handovers and ping-pong

switching rate, but the user will reluctant to change serving satellite even know other satellite may offer more free channels. In an LEO satellite network with a certain constellation, the long coverage time means that in this serving period, the user will get bigger maximum communication elevation angle [17], so the overall signal quality will be higher than other baseline algorithms.

- **Maximum instantaneous signal strength scheme (MIS)** [10]: This criterion is almost equivalent to the maximum elevation angle, because, for satellite communication, small elevation angle between the mobile terminal and the satellite leads to frequent shadowing and blockage events for the signal due to trees, buildings, hills, etc. [5]. Thus, it will avoid link failures and get relative high signal quality, but due to the dynamic characteristics of the channel, it is only able to guarantee the instantaneous signal quality. Once the channel changes rapidly, the handover decision will be unstable.

- **Random handover scheme (RH)**: According to this criterion, the user will randomly choose an available satellite in the slot of handover decision-making.

It can be seen that MCSQ, after iterations, gets the greatest signal quality and obtains almost the same low average handover number as MST. Since the cumulative signal quality combines the effect of elevation angle and remaining serving time, according to the criterion, the MCSQ combines the advantages of MIS scheme and MST scheme in data representation. Besides, the reinforcement learning framework makes the proposed scheme more telescopic in handover decision, so the MCSQ achieves the overall stable improvement in the whole satellite cycle.

5 Conclusion

In this paper, we have investigated a handover problem for LEO satellite networks, and we further formulated it as a stochastic optimization problem to get optimal signal quality of users. To solve this problem, we model the channel by the O-U process and introduce a criterion that combines signal quality and remaining serving time. Then, we propose a reinforcement learning handover strategy, which has shown that it has a good effect on improving the overall signal quality between satellites and users in a period, also greatly reduces the average handover number of users. In reality, the distribution of signal received cannot be predicted by a simple model, but the Q table after training could be a good reference when users are choosing the next hop satellites. Therefore, it can be generalized into a different network model easily, ensuring the flexibility of the algorithm. In further work, some novel reinforcement learning methods can be applied in this network model and handover procedure to improve the efficiency of this scheme.

References

1. Al-Rawi, H.A., Ng, M.A., Yau, K.L.A.: Application of reinforcement learning to routing in distributed wireless networks: a review. Artif. Intell. Rev. **43**(3), 381–416 (2015)
2. Ali, I., Al-Dhahir, N., Hershey, J.E.: Predicting the visibility of LEO satellites. IEEE Trans. Aerosp. Electron. Syst. **35**(4), 1183–1190 (1999)
3. Bertiger, B.R., Leopold, R.J., Peterson, K.M.: Method of predicting cell-to-cell hand-offs for a satellite cellular communications system, uS Patent 5,161,248, 3 Nov 1992
4. Charalambous, C.D., Menemenlis, N.: Stochastic models for short-term multipath fading channels: chi-square and Ornstein-Uhlenbeck processes. In: Proceedings of the 38th IEEE Conference on Decision and Control (Cat. No. 99CH36304), vol. 5, pp. 4959–4964. IEEE (1999)
5. Chini, P., Giambene, G., Kota, S.: A survey on mobile satellite systems. Int. J. Satell. Commun. Netw. **28**(1), 29–57 (2010)
6. Chowdhury, P.K., Atiquzzaman, M., Ivancic, W.: Handover schemes in satellite networks: state-of-the-art and future research directions. IEEE Commun. Surv. Tutorials **8**(4), 2–14 (2006)
7. Del Re, E., Fantacci, R., Giambene, G.: Efficient dynamic channel allocation techniques with handover queuing for mobile satellite networks. IEEE J. Sel. Areas Commun. **13**(2), 397–405 (1995)
8. Del Re, E., Fantacci, R., Giambene, G.: Handover queuing strategies with dynamic and fixed channel allocation techniques in low earth orbit mobile satellite systems. IEEE Trans. Commun. **47**(1), 89–102 (1999)
9. Duan, C., Feng, J., Chang, H., Song, B., Xu, Z.: A novel handover control strategy combined with multi-hop routing in LEO satellite networks. In: 2018 IEEE International Parallel and Distributed Processing Symposium Workshops (IPDPSW), pp. 845–851. IEEE (2018)
10. Gkizeli, M., Tafazolli, R., Evans, B.: Modeling handover in mobile satellite diversity based systems. In: Proceedings of the IEEE 54th Vehicular Technology Conference on VTC Fall 2001, (Cat. No. 01CH37211), vol. 1, pp. 131–135. IEEE (2001)
11. Li, R., Zhao, Z., Chen, X., Palicot, J., Zhang, H.: TACT: a transfer actor-critic learning framework for energy saving in cellular radio access networks. IEEE Trans. Wireless Commun. **13**(4), 2000–2011 (2014)
12. Liu, Y.J., Tang, L., Tong, S., Chen, C.P., Li, D.J.: Reinforcement learning design-based adaptive tracking control with less learning parameters for nonlinear discrete-time mimo systems. IEEE Trans. Neural Netw. Learn. Syst. **26**(1), 165–176 (2015)
13. Moura, J., Edwards, C.: Future trends and challenges for mobile and convergent networks. arXiv preprint arXiv:1601.06202 (2016)
14. Papapetrou, E., Karapantazis, S., Dimitriadis, G., Pavlidou, F.N.: Satellite handover techniques for LEO networks. Int. J. Satell. Commun. Netw. **22**(2), 231–245 (2004)
15. Papapetrou, E., Pavlidou, F.N.: QoS handover management in LEO/MEO satellite systems. Wireless Pers. Commun. **24**(2), 189–204 (2003)
16. Sadek, M., Aissa, S.: Personal satellite communication: technologies and challenges. IEEE Wirel. Commun. **19**(6), 28–35 (2012)
17. Seyedi, Y., Safavi, S.M.: On the analysis of random coverage time in mobile LEO satellite communications. IEEE Commun. Lett. **16**(5), 612–615 (2012)

18. Sweeting, M.N.: Modern small satellites-changing the economics of space. Proc. IEEE **106**(3), 343–361 (2018)
19. Taleb, T., Hadjadj-Aoul, Y., Ahmed, T.: Challenges, opportunities, and solutions for converged satellite and terrestrial networks. IEEE Wirel. Commun. **18**(1), 46–52 (2011)
20. Wu, Z., Jin, F., Luo, J., Fu, Y., Shan, J., Hu, G.: A graph-based satellite handover framework for LEO satellite communication networks. IEEE Commun. Lett. **20**(8), 1547–1550 (2016)
21. Yue, P.C., Qu, H., Zhao, J.H., Wang, M., Wang, K., Liu, X.: An inter satellite link handover management scheme based on link remaining time. In: 2016 2nd IEEE International Conference on Computer and Communications (ICCC), pp. 1799–1803. IEEE (2016)
22. Zhang, C., Patras, P., Haddadi, H.: Deep learning in mobile and wireless networking: a survey. arXiv preprint arXiv:1803.04311 (2018)
23. Zhang, Y., Kang, C., Ma, T., Teng, Y., Guo, D.: Power allocation in multi-cell networks using deep reinforcement learning. In: 2018 Technology Conference (VTC-Fall). IEEE (2018)
24. Zhao, W., Tafazolli, R., Evans, B.: Combined handover algorithm for dynamic satellite constellations. Electron. Lett. **32**(7), 622–624 (1996)

A Non-contact and Unconstrained Sleep Health Monitoring System

Zeyu Chen, Fuze Tian, Qinglin Zhao[✉], and Bin Hu[✉]

Gansu Provincial Key Laboratory of Wearable Computing,
School of Information Science and Engineering, Lanzhou University,
Lanzhou, China
{chenzy17, tianfz17, qlzhao, bh}@lzu.edu.cn

Abstract. Clinically, polysomnography (PSG) is used to assess sleep quality by monitoring various parameters, such as Electroencephalogram (EEG), electrocardiogram (ECG), Electrooculography (EOG), Electromyography (EMG), pulse, oxygen saturation, and respiratory rate. However, in order to assess these parameters, PSG requires a variety of sensors that must make direct contact with patients' bodies, which can affect patients' quality of sleep during testing. Thus, the use of PSG to assess sleep quality can yield invalid and inaccurate results. To address this gap, this paper proposes a sleep health monitoring system that has no restraints and does not interfere with sleep. This method collects ballistocardiogram (BCG) due to ejection by placing a piezoelectric film sensor under a sleeping cushion and evaluates three indicators: heart rate variability (HRV), respiration, and body movements. The ECG and BCG of 10 subjects were collected synchronously while the subjects were lying flat. Specifically, power-line interference was eliminated by adaptive digital filtering with a minimum mean square. A paired t-test revealed that there were no significant differences between BCG and standard ECG signals in the time-domain, frequency-domain, and nonlinear parameters of HRV. Respiration and body motion were extracted from the BCG in order to effectively monitoring of sleep apnea and nighttime bed-off times. Compared with other non-contact monitoring methods, such as acceleration sensors, coupling electrodes, Doppler radar and camera, the system presented in this paper is superior, as it has high signal quality, strong anti-interference ability, and low cost. Moreover, it does not interfere with normal sleep.

Keywords: Sleep · BCG · HRV · Non-contact

1 Introduction

Sleep quality affects humans' mental and physical states. In order to address and resolve sleep issues, polysomnography (PSG) is used to monitor sleep clinically [1]. By simultaneously collecting parameters such as EEG, EOG, ECG, EMG, respiration, and oxygen saturation, PSG can assess whether snoring severity or apnea occur during sleep or whether blood oxygen saturation is reduced (obstructive sleep). Moreover, PSG can determine apnea syndrome, abnormal heart rate, and sleep stage (awakening period, rapid eye movement period, non-rapid eye movement period). However, using

© Springer Nature Switzerland AG 2019
D. Milošević et al. (Eds.): HCC 2019, LNCS 11956, pp. 56–66, 2019.
https://doi.org/10.1007/978-3-030-37429-7_6

PSG to clinically measure sleep parameters can interfere with normal sleep, patients must fully cooperate with the requirements of the procedure to yield accurate results (such as wearing various complicated electrodes, adapting to the sleeping environment in advance, limited drinking water, not engaging in vigorous exercise, etc.).

Current research has been focusing on how to accurately monitor sleep parameters without using contact (i.e., non-contact methods) [2]. Nakajima [3], Deng [4], and other researchers used charge coupled device (CCD) cameras to analyze the sleeping position and respiration of subjects in bed in real time. However, the camera imposed a burden on the subjects and indirectly affected sleep. Currently, it is possible to accurately measure sleep-related indicators without disturbing normal sleep. Some relatively novel non-contact monitoring methods include coupled electrodes, Doppler radar, accelera-tion sensors, and piezoelectric films. Lee [5] and Wu [6] used two coupling electrodes instead of the traditional Ag/AgCl patch electrodes to determine ECG. The advantage of such a method is that the ECG signal is directly collected. However, the signal quality of the coupled electrode is easily affected by contact parts, clothing thickness, the environment, and other factors. Moreover, the electrode is difficult and expensive to manufacture. Zhincheng Yang used a 60 GHz mm-wave radar to identify sleep pos-tures and determine respiration and heart rate [7]. The advantage of this approach is that it can effectively monitor multiple people at the same time, even behind a wall. However, it is difficult to determine heart rate from the signal. Zhang [8] and Nuksawn [9] used an accelerometer attached to the chest to monitor sleep postures, heart rate, and respiration. An obvious advantage of this method is its low cost. However, in terms of signal quality, this method did not achieve good results. The accelerometer also needs to be worn. Piezoelectric film is designed to collect signals using the piezoelectric effect of a polymer material. When pressed by a human body, the piezoelectric film senses a slight change in the body's periodic pressure caused by cardiac ejection. The signal obtained by this method is the BCG, whose main features are H, I, J, K, L, M, and N bands [10]. By extracting the J wave from the BCG, the heart rate can be obtained, and the heart rate variability can be calculated [11–13]. Because respiration causes a slight change in the body's pressure, the BCG signal also contains a respiratory signal that overlaps with heart beats in frequency [14]. The respiratory signal can be separated by a Finite impulse response (FIR) band-pass filter of 0.1–0.5 Hz [15]. Jiang et al. [16] found that the quality of a BCG signal is related to the compression site of the human body by studying the quality of BCG signals (sitting, standing, lying flat). Liu et al. [17] studied the amplitude characteristics of heart-impact signals under different sleeping positions. Jose [18], Sadek [19] successfully obtained heart rate by placing a piezoelectric film on the patients' wrist and headrest. Compared with other non-contact monitoring methods, piezoelectric film does not disturb normal sleep, the signal quality is excellent, and heart rate characteristics are apparent.

In this paper, the piezoelectric film (PVDF) was placed under a sleeping cushion to obtain the BCG and body motion signal with obvious J-wave characteristics. The ECG and BCG of 10 subjects lying flat were collected synchronously in experiments. After the application of the adaptive filtering algorithm, the power frequency noise is effectively reduced. A high-order FIR band-pass filter was used to extract the respi-ratory signal, and the J-wave was identified by the peak identification method. The J-J intervals were analyzed instead of the R-R intervals, which was important in analyzing heart rate variability. Then, a paired t-test was used to analyze multiple indicators of heart rate variability. It was determined that there were no significant differences in analyzing HRV between BCG and ECG. Moreover, the method presented in this paper can accurately determine if apnea has occurred and accurately detect nighttime bed-off times.

2 Method

2.1 Signal Acquisition

The PVDF sensor is made of polymer material polyvinylidene difluoride (PVDF). When the α-type crystal structure of the PVDF film is rolled and stretched to become a β-type crystal structure, the PVDF film has a strong piezoelectric effect [20]. When the PVDF after stretching and polarization is subjected to a certain direction of pressure, the polarization plane of the material will generate a certain charge due to the piezo-electric effect.

The system converts the charge into a voltage through a charge integrator amplifier; then, a weak periodic pressure change signal (BCG signals) is obtained through alternating current (AC) amplifier and an anti-alias filter. Compared with the BCG, the amplitude of the body motion signal is extremely large. When the body motion is severe, the BCG is distorted due to the limited power rail. Therefore, the body motion signal needs to be acquired from the output of the charge integrator amplifier. The analog to digital converter (ADC) acquisition part is our self-developed, eight-lead EEG sensor with 24-bit resolution, a sampling rate of 250SPS, and 2 µVpp noise (0.01 Hz to 70 Hz) from our laboratory, which can collect BCG and body motion signals. The signal acquisition process is shown in Fig. 1. The collected data is transmitted to the host computer via Bluetooth for further processing.

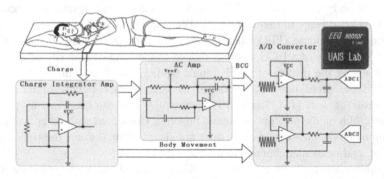

Fig. 1. BCG hardware acquisition part

2.2 Digital Filter

Digital filtering mainly considers the power frequency noise (50 Hz/60 Hz). In a traditional digital notch design, in order to ensure enough attenuation, the order of the filter should be set high. However, the calculation therefore becomes massive, and the design is complicated and not conducive to dynamic adjustment. An adaptive filter can be used to achieve better effects than a traditional notch filter [21] and can effectively extract a useful signal when the signal-to-noise ratio (SNR) is −100 dB. The principle of the least mean square filter (LMS) adaptive filter is as follows:

$$y(n) = \sum\nolimits_{i=0}^{M-1} \omega_i(n)x(n-i) \tag{1}$$

$$e(n) = d(n) - y(n) \tag{2}$$

$$\omega_i(n+1) = \omega_i(n) + 2\mu e(n)x(n-i), i = 0, 1, \cdots, M-1 \tag{3}$$

where μ is the convergence factor.

As shown in Fig. 2, where x(k) is the signal with a certain frequency interference, which means the input original signal as well, it is input from the desired signal end of the adaptive filter, $\sin(2\pi k f_0/f_s)$ and $\cos(2\pi k f_0/f_s)$ are interference signals with frequency f_0 (where fs is the sampling rate). Interference signals are multiplied by W_1 and W_2 respectively for proper linear combination so that the output y(k) can be closer to the actual interference, and the final output error e(k) is the signal that matters [22]. The algorithm can be used to determine how to choose W_1 and W_2. Specifically, the algorithm uses some criterion to construct a function about the error e(k), which is obtained by minimizing e(k). The most common criterion for W_1 and W_2 is to minimize the sum of the least mean square to limit the error. The most common criterion for W_1 and W_2 is to minimize the sum of the least mean square of e(k).

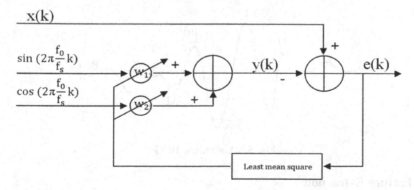

Fig. 2. Adaptive filter

The ECG signal with power frequency interference is generated by the standard signal source. The input signal x(k) = S_x (k) + S_y (k), where x(k) is the ECG signal contaminated with the power frequency noise, S_y (k) is the ECG signal, and S_x (k) is the power frequency signal. When x(k) is treated as the desired output signal and S_x (k) as the input signal, the actual required S_y (k) is obtained by the adaptive filter. The effect is shown in Fig. 3.

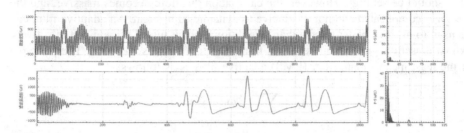

Fig. 3. Comparison of standard ECG signals before and after adaptive filtering.

As can be seen, when the adaptive filter is applied to the processing of the collected BCG, the power frequency interference can be removed effectively. The processed heart impact map signal is shown in Fig. 4. Compared with a standard signal, the peaks are obvious and the signal quality is verified (Fig. 5).

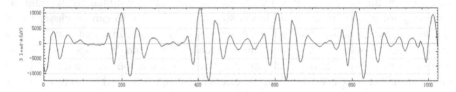

Fig. 4. Actual collected BCG

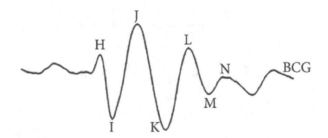

Fig. 5. A standard BCG

2.3   Feature Extraction

Heart Rate Variability. The J wave is the largest of all bands, and the peak detection makes it easy to identify the J wave and obtain the heart rate [23]. We analyzed heart rate variability by replacing the traditional R-R interval with a J-J interval. There are three methods of heart rate variability analysis: Time-Domain, Frequency-Domain, and Nonlinear. Each method has multiple analytical indicators. The main measurement indicators of time domain analysis in this paper include the average of the time intervals between successive BCG J-waves (AVNN), the standard deviation of normal-to-normal J-J intervals (SDNN), the root mean square of successive J-J interval differences (RMSSD), and the relative percent of successive J-J interval pairs that differ by more than 50 ms (pNN50). The frequency domain analysis mainly analyzes the energy distribution of each frequency band through FFT or autoregressive (AR) models. This paper uses an AR model for spectral estimation. The spectrum can be divided into very low frequency (VLF < 0.04 Hz), low frequency (LF is 0.04 to 0.15 Hz), and high frequency (HF is 0.15 to 0.4 Hz). The main indicators in the frequency domain analysis are low frequency energy percentage (nLF), high frequency energy percentage (nHF), and low frequency/high frequency energy ratio (LF/HF). HRV nonlinear analysis methods mainly include Poincare plot, Detrended fluctuations (DFA), etc. [24]. This study calculates the root mean square of the short axis (SD1), the root mean square of the long axis (SD2) and the ratio, SD2/SD1, from Poincare plot.

SD1 reflects the instantaneous HRV. We can compute it by

$$SD1 = \sqrt{Var(\frac{JJ_i - JJ_{i+1}}{\sqrt{2}})} \tag{4}$$

SD2 reflects the instantaneous HRV. We can compute it by

$$SD2 = \sqrt{Var(\frac{JJ_i + JJ_{i+1}}{\sqrt{2}})} \tag{5}$$

where Var represents the variance and JJ_i represents the i-th heartbeat interval.

Respiration. The bandwidth of the respiratory signal is 0.1 to 0.5 Hz. A 500th-order FIR bandpass filter is used. The upper 3 dB cutoff frequency is set to 0.5 Hz, the lower 3 dB cutoff frequency is 0.1 Hz, and the sampling rate is 250SPS. The signal after removing the power frequency is subjected to FIR bandpass filtering to obtain a respiratory signal, and the signal is as shown in Fig. 6.

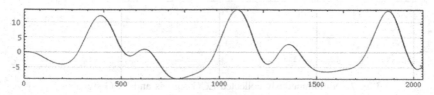

Fig. 6. Respiratory signal from the BCG

3 Experiment

3.1 Experimental Setup and Data Collection

To validate the reliability of the BCG signal, we placed the piezoelectric film sensor under a mattress and equipped subjects with ECG electrodes. Then, both the ECG signals and BCG signals of the 10 subjects were collected simultaneously. The characteristics of the research subjects are shown in Table 1. The time for collection was 11 min, and all of subjects completed a round of six movements, which were getting to bed, lying flat, getting off bed, getting to bed, suffocating for 15 s, and getting off bed. During the period of this pipeline, the length of time for lying flat was 10 min, and the interval between getting off bed and getting to bed was 1 min.

Table 1. Characteristics of research subjects

Index	Mean ± std	Unit
Age	25.24 ± 2.84	Years
Weight	75.11 ± 10.21	Kg
Height	1.78 ± 0.04	M
BMI	23.65 ± 2.70	Kg/m^2

Because the salient body motion signals produced by the two actions of getting off bed and getting to bed would have seriously affected the BCG signal, we discarded this part of the ECG and BCG signals, and only the body motion was considered. Apart from the above two actions, ECG signals and BCG signals were clearly collected during lying and suffocating (Fig. 7).

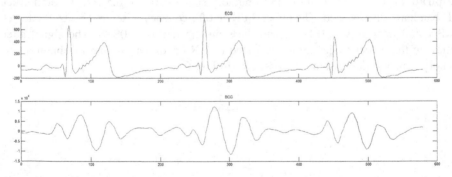

Fig. 7. Simultaneously collected ECG signals and BCG signals

3.2 Analysis of Significant Difference Between BCG Signal and ECG Signal

By measuring the R wave and J wave of the BCG and ECG signals in the lying state, the measured values of the HRV_{ECG} and HRV_{BCG} in the time domain were obtained (i.e., AVNN, SDNN, RMSSD, nPNN50, respectively), as well as the measured value of the low frequency energy ratio, nLF; the high frequency energy ratio in the frequency domain, nHF; the low frequency/high frequency energy ratio, LF/HF, in the frequency domain; the root mean square value, SD1, of the short axis of the nonlinear measured value; the root mean square value, SD2, of the long axis; and the ratio SD2/SD1. Table 2 shows that the sample statistics do not exceed $t_{0.05/2,9}$, and there is no significant divergence between the HRV indicators of 10 types of BCG signals and ECG signals. Therefore, the analysis confirmed that the collected BCG signals were reliable in analyzing the HRV index.

Table 2. Mean, Standard deviation, T-value, and P-value of ECG and BCG during static state

Parameter	HRV_{ECG}	HRV_{BCG}	t	p
AVNN/ms	764.86 ± 30.86	764.55 ± 30.77	1.396	0.196
SDNN/ms	26.17 ± 8.29	26.99 ± 6.95	−1.385	0.199
RMSSD/ms	29.13 ± 10.82	30.14 ± 7.48	−0.698	0.503
PNN50/%	9.24 ± 10.88	9.35 ± 8.55	−0.090	0.930
LF/HF	0.95 ± 0.96	0.89 ± 1.07	0.764	0.464
nLF/%	38.48 ± 18.48	36.50 ± 17.65	1.026	0.332
nHF/%	56.88 ± 20.32	59.53 ± 19.28	−1.292	0.229
SD1/ms	20.76 ± 7.71	21.48 ± 5.33	0.702	0.501
SD2/ms	30.60 ± 9.39	31.32 ± 8.92	−1.320	0.219
SD2/SD1	1.51 ± 0.27	1.47 ± 0.31	0.811	0.438

3.3 Recognition of Sleep Apnea

This experiment simulated apnea in patients with sleep apnea syndrome by suffocating the subjects for 15 s. By analyzing the waveform, we found that the regular breathing has a law of periodicity, and thus the breathing rate was easily obtained. But if the patient were in the apnea state, the waveform barely fluctuated. The experiment suggested that the BCG signal could be very convenient and effective to monitor the breathing of patients with obstructive sleep apnea (Fig. 8).

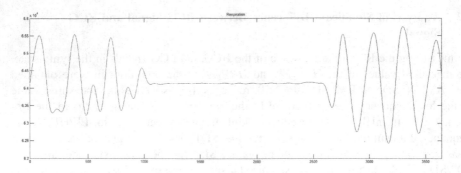

Fig. 8. 15-second apnea signal

3.4   Nighttime Bed-Off Identification

The salient feature of bed-off movements at night is that there are two abnormal, violent body shaking signals in the continuous BCG signal collected (getting on bed, getting off bed), and with no BCG signal between the two body shaking signals, as shown in Fig. 9. Given the experimental paradigm, besides the initial getting to bed and the last getting off bed, a pair of bed-off and bed-on motions with 30 s' intervals was included to imitate the actual sleep scenario. These preparations yielded an experimental accuracy of 100% during testing. The results showed that the BCG can accurately identify the bed-off action during nights. Thus, the BCG provides the possibility of long-term monitoring of the number of bed-off movements per night and analysis of personal sleep patterns.

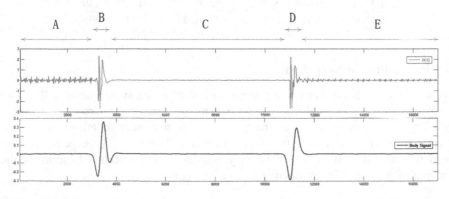

Fig. 9. A complete process of getting off the bed; the A and E phases are lying flat, B is the bed-off motion, and D is the bed-on motion. The C stage represents the time of leaving bed.

4 Results and Discussion

This paper proposed a sleep health monitoring system based on BCG signal. By adopting a non-contact and non-interfering acquisition method, the BCG signals with obvious J-wave characteristics are collected in real time. Minimum mean square adaptive filtering was used to eliminate power frequency interference. The ECG and ECG were collected synchronously during experiments. The HRV was not significantly different among the indicators of time domain, frequency domain, and nonlinear, indicating that the BCG is reliable in analyzing HRV indicators. At the same time, this method can accurately determine apnea and nighttime bed-off motion characteristics, indicating that the BCG can be used as a non-contact sleep quality monitoring method. Compared to other non-contact monitoring solutions, this solution is more convenient and accurate and helps to achieve real-time monitoring of long-term sleep quality.

Acknowledgment. The project was funded by the National Basic Research Program of China (973 Program) (No. 2014CB744600), the Program of International S&T Cooperation of MOST (No.2013DFA11140), the National Natural Science Foundation of China (grant No. 61210010, No. 61632014), the National key foundation for developing scientific instruments (No.61627808), Program of Beijing Municipal Science & Technology Commission (No. Z171100000117005).

References

1. Mador, M.J., Kufel, T.J., Magalang, U.J., et al.: Prevalence of positional sleep apnea in patients undergoing polysomnography. Chest **128**(4), 2130–2137 (2005)
2. Ma, Y., Tian, F., Zhao, Q., et al.: Design and application of mental fatigue detection system using non-contact ECG and BCG measurement. In: 2018 IEEE International Conference on Bioinformatics and Biomedicine (BIBM), pp. 1508–1513. IEEE (2018)
3. Nakajima, K., Matsumoto, Y., Tamura, T.: A monitor for posture changes and respiration in bed using real time image sequence analysis. In: Proceedings of the 22nd Annual International Conference of the IEEE Engineering in Medicine and Biology Society (Cat. No. 00CH37143), vol. 1, pp. 51–54. IEEE (2000)
4. Deng, F., Dong, J., Wang, X., et al.: Design and implementation of a noncontact sleep monitoring system using infrared cameras and motion sensor. IEEE Trans. Instrum. Measur. **67**(7), 1555–1563 (2018)
5. Lee, H.J., Hwang, S.H., Yoon, H.N., et al.: Heart rate variability monitoring during sleep based on capacitively coupled textile electrodes on a bed. Sensors **15**(5), 11295–11311 (2015)
6. Wu, K., Zhang, Y.: Contactless and continuous monitoring of heart electric activities through clothes on a sleeping bed. In: 2008 International Conference on Information Technology and Applications in Biomedicine, pp. 282–285. IEEE (2008)
7. Yang, Z., Pathak, P.H., Zeng, Y., et al.: Vital sign and sleep monitoring using millimeter wave. ACM Trans. Sens. Netw. (TOSN) **13**(2), 14 (2017)
8. Zhang, Z., Yang, G.Z.: Monitoring cardio-respiratory and posture movements during sleep: what can be achieved by a single motion sensor. In: 2015 IEEE 12th International Conference on Wearable and Implantable Body Sensor Networks (BSN), pp. 1–6. IEEE (2015)

9. Nuksawn, L., Nantajeewarawat, E., Thiemjarus, S.: Real-time sensor-and camera-based logging of sleep postures. In: 2015 International Computer Science and Engineering Conference (ICSEC), pp. 1–6. IEEE (2015)
10. Dock, W., Taubman, F.: Some technics for recording the ballistocardiogram directly from the body. Am. J. Med. **7**(6), 751–755 (1949)
11. Jose, S.K., Shambharkar, C.M., Chunkath, J.: HRV analysis using ballistocardiogram with LabVIEW. In: 2015 International Conference on Computing and Communications Technologies (ICCCT), pp. 128–132. IEEE (2015)
12. Anoop, K., Ahamed, V.I.: Heart rate estimation in BCG. In: Proceedings of National Conference CISP (2013)
13. Acharya, U.R., Joseph, K.P., Kannathal, N., et al.: Heart rate variability: a review. Med. Biol. Eng. Comput. **44**(12), 1031–1051 (2006)
14. Zhao, W., Ni, H., Zhou, X., et al.: Identifying sleep apnea syndrome using heart rate and breathing effort variation analysis based on ballistocardiography. In: 2015 37th Annual International Conference of the IEEE Engineering in Medicine and Biology Society (EMBC), pp. 4536–4539. IEEE (2015)
15. Koivuluoma, M., Barna, L., Koivistoinen, T., et al.: Influences of digital band-pass filtering on the BCG waveform. In: BIOSIGNALS, vol. 2, pp. 84–89 (2008)
16. Jiang, F., Song, S., Cheng, J., et al.: A research based on BCG signal detection device. In: 2014 International Conference on Mechatronics, Electronic, Industrial and Control Engineering (MEIC-2014). Atlantis Press (2014)
17. Liu, M., Ye, S.: A novel body posture recognition system on bed. In: 2018 IEEE 3rd International Conference on Signal and Image Processing (ICSIP). IEEE (2018)
18. Jose, S.K., Shambharkar, C.M., Chunkath, J.: Cardiac arrhythmia detection using ballistocardiogram signal. In: 2015 IEEE International Conference on Signal Processing, Informatics, Communication and Energy Systems (SPICES), pp. 1–5. IEEE (2015)
19. Sadek, I., Biswas, J., Abdulrazak, B., et al.: Continuous and unconstrained vital signs monitoring with ballistocardiogram sensors in headrest position. In: 2017 IEEE EMBS International Conference on Biomedical & Health Informatics (BHI), pp. 289–292. IEEE (2017)
20. Zheng, J., He, A., Li, J., et al.: Polymorphism control of poly (vinylidene fluoride) through electrospinning. Macromol. Rapid Commun. **28**(22), 2159–2162 (2007)
21. Haykin, S.: Adaptive Filter Theory. Prentice-Hall, Englewood Cliffs (1991)
22. Allred, D.J., Yoo, H., Krishnan, V., et al.: LMS adaptive filters using distributed arithmetic for high throughput. IEEE Trans. Circuits Syst. I Regul. Pap. **52**(7), 1327–1337 (2005)
23. Pino, E.J., Chávez, J.A.P., Aqueveque, P.: BCG algorithm for unobtrusive heart rate monitoring. In: 2017 IEEE Healthcare Innovations and Point of Care Technologies (HI-POCT), pp. 180–183. IEEE (2017)
24. Kamen, P.W., Tonkin, A.M.: Application of the Poincare plot to heart rate variability: a new measure of functional status in heart failure. Aust. N. Z. J. Med. **25**(1), 18–26 (1995)

Sentiment Analysis of Social Networks' Comments to Predict Stock Return

Juan Cheng[1], Jiaolong Fu[2], Yan Kang[3], Hua Zhu[4],
and Weihui Dai[5(✉)]

[1] Xianda College of Economics, Shanghai International Studies University,
Shanghai, China
[2] Law School, Hunan University of Arts and Science, Changde, China
[3] School of Humanities and Management Sciences, Southwest Medical
University, Luzhou, China
[4] School of Software, Fudan University, Shanghai, China
[5] School of Management, Fudan University, Shanghai, China
whdai@fudan.edu.cn

Abstract. Financial intelligence has become a research hotspot in recent years
with the development of behavioral finance which introduces the social emotion
and behavior factors in the decision-making. The data mining technology is
widely used in the research on financial intelligence. This paper collected the
investors' comments from Social Network Sites (SNS) by crawler technology
and segmented each piece of comment into words by Chinese text processing
technology to build a financial sentiment lexicon. Applying the sentiment lexicon, a sentiment computing model based on SO-PMI algorithm was designed to
compute the sentiment indices of the investors. Finally, the paper made an
empirical analysis through linear regression between the return of the stock and
its investors' sentiment index. The result proved that the sentiment indices based
on the investors' comments are better to measure the investors' sentiment and
can be used to predict the stock return.

Keywords: Financial intelligence · Social network sites (SNS) · Sentiment
analysis · Stock return prediction

1 Introduction

Behavioral finance thinks that the returns of the stocks are influenced by the investors'
sentiment. This returns of the stocks can be predicted according to the investors'
sentiment (Hui and Li 2014, Wang and Sun 2004, Yu and Meng 2010). How to
measure the investors' sentiment is a key point of the prediction since it determines the
accuracy of the prediction result. Most existing approaches have used some proxy
variables such as the turnover rate, the discount of close-end fund, the growth rate of
new investor account and the consumer confidence index to measure the investors'
sentiment (Yi et al. 2017; Jing and He 2016). But the predicting result is not good since
these proxy variables measure the investors' sentiment indirectly. Collecting data and
mining the information that reflects the investors' sentiment from the comments of the

© Springer Nature Switzerland AG 2019
D. Milošević et al. (Eds.): HCC 2019, LNCS 11956, pp. 67–74, 2019.
https://doi.org/10.1007/978-3-030-37429-7_7

investors on the financial social network sites is gradually adopted, and has become a research hotspot of financial intelligence in recent years (Liu and Lu 2007, Dickinson and Hu 2015, Huang 2015, Lee et al. 1991, Huang et al. 2017, Lan 2011). This approach has been proved to be better than the proxy variables in computing the investors' sentiment since it reflects the investors' sentiment directly (Zhu and Zhao 2016). Building a sentiment lexicon and a sentiment computing model is the main work of the approach (Zhu 2015). These two have great influence on the measurement of the investors' sentiment and are improved constantly. However, the data mining and sentiment computing methods have much room for improvement in terms of accuracy.

The paper focuses on the method to measure investor sentiment most intuitively. We used Scrapy crawler framework to collect the comments on the financial social network site and HanLP (Chinese text processing) to mine the texts to build a sentiment lexicon. Then, a sentiment computing model based on SO-PMI algorithm was proposed. In order to evaluate the effectiveness of the sentiment computing model, the paper chose the stock "Letv" to analyze the relationship between the return of the stock and the sentiment index by linear regression. The result supported that the sentiment computing model proposed in this paper performs well by comparing the return of regression with the real return of the stock.

2   Data Collecting and Processing

2.1   Data Collecting and Segmentation

Snowball website is a Chinese SNS focused on stock investment which was established in 2010. Every stock has a forum on Snowball. The investors interested in a specific stock can read the posts and replies to the posts of the stock. More than 10 million users on Snowball made over 30 thousand's pieces of posts and replies every day. The paper just chose one stock named "Letv" as an example. Its web site is http://xueqiu.com/S/SZ300104. We used "Scrapy" to collect the data from the web site.

The data collected includes 450 thousands user's ID, 130 thousands posts, 560 thousands replies and mutual following relationship of the users. Some data cleaning tool such as "Re package", "Json package" and "Response object" were used to clean the data according to the data structure features. The symbols without actual meaning like "@" and emoticon are deleted in the cleaning. The data after cleaning was saved on MongoDB (a storage platform).

We used HanLP as the Chinese text processing tool to process the posts and replies. First, we used Chinese full-pitch punctuations comma, dot, colon, semicolon, question mark, exclamation point as the separators to segment the sentences into the clauses. For example, Fig. 1 is the original post in Chinese.

回复@老缠游记: 出掉$乐视网(SZ300104)$先，转战长春高新了 89.6{[俏皮]}///@老缠游记:回复@stzp:你赚变得比老缠还快，纯粹的投机者！握个手！

Fig. 1. The original post in Chinese

The above post and its reply can be segmented into clauses and a group of words as shown in Table 1. In Table 1, left column is the clause ID, middle column is the segmented clauses, and right column is the segmented words. All the words except those without emotion are saved in the database XqWordList.

Table 1. The segmentation of the sentences

Clause ID	The text(in Chinese)	Words(with translation in English)
21816	出掉乐视网先	出/vf(sell), 掉/v(out), 乐视网/nt(Letv), 先/d(first)
21817	转战长春高新了 89.6	转战/v(transfer to), 长春高新/nt(ChangChunGaoXin), 了/ule, /w, 89.6/m
21818	你转变得比老缠还快	你/rr(You), 转变/vi (change), 比/p(than), 老/a, 缠/v, 还/d(much), 快/a(fast)
21819	纯梓的投机者	纯梓/b(absolute), 的/ude1, 投机者/n(speculator)
21820	握个手	握/v(shake), 个/q, 手/n(hand)

2.2 Modeling Investor's Sentiment

15 emotional symbols were selected from these words to construct a basic financial sentiment lexicon through voting by 5 volunteers. For each word with high frequency, we can compute its sentiment index by SO-PMI algorithm on the base of the basic financial sentiment lexicon as shown in Formulas (1) and (2).

$$SO - PMI(Word) = \sum_{PWord \in P-Words} PMI(PWord, Word)$$
$$- \sum_{PWord \in N-Words} PMI(PWord, Word) \tag{1}$$

$$\text{Here,} \quad PMI(word1, word2) = \log_2^{(N \cdot \frac{p(word1 \& word2)}{p(word1) \cdot p(wordd2)})} \tag{2}$$

Where, N is the number of words in the lexicon, p(Word1&Word2)is the frequency of word1 and word2 which appear together. p(Word1) is the frequency of word1, p (Word2) is the frequency of word2. The greater PMI means the stronger correlation between the two words.

Then, we set the thresholds for the positive or negative words in order to recognize whether a word is positive or negative. The test group consisted of 100 words with distinct emotional orientation voted by five volunteers. The result showed that the reasonable threshold of positive word was 3 and that of negative word was −24. With the thresholds, the correct rate of recognition was more than 80%.Each word can be classified according to the thresholds as Formula (3).

$$SO - PMI(Word) = \begin{cases} \geq 3, positive \\ else, neutral \\ \leq -24, negative \end{cases} \tag{3}$$

There are 1974 positive words, 1142 negative words and 6356 neutral words after the classification. The intension of the sentiment of a word is also influenced by the modifiers such as adverb of degree and negative words. So, the sentiment index of a clause is computed by Formula (4)

$$SO_{clause} = SO\_PMI(Word) \times F_{w_d} \times F_{w_n} \tag{4}$$

F_{w_d} is the weight of the adverb, F_{w_n} is the weight of the negative word.

Add all the sentiment indices of the clauses in a sentence to get the sentiment index of the sentence as Formula (5).

$$SO_{sentence} = \sum_{i=1}^{n} SO_{clause} \tag{5}$$

Add all the sentiment indices of all posts of an investor, then divide the sum by the number of posts of the investor to get the investor's sentiment index as Formula (6) (Table 2).

$$SO_{investor} = \sum_{i=1}^{n} SO_{sentence} \Big/ n \tag{6}$$

Table 2. Demonstration of computing the sentiment index

Computing Process	Initial Data	Clauses Process	Words Process	The sentiment of words	The sentiment of Clause: SO_{clause}	The sentiment of a piece of post or reply: $SO_{sentence}$
A piece of post or reply	You suppo rt Letv at t his time.It is worth a greement! (老跟能在 这个时候 力挺乐 视，值得 称{[很 赞]})	The first clause: You support Letv at this time.(老跟 能在这个时候 力挺乐视) The second clause: It is worth agree-ment!(值得称 {[很赞]})	The first grou p:You(老 能)，at this ti me(时候)，ag reement(力 挺)，Letv(乐 视). The second group: worth(值 得)， agreement(很 赞)	The first group:You（0），at this time(0),support(1),let v(0). The second group:worth(1),agreement(1)	The first group:1. The second group:2	1.5

3   Sketching the Sentiment Map

There is a complex investors' relationship network due to the investors' mutual following. The relationship network is like the page link structure. So the PageRank algorithm for evaluating the influence of the web page can be used to evaluate the influence of an investor. The influence of an investor is calculated by Formula (7).

$$UserRank(v) = (1 - d) + d \sum_{u \in U(v)} \frac{UserRank(u)}{N(u)} \tag{7}$$

UserRank(v) is the value of the influence of the user v. u is another user who follows the user v. N(v) is the number of the users who follow the user v. U(v) is the set of the users who follow the user v. d is the damping coefficient.

The algorithm run the computation of UserRank iteratively until all the UserRanks are unchanged. The more the following users, the greater the UserRank will be. After 15 iterations, we get the UserRanks of all users. In order to visually display the sentiment of the users, we drawn a sentiment map according to the sentiment indices of the investors. A map is drawn every 5 min since fifteen past nine in the morning. In the map, each point represents a user. The influence of the user is represented by the size of

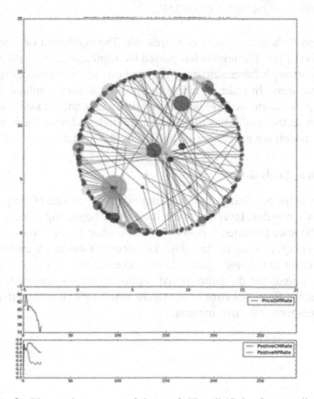

Fig. 2. The sentiment map of the stock "Letv" (Color figure online)

the point. The colour of the points represents the sentiment of the investors. The red point means the investor's sentiment is positive and the green point means the opposite. The connections between points represent the mutual following relationship. There are 288 maps each day and they are made into an animation. The investors can observe the changing of the sentiment by watching the animation. Figure 2 shows the sentiment map of the stock "Letv".

4   Regression Test

4.1   Regression Model

We used all-day data during 2015 to analyze the relationship between the sentiment and the return of the stock respectively. There are 222 pieces of data in 2015. The regression factors are the number of the posts (F_1), the ratio of the positive posts (F_2), the ratio of the negative posts (F_3), the sum of the sentiment indices of all posts (F_4), the number of the replies (F_5), the ratio of the positive replies (F_6), the ratio of the negative replies (F_7), the sum of the sentiment indices of all replies (F_8). We obtained the following regression formula.

$$R = 0.00249 + 0.00251F_1 + 0.0579F_2 - 0.0398F_3 + 0.00624F_4 + 0.007135F_5 \\ + 0.1209F_6 - 1.0438F_7 + 0.00952F_8 \tag{8}$$

Tables 3 and 4 show the results of regression. The coefficient of determination of the regression is 0.4016. The regress has passed the significance test with 5% level. The users who have strong influence are more accurate on the prediction of the stock return than the general users. In order to reflect the impact of user's influence on the prediction of the stock return, we have done many times tests and introduce the cubic root of the UserRank as the coefficients of the sentiment indices finally. In order to evaluate the regression model, we made a mock trading by the model.

4.2   Regression Analysis

First, we buy in at the beginning of the next day when the change of the price predicted is more than 4.4% nowday. Inversely, we sell out at the beginning of the next day when the change of the price predicted is below −4.4% nowday. Second, we sell out all of the stock at the opening price on the last day. The return of the mock trading is 77.96%, which is very close to the real return 74.48%. The fractional error of the return was 4.67%. We also predicted the return based on the investor's sentiment measured by proxy variables, the fractional error of the return was 8.19%. It proved that the method in this paper was more effective method.

Table 3. Analysis of variance

Analysis of variance					
Source	HF	Sum of squares	Mean square	FValue	Pr > F
Model	8	12289.94	1536.243	2.01	0.046597
Error	213	162794.1	764.3		
Corrected total	221	172459.61			

Table 4. Parameter estimates

Parameter estimates					
Variable	DF	Parameter estimates	Standard error	t Value	Pr > |t|
F1	213	0.000251	0.00016	1.5688	0.1132
F2	213	0.0579	0.02718	2.1302	0.0343
F3	213	−0.0398	0.01638	−2.4298	0.0159
F4	213	0.000624	0.00043	1.4512	0.1482
F5	213	0.007135	0.00246	2.9004	0.0041
F6	213	0.1209	0.05079	4.1150	0.0006
F7	213	−1.0438	0.34678	−3.0099	0.0029
F8	213	0.00952	0.00480	1.9833	0.0486

5 Conclusions

This paper applied the data mining technology into the measurement of the sentiment of the investors in order to predict the stock return. Compared with the traditional methods which choose some proxy variables to measure the sentiment, it can measure the sentiment of the investors more directly and accurately. Besides, the paper visualized the sentiment by drawing the sentiment map. The investors can observe the change of the sentiment. However, the grammatical structure and the modification of the sentences are not considered in the sentiment computing method. It is expected to be improved in the future research. Besides, the regression model, which is multiple regression, also can be improved in order to obtain more accurate prediction.

Acknowledgments. This work was supported by International Cooperation and Exchange Program, Ministry of Science and Technology of China (No. 4-9/2018), and Major Project of Philosophy and Social Science Research, Ministry of Education of China (No. 19JZD010). Juan Cheng, Jiaolong Fu, Yan Kang are the joint first authors, and Weihui Dai is the corresponding author of this paper.

References

Hui, B., Li, P.: Does investor sentiment predict stock returns? The evidence from Chinese stock market. J. Syst. Sci. Complex. **27**(1), 130–143 (2014)
Zhu, B., Zhao, S.: The PSM-black-litterman model based on the sentiment mining. Times Finan. **36**(12), 224–226, 229 (2016)

Dickinson, B., Hu, W.: Sentiment Analysis of Investor Opinions on Twitter (2015). http://dx.doi.org/10.4236/sn.2015.43008. Published Online July 2015

Lee, C.M., Shleifer, A., Thaler, R.H.: Investor sentiment and the closed-end fund puzzle. J. Finan. **46**(1), 75–109 (1991)

Zhu, H.: Research and Implementation based on SNS Emotional Information of Stock Price Change Trend. Fudan University, Shanghai (2015)

Huang, F., Feng, S., Wang, D.: Mining topic sentiment in microblogging based on multi-feature fusion. Chin. J. Comput. **40**(4), 872–888 (2017)

Yi, H., Cai, Y., Dong, D.: The research on construction and effect of stock market investor sentiment index in China. Price Theory Pract. **28**(10), 130–133 (2017)

Huang, J.: The construction of the investors' sentiment index based on social media. Contemp. Acc. **2**(18), 18–20 (2015)

Liu, J., Lu, Y.: Survey on topic-focused web crawler. Appl. Res. Comp. **24**(10), 26–29 (2007)

Wang, M., Sun, J.: Stock market returns, volatility and the role of investor sentiment in China. Econ. Res. J. **50**(10), 75–83 (2004)

Lan, Q.: Research of collecting financial data from internet. Comput. Eng. Des. **32**(5), 1829–1832 (2011)

Yu, Q.H., Meng, W.: Cointegration analysis on relation of investor sentiment and the composite index of shanghai stock exchange. Forecasting **29**(5), 53–57 (2010)

Jing, R., He, X.: The construction of the investor's sentiment in China. Contemp. Econ. **31**(24), 43–47 (2016)

Tetlock, P.C.: Giving content to investor sentiment: the role of media in the stock market. J. Finan. **62**(3), 1139–1168 (2007)

SC-RBAC: A Smart Contract based RBAC Model for DApps

Yi Ding[1], Jun Jin[1(✉)], Jinglun Zhang[1], Zhongyi Wu[2], and Kai Hu[3]

[1] School of Information, Beijing Wuzi University, Beijing, China
jinjun@bwu.edu.cn
[2] Key Laboratory of Advanced Public Transportation Science, Beijing, China
[3] School of Computer Science and Engineering,
Beihang University, Beijing, China

Abstract. Blockchain technology with its non-centralized, transparent, trustful, traceable and tamper-resistant features draws more and more attention both in commercial and scientific area. Smart contracts and DApps (Decentralized Applications) are programs naturally running automatically on blockchain. Access control is a principle that regulates the access to critical resources. RBAC (Role based Access Control) is one of access control mechanisms and it involves three parts: user, role and permission, with their relations, corresponding to real business. However, traditional implementation of RBAC relies on centralized server which is in danger of being modified, invaded or a single point of failure. The paper proposes a decentralized and smart contract based RBAC model named SC-RBAC for DApps. It is developed by Ethereum's Solidity and offers a strong compatibility with different DApps. The features of SC-RBAC associated with flexible interfaces, traceability and security enrich the community of DApps. The results of two experiments are discussed to evaluate the overheads of SC-RBAC model.

Keywords: Access control · RBAC · Blockchain · Smart contract · DApps

1 Introduction

Nowadays, one viewpoint has been wildly adopted that Internet, especially mobile Internet changes the human's life style and society dramatically. Advances in technology are often accompanied by security threats as well. Then reliable measures are required to prevent data misuse, data vandalism and data leakage, etc.

Roles are frequently exploited to design the permissions of users to access certain services or operations. RBAC [1] (Role-based Access Control), one mechanism to describe the access control relations between users and permissions, is welcomed in industrial and scientific areas for their resources' management. The mechanism is often maintained in a centralized database, which may lead to data tampering and bring potential dangers into companies' business. Obviously, role disguise is more difficult to be prevented in the Internet world, comparing to face-to-face communication. How can we effectively solve the potential security problems for RBAC mechanism is a challenge.

© Springer Nature Switzerland AG 2019
D. Milošević et al. (Eds.): HCC 2019, LNCS 11956, pp. 75–85, 2019.
https://doi.org/10.1007/978-3-030-37429-7_8

Blockchain technology was introduced in 2008 [2] and is generally called a decentralized ledger. Blockchain, as one distributed system without central authority, is composed of block (i.e. a time-stamped series of immutable transaction data) and chain (i.e. blocks are secured and bound to each other with cryptographic techniques). The nodes in the blockchain, owning all the records individually and being connected by point-to-point network, don't have to trust each other in advance and rely on the mechanism to maintain the relations.

Blockchain technology has been developed rapidly in recent years, from 1st generation to 2nd generation. 1st generation primarily focuses on cryptocurrency such as Bitcoin, while 2nd generation introduces more flexible program interface like smart contracts on Ethereum [3, 4]. Smart contract, one digital version of contract generated in 1997, is not a new concept. However, it does not play a significant role in current business without blockchain. The generation of blockchain brings new opportunity to smart contract, and they are natural allies. Smart contract running on blockchain supports credible transactions without third parties. Once deployed, smart contracts will be executed on all the blockchain's nodes, which cannot be modified, and thus trusted.

DApps are distributed and innovative applications using blockchain technology. Smart contract is an essential part for DApps, providing some specified functions or useful services. Security access control is indispensable for DApps. In this research, we try to use the technologies of blockchain and smart contracts to solve the problems stated above.

In this paper, we present a smart contract based RBAC(SC-RBAC) model on blockchain. It is designed to be decentralized (blockchain based on consensus mechanism is centerless), traceable (any accesses to the system are recorded on blockchain and it is suitable for audit), securable (private key or biometric features like fingerprint is provided to verify user identity and prevents role disguise), transparent (audit criteria and smart contracts are visible among nodes), full automatic (smart contracts perform automatically without any human interventions), generic (useful and flexible RBAC interfaces can be used for different DApps' integration) and practical (the overheads of access control mechanism is comparable low). The major idea is to preserve the information of users, roles, resources and their relations on the blockchain and let SC-based access control self-execute. A prototype has been implemented by Solidity programming language [5] and deployed on Entereum.

Main contributions of this paper are listed in the following:

(1) A smart contract based model named SC-RBAC is provided and its feasibility is verified by experiments.
(2) A prototype of SC-RBAC is developed and deployed on Ethereum.
(3) A generic template for fast integrating RBAC to DApps are described.

The rest of the paper is organized as follows: Related works about RBAC on blockchain are mentioned in Sect. 2. Section 3 presents our SC-RBAC model and the generic template for DApps. Experiments and performance analysis about SC-RBAC prototype are discussed in Sect. 4. Finally, Sect. 5 summaries the conclusion and outlines future work.

2 Related Work

RBAC [1, 6, 7] is a typical access control model. Traditionally, access control policies and related data are stored on a central server (or server cluster). The server undertakes all the task of controlling access for an application system. In this paradigm, the data and programs may be in danger of being attacked or modified in an abnormal way. Blockchain provides a trustful platform for storage and computing, which will bring more security and transparency for access control mechanism. Many approaches have been studied in this area, and we will discuss previous researches of access control using blockchain.

Bitcoin Oriented
These research works [8–12] are primarily concentrated on Bitcoin blockchain. The key idea is to use customized script to realize access control policies and to embed the scripts into the transactions. Since Bitcoin framework does not support universal programming language, customized scripts cannot perform flexible and powerful logics.

Smart Contract Oriented
Unlike above works, these researches [13–15] make use of smart contract running on blockchain. The objective of the paper [13] is to study access control of Internet of things. It contains multiple access control contracts, such as judge contract and register contract. [14] focuses on Hyperledger Fabric to explore access control management. Compared to these studies, we try to do some work for DApps and implements one prototype with smart contract language—Solidity. [15] is quite similar to our work, which is also tested on Ethereum platform and related to RABC mechanism. However, its goal is to cope with trans-organizational RBAC problem, while our model could be suitable for both intra-organizational and trans-organizational problems. The role in our model isn't limited in one organization.

3 The Principle of Smart Contract Based RBAC (SC-RBAC)

3.1 Overview

Typically, blockchain technology runs a consensus protocol to finish the data synchronization as well as storage, and performs automated script codes - smart contracts to complete business logic and to process related data. It has the features of decentralization, security and tamper-resistance. Smart contracts running on blockchains can reduce human interventions and enhance data security.

The destination of this research is to provide a credible and auditable RBAC model for DApps. To achieve this, we present a generic smart contract based RBAC model called SC-RBAC. In this model, the relevant data of users, roles and permissions, as well as the codes of smart contracts, are all recorded and stored on the blockchain. Figure 1 gives an overview of SC-RBAC and its typical authorization process.

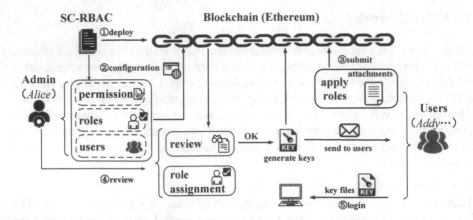

Fig. 1. Overview of SC-RBAC and its typical authorization process

Initialization Phase

Step1: SC-RBAC is in form of smart contracts and deployed on Ethereum.

Step2: Administrator *Alice* configures roles, permissions and users (no accounts or users at the initial stage) according to the services and managements *DAppx* (*DAppx* here is one representative of DApps) brings. The related data is stored on Ethereum.

User Registration Phase

Step3: If a user *Addy* chooses to use *DAppx*, then he will submit application with some attachments regulated by *Alice*. *DAppx* creates a new account and generates his public and private keys.

Step4: *Alice* reviews *Addy*'s application and assigns a role to him. The key pair was downloaded and sent to *Addy*. After that, the key pair will be deleted immediately to avoid illegal usage.

User Login Phase

Step5: After received his key pair, *Addy* uses them to login and access the authorized services.

3.2   SC-RBAC Smart Contract

Ethereum supports various high-level programming languages, such as Solidity whose grammar regulation is similar to JavaScript, Serpent whose grammar regulation is similar to Python, and LLL whose grammar regulation is similar to Lisp, etc. Nowadays, the widely adopted and known language is Solidity. In this paper, we use Solidity to create smart contracts, which are compiled into bytecodes and then deployed to execute on Ethereum virtual machines.

Smart contracts implemented with Solidity are similar to classes in object-oriented programming languages, where a smart contract can declare multiple members such as state variables, functions, events, and so on. One contract can inherit other contracts to increase reusability. As described in Fig. 2, SC-RBAC is composed of four parts: Ownable.sol, Permission.sol, Role.sol and User.sol. Ownable is a basic contract, and

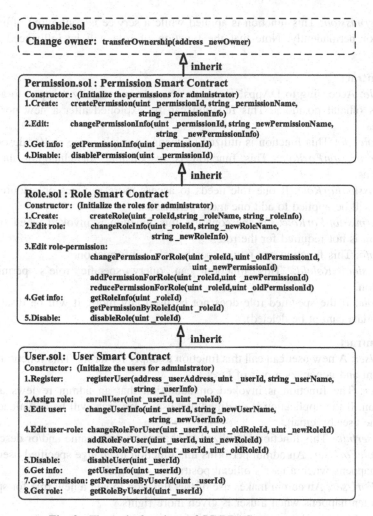

Fig. 2. The main functions of SC-RBAC smart contract

almost all the self-defined contracts inherit from it. In order to reuse the contracts and functions, role contract inherits permission contract while user contract inherits role contract.

Permission Contract

createPermission: When a new accessible service is deployed online, this function could be called. Here the permissionId and permissionName map to a service unit to identify the service, while. permissionInfo is the service's description.

changePermissionInfo: This function is utilized to change a service's name and/or description.

getPermissionInfo: This function is employed to acquire a specified service's description.

disablePermission: This function is applied while a service is configured offline temporarily or permanently. Note that the function of role will be influenced by this operation.

Role Contract

createRole: According to DApps' requirements, role is normally mapped to an organization's official positions. This function can be exploited after a new position is proposed.

changeRoleInfo: This function is utilized to change a role's name and/or description.

changePermissionForRole: This function is responsible for changing the role's permissions.

addPermissionForRole: If one role needs to access more services or resources, this function will be applied to add one more permission.

reducePermissionForRole: On the contrary, this function is invoked when specified permission is not required for the role.

getRoleInfo: This function can return specific role's information.

getPermissionByRoleId: This function can query specific role's permissions' information.

disableRole: If the specified role does not be used any more, it will be disabled(the recorded data cannot be deleted).

User Contract

registerUser: A new user can call this function to apply for a role. This function creates an account and generates a pair of keys.

enrollUser: The function is invoked on condition that an admin reviews a user's application. If the application is qualified, the user's key pair will be downloaded and sent to the user by email.

changeUserInfo: This function is utilized to change a user's name and/or description.

changeRoleForUser: An admin performs this function to change specified user's role. It often happens when a user's official position is changed.

addRoleForUser: An admin makes use of this function to add a new role for specified user. It often happens when a user is given more rights.

reduceRoleForUser: An admin utilizes this function to reduce a specified user's role. It often happens when a user is deprived of rights.

disableUser: If a user does not want to use this DApp any more, the administrator will invoke this function to disable the user's account.

getUserInfo: This function provides the specified user's name and other description.

getPermissionByUserId: This function can query specific user's permissions' information.

getRoleByUserId: This function can return specified user's role.

Note that all the modification functions are like *createXXX*, *changeXXX* or *disableXXX*. Since there is no actual deletion operation on blockchain, what the functions do is changing the state called the disabled operation. After these operations are returned, the newest states are recorded.

3.3 Generic Template for DApps

It is well known that security is essential requirements for DApps on the blockchain platform. We aim to propose an effective method for construction of security DApps. Figure 3 shows the hierarchical architecture of the new DApps that integrate SC-RBAC model. For real application, it provides rich access control services displayed in Fig. 2. Web application can easily call SC-RBAC model through Web3.js or JSON-RPC. SC-RBAC model plays the role of access control and filters illegal accesses. The smart contracts in Fig. 3 are the elementary functions for the specific application.

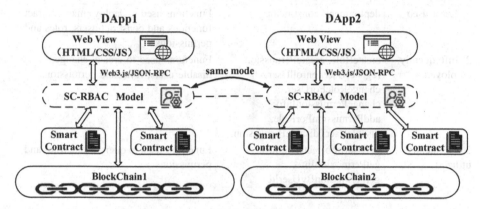

Fig. 3. SC-RBAC generic template for DApps

4 Experiment and Evaluation

To verify the practicability of our SC-RBAC prototype, two groups of experiments are designed in this section. One group is about execution time, and the other one is about gas [4, 16] cost.

4.1 Environment Setting

We use Ethereum Geth [17] developed by programming language Go [18] to deploy SC-RBAC prototype. Geth is a kind of Ethereum private blockchain and can be launched on various platforms, such as Windows, Linux/Unix, Mac and so on. The experimental platform is composed of five personal computers (Intel Core i3-4160 @ 3.60 GHz, 4G memory capacity). The operating system and software version are CentOS7(64bit), Geth1.8.9 and GO1.10.1. We design two groups of trials, one group is executed on five nodes, while the other one only uses one node.

4.2 Results and Evaluation

Considering the frequency of invoking functions, three types can be summarized as shown in Table 1. The first category refers the functions least used, including deployment and constructor functions, which are invoked only once during the initialization phase when SC-RBAC was deployed on the Ethereum. The second category relates to the functions infrequently employed. In the execution stage, a DApp's permission operation frequency is relatively stable. The contents of permissions, roles and users are occasionally modified. This category involves all the creating, changing and disabling operations. The third category points to the functions frequently utilized. Some typical querying operations are listed here.

Table 1. Functions invoked in our experiment.

Category	Operation or function name	Illustration
1. Least used	deployment, constructor	Functions used to deploy smart contract functions and construct users, roles and permissions
2. Infrequently employed	createRole, createPermission, registerUser, enrollUser, changeRoleForUser, changePermissionInfo, addPermissionForRole, disableUser, disablePermission	Functions used to create, change, disable users, roles and permissions
3. Frequently utilized	getUserInfo, getRoleInfo, getPermissionInfo, getPermissionByUserId	Functions used to query users, roles and permissions

In our experiments, we designed the tests on Ethereum platform with default configurations. With regards to the scenarios of a DApp's permission management, SC-RBAC functions may be utilized rarely while a role's permissions are changed, or may be employed frequently while some users register intensively. For the former one, there is a strong possibility that only one transaction is packed in a block. For the latter one, more than one transaction is probably packed in a block together. [19] also illustrates the similar situation. The average time of these transactions is comparable less. Nonetheless, the DApp's users have to wait for the average block generation time regularly to complete the operations. There is almost no difference between the two situations related above.

Figure 4 displays the time costs of these smart contract functions both in one and five node environments (the data is based on the assumption that all the blocks are generated correctly at one time). The procedure has been repeated 30 times and the average data is adopted. The execution time of 1st and 2rd category functions is approximate 15 s. Noticeably, the four querying functions on the right of the horizontal axis belong to the 3rd category, whose execution times are comparatively lower than the left ones. In order to be illustrated more clearly, they are also amplified on the right with the precision of milliseconds. In addition, there is no significant different between one node and five environments.

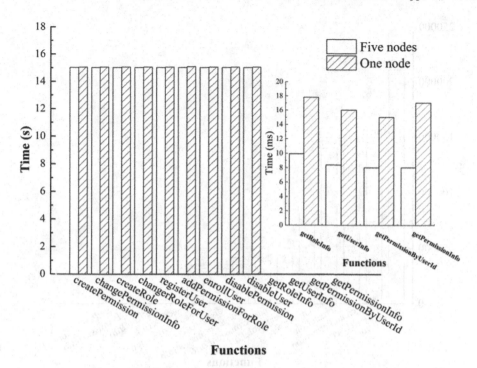

Functions

Fig. 4. Time cost of SC-RBAC smart contract functions

Gas is a fraction of an Ethereum token, and is employed by the contracts to pay for the transaction on the blockchain. The gas costs of SC-RBAC functions were measured and shown in Fig. 5. It is well known that the gas costs are in proportion to the number of function's instructions. The highest cost originates from the deployment of SC-RBAC prototype on blockchain with 6719667 gas (this quantity is too high to be displayed normally in Fig. 5). Once deployed, SC-RBAC model will be executed continually. The deployment function cost, as well as the constructor function, can be acceptable because it only occurs one time. Other functions (the second category), like *registerUser*, *createPermission*, *createRole*, *enrollUser*, are often called infrequently, so their gas costs are reasonable as well. Moreover, it is unequivocally clear that querying functions (the third category) are utilized frequently without gas cost.

The above experiments' results illustrate that both the time and gas costs appear a downward trend from first category to third category. Although the functions in first or second category cost much, they are invoked rarely. Apparently, the functions in third category are called quite frequently. They are inexpensive, and thus can be utilized freely. Consequently, it is proved that the SC-RBAC costs are acceptable at most time with its effective security features.

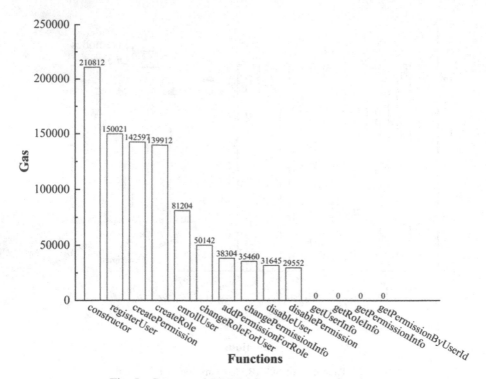

Fig. 5. Gas cost of SC-RBAC smart contract functions

5   Conclusion

Security access control is prominent for DApps. This paper presents one smart contract based RBAC model called SC-RBAC. The advantages of this model are that it not only provides secure, auditable and transparent access control, but also one generic template for integration into DApps. No extra work is required for DApps to design the access control module, and customized SC-RBAC is an efficient solution. In order to verify the feasibility of SC-RBAC, a prototype was implemented using Solidity and deployed on Ethereum. The results of the performance experiment indicate that the overheads of the most frequently called functions, such as *getXXX*, are negligible, while the costs of rarely called functions, such as *createXXX*, *changeXXX* and *disableXXX,* are relatively higher due to the mechanism of blockchain. In a word, the performance of the model is acceptable to DApps.

In the future, more work can be done to perfect the model with greater reliability and to provide more intelligence. Furthermore, better performance is also one destination for our work.

Acknowledgment. This research presented is supported by Research Base Project of Beijing Municipal Social Science Foundation (No. 18JDGLB026), Science and Technique General Program of Beijing Municipal Commission of Education (No. KM201910037003), Project of 2018 "Shipei plan" of Beijing Wuzi University, and Beijing Intelligent Logistics System Collaborative Innovation Center (PXM2018_014214_000009).

References

1. Sandhu, R.S., Coyne, E.J., Feinstein, H.L., et al.: Role-based access control models. IEEE Comput. **29**(2), 38–47 (1996)
2. Nakamoto, S.: Bitcoin: A peer-to-peer electronic cash system. https://bitcoin.org/bitcoin.pdf. Accessed 11 May 2019
3. Ethereum Blockchain App Platform. https://ethereum.org/. Accessed 27 May 2019
4. Wood, G.: Ethereum: A secure decentralised generalised transaction ledger, Yellow Paper. https://ethereum.github.io/yellowpaper/paper.pdf. Accessed 27 May 2019
5. Solidity. https://solidity.readthedocs.io/en/develop/. Accessed 30 May 2019
6. Ferraiolo, D.F., Kuhn, D.R.: Role-based access controls. Computer **4**(3), 554–563 (1992)
7. Moyer, M.J., Abamad, M.: Generalized role-based access control. In: Proceedings 21st International Conference on Distributed Computing Systems, pp. 391–398. IEEE (2001)
8. Ouaddah, A., Abou Elkalam, A., Ait, O.A.: FairAccess: a new blockchain based access control framework for the internet of things. Secur. Commun. Netw. **9**(18), 5943–5964 (2016)
9. Ramachandran, A., Kantarcioglu, D.: Using Blockchain and Smart Contracts for Secure Data Provenance Management. arXiv preprint arXiv:1709.10000 (2017)
10. Ouaddah, A., Elkalam, A.A., Ouahman, A.A.: Towards a novel privacy-preserving access control model based on blockchain technology in IoT. In: Rocha, Á., et al. (eds.) Europe and Mena Cooperation Advances in Information and Communication Technologies. AISC, vol. 520, pp. 523–533. Springer, Cham (2017). https://doi.org/10.1007/978-3-319-46568-5_53
11. Outchakoucht, A., Hamza, E.S.S., Leroy, J.P.: Dynamic access control policy based on blockchain and machine learning for the internet of things. Int. J. Adv. Comput. Sci. Appl. **8** (7), 417–424 (2017)
12. Di Francesco Macsa, D., Mori, P., Ricci, L.: Blockchain based access control. In: Chen, Lydia Y., Reiser, Hans P. (eds.) DAIS 2017. LNCS, vol. 10320, pp. 206–220. Springer, Cham (2017). https://doi.org/10.1007/978-3-319-59665-5_15
13. Zhang, Y., Kasahara, S., Shen, Y., Jiang, X., Wan, J.: Smart contract-based access control for the internet of things. IEEE Internet Things J. 1–11 (2018)
14. Ihle, C., Sanchez, O.: Smart contract-based role management on the blockchain. In: Abramowicz, W., Paschke, A. (eds.) BIS 2018. LNBIP, vol. 339, pp. 335–343. Springer, Cham (2019). https://doi.org/10.1007/978-3-030-04849-5_30
15. Cruz, J.P., Kaji, Y., Yanai, N.: RBAC-SC: role-based access control using smart contract. IEEE Access **6**, 12240–12251 (2018)
16. Ethereum Homestead Documentation. https://ethereum-homestead.readthedocs.io/en/latest/index.html. Accessed 30 May 2019
17. Go Ethereum. https://geth.ethereum.org/. Accessed 27 May 2019
18. The Go Programming Language. https://golang.google.cn/. Accessed 30 May 2019
19. Ethereum (ETH) Blockchain Explorer. https://etherscan.io/. Accessed 30 May 2019

Customer Classification-Based Pre-sale Multi-value-chain Collaborative Mechanism Verification

Lei Duan[1,2], Wen Bo[1,3], Changmao Wu[1], Huansheng Ning[2],
and Changyou Zhang[1(✉)]

[1] Laboratory of Parallel Software and Computational Science,
Institute of Software, Chinese Academy of Sciences, Beijing,
People's Republic of China
changyou@iscas.ac.cn
[2] School of Computer and Communication Engineering,
University of Science and Technology Beijing, Beijing,
People's Republic of China
[3] School of Computer and Information Technology, Shanxi University,
Taiyuan, Shanxi, People's Republic of China

Abstract. During the automobile pre-sale service, understanding customer preferences and segmenting customer purchasing power can provide a foundation for automobile dealers' vehicle allocation strategy to manufacturers. It can improve the service ability of dealers and increase marketing revenue. This paper comes up with a verification method based on customer segmentation for multi-value-chain collaborative mechanism. First, we studied the process and inner accounting mechanism of the automobile marketing value chain and production value chain. We can explore the value-added of customer segmentation in each value chain and examine the extent to which customer segmentation results affect the value-added. Then, we construct a colored Petri net model for the value chain of each accounting unit. The sales service is mapped to transition. The sales resource management department is mapped to the place. The sales resource is mapped to the Token in the place. And we simulate the value-added process of the automobile marketing value chain and production value chain. Finally, we offer differentiated sales service processes for different categories of customers based on customer segmentation and design a multi-value-chain collaborative mechanism. On the CPN Tools simulation platform, we simulate the value-added quantitative effects in multi-value-chain collaborative state. We verified the correctness of the collaborative mechanism. In the simulation experiment, we monitored the number of automobile sales, the number of backlogs. The simulation result shows that the multi-value-chain collaborative mechanism based on customer segmentation designed in this paper can effectively increase the number of automobile sales by dealers and manufacturers. It provides a quantifiable basis for the optimization solutions of multi-value-chain.

Keywords: Customer segmentation · Multi-value-chain collaboration ·
Automobile Pre-sale Service · Colored Petri net · CPN Tools

© Springer Nature Switzerland AG 2019
D. Milošević et al. (Eds.): HCC 2019, LNCS 11956, pp. 86–97, 2019.
https://doi.org/10.1007/978-3-030-37429-7_9

1 Introduction

In recent years, China has a huge automobile consumer and the demand in the automobile market continues to grow. It not only promotes the rapid development of the domestic automobile manufacturing, but also attracts the world's major automobile multinationals to enter China. Domestic and foreign auto companies have launched fierce competition in this emerging market. The automobile value chain includes upstream procurement, midstream machining and downstream sales and services. The value of the collaborative mechanism is mainly reflected in the value-added by the services that it provides for the upstream and downstream operations of the multi-value-chain. Its purpose is to enhance the marketing revenue and service capabilities of upstream and downstream companies in the value chain. Customer segmentation is the product of modern marketing concepts [1]. It is the basis for better implementation of multi-value-chain collaboration. We divide customers into categories based on factors such as their natural characteristics, value, buying behavior, needs and preferences. The customer value and needs in each category are similar and we offer similar marketing strategies. We can take appropriate marketing tools for each customer group and provide products and services that meet the customer group. At the same time, we can also provide differentiated products and services for different customer groups. Enterprises have improved marketing efficiency within a limited range of resources [2].

In summary, customer segmentation is an important component of better implementation of pre-sales services. Pre-sales service multi-value-chain collaborative mechanism based on customer segmentation is the basis for enterprises to grasp customer needs and seek long-term development. This paper constructs a colored Petri net model of the dealer marketing value chain and the manufacturer production value chain. We provide differentiated service for different categories of customers and design multi-value-chain collaborative mechanism based on customer segmentation. Through the CPN Tools simulation platform, we simulate the gain effects of multi-value-chain collaboration and provide a quantifiable basis for multi-value-chain optimization.

2 Related Work

2.1 The Conception of Value Chain

The conception of value chain was first proposed by Michael E. Porter in 1985. Porter believes that "Every business is a collection of activities in the process of designing, producing, selling, sending and assisting its products. All of these activities can be represented by a value chain." The value creation of a company is made up of a series of activities. These activities can be divided into basic activities and auxiliary activities. Basic activities include inner logistics, production, external logistics, marketing and sales, services, etc.; Auxiliary activities include procurement, technology development, human resource management, and enterprise infrastructure. These production and management activities are different and related to each other and they constitute a dynamic process of creating value. This is the value chain. Every value activity in the value chain has an impact on how much value the company can achieve. Value chains play a major role in revenue, division of labor, and business strategy.

2.2 Petri Net Model

Petri Net was introduced in 1962 by Carl Adam Petri as a process modeling and analysis tool [3, 4]. Petri nets have an intuitive graphical representation and a solid mathematical foundation. It can effectively describe and model information systems, and it has strong dynamic analysis capabilities for system concurrency, asynchrony and uncertainty [5]. The system behavior characteristics of Petri net research mainly include reachability of state, location boundedness, transition activity, reversible initial state, and reachability between identifiers. The definition of Petri net is as follows.

Definition 1. A triplet N = (S, T, F) is a Petri net, if and only if:

① $S \cup T \neq \varnothing. S \cap U = \varnothing$;

② $F \cap (S \times T) \cup (T \times S)$;

③ $dom(F) \cup cod(F) = S \cup T, dom(F) = \{x | \exists y : (x, y) \in F\}, cod(F) = \{y | \exists x : (x, y) \in F\}$. They are the domain and range of F.

S is the place set of N; T is the transition set of N; F is the flow relationship.

2.3 The Conception of Customer Segmentation

The customer segmentation theory was first proposed by an American scholar, Wendell Smith, in the 1950s. It means that companies develop precise strategic strategies based on their business operations models. In a certain marketing, the customer is classified according to various factors such as the customer's natural characteristics, value, purchase behavior, demand and preferences. And provides targeted products, services and marketing within the limited resources of the company. The research on customer segmentation is a requirement for establishing a collaborative mechanism between marketing value chain and production value chain. If you divide your customers into multiple categories and the customer value or demand in each category is similar, its corresponding marketing strategy will be similar. Then we can take the appropriate marketing tools for each customer group and provide products and services that match the customer base. It can improve marketing efficiency [6]. Through customer segmentation, we can grasp the characteristics of different customer groups and provide customers with satisfactory products and services. It can maintain a good supply and demand relationship with customers and enhance their core competitiveness. The pre-sales service marketing model is to meet the needs of customers and establish a customer-centric auto marketing model [7, 8].

3 Petri Net Modeling of Pre-sales Service Multi-value-chain

3.1 Automobile Pre-sales Service Multi-value-chain

Automobile pre-sales service refers to a series of services that the company launches before the customer does not touch the product, which stimulates the customer's desire to purchase. Its main purpose is to accurately grasp the needs of customers and try to satisfy them [9]. The pre-sales service multi-value-chain includes the dealer marketing value chain and the manufacturer production value chain [10]. The dealer marketing value chain starts from the customer to the store, enters the customer information, and understands the customer's intention. It develops a tracking sales plan based on customer preferences and needs. And it provides a series of marketing services in the process of customers purchasing cars, handling customer objections and handling delivery business. The dealer marketing value chain is shown in Fig. 1. The manufacturer production value chain begins with the receipt of customer orders. It is approved by financial review and then develops a production plan. The vehicle is in storage after the quality inspection is completed [11]. The manufacturer production value chain is shown in Fig. 2.

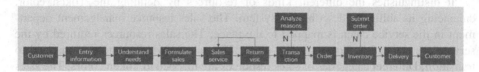

Fig. 1. The dealer marketing value chain.

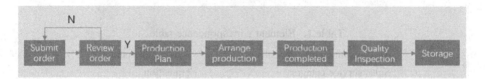

Fig. 2. The manufacturer production value chain.

After analyzing the dealer marketing value chain and the manufacturer production value chain, we understand its workflow and inner accounting mechanism, clarify the gain link and gain factors, and lay the foundation for the design of multi-value chain coordination mechanism and clarify the gain link and gain factors. It can lay the foundation for the design of multi-value-chain collaborative mechanism. The main factor affecting the revenue of dealers and manufacturers is the number of automobile sales.

3.2 Multi-value-chain Modeling Based on Colored Petri Net

In the process of establishing a multi-value-chain network model, we must consider the value-added factors of each link. This paper designs and verifies the multi-value-chain of automobile pre-sale service by using colored Petri net system. The colored Petri net defines a Token color set for each place in the network and defines an action color set for each transition. It can not only describe more complex systems, but also make the Petri net system simple and clear [12]. The definition of a colored Petri net is as follows.

Definition 2. A colored Petri net can be represented as a seven-tuple: CTPN = (P, T, I, O, C, TF, M_0). Among them:

P = {p1, p2, ..., pn} is a limited place set;

T = {t1, t2, ..., tm} is a finite set of transitions, and $P \cap T \neq \Phi$, $P \cup T \neq \Phi$;

I(p, t) is the input function from the place p to the transition t: $C(p) \times C(t) \rightarrow N0$ (Non-negative integer). It corresponds to the colored directed arc from p to t;

O(p, t) is the output function from the transition t to the place p: $C(t) \times C(p) \rightarrow N0$ (Non-negative integer). It corresponds to the colored directed arc from t to p;

C is a color set associated with place and transitions;

TF is the set of delays for all transitions;

M_0: $P \rightarrow N0$ is the initial identification. In the graphical representation, the place is represented by a circle; the transition is represented by a rectangle; the flow relationship between the elements is represented by an arc with an arrow.

It distinguishes the different kinds of resources by defining the Token color, enhancing its ability to describe the system. The sales resource management department in the service chain is mapped to the place; The sales resources required by the service are mapped to Tokens in the place; The sales service category is mapped to transition; Different categories of sales resources are mapped to Token colors; The sales resource information is mapped to a Token value. They are important factors influencing the gains of the value chain. The corresponding relationship of elements is shown in Table 1.

Table 1. Element correspondence table

Multi-value-chain element	Petri net element
Sales service category	Transition
Sales resource management department	Place
Sales resources	Token
Categories of sales resources	Token color
Sales resource information	Token value
Service control	Directed arc

We add control elements based on the colored Petri nets to represent differentiated services, control service occurrence probability, and resource constraints for different customers. The automobile pre-sales service network system is as follows.

Definition 3: Automobile Pre-sale Service Network System

Π = (P, T; I, O, C, K, D, M_0) The necessary and sufficient condition for it to be called the pre-sale service network system is:

$\sum = (P, T; I, O, C, M0)$ is a colored network system;

For $t \in T$, D(t) is the control function of transition t;

For $p \in P$, K(p) is the capacity limit function of the place p.

3.3   Constructing a Single Value Chain Petri Net Model

Building a dealer marketing value chain Petri net is shown in Fig. 3. First, after the customer comes to the store, we confirm whether the customer is coming to the store for the first time and register the customer information. We can understand customer preferences and needs based on customer information. Then we develop a tracking sales plan. In the first phase (P1), the salesperson provides the vehicle information to the customer and displays the appearance and function of the vehicle. Then he introduces the advantages and benefits of the customer's interesting configuration and invites the customer to feel inside the vehicle; In the second stage (P2), the salesperson provides the test drive service for the customer after the static display of the vehicle. This will give the customer a sensible understanding of the vehicle through a direct driving experience. It will enhance the customer's confidence in the actual driving impression of the vehicle's various functions. This will enhance the customer's recognition of the product and help to further stimulate the customer's desire to purchase; In the third stage (P3), after the customer has not decided to purchase the vehicle to leave the store, the salesperson should analyze the reason and actively contacts the customer to handle the customer objection. When the customer intends to purchase a vehicle, the salesperson checks the inventory to confirm whether there is a vehicle. The customer can pick up the car at the first time after signing the order, which is also one of the important factors affecting the customer's order. It will directly affect the number of vehicles sold. Finally, we will handle the delivery procedures for the customers and make follow-up visits. It reflects the continuity of the service. It can enhance customer satisfaction and win business opportunities in the aftermarket [13]. Building a manufacturer production value chain Petri net is shown in Fig. 4. After receiving the order, the manufacturer generates a production plan and arranges production. The produced vehicles undergo a series of quality and safety tests [14]. The vehicles passing the inspection are arranged into the warehouse.

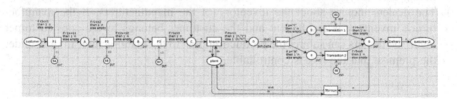

Fig. 3. Dealer marketing value chain Petri net model.

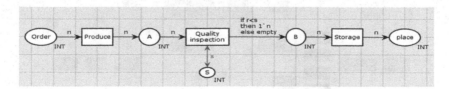

Fig. 4. Manufacturer production value chain Petri net model.

4 Pre-sale Service Multi-value-chain Collaboration

4.1 Multi-value-chain Collaborative Mechanism

Multi-value-chain collaborative mechanism provides a way for manufacturers and dealers to work together. For the manufacturers, it realizes the management of informationization in the sales process. It optimizes the inventory structure; Dealers can use the collaborative mechanism to get some basic information about the vehicles in the factory. It can complete the operation of querying, transferring and selling vehicles. In addition, the manufacturer can keep abreast of the sales of the dealers. It can improve the business production efficiency and operational efficiency between the two, and increase the market competitiveness of the enterprise [15].

The collaborative mechanism between the dealer marketing value chain and the manufacturer production value chain is shown in Fig. 5. First, the customer is divided according to various factors such as customer basic characteristics, customer value, purchase behavior, demand and preferences. Customers are divided into four categories: core customers, senior customers, general customers and potential customers. It provides differentiated services to different categories of customers based on customer segmentation results and establishes collaborative between dealer and manufacturer. The collaborative points are:

(1) For core customers and senior customers, dealers and manufacturers will provide personalized services for them. (Such as: automobile surface color, leather seats, seat ventilation and heating and performance kits, etc.) It increases the interests of manufacturers and dealers while meeting customer preferences.

(2) The dealers can check the inventory of the manufacturers when the inventory is insufficient. It can cooperate with the manufacturer to make the whole vehicle transfer. It guarantees that the high-level customers will handle the delivery business the first time after placing the order. It not only improves dealer marketing efficiency, but also improves customer satisfaction.

(3) The dealers inquire about the inventory of the manufacturers. If the manufacturer's inventory is insufficient and the customer belongs to the core customer level, the dealers can submit the production plan to the manufacturer in time under the premise that the risk is controllable.

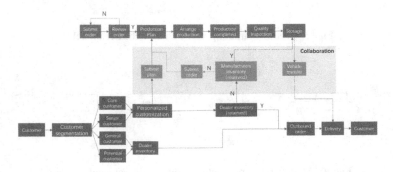

Fig. 5. Marketing value chain and production value chain collaborative Petri Net Model.

4.2 Verification of Collaborative Mechanism

Through the CPN Tools simulation platform, we validate the collaborative mechanism proposed in the previous section. After customer segmentation, customers are divided into four categories: core customers (Customer A), senior customers (Customer B), general customers (Customer C) and potential customers (Customer D). We provide differentiated marketing services for different categories of customers as shown in Fig. 6.

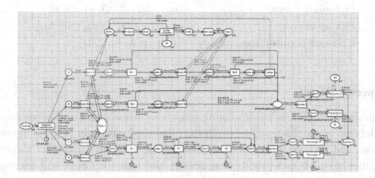

Fig. 6. Marketing value chain and production value chain collaborative simulation model.

5 Simulation Experiment

5.1 Experimental Environment and Parameter Settings

The Petri net models of the marketing value chain and production value chain, and the Petri net model of the multi-value-chain collaborative mechanism based on customer segmentation were tested on the CPN Tools simulation platform. We analyze the historical data of corporate customers through customer segmentation, and summarize the proportion of core customers, senior customers, general customers and potential customers.

We set the random number to control the customer classification ratio in the Petri net model. We take the number of customers and the number of automobiles purchased as input. We control the probability of each service link by setting a random number. We monitor the value-added benefits of each service link and verify the correctness of multi-value-chain collaboration based on customer segmentation. CPN Tools uses the ML language to define specifications and describe the network. It controls the data type by defining a color set and describes the color set by predefining a color set function. The main parameter types defined in this paper are shown in Table 2.

Table 2. Main parameter description table.

Name	Type	Description
Number of automobiles purchased	int	n
Dealer inventory	int	m1
Manufacturer inventory	int	m2
Reserve dealer inventory	int	a
Reserve manufacturer inventory	int	b
Insufficient inventory	int	c
Probability control	int	s1, s2, s3, s4, s5, s6
Pick up the car	string	q ("Y": Yes, "N": No)
Customer segmentation category	string	p(A, B, C, D)

5.2   Expected Result

In the experimental process of simulating marketing value chain and production value chain collaborative mechanism, we input the quantified customer automobile purchase information, and provide differentiated services for different customers through customer segmentation. It can predict that dealers and manufacturers will increase their automobile sales after adopting the collaborative mechanism.

5.3   Result Analysis

In the same experimental environment, the multi-value-chain Petri net model with collaborative mechanism and the multi-value-chain Petri net model without collaborative mechanism were simulated respectively. We control the input of the customer's car purchase information in the two sets of models. The same service uses the same transaction probability. After 10 simulation experiments, we monitor the number of automobile sales and the amount of backlog inventory as shown in Table 3.

Table 3. Simulation experiment result table.

Simulation experiment number	The multi-value-chain Petri net model with collaborative mechanism		The multi-value-chain Petri net model without collaborative mechanism	
	Automobile sales	Backlog inventory	Automobile sales	Backlog inventory
1	1681	368	1179	311
2	1690	360	1212	305
3	1594	410	1146	304
4	1766	344	1308	269
5	1646	351	1197	288
6	1672	379	1069	311
7	1730	341	1320	273
8	1591	389	1185	284
9	1709	340	1167	293
10	1556	327	1201	267
Average	1663	360	1198	290

In order to more clearly compare the experimental results of the multi-value-chain Petri net model with collaborative mechanism and the multi-value-chain Petri net model without collaborative mechanism, according to the data in the table, we draw as shown in Fig. 7.

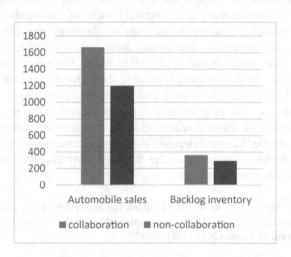

Fig. 7. Comparison of simulation results.

After tidying up the results of ten experimental results, we found that in the multi-value-chain collaborative network model, the average number of automobile sales is 1663, and the number of backlogs is 360. In the multi-value-chain non-cooperative network model, the average number of automobile sales is 998, and the number of backlogs is 110. According to the experimental results, the dealers and manufacturers adopt the multi-value-chain collaborative mechanism. Although the number of back-logs in the automobile has increased slightly, the benefits are more objective. Through simulation experiments, we prove the correctness of the pre-sale multi-value-chain collaborative mechanism based on customer segmentation. It is effective in increasing the sales of dealers and manufacturers, and it brings them greater benefits.

6 Conclusions and Prospects

This paper offered a pre-sale multi-value-chain collaborative mechanism based on customer segmentation, and simulates the multi-value-chain collaborative service process. After analyzing the income situation, we ultimately verify the correctness and practicability of the collaborative mechanism and provide a quantitative reference for the optimization of multi-value-chain. The research content the following aspects: First, we analyzed the research status of customer segmentation, multi-value-chain collaboration, and Petri net application. We analyzed the composition of the dealer marketing value chain, the manufacturer production value chain, and explore the value-added role of customer segmentation in each value chain. We examined the extent to which customer segmentation results affect the multi-value-chain. Then we built two colored Petri net models of the value chains through the CPN Tools platform. Finally, based on the gain effect of customer segmentation in each value chain, we established intact multi-value-chain collaborative Petri net model and conduct simulation experiments. The experimental result shows that the multi-value-chain collaborative mechanism based on customer segmentation can effectively increase the number of automobile sales by dealers and manufacturers. It can improve the marketing efficiency of enterprises and bring more benefits to dealers and manufacturers. However, there are still some problems in this research that need further study: We select customer segmentation variables based on the company's existing data. Due to the continuous changes of enterprises and customers, the selection of customer segmentation variables should be sustainable adjusted. The proportion of customer classification in the collaborative mechanism also should be adjusted according to the customer segmentation results, and adapt the market in time. The next step is optimizing customer segmentation based on a large amount of customer data, improving the rationality and accuracy of customer segmentation in further research work.

Acknowledgment. The authors would like to thank the anonymous referees for their valuable comments and helpful suggestions. This paper is supported by The National Key Research and Development Program of China (2017YFB1400902).

References

1. Viegas, J.L., Vieira, S.M., Melicio, R., Mendesjoao, V.M.F., Sousa, M.C.: Classification of new electricity customers based on surveys and smart metering data. Energy (09), 123–125 (2016)
2. Parsell, R.D., Wang, J., Kapoor, C.: CUSTOMERSEGMENTATION: WO/2014/099928 (2014)
3. Murata, T.: Petri nets: properties, analysis and applications. Proc. IEEE **77**(4), 541–580 (1989)
4. Yuan, C.Y.: Principle and Application of Petri Nets. Publishing House of Electronics Industry, Beijing (2005). (in Chinese)
5. Zhou, M., Wu, N.: System modeling and control with resource-oriented Petri nets. Computer Science, Engineering & Technology (2010)
6. Bao, Z., Zhao, Y., Zhao, Y.: Segmentation of Baidu takeaway customer based on RFA model and cluster analysis. Comput. Sci. **45**(S2), 436–438 (2018)
7. Yu, Q., Cai, L., Tan, X.: Airport emergency rescue model establishment and performance analysis using colored Petri Nets and CPN tools. Int. J. Aerosp. Eng. **2018** (2018). Article ID 2858375, 8 pages
8. Jiang, S., Cai, S., Olle Olle, G., Qin, Z.: Durable product review mining for customer segmentation. Kybernetes **44**(1), 12–17 (2015)
9. Yang, Z., Zhang, J., Chen, X.: The pricing strategy for the dual channel supply chain with pre-sale service. In: 2018 15th International Conference on Service Systems and Service Management (ICSSSM) (2018)
10. Xin, Z., Minghai, Y., Mingdi, J.: Study on designing modern auto service marketing mode based on value chain theory. In: The 3rd International Conference on Information Sciences and Interaction Sciences, 23–25 June 2010

11. Arciniegas, D., Herrera, M.A., Táutiva, K.: Automation process modeling of a electric cars production line through Petri nets and GRAFCET. In: 2017 IEEE 3rd Colombian Conference on Automatic Control (CCAC), 18–20 October 2017

12. Chen, Z.-Y., Fan, Z.-P., Sun, M.: A multi-kernel support tensor machine for classification with multitype multiway data and an application to cross-selling recommendations. Eur. J. Oper. Res. **11**, 34–37 (2016)

13. Wei, H.: Empirical study on Chinese car enterprise upgrading based on global value chain. In: 2011 International Conference on Product Innovation Management (ICPIM 2011) (2011)

14. Nauhria, Y., Kulkarni, M.S., Pandey, S.: Development of strategic value chain framework for indian car manufacturing industry. Glob. J. Flex. Syst. Manage. **19**(1), 21–40 (2018)

15. Yuxin, Y., Hongyan, Z.: The research of auto-adaptive production strategy based on supply chain coordination. In: 2010 3rd International Conference on Information Management, Innovation Management and Industrial Engineering, 26–28 November 2010

A Real-Time Update Approach for Visualizing Multidimensional Big Data

E. Haihong$^{(\boxtimes)}$, Huihui Kong, Yunfeng Liu, Meina Song,
and Zhonghong Ou

Beijing University of Posts and Telecommunications, Beijing 100876, China
{ehaihong, hhkong, yfliu, mnsong,
zhonghong. ou}@bupt. edu. cn

Abstract. Multidimensional data in different fields (e.g., science and technology resources, medicine, finance and transportation) have emerged in the big data era. Technology of visualization is critical to analyze multidimensional data which is large-scale. Thus, the highly variability of multidimensional data leads to the necessity of improving data management efficiency by updating data in real time. Nevertheless, traditional real-time communication technology suffers from low efficiency and large bandwidth consumption. Based on WebSocket, this paper proposes a multidimensional data service platform architecture that is scalable, real-time and versatile. First, on the platform, we compare the network load and network latency above Ajax polling and WebSocket. Secondly, for multidimensional data, we further validate the effectiveness of WebSocket in the application scenarios of frequent updating and multi-user concurrent requests. WebSocket-based platform greatly shortens the time for people to understand and analyze massive and complex multidimensional data. The application results show that the design is effective for real-time updating of multidimensional data.

Keywords: WebSocket · Multidimensional data · Real-time updating

1 Introduction

With the rapid development of computer networks, multidimensional data generated in different fields are exponentially growing. These data are not only large-scale and complex, but also increasingly diversified.

Since web technology is widely used in different fields, it is important to visually display multidimensional data in the procession of analysis. Visualization technology is widely used to analyze the multidimensional data, which presents data through charts. However, there are two major problems for visualization technology. First, the huge amount of multidimensional data and the concurrent requests of users lead to the decline in the performance of visualization. Second, the displayed chart needs to be redrawn when the user changes the data, which reduces the efficiency of work for user. The real-time updating is the core part in visualization technology, which not only solves the problem of synchronous updating for multidimensional data and charts, but also supports the concurrent requests of multi-user by WebSocket. Through monitoring

D. Milošević et al. (Eds.): HCC 2019, LNCS 11956, pp. 98–104, 2019.
https://doi.org/10.1007/978-3-030-37429-7_10

and analyzing the multidimensional data in real-time, it not only increases the management efficiency of the diversified multidimensional data, but also helps to observe the changing status of systems, devices, networks and so on, which ensure that it runs stably.

For real-time updating of multidimensional data, Teng et al. [3] collected and transformed the real-time operational data which finally completed the real-time push of the monitoring data. Lin et al. [4] designed a real-time system, which completed the dynamic adjustment of the real-time graph for the log multidimensional data in the data warehouse. By combining OpenLayers and WebSocket, Bao et al. [5] designed a platform architecture which supported real-time update for data of tourism geographic. However, there were two defects in these systems. First, the aforementioned schemes were implemented on the basic of multidimensional data in the specific field, which leaded to the application limitations. Second, they did not visualize the multidimensional data and update the charts of data in real-time, which leaded to the decline in the rendering effect and analysis efficiency for the multidimensional data.

In order to solve the above problems, this paper proposes a general platform architecture in real-time updating for multidimensional data. The main idea is to maintain a real-time connection between multidimensional data and charts through the WebSocket.

In general, this paper mainly solves the three problems as follows:

(1) We implemented a stable system to visualize the multidimensional data into charts, which was suitable for different areas.
(2) We achieved real-time update for visualizing multidimensional data by WebSocket, which conquered high concurrent requests in different scenarios.
(3) We collected the multidimensional data in different fields and examined the stability and error rate for the system. The experimental results demonstrated the superiority of our approach.

2 Related Work

The visualization of multidimensional data is becoming more and more mature when the multidimensional data has continuously generated in different fields. Jin et al. [1] proposed a visualization system of infectious disease based on infectious disease data in the medical field, which effectively helped doctors to explore the pattern of infectious diseases. Song et al. [2] proposed an improved mode of book interview based on multidimensional data analysis, which synthesized feedbacks after article researches. Zhao et al. [17] attempted to calculate and analyze the functional strength of various types of urban development land, and then achieved the comprehensive assessment of various types of development land by the network address analysis data. In conclusion, the multidimensional data is widely used in the visualization system described above. However, they lack the real-time analysis for multidimensional data.

The technology of real-time updating is widely used in various types of visualization systems, which has reached significant achievement. Zheng et al. [9] provided a solution based on WebSocket which achieved two-way communication to really accomplish the real-time data updating. It greatly improved the management level of system and reduced the network latency and the burden of the server. Zhong et al. [6] used cache database to save the real-time GPS information from the interactive terminal server. And then the pushing service pushed the newest GPS information to Web application client. In this way, some more serious disasters would be avoided. By combining Node.js and Socket.io, Qi et al. [7] designed and implemented a message push platform in real-time based on WebSocket. Pimentel et al. [8] built a Web application to compare the latency of WebSocket to HTTP polling and long polling. Jin et al. [11] designed and implemented a WebSocket server which can provide real-time message push service to a large number of different users' subscription requests.

3   A Visual Analysis Platform for Multidimensional Data

The continuous accumulation of the enterprise data has brought great pressure to the database. In order to analyze multidimensional data efficiently, the analysis platform built in this paper mainly achieves the following aspects. On the one hand, we define the data model for the multidimensional data sets and build the Cubes by KYLIN. Finally, we cache the multidimensional Cubes which are set by users into HBase and wait for the analysis. On the other hand, the multidimensional data is usually growing over time. KYLIN is used to achieve the incremental calculation and store the Cube for avoiding the full calculation of Cube, which increases the speed of querying the multidimensional data in real-time. In the end, we apply Echarts technology to visualize and analyze the multidimensional data by dragging dimensions and measures. The Echarts provides a variety of charts, which explore the value of multidimensional data from different perspectives. We push the new multidimensional data in real time by WebSocket when the user changes the data. The pipeline of our scheme is shown in Fig. 1.

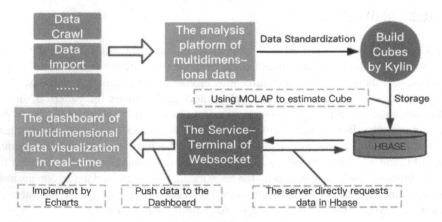

Fig. 1. The pipeline of our scheme.

4 Design and Implementation of Real-Time Updating

4.1 The Introduction of WebSocket

WebSocket is a new communication protocol based on HTML5. WebSocket provides full-duplex and two-way communication on a single persistent connection, which implements the communication in real-time for big data between the web page and the server. The two-way connection channel is mainly established through the TCP socket. Then, the real-time verification between the client and the server is implemented through the handshake protocol. The server can push the updating information to the client after successful verification. The request header in each client contains an Upgrade field. The browser connects to different protocols based on this field. After upgrading to the WebSocket protocol, the browser sends data to the server through the method of send, and receives the data returned by the server through the event of onmessage.

4.2 Experimental Result

The Comparison Between WebSocket and Ajax Polling. Based on multidimensional analysis platform, this paper verified the real-time and high efficiency of WebSocket in real-time updating. WebSocket and Ajax polling are respectively applied to meet the demand of dashboard in real-time updating. Through the displayed data of the development test tool in Chrome browser, we compare the request and response information in the procession of communication. As shown in Table 1, we compare the average resource cost which required for each connection between AJAX polling and WebSocket.

Table 1. The comparison of average resource cost between AJAX polling and WebSocket.

Method of implementation	Need to transmit data frames	Average data frame size (Byte)
	Establish connection	Request connection
AJAX polling	8	871
WebSocket	1	54

In Table 1, Ajax polling has a large number of header information and requests when it establishes and maintains a connection. In contrast, WebSocket only needs one handshake to establish a stable connection with less header information, which is only generated during the handshake. At the same time, the multiple delays caused by frequent requests and responses results from the Ajax polling requiring a persistent connection to be maintained. And the server is unable to send any message to the browser during the delay, which results in the waste of resources. On contrary, in WebSocket, the message will immediately return to the browser when it arrives at the server. After a delay, the channel will always remain open and there is no necessary to send requests to the server repeatedly. Therefore, the network delay and network congestion are greatly reduced, so that the performance in real-time is improved.

We verified the network overhead of each request in Ajax polling and WebSocket. Three different sets of examples were used. (A: 1000 clients, polled once per second; B: 10000 clients, polled once per second; C: 100000 clients, polled once per second). Ajax polling sends full header information every request. Therefore, the network overhead of the Ajax polling request is 871 bytes as shown in Table 1, while WebSocket sends header information only for first time request. Therefore, each message in WebSocket is a WebSocket frame with an overhead of only 2 bytes. The network throughput of the two methods is compared as shown in Fig. 2.

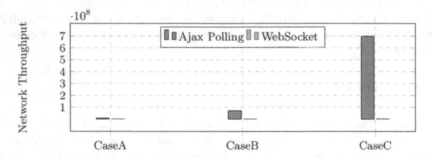

Fig. 2. The network throughput of the two methods.

It can be concluded from the Fig. 2 that WebSocket has the great performance advantage over Ajax Polling in the case of a large number of requests.

The Test of High Concurrent Transmission in WebSocket. There are some problems in WebSocket for the multidimensional analysis platform. On the one hand, the analysis platform employs multidimensional data, which is continuously updated as the time goes. That leads to high concurrent transmission. On the other hand, different users analyze the multidimensional data on this platform, which leads to high concurrent requests. Thus, we verified whether the high concurrency and stability of the WebSocket server was up to standard in the case of frequently requesting. The number of different concurrent connections was respectively exercised. Finally, we summarized the performance of WebSocket which was shown in Fig. 3.

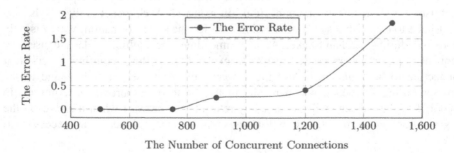

Fig. 3. With the increase of concurrent connection, we demonstrated the error rate of platform.

As shown in Fig. 3, when the number of concurrent connections is less than 750, the error rate is 0. When the number of connections reaches 1500, the error rate is about 1.83%, which means that the WebSocket can meet the requirement of the multidimensional data in real-time. It reaches good stability and practical applicability under the mode of multi-application and multi-tenant.

The Result of Visualization. Finally, in order to show real-time effects better, we capture the real-time changes of data by visualizing the data into indicator cards. The final result of visualization is shown in Fig. 4.

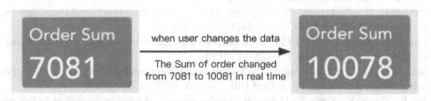

Fig. 4. The result of visualization.

5   Conclusion

This paper mainly proposes a real-time update technology of data and graphs on the multidimensional data visualization platform. We compared the difference between Ajax polling and WebSocket scheme in network load and network delay. Finally, we choose WebSocket as the technology solution of real-time updating with relatively high performance. At the same time, we tested the high concurrency performance of WebSocket due to the rapid change of the multidimensional data, which is great for the scenarios of multi-user in practical applications. In the end, we give the result of visualization. The experimental results show that WebSocket achieve the intended design goals in terms of functionality and performance, which not only reduces the number of network requests and network delay, but also maintains the system stability in high concurrent requests.

In the future work, we will combine WebSocket with the front-end framework (e.g., VUE, React, Angular) to provide a more effective solution for application systems.

Acknowledgements. This work is supported by the National Key R&D Program of China (Grant No. 2017YFB1401501) and Engineering Research Center of Information Networks, Ministry of Education.

References

1. Jin, S., Tao, Y., Yan, Y., Dai, H.: Disease patterns analysis based on visualization of multidimensional space-time data. J. Comput. Aided Des. Comput. Graph. **31**(2), 241–255 (2019)
2. Song, A., Zhu, Y., Yuan, S.: An empirical study of library acquisition improvement based on multidimensional data analysis, pp. 1–8. Library Forum (2019)

3. Teng, H., Teng, F.: Multidimensional database technology in real time monitor and control system. Ph.D. dissertation (1999)
4. Lin, Z., Yang, D., Song, G.: Materialized views selection of multi-dimensional data in real-time active data warehouses. Ph.D. dissertation (2008)
5. Bao, S., Zhou, W., Zhu, R.: Design and implementation of real - time visualization of tourism resources configuration based on OpenLayers and WebSocket. Geomatics Spat. Inf. Technol. **40**(12), 122–125 (2017)
6. Zhong, B., Tao, Z.: Implementation for system of real-time pushing vehicle's GPS information based on WebSocket. Electron. Des. Eng. **13**, 48–52 (2017)
7. Hua, Q., Jia, L., Jun, L.: Design and implementation of real-time server message push based on Websocket. Microprocessors **37**(3), 36–39 (2016)
8. Pimentel, V., Nickerson, B.G.: Communicating and displaying real-time data with WebSocket. IEEE Internet Comput. **16**(4), 45–53 (2012)
9. Zhend, L., Zheng, X., et al.: Research on power system real-time data updating based on WebSocket. Comput. Mod. **1**(1), 85–87 (2013)
10. Ma, K., Zhang, W.: Introducing browser-based high-frequency cloud monitoring system using WebSocket proxy. Int. J. Grid Util. Comput. **6**(1), 21–29 (2014)
11. Jin, F., Zhang, Y., Yong, P.: Design and implementation of monitoring system in B/S mode base on two servers. Comput. Simul. **31**(2), 201–205 (2014)
12. Chen, L., Xiang-yang, F., et al.: Study and implementation of html5 WebSocket handshake protocol. Comput. Appl. Softw. **32**(1), 128–131 (2015)
13. Liu, J., et al.: Service impact analysis and visualization display based on massive real-time monitoring data. Telecommun. Sci. **34**(3), 192–196 (2018)
14. Heer, J., Kong, N., Agrawala, M.: Sizing the horizon: the effects of chart size and layering on the graphical perception of time series visualizations. In: Proceedings of the SIGCHI Conference on Human Factors in Computing Systems, pp. 1303–1312. ACM (2009)
15. Qin, J.: Research and performance analysis on instant messaging based on WebSocket. Mob. Commun. **41**(12), 44–48 (2017)
16. Xue, L., Liu, Z.: Network real-time communication based on WebSocket. Ph.D. dissertation (2014)
17. Zhao, M., Liang, J., Guo, Z.: Classifying development-land type of the megacity through the lens of multisource data. Urban Res. **5**, 72–77 (2018)
18. Zhao, H.: A platform for real time multidimensional statistical analytics of large-scale log data. Comput. Sci. Appl. **7**(04), 351 (2017)
19. Zhao, W., et al.: A real-time identification method of highway service level based on multidimension data. J. Transp. Syst. Eng. Inf. Technol. **18**(4), 240–246 (2018)
20. Fan, T.: Development of online configuration monitoring system based on WebSocket and SVG. Master's thesis, Nanjing University (2018)
21. Zhang, H.: Application of real time web monitoring technology based on WebSocket and Redis in WiFi management. China Comput. Commun. (6), 43–45 (2017)

Cultural Psychology Analysis on Ancient Chinese Archive

Miaorong Fan[1,2], Fugui Xing[1,2], Hua Li[3(✉)], and Tingshao Zhu[1(✉)]

[1] Institute of Psychology, Chinese Academy of Sciences,
Beijing 100101, People's Republic of China
tszhu@psych.ac.cn
[2] Department of Psychology, University of Chinese Academy of Sciences,
Beijing 100049, People's Republic of China
[3] Institute of Qilu-Culture Studies, Shandong Normal University,
Jinan 250014, China
lh999ren@163.com

Abstract. With the developing of big data analysis technology and the digitization of ancient Chinese books, we are able to do more thoroughly psychology-related analysis on Chinese history, culture and people. It is an opportunity that we can have a glance at Chinese behavioral and psychological variation from 2000 years ago to the beginning of the nineteenth century. We aim at figuring out the change of the mental process of Chinese people through the past 2000 years and how the mind and behavior are shaped by the specific culture at each stage, based on the analysis of behavior and psychology, with the usage of big data and artificial intelligence techniques. In this paper, we construct the Ancient Chinese Archive (ACA), beginning with twenty-five official dynastic history books. The timeline begins with the pre-Qin period and ends with Qing dynastic. Our selection contains about 27 million Chinese characters (except punctuations). For better observation of the variation tendencies through every period, we merge some books into one period and divide all the dynasties into 9 parts. In each period, we count the frequency of some keywords such as Xiào (孝, filial piety), Lǐ (禮/礼, proper rite), etc., which represent Chinese social morality and culture ethics. We analyze the change process of the word frequency through time and areas and try to find out the explanation of these trends.

Keywords: Big data · Culture psychology · Chinese history · Twenty-Five Histories · Ancient Chinese Archive

1 Introduction

China is one of the world's oldest civilizations [1], with thousands of years of history. With the developing of big data analysis technology, and digitization of ancient Chinese books, we are able to do more thoroughly psychology-related researches on Chinese history, culture and people. A history book is a microcosm of a dynasty. We can learn a lot from the history of ancient China. The inner world of many history figures is worth exploring and analyzing. It is an opportunity that we can have a glance

© Springer Nature Switzerland AG 2019
D. Milošević et al. (Eds.): HCC 2019, LNCS 11956, pp. 105–110, 2019.
https://doi.org/10.1007/978-3-030-37429-7_11

at Chinese people's behavioral and psychological variation from 2000 years ago to the beginning of the nineteenth century.

In this paper, we aim at figuring out the changing of mental process of Chinese people through the past 2000 years, based on the analysis of behavior and psychology, with the usage of big data and artificial intelligence techniques. We are transferring modern psychological prediction models (such as personality prediction, etc.) to ancient context.

In this paper, we construct the Ancient Chinese Archive (ACA), beginning with twenty-five official dynastic history books. Twenty-Five Histories are also known as the Twenty-Five Official Dynastic Histories (Chinese: 二十五史), including 25 orthodox histories books, covering a period from 3000BC to Qing dynasty. They are the most important bibliographies and are of great reference value when we research on China's history.

2 Data Sources

To conduct the analysis, we acquire data from twenty-five official dynastic history books. The timeline begins with the pre-Qin period and ends with Qing dynastic. Selection contains 25 books and about 27 million Chinese characters (except punctuations).

For better understanding of the variation tendency through every period, we merge some books and divide all the dynasties into eight parts. For Records of the Grand Historian, it corresponds from the age of the legendary Yellow Emperor to the reign of Emperor Wu of Han. The part of Han dynasty is coincided with Book of Han, so we remove the part of western Han dynasty and select biographies from the pre-Qin and Qin dynasty period, including Basic Annals (No. 1 to No. 7), Hereditary Houses (No. 1 to No. 21), Ranked Biographies (No. 1 to No. 31). The whole collection and dynasties merging are as Table 1.

Table 1. Data collection and dynasties merging.

Merged dynasties	Book title	Number of words in Chinese (except punctuations)
Pre-Qin period and Qin dynasty	*Records of the Grand Historian*	253,301
	Basic Annals (No. 1 to No. 7)	
	Hereditary Houses (No. 1 to No. 21)	
	Ranked Biographies (No. 1 to No. 31)	
Han dynasty	*Book of Han*	739,695
	Book of the Later Han	1,372,655

(*continued*)

Table 1. (*continued*)

Merged dynasties	Book title	Number of words in Chinese (except punctuations)
Wei, Jin, Southern and Northern dynasties	*Records of the Three Kingdoms*	452,674
	Book of Jin	1,156,143
	Book of Song	810,379
	Book of Southern Qi	293,016
	Book of Liang	295,500
	Book of Chen	164,117
	Book of Wei	998,411
	Book of Northern Qi	128,295
	Book of Zhou	259,628
Sui, Tang and Five dynasties	*Book of Sui*	673,473
	History of the Southern Dynasties	662,256
	History of the Northern Dynasties	1,092,522
	Old Book of Tang	1,989,653
	New Book of Tang	1,634,578
	Old History of the Five Dynasties	312,277
	Historical Records of the Five Dynasties	290,155
Song dynasty	*History of Song*	3,911,239
Liao and Jin dynasty	*History of Liao*	354,130
	History of Jin	921,984
Yuan dynasty	*History of Yuan*	1,557,751
Ming dynasty	*History of Ming*	2,750,857
Qing dynasty	*Draft History of Qing*	4,405,553

The total amount of text for each period is shown in Table 2.

Table 2. The total amount of text for each period.

Merged dynasties	Total amount of text
Pre-Qin period and Qin dynasty	253,301
Han dynasty	2,112,350
Wei, Jin, Southern and Northern dynasties	4,558,163
Sui, Tang and Five dynasties	6,654,914
Song dynasty	3,911,239
Liao and Jin dynasty	1,276,114
Yuan dynasty	1,557,751
Ming dynasty	2,750,857
Qing dynasty	4,405,553

3   Data Analysis

We have conducted multiple analysis on this dataset, such as the frequency changes of some keywords inside the history books and do cultural psychology research. We also try to transfer psychology prediction models such as personality model into ACA, and do computation using the models after transfer.

3.1   Cultural Psychology Analysis

Culture is created by people, and people are influenced unconsciously by the cultural around them. In Chinese culture, Confucianism is the most influential genre and the mainstream consciousness in ancient China.

In Confucian culture, the three cardinal guides and the five constant virtues are very important. Confucianism maintains the ethics and moral system of the society through the education of the three cardinal guides and the five constant virtues and plays an extremely important role in the development of Chinese civilization.

The Three Cardinal Guides include: ruler guides subject, father guides son and husband guides wife. Five Constant Virtues include Rén (仁, benevolence, humaneness); Yì (義/义, righteousness or justice); Lǐ (禮/礼, proper rite); Zhì (智, knowledge); Xìn (信, integrity). Xiào (孝, filial piety) is a core concept inside the three cardinal guides and is advocated by Confucianism, so we choose Xiào (孝, filial piety) for our analysis.

We count the frequency statistics for Xiào (孝, filial piety) in each period. Here, frequency of a specific word is defined as below:

$$frequency\ of\ word\ x = \frac{number\ of\ word\ x}{total\ amount\ of\ text} \times 100\% \qquad (1)$$

For example:

$$frequency\ of\ word\ Xiào = \frac{number\ of\ Xiào}{total\ amount\ of\ text} \times 100\% \qquad (2)$$

Table 3. Statistical result for frequency of word Xiào (孝, filial piety).

Merged dynasties	Number of word Xiào	Frequency of word Xiào
Pre-Qin period and Qin dynasty	191	0.0754%
Han dynasty	2312	0.1095%
Wei, Jin, Southern and Northern dynasties	4797	0.1052%
Sui, Tang and Five dynasties	7480	0.1124%
Song dynasty	2251	0.0576%
Liao and Jin dynasty	623	0.0488%
Yuan dynasty	427	0.0274%
Ming dynasty	1730	0.0629%
Qing dynasty	1736	0.0394%

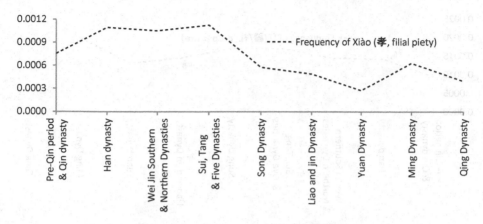

Fig. 1. Frequency of word Xiào (孝, filial piety)

From Fig. 1, we can see that from pre-Qin and Qin Dynasty to Han dynasty, frequency of 'Xiao' rose up by 45.2% to 0.11%. It is positively correlated with that the emperors of Han dynasty valued filial piety highly and honored Confucianism.

The Han dynasty is a dynasty that China's feudal imperial society fully stereotyped. At the same time, Han is a very important stage in the development of filial piety. It is since Han Dynasty that "以孝治天下(yǐ xiào zhì tiān xià)", which means "governing the world with filial piety", is advocated by the rulers, and for the first time, filial piety went on the political stage. Especially during the period of Emperor Wu of the Han Dynasty, after Dong Zhongshu put forward the suggestion to honor Confucianism alone, Confucianism became a feudal orthodoxy. Since Confucianism attached great importance to Filial piety, it became the traditional ethical norms of China. Filial piety and loyalty complemented each other and became the core of the social ideological and moral system.

When it came to Song Dynasty, frequency rapidly dropped by 48.8% to 0.06%. This is because loyalty is more respected in Song society and culture, rather than filial piety. One man can't have filial piety and loyalty at the same time. At first, loyalty is a part of filial piety culture. but in Song Dynasty, loyalty became more important than filial piety. To serve the country and repay the country with supreme loyalty is more honored by society and the emperors.

In Ming Dynasty, filial piety rose up again. The political system of the Ming Dynasty and the Confucian concepts are mutually based and supported, forming a unified whole, and at this time filial piety became a legal support to the national political system. By contrast, Qing Dynasty's national system and Confucian ethics are not so unified. Confucian filial ethics is not a norm of behavior for the entire country. Qing dynasty is more like a pluralistic complex situation. So frequency of 'Xiao' went down in Qing dynasty is reasonable.

For frequency of Lǐ (禮/礼, proper rite) (See Fig. 2), it also rose up in Han dynasty and in Ming Dynasty, when Confucianism was more popular and welcomed by the rulers at that time and was used as powerful and efficient vehicles for governing.

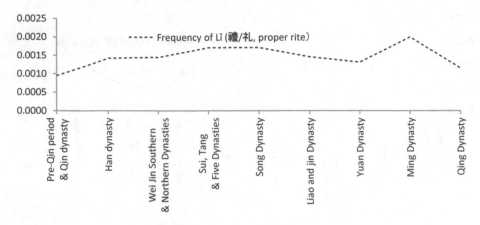

Fig. 2. Frequency of Lǐ (禮/礼, proper rite).

4 Conclusion

Culture is created by people, and people are influenced unconsciously by the cultural around them. In this paper, we analyze the frequency changes of keywords that represent some certain aspects of Chinese culture, especially Confucian culture. We also explain the reason for the changes. The contributing factors of the changes in each dynasty include ruling style and favorite ideas of the monarch, the nation that the ruling class is from (whether they are from the Han nationality or from national minority), the region that the dynasty is established and so on. Finally, in our further research, we will try to do more deep-going analysis such as modeling and transfer learning on ACA.

Acknowledgement. This project was partly supported by the National Social Science Foundation of China (Grant No. 14CZW019) and supported by Taishan Scholars Foundation of Shandong Province, China (Grant No. tsqn20161024).

References

1. China country profile. BBC News, 18 October 2010. Accessed 07 Nov 2010
2. Yang, Z.H.J.J.o.N.I.o.A.T.: On filial piety culture characteristics in the Song Dynasty (2005)
3. Wiki, Twenty-Four Histories. https://en.wikipedia.org/wiki/Twenty-Four_Histories. Accessed 14 Mar 2019
4. Wiki, Records of the Grand Historian. https://en.wikipedia.org/wiki/Records_of_the_Grand_Historian. Accessed 06 June 2019
5. Wiki, Dynasties in Chinese history. https://en.wikipedia.org/wiki/Dynasties_in_Chinese_history. Accessed 25 May 2019
6. ChinaKnowledge.de, The Twenty-Five Official Dynastic Histories (*ershiwushi* 二十五史), http://www.chinaknowledge.de/Literature/Historiography/ershiwushi.html. Accessed 08 July 2010

Real-Time Collaborative Annotation System Supporting Separation of Content and Annotation

Lili Gao[1(✉)], Tian Cheng[2], Liping Gao[2], and Dongbin Guo[2]

[1] College of Special and Preschool Education, Weifang University,
Weifang 261061, China
snowgao@163.com

[2] School of Optical-Electrical and Computer Engineering,
University of Shanghai for Science and Technology, Shanghai 200093, China

Abstract. In earlier collaborative annotation system, the details of all comments were presented in the form of images and were marked directly on the contents. This approach may cause visual confusion no matter what style people choose to display annotations by overlaying contents or by overlapping other comments. This article introduces a new annotation display model to separate contents and annotations, and the MPSAC (Multi-processing and Separation of Annotation and Content) strategy was presented to achieve the consistency maintenance of different collaborative sites based on this model. With the foundation of controlling executive operations' effect and maintaining the consistency of annotations, the strategy discovers and resolves the collision problem among overlapped annotations and provide a better interactive experience for users. The feasibility and correctness of this strategy were verified by case analysis and CoNote model system at the end of this paper.

Keywords: Collaborative annotation · Conflict resolution · Consistency maintenance · MPSAC

1 Introduction

With the development of cooperative computing and network communication technology, teamwork has been broadly applied to handle complex transactions. In collaborative work, people have a higher demand for interactivity. Being a method to improve interactivity, collaborative annotation has been widely used in various fields such as decision making, product design, doctor consultation, online classroom discussion, etc. [1–11]. Texts, tables and other comments were displayed in the form of images and the previous collaborative annotation system distinguished comments by using different colors, increased accuracy by using a stylus and stored comments according to their classification to enhance the display effect of comments. However, none of these methods can absolutely solve the problem of comments' overlaying of the contents or the overlapping among themselves. For example, Eppler and Pfister [1, 2] has proposed using different colors to represent different identities and replacing different comments with different symbols during the discussion of a meeting. This method improves interactivity in some degree, but it can only be used in a small number of users with low mobility. The reasons

© Springer Nature Switzerland AG 2019
D. Milošević et al. (Eds.): HCC 2019, LNCS 11956, pp. 111–122, 2019.
https://doi.org/10.1007/978-3-030-37429-7_12

are as follows: (1) The amounts of color limits the multi-level division of identities; (2) Too many symbols may not be quickly accepted by the users which lead to bad experience. Gorgan [3, 4] advanced using a stylus but not a mouse to select the part accurately which is to be annotated and to comment on the original image. But when there are too many comments, this method may affect the latecomers' view of the original content. Camba [6] described the collaborative annotations in the process of 3D design and proposed to store annotations according to the classifications of the comments but displayed them by hashing them in the original image. This raised a new problem—the overlapping among comments. Coustaty [11] described the process of taking electronic notes on the printed documents. Since different users use different sizes of the symbol when annotating the same object, they may be ambiguous when selecting the annotated object. Although there are varieties of collaborative annotation systems in the past, the user experience was not perfect. The reasons are as follows: (1) All comments are showed in the form of images. (2) All comments are stacked on the original image which leads to the overlapping among themselves.

This article introduces a new model to show annotations which separates contents and comments. And the content in a text form is taken as an example to be separated into keywords and invalid parts. The keywords are chosen and annotated by users as an annotation object and the consistency maintenance strategy MPSAC (Multi-processing and Separation of Annotation and Content) is proposed based on the new model that separates content and annotation. In MPSAC, each annotation added by the user will be uniquely identified and recorded by a timestamp and site priority. Only the valid annotation associated with a keyword which was generated earliest will be displayed and other relevant annotations will be hidden. In order to ensure that annotations from different sites are consistent and non-overlapping, algorithms like DeleteObject, UndoObject and SetAL are proposed in this paper. The DeleteObject and InsertObject algorithms ensure the consistency of all sites by controlling the executive effort of delete and insert operations and SetAL algorithm is used to determine and resolve the conflicts among annotations.

2   Annotation Document Model Supporting Separation of Content and Annotation

The participants of the collaborative annotations such as annotated objects, annotations, operations (Insert, Delete and Undo) and presentative annotations are re-modeled. The representation of annotation is a new concept which is put forward in this article and will be analyzed with text content in this article. The following gives the corresponding definitions.

Definition 1: Content Column(Wd): The content column is used to store various forms of content and the text content in different numbers of lines is presented according to Wd.

Definition 2: Annotation Column(Zd): The annotation column is used to store comments edited by users with a width of Zd. The annotation column provides a separated comment area for each line of the text content and each area only displays annotations of the corresponding line.

Definition 3: Annotated Object(AO): AO_{ij} represents the j-th annotated object in the i-th line. The line number and sequence number of one annotated object are unique, even if an annotation object group which contains keywords in different lines can only have one-line number.

Definition 4: Annotation(A): The annotation object is a square area containing descriptive content defined by users. It can be described by A = [Lu, Rd, C], where C is the content of the annotation, and Lu and Rd are both two-dimensional coordinates. Lu Indicates the annotation coordinate which is described as [Lu.x, Lu.y] located in the upper left and Rd indicates the annotation coordinate [Rd.x, Rd.y] located in the lower right. The position of the annotation can be judged by Lu and Rd. There are four conditions for the two annotation objects A1 and A2:

> **Condition 1**(A1, A2): Indicates that there is an overlapping Angle between A1 and A2.
> **Condition 2**(A1, A2): Indicates that two angles of A2 are contained in A1.
> **Condition3**(A1, A2): Indicates that four angles of A2 are all contained in A1.
> **Condition4**(A1, A2): Indicates A1 and A2 are not overlapped.

Definition 5. Operation Definition: The operations in this model is defined as: Ins(Ao, A, T, S), Del (AO, A, T, S), and Undo (O, S), where AO indicates the annotated object, A indicates the annotation, T indicates the generation time of the operation, and S indicates the status to record whether the execution effect of the operation is presented in the annotation column. It should be noted that when Del and Undo operate on the same annotated object after Ins, the Del operation has higher priority than the Undo operation, that is, the Del's status is stored in the operation sequence with the form of 1, and the Undo's is 0.

Definition 6. Active Width of the Annotation(Md): Based on the upper left and lower right vertices of the annotation, the left and right annotations which are the closest to the annotation but not overlapped with each other are found and indicated as A1 and A2. Then Md is defined as A2.Lu.x-A1.Rd.x. If there is no comment in the right of the annotation, that's is to say A2 doesn't exist, Md is defined as the distance from the right side of A1 to the right side of the annotation column, i.e. Md = Zd-A1.Rd.x.

Definition 7. Show Annotation(SA): Each annotated object AO has a series of comments. In which the earliest comment is generated as a comment presented to the user and this is the presentation annotation of the AO. There is at most one show annotation for each comment.

Definition 8. Comment Sequence(LA): The SA for each annotated object was held in the comment sequence. These comments are the earliest valid comments for the object being annotated. LAi is an array whose length is the same as the number of annotated objects in the ith line of the content column. When LAi[k] is not empty, it indicates the show annotation of the kth annotated object in the ith line of the content column.

3 Consistency Maintenance Strategy

The new model puts forwards new requirement for the consistency of the collaborative annotation process. It not only requires the same annotations to be displayed at each site, but also requires annotations to be presented without overlapping. Therefore, the

MPSAC strategy divides the process of consistency maintenance of each site into two steps: 1. Control the execution process of various operations to ensure that annotation sequence of each station and the status of the operation in the history sequence are the same even if they are executed in different orders. 2. According to the same annotation sequence provided in step one, the SetAL algorithm will be used to discover and resolve the position conflict among annotations.

3.1   Processing of the Insert Operation

The process of the insert operation is relatively simple. When the operation $O = $ Ins (AOij, A, T, S) comes, only the earliest valid comment will be found in the relevant annotation of the jth in the i-th line: A1. If the generation time of A1 is later than that of A, A is selected as the show annotation of AOij and will be placed in LA, that is, LAi [j] = A. While if the generation of A1 is earlier than the generation of A, there is no effect on the A's comment list. Finally, if the operation O is placed in the HB in the order of generation time and placed in the HQ according to the execution time.

3.2   Processing of Delete Operation

When the delete operation $O = $ Del(AOij, A, T, S) comes, the first two earliest valid annotations of the j-th annotated object in the i line should be found: A1 and A2. If A1 is the same as A2, it means that the show annotation is deleted, then we should adjust LAi[j] = A2 and set the flag S of the corresponding insert operation in HB to 0 at the same time. If A1 and A2 are different, we only need is to set the S of the corresponding operation to 0 and put the operation O into HQ. The corresponding algorithm is detailed as follows.

3.3   Processing of Undo Operation

When an Undo operation Undo(O, S) comes, the execution of it is determined by whether the target operation O is an Insert operation or a Delete operation. If O is an Insert operation, HB is traversed to find the corresponding insert operation O1 and the first two valid comments A1 and A2 which are corresponding to O.AOij. Then two conditions may occur: (1) If the corresponding insert operation is valid and the corresponding comment is a show annotation, that is, O1.S = 1 and O1.A = A1, the status of operation O is set to 0 and A2 is set to show annotation of AOij and be put in LAi[j]. After that, Undo(O, S) is stored in HB and HQ; (2) If the corresponding insert operation is valid but the corresponding comment is not a show annotation, ie O1.S = 1 and O1. A $\neq$ A1, change status of O1 to 0 and set HB and HQ. Then the operation O and Undo (O, S) will comprise a do-undo-pair; (3) If the concurrent operation is executed firstly which leads to the corresponding insertion operation to be invalid, that is, O1.S = 0. It only needs to maintain the relationships of the do-undo-pair in the HB and save the operation in the HQ. If O is a delete operation, it is necessary to find the corresponding delete operation del_1 and insert operation ins_1 in HB, where del_1 and ins_1 are operations aiming at annotation A in AOij, and the earliest valid annotation A1 which is corresponding to AOij will be obtained. (1) If the executed comment is the previous show annotation, ie A.T > A1.T, ins1.T is set to valid and the position of undo in HB is adjusted to guarantee the relationship of do-undo-pair and afterwards undo operation is stored in HQ. (2) Otherwise if A == A1 or AT < A1.T, ins1.T is set to valid and A is changed to the show annotation of AOij. At last the position of undo in HB is adjusted to ensure the relationship of do-undo-pair and save the undo operation in HQ.

3.4 Case Analysis

Now we use the above control algorithm to illustrate the execution effect of each site. When the HB and AL of all sites are the same, each site implements the consistency of the show annotation. Assume that the initial state of the i-th line of the comment column is ALi = {null, null ...} and that $O1 \sim O5$ all process the keywords of the i-th line, where $O1$ = Ins(AOi1, A1, T1, 1), $O2$ = Ins(AOi1, A2, T2, 1), $O3$ = Ins(AOi1, A3, T3, 1), $O4$ = Del(AOi1, A1, T4, 1), $O5$ = Undo(O2, 1), and $T1 < T2 < T3 < T4$. The relationship among operations is shown in Fig. 1.

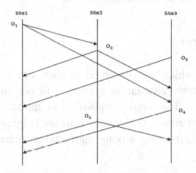

Fig. 1. Operation relationship

At Site 1, operation O1 is generated and executed immediately. Since A1 is the first comment of AOi1, A1 is placed into AL as a show annotation, that is, $ALi[0]$ = A1. And O1 is placed into HB according to the order of generation time; When O2 arrives, since the generation time of A2 is later than the generation time of A1, O2 is directly put into HB in the order of generation time; the state and the execution mode of O3 and O2 are the same, so the operation sequence of the i-th line is HBi = [O1, O2, O3], and the comment sequence is ALi = [A1, null, ...]; When O5 is executed, it is found that the comment carried by the operation is not the show annotation A1 of the annotated object AOi1, so it only needs to maintain the relationship between O5 and O2 of the do-undo-pair in HBi. At this time, HBi = [O1, O2, O5, O3], ALi = [A3, null, ...]; The effect of O4 is to delete the show annotation of AOi1. Thus it is necessary to set the second earliest annotation A3 to show annotation, and set the status of O1 to 0 in the HBi to maintain the ins-del-pair relationship of O4 and O1. At this time in the annotation column HBi = [O1, O4, O2, O5, O3], ALi = [A3, null, ...].

At site 3, the operation O3 is generated and executed immediately, so A3 is set as the show annotation of AOi1 and be put into the annotation sequence. When O2 arrives, since the generation time of the corresponding annotation A2 is earlier than A3, the show annotation of AOi1 is replaced by A2. Since A1 which carried by O1 is generated earlier than the valid annotations in the AOi1, when O1 arrives at station 1, A1 will be regarded as the show annotation and the operations in the HBi are maintained in chronological order; When O4 is generated, the two earliest comments A1 and A2, which contain the comment A1 that is to be deleted by O4 should be found. Therefore, it is necessary to set A2 as the show annotation of AOi1 and set the state of O1 to invalid and maintain the ins-del-pair relationship in HBi. The status of the comment column is HBi = [O1, O4, O2, O3], ALi = [A2, null, ...]; When O5 arrives, the comment A2 carried by the operation to undo O2 is a show annotation, so the following valid annotation A3 is set to the show annotation, and the relationship of do-undo-pair is maintained. Finally, the status of the comment column is HBi = [O1, O4, O2, O5, O3], ALi = [A3, null, ...].

By analyzing the execution status of the operations, it is found that each site has maintained the consistency of the annotations presented in the annotation column. The first part of the consistency in the collaborative annotation system is completed. The second part of the work will be introduced below.

4   Conflict Resolution

When the annotations of which are presented at each site need to be consistent, the follow-up of the collaborative annotation system is to present the annotations to the user in a friendly manner. The SetAL algorithm is introduced to resolve conflicts among show annotations to ensure that annotations are displayed in a non-overlapping way. In order to handle the overlap among annotations, there are two situations that need to be handled.

4.1   Conflict Analysis

4.1.1   Case 1: The Width of the Annotation Column Is Shorter Than Md

When the width of the annotation column is set as Zd = 945, there is an comment of AOi4, that is A1 = [[10, 10], [90, 90]] in the annotation column. At this time, there is an operation O1 which carriers the comment of AOi5, that is A2 = [[50, 50], [120, 120]]. Since A2 and A1 satisfy Condition1 (A2, A1), the result of directly executing O1 is that A2 and A1 are overlapped—as shown below. By calculation, A3.Rd.x-A3.Lu.x = 70, that is, the width of the annotation is 70 is shorter than Md, so simple translations can avoid overlapping of two annotations, as shown in Fig. 2. Here we resolve the conflict by translating the post-inserted comment horizontally to the right.

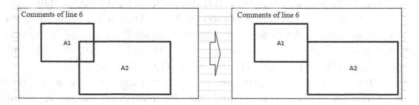

Fig. 2. Conflict resolution strategy in case that the width of the annotation column is shorter than Md

4.1.2   Case 2: The Width of the Annotation Column Is Longer Than Md

Suppose that the width of the annotation column is set to Zd = 945, and there are two annotations in the comment column: A1 and A2, which belong to different keywords, where A1 = [[10, 10], [90, 90]], A2 = [[150, 20], [230, 10]], there is a comment A3 = [[50, 50], [120, 120]] for ALi6. Since A3 and A1 satisfy Condition1 (A3, A1), we know that if A3 is inserted directly, overlap will be caused between A3 and A1. It is found by calculation that A3.Rd.x-A3.Lu.x = 70 > Md, so the lateral translation cannot eliminate the conflict. At this time the adjustment of width should be taken into consideration. In the adjust process, we have to ensure that the area of the annotation being adjusted does not change, so that it will not affect the content of the show annotation and eliminates the overlap problem among annotations (Fig. 3).

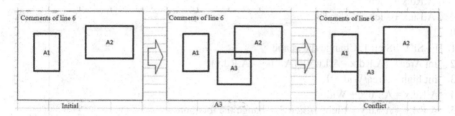

Fig. 3. Conflict resolution strategy in case that the width of the annotation column is longer than Md

4.2   Conflict Resolution Scheme

In order to avoid overlapping among annotations, the calculation of an annotation's Md should be done firstly before A is put into the annotation column. If Md is longer than the width of A and there is no overlap, A is placed directly; if the annotation placed previously has a position overlap with A and Md is longer than the width of A, the placement position of the annotation A will be adjusted; if the annotation placed previously has a position overlap with A and Md is shorter than the width of A, the shape of A is adjusted under the condition that the area of annotation A will not be unchanged. The corresponding implementation algorithm is as follows:

```
Function SetAL(ALi){
1   int Flag;
2   Al = Ar = null;
3   For(int i; i<ALi.size; i++){
4       Md = GetMd(ALi[i],ALi,Al,Ar);
5       IF(A = = null &&Ar = = null){
6           Set(AL[i]);
7       }ELSE{
8           Set(A` = TranAnnotation(ALi[i],Md));    //set transformed annotation A
```

```
9    }
10 }
11}
//adjust the location and shape according to A's Md and conflict annotation
Function TranAnnotation(A,Wid,Al,Ar){
1  IF(Wid > (A.Rd.x - A.Lu.x)){
2     IF(Al = = null){
3       int len =A.Rd.x - Ar.Lu.x;
4       A.Rd.x += len;
5       A.Lu.x +=len;
6     }ELSE{
7       int len = A.Lu.x - Al.Rd.x;
8       A.Rd.x += len;
9       A.Lu.x += len;
10    }
11 }ELSE{  //Step 1. Change the shape
12    int Area = (A.Rd.x - A.Lu.x)*(A.Lu.y - A.Rd.y);
13    int high = Area/Wid + 1;
14    A.Rd.x = A.Lu.x + Wid;
15    A.Rd.y = A.Lu.y + high;  //Step 2. Change the location
16    int len = A.Lu.x - Al.Rd.x;
17    A.Rd.x += len;
18    A.Lu.x += len;
19 }
20 Return A;
21}
//Find annotation A's Md
Function GetMd(A,ALi,Al,Ar){
1   int left = 0;
2   int right = Zd;
3   For(int i; i < ALi.size; i++){
4     IF(A≠ALi[i]){
5     IF(ALi[i].Rd.x < A.Rd.x && Al.Lu.x < ALi[i].Lu.x){
6       left = ALi[i].Rd.x;
7       Al = ALi[i];
8       IF(ALi[i].Lu.x > A.Lu.x && Ar.Lu.x > ALi[i].Lu.x){
9         right = ALi[i].Lu.x;
10        Ar = ALi[i];
11      }
12    }ELSE{
13      Break;
14    }
15   }
16 Return right - left;
17 }
18}
```

4.3   Control Algorithm

As mentioned above, the MPSAC policy requires that each operation is performed firstly. And then, under the premise of the consistency maintenance of the annotation sequence, the selected annotations are placed without overlap, and finally the sites are displayed identically.

```
Function Execute(O){
1  IF(TypeOf(O) = = ins){
2    InsertObject(O);
3  }ELSE IF(TypeOf(O) = = del){
4    DeleteObject(O);
5  }ELSE{
6    UndoObject(O);
7  }
8  SetAL[ALi];
9 }
```

5   CoNote Model System

In order to verify the correctness and feasibility of the proposed algorithm, a collaborative annotation system CoNote that supports separation of content and annotation has been proposed. The goal of the system is to allow different users in distributed locations to operate comments on the text content in real time. After performing the same operation, the annotations of each site are not overlapped and displayed identically.

The CoNote system is based on the full-duplex communication WebSocket protocol and is developed by using HTML, JavaScript, CSS, and Java. The CoNote system is not a complete replication architecture. Each client in the CoNote system keeps a copy of the document, and the operations generated by each client are sent to the server. The server is regarded as a special client node, that is, the server will only synchronize and transmit operations sent by other clients without generating new operations.

On the client side, the system needs to complete three tasks. Firstly, it accepts operations generated by the user and execute them. Secondly, it establishes a connection to the server and receives the operation transmitted by the server and send to the local operation to achieve data synchronization. Thirdly, it performs operations synchronized from the server. The server serves as the end point of a data summary to synchronize the operations between the local and current connected clients' data.

The following figure shows the user interface of the CoNote system. Figure 4 is a case where the user does not select the annotation space, where in the upper part is the display space and the lower part is the user operation space; Fig. 5 is a show annotation image in which the user selected the comment space.

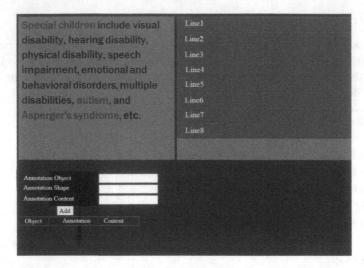

Fig. 4. System status before user select

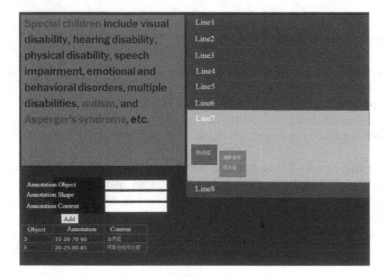

Fig. 5. System status when user select

6   Conclusion

There are a lot of collaborative annotation systems in the past, but they don't bring a good user experience. This is because these systems didn't notice the overlap problem between content and comments and overlap among comments. In addition, Undo functions are not supported by those collaborative systems. In response to the above three questions, the participants of collaborative annotation systems are re-modeled and separation between content and annotation is put forward as well as the consistency strategy MPSAC. To store content and annotation separately is to store content and comments in two separate spaces and divide the content into keywords and irrelevant content. The MPSAC strategy decomposes the process of the latter two problems into maintain annotation sequence consistency and resolve annotation conflict. Maintaining annotation consistency is ensuring the consistency of each site by preserve the consistency by using generation time, insert, delete and undo algorithms under the condition of only one show annotation of each annotation object is represent to the user. Ins-del-pair, do-undo-pair and their priorities' relationships are introduced in this paper to simplify the maintenance process of the operation sequence. With the new annotation model, the conflicts between annotations are mainly in two-dimensional space. The conflict resolution among annotations is to use the SetAL algorithm to handle the position conflict among annotations on the premise that the annotation sequences selected by each site are consistent.

In the collaborative annotation environment, the MPSAC strategy can achieve the site consistency within three types of operations and ensure the non-overlapping placement among the annotations. However, the SetAL algorithm proposed in this paper can only solve the problem of horizontal conflict among annotations. This method of resolution does not guarantee that the annotations are compactly placed in the comment column. In other words, there may be a vertical position between the annotations presented wasted.

Collaborative annotations have brought great convenience to users in online discussions, but there are still many problems in collaborative annotation. It is believed that with these problems optimized, the collaborative annotation system will play a more important role in people's work.

References

1. Eppler, M.J., Pfister, R.A.: Drawing conclusions: supporting decision making through collaborative graphic annotations. In: 14th International Conference Information Visualisation, London, UK, pp. 369–374. IEEE Computer Society (2010)
2. Eppler, M.J., Kernbach, S.: Dynagrams: enhancing design thinking through dynamic diagrams. Des. Stud. **47**(11), 91–117 (2016)
3. Gorgan, D., Stefanut, T., Gavrea, B.: Pen based graphical annotation in medical education. In: 12th IEEE International Symposium on Computer-Based Medical Systems, Maribor, Slovenia, pp. 681–686. IEEE (2007)
4. Smit, N., Hofstede, C.-W., et al.: The online anatomical human: web-based anatomy education. In: 37th Annual Conference of the European Association for Computer Graphics, Lisbon, Portugal, pp. 37–40. Eurographics Association, Aire-la-Ville (2016)
5. Gao, L., Xu, X.: A new algorithm for real-time collaborative graphical editing system based on CRDT. In: Sun, Y., Lu, T., Xie, X., Gao, L., Fan, H. (eds.) ChineseCSCW 2018. CCIS, vol. 917, pp. 201–212. Springer, Singapore (2019). https://doi.org/10.1007/978-981-13-3044-5_15
6. Camba, J.D., Contero, M., et al.: On the integration of model-based feature information in Product Lifecycle Management systems. Int. J. Inf. Manage. **37**(6), 611–621 (2017)

7. Coburn, J.Q., Salmon, J.L., Freeman, I.: Effectiveness of an immersive virtual environment for collaboration with gesture support using low-cost hardware. J. Mech. Des. **140**(4), 1–9 (2018)
8. Goy, A., Magro, D., Petrone, G., Picardi, C., Segnan, M.: Ontology-driven collaborative annotation in shared workspaces. Future Gener. Comput. Syst. **54**(1), 435–449 (2016)
9. CheeWyai, L., Cheah, W., Chowdhury, A.K., Gulden, C.: Engineering sustainable software: a case study from offline computer support collaborative annotation system. In: 9th Malaysian Software Engineering Conference (MySEC), Kuala Lumpur, pp. 272–277 (2015)
10. Poster, S.R.: Interactive and collaborative source code annotation. In: IEEE International Conference on Software Engineering, Florence, Italy, pp. 799–800. IEEE (2015)
11. Coustaty, M., Sidere, N., Ogier, J.: Augmented documents for research contact management. In: 4th International Forum on Research and Technology for Society and Industry (RTSI), Palermo, pp. 1–6. IEEE (2018)

An Advanced Membrane Evolutionary Algorithm for Constrained Engineering Design Problems

Wenxiang Guo, Laisheng Xiang, and Xiyu Liu[(⊠)]

College of Business, Shandong Normal University, Jinan, China
18862003616@163.com, xls3366@163.com, sdxyliu@163.com

Abstract. The main goal of this paper is to propose an advanced dynamic membrane algorithm (ADMA-PSO/GA) based on the particle swarm optimization (PSO) and genetic algorithm (GA) for solving the constrained problems in engineering design. The proposed algorithm combines membrane computing with the particle swarm optimization and genetic algorithm. In the PSO phase, the worst solutions are enhanced by the global-local best inertia weight and acceleration coefficients. In the second phase, crossover and mutation operators have been used to further improve the balance between the exploration and exploitation ability in PSO. The constraints are handled using a parameter-free penalty function. Several well-known engineering design problems have been used to measure the performance of the proposed algorithm in this paper. The simulation results show that the proposed algorithm is superior to the state-of-the-art algorithms.

Keywords: P system · Particle swarm optimization · Genetic algorithm · Engineering design problem

1 Introduction

Constrained optimization is an important part of engineering design. Most of these problems in engineering have various types of constraints. Due to these constraints and their interrelationship between the objective functions, constrained optimization problems are not easy to deal with. In order to solve such problems, mathematical programming and meta-heuristic methods are two main solutions. Traditional mathematical methods can't deal with all these problems. While the meta-heuristic optimization techniques (such as PSO [1], ABC [2], ACO [3]) can obtain global or near-global optimum solutions. Behnam combines genetic algorithm, particle swarm optimization, and symbiotic organisms search to find a better solution of problems in complex design space and to control the feasibility of finding solutions using the penalty function method [4]. Bahriye uses ABC algorithm to solve large scale optimization problems and apply it to engineering design problems by extending the basic ABC algorithm [5]. These methods are quite suitable for global searches due to their capability of exploring and detecting promising regions in the search space at an

© Springer Nature Switzerland AG 2019
D. Milošević et al. (Eds.): HCC 2019, LNCS 11956, pp. 123–132, 2019.
https://doi.org/10.1007/978-3-030-37429-7_13

accurate time. A general non-linear constrained optimization problem can be formulated as follows [9]:

$$
\begin{aligned}
& Minimize\ f(x) \\
& subject\ to \\
& g_j(x) \le 0 \quad ; \quad j = 1, 2, \cdots, M \\
& l_i \le x_i \le u_i \quad ; \quad i = 1, 2, \cdots, n
\end{aligned}
\tag{1}
$$

where $x = \left[x_1, x_2, \ldots, x_n\right]^T$ denotes the n-dimensional vector of variables; $f(x)$ is the objective function; l_i and u_i are the minimum and maximum permissible values for the i_{th} variable respectively; p and q refer to the number of equality constraints and inequality constraints, $S = \left\{x \in R \mid g_j(x); j = 1, 2, \ldots, p + q = (M)\right\}$ be the set of feasible solutions.

Membrane computing (P system) is a computing model constructed by simulating the structure of cells and was proposed by Gheroghe paun [10]. For its hierarchical structure and intrinsic parallelism, membrane computing has been widely used, especially in solving the NP-complete problems. The membrane evolutionary algorithm was developed by combing the network structure of the membrane with the evolutionary algorithm. Nishida developed the first membrane algorithms based on the nest like membrane structure to solve the traveling salesman problem [6]. He et al. developed an adaptive membrane algorithm for solving combinatorial optimization problems [7]. Niu et al. developed a novel membrane algorithm based on ant colony optimization to solve the capacitated vehicle routing problem [8]. These aforementioned techniques have been successfully applied to various constrained optimization problems. However, the performance while getting optimal or near to optimal solution shows a significant difference. Particle swarm optimization and genetic algorithm are two typical evolutionary algorithms, both the two algorithms have strengths and weaknesses. This paper combines membrane computing with the two algorithms and proposed an improved dynamic membrane evolutionary algorithm for solving the engineering design problems.

The remaining contents of this paper are organized as follows. Section 2 gives a brief description of the P-system, PSO and GA. The proposed dynamic membrane evolutionary algorithm is described in Sect. 3. Section 4 shows the simulation results and analysis. Conclusion and prospect are given in Sect. 5.

2   Preliminaries

2.1   Design of Membrane System

Based on many years of research on DNA computing and inspired by biological cells, Paun proposed membrane computing model. According to the membrane structure, the P system can be classified into three main families: cell-like P system, tissue-like P system and neural-like P system [11]. A common cell-like membrane system based on evolution and communication rules is defined as follows [12] and is showed in Fig. 1.

$$\Pi = (O, u, H, \omega_1, \ldots, \omega_m, R_1, \ldots, R_m, i_o) \qquad (2)$$

In which,

(1) $O \{a, b, c\}$ is the alphabet, and its elements are called objects.
(2) μ is the membrane structure.
(3) H is the number of membrane structure.
(4) ω_i are strings over O, describing multisets of objects in the region i.
(5) R_i $(i = 1, 2 \ldots m)$ are different rules over O^* in each cell.

   (i) $[_iS_1 \rightarrow S_2]_i$, $i \in \{1, 2, \cdots, m\}$ and $S_1, S_2 \in O^*$ (Evolution rules)
   (ii) $S_1[_i]_i \rightarrow [_iS_2]$, $i \in \{1, 2, \cdots, m\}$ (In-communication rules)
   (iii) $[_iS_1]_i \rightarrow [_i]_iS_2$, $i \in \{1, 2, \cdots, m\}$ (Out-communication rules)
   (iv) $[_iW] \rightarrow [_iU]_i[W - U]_i$, $W, U \in O$ (Division rules).

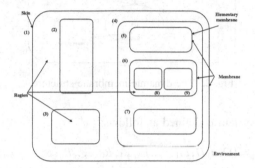

Fig. 1. A common cell-like membrane system

2.2 Particle Swarm Optimization and Genetic Algorithm

Particle swarm optimization (PSO) was proposed by Kennedy and Eberhart, and is a kind of heuristic evolutionary algorithm based on swarm intelligence. Each particle decides its path on the basis of its previous best position (*pbest*) and global best position (*gbest*) among all the particles.

Genetic algorithm (GA) is a computational model that simulates the natural evolution of Darwin's biological evolution theory and the evolutionary process of genetics [13]. GA is modeled loosely on the principles of the evolution through natural selection, employing a population of individuals that undergo selection in the presence of variability-inducing operators such as mutation and crossover. A fitness function is used to evaluate individuals, and reproductive success varies with fitness [14].

3 An Advanced Dynamic Membrane Evolutionary Algorithm for CEDPs

3.1 Improved Dynamic Membrane Structure

Inspired by PSO, GA and P system, a new dynamic membrane evolutionary algorithm called AMDEA-PSO/GA was proposed in this paper, which combines the P system, particle swarm optimization, and genetic algorithm. ADMEA-PSO/GA uses an improved dynamic cell-like P system framework [15]. Generally, the P system consists of three main parts: the dynamic membrane structure, the multisets that represent objects and evolution rules. The structure of the dynamic membrane with rules is shown in Fig. 2.

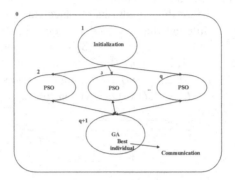

Fig. 2. The dynamic membrane structure.

This cell-like P system is defined as follows:

$$\Pi = (O, u, H, \omega_1, \ldots, \omega_{q+1}, R_0, \ldots, R_q, R_{q+1}, i_{out}) \tag{3}$$

where, ω_1 and ω_{q+1} represents the initial objects in membrane 1 and the objects gathered from $q - 1$ subpopulations separately. R_0 is the communication rule, R_1 is division rule that separates the membrane 1 to $q - 1$ elementary membranes, $R_2, \ldots, R_{q+1}$ are the evolution rules in membrane $2, \cdots, q + 1$ separately, i_{out} is the output result in membrane 0.

In the cell-like P system, membrane rules mainly contain evolutionary rules and communication rules. In this paper, evolution rules include PSO rules, selection rule, crossover rule, and mutation rule which are used to evolve the objects in the membrane $2, \cdots, q + 1$. Selection rule and crossover rule are used to select objects to evolve and improve the global convergence separately. Communication rules are used to communicate and share information between two regions, exchange superior objects.

3.2   Procedure of AMDEA-PSO/GA

The steps of the improved membrane evolutionary algorithm can be described as follows:

Step1: Initialize membrane structure. An improved dynamic membrane structure is constructed, consists of elementary membranes 1, $q + 1$ and a skin membrane 0 at iteration $i = 1$. $m * (q - 1)$ objects are randomly generated by the rule in the membrane 1.

Step2: Division rule is implemented. The elementary membrane 1 separates to $q-1$ elementary membranes, each elementary membrane consists of m objects.
Step3: During the computation, update the individuals in each elementary membrane according to PSO rule separately.
Step4: Transport updated individuals in each elementary membrane above to the elementary membrane $q+1$.
Step5: Update the objects inside the membrane $q+1$ with specific evolutionary rules, calculate the fitness and transport the best object to membrane 0, replace the present one if its fitness value is larger. The communication rule is used to communicate its best object to the environment. If the termination conditions met, output the best individual. Otherwise, transport the objects in the membrane $q+1$ to the membrane $2, \cdots, q$ and go to step3. In this paper, the maximum iteration is used as the halting condition. After the system halts, the best object in the environment is the best solution.

4 Simulation Results

In order to measure the performance of the proposed algorithm, this paper selects three benchmarks which have already been solved by other algorithms [16]. These problems all have constraints and could evaluate how the proposed algorithm in this paper works. In our experiments, the present algorithm is implemented in Matlab 7.0a. For eliminating stochastic discrepancy, 30 independent runs are performed in each example.

4.1 Welded Beam Design Problem (Example 1)

The welded beam design problem was first proposed by Coello [17]. The structure of the welded beam is shown in Fig. 3. The goal is to find the minimum fabricating cost of the welded beam subject to several constraints. Four design variables are presented here, the thickness of the weld $h = x_1$, length of the welded joint $l = x_2$, the width of the beam $t = x_3$, and the thickness of the beam $b = x_4$ i.e. the decision vector is $X = (h, l, t, b) = (x_1, x_2, x_3, x_4)$. Here we define the mathematical formulation of the objective function $f(x)$ as follows:

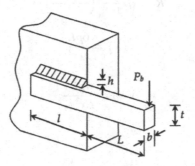

Fig. 3. Welded beam design problem.

$$Minimize\ f(x) = 1.10471x_1^2x_2 + 0.04811x_3x_4(14+x_2) \tag{4}$$

$$s.t.\ g_1(X) = \tau(X) - \tau_{max} \leq 0 \tag{5}$$

$$g_2(X) = \sigma(X) - \sigma_{max} \leq 0 \tag{6}$$

$$g_3(X) = x_1 - x_4 \leq 0 \tag{7}$$

$$g_4(X) = 0.125 - x_1 \leq 0 \tag{8}$$

$$g_5(X) = \delta(X) - 0.25 \leq 0 \tag{9}$$

$$g_6(X) = P - P_c(X) \leq 0 \tag{10}$$

$$g_7(X) = 0.10471x_1^2 + 0.04811x_3x_4(14+x_2) - 5 \leq 0 \tag{11}$$

$$0.1 \leq x_1 \leq 2, \quad 0.1 \leq x_2 \leq 10, \quad 0.1 \leq x_3 \leq 10, \quad 0.1 \leq x_4 \leq 2 \tag{12}$$

Where,

$$\tau(X) = \sqrt{(\tau')^2 + 2\tau'\tau''\frac{x_2}{2R} + (\tau'')^2} \tag{13}$$

$$\tau' = \frac{p}{\sqrt{2}x_1x_2} \tag{14}$$

$$\tau'' = \frac{MR}{J} \tag{15}$$

$$M = p(L + \frac{x_2}{2}) \tag{16}$$

$$J = 2\left\{\sqrt{2}x_1x_2\left[\frac{x_2^2}{12} + \left(\frac{x_1+x_3}{2}\right)^2\right]\right\} \tag{17}$$

$$R = \sqrt{\frac{x_2^2}{4} + \left(\frac{x_1+x_2}{2}\right)^2} \tag{18}$$

$$\sigma(X) = \frac{6pL}{x_4x_3^2} \tag{19}$$

$$\delta(X) = \frac{4pL^3}{Ex_3^3x_4} \tag{20}$$

$$p_c(X) = \frac{4.013E\sqrt{(x_3^2 x_4^6/36)}}{L^2} * \left(1 - \frac{x_3}{2L}\sqrt{\frac{E}{4G}}\right) \qquad (21)$$

$$P = 6000 \; lb, \quad L = 14 \; in, \quad E = 30 * 10^6 psi, \quad G = 12 * 10^6 psi \qquad (22)$$

A variety of methods has been applied to this problem, including traditional cultural algorithms with evolutionary programming (CAEP), co-evolutionary particle swarm optimization (CPSO). The results they got are presented along with the proposed algorithm in Table 1 where we can easily find that the best solution through the proposed algorithm is $f(x) = 1.7248523$, which is better than the other algorithms. The mean solution given by the algorithm is $f(x) = 1.7248523$ with a standard deviation of 1.11E−07, it is obvious that the algorithm possesses the best performance.

Table 1. Comparison of the best solution for the above problem found by different methods.

	Proposed	CAEP	MBA	MEA	CPSO
x_1	1.9890848	0.20577	0.205729	0.205675	0.202369
x_2	5.9059576	3.4705	3.470493	3.470993	3.544214
x_3	1.5362705	9.0366	9.036626	9.040587	9.04821
x_4	0.3912233	0.2057	0.205729	0.205728	0.205723
Best	**1.7248523**	**1.724852**	1.724853	1.725507	1.728024
Mean	**1.7248523**	1.748831	1.724853	1.726594	1.748831
Worst	**1.72485278**	1.7482143	1.724853	1.728121	1.782143
Std.	**1.11E−07**	4.43E−01	**6.94E−19**	7.25E−04	1.29E−02

4.2   Design of Pressure Vessel (Example 2)

The pressure vessel design problem was proposed by Kannan and Kramer [18]. The objective is to minimize the total cost, including the cost of the material, forming, and welding. A cylindrical vessel is capped at both ends by hemispherical heads as shown in Fig. 4. There are four variables related to the problem as shown in the figure below, the variable vectors are given by $X = (T_s, T_h, R, L) = (x_1, x_2, x_3, x_4)$. The mathematical model of the problem can be defined as follows:

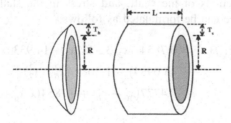

Fig. 4. Pressure vessel design problem.

$$\text{Minimize } f(x) = 0.0224x_1x_3x_4 + 1.7781x_2x_3^2 + 3.1661x_1^2x_4 + 19.84x_1^2x_3 \qquad (23)$$

$$s.t. \, g_1(X) = -x_1 + 0.0193x_3 \le 0 \qquad (24)$$

$$g_2(X) = -x_2 + 0.00935x_3 \le 0 \qquad (25)$$

$$g_3(X) = -\pi x_3^2 x_4 - \frac{4}{3}\pi x_3^3 + 1296000 \le 0 \qquad (26)$$

$$g_4(X) = x_4 - 240 \le 0 \qquad (27)$$

$$0 \le x_1, x_2 \le 100 \qquad (28)$$

$$10 \le x_3, x_4 \le 200 \qquad (29)$$

The comparison with the existing algorithms is shown in Table 2, the best solution, mean solution and the standard deviation given by the present algorithm are $f(x) = 5885.3327, f(x) = 5885.3353, 7.89\text{E}{-}03$, respectively. All these indexes show that the proposed algorithm possesses the best performance.

Table 2. Comparison of the best solution for the above problem found by different methods

	Proposed	CDE	CPSO	MBA	MEA
x_1	0.7781686	0.8125	0.8125	0.7802	0.827599
x_2	0.3846491	0.4375	0.4375	0.3856	0.413794
x_3	40.319618	42.098411	42.0913	40.4292	42.703137
x_4	200	176.63769	176.7465	198.4964	169.965254
Best	**5885.3327**	6059.734	6061.0777	5889.3216	6029.1811
Mean	**5885.3353**	6085.2303	6147.1332	6200.64765	6136.7807
Worst	**5885.3684**	6371.0455	6363.8041	6392.5062	6288.263
Std.	**7.89E−03**	43.013	86.45	160.34	56.76628

4.3   Speed Reducer Design Problem (Example 3)

Another classic constrained design problem is Speed reducer design problem, the problem was developed by Mezura-Montes and Coello [17]. The objective is to minimize the weight subject constraints on the bending stress of the gear teeth, surface stress, transverse detections of the shaft, and stress in the shafts. There are seven variables. The objective can be formulated as follows:

$$\begin{aligned} \text{Minimize } f(x) = \; & 0.7854x_1x_2^2(3.3333x_3^2 + 14.9334x_3 \\ & - 43.0934) - 1.508x_1(x_6^2 + x_7^2) \\ & + 7.4777(x_6^3 + x_7^3) + 0.7854(x_4x_6^2 + x_5x_7^2) \end{aligned} \qquad (30)$$

After 30 independent runs, the results they got are presented along with the proposed algorithm in Table 3. It shows that the performance of the proposed algorithm performs better than the current algorithms in this problem.

Table 3. Comparison of the best solution for the above problem found by different methods

	Proposed	MDE	HEAA	MBA	PSO-DE
Best	**2994.471066**	2996.356689	2994.499107	2996.348167	2996.348167
Mean	**2994.472491**	2996.36722	2994.613368	2996.769019	2996.348174
Worst	**2994.479811**	NA	2994.752311	2999.652444	2996.348204
Std.	**2.65E−03**	8.20E−03	7.00E−02	1.56	**6.40E−06**

5 Conclusion

This paper presents an advanced membrane evolutionary algorithm ADMA-PSO/GA to solve constrained engineering design problems. First, the solution was found by the PSO, and then the solution in the current population will perform genetic algorithm operators to increase the exploration of the solution space for the next generation. Several constrained problems prove that the proposed algorithm shows more effectiveness and robustness and is reliable in solving different engineering design optimization problems. In the future, the advanced dynamic membrane structure can be combined with the other swarm intelligence algorithms, such as the Artificial Bee Colony algorithm.

Acknowledgment. Project is supported by National Natural Science Foundation of China (61472231, 61502283, 61876101, 61802234, 61806114), Social Science Fund Project of Shandong Province, China (16BGLJ06, 11CGLJ22), Postdoctoral Project, China (2017M612339).

References

1. Kennedy, J.: Particle swarm optimization. In: ICNN 95-International Conference on Neural Networks. IEEE (2002)
2. Liu, F., Sun, Y., Wang, G.G., et al.: An artificial bee colony algorithm based on dynamic penalty and Lévy flight for constrained optimization problems. Arab. J. Sci. Eng. **1**, 1–20 (2018)
3. Blum, C.: Beam-ACO: hybridizing ant colony optimization with beam search: an application to open shop scheduling. Comput. Oper. Res. **32**(6), 1565–1591 (2005)
4. Farnad, B., Jafarian, A.: A new nature-inspired hybrid algorithm with a penalty method to solve constrained problem. Int. J. Comput. Methods **15**, 1850069 (2018)
5. Akay, B., Karaboga, D.: Artificial bee colony algorithm for large-scale problems and engineering design optimization. J. Intell. Manuf. **23**(4), 1001–1014 (2012)
6. Nishida, T.: An approximate algorithm for NP-complete optimization problems exploiting P-systems. In: The Brainstorming Workshop on Uncertainty in Membrane Computing, pp. 185–192 (2004)

7. He, J., Xiao, J., Shao, Z.: An adaptive membrane algorithm for solving combinatorial optimization problems. Acta Mathematica Scientia **34**(5), 1377–1394 (2014)
8. Niu, Y., Wang, S., He, J., Xia, J.: A novel membrane algorithm for capacitated vehicle routing problem. Soft. Comput. **19**(2), 471–482 (2015)
9. Xiao, J., He, J.J., Chen, P., et al.: An improved dynamic membrane evolutionary algorithm for constrained engineering design problems. Nat. Comput. **15**(4), 579–589 (2016)
10. Gheorghe, P.: Membrane computing: an introduction. Theoret. Comput. Sci. **287**(1), 73–100 (2002)
11. Pan, L., Wu, T., Su, Y., et al.: Cell-like spiking neural P systems with request rules. IEEE Trans. Nano Biosci. **99**, 1 (2017)
12. Song, B., Pan, L., Perez-Jimenez, M.J.: Cell-Like P systems with channel states and symport/antiport rules. IEEE Trans. Nano Biosci. **15**(6), 555–566 (2016)
13. Deb, K., Pratap, A., Agarwal, S., et al.: A fast and elitist multi-objective genetic algorithm: NSGAII. IEEE Trans. Evol. Comput. **6**(2), 182–197 (2002)
14. Garg, H.: A hybrid PSO-GA algorithm for constrained optimization problems. Appl. Math. Comput. **274**, 292–305 (2016)
15. Xiao, J., Liu, Y., Zhang, S., et al.: An adaptive membrane evolutionary algorithm for solving constrained engineering optimization problems. J. Univers. Comput. Sci. **23**(7), 652–672 (2017)
16. Zhang, G., Cheng, J., Gheorghe, M., et al.: A hybrid approach based on differential evolution and tissue membrane systems for solving constrained manufacturing parameter optimization problems. Appl. Soft Comput. **13**(3), 1528–1542 (2013)
17. Mezura-Montes, E., Coello, C.A.C.: Useful infeasible solutions in engineering optimization with evolutionary algorithms. In: Gelbukh, A., de Albornoz, Á., Terashima-Marín, H. (eds.) MICAI 2005. LNCS (LNAI), vol. 3789, pp. 652–662. Springer, Heidelberg (2005). https://doi.org/10.1007/11579427_66
18. Kannan, B.K., Kramer, S.N.: An augmented lagrange multiplier based method for mixed integer discrete continuous optimization and its applications to mechanical design. J. Mech. Des. **116**(2), 405 (1994)

Aspect Sentiment Classification Based on Sequence to Sequence Reinforced Learning

Hanlu Chu, Yaoxing Wu, Yong Tang$^{(\boxtimes)}$, and Chengjie Mao

School of Computer Science, South China Normal University,
Guangzhou, China
{hlchu,yxwu,ytang}@m.scnu.edu.cn, maochj@qq.com

Abstract. The task of aspect sentiment classification (ASC) is a fundamental task in sentiment analysis. Given an aspect and a sentence, the task classifies the sentiment polarity expressed on the target in the sentence. Previous work usually distinguish the sentiment based on one-way LSTM, which are often complicated and need more training time. In this paper, motivated by the BERT from Google AI Language, we propose a novel two-way encoder-decoder framework that automatically extracts appropriate sentiment information according to sequence to sequence reinforced learning. We use reinforcement learning to explore the space of possible extractive targets, where useful information provided by earlier predicted antecedents could be utilized for making later coreference decisions. The experiments on SemEval datasets demonstrate the efficiency and effectiveness of our models.

Keywords: Aspect sentiment classification · Sequence to sequence editing · Reinforced learning

1 Introduction

Typically, aspect sentiment classification (ASC) is a classification task of sentiment analysis [1]. Sequence to Sequence Reinforced Learning (SSRL) based approaches are becoming popular for this task, which aim to editing an input sequence to 3 sentiment level. Given the recent success of deep learning models, various neural network models and advanced attention mechanisms have been proposed for this task and have achieved highly competitive results on several benchmark datasets [2, 3, 6].

However, the models must overcome a heavy reliance on a large amounts of annotated data in order to learn a robust feature representation for multi-aspect emotion classification [7]. In reality, large-scale datasets are usually not readily available and costly to obtain, partly due to the ambiguity of many informal expressions in user-generated comments. Conversely, it is easier to find datasets associated with another closely related task: sentiment classification, which aims to classify the sentiment polarity of a given piece of text (i.e., positive, neural, negative). We expect that these resources may allow us to improve target-sensitive representations and thus more accurately identify emotions in business reviews [10]. To achieve these goals, we propose a two-way SSRL approach in this paper.

© Springer Nature Switzerland AG 2019
D. Milošević et al. (Eds.): HCC 2019, LNCS 11956, pp. 133–142, 2019.
https://doi.org/10.1007/978-3-030-37429-7_14

Attention mechanisms has been successfully used in many ASC tasks. It first computes the alignment scores between context vectors and target vector; then carry out a weighted sum with the scores and the context vectors. However, both LSTM and attention layer are very time consuming during training. LSTM processes one token in a step. Attention layer involves exponential operation and normalization of all alignment scores of all the words in the sentence [9]. Moreover, some models needs the positional information between words and targets to produce weighted LSTM [11], which can be unreliable in noisy review text. In this paper, we propose a fast and effective neural network based on convolutions and gating mechanisms, which has much less training time than LSTM based networks, but with better accuracy.

To evaluate the efficacy of our framework, we extracted divided restaurant and laptop datasets from SemEval datasets to verify the efficiency and effectiveness of our models. Specifically, in the training stage, each input sentence is associated with a numeric outcome. a model needs to edit the input to generate a new sentence that will satisfy the outcome target with high probability.

Our contributions are concluded as follows:

(1) We provide a two-way SSRL methods which far beyond the previous results.
(2) The outcome here is numerical, and it can be regarded as a generalization of the categorical outcome. We can specify a particular target rating.
(3) We summarize the challenges, advantages, and disadvantages of using different RL methods for seq2seq training.

2 Proposed Model

We use a variant of self-attention mechanism to distinguish the emotional and non-emotional words. Then the positive words and negative words are used to update the corresponding memory modules. Finally, the decoder uses the target sentiment information extracted from the memory and the content representation to perform decoding. The definitions of related notations are given in Table 1.

In this paper, a method of dividing regions based on specific target words is introduced. this method can expand the same sentence based on different aspects, so that each independent region can contain important characteristic information from different aspects, so that the parameter information can be adjusted according to the region in which different aspects are located in the training process of network model, and the important emotional information of specific aspects can be effectively preserved.

Table 1. SSRL model parameters

SSRL model parameters	
t	A target word, $t \in R^{V \times 1}$
K	Number of sentiment classes
s	Sentiment score, $s \in R^{K \times 1}$
y	Sentiment probability
H_t	Encoder hidden state at time t
$\hat{Y}$	Sets of actions that the agent is taking for a period of time t
π	The policy that the agent uses to take the action
π_θ	SSRL models use RNNs with parameter for the policy
$A(s_t; y_t)$	Advantage function which defines how good a state-action pair is w.r.t the expected reward that can be received at this state
$Q\pi$	The Q-value (that shows the estimated reward of taking action y_t when at state s_t)

2.1 Modeling Input Sentences

Given a target sentence, an embedding matrix A is used to convert t into a vector representation, the values in the embeddings st are parameters that are optimised during training [21] (Fig. 1).

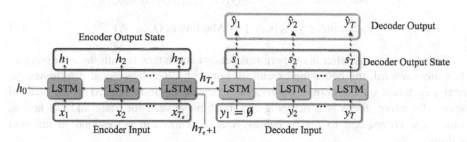

Fig. 1. A simple SSRL model. The blue boxes correspond to the encoder part which has Te units. The green boxes correspond to the decoder part which has T units. (Color figure online)

2.2 Modeling Output Sentences

We modify the LSTM model as follows: the aspect detection is integrated into the neural network architecture permitting an end-to-end optimization of the whole model during training. This is achieved by formatting the classifier output as a vector. This corresponds to predicting one of the three classes positive, negative and neutral for each aspect. Specifically, we transform the feature vector v extracted from the sentence to a score vector $\hat{y}^{(a)}$ for each aspect and apply softmax normalization:

Where

$$\hat{y}^{(a)} = soft\ \max(W^{(a)}v + b^{(a)}) \tag{1}$$

$$\text{softmax}(x)_i = \frac{\exp(x^i)}{\sum_{k=0}^{3} \exp(x^k)} \tag{2}$$

The loss is simply the cross entropy summed over all aspects:

$$L(\theta) = \sum_{a \in A} H(y^{(a)}, \hat{y}^{(a)}) \tag{3}$$

$$H(y, \hat{y}) = -\sum_i y_i \cdot \log(\hat{y}_i) \tag{4}$$

2.3 Reinforcement Learning Model

In reinforcement learning, the goal of an agent interacting with an environment is to maximize the expectation of the reward that it receives from the actions. Therefore, the focus is on maximizing one of the following objectives:

$$\text{Maximize } E_{\hat{y}_1,\dots,\hat{y}_T} \sim \pi\theta(\hat{y}_1,\dots,\hat{y}_T)[r(\hat{y}_1,\dots,\hat{y}_T)] \tag{5}$$

$$\text{maximize } y\ A_\pi(s_t, y_t) \rightarrow \text{Maximize}_y\ Q_\pi(s_t, y_t) \tag{6}$$

The reward is that after agent performs actions and interacts with the environment. The more reward, the better the execution. For each time slice, agent determines the next step based on the current observation. each observation is used as the state of agent. Therefore, there is a mapping relationship between the state and the action, which can correspond to the probability that a action can correspond to different actions.

2.4 Sentiment Score

The aspect sentiment scores for positive, neutral, and negative classes are then calculated, where a sentiment-specific weight matrix C is used. The sentiment scores are represented in a vector H, where K (K = 3) is the number of classes. The final sentiment probability y is produced with a softmax operation, i.e., y = softmax(s).

3 Related Work

For aspect-based sentiment analysis, past approaches use classifiers with expensive hand-crafted features based on n-grams, parts-of-speech, negation words, and sentiment lexica [9, 10]. However, it only extracts features from the preceding and subsequent sentence without any notion of structure. Neural network-based approaches include an LSTM that determines sentiment towards a target word based on its position [4, 5] as well as a recursive neural network that requires parse trees [14, 16]. Which capture the importance of different context word to a particular aspect and use this information as a semantic representation of the sentence.

For aspect sentiment classification, there are two main methods: supervised learning and dictionary-based supervised learning [11, 13]: the key problem for known aspect, is how to determine the scope. for each sentiment. The mainstream method is to use dependency, weighted weight according to dependency. Dictionary-based method sentiment shifters has some words that change the polarity of emotion, such as not, never, none, nobody, nowhere, neither. In addition to the above methods, there are many emotions that are implicit and difficult to extract, using Basic rules of opinions [21].

Sequence to sequence learning [12] is a challenging task in artificial intelligence, and covers a variety of language generation applications such as NMT [1, 2, 8] text summarization [17, 18], question answering [1] and conversational response generation [9, 10]. Sequence to sequence learning has attracted much attention in recent years due to the advance of deep learning. However, many language generations tasks such as AMN lack paired data but have plenty of unpaired data. Therefore, the pre-training on unpaired data and fine-tuning with small scale paired data will be helpful for these tasks, which is exactly the focus of this work.

LSTM is a neural network model for processing sequence data. It can be used in language model, machine translation, picture tagging, music automatic generation and so on, and plays an indispensable role in deep learning and big data [8]. Both on the review and on the sentence level, sentiment is dependent not only on preceding but also successive words and sentences. A Bidirectional LSTM (Bi-LSTM) [13] allows us to look ahead by employing a forward LSTM, which processes the sequence in chronological order, and a backward LSTM, which processes the sequence in reverse order. The output h_t at a given time step is then the concatenation of the corresponding states of the forward and backward LSTM [20].

4   Experiments and Discussion

4.1   Datasets Preparation

We perform experiments on the datasets of SemEval dataset task 2014, which contain online reviews from domain Laptop and Restaurant [20]. In these datasets, aspect sentiment polarities are labeled. The training and test sets have also been provided. Full statistics of the datasets are given in Table 2.

Table 2. Statistics of datasets

Dataset	Positive		Neutral		Negative	
	Train	Test	Train	Test	Train	Test
Restaurant	2164	728	637	196	807	196
Laptop	994	341	464	169	870	128

The sentences which have different sentiment labels for different aspects or targets in the sentence are more common in review data than in standard sentiment classification benchmark. Therefore, to access how the models perform on review sentences more accurately, we create small but difficult datasets, which are made up of the sentences having opposite or different sentiments on different targets.

4.2   Models for Comparison

AMN: A state-of-the-art memory network used for ASC. The main difference from MN is in its attention alignment function.

ATAE-LSTM: ATAE-LSTM is an attention-based LSTM for ACSA task. It appends the given aspect embedding with each word embedding as the input of LSTM, and has an attention layer above the LSTM layer.

Sequence to Better Sequence (S2BS) [9]: For training, S2BS also requires each sentence is associated with an outcome. For test, S2BS only revises a sentence such that the output is associated with a higher outcome, which is not a specified value. For comparison, we adapt our revision method for S2BS, by which their trained model is able to conduct quantifiable sentence revision.

RRSL: a novel two-way encoder-decoder framework that automatically extracts appropriate sentiment information according to sequence to sequence reinforced learning. We use reinforcement learning to explore the space of possible extractive targets, where useful information provided by earlier predicted antecedents could be utilized for making later coreference decisions.

4.3   Training the Models

We use the BERT for the initialization of word vectors. We initialize

The dimension of the word embeddings and the size of the hidden layers are 300. The learning rate is set to 0.01 and the dropout rate is set to 0.1. Stochastic gradient descent is used as our optimizer. The position encoding is also used.

We implement all models in the TensorFlow environment using same input, embedding size, dropout rate, optimizer, etc. Target aspect as x and t respectively. The appear together in a pair format(x, t) as input and all such pairs construct the dataset H (Table 3).

Table 3. The algorithm based on sequence to sequence reinforced learning

Algorithm 1 Training a simple seq2seq model	Algorithm 2 Reinforce Learning algorithm

Input: Input sequences (X) and ground-truth output sequences (Y).
Output: Trained seq2seq model.
Training Steps:
for batch of input and output sequences X and Y do
Run encoding on X and get the last encoder state h_t.
Run decoding by feeding h_t to the first decoder and obtain the
sampled output sequence $\hat{Y}$.
Calculate the loss according to Eq. (3) and update the parameters
of the model.
end for
Testing Steps:
for batch of input and output sequences X and Y do
Use the trained model and Eq. (4) to sample the output $\hat{Y}$
Evaluate the model using a performance measure
end for
	Input: Input sequences (X), ground-truth output sequences (Y),and (preferably) a pre-trained policy
	Output: Trained policy with REINFORCE.
	Training Steps:
	while not converged do
	Select a batch of size N from X and Y.
	Sample N full sequence of actions:
	$f(\hat{y}_1)$;
	Observe the sequence reward and calculate the baseline.
	Calculate the loss according to Eq. (5).
	Update the parameters of network.
	end while
	Testing Steps:
	for batch of input and output sequences X and Y do
	Use the trained model and Eq. (6) to sample the output $\hat{Y}$.
	Evaluate the model using a performance metric.
	end for

4.4 Evaluation Measure

We use F1-score as our evaluation measure. We report both F1-Macro over all classes and all individual classes based F1 Scores. F1-score is calculated based on recall rate and precision rate.

We use the open-domain word embeddings 1 for the initialization of word vectors. We initialize other model parameters from a uniform distribution $U(-0.05, 0.05)$. The dimension of the word embedding and the size of the hidden layers are 300. The learning rate is set to 0.01 and the dropout rate is set to 0.1. Stochastic gradient descent is used as our optimizer. When we do binary classification, we think that if h (x) $>=$ 0.5, then predict = 1; if h (x) < 0.5, then predict = 0. Here 0.5 is the classification threshold. By increasing the threshold, we will be more confident in the forecast value, we will increase the precision rate. However, this will reduce the recall rate. so the model will not let go of the real example, and the recall rate will increase the. We implemented all models in the TensorFlow environment using same input, embedding size, dropout rate, optimizer, etc.

4.5   Result Analysis

The classification results are shown in Table 4. Note that the all mentioned above models.

Table 4. Results of all models on two datasets of F1 score.

Model	Restaurant			Laptop		
	Neg.	Neu.	Pos.	Neg.	Neu.	Pos.
AMN + word2vec	57.07	36.81	82.86	52.67	47.89	75.63
ATAE-LSTM + word2vec	64.22	49.40	85.73	55.27	42.15	80.81
SSRL + word2vec	**67.11**	**50.47**	**87.90**	**57.73**	**48.12**	**81.96**
AMN + Bert	64.18	47.86	85.90	60.00	52.56	82.91
ATAE-LSTM + Bert	67.89	51.47	88.19	61.08	57.49	83.11
SSRL + Bert	**68.72**	**54.03**	**89.51**	**63.17**	**59.11**	**84.39**

Compared to LSTM, our two model variants could successfully distinguish the describing words for corresponding aspect targets. In addition, it can also be seen from the laboratory results that the model proposed in this paper, by accepting the attention mechanism of the word layer and the sentence layer, can pay high attention to the internal dependence of specific aspects on the sentence to be classified, and reduce the dependence on the region division, and the classification accuracy still maintains a good distribution when the region length is reduced. In the sentence "the food are delicious, but the service is slow", a LSTM outputs the same negative sentiment label for both aspect terms "food" and "service", while SSRL + word2vec and SSRL + Bert successfully recognize that "slow" is only used for describing "service" and output neutral and negative labels for aspects "food" and "service" respectively. In another example "the staff members are extremely friendly and even replaced my drink once when i dropped it outside", our models also find out that positive and neutral sentiment for "staff" and "drink" respectively.

5   Conclusion and Future Work

In this paper, we integrated the advantage of the enhanced learning and Sequence editing, RRSL creatively and put forward a model used to solve the fine-grained aspect sentiment classification. Two classical classification experiments show that the method used in this paper can faster training data improve classification accuracy. In this paper, users' comments on unclear expression of emotion polarity, long or short sentence length, omission of content, vague reference, and disordered expression of emotion are also prominent. Our work reports improvements the training speed and bear a great potential for future applications on larger scale. In the future, we prepare used the masked sequence to sequence model to improve the results based on the RRSL model, and the results can be quantitatively precious.

References

1. Zhang, Y., Xu, J., Yang, P., Sun, X.: Learning sentiment memories for sentiment modification without parallel data. In: Empirical Methods in Natural Language Processing (EMNLP), pp. 1097–1102 (2018)
2. Yi, L., Li, D., Li, P., Shi, S., Lam, W., Zhang, T.: Learning sentiment memories for sentiment modification without parallel data. In Empirical Methods in Natural Language Processing (EMNLP), pp. 3855–3864 (2018)
3. Keneshloo, Y., Shi, T., Ramakrishnan, N., Reddy, C.K.: Deep Reinforcement Learning for Sequence-to-Sequence Models. IEEE (2018)
4. Dahou, A., Elaziz, M.E.A., Zhou, J., Xiong, S.: Arabic sentiment classification using convolutional neural network and differential evolution algorithm. Comput. Int. Neurosc. 2537689:1–2537689:16 (2019)
5. Ma, R., Wang, K., Qiu, T., Sangaiah, A.K., Lin, D., Liaqat, H.B.: Feature-based compositing memory networks for aspect-based sentiment classification in social Internet of Things. Future Gener. Comput. Syst. **92**, 879–888 (2019)
6. Zhang, T., Wu, X., Lin, M., Han, J., Hu, S.: Imbalanced sentiment classification enhanced with discourse marker. CoRR abs/1903.11919
7. Sharma, R., Bhattacharyya, P., Dandapat, S., Bhatty, H.S.: Identifying transferable information across domains for cross-domain sentiment classification. In: ACL, pp. 968–978 (2018)
8. Zhu, P., Qian, T.: Enhanced aspect level sentiment classification with auxiliary memory. In: COLING, pp. 1077–1087 (2018)
9. Lv, G., Wang, S., Liu, B., Chen, E., Zhang, K.: Sentiment classification by leveraging the shared knowledge from a sequence of domains. In: Li, G., Yang, J., Gama, J., Natwichai, J., Tong, Y. (eds.) DASFAA 2019. LNCS, vol. 11446, pp. 795–811. Springer, Cham (2019). https://doi.org/10.1007/978-3-030-18576-3_47
10. Liu, S., Lee, I.: Sentiment classification with medical word embeddings and sequence representation for drug reviews. In: Siuly, S., Lee, I., Huang, Z., Zhou, R., Wang, H., Xiang, W. (eds.) HIS 2018. LNCS, vol. 11148, pp. 75–86. Springer, Cham (2018). https://doi.org/10.1007/978-3-030-01078-2_7
11. Iqbal, F., et al.: A hybrid framework for sentiment analysis using genetic algorithm based feature reduction. IEEE Access **7**, 14637–14652 (2019)
12. Xu, G., Meng, Y., Qiu, X., Yu, Z., Wu, X.: Sentiment analysis of comment texts based on BiLSTM. IEEE Access **7**, 51522–51532 (2019)
13. Seifollahi, S., Shajari, M.: Word sense disambiguation application in sentiment analysis of news headlines: an applied approach to FOREX market prediction. J. Intell. Inf. Syst. **52**(1), 57–83 (2019)
14. Pang, B., Lee, L., Vaithyanathan, S.: Thumbs: sentiment classification using machine learning techniques. In: Empirical Methods in Natural Language Processing, pp. 79–86 (2002)
15. Polanyi, L., Zaenen, A.: Contextual valence shifters. In: Shanahan, J.G., Qu, Y., Wiebe, J. (eds.) Computing Attitude and Affect in Text: Theory and Applications, pp. 1–10. Springer, Dordrecht (2006). https://doi.org/10.1007/1-4020-4102-0_1
16. Pontiki, M., Galanis, D., Pavlopoulos, J., Papageorgiou, H., Androutsopoulos, I., Manandhar, S.: SemEval-2014 task4: aspect based sentiment analysis. In: ProWorkshop on Semantic Evaluation (SemEval-2014). Association for Computational Linguistics (2014)

17. Socher, R., Perelygin, A., Wu, J., Chuang, J., Manning, C.D., Ng, A., Potts, C.: Recursive deep models for semantic compositionality over a sentiment treebank. In: Empirical Methods in Natural Language Processing, pp. 1631–1642 (2016)
18. Chen, P., Sun, Z., Bing, L., Yang, W.: Recurrent attention network on memory for aspect sentiment analysis. In: EMNLP, pp. 463–472 (2017)
19. Chung, J., Gulcehre, C., Cho, K., Bengio, Y.: Empirical evaluation of gated recurrent neural networks on sequence modeling. In: NIPS (2014)
20. Collobert, R., Weston, J., Bottou, L., Karlen, M., Kavukcuoglu, K., Kuksa, P.P.: Natural language processing (almost) from scratch. J. Mach. Learn. Res. **12**, 2493–2537 (2011)
21. Bowman, S.R., Vilnis, L., Vinyals, O., Dai, A.M., Jozefowicz, R., Bengio, S.: Generating sentences from a continuous space. In: Proceedings of the 20th SIGNLL Conference on Computational Natural Language Learning, pp. 10–21 (2016)

Performance-Improved UCD-Based Hybrid Precoding for Millimeter-Wave Massive MIMO Single User Systems

Shuxiang Heng[✉] and Yang Liu

Beijing University of Posts and Telecommunications,
No. 10 Xitucheng Road, Beijing, China
{hengshuxiang, ly}@bupt.edu.cn

Abstract. Hybrid precoding scheme can significantly reduce the number of radio frequency (RF) chains and the huge energy consumption in mmWave massive MIMO systems. Most existing hybrid precoding papers are based on singular value decomposition (SVD) or geometric mean decomposition (GMD). The GMD-based hybrid precoding scheme can avoid the complicated bit allocations in SVD-based method, because it decomposes the channel into several identical SNR subchannels. However, the GMD-based method will suffer from considerable capacity loss at low SNR in mmWave systems. To overcome this shortcoming, we can combine the hybrid precoding structure with uniform channel decomposition (UCD). In this paper, we propose the UCD-based hybrid precoding for mmWave massive MIMO single user systems. Simulation results verify that the proposed UCD-based hybrid precoding scheme outperforms the conventional SVD-based and GMD-based hybrid precoding schemes, because it achieves much higher spectral efficiency and lower bit error rate (BER) at low SNR systems.

Keywords: mmWave · Hybrid precoding · SVD · GMD · UCD

1 Introduction

Nowadays, millimeter wave (mmWave) communications are expected to be a promising key technique of the 5G wireless network [1–5]. The large available bandwidth provides gigabit-per-second data rates and the short wavelength of the mmWave enables much more antennas to be packed into the small size, which allows for large scale spatial multiplexing and highly directional beamforming. To compensate for the severe path loss of mmWave channel, we should combine massive multiple-input multiple-output (MIMO) with mmWave to provide higher data rates, higher spectral efficiency and lower latency [6–8].

In traditional low-frequency MIMO systems, precoding is implemented in the digital domain, which antenna is connected to one RF chain (including amplifier and mixer) [6]. But in mmWave systems, it is unrealistic to assign a RF chain for each antenna, which will add extra power consumption and implement complexity. Therefore, hybrid precoding architecture has been proposed [9–14]. The analog beamforming

© Springer Nature Switzerland AG 2019
D. Milošević et al. (Eds.): HCC 2019, LNCS 11956, pp. 143–153, 2019.
https://doi.org/10.1007/978-3-030-37429-7_15

is implemented with phase shifters to take advantage of beamforming gains, while the digital beamforming works at the baseband to take advantage of multiplexing gains.

Leveraging the sparse scattering characteristic of mmWave, we can formulate the hybrid precoding design problem as a sparse reconstruction problem, and use the orthogonal matching pursuit (OMP) algorithm [10–14]. In [14], the author transforms the problem of maximizing the mutual information of hybrid precoder into minimizing the Euclidean distance between the optimal unconstrained digital precoder and hybrid precoder. Most existing papers utilize the singular value decomposition (SVD) of the channel matrix to obtain the optimal unconstrained precoding matrix. Nevertheless, the very different SNR for each subchannel results in the complicated bit allocations and high order modulation [15]. [16–18] use a mmWave massive MIMO hybrid precoding scheme based on the geometric mean decomposition (GMD). It converts the mmWave channel into identical SNR subchannels and has lower BER compared with SVD-based method [19–21].

However, because of the zero-forcing (ZF) detection method, GMD-based hybrid precoding scheme will suffer from considerable capacity loss at low SNR [15], while low SNR is a typical characteristic in mmWave systems [2, 10]. Therefore, we need to find a better hybrid precoding method in mmWave systems. Luckily, [19] propose a digital precoding method based on the uniform channel decomposition (UCD) matrix decomposition, which has better capacity performance and can achieve high data rates transmission using small symbol constellations. In this paper, we combine the uniform channel decomposition (UCD) scheme with mmWave hybrid precoding structure, and propose a UCD-based performance-improved hybrid precoding method for mmWave massive MIMO single user systems. Two remarkable merits of UCD-based hybrid precoding make it outperform the SVD-based and GMD-based hybrid precoding. Firstly, it doesn't make a tradeoff between spectral efficiency and BER performance like GMD-based method, instead, it achieves best of both at the same time. Secondly, UCD scheme has the maximal diversity gain. It can make a best balance between the diversity gain and multiplexing gain [19]. As a result, the proposed UCD-based hybrid precoding method is the more suitable than other methods in mmWave systems.

Notation: Lower-case and upper-case boldface letters a and A denote a vector and a matrix. $tr\{\mathbf{A}\}$, $\{\mathbf{A}\}^{\mathrm{T}}$, $\{\mathbf{A}\}^{\mathrm{H}}$, $\|\mathbf{A}\|_{\mathrm{F}}$ and $\{A\}_{i,j}$ denote the trace, transpose, conjugate transpose, Frobenius norm, and the element in the ith row and the jth column of A, respectively. $\mathbf{A}(:, 1)$ and $\mathbf{A}(:, 1 : \mathrm{N_s})$ means the first column and the first Ns columns of matrix A, respectively.

2 System Model

2.1 Fundamental model

We consider a typical narrowband mmWave massive MIMO system with a hybrid precoding structure [7–10]. A base station (BS) with N_{RF} RF chains and Nt transmit antennas sends Ns independent data streams to the mobile station (MS) with N_{RF} RF chains and Nr receiving antennas as shown in Fig. 1.

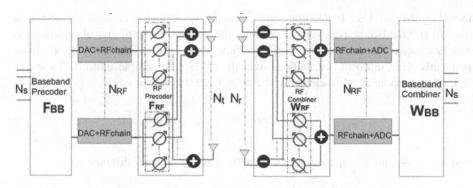

Fig. 1. Hybrid analog/digital architectures divide the precoding/combining processing between analog and digital domains.

We assume that Ns $\leq$ N_{RF} $\leq$ Nt and Ns $\leq$ N_{RF} $\leq$ Nr. In the hybrid precoding structure, the precoding matrix $\mathbf{F} \in \mathbb{C}^{N_t \times N_s}$ can be divided as $\mathbf{F} = \mathbf{F}_{RF}\mathbf{F}_{BB}$: the low-dimension digital precoding matrix $\mathbf{F}_{BB} \in \mathbb{C}^{N_{RF} \times N_s}$ and the high-dimension analog precoding matrix $\mathbf{F}_{RF} \in \mathbb{C}^{N_t \times N_{RF}}$. The transmitted signal symbol is $\mathbf{s} \in \mathbb{C}^{N_s \times 1}$, and the transmitted signal vector can be expressed as $\mathbf{x} = \mathbf{F_s} = \mathbf{F}_{RF}\mathbf{F}_{BB}\mathbf{s}$. And $\mathbf{E}[\mathbf{ss}^H] = \frac{P_s}{N_s}\mathbf{I}_{N_s}$, $\mathbf{P_s}$ is the average total transmitted power. Since $\mathbf{F}_{RF}$ is implemented using analog phase shifters, its entries are of constant modulus, so $\left|[\mathbf{F}_{RF}]_{m,n}\right|^2 = \mathbf{N_t^{-1}}$. To meet the power constraint, we bound the transmit power at the BS as $\|\mathbf{F}_{RF}\mathbf{F}_{BB}\|_F^2 = N_S$. Similarly, the hybrid combing matrix $\mathbf{W} \in \mathbb{C}^{N_r \times N_s}$ can be expressed as $\mathbf{W} = \mathbf{W}_{RF}\mathbf{W}_{BB}$. After the combining matrix, the signal vector is

$$y = \sqrt{\rho}\mathbf{W}^H\mathbf{H}\mathbf{F}\mathbf{s} + \mathbf{W}^H\mathbf{n} = \sqrt{\rho}\mathbf{W}_{BB}{}^H\mathbf{W}_{RF}{}^H\mathbf{H}\mathbf{F}_{RF}\mathbf{F}_{BB}\mathbf{s} + \mathbf{W}_{BB}{}^H\mathbf{W}_{RF}{}^H\mathbf{n}. \quad (1)$$

The ρ is the averaged received power. $H \in \mathbb{C}^{N_r \times N_t}$ is the channel matrix between BS and the MS, and $n \in \mathbb{C}^{N_r \times 1}$ is the additive white Gaussian noise (AWGN) vector at the MS following the circularly symmetric complex Gaussian distribution with zero mean vector and covariance matrix, i.e.CN (0, σ^2 INr). We assume that perfect channel state information (CSI) is known at both the transmitter and receiver.

2.2 Channel Model

Since mmWave channels are expected to have limited scattering, we adopt the modified Saleh-Valenzuela (SV) model in this paper [10–13]

$$\mathbf{H} = \sqrt{\frac{N_tN_r}{N_{cl}N_{ray}}}\sum_{i=1}^{N_{cl}}\sum_{l=1}^{N_{ray}}\beta_{il}\boldsymbol{\alpha}_r(\varphi_{il}^r)\boldsymbol{\alpha}_t(\varphi_{il}^t)^H \quad (2)$$

where N_{cl} and N_{ray} represent the number of clusters and the number of paths in each cluster. $\beta_{il} \sim CN(0,1)$ is the complex gain of l th ray in the i th propagation cluster, including the path loss. In addition, $\alpha_r(\varphi_{il}^r)$ and $\alpha_i(\varphi_{il}^t)$ are the array response vectors at the MS and the BS, respectively. φ_{il}^r denotes the azimuth angle of arrival (AoA) at the MS, and φ_{il}^t is the azimuth angle of departure (AoD) at the BS, and both are independently from uniform distributions over $[0, \pi]$. For the simple uniform linear line (ULA) antenna array of N elements, the array response vector is

$$\alpha_{ULA} = \sqrt{\frac{1}{N}}\left[1, e^{jkd\sin(\varphi)}, \cdots e^{j(N-1)kd\sin(\varphi)}\right]^T \tag{3}$$

where $k = \frac{2\pi}{\lambda}$, and λ is the signal wavelength, d is the antenna distance.

3 The Optimal Digital Unconstrained Precoder

3.1 Fully Digital SVD-Based and GMD-Based Precoding

In the traditional optimal digital precoding mmWave schemes, most existing papers utilize the singular value decomposition (SVD) of the channel matrix: $\mathbf{H} = \mathbf{U\Sigma V^H}$ to obtain the optimal unconstrained precoding and combining matrix [14], which means we let the first Ns columns of $\mathbf{V}$ and $\mathbf{U}$ be the optimal unconstrained precoding and combining matrix, respectively.

But due to the limited spatial scattering in mmWave propagation, the singular values of channel matrix H vary a lot, which results in the very different SNR subchannels. To solve this problem, GMD-based precoding has been proposed [15], where the channel matrix H is decomposed as $\mathbf{H} = \mathbf{GRQ^H}$. R1 $\in \mathbb{C}^{N_s \times N_s}$ is the first Ns rows and Ns columns of R. It's an upper triangular matrix with identical diagonal elements, which is the geometric mean of the first Ns singular values of H, and the diagonal element is $r_{ii} = \bar{r} = (\sigma_1\sigma_2...\sigma_{N_s})^{\frac{1}{N_s}}$. If we let the first Ns columns of $\mathbf{Q}$ and $\mathbf{G}$ be the optimal unconstrained precoding and combining matrix, respectively, then (1) will be

$$y = \sqrt{\rho}\mathbf{R}_1 s + \mathbf{G}_1{}^H n \tag{4}$$

Then, by utilizing successive interference cancellation (SIC) at the receiver, we can acquire Ns sub-channels with equal SNR and sub-channel gain r_{ii} [15].

3.2 Fully Digital UCD-Based Precoding

However, GMD-based hybrid precoding scheme will suffer from considerable capacity loss at low SNR systems. Based on GMD matrix decomposition method, [19] propose a uniform channel decomposition (UCD) matrix decomposition scheme, which can decompose a MIMO channel into multiple subchannels with identical SNR and capacity. We modify the optimal unconstrained digital precoder is $\mathbf{F}_{opt} = \mathbf{V}_1\theta^{\frac{1}{2}}\mathbf{\Omega^H}$, where $\mathbf{V}_1$ is the first Ns column of V. θ is a diagonal matrix whose diagonal element θ_k

determines the power loaded to the k th subchannel and is found via "water-filling" power allocation [22]:

$$\theta_k = \max\left\{\mu - \frac{\alpha}{\lambda_{H,k}^2}, 0\right\}, \sum_{k=1}^{K}\theta_k = 1 \tag{5}$$

where $\lambda_{H,k}$ is the nonzero singular values of H, μ is a constant and $\alpha = N_s\sigma^2$.

$\Omega \in \mathbb{C}^{N_s \times N_s}$ is a unitary matrix. The introduction of Ω makes the precoding matrix design more flexible than the SVD. According to [19], the virtual channel is

$$\mathbf{G} \triangleq \mathbf{HF} = \mathbf{U}\Sigma\mathbf{V}^{\mathbf{H}}\mathbf{V}\theta^{\frac{1}{2}}\Omega^{\mathbf{H}} \triangleq \mathbf{U}\Lambda\Omega^{\mathbf{H}} \tag{6}$$

where $\Lambda = \Sigma\theta^{\frac{1}{2}}$ is a diagonal matrix with diagonal elements for $\{\sigma_i\}_{i=1}^{Ns}$.

Then we can have the following GMD

$$\mathbf{J} \triangleq \begin{bmatrix} \mathbf{U}\Lambda \\ \sqrt{\alpha}\,\mathbf{I}_{N_s} \end{bmatrix} = \mathbf{G}_J\mathbf{R}_J\mathbf{Q}_J^{\mathbf{H}} \tag{7}$$

where $\mathbf{G}_J \in \mathbb{C}^{(N_r + N_s) \times N_s}$ is semi-unitary matrix. $\mathbf{R}_J \in \mathbb{C}^{N_s \times N_s}$ is an upper triangular matrix with equal diagonal elements, and $\mathbf{Q}_J \in \mathbb{C}^{N_s \times N_s}$ is unitary matrix. [19] shows that the SVD decomposition of $\mathbf{J}$ is

$$\mathbf{J} \triangleq \begin{bmatrix} \mathbf{U}\Lambda\tilde{\Sigma}^{-1} \\ \sqrt{\alpha}\tilde{\Sigma}^{-1} \end{bmatrix}\tilde{\Sigma}\mathbf{I}_{N_s} \tag{8}$$

$\tilde{\Sigma}$ is an $N_s \times N_s$ diagonal matrix with the diagonal elements $\tilde{\sigma}_i = \sqrt{\sigma_i^2 + \alpha}$. According to the relationship between the SVD decomposition and GMD decomposition [18, 19], we have the following collusion

$$\mathbf{G}_J = \begin{bmatrix} \mathbf{U}\Lambda\tilde{\Sigma}^{-1} \\ \sqrt{\alpha}\tilde{\Sigma}^{-1} \end{bmatrix}\mathbf{S}_L, \mathbf{Q}_J = \mathbf{I}_{N_s}\mathbf{S}_R, \mathbf{R}_J = \mathbf{S}_L^H\tilde{\Sigma}\mathbf{S}_R \tag{9}$$

where $\mathbf{S}_R \in \mathbb{C}^{N_s \times N_s}$ and $\mathbf{S}_L \in \mathbb{C}^{N_s \times N_s}$ are unitary matrix depended on $\tilde{\Sigma}$. Let $\Omega = \mathbf{Q}_J^{\mathbf{H}}$, and

$$\mathbf{Q}_{\mathbf{Ga}} = \begin{bmatrix} \mathbf{I}_{N_r} & \mathbf{0} \\ \mathbf{0} & \Omega \end{bmatrix}\mathbf{G}_J \tag{10}$$

we let $\mathbf{Q}_{\mathbf{Ga}}^{\mathbf{u}}$ denotes the first N_r rows of $\mathbf{Q}_{\mathbf{Ga}}$, or equivalently, the first N_r rows of $\mathbf{G}_J$. According to [19], the nulling vectors are

$$\mathbf{w}_i = r_{J,ii}^{-1}\mathbf{q}_{\mathbf{Ga},i} \tag{11}$$

where $r_{J,ii}$ is the i th diagonal element of $\mathbf{R}_J$, and $\mathbf{q}_{\mathrm{Ga},i}$ is the i th column of $\mathbf{Q}^{\mathrm{u}}_{\mathrm{Ga}}$.

$$\mathbf{Q}^{\mathrm{u}}_{\mathrm{Ga}} = \mathbf{U}\Lambda\tilde{\Sigma}^{-1}\mathbf{G}_1 \tag{12}$$

$\mathbf{G}_1$ is the first N_s column of $\mathbf{G}_J$. Considering Eq. (9), we have the following collusion: the optimal digital combining and precoding matrix are

$$\mathbf{W}_{\mathrm{opt}} = r^{-1} * \mathbf{U}_1\Lambda\tilde{\Sigma}^{-1}\mathbf{S}_L \tag{13}$$

$$\mathbf{F}_{\mathrm{opt}} = \mathbf{V}_1\theta^{\frac{1}{2}}\mathbf{S}_R \tag{14}$$

4 The Constrained UCD-Based Hybrid Precoding Design

According to [14], our purpose is maximizing the spectral efficiency in the hybrid precoding case:

$$\underset{\mathbf{F}_{RF}\mathbf{F}_{BB}}{\arg\max}\log_2\left|I + \frac{\rho}{\mathrm{N}_s}\mathbf{R}_{\mathrm{n}}^{-1}\mathbf{W}_{\mathrm{BB}}^{\mathrm{H}}\mathbf{W}_{\mathrm{RF}}^{\mathrm{H}}\mathbf{H}\mathbf{F}_{RF}\mathbf{F}_{BB}\mathbf{F}_{BB}^{\mathrm{H}}\mathbf{F}_{RF}^{\mathrm{H}}\mathbf{H}^{\mathrm{H}}\mathbf{W}_{\mathrm{RF}}\mathbf{W}_{\mathrm{BB}}\right| \tag{15}$$

$\mathbf{R}_{\mathrm{n}} = \sigma_n^2\mathbf{W}_{\mathrm{BB}}^{\mathrm{H}}\mathbf{W}_{\mathrm{RF}}^{\mathrm{H}}\mathbf{W}_{\mathrm{RF}}\mathbf{W}_{\mathrm{BB}}$, and is the noise covariance matrix after combining. To simplify joint optimization design, we temporarily decouple the joint transmitter receiver optimization problem and focus on the transmitter design of the hybrid precoders FRF, FBB, assuming that the receiver can perform optimal nearest-neighbor decoding based on the Nr-dimensional received signal y, which means we let $\mathbf{W}_{\mathrm{RF}} = \mathbf{W}_{\mathrm{BB}} = \mathbf{I}_{\mathrm{Nr}}$.

The objective function is converted to maximize the mutual information:

$$\underset{\mathbf{F}_{RF}\mathbf{F}_{BB}}{\arg\max}\log_2\left|I + \frac{\rho}{\mathrm{N}_s\sigma_n^2}\mathbf{H}\mathbf{F}_{\mathrm{RF}}\mathbf{F}_{\mathrm{BB}}\mathbf{F}_{\mathrm{BB}}^{\mathrm{H}}\mathbf{F}_{\mathrm{RF}}^{\mathrm{H}}\mathbf{H}^{\mathrm{H}}\right|$$

$$\text{s.t. } \mathbf{F}_{\mathrm{RF}} \in \psi, \ \|\mathbf{F}_{\mathrm{RF}}\mathbf{F}_{\mathrm{BB}}\|_{\mathrm{F}}^2 = \mathrm{N}_S \tag{16}$$

where ψ is the a candidate set of $\mathbf{F}_{RF}$ matrix column vectors, and [14] transforms it as

$$\underset{\mathbf{F}_{RF}\mathbf{F}_{BB}}{\arg\min}\left\|\mathbf{F}_{\mathrm{opt}} - \mathbf{F}_{\mathrm{RF}}\mathbf{F}_{\mathrm{BB}}\right\|_{\mathrm{F}}$$

$$\text{s.t. } \mathbf{F}_{\mathrm{RF}} \in \psi, \ \|\mathbf{F}_{\mathrm{RF}}\mathbf{F}_{\mathrm{BB}}\|_{\mathrm{F}}^2 = \mathrm{N}_S \tag{17}$$

where $\mathbf{F}_{\mathrm{opt}}$ is the optimal unconstrained digital precoder, and we let the $\mathbf{F}_{\mathrm{opt}}$ be $\mathbf{V}_1\theta^{\frac{1}{2}}\mathbf{S}_R$. However, solving (17) is still difficult, since $\mathbf{F}_{\mathrm{RF}}$ and $\mathbf{F}_{\mathrm{BB}}$ are coupled, and the $\mathbf{F}_{\mathrm{RF}} \in \psi$

constraints is non-convex, so we design $\mathbf{F}_{RF}$ and $\mathbf{F}_{BB}$ separately. When we design $\mathbf{F}_{RF}$, we think $\mathbf{F}_{BB}$ is fixed and vice versa. Therefore,

$$\left\| \mathbf{V}_1 \boldsymbol{\theta}^{\frac{1}{2}} \mathbf{S}_{R1} - \mathbf{F}_{RF}\mathbf{F}_{BB} \right\|_F \overset{a}{=} \left\| \mathbf{V}_1 \boldsymbol{\theta}^{\frac{1}{2}} - \mathbf{F}_{RF}\mathbf{F}_{BB}\mathbf{S}_{R1}{}^H \right\|_F = \left\| \mathbf{V}_1 \boldsymbol{\theta}^{\frac{1}{2}} - \mathbf{F}_{RF}\tilde{\mathbf{F}}_{BB} \right\|_F \quad (18)$$

where (a) holds because the Frobenius norm is invariant under rotations, and $\tilde{\mathbf{F}}_{BB} = \mathbf{F}_{BB}\mathbf{S}_{R1}^H$. Now we design the $\mathbf{F}_{RF}$ and $\tilde{\mathbf{F}}_{BB}$ separately.

This paper we use a new method based on SVD decomposition and residual matrix iterative algorithm to design transmitter hybrid precoding matrix [18]. The proposed algorithm is as follows:

Require: H	(7) $\mathbf{F}_{res} = \dfrac{\mathbf{F} - \mathbf{F}_{RF}\tilde{\mathbf{F}}_{BB}}{\left\| \mathbf{F} - \mathbf{F}_{RF}\tilde{\mathbf{F}}_{BB} \right\|_F}$
(1) $\text{SVD}(\mathbf{H}) = \mathbf{U}\boldsymbol{\Sigma}\mathbf{V}^H$, $\mathbf{V}_1 = \mathbf{V}(:,1:N_s)$	
(2) according to (5)~(14), we can obtain:	(8) $\text{SVD}(\mathbf{F}_{res}) = \mathbf{U}_1\mathbf{S}_1\mathbf{V}_1^H$
$\boldsymbol{\theta}$、 $\mathbf{S}_R$ and let $\mathbf{F} = \mathbf{V}_1\boldsymbol{\theta}^{\frac{1}{2}}$	(9) $\mathbf{n} = \dfrac{1}{\sqrt{N_t}} * e\{j * angle(\mathbf{U}_1(:,1))\}$
(3) $\text{SVD}(\mathbf{F}) = \tilde{\mathbf{U}}\tilde{\mathbf{S}}\tilde{\mathbf{V}}^H$	
(4) $\mathbf{A} = \dfrac{1}{\sqrt{N_t}} * e\{j * angle(\tilde{\mathbf{U}})\}$	(10) $\mathbf{F}_{RF} = \left[\mathbf{F}_{RF}, \mathbf{n}\right]$
	End for
(5) $\mathbf{F}_{RF} = \mathbf{A}(:,1)$	(11) $\mathbf{F}_{BB} = pinv(\mathbf{F}_{RF}) * \mathbf{F} * \mathbf{S}_R$
For i =1: $N_{RF} - 1$	
(6) $\tilde{\mathbf{F}}_{BB} = pinv(\mathbf{F}_{RF}) * \mathbf{F}$	(12) $\mathbf{F}_{BB} = \dfrac{\sqrt{N_s}\mathbf{F}_{BB}}{\left\| \mathbf{F}_{RF}\mathbf{F}_{BB} \right\|_F}$
	(13) Return: $\mathbf{F}_{RF}$ and $\mathbf{F}_{BB}$

Step (1)–(3) are to obtain left unitary matrix of the optimal matrix F by SVD as well as unitary matrix SR. Step (4) is to meet the restrictions of the elements of $\mathbf{F}_{RF}$. So we keep the phase of all the elements in the $\tilde{\mathbf{U}}$, and then forcibly change the phase as $1/\sqrt{N_t}$. We select the first column vector of A as the first column of $\mathbf{F}_{RF}$ in step (5), because the first singular value is often much larger than other singular values. We select the corresponding left singular column vector with the largest correlation with the original matrix. And then We use the least squares (LS) method to calculate pseudo digital precoding matrix $\tilde{\mathbf{F}}_{BB}$ in step (6).

In step (7), we obtain the residual matrix Fres, and then continue to do the SVD decomposition on the residual matrix to obtain the largest left singular column vector. And then transform it into a column vector with the magnitude of $1/\sqrt{N_t}$, adding it as a second column of the original analog precoding matrix (8)–(10). So we update to obtain a new analog precoding matrix, and then repeat the following process until the final complete analog precoding matrix is obtained. Step (11) is to obtain digital precoding matrix by least squares (LS) method and **SR**. Step (12) is to meet the constraint of $\|\mathbf{F}_{RF}\mathbf{F}_{BB}\|_F^2 = N_S$.

At the receiver side, similarly, we let the full digital combiner be $\mathbf{W}=r_{J,ii}^{-1}\mathbf{U}_1\Lambda\tilde{\Sigma}^{-1}$, and $\mathbf{W_{BB}} = pinv(\mathbf{W_{RF}}) * \mathbf{W} * \mathbf{S_L}$. Then we use the same algorithm as transmitter side to obtain the RF combiner WRF and baseband combiner $\mathbf{WBB}$.

5 Simulations

In this section, we will evaluate the performance of the proposed UCD-based hybrid precoding method for mmWave systems by MATLAB. The parameters are set as Nt = 128 or 256-element ULAs and Nr = 36-element ULAs with the antenna spacing d = $\lambda/2$ antennas at the BS and MS. The BS and MS both adopt N_{RF} = 4 RF chains, Ns = 4 data streams and the carrier frequency is 28 GHz. Besides, N_{C1} = 5 and each cluster includes N_{ray} = 10 paths. The complex gain of each path follows the distribution CN (0, 1), and the AoAs and AoDs are uniformly distributed in $[0, \pi]$. The modulation scheme is 16QAM. We will give comparison of performance of spectral efficiency and bite error rate (BER) between different schemes.

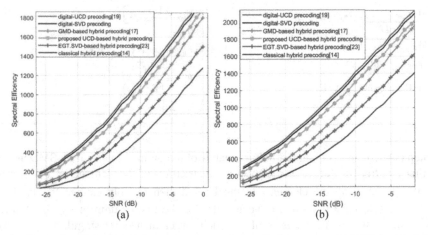

Fig. 2. Spectral efficiency performance comparison of the proposed UCD-based hybrid precoding in a 128 × 36 (a) and 256 × 36 (b) mmWave massive MIMO system.

Figure 2(a) shows the spectral efficiency performance comparison of the proposed UCD-based hybrid precoding scheme and other schemes in a 128 × 36 mmWave massive MIMO system. We can observe that the proposed UCD-based precoding schemes can achieve a better spectral efficiency performance than the classical hybrid precoding [14], the EGT. SVD-based hybrid precoding [23] and the GMD-based precoding [17], and is much closer to the optimal digital-UCD and digital-SVD precoding scheme. The reason is that UCD-based hybrid precoding schemes adopt the MMSE-VBLAST detecting method, it won't cause capacity loss at low SNR systems. Besides, with the help of water-filling algorithm, UCD-based precoding schemes can easily achieve better spectral efficiency than others.

Figure 2(b) shows the spectral efficiency performance in a 256 × 36 mmWave massive MIMO system. It seems that the proposed UCD-based hybrid precoding scheme has better performance when the number of transmitter antennas is larger and closer to the optimal full digital-UCD and digital-SVD precoding method. This is because systems can utilize more beamforming gains as transmitter antennas increases.

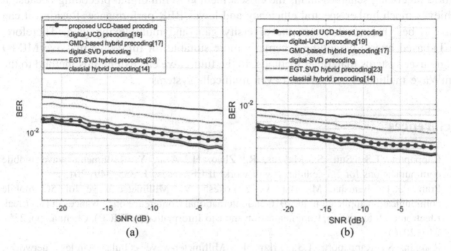

Fig. 3. BER performance comparison of the proposed UCD-based hybrid precoding in a 128 × 36 (a) and 256 × 36 (b) mmWave massive MIMO system.

Figure 3(a) and (b) show the BER performance of the proposed UCD-based hybrid precoding and others in a 128 × 36 and 256 × 36 mmWave systems. we can find the proposed UCD-based precoding scheme has a lower BER result and is closer to the full digital precoding system than other methods. Because the UCD-based precoding schemes is based on the GMD-based precoding method, it can convert the mmWave MIMO channel into multiple sub-channels with the identical SNR and avoid complicated bit allocation. Besides, as the number of antennas increases, the bit error rate (BER) decreases significantly. The reason is that more transmitted antennas can provide more array gains.

6 Conclusions

In this paper, we combine the uniform channel decomposition (UCD) scheme with mmWave hybrid precoding structure, and innovatively propose a UCD-based performance-improved hybrid precoding method for mmWave massive MIMO single user systems. It avoids the complicated bit allocations in the conventional SVD-based hybrid precoding and capacity loss in GMD-based hybrid precoding system. Firstly, the precoding matrix is designed by using the SVD decomposition of the residual matrix to avoid the design of the candidate vector set [18, 23]. Then the digital precoding is designed using the GMD and UCD decomposition of the channel matrix. We can convert the mmWave MIMO channel into multiple identical SNR and capacity subchannels, which naturally doesn't need complicated bit allocations. Combining with "water-filling" power allocation scheme, we make UCD-based hybrid capacity improved at low SNR systems. Simulation results verify that the proposed UCD-based hybrid precoding scheme outperforms conventional SVD-based and GMD-based hybrid precoding schemes and is the closest method to full digital precoding because it achieves much higher spectral efficiency and lower BER performance. Besides, it can make a best balance between the diversity gain and multiplexing gain. Therefore, UCD-based hybrid precoding scheme is more suitable for mmWave massive MIMO single user systems than other methods. In the future, we can extend this method to the mmWave multiusers systems and even multicells systems.

References

1. Rappaport, T.S., Sun, S., Mayzus, R., Zhao, H., Azar, Y.: Millimeter wave mobile communications for 5G cellular: it will work! IEEE Access 1, 335–349 (2013)
2. Prabu, R.T., Benisha, M., Bai, V.T., Yokesh, V.: Millimeter wave for 5G mobile communication application. In: 2016 2nd International Conference on Advances in Electrical, Electronics, Information, Communication and Bio-Informatics (AEEICB), Chennai, pp. 236–240 (2016)
3. Rangan, S., Rappaport, T.S., Erkip, E.: Millimeter-wave cellular wireless networks: Potentials and challenges. Proc. IEEE 102(3), 366–385 (2014)
4. Niu, Y., Li, Y., Jin, D., Su, L., Vasilakos, A.V.: A survey of millimeter wave communications (Mmwave) for 5G: opportunities and challenges. Wireless Netw. 21(8), 2657–2676 (2015)
5. Roh, W.: Millimeter-wave beamforming as an enabling technology for 5G cellular communications: theoretical feasibility and prototype results. IEEE Commun. Mag. 52(2), 106–113 (2014)
6. Venkateswaran, V., Van, A.: Analog beamforming in MIMO communications with phase shift networks and online channel estimation. IEEE Trans. Signal Process. 58(8), 4131–4143 (2010)
7. Sohrabi, F., Yu, W.: Hybrid digital and analog beamforming design for large-scale MIMO systems. In: Proceedings of IEEE International Conference on Acoustics, Speech Signal Process. (ICASSP), South Brisbane, QLD, Australia, pp. 2929–2933 (2015)
8. Alkhateeb, A., Mo, J., González-Prelcic, N.: MIMO precoding and combining solutions for millimeter-wave systems. IEEE Commun. Mag. 52(12), 122–131 (2014)
9. Han, S., Xu, Z., Rowell, C.: Large-scale antenna systems with hybrid analog and digital beamforming for millimeter wave 5G. IEEE Commun. Mag. 53(1), 186–194 (2015)
10. Alkhateeb, A., Ayach, O.E., Leus, G.: Channel estimation and hybrid precoding for millimeter wave cellular systems. IEEE J. Sel. Topics Signal Process. 8(5), 831–846 (2014)
11. Tropp, J.A., Gilbert, A.C.: Signal recovery from random measurements via orthogonal matching pursuit. IEEE Trans. Inf. Theory 53(12), 4655–4666 (2007)
12. Rusu, C., Mendez-Rial, R., Gonzalez-Prelcicy, N., Heath, R.W.: Low complexity hybrid sparse precoding and combining in millimeter wave MIMO systems. In: Proceedings of 2015 IEEE International Conference on Communication (ICC), London, pp. 1340–1345 (2015)

13. Yu, X., Shen, J., Zhang, J., Letaief, K.B.: Alternating minimization algorithms for hybrid precoding in millimeter wave MIMO systems. IEEE J. Sel. Topics Sig. Process. **10**(3), 485–500 (2016)
14. Ayach, O.E., Rajagopal, S., Abu-Surra, S.: Spatially sparse precoding in millimeter wave MIMO systems. IEEE Trans. Wireless Commun. **13**(3), 1499–1513 (2014)
15. Jiang, Y., Li, J.: Joint transceiver design for MIMO communications using geometric mean decomposition. IEEE Trans. Signal Process. **53**, 3791–3803 (2005)
16. Sun, Y., Zhan, J.L., Zhao, M.: GMD-based hybrid precoding algorithm. Video Eng. **42**(1), 63–66 (2018)
17. Tian, X., Linglong, D., Xinyu, G.: GMD-based hybrid precoding for millimeter-wave massive MIMO systems. China Commun. **15**(5), 229–238 (2018)
18. Yichun, Z., Yaoming, H.: Hybrid precoding method for millimeter - wave MIMO systems based on SVD and GMD. Information & Communications Sum. No 180 (2017)
19. Jiang, Y., Li, J., Hager, W.W.: Uniform channel decomposition for MIMO communications. IEEE Trans. Signal Process. **53**(11), 4283–4294 (2005)
20. Chao, C., Tsai, S., Hsu, T.: Bit allocation schemes for MIMO equal gain precoding. IEEE Trans. Wireless Commun. **10**(5), 1345–1350 (2011)
21. Pei, L., Yanwen, W.: Research on precoding for MIMO system with SVD, GMD and UCD. Mod Electr Tech. **7**, 78–79 (2010)
22. Daniel, P.P.: Practical algorithms for a family of water filling solutions. IEEE Trans. Signal Process. **53**(2), 3–8 (2005)
23. Shasha, L., Xun, Z.: Hybrid precoding method of millimeter-wave MIMO system based on SVD and EGT. Softw. Guide **5**, 56 (2018)

Evaluation of Augmented Reality Occlusion Scheme Based on Analytic Hierarchy Process

Ce Wang$^{(\boxtimes)}$ and Wenjun Hou$^{(\boxtimes)}$

Beijing University of Posts and Telecommunications, Beijing 100876, China
wangce0518@126.com, hou1505@163.com

Abstract. In this paper, the metrics of virtual and real occlusion effects are proposed for different augmented reality scenarios, and seven occlusion schemes are proposed correspondingly. In order to evaluate the adaptability of various virtual and real occlusion schemes in different scenarios, the augmented reality scenes with label information are used to construct the evaluation model of the virtual and real occlusion schemes by analytic hierarchy process. The two-step experimental calculation is carried out to find the most suitable occlusion scheme in the scene.

Keywords: Augmented reality · Analytic Hierarchy Process · Occlusion scheme evaluation

1 Introduction

1.1 Research Background

Augmented reality can superimpose computer generated virtual objects or system prompt information into real scenes, so as to effectively expand and enhance the real scene [1], and also support users to interact with them. Synthetic scenes make the user feel the strength of reality, depending on the fusion effect of the virtual object and the real scene. Whether the correct or appropriate occlusion of the virtual and real objects in the scene is the key factor affecting the fusion effect.

With the development of technology, the upgrading of hardware devices and the growing sci-fi cognition, augmented reality scenes are no longer satisfied that virtual objects can block the background, but require that they can also be obscured by foreground objects so that users can synthesize scenes. Get more natural spatial perception. It can be said that the augmented reality scene has higher and higher requirements for realism.

1.2 Main Problem and the Function to Solve

Augmented reality occlusion is a relatively new area, and there is no metric for the occlusion effect in the industry. This paper will combine the research of the predecessors and my own exploration to summarize and classify the functions of the augmented reality scene and the augmented reality product, so as to realize the division of the metrics for the augmented reality occlusion effect. Then, the evaluation experiment

© Springer Nature Switzerland AG 2019
D. Milošević et al. (Eds.): HCC 2019, LNCS 11956, pp. 154–165, 2019.
https://doi.org/10.1007/978-3-030-37429-7_16

is carried out, and the weights of each metrics in different scenarios are determined by the analytic hierarchy process, and then the occlusion schemes are scored, and the appropriate occlusion schemes in various scenarios are calculated.

2   Augmented Reality Scene Division

Augmented reality scenes or today's augmented reality products can be divided into the following categories: information extension and presentation products, indicator mark products, object modeling reproduction products, and game entertainment products. Different augmented reality products have different metric dimensions.

2.1   Information Extension and Presentation Products

Such augmented reality products are intended to maximize the small screen cognition by augmenting reality scenes by expanding information carried by a smaller range through virtual icons or labels, as shown in Fig. 1. Display additional content, such as washing method precautions; as shown in Fig. 2, by scanning the price tag of an item

Fig. 1. Augmented reality extended clothing label (add footnotes to objects)

Fig. 2. Augmented reality supermarket price tag changes

Fig. 3. Augmented reality looking for an online evaluation of the same item

in the supermarket to view the preferential price or price change; as shown in Fig. 3, you can scan the offline shelf A product to find the evaluation of the same item on the Internet.

2.2 Indicator Mark Products

Such augmented reality products mainly guide users by adding augmented reality signs or graphics, which are commonly found in augmented reality navigation products. For such products, the clarity and legibility of the logo is an important dimension.

2.3 Object Modeling Reproduction Products

Such augmented reality products are mainly reproduced by combining three-dimensional modeling of objects that have been damaged or objects that cannot be presented in a real scene due to distance and security factors, and then combined with real application scenarios. Common in exhibition fields such as ancient cultural relics reproduction, medical fields such as simulated anatomy, pre-operative modeling, military fields such as simulated sand table, industrial fields such as virtual assembly, virtual repair, etc.

2.4 Game Entertainment Products

Such products cover a wide range and are involved in many dimensions. The user's focus is mainly on the realism of the dynamics.

2.5 Summary of Metrics

Combining the augmented reality scenes and products listed in the above subsections, we can summarize the seven dimensions of easyness, readability, clarity, positional realism, level of realism, dynamic realism, and comprehensive experience, as shown in Table 1.

Table 1. Augmented reality occlusion effect metrics and their description

	Description
Obvious	Easy to find virtual objects in this augmented reality scene
Readability	I can clearly understand the information contained in the virtual object
Sharpness	The virtual objects in the augmented reality scene are clear and will not be confused with the scene
Positional realism	I know the position of this virtual object in the real coordinate system
Hierarchical realism	I can tell which object is in front and which object is in the back
Mobility realism	The realism of this virtual object makes me feel realistic and has a low sense of violation
Comprehensive experience	I am subjectively willing to accept that this is an augmented reality scene

3 Augmented Reality Occlusion Visualization

Several occlusion effect metrics are extracted from the previous section, which is instructive for the visualization of augmented reality stagnation. Predecessors have little research in related fields. This paper defines seven visual and real occlusion visualization schemes through research on design psychology and emotional expression, combined with the specific characteristics of augmented reality scenes. When occlusion occurs, how to integrate the virtual and real scenes to adapt to different functions to maximize the measurement index is the fundamental purpose of the design visualization scheme.

3.1 Contour Marking Method

The line is the trajectory of the point movement, which is the result of the extremely thin planes touching each other. The line created by the point breaking its quiescent state has motion and direction. In the category of geometry, the line exists to the position, length and direction [4], but the line elements in the virtual and real occlusion visualization not only have position, length and direction, but also have attributes such as thickness, shading, width, color and virtual reality.

All forms and representations and foundations of the line, in the online element notation, extract the outline by the shape of the line, representing the difference between the most basic virtual and real object boundaries. At the same time, the line elements also have rich and strong emotional expression. For example, the straight line gives a simple, rigid and rational feeling, and has a strong masculine character. The opposite curve shows a soft, gentle and elastic feeling. Feminine character [5]. The thickness of the line elements affects the expressiveness of the line. Thin lines often mean lightness, simplicity, and speed; thick lines are rough, steady, and sluggish.

Open lines are called lines, and closed lines are outlines. Most of the virtual objects in the augmented reality scene are pre-created three-dimensional models or two-dimensional images, etc., which contain relatively clear outlines. The outline represents the true face of an object, which is important in augmented reality scenes.

Reasonable use of line elements to express emotions has great significance for the expressiveness of false and real occlusion in augmented reality. The conclusions obtained from the analysis of the reference line elements, this section will introduce the specific application of the contour marking method in the augmented reality virtual reality occlusion effect.

3.2 Thick Outline Solid Line Marking Method

The outline marking methods are shown as Fig. 4:

Fig. 4. Outline marking methods

3.3 Opacity Method

In real-life scenes, few objects have the property of opacity, and most real objects are opaque. Therefore, the opacity of the virtual object or the change of the opacity of the reconstructed model can achieve the effect of real occlusion. As shown in Fig. 5:

Fig. 5. Example of opacity method

3.4 Color Modification Method

Color is the "expression" of nature and human life. Because of different cultural values, people have different geographical environments, different stages of life, different moods when facing specific colors, and different views on color.

Color has a sensible form that visually perceives the form of a hue, lightness, and chroma that it exhibits [6]. This perceptible color becomes a symbol that conveys some abstract content. Human emotions about color are the result of evolution. Because the human eye is a very delicate structure, color is a visual effect on light produced by light waves of different lengths through the eyes, brain and our life experience.

Visuality, like touch and hearing, is one of the most primitive and profound sensory cognitions of human cognition. The color is cold and warm, but the color itself has no temperature difference between cold and warm. It is the psychological association that causes people to feel cold and warm.

For example, seeing blue is easy to think of space, ice, snow, ocean and other five, it will produce a feeling of cold, rational, calm.

When you see red, you will think of the sun, flame, magma and other objects, creating a warm, warm, dangerous feeling, making people feel impulsive.

Assigning an obstruction or an occluded object to a specific color to distinguish objects of different levels in an augmented reality scene is defined as a color modification method. As shown in Fig. 6:

Fig. 6. Example of color modification

3.5 Model Flicker Method

For a specific game or entertainment scene, proper flickering of the model can improve the dynamism. As shown in Fig. 7:

Fig. 7. Example of model flicker

4 Evaluation Experiment of Augmented Reality Occlusion Effect

4.1 Overview of Analytic Hierarchy

In this paper, the analytic hierarchy process is used to evaluate the effect of augmented reality occlusion.

The Analytic Hierarchy Process (AHP) was conducted by the American operations researcher Pittsburgh University professor Satti (TLSaaty) in the early 1970s [7]. This method is characterized by the in-depth analysis of the nature, influencing factors and internal relations of complex decision problems, using less quantitative information to mathematicalize the thinking process of decision making, thus multi-objective and multi-criteria Or complex decision problems with no structural characteristics provide a simple decision-making method. It is a model and method for making decisions on complex systems that are difficult to quantify completely.

The basic principle of the analytic hierarchy process is to decompose the problem into different components according to the nature of the problem and the overall goal to be achieved, and to combine the factors according to the interrelated influences and affiliation of the factors to form a multi-level. The analysis of the structural model ultimately leads to the problem being attributed to the determination of the relative importance of the lowest level (for decision-making schemes, measures, etc.) relative to the highest level (total target) or the ranking of relative merit sequences.

In this paper, when using the AHP to construct the system model, it is divided into the following four steps:

1. Establish a hierarchical model
2. Construction judgment (pairwise comparison) matrix
3. Hierarchical single sorting and consistency check
4. Hierarchical total ordering and consistency test

Firstly, the construction of the hierarchical model is carried out, and the goals of the decision, the factors considered (decision criteria) and the decision objects are divided into the highest layer, the middle layer and the bottom layer according to the mutual relationship between them, and the hierarchical structure diagram is drawn. The highest level is the purpose of decision-making and the problem to be solved. The lowest level is the alternative when making decisions, and the middle layer is the consideration factor and the criterion of decision-making. In the experimental scenario of this paper, the target layer is the occlusion scheme selection under different augmented reality scenes. The criterion layer is aptitude, readability, definition, positional realism, level of realism, dynamic reality, and comprehensive experience. The seven metrics, the scheme layer are seven occlusion methods to choose from, namely, the thick contour solid line marking method, the thin contour implementation marking method, the thick contour dotted line marking method, the thin contour dotted line marking method, the opacity method, and the color decoration. Method and model flicker method.

Then construct a judgment (pairwise comparison) matrix. When determining the weights between the various factors at each level, if it is only a qualitative result, it is often not easy to be accepted by others. Therefore, Saaty et al. propose a uniform matrix method, that is, not putting all the factors together, but two In comparison, the relative scale is used in this case to minimize the difficulty of comparing factors with different properties to improve the accuracy [8]. For a certain criterion, the schemes under it are compared in pairs, and the grades are rated according to their importance. a_{ij} is the result of comparing the importance of factor i with factor j, Table 2 lists the nine importance levels and their assignments given by Saaty. A matrix composed of two or two comparison results is called a judgment matrix. The judgment matrix has the following properties: $a_{ij} = \frac{1}{a_{ji}}$.

Then, the hierarchical single ordering and its consistency test are performed, and the eigenvector corresponding to the largest eigenvalue as λ_{max} of the judgment matrix is normalized (so that the sum of the elements in the vector is equal to 1) and then denoted as W. The element of W is the ranking weight of the same level factor for the

Table 2. The level of the judgment matrix

Scales	Meaning
1	Expressing equal importance compared to two factors
3	One factor is slightly more important than the other
5	One factor is significantly more important than the other
7	One factor is strongly important than the other
9	One factor is extremely important than the other
2, 4, 6, 8	Median value of the above two adjacent judgments
Reciprocal	The judgment of the factor i compared with the factor j is a_{ij}, then the comparison of the factor j and factor i is judged as the reciprocal

relative importance of a factor of the previous level factor. This process is called hierarchical single ordering. If it is possible to confirm the ordering of the hierarchical order, a consistency check is required. The so-called consistency check refers to the allowable range for determining the inconsistency for A. Wherein, the unique non-zero eigenvalue of the n-order uniform matrix is n; the largest eigenvalue of the n-order positive reciprocal matrix A as $\lambda \geq n$, if and only if $\lambda = n$, A is a uniform matrix.

Since λ is continuously depends on a_{ij}, the more λ is larger than n, the more serious the inconsistency of A, and the consistency index is calculated by CI. The smaller the CI, the greater the consistency. The feature vector corresponding to the maximum eigenvalue is used as the weight vector of the degree of influence of the comparison factor on a certain factor of the upper layer. The greater the degree of inconsistency, the larger the judgment error caused. Therefore, the magnitude of the λ-n value can be used to measure the degree of inconsistency of A. The definition of the consistency indicator is CI: $CI = \frac{\lambda - r}{n - 1}$, there is complete consistency if CI = 0, there is satisfactory consistency if CI is close to 0, the larger the CI, the more serious the inconsistency. To measure the size of CI, a random consistency indicator RI: $RI = \frac{CI_1 + CI_2 + \ldots + CIr}{n}$ was introduced.

Among them, the random consistency index RI is related to the order of the judgment matrix. In general, the larger the matrix order, the greater the probability of a uniform random deviation. The corresponding relationship is shown in Table 3:

Table 3. Average random consistency index RI standard value (different standards, RI values will also have slight differences)

Matrix order	1	2	3	4	5	6	7	8	9	10
RI	0	0	0.58	0.90	1.12	1.24	1.32	1.41	1.45	1.49

Considering that the deviation of consistency may be due to random reasons, when testing whether the judgment matrix has satisfactory consistency, it is also necessary to compare the CI with the random consistency index RI to obtain the test coefficient CR, CR = CI/RI. Generally, if CR < 0.1, the judgment matrix is considered to pass the consistency check, otherwise there is no satisfactory consistency. Finally, the total ranking of the hierarchy and its consistency test are performed. The weight of all factors of a certain level for the relative importance of the highest level (total target) is calculated, which is called the total ranking of the hierarchy. This process is carried out in order from the highest level to the lowest level.

4.2 Evaluation Experiments in This Paper (Tag Information Type Scenario as an Example)

Experimental Design

The experimental questionnaire in this paper is divided into two stages. The first stage is defined as the expert user review stage. The purpose is to obtain the weight of the unacceptable evaluation elements in each augmented reality scenario. The second stage is the augmented reality experiencer evaluation stage. It is to find the performance of different augmented reality occlusion schemes on each evaluation element.

In the first stage, six augmented reality research workers were invited, each of whom experienced eight representative scenes:

1. Label information scene - online product same paragraph
2. Model reproduction scene - damage Relics Reproduction
3. Scenes that Cannot Touch Objects - Preoperative Skull Modeling
4. Mobility Scenes - AR Electronic Pets
5. Displacement Scenes - Virtual Home Renovation
6. Virtual Object Creation Scenes - AR Stereo Paintings
7. Navigation Scenes - Hotstpper navigation
8. Self-timer scene - Animoji expression.

The experience time of each scene is 3–10 min. During the period, fill in the assessment elements of the current scene and score a questionnaire, and rest for 5 min after performing a scene to prevent confusion between different scenes.

In the second stage, 15 subjects were invited to score the evaluation. Experimental process: Install the application of eight experimental scenarios on iphoneX in advance, and then invite 15 testees to experience 3–10 min for each of the eight experimental scenarios, and occlude the different metrics after each experience. The scoring of the program still uses the consistent matrix method. The scoring topic is as follows: In this scenario, from the perspective of legibility, you think that the thick line contour method: the thin line contour method can make a few points. Through the 15 scores tested, the judgment matrix of the second stage is obtained.

Establishing a Hierarchical Model

There are 8 different hierarchical models in this experiment. The difference between them is the difference of the target layer. Taking the tag information class augmented reality occlusion scene as an example, the hierarchical structure model can be drawn as shown in Fig. 8.

Fig. 8. Label information class augmented reality occlusion scheme hierarchy mode

Construction Judgment Matrix and Hierarchical Ordering

Through the questionnaire data of the first stage, the judgment matrix of the target layer A can be obtained as follows:

$$A = \begin{pmatrix} 1 & 1 & 3 & 5 & 6 & 9 & 3 \\ 1 & 1 & 3 & 3 & 7 & 7 & 3 \\ 1/3 & 1/3 & 1 & 5 & 6 & 7 & 3 \\ 1/5 & 1/3 & 1/5 & 1 & 3 & 5 & 1/3 \\ 1/6 & 1/7 & 1/6 & 1/3 & 1 & 1 & 1/4 \\ 1/9 & 1/7 & 1/7 & 1/5 & 1 & 1 & 1/3 \\ 1/3 & 1/6 & 1/3 & 3 & 4 & 3 & 1 \end{pmatrix}$$

The maximum eigenvalue as λ_{max} of A can be found to be 7.44, and the corresponding normalized eigenvector is: $W_A = (0.30, 0.28, 0.19, 0.08, 0.03, 0.03, 0.10)^T$

According to $CI = \frac{\lambda - r}{n - 1}$, find CI = 0.073, look up the table to know RI = 1.32, so CR = 0.055 < 0.1, passed the consistency test.

Through the questionnaire data of the second stage, the judgment matrix of the criterion layer B relative to the scheme layer C can be obtained, taking the legibility (B_1) as an example, as follows:

$$B_1 = \begin{pmatrix} 1 & 4 & 3 & 4 & 6 & 3 & 3 \\ 1/4 & 1 & 1/3 & 1 & 3 & 1 & 1/3 \\ 1/3 & 3 & 1 & 3 & 5 & 1 & 1 \\ 1/4 & 1 & 1/3 & 1 & 1 & 1/3 & 1/5 \\ 1/6 & 1/3 & 1/5 & 1 & 1 & 1/3 & 1/5 \\ 1/3 & 1 & 1 & 3 & 3 & 1 & 1 \\ 1/3 & 3 & 1 & 5 & 5 & 1 & 1 \end{pmatrix}$$

CR = 0.037 < 0.1, passed the consistency test. According to the same method, the normalized feature matrix of the discernibility (B_1) to the comprehensive experience (B_7) is obtained, and all pass the consistency test. Organize as W_B, as shown below:

$$W_B = \begin{pmatrix} 0.35 & 0.18 & 0.35 & 0.13 & 0.30 & 0.21 & 0.42 \\ 0.08 & 0.22 & 0.16 & 0.32 & 0.17 & 0.08 & 0.11 \\ 0.16 & 0.13 & 0.16 & 0.13 & 0.08 & 0.02 & 0.05 \\ 0.16 & 0.13 & 0.16 & 0.13 & 0.08 & 0.02 & 0.05 \\ 0.04 & 0.04 & 0.03 & 0.04 & 0.05 & 0.18 & 0.20 \\ 0.13 & 0.23 & 0.18 & 0.04 & 0.02 & 0.09 & 0.03 \\ 0.18 & 0.04 & 0.02 & 0.28 & 0.25 & 0.39 & 0.06 \end{pmatrix}$$

Level Total Ordering

Based on the results of and, the resulting matrix can be found as follows:

$$W = W_B W_A = \begin{pmatrix} 0.35 & 0.18 & 0.35 & 0.13 & 0.30 & 0.21 & 0.42 \\ 0.08 & 0.22 & 0.16 & 0.32 & 0.17 & 0.08 & 0.11 \\ 0.16 & 0.13 & 0.16 & 0.13 & 0.08 & 0.02 & 0.05 \\ 0.16 & 0.13 & 0.16 & 0.13 & 0.08 & 0.02 & 0.05 \\ 0.04 & 0.04 & 0.03 & 0.04 & 0.05 & 0.18 & 0.20 \\ 0.13 & 0.23 & 0.18 & 0.04 & 0.02 & 0.09 & 0.03 \\ 0.18 & 0.04 & 0.02 & 0.28 & 0.25 & 0.39 & 0.06 \end{pmatrix} \begin{pmatrix} 0.30 \\ 0.28 \\ 0.19 \\ 0.19 \\ 0.03 \\ 0.03 \\ 0.10 \end{pmatrix}$$

$$= \begin{pmatrix} 0.29 \\ 0.16 \\ 0.13 \\ 0.10 \\ 0.06 \\ 0.15 \\ 0.12 \end{pmatrix}$$

It is concluded that the thick outline marking method is most suitable in the augmented reality scene of the label information type as 0.29, and is obviously superior to other methods. The maximum one of others is 0.16.

5   Conclusion

By analyzing the characteristics of different augmented reality scenes, this paper summarizes the seven elements that affect the virtual and real occlusion effects in augmented reality, namely, legibility, readability, clarity, positional realism, level of realism, dynamic reality, Comprehensive experience. Then through the study of color, line and design psychology, the contour marking method (subdivided according to thickness and virtual reality), different brightness method, color modification method and model flickering method are proposed.

The hierarchical structure model of different scenarios is constructed by using seven elements as the metric and seven occlusion schemes. The judgment matrix of the hierarchical model is determined by scoring by expert users and other users' experience, and the augmented reality scene of the tag information class is For example, it is concluded that the thick outline marking method is the most suitable occlusion scheme in this scene.

References

1. Tian, Y.: Research on virtual and real occlusion processing methods in augmented reality. Huazhong University of Science and Technology (2010)
2. Li, M., Wang, Y., Yuan, B., et al.: Research on the database of cultural relics protection of terracotta warriors and horses of Qin Shihuang mausoleum based on GIS. J. Baoji Univ. Arts Sci. (Nat. Sci. Ed.) **27**(3), 246–248 (2007)
3. Zhang, J., Luo, S.: Research on medical augmented reality modeling method. J. Syst. Simul. **21**(12), 3658–3661 (2009)
4. Kandinsky, Shiping, L.: Kandinsky Argument Line. Renmin University of China Press (2003)
5. He, L., Liu, H., Ke, S.: Application of line elements in product shape design. Packag. Eng. **18**, 100–103 (2012)
6. Zhu, Q., Li, W.: The significance of color and its application in industrial design. J. Chongqing Univ. Sci. Technol. Soc. Sci. Ed. **4**, 151–152 (2008)
7. Xu, C.: Application of analytic hierarchy process in the selection of raw material suppliers in enterprises. Orporate Her. (7), 262 (2010)
8. Chen, J.: Fuzzy consistent matrix method for multi-scenario evaluation of highway routes. J. Highw. Transp. Res. Dev. **19**(5), 42–44 (2002)

Suicidal Ideation Detection via Social Media Analytics

Yan Huang[1,2], Xiaoqian Liu[1], and Tingshao Zhu[1(✉)]

[1] Institute of Psychology, Chinese Academy of Sciences, Beijing, China
hihihey@qq.com, tszhu@psych.ac.cn
[2] University of Chinese Academy of Sciences, Beijing, China

Abstract. Suicide is one of the increasingly serious public health problems in modern society. Traditional suicidal ideation detection using questionnaires or patients' self-report about their feelings and experiences is normally considered insufficient, passive, and untimely. With the advancement of Internet technology, social networking platforms are becoming increasingly popular. In this paper, we propose a suicidal ideation detection method based on multi-feature weighted fusion. We extracted linguistic features set that related to suicide by three different dictionaries, which are data-driven dictionary, Chinese suicide dictionary, and Language Inquiry and Word Count (LIWC). Two machine learning algorithms are utilized to build weak classification model with these three feature sets separately to generate six detection results. And after logistic regression, to get the final weighted results. In such a scheme, the results of model evaluation reveal that the proposed detection method achieves significantly better performance than that use existing feature selection methods.

Keywords: Suicidal ideation · Detection · Social media

1 Introduction

1.1 Suicidal Ideation and Social Media

Traditionally, mental health professionals usually rely on mental status examination through specialized questionnaires and the Suicidal Behaviors Questionnaire Revised (SBQ-R) (Osman et al. 2001) or patients' self-reported feelings and experiences to detect suicidal ideation, which have been considered insufficient, passive, and untimely (McCarthy 2010; Choudhury et al. 2013). In fact, patients with mental diseases do not often go to health care providers. Given the high prevalence and devastating consequences of suicide, it is critical to explore proactive and cost-effective methods for real-time detection of suicidal ideation to perform timely intervention.

In 2012, a 22-year-old girl nicknamed Zoufan left her last words on Weibo that she was suffering from depression and decided to die. She hanged herself in the dormitory, the police and medical staff could not save her life even though they did their best to

---

Y. Huang and X. Liu—These authors contributed equally to this work.

© Springer Nature Switzerland AG 2019
D. Milošević et al. (Eds.): HCC 2019, LNCS 11956, pp. 166–174, 2019.
https://doi.org/10.1007/978-3-030-37429-7_17

rescue. Netizens were generally saddened and regretted about this news, and more and more people are calling for greater attention to people with depression and preventing them from being suicidal.

Previous studies have found that suicidal ideation of a person can be discovered from his/her communication through analysis of his/her self-expression (Stirman and Pennebaker 2001; Li et al. 2014). With the advancement of Internet technology, more and more people share their lives, experiences, views, and emotions in those online venues (Kaplan and Haenlein 2010; Robinson et al. 2015). There has been an increase of research on the effect of social media use on mental health (Pantic 2014). Research has shown that people with mental health problems, especially those with depression, have high social media use (Davila et al. 2012; Rosen et al. 2013). Analyzing people's behavior in social media may provide a genuine, proactive, and cost-effective way to discover the intention or risk of an individual to commit suicide.

Wang et al. (2013) developed a depression detection model with characteristics of users' language and behavior as input features. 180 users' social media data were used to verify the models. These results demonstrate that it is feasible to detect individuals' suicidal ideation through social media.

1.2   Related Work

In the past decade, there has been increasing efforts toward automatic identification of suicidal ideation from social media data (O'Dea et al. 2015; Paul and Dredze 2014; Wang et al. 2015). Suicidal ideation identification is a binary classification problem – to determine if an individual has a suicidal ideation or not. As a result, the majority of studies in this field use a classification machine learning approach to build suicidal ideation detection models, in which feature selection is critical. Existing models use different predictive features, including LIWC-based (Pestian et al. 2012; Li et al. 2014), dictionary-based (Lv et al. 2015), and social media behavior based (e.g., the total number of followers, the number of blogs published after midnight) (Guan et al. 2015). Each has its advantages and disadvantages. The knowledge-experience-driven approach is gradually accumulated in practice, and the data-driven approach is derived from factual analysis under a large number of data records. The former is more targeted and flexible than the latter, but experience is easily limited by subjective judgment and data insight is more objective. However, there is a lack of studies that investigate which feature selection method might be more effective to select predictive features.

This study is aimed to propose a new suicidal ideation detection method based on multi-feature weighted fusion (Called MFWF method hereafter), to verify if this new method provides better model performance than existing suicidal prediction model. If this assumption is true, we will get higher accuracy of suicidal ideation. Based on the method of big data analysis, we can automatically identify the blogger's suicidal ideation by analyzing their blog posts, which can also be used to assist manual screening, and ultimately improve the efficiency of mental health services.

2 Methods

In this research, we aim to advance knowledge about correlations between linguistic features of textual social media content and suicidal ideation, and provide an optimized weighting method that combines multiple features and logistic regression to detect suicidal ideation. The proposed MFWF method works as follows.

First and foremost, collect the data from blog posts with suicidal ideation. Then linguistic features are extracted through three ways, including data-driven dictionary that automatically generated by n-gram (Liu et al. 2019), Chinese suicide dictionary, and Language Inquiry and Word Count (LIWC) (Pennebaker et al. 2001). Two machine learning algorithms are utilized to build weak classification model with these three features sets separately to generate six detection results. Input them into logistic regression model to get different parameters that corresponded to weighting results of different linguistic feature. Finally, get the judgment on whether there is suicidal ideation (Fig. 1).

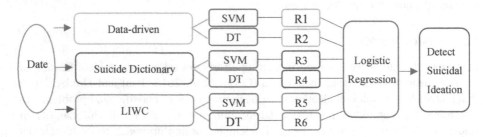

Fig. 1. The multi-feature weighted fusion method

2.1 Data Collection

The data source for this study is from Sina Weibo, China's largest open social media platform, with current monthly active users reaching 465 million.

After Zoufan's death, people still paid attention and left messages below her last blog, and some of the messages revealed suicidal thoughts. Our research team crawled 65,352 messages from there through Sina Weibo API. They were posted by 23,000 bloggers. Considering that extremely short blogs may contain little useful information, we removed the blogs containing fewer than 5 words (excluding punctuation and Arabic numerals). The average length of the remaining blogs was 29.63 words.

We established a team of three experts who specialized in psychology and suicidal behavior, to assess whether there is suicidal ideation and manually labelled each of the collected blogs. In order to ensure the accuracy of the results, each expert was trained before the assessment, and the consistency of the assessment results was verified on a small sample. Then multiple rounds of assessment began, and the label would be confirmed only if the three experts make a consistent judgment, otherwise re-determine.

The labeling results of individual coders were examined for consistency. Finally, we had 8,548 blogs labeled as "suicide" (i.e., with suicidal ideation). To avoid the common data imbalance problem in machine learning, we randomly selected 10,000 blogs from the ones labelled as non-suicide as negative training samples.

2.2 Feature Extraction

We used three kinds of dictionaries for linguistic feature extraction, including (1) automated machine learning dictionary driven by N-gram. N-gram is an algorithm based on statistical language model. Its basic idea is to operate the contents of the text in a sliding window of size N in bytes to form a sequence of byte segments of length N. Each byte segment is called gram, and the frequency of occurrence of all grams is counted and filtered according to a preset threshold to form a key gram list, which is the vector feature space of the text. N-gram is based on the assumption that the occurrence of the nth word is related to the first n-1 words, and is not related to any other words. The probability of each word can be calculated by statistical calculations in the corpus. The probability that the entire sentence appears is equal to the probability product of the occurrence of each word. A 3-gram algorithm was utilized to process blog content and identify one-character, two-character, and three-character terms in the blog corpus. We removed stop words and considered the remaining terms as candidate terms. (2) Chinese suicide dictionary, a suicide-related lexicon that developed by some Chinese experts, it was based on knowledge and experience. (3) SCMBWC, the Simplified Chinese Micro-Blog Word Count tool (Gao et al. 2013), it is a Chinese version of the commonly used Language Inquiry and Word Count (LIWC) (Pennebaker et al. 2001), a classic tool for sentiment analysis. The set of selected terms using the above three kinds of dictionaries were called feature set A, B and C.

To constitute feature set A, a Chi-square test was then performed on the candidate terms to identify those that are mostly related to the blogs labelled as suicide. Chi-square test is a widely used hypothesis testing method that examines whether two categorical variables are independent of each other. In order to find terms closely related to suicide, we first propose a null hypothesis that each candidate term is independent of the "suicide" class. The Chi-square value of a candidate term is calculated by Eq. (1), P value <0.05.

$$X2_t = \frac{N(A_t D_t - B_t C_t)^2}{(A_t + C_t)(A_t + B_t)(B_t + D_t)(B_t + C_t)} \tag{1}$$

Where N denotes the total number of blogs in the corpus. For each term t, A_i is the number of blogs that contain t and are labeled as "suicide"; B_t is the number of blogs that contain t and are labeled as "non-suicide"; C_t is the number of blogs that do not contain t and are labeled as "suicide"; and D_t is the number of blogs that do not contain t and are labeled as "non-suicide". A larger Chi square value $X2_t$ implies that the term t is more related to suicidal ideation.

After calculating the Chi-square value of each candidate term, we selected the top 2,000 terms with the highest Chi-square values to constitute feature set A, such as "negative energy", "miserable", and "rubbish", as the predictive features for our proposed suicidal ideation detection model. Quite some of those feature terms do not exist in any existing emotion- or suicide-related dictionaries, which shows that the proposed feature selection method can identify new suicide-related terms from social media content, which may help with the expansion of existing dictionaries.

We use Term Frequency–Inverse Document Frequency (TF-IDF) (Salton and McGill 1986) for feature extraction, which is a statistical method used to assess the importance of a word for a file set or one of the files in a corpus. The importance of a word increases proportionally with the number of times it appears in the file, but it also decreases inversely with the frequency it appears in the corpus. The main idea of using TF-IDF is that if a word or phrase appears in a class of blog posts with a high frequency TF and rarely appears in other categories, the word or phrase is considered to be suitable for classification.

Each blog f_j, after word segmentation, would be represented as a feature vector with 2,000 dimensions corresponding to 2,000 feature terms. The TF-IDF value of each feature term s_i was calculated as the feature value by Eqs. (2)–(4).

$$\text{tfidf}_{i,j} = tf_{i,j} \times idf_i \tag{2}$$

$$\text{tf}_{i,j} = \frac{n_{i,j}}{\sum_k n_{k,j}} \tag{3}$$

$$\text{idf}_i = \log \frac{|D|}{|j : s_i \in d_i|} \tag{4}$$

Where $n_{i,j}$ in Eq. (3) denotes the number of times a feature term s_i appeared in the blog f_j, and the denominator is the sum of occurrences of all feature terms appeared in f_j. In Eq. (4), D is the total number of blogs in the corpus, and the denominator is the number of blogs containing s_i.

Feature set B contains 295 terms, which were manually selected by a panel of domain experts based on their correlations with suicide. Feature set C contains 102 linguistic features, including emotional word categories such as positive words, sad words, and angry words. Among the 102 linguistic features extracted, 88 corresponded to the 88 word categories of SCMBWC; 3 features were related to the categories indicating past, now and future, respectively, such as "yesterday", "today" and "to-morrow"; and the remaining 11 features were related to different kinds of punctuations, including period, comma, colon, semicolon, question mark, exclamation point, dash, quotation mark, apostrophe, parent, and other punctuations. The three original different predictive feature sets are presented in Table 1.

Table 1. The original predictive feature sets

	Number of features	Selection methods
Feature set A	2,000	Learned from real blog data
Feature set B	295	Chinese suicide dictionary
Feature set C	102	SCMBWC

2.3 Model Construction with Machine Learning Algorithms

We selected two algorithms, SVM (Support Vector Machine) and DT (Decision Tree), which not only the most commonly used classification machine learning algorithms, but also frequently used in the psychological character, emotion, and suicidal ideation detection (Gamon et al. 2013; Huang et al. 2014; Guan et al. 2015; Iliou et al. 2016).

SVM is an important example of "kernel methods" in statistical learning, one of the key areas in machine learning. It is a discriminative classifier for binary classification formally defined by a separating hyperplane. The operation of the SVM algorithm is based on finding the hyperplane that separates m-dimensional data into two classes and gives the largest distance to the nearest training samples of the two classes.

DT uses a treelike graph to model the reasoning process of mapping an input feature set to one of the predefined class labels. We chose the CART (classification and regression trees) algorithm in this study. CART constructs a binary decision tree by splitting a node into two child nodes repeatedly, beginning with the root node that contains the whole learning sample.

2.4 Measures of Model Performance

The performances of the constructed suicidal ideation detection models were evaluated using four metrics that have been commonly used in evaluating classification models, including precision (P), recall (R), F-measure (F), and accuracy (A), as defined by Eqs. (5)–(8). Precision is the percentage of all blogs detected as suicidal ideation that are indeed suicide ones; recall is the percentage of blogs in the entire corpus with suicidal ideation that are correctly identified; F-measure is a harmonic mean of precision and recall; and accuracy is the percentage of correctly detected blogs with and without suicidal ideation among the total number of blogs examined.

$$P = \frac{TP}{TP + FP} \tag{5}$$

$$R = \frac{TP}{TP + FN} \tag{6}$$

$$F = \frac{2 \times P \times R}{P + R} \tag{7}$$

$$A = \frac{TP + TN}{TP + FP + TN + FN} \tag{8}$$

Where TP denotes the number of blogs that are correctly classified as suicidal ideation; TN denotes the number of blogs that are correctly classified as non-suicidal ideation; FN denotes the number of blogs with suicidal ideation but classified as non-suicidal ideation incorrectly; and FP denotes the number of blogs without suicidal ideation but classified as suicidal ideation incorrectly.

3   Suicidal Ideation Detection Results

After calculation, we found the model constructed by SVM achieved better performances than DT, and feature set A made the best predictive performance. Therefore, the four metrics value of SVM from feature set A were chosen as the result of MFWF method.

We then compared the performances with the existing data-driven results (Liu et al. 2019) to draw final conclusion of this study. Table 2 shows the performances of the two different detection method.

Table 2.   Precision, Recall, F-measure, and Accuracy of two detection methods

Detection method	Precision	Recall	F-measures	Accuracy
MFWF method	**0.89**	**0.88**	**0.88**	**0.89**
Data-driven method	0.88	0.78	0.83	0.85

The results show that the suicidal ideation detection based on MFWF method ($p < 0.01$), achieved better performances than data-driven method. It represented that the multi-feature fusion weighting method proposed in this study is an improvement for the data-driven method with single feature extraction. It's worth noting that recall of MFWF method increased from 0.78 to 0.88 by 12.8%, which also indicated that the rate of missed detection decreased. It's of great significance for suicidal ideation detection.

4   Discussion

There are two major findings and contributions of this research. First, on suicidal ideation detection, the multi-feature fusion method is better than the single feature extraction method, as it effectively integrates the existing methods. not only considered data and expert opinion (Chinese suicide dictionary), but also combines the features of emotional expression (LIWC). Second, general linguistic features extracted by or traditional suicide dictionary or natural language processing tool such as SCMBWC, may not be very effective. People may use some suicide-related terms that do not exist in any existing suicide lexicons, as we observed in this study. It suggests that the predictive features for a suicidal ideation detection model for social media should take into consideration suicide-related terms used in social media.

The results of experiments show that a suicidal ideation detection system can successfully discover suicidal ideations from social media content with a reasonably high accuracy. The promising results demonstrate the practical feasibility and value of real-time detection of suicidal ideation and prevention on social media, which may help reduce suicidal intentions and attempts.

Based on the conclusions of this study, we can better use social media content analysis to prevent suicide in practice. For example, evaluating blog posts on social media on a daily basis by machine. If machine screening detected suicidal ideation, we would conduct a manual assessment. Once we confirmed the same conclusion, early warning and intervention would be implemented in a timely manner. Which is also the reason why we want to achieve more accurate and efficient machine screening result.

There are some limitations of our study. First, we used blogs collected from a single social media site. There are other types of textual social media content, such as discussion messages in online communities that may possess different characteristics. Therefore, the generalizability of the findings of this research needs to be taken with caution. Second, given the scope of this study, we did not include other potential feature selection methods, such as feature selection through deep learning, which is worth to be explored. Third, the units of the analysis in this study are individual blogs. We did not consider any significant changes in emotions, self-expressions, and other relevant features across multiple contiguous blogs, which may provide additional helpful cues for assessment. These limitations provide us opportunities for future research.

Acknowledgement. This study was partially supported by the Key Research Program of the Chinese Academy of Sciences (No. ZDRW-XH-2019-4).

References

Choudhury, M.D., Gamon, M., Counts, S., Horvitz, E.: Predicting depression via social media. In: Proceedings of the Seventh International AAAI Conference on Weblogs and Social Media, Cambridge, Massachusetts, USA, 8–11 July, pp. 128–137 (2013)

Davila, J., Hershenberg, R., Feinstein, B.A., Gorman, K., Bhatia, V., Starr, L.R.: Frequency and quality of social networking among young adults: associations with depressive symptoms, rumination, and co-rumination. Psychol. Popul. Media Cult. 1(2), 72–86 (2012)

Gao, R., Hao, B., Li, H., Gao, Y., Zhu, T.: Developing simplified Chinese psychological linguistic analysis dictionary for microblog. In: Imamura, K., Usui, S., Shirao, T., Kasamatsu, T., Schwabe, L., Zhong, N. (eds.) BHI 2013. LNCS (LNAI), vol. 8211, pp. 359–368. Springer, Cham (2013). https://doi.org/10.1007/978-3-319-02753-1_36

Gamon, M., Choudhury, M.D., Counts, S., Horvitz, E.: Predicting depression via social media. AAAI (2013)

Guan, L., Hao, B., Cheng, Q., Yip Paul, S.F., Zhu, T.: Identifying Chinese microblog users with high suicide probability using internet-based profile and linguistic features: classification model. Jmir Mental Health 2(2), e17 (2015)

Huang, X., Zhang, L., Liu, T., Chiu, D.: Detecting suicidal ideation in Chinese microblogs with psychological lexicons. In: Proceedings of Intl Conf on Ubiquitous Intelligence and Computing, and Intl Conf on Autonomic and Trusted Computing, and Intl Conf on Scalable Computing and Communications and ITS Associated Workshops, pp. 844–849 (2014)

Iliou, T., et al.: Machine Learning Preprocessing Method for Suicide Prediction. In: Iliadis, L., Maglogiannis, I. (eds.) AIAI 2016. IAICT, vol. 475, pp. 53–60. Springer, Cham (2016). https://doi.org/10.1007/978-3-319-44944-9_5

Kaplan, A.M., Haenlein, M.: Users of the world, unite! The challenges and opportunities of social media. Bus. Horiz. 53(1), 59–68 (2010)

Li, T.M.H., Chau, M., Yip, P.S.F., Wong, P.W.C.: Temporal and computerized psycholinguistic analysis of the blog of a Chinese adolescent suicide. Crisis 35(3), 168 (2014)

Lv, M., Li, A., Liu, T., Zhu, T.: Creating a Chinese suicide dictionary for identifying suicidal ideation on social media. Peerj **3**(10.7177), e1455 (2015)

McCarthy, M.J.: Internet monitoring of suicidal ideation in the population. J. Affect. Disord. **122** (3), 277–279 (2010)

O'Dea, B., Wan, S., Batterham, P.J., Calear, A.L., Paris, C., Christensen, H.: Detecting suicidality on Twitter. Internet Interv. **2**, 183–188 (2015)

Osman, A., Bagge, C.L., Guitierrez, P.M., Konick, L.C., Kooper, B.A., Barrios, F.X.: The suicidal behaviors questionnaire-revised (SBQ-R): validation with clinical and nonclinical samples. Assessment **5**, 443–454 (2001)

Pantic, I.: Online social networking and mental health. Cyberpsychology Behav. Soc. Netw. **17** (10), 652–657 (2014)

Paul, M.J., Dredze, M.: Discovering health topics in social media using topic models. Plos One **9** (8), e103408 (2014)

Pennebaker, J.W., Francis, L.E., Booth, R.J.: Linguistic inquiry and word count: LIWC2001. Lawrence Erlbaum Associates, Mahwah (2001)

Pestian, J.P., et al.: Sentiment analysis of suicide notes: a shared task. Biomed. Inform. Insights **5** (Suppl 1), 3–16 (2012)

Rosen, L.D., Whaling, K., Rab, S., Carrier, L.M., Cheever, N.A.: Is Facebook creating "iDisorders"? The link between clinical symptoms of psychiatric disorders and technology use, attitudes and anxiety. Comput. Hum. Behav. **29**, 1243–1254 (2013)

Salton, G., McGill, M.J.: Introduction to Modern Information Retrieval. McGraw-Hill, New York (1986)

Stirman, S.W., Pennebaker, J.W.: Word use in the poetry of suicidal and nonsuicidal poets. Psychosom. Med. **63**(4), 517–522 (2001)

Wang, X., Zhang, C., Ji, Y., Sun, L., Wu, L., Bao, Z.: A depression detection model based on sentiment analysis in micro-blog social network. In: Li, J., et al. (eds.) PAKDD 2013. LNCS (LNAI), vol. 7867, pp. 201–213. Springer, Heidelberg (2013). https://doi.org/10.1007/978-3-642-40319-4_18

Wang, X., Li, A., Zhu, T.: Digital detection of suicidal ideation on social media. Int. J. Emerg. Ment. Health Hum. Resil. **17**(3), 661–663 (2015)

Liu, X., et al.: Proactive Suicide Prevention Online (PSPO): machine identification and crisis management for Chinese social media users with suicidal thoughts and behaviors. J. Med. Internet Res. (2019). https://doi.org/10.2196/11705

The Latent Semantic Power of Labels: Improving Image Classification via Natural Language Semantic

Haosen Jia[1], Hong Yao[1,2(✉)], Tian Tian[1,2], Cheng Yan[1], and Shengwen Li[3]

[1] School of Computer Science, China University of Geosciences,
Wuhan 430074, China
yaohong@cug.edu.cn
[2] Hubei Key Laboratory of Intelligent Geo-Information Processing,
China University of Geosciences, Wuhan 430074, China
[3] Faculty of Information Engineering,
China University of Geosciences, Wuhan 430074, China

Abstract. In order to address the problem that numerical labels are difficult to optimize, one-hot encoding is introduced into image classification tasks, and has been widely used in current models based on CNNs. However, one-hot encoding neglects the textual semantics of class labels, which closely relate to image characteristics and contain latent connections between images. Inspired by distributional similarity based representations in Natural Language Processing society, we propose a framework by introducing Word2Vec into classic CNN models to improve image classification performance. By mining the latent semantic power of classes labels, word vector representations participate in the classification model instead of the traditional one-hot encoding. In the evaluation experiments implemented on data sets of CIFAR-10 and CIFAR-100, a series of representative CNNs have been tested as the feature extraction component for our framework. Experimental results show that the proposed method has revealed compelling ability to improve the classification accuracy.

Keywords: Image classification · Convolutional Neural Networks · Natural language semantic · Word vectors

1 Introduction

In recent years, a surge in demand for visual object recognition applications has arisen, such as autonomous vehicles [3] and content-based image searches [35]. To address the basic problem of image classification, a large amount of methods

This research was supported by Project 61672474, 61501412 supported by NSFC. National Science and Technology Major Project (No. 2017ZX05036-001-010). Science and Technology Planning Project of Guangdong Province, China. (No. 2018B020207012).

have been proposed. Among these researches, Convolutional Neural Networks (CNNs) have achieved a great success on many computer visual tasks by imitating human visual hierarchical structure. Since CNN has been used for recognizing the handwriting numbers, some advanced models based on CNNs [10] have even exceeded human's performance on certain image classification data sets.

The emergence of image object detection and recognition challenges such as ImageNet [7] and COCO [20] further promotes the development of CNNs. A CNN structure called AlexNet [14], winner of ImageNet competition in 2012, has started a new era of deep learning and CNNs. Since then, researchers began to make various efforts to improve the performance of deep networks: VGGNet [28] and MSRANet [9] deepen the layers of networks, NIN [19] and GoogLeNet [31] enhance the convolution module function, and ResNet [10] adopts the above two means in a uniform model. Although network structures are quite different, the encoding scheme of labels are the same. During the supervised training of networks, it requires the participation of labels. Studies mentioned above invariably adopt the one-hot encoding of class labels.

In spite of the significant improvements made by previous models based on CNN, the discussion on the one-hot encoding is not sufficient. One-hot encoding removes the integer values and assigns an exclusive binary value for each unique label. Combined with Cross-Entropy as the loss function, it solves the problem that numerical labels are difficult to optimize. Although it is better than natural numerical encoding of labels, the one-hot encoding neglects and losses the textual semantics of class labels. The different mappings from images to labels with different label encodings are shown in Fig. 1. In fact, semantic relations play a very important role. For example, in the field of natural language processing (NLP), each word was first represented as a one-hot encoding in the early 2000 s. Afterwards, word vectors have been proved to be more effective in NLP tasks, such as parsing, language modeling, Named-Entity Recognition (NER) [24] and so on [32]. In image classification, it is easy to understand that similar objects should have similar semantic relationships on their labels. Moreover, they should share some similar visual features at lower or higher levels as well. We believe that the latent semantic power of labels will significantly contribute to image classification tasks, as long as we take good advantage of it.

In order to verify our conjecture, we propose a new framework. Low dimensional label semantic vectors are learned from natural languages and introduced into classical CNN models, and the deep network will be trained to learn the mapping from image features to label semantic representations. We believe this study provides new insights into integrations of image classification and natural language semantic, and makes an original contribution to image classification with the help of external knowledge. It is hoped that this research will lead to a deeper understanding of CNN models for image classification, and provide an alternative choice to speed up CNN model's training.

After the motivation and purpose of this paper we've stated above, the remaining part of it proceeds as follows: related work is reviewed in Sects. 2, and 3 describes the method. Experimental results and discussions are detailed in Sect. 4. And finally we draw the conclusion and further work.

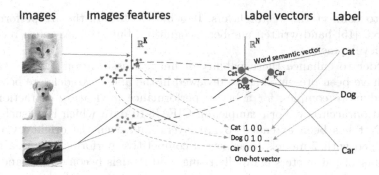

Fig. 1. The mapping of images to labels.

2   Related Work

2.1   CNN Models

The well-known approaches to improve CNNs mainly make efforts on the following aspects: deepening the depth of the network [15,28], enhancing the convolution module function [19,31] and combining the above two kinds of method [10]. During the past decade, the most outstanding and widely used CNN models include LeNet [15], AlexNet [14], VGGNet [28], GoogLeNet [31] and ResNet [10].

LeNet is a concise convolutional neural network which has three convolution layers, two pooling layers, and two fully connected layers including a Softmax output layer. AlexNet adopts the basic idea of LeNet and expands the fundamental architecture of CNNs. Moreover, several advanced technologies such as ReLU, Dropout and LRN are successfully applied in the model training. VGGNet makes the improvement over AlexNet by replacing large kernel-sized filters with successive 3×3 kernel-sized filters, and validates multiple stacked kernels of smaller size perform better than one with a larger size kernel. After that, to improve utilization of computing resources inside the network, Google proposed GoogLeNet, which is achieved by a carefully crafted design that allows increasing the depth and width of network while keeping the computational budget constant. As the network architecture goes deeper and deeper, vanishing gradient problem occurs which makes the network difficult to train. For solving this problem, ResNet creates a direct path between the input and output to the module, which can effectively propagate gradient backwards during the training of deep networks. By means of this contribution, deep networks up to hundreds of layers are able to be trained and used in practical tasks.

To some extent, all of current deep network models can be seen as a nonlinear mapping function from images to labels. In this black box, only the input and output layers are clear, while the mechanism of hidden layers remains difficult to interpret. Recently, for the purpose of developing explicable network models, Hinton proposed CapsNet [27]. CapsNet takes the length of the vector to represent the existence probability of the instance, and adopts dynamic routing to connect the vector layers. It achieves state-of-the-art performance on MNIST [15] hand-written number recognition, but still has much room for further development.

In order to enhance the capability of image classification models, various designs have been presented as we reviewed in the above. Models are becoming more and more complex for a better performance on feature extraction and classification accuracy. For example, on CIFAR-10 data which isn't such a big set, ResNet has been extended to 110 layers, and even the original CapsNet is assembled with 7 models to achieve a competitive performance. As a result, the number of parameters is rapidly rising, and models become more and more difficult to train. In our point of view, making improvement on feature extraction layers should not be the only way to develop the classification framework. The output classification layer and the sample labels are worthy to be reconsidered and investigated.

2.2   One-Hot Encoding

In the classical CNN models for image classification, category literal labels are considered not helpful in the tasks. So images are marked with numerical values, such as integer values 0, 1, 2 in most available data sets. The integer values have a natural ordered relationship between each other, and machine learning algorithms may spontaneously understand and harness this incorrect relationship. In fact, using this encoding and allowing the model to assume a natural ordering between categories may result in poor performances or unexpected results.

An alternative model widely adopted is the one-hot encoding, which removes the integer values and adds a new binary variable for each unique integer value [32]. One-hot origins from an M-bit status register to encode M states. Under this encoding, each class label has its own separate register bits. At any moment, only one of them is valid. For data set label encoding, if this set has K possible class labels, there will be K binary class labels after being uniquely encoded. Also, these class labels are mutually exclusive, with only one activated at a time. Combined with Cross-Entropy loss and Softmax function, One-hot encoding achieves the fair losses of different types of errors, and solves the problem that numerical labels are difficult to optimize. CNN models also benefit from it to process multi-classification tasks.

However, one-hot encoding is premised on the assumption of flat label space. Thus, rich relationships existing among labels are all ignored. Don't these relationships matter in image classification? We believe the answer is yes.

2.3   Image Classification and Semantics

Since many studies refer to the statement of image classification or image semantic, we introduce some representatives here in order to clarify the problem we will focus on in this paper. In general, studies related to semantic and images can be concluded into several categories. The first category is scene classification based on scene semantics. The models include statistics of local features [30], histograms of textons [16], bag-of-words(BoW) model [5,29], and bag of semantics (BoS) [8], etc. These researches study the semantics between objects in an image, rather than the natural language semantics.

Another category is image captioning that focuses on how to automatically generate a natural language description for an image or a video. It is a fundamental problem in artificial intelligence that connects computer vision and natural language processing [34]. Furthermore, to achieve success on cognitive tasks, models are trying to understand the interactions and relationships between objects in an image [12].

Other relevant work includes some image classification methods via label embedding. We take One-Shot Learning [17] or Zero-Shot Learning as an example. One-Shot or Zero-Shot Learning aims to predict the labels of an object which has very few training samples or no sample at all. To recognize an unknown object based on some known samples, the major method is to mine some nonvisual source of information, e.g., attributes or labels between the known and unknown. Specific ways include embedding tag attributes [1], hierarchical semantic embedding (HierSE) [18], and improving label embedding by semantic of features [23,36], etc. These efforts mainly improve Zero-Shot Learning by the semantic similarity among different images, and their objective and method are different from this work. Our work only focuses on classical image classification tasks, and our objective is to improve the classification accuracy with the help of class label semantics from natural language.

3   Method

Similar objects, which are similar in semantics, are considered to have similar word vector representations in the research of natural language processing. Therefore, using word vectors instead of one-hot vector to encode the sample labels will take the similarities between objects into account, which is expected

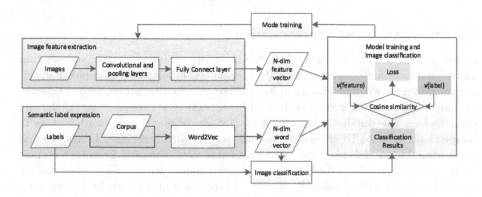

Fig. 2. The architecture of our proposed method.

to be helpful for fitting or training. To be specific, external knowledge is utilized to generate the text vectors of labels required in image classification. The architecture of the proposed method, as shown in Fig. 2, consists of three components. The first one aims to extract image features by an arbitrary CNN. The second one is a semantic embedding component which extracts semantic vectors from natural language corpus according to class labels. And the third component concerns the integration of the above two into one whole training and test model, where similarity between the encodings of visual features and label semantic representations will be a key point.

3.1   Image Feature Extraction

The image semantic component of the model is used to extract image features, where any of a classical CNN model can be used here. It takes an image as the input and outputs a feature vector after processed by several layers, usually including convolutional layers, pooling layers, and fully connected layers. The space where this visual feature lies in can be called as the image feature space, in which feature representation of image contents is embedded.

Any popular CNN models, even customized networks are qualified in this framework, on the condition of a slight modification. And the adjustment should be performed on the last fully connected layers of the employed CNN model. Original output layer is equivalent to a Softmax classification layer, which contains K neurons corresponding to the K classes. Here, we change this layer to output N-dimensional vector instead of K according to the dimension of word vectors that we will discuss in the next subsection.

As for the particular CapsNet, whose output layer is a class capsule layer where each capsule corresponds to a class label, idea of the change is the same. In the original design of CapsNet, Hinton suppose the length of each class capsule vector indicating the probability of an instance's existence. Although this change may violate the above hypothesis, we can give a new explanation of the modified output capsule layer: it is still a feature layer, carrying higher level descriptions of an instance, on which dynamic routing is applicable as usual.

3.2   Semantic Label Representation

In the proposed method, class labels are transformed into semantic vectors that are related to image contents. Actually, literal class labels, such as apple, dog, cat, are meaningless without empirical knowledge or semantic scene. In our scheme, we employ the semantic vectors derived from texts to represent class labels, and then replace the one-hot encoding with the above mentioned word vectors in image classification. To generate semantic vectors for literal class labels, we find some available models that are capable of extracting semantic from text corpus, such as Word2Vec [21], GLoVe [25], etc. In this paper, we choose the benchmark Word2Vec method to model the context information and generate the semantic vectors, and more experiments with other algorithms will be provided in the experiment section.

Word2Vec [21], made open source by Google in 2013, is a deep-learning tool for transforming words into vectors. It applies a sliding window moving on a corpus to generate training samples. The word in the center of the window is the target word and the others are context words. By building a Neural Network Language Model (NNLM) [2] using the input training corpus, Word2Vec model can map each word to characterize real valued vectors via its contextual contents. There are two Word2Vec implements, i.e., CBOW and Skip-gram [21]. The CBOW model uses the average or sum of context words as input to predict the target, and the Skip-gram model uses the target word as input to predict each context word. To simplify, we represent the objective of each prediction as:

$$L_{context} = Pr(w, c) = \frac{exp(w \cdot c)}{\sum_{w' \in V} exp(w' \cdot c)} \qquad (1)$$

In CBOW, w is the target word, and c is the vector of the context words; whereas in Skip-gram, w is each word in the context, and c is the vector of the target word.

In the training process of Word2Vec, cosine similarity is a basic measurement which evaluates the distance of two word vectors. The closer in semantics two words are, the closer the distance between their word vectors is. It should be noted that, the training of CBOW of Skip-gram is pre-done separately from our image classification task, which means this natural language semantic component doesn't train and update together with the feature extraction networks. Pre-trained models provided by Mikolov [21] offer us a ready-made tool to generate word vectors for each class labels.

3.3   Model Training and Image Classification

Since deep network model has output an N dimensional feature and another N dimensional vector of label is prepared, the next point should focus on the model training and test. The connection of two parts operates at the last layer of the CNN for feature extraction, which actually functions as a Softmax classifier. However, in traditional one-hot deep models, the Softmax classifier predicts the probability of each class, and it is usually trained according to the Cross-Entropy loss [6]. In our framework, things are somewhat different: this classifier no longer outputs the probabilities, but the results that fit the word vectors. Because word vector does not represent a probability distribution anymore, Cross-Entropy function is no longer applicable in our framework. Inspired by the similarity measure in Word2Vec, here we choose cosine proximity as the loss function. Cosine similarity is a measure of similarity between two non-zero vectors of an inner product space, therefore our loss function can be indicated by Eq. 2

$$Loss = -\frac{1}{n} \sum_{i=1}^{n} \frac{v^i_{image} \cdot v^i_{word}}{|v^i_{word}| \cdot |v^i_{image}|} \qquad (2)$$

where v_{image} is the N-dim feature vector, and v_{word} is the N-dim word vector.

We adopt two optimization functions, Adam [11] and RMSProp [33], to optimize our model. Both RMSprop and Adam are adaptive learning rate methods, which are frequently used in deep network training.

In the image classification tests, test samples are put into the CNN and features are output as a result of feature extraction. Cosine similarity is computed again between this feature vector and each word vector of labels, and then the largest result will indicate the category this sample belongs.

4   Experiments

4.1   Datasets

Two types of data sets participate in our experiments. One type is used for carrying out multi-class image classification task, and the other one is the text corpus used for attaching semantics to class labels.

As for image classification, we select CIFAR-10 [13] as the first set to verify classification accuracy. CIFAR-10 has a total of 60,000 color images with the size of 32×32 pixels, which are divided into 10 classes. We randomly take 50,000 images (5,000 images in each class) as the training samples, and the other 10,000 ones are left as the validation part. The training process is carried out in 5 batches, where 10,000 images are trained for each batch. Note that the images in 5 batches are divided randomly, which means the sample numbers of each class in one batch are not necessarily the same.

In addition, we supplement dataset to evaluate our method on the condition of greater number of categories. CIFAR-100 [13] is basically similar as the CIFAR-10 dataset, except for the number of classes it contains. CIFAR-100 divides more superclasses on the original CIFAR data, therefore, 100 classes containing 600 images each can be viewed in this data set. Similarly, we randomly choose 500 training images and 100 validation images per class when using CIFAR-100.

We represent semantic of class labels based on the Word2Vec model by taking the Text8 corpus [22] and Wiki-en-text corpus as data sources. Text8 corpus derives from Enwiki8, which is originally used for text compression and has become a standard data set for NLP. By collecting the first 100,000,000 characters from Wikipedia, it is obtained after processing on these characters: all the strange symbols are removed, including non-English characters; and some conversions are performed, such as upper case to lower case, numeral characters to the corresponding English words, etc. Text8 contains only 27 characters: lower case letters from a to z, as well as a space character. Actually, it looks like Wikipedia with the removal of all the punctuations, and the size of it is 95.3M. Since some label words of CIFAR-100 are not included by Text8 corpus, we further crawl a corpus by ourselves from Wikimedia using a tool (https:// dumps.wikimedia.org/) after running 24 h, which is called Wiki-en-text and has the size of 10 GB.

We implement our method in the deep learning framework Keras [4], and evaluate the proposed method by a series of experiments on CIFAR-10 and CIFAR-100 data sets. Class label semantic vectors based on Word2Vec are trained with corpus by the python gensim library [26].

4.2   Results and Discussions

Results with Classical CNN Models. To evaluate the effectiveness of our method, we consider a wide variety of convolutional neural network models including LeNet, AlexNet, VGGNet (VGG-13 and VGG-16), GoogLeNet, and ResNet. We set the loss function to cosine function and optimizer to Adam, and test different CNN models to observe their losses and accuracy rates. We first run experiments for 12 epochs, which are default settings of training epochs in the implements of those CNN networks. Since the default training of 12 epochs may be not enough to show the potential of networks, experiments with 200 epochs are also conducted with the same settings. Results of experiments are listed in Table 1. Here the accuracy means the top-1 prediction accuracy rate. It is necessary to note that, the test loss values in Table 1 are listed for longitudinal reference only, because the loss function in our method is different from the original one in classical CNN-One-hot models.

It can be seen from the table that results of all network models have been improved by approximate 10% after bringing in word semantic of labels. Under the training of 12 epochs, the proposed method in LeNet has achieved the greatest increase of 17.8% on classification accuracy, and even the smallest improvement can be found on VGGNet with about 7% promotion. Apparently, better results are obtained with a training epoch number of 200. More sufficient training produces higher accuracy for each classification models, and the improvement brought by our scheme remains obvious: ranging from 8% to 14%.

Since distinctions and relations of class label semantic representations may be more visible with the increase of category numbers, we further test our framework on CIFAR-100 data set. Because some of class label words in CIFAR-100 cannot be found in Text8 corpus, the word vector for CIFAR-100 in this paper is trained with Wiki-en-text corpus that created by ourselves. Since the results after 200-epoch training are superior and able to reflect the effect adequately in the above experiments, to save running time and paper layout, we just list the results of top-1 accuracy after 200-epoch training in Table 2.

Similar findings can be discovered from Table 2 for our method. Except for the inconspicuous increase on GoogLeNet, all models have been greatly improved by an accuracy enhancement from 14% to 34%. As observed from Tables 1 and 2, we believe that the natural language semantics can contribute even exceed our imagination, which verify the effectiveness of our proposed method.

Table 1. The classification results of different networks with one-hot labels and our Word2vec labels respectively on CIFAR-10 data set.

epoch	Networks	Loss		Accuracy		
		One-hot	Word2vec	One-hot	Word2vec	Improvement
12	LeNet	−0.7364	−0.724	0.6937	0.8717	0.1780
	AlexNet	−0.7871	−0.7349	0.749	0.8948	0.1458
	VGG13	−0.7810	−0.7114	0.7681	0.8400	0.0719
	VGG16	−0.8074	−0.7240	0.7830	0.8612	0.0782
	GoogLeNet	−0.8343	−0.7373	0.8073	0.9205	0.1132
	ResNet20	−0.6732	−0.6861	0.6793	0.8074	0.1281
200	LeNet	−0.7411	−0.7211	0.7036	0.8836	0.1800
	AlexNet	−0.8186	−0.7440	0.8059	0.9284	0.1225
	VGG13	−0.6887	−0.7192	0.7964	0.9067	0.1103
	VGG16	−0.6120	−0.7086	0.7316	0.8435	0.1119
	GoogLeNet	−0.7534	−0.7536	0.8737	0.9639	0.0902
	ResNet20	−0.8249	−0.7357	0.8221	0.9304	0.1083

Table 2. The classification results of different networks on CIFAR-100 data set with 200-epoch training.

Networks	Loss		Accuracy		
	One-hot	Word2vec	One-hot	Word2vec	Improvement
LeNet	−0.3940	−0.5049	0.3425	0.5475	0.2050
AlexNet	−0.4877	−0.5432	0.4627	0.6105	0.1478
VGG13	0.3940	−0.4818	0.3425	0.5735	0.2310
VGG16	−0.2284	−0.4761	0.2282	0.5746	0.3464
GoogLeNet	−0.8825	−0.8854	0.8739	0.8742	0.0003
ResNet20	−0.3928	−0.5115	0.3943	0.5811	0.1868

More Details About Training Settings. More details involved in our method and the classification experiments will be provided and discussed in this subsection. First, as matters of experience, different optimization functions in CNN model training may lead to different results. To examine the impact of optimization functions, we conduct experiments on CIFAR-10 by taking LeNet-5 as an example. We test two frequently used optimizers Adam and RMSProp which have been discussed in Sect. 3.3 to observe the change of results in image classification. And results are listed in Table 3. Moreover, although the accuracy rates of the proposed method are various over different optimizers, the improvements of utilizing our method remain above 16%.

Table 3. Impact of optimization functions with 12-epoch and 200-epoch training.

Dataset	epoch	Optimizer	Loss		Accuracy		
			One-hot	Word2vec	One-hot	Word2vec	Improvement
Cifar10	12	Adam	−0.7294	−0.7233	0.6823	0.8454	0.1631
		RMSprop	−0.7293	−0.7272	0.7027	0.8943	0.1916
	200	Adam	−0.7411	−0.7211	0.7036	0.8836	0.1800
		RMSprop	−0.7353	−0.7230	0.6917	0.8793	0.1876
Cifar100	12	Adam	−0.3986	−0.5032	0.3511	0.5427	0.1916
		RMSprop	−0.3976	−0.5009	0.3556	0.5459	0.1903
	200	Adam	−0.3940	−0.5049	0.3425	0.5475	0.2050
		RMSprop	−0.3645	−0.5020	0.3606	0.5473	0.1867

An overview of Table 3 shows that two optimizes perform with small fluctuations on various training epochs. Since no absolute advantage of any optimizer is observed, we employ Adam as the uniform optimizer for all network models in Sect. 4.2. Moreover, although the accuracy rates of the proposed method are various over different optimizers, the improvements of utilizing our method remain above 16%.

In order to better observe how the classification accuracy changes with the number of training epoch, classification results of LeNet on CIFAR-10 and CIFAR-100 when models are trained from 1 epoch to 200 epochs are recorded and plotted in Fig. 3. From this figure, we can see the stability of convergence as well as the superiority of accuracy rates of our proposed method.

Results with CapsNet Model. Since the CapsNet uses capsules and vectors to represent features, it has a lot of differences on model structure and training method from classical convolutional neural networks. As a remarkable and brannew design of image classification model, CapsNet is worthy to be examined under our proposed framework. As we have mentioned above, we adopt cosine proximity as the training loss, and perform comparative experiments on CIFAR-10 data set. The results of classifying CIFAR-10 images using CapsNet model are as plotted in Fig. 4.

Some conclusions can be draw from Fig. 4: (1) The training accuracy of the original CapsNet model shown in Fig. 4(a) is significantly greater than the verification accuracy, which might be caused by over-fitting problem. (2) Compared with Fig. 4(a), the method we display in Fig. 4(b) basically shows less over-fitting phenomenon, which improves the accuracy by 15%. (3) The proposed method achieves better convergence speed than the original CapsNet, because the right curve takes less than 10 epochs to approach flat and the left one takes about 20 epochs shown in Fig. 4.

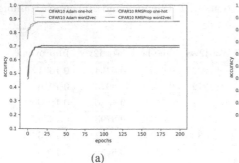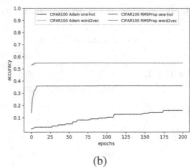

Fig. 3. Classification accuracy curves for LeNet-5 on (a) CIFAR-10 and (b) CIFAR-100 with different numbers of training epoch.

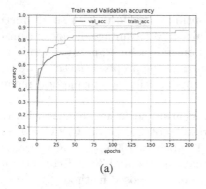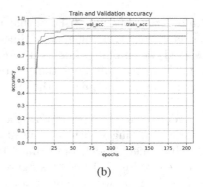

Fig. 4. The training and validation accuracy curves of CapsNet with (a) one-hot label vector, (b) word semantic label vector.

Dimension of Word Vectors. There are some differences between the dimension of one-hot vectors and natural language vectors. One-hot encoding transforms each label of K classes into one K-dimensional vector, where the dimension of encoding vectors is fixed according to the number of categories. However, the dimension of natural language vector is not limited to the number of K classes, which can take any value more than zero. The previous results are presented with 10-dimensional word vectors, which are implemented based on experiments we have done to evaluate the impact of word vector dimension. Actually, in the applications of Word2Vec in the field of natural language processing, dimensions of word vector usually vary from 100 to 300. In our application, we test the impact of word vector dimension and results are plotted in Fig. 5. As shown in Fig. 5, accuracy of classification is generally decreasing with the increase of word vector dimension. This phenomenon may be caused by the noise introduced by higher dimension vector representation. As a result, the above experiments suggest that we'd better choose a lower dimension of word vectors carefully.

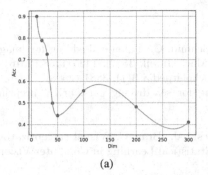

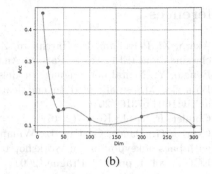

(a)   (b)

Fig. 5. Classification results employing different dimension vectors of (a) CIFAR-10 with text8 corpus and (b) CIFAR-100 with wiki-en-text corpus.

As shown in Fig. 5, accuracy of classification is generally reducing with the increase of word vector dimension. This phenomenon may be caused by the noise introduced by higher dimension vector expression, indicating the importance of appropriate vector expressions.

5   Conclusion

In this paper, we propose a novel method to improve image classification by introducing word semantics into CNN models and CapsNet. The method extracts natural language semantic of the labels before performing image classification, and replaces basic one-hot classification labels with word vectors derived from external knowledge. Experiments on the CIFAR-10 and CIFAR-100 data sets are conducted for evaluating the performance. Compared to original models with one-hot label encoding, all of CNN models and CapsNet integrated in our framework yield better accuracy rates in image classification tasks with a significant improvement. And some experiments show better convergence efficiency and less over-fitting, which benefit the training of models. This study provides new approaches of integrating image classification and natural language semantic, and will contribute to the development of classification model construction and training.

The proposed method still has some limitations. When class labels' semantics have less distinct characteristics, such as t-shirt and shirt in the Fashion MNIST dataset, the method is expected to be less effective or fail to distinguish these categories. In addition, model performance is closely connected with the dimension of natural language word vector, and the inherent reasons need to be further explored. Finally, limited by computational power, we choose CIFAR-10 and CIFAR-100 which are relatively small to perform our experiments. Results of our proposed method on big datasets, such as ImageNet, are not investigated so far. In the future work, we will experiment on these greater dataset to further validate and improve our approach.

References

1. Akata, Z., Perronnin, F., Harchaoui, Z., Schmid, C.: Label-embedding for image classification. IEEE Trans. Pattern Anal. Mach. Intell. **38**(7), 1425–1438 (2016)
2. Bengio, Y.: Neural net language models. Scholarpedia **3**(1), 3881 (2008)
3. Bojarski, M., et al.: End to end learning for self-driving cars. arXiv preprint arXiv:1604.07316 (2016)
4. Chollet, F., et al.: Keras (2015)
5. Csurka, G., Dance, C., Fan, L., Willamowski, J., Bray, C.: Visual categorization with bags of keypoints. In: Workshop on Statistical Learning in Computer Vision, ECCV, vol. 1, pp. 1–2, Prague (2004)
6. De Boer, P.T., Kroese, D.P., Mannor, S., Rubinstein, R.Y.: A tutorial on the cross-entropy method. Ann. Oper. Res. **134**(1), 19–67 (2005)
7. Deng, J., Dong, W., Socher, R., Li, L.J., Li, K., Fei-Fei, L.: ImageNet: a large-scale hierarchical image database. In: 2009 IEEE Conference on Computer Vision and Pattern Recognition, CVPR 2009, pp. 248–255. IEEE (2009)
8. Dixit, M., Chen, S., Gao, D., Rasiwasia, N., Vasconcelos, N.: Scene classification with semantic fisher vectors. In: Proceedings of the IEEE Conference on Computer Vision and Pattern Recognition, pp. 2974–2983 (2015)
9. He, K., Zhang, X., Ren, S., Sun, J.: Delving Deep into Rectifiers: Surpassing Human-Level Performance on ImageNet Classification. CoRR abs/1502.01852 (2015)
10. He, K., Zhang, X., Ren, S., Sun, J.: Deep residual learning for image recognition. In: Proceedings of the IEEE Conference on Computer Vision and Pattern Recognition, pp. 770–778 (2016)
11. Kingma, D.P., Ba, J.L.: Adam: a method for stochastic optimization (2015)
12. Krishna, R., et al.: Visual genome: connecting language and vision using crowd-sourced dense image annotations. Int. J. Comput. Vis. **123**(1), 32–73 (2017)
13. Krizhevsky, A., Nair, V., Hinton, G.: The CIFAR-10 dataset (2014). http://www.cs.toronto.edu/kriz/cifar.html
14. Krizhevsky, A., Sutskever, I., Hinton, G.E.: ImageNet classification with deep convolutional neural networks. In: Advances in Neural Information Processing Systems, pp. 1097–1105 (2012)
15. LeCun, Y., Bottou, L., Bengio, Y., Haffner, P.: Gradient-based learning applied to document recognition. Proc. IEEE **86**(11), 2278–2324 (1998)
16. Leung, T., Malik, J.: Representing and recognizing the visual appearance of materials using three-dimensional textons. Int. J. Comput. Vis. **43**(1), 29–44 (2001)
17. Li, F.F., Fergus, R., Perona, P.: One-shot learning of object categories. IEEE Trans. Pattern Anal. Mach. Intell. **28**(4), 594–611 (2006)
18. Li, X., Liao, S., Lan, W., Du, X., Yang, G.: Zero-shot image tagging by hierarchical semantic embedding. In: Proceedings of the 38th International ACM SIGIR Conference on Research and Development in Information Retrieval, pp. 879–882. ACM (2015)
19. Lin, M., Chen, Q., Yan, S.: Network in network. CoRR abs/1312.4400 (2013)
20. Lin, T.-Y., et al.: Microsoft COCO: common objects in context. In: Fleet, D., Pajdla, T., Schiele, B., Tuytelaars, T. (eds.) ECCV 2014. LNCS, vol. 8693, pp. 740–755. Springer, Cham (2014). https://doi.org/10.1007/978-3-319-10602-1_48
21. Mikolov, T., Chen, K., Corrado, G., Dean, J.: Efficient estimation of word representations in vector space. arXiv preprint arXiv:1301.3781 (2013)

22. Mikolov, T., Joulin, A., Chopra, S., Mathieu, M., Ranzato, M.: Learning longer memory in recurrent neural networks. arXiv preprint arXiv:1412.7753 (2014)
23. Morgado, P., Vasconcelos, N.: Semantically consistent regularization for zero-shot recognition. In: CVPR, vol. 9, p. 10 (2017)
24. Nadeau, D., Sekine, S.: A survey of named entity recognition and classification. Lingvisticae Invest. **30**(1), 3–26 (2007)
25. Pennington, J., Socher, R., Manning, C.: Glove: global vectors for word representation. In: Proceedings of the 2014 Conference on Empirical Methods in Natural Language Processing (EMNLP), pp. 1532–1543 (2014)
26. Řehůřek, R., Sojka, P.: Software framework for topic modelling with large corpora. In: Proceedings of the LREC 2010 Workshop on New Challenges for NLP Frameworks, pp. 45–50. ELRA, Valletta, May 2010. http://is.muni.cz/publication/884893/en
27. Sabour, S., Frosst, N., Hinton, G.E.: Dynamic routing between capsules. In: Advances in Neural Information Processing Systems, pp. 3856–3866 (2017)
28. Simonyan, K., Zisserman, A.: Very deep convolutional networks for large-scale image recognition. arXiv preprint arXiv:1409.1556 (2014)
29. Sivic, J., Zisserman, A.: Video Google: a text retrieval approach to object matching in videos. In: Null, p. 1470. IEEE (2003)
30. Su, Y., Jurie, F.: Improving image classification using semantic attributes. Int. J. Comput. Vis. **100**(1), 59–77 (2012)
31. Szegedy, C., et al.: Going deeper with convolutions. In: Proceedings of the IEEE Conference on Computer Vision and Pattern Recognition, pp. 1–9 (2015)
32. Tang, D., Wei, F., Yang, N., Zhou, M., Liu, T., Qin, B.: Learning sentiment-specific word embedding for Twitter sentiment classification. In: Proceedings of the 52nd Annual Meeting of the Association for Computational Linguistics (Volume 1: Long Papers), vol. 1, pp. 1555–1565 (2014)
33. Tieleman, T., Hinton, G.: Rmsprop. Lecture, COURSERA (2012)
34. Vinyals, O., Toshev, A., Bengio, S., Erhan, D.: Show and tell: lessons learned from the 2015 MSCOCO image captioning challenge. IEEE Trans. Pattern Anal. Mach. Intell. **39**(4), 652–663 (2017)
35. Wan, J., et al.: Deep learning for content-based image retrieval: a comprehensive study. In: Proceedings of the 22nd ACM International Conference on Multimedia, pp. 157–166. ACM (2014)
36. Zhang, L., Xiang, T., Gong, S., et al.: Learning a deep embedding model for zero-shot learning (2017)

Clustering Method for Low Voltage Substation-Area Users Based on Edge Computing Architecture

Jing Jiang[1]([⊠]), Yudong Wang[1]([⊠]), Weijun Zheng[2]([⊠]),
Zhepei Xin[1]([⊠]), Lirong Liu[1]([⊠]), and Bin Hou[3]([⊠])

[1] State Grid Economic and Technological Research Institute Co., Ltd., Beijing,
People's Republic of China
{jiangjing,wangyudong,xinpeizhe,liulirong}
@chinasperi.sgcc.com.cn
[2] Jiaxing Power Supply Company, State Grid Zhejiang Electric Power Co., Ltd.,
Jiaxing, People's Republic of China
zhengweijun@zj.sgcc.com.cn
[3] School of Electronic Engineering, Beijing University of Posts
and Telecommunications, Beijing, People's Republic of China
robinhou@bupt.edu.cn

Abstract. Measurement and monitoring of data in power networks are of great value. A hierarchical clustering analysis method based on edge computing architecture for low-voltage power consumption data is proposed. Based on edge computing architecture, edge nodes are used to collect data and cluster locally, a cloud computing platform is used to aggregate and optimize clustering results. The algorithm flow and optimization strategy are described. The simulation results show that the method has certain advantages in improving computational efficiency and reducing time delay.

Keywords: Edge computing · Hierarchical clustering hierarchical · AMI data analysis · FCM

1 Introduction

In the process of building a unified and strong smart grid with the characteristics of information, digitalization, automation and interaction, a large amount of measurement data and SCADA (Supervisory Control and Data Acquisition) data has been accumulated [1]. These data have the characteristics of mass, high frequency and positional, and have great analytical values. Especially the measurement data implies many kinds of information, such as users' electricity consumption behavior habits. Mining and clustering analysis of these data and user behavior can help power grid companies to understand the personalized and differentiated service needs of users. It can be applied to load forecasting, time-sharing tariff, anomaly detection, and other scenarios, so that grid companies can further expand the depth and breadth of service, and make corresponding policies for future power supply [2].

D. Milošević et al. (Eds.): HCC 2019, LNCS 11956, pp. 190–199, 2019.
https://doi.org/10.1007/978-3-030-37429-7_19

Edge computing is a distributed open platform that integrates network, computing, storage and application core competence on the edge of the network near to the source of objects or data. It provides devices access, data caching and real-time analysis services near the edge of the network, to meet the key requirements of industry informatization and intellectualization, in such aspects as agile connection, real-time business, data optimization, application intelligence, security and privacy protection.

If the data analysis and control logic are all implemented in the cloud side, it may generate huge data transmission requirements, and it is difficult to meet the real-time requirements of business, and it is unable to process sensitive data, so it can't well realize real-time analysis and privacy protection of massive data.

Comparatively speaking, cloud computing is suitable for large data analysis with offline and long-period. Edge computing is suitable for real-time and short-period data analysis and local decision-making scenarios.

The combination of edge and cloud can provide hierarchical data aggregation and utilization, as well as achieve unified data management and distributed data analysis, and achieve more flexible and efficient analysis and processing of massive power metering data.

The clustering analysis of users' electricity consumption characteristics based on power metering data is a popular issue in recent years. Typical research includes:

Cloud computing can realize the unified management and analysis of massive heterogeneous data, which provides a new method for data mining and effectively solves the storage, management and utilization of massive data. In [3], they introduced the data mining method based on cloud computing, and the problems and prospects of data mining based on cloud computing are described. In [4], they developed a parallel distributed data mining toolkit platform based on distributed data processing platform Hadoop, support association rule analysis, classification and clustering algorithms. In [5], a specific method and strategy for realizing hierarchical and virtualized K-means clustering in a cloud environment is proposed. This method establishes a cloud data mining system and describes the specific process of cloud data mining. In [6, 7], they introduced the concept of seamless computing, weakened the role of cloud platforms in the network, and emphasized the ability of edge devices to participate in the business. Seamless computing environment is built through applications to support workload movement between cloud and edge, and to transfer standard cloud computing capability to edge resource-constrained computing environment.

Advanced metering infrastructure (AMI) offers utilities new ways to model and analyze distribution circuits [8] and [9]. In [10], a distributed FCM method is described, which distributes the subset of data sets to multiple clusters, so as to achieve efficient distributed clustering. The FCM algorithm is based on minimizing a c-means objective function to determine an optimal classification. Some validity index measurements were carried out to estimate the compactness of the resulting clusters or to find the optimal number of clusters for a data set. In [11], fuzzy C-means (FCM) clustering and probabilistic neural network (PNN) classification are used to determine the typical load profile (TLP). They propose an optimal number of cluster determination by using the Davies-Bouldin index. The load curve provides users and suppliers with useful information about electricity consumption. With this approach, secondary connectivity and impedance models can be auto generated.

By introducing edge computing model into the analysis of power metering data, a hierarchical clustering method for transformer users based on edge computing architecture is proposed.

Constructing the edge computing system, the efficient and fast clustering analysis of massive user data can be realized. The user data is cached to the nearby edge computing nodes, and the edge computing nodes are used to analyze and mine the data to acquire local knowledge. Then the global knowledge is acquired through interaction with a cloud platform, and the user's electricity consumption behavior is analyzed according to global knowledge, to realize anomaly detection and prediction.

The rest of this paper is organized as follows. Section 2 presents the architecture of the edge computing system applied to the analysis of AMI data in low-voltage stations, introduces the main modules and characteristics. Section 3 introduces the distributed clustering scheme (FCM) of AMI data in low-voltage stations based on edge computing system, introduces the algorithm flow, key formulas and the global optimization mechanism based on cloud computing center. The conclusions of this study are given in Sect. 4.

2 Hierarchical Clustering System Architecture of Measurement Data Based on Edge Computing

2.1 System Architecture

The Architecture of Power Metering Data Analysis System Based on Edge Computing is shown in Fig. 1. The architecture of this edge computing system is divided into three layers: data acquisition layer, edge computing layer and cloud computing layer.

Data Acquisition Layer. Data Acquisition Layer relies on traditional AMI and SCADA systems in low-voltage stations to collect high-frequency measurement data and monitoring data. Collected data would be aggregated into the edge computing layer for processing and analysis.

Edge Computing Layer. This layer consists of a large number of distributed edge computing nodes, which are deployed based on geographical location (low-voltage station area). Data acquisition layer can cache data to nearby nodes according to geographical location or link relationship. Analysis methods and strategies can be obtained from the cloud computing layer, and local data can be analyzed independently according to these strategies, and local knowledge can be generated.

Computing is mainly carried out on distributed edge device nodes, which can reduce the total amount of data transmission and improve the analysis speed and effect.

Cloud Computing Layer. Establish a cloud computing platform to implement the analysis strategy of edge computing nodes, data aggregation and node behavior management. After analyzing the local knowledge of the edge computing layer, the analysis conclusions can be summarized to the cloud computing layer. On the one hand, cloud computing layer constitutes global knowledge. On the other hand, the cloud computing layer can construct new analysis strategies based on global knowledge and send them to edge computing layer to optimize the analysis process of local knowledge.

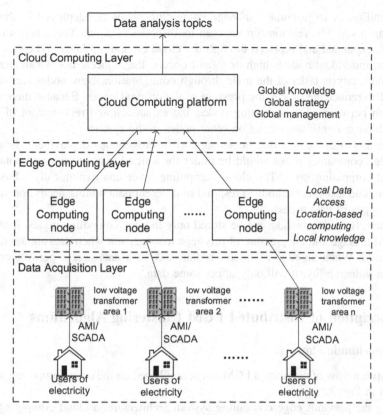

Fig. 1. The Architecture of Power Metering Data Analysis System Based on Edge Computing.

Compared with the centralized data analysis method which directly collects data to one cloud platform, the edge computing architecture can reduce the data access delay and improve the efficiency of data analysis.

2.2   A Subsection Sample

The following advantages can be gained by using the distributed architecture based on edge computing in power metrology analysis:

Communication efficiency: Because of the geographic dispersion of edge computing, the efficiency of data acquisition and transmission is improved. The edge computing node can quickly process the information of low-voltage stations nearby, and transmit the calculation results and a small amount of necessary data to the cloud so that the edge computing layer and cloud computing layer can interact more efficiently.

Scalability: A large number of edge computing nodes are deployed in the edge computing layer. The relationship between these nodes is equal. These nodes can be managed, expanded and replaced through the cloud computing layer. When the pressure of an individual node is high or a fault occurs, the adjacent nodes can share the access and analysis tasks of the node through configuration. New nodes can also be deployed to reduce the processing pressure of the original nodes. Because there is less interaction between edge computing nodes, the expansion and replacement of nodes will not have a significant impact on other nodes in the system.

Generality: Edge computing nodes support heterogeneous deployment, heterogeneous edge computing nodes would be under the unified management and control of the cloud computing layer. The cloud computing layer can dynamically adjust data processing and analysis methods to respond to different data mining needs and support different types of data access.

Privacy: Because raw data can be stored only in edge computing nodes, it reduces the risk of leakage and the pressure of privilege management when data are aggregated to the cloud. When an edge calculates a data leak or a data access permission configuration vulnerability, it will only affect some data.

3   Description of Distributed FCM Clustering Algorithms

3.1   Algorithmic Flow

The algorithm flow of distributed FCM clustering based on the edge computing system is shown in Fig. 2.

Using the previous edge computing system architecture, a cloud computing platform is used as the master station, and edge computing node is used as the slave station. Local AMI data are obtained from low-voltage stations and clustered by FCM to obtain local knowledge. The master station gathers the local knowledge of each slave station to form global knowledge. According to global knowledge, the optimal clustering number and cluster center are obtained and sent to the slave node. The specific process is as follows:

The master station uses FCM clustering algorithm to obtain the number of clusters and the global initial clustering center. The master station broadcasts the global cluster center to all slave stations. Each site classifies each data object in its local data set and calculates the local cluster information. Each slave station divides the cluster information and sends the spit-out sub-information to the other slave stations. Each slave station receives the residual sub-information sent from the slave station, recalculates the cluster information, and sends the calculation results to the master station. The master station receives the cluster information sent from the slave station and calculates the global cluster center. Iterate iteratively until the global cluster center remains unchanged.

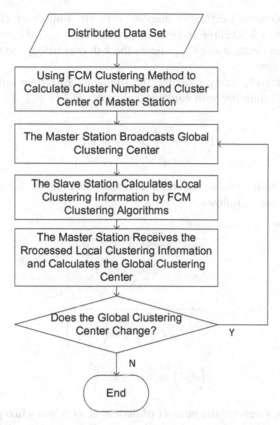

Fig. 2. Algorithm flow of distributed FCM clustering.

3.2 The Core Process of FCM

In the fuzzy C-means clustering method, each data point belongs to a clustering center according to a certain degree of fuzzy membership. Clustering centers are continuously modified using an iteration method. The iteration process aims at minimizing the weighted sum of jump and membership from all data points to each clustering center. Membership matrix μ is allowed to be between [0, 1]. With standardization, the sum of membership degrees of a user is equal to 1, that is:

$$\sum_{i=1}^{c} \mu_{ik} = 1, \quad k = 1, 2, \ldots, n \tag{1}$$

The objective function is:

$$J = \sum_{k=1}^{n} \sum_{i=1}^{c} \mu_{ik}^{m} \|x_k - v_i\|^2 \tag{2}$$

In Eq. (2): n denotes load curve number, c is the number of clusters specified beforehand. x_k is the k load curve and expressed as a vector. v_i is the center of type, and i is load curve when iteration stops. μ_{ik} means the k-th user belongs to the membership degree of the i-th user.

According to the clustering criterion, the following Lagrange functions are constructed, Lagrange multiplier with equality constraints is $I_k(k = 1, 2, \ldots, n)$.

$$F = J + I_k \left(\sum_{i=1}^{c} \mu_{ik} - 1 \right) \tag{3}$$

According to Kuhn-Tucker theorem, necessary conditions for minimizing the objective function are as follows.

$$v_i^{(r+1)} = \frac{\sum_{k=1}^{n} \left(\mu_{ik}^{(r)} \right)^m x_k}{\sum_{k=1}^{n} \left(\mu_{ik}^{(r)} \right)^m} \tag{4}$$

$$\mu_{ik}^{(r+1)} = \frac{1}{\left[\sum_{j=1}^{c} \left(d_{ik}^{(r)} \right)^2 / \left(d_{jk}^{(r)} \right)^2 \right]^{\frac{1}{m-1}}} \tag{5}$$

$$\left(d_{ik}^{(r)} \right)^2 = \left\| x_k - v_i^{(r)} \right\|^2 \tag{6}$$

In Eqs. (4)–(6): r denotes the number of iterations, m denotes fuzzy variable. d_{ik} is the distance between the center of the load curve of type i obtained from the second iteration of the load curve of section k. In each iteration, the iteration center and membership degree are updated by formula (4) and formula (5), respectively. The clustering process is as follows:

Take initial values: $m = 2, 2 \leq c \leq n$. Calculate v_i^r and $\left(d_{ik}^{(r)} \right)^2$ according to formula (4) and formula (6), and then calculating $\mu_{ik}^{(r+1)}$ from Eq. (5). For a given threshold ρ, if there is $\left[J^{(r+1)} - J^{(r)} \right] < \rho$, then the clustering process stops or returns to (4).

Fuzzy C-means clustering generates some clustering centers and membership matrix at the end of clustering. The matrix represents the degree of membership of each curve relative to each classification. According to the principle of maximum membership degree, the classification of each curve can be judged, and the purpose of clustering can be realized.

3.3   Clustering Strategy Optimization Based on Global Knowledge

When the master station acquires the local knowledge of each slave station, it includes information such as the cluster and the number of instances of each slave station. The following formula is used to recalculate the global cluster center and broadcast it to subordinate sites to obtain globally optimized clustering results. Before the global clustering center is stable, the clustering process can be repeated many times.

$$c = \frac{1}{n} \sum_{i=1}^{j} c_i n_i \tag{7}$$

$$n = \sum_{i=1}^{j} n_i \tag{8}$$

In Eqs. (7) and (8), n_i is the number of users of the slave station i, c_i is the number of clusters from site i, n is the sum of all users from all station.

4 Simulation Results

The original data set is from the UCI Machine Learning Repository. The selected data set consists of 370 users intercepted in one month (AMI data). The users are divided into three groups: 111 users in group 1, 111 users in group 2 and 148 users in group 3. Data is collected every 15 min, so each user has 672 attributes or dimensions. Before analysis, these data will be processed and merged.

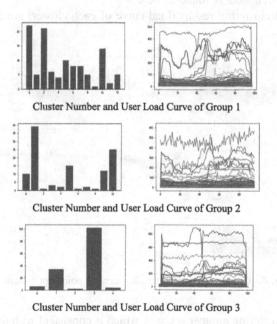

Cluster Number and User Load Curve of Group 1

Cluster Number and User Load Curve of Group 2

Cluster Number and User Load Curve of Group 3

Fig. 3. Number and load curve of slave stations clustering.

Put dataset group 1, group 2 and group 3 in slave stations S1, S2 and S3 respectively for FCM clustering analysis. The clustering results are as follows.

S1 set clustering number $c_1 = 13$, then 111 users in group 1 are divided into 13 categories. The number of clustering users per category is:

$$n_1 = [22, 5, 21, 6, 4, 10, 8, 8, 5, 1, 14, 2, 5]$$

S2 set clustering number $c_2 = 11$, then 111 users in group 2 are divided into 11 categories. The number of clustering users per category is:

$$n_2 = [10, 39, 1, 3, 2, 15, 1, 2, 1, 12, 25]$$

S2 set clustering number $c_3 = 5$, then 148 users in group 3 are divided into 5 categories. The number of users in each category is:

$$n_3 = [6, 34, 2, 102, 4]$$

The results of clustering number and users load curve of slave stations are shown in Fig. 3.

Local clustering results of each slave station are transmitted to the master station, including the clustering number and data distribution. The master station adjusts the number of clustering according to formula (7) and (8). After one or more rounds, the optimal number of clusters is found to be c = 9.

The optimized clustering results (load curve of each cluster) are shown in Fig. 4.

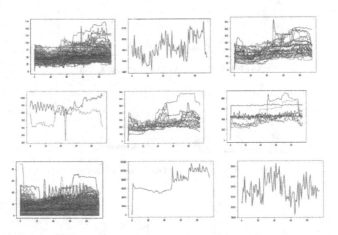

Fig. 4. Clustering load curve after central point optimization.

The optimal clustering number is c = 9, which is consistent with the above process.

The method proposed is based on the edge computing model, which can process local data in parallel. When the optimal clustering results are obtained, the transmission time and processing analysis time are greatly reduced. When the composition of power users in low-voltage stations changes, the new clustering results can be obtained more quickly by this method.

5   Conclusion

The collection and analysis of AMI and SCADA data in low voltage stations can produce great value. Clustering of these data plays a supporting role in the fields of power forecasting and abnormal power consumption analysis.

Traditional data clustering methods based on cloud computing have shortcomings in efficiency, while traditional distributed clustering methods lack the ability of unified adjustment. Based on edge computing architecture, a general and dynamic distributed clustering analysis method can be realized. The simulation results show that this method can reduce the data transmission overhead, reduce the time of data analysis and the number of iterations to obtain the optimal clustering effect.

The hierarchical FCM method proposed in this paper can be further extended, such as optimizing and adjusting the clustering centers by the main station, and weighting the local clustering results and the optimization information of the main station.

References

1. Gupta, R., Moinuddin, Kumar, P.: Cloud computing data mining to SCADA for energy management. In: 2015 Annual IEEE India Conference (INDICON), pp. 1–6. IEEE, India (2015)
2. Bergman, D.C., Jin, D., Juen, J.P., Tanaka, N., Gunter, C.A., Wright, A.: Nonintrusive load-shed verification. IEEE Pervasive Comput. **10**(1), 49–57 (2011)
3. Hu, T., Chen, H., Huang, L., Zhu, X.: A survey of mass data mining based on cloud-computing. In: Anti-counterfeiting, Security, and Identification, pp. 1–4. IEEE, Taiwan (2012)
4. He, Q., Zhuang, F., Zeng, L., Zhao, W., Tan, Q.: PDMiner:a cloud computing based parallel and distributed data mining toolkit platform. Scientia Sinica Informationis **44**(7), 871–885 (2014)
5. Nair, M.: Data mining using hierarchical virtual k-means approach integrating data fragments in cloud computing environment. In: 2011 IEEE International Conference on Cloud Computing and Intelligence Systems, pp. 230–234. IEEE, China (2011)
6. Mueller, H., Gogouvitis, S.V., Haitof, H., Seitz, A., Bruegge, B.: Seamless computing for industrial systems spanning cloud and edge. In: 2017 International Conference on High Performance Computing and Simulation (HPCS), pp. 209–216. IEEE, Italy (2017)
7. Mueller, H., Gogouvitis, S.V., Haitof, H., Seitz, A., Bruegge, B.: Poster abstract: continuous computing from cloud to edge. In: 2016 IEEE/ACM Symposium on Edge Computing (SEC), pp. 97–98. IEEE, USA (2016)
8. Short, T.A.: Advanced metering for phase identification, transformer identification, and secondary modeling. IEEE Trans. Smart Grid **4**(2), 651–658 (2013)
9. Berrisford, A.J.: A tale of two transformers: an algorithm for estimating distribution secondary electric parameters using smart meter data. In: 2013 26th IEEE Canadian Conference on Electrical and Computer Engineering (CCECE), pp. 1–6. IEEE, Canada (2013)
10. Prahastono, I., King, D.J., Ozveren, C.S., Bradley, D.: Electricity load profile classification using Fuzzy C-Means method. In: 2008 43rd International Universities Power Engineering Conference, pp. 1–5. IEEE, Italy (2008)
11. Anuar, N., Zakaria, Z.: Electricity Load Profile Determination by Using Fuzzy CMeans and Probability Neural Network, 1st edn. Elsevier, Amsterdam (2012)

Different Performances of Speech and Natural Gait in Identifying Anxiety and Depression

Chunke Jing[1,2], Xiaoqian Liu[1(✉)], Nan Zhao[1], and Tingshao Zhu[1(✉)]

[1] Institution of Psychology, Chinese Academy of Sciences, Beijing, China
{liuxiaoqian, tszhu}@psych.ac.cn
[2] Department of Psychology, University of Chinese Academy of Science, Beijing, China

Abstract. Recognition of individual's emotions using the data of speech and natural gait has been proved effective by previous studies. This study establishes a supervised regression model based on individual's speech and natural gait data to predict the scores of depression and anxiety. The results show that the performance of the predictive model based on gait features is better than that of speech features. The basic processes are as follows: First, we recruited 88 participants and collected their speech data and natural gait data in laboratory; Second, we required each participant to finish the anxiety and depression scales (PHQ-9, CES-D, T-AI, and GAD-7), to get the anxiety and depression scores as the tagging data for modeling; Third, we extracted 6125 features from each participant's speech data and 1602 features from the gait data; Finally, we used a variety of machine learning algorithms to train and test the predictive models. The results show that the max Pearson Correlation Coefficient between the predicted scores and the questionnaire scores of the speech-based model is 0.43, and the max Pearson Correlation Coefficient of the gait-based model is 0.73, which is much better than the speech-based model.

Keywords: Speech · Gait · Anxiety detection · Depression detection

1 Introduction

Depression and anxiety disorder are two common mental disorders and important public health issues worldwide. The lifetime prevalence of depression is 2%–15% [1], and the 12-month prevalence of anxiety is 1.9%–5.1% [2]. Moreover, both these two disorders usually show long duration and are easy to relapse, accompanied by other diseases and suicide risk [3, 4], which not only seriously damage the health, social roles and social interaction of the patients [5], but also bring tremendous disease burden to the society. According to the statistics, depression and anxiety disorder ranked first and second respectively in the burden of mental illness, accounting for 40.5% and 14.6% (calculated according to DALYs) [6].

Depression and anxiety disorders are often screened and diagnosed using psychological scales, such as the PHQ-9 [7] and CES-D [8] scales to assess the degree of depression, and the T-AI [9] and GAD-7 [10] scales to assess the degree of anxiety.

The PHQ-9 is the 9-item depression module of the full PHQ (Patient Health Questionnaire). It includes nine items according to nine diagnostic criteria of DSM-IV. As a severity measure, the PHQ-9 score can range from 0 to 27, since each of the 9 items can be scored from 0 (not at all) to 3 (nearly every day). The higher the score, the higher the degree of depression [7]. The CES-D consists of 20 questions. The score of each question ranges from 0 to 3, and the score range of the CES-D scale is 0–60. The larger the score, the more depressive symptoms appear [8]. The T-AI is a trait scale in STAI. STAI uses self-report method to measure and distinguish anxiety as a trait or different aspects of a state. Trait anxiety refers to the difference in frequency and intensity of individual anxiety performance, and tends to be relatively stable personality characteristics, which is tested by T-AI subscale ranging from 20 to 80 [9]. GAD-7 is a concise patient self-rating scale to test the risk and severity of generalized anxiety disorder. The scale contains seven questions, and the total score is between 0 and 21. The higher score indicates the higher anxiety level [10].

Although the four scales are widely used and proved to have good reliability and validity [7–10], they still have some limitations as self-report measurements. Restricted by the ability to perceive and express, in practice people are not always able to accurately assess and report their symptoms and severities. In addition, because the self-report scale is purely subjective, it is easy to conceal, so using objective indicators to assess anxiety and depression as supplementary means has a certain significance [11].

Previous studies have shown that psychomotor disturbance is one of the basic characteristics of depression [12, 13]. Speech and motor characteristics are also objective criteria for measuring psychological activity in depressive state [14]. For example, the general changes of the voice of depressive patients included slow speech speed, more pauses, longer pauses, shorter pronunciation time, etc. [15, 16]. Meanwhile, there are also significant correlations between pitch and depression score in the prosodic features of depressive patients' voice, and the decrease of pitch in bandwidth, amplitude, energy, etc. [17, 18], which provides a basis for evaluating depression degree by using phonetic features. On the other hand, it was found that compared with healthy people, depression patients have significantly slower pace, shorter pace, less time for both feet support and circulation [19], indicating that it is feasible to assess the degree of depression based on gait. For anxiety, studies have shown that anxiety can be reflected in facial and behavioral characteristics, including the behavior of eyes, mouth, cheeks and the whole head (head movement, head speed, etc.) [20]. At the same time, anxiety can also be reflected in speech characteristics, such as the average maximum pitch, the high frequency components of the spectrum and the proportion of silent pause [21]. These studies also show that it is possible to recognize anxiety by speech and gait features.

Although both depression and anxiety can leave clues on speech and gait characteristics, and studies have used them to recognize depression and anxiety [22, 23], the recognition performances of speech and gait has not been compared on one data set of normal subjects. So we have two main issues discussed in this paper. The first is to further verify the effectiveness of normal human speech and gait (especially two-dimensional gait information) in identifying depression and anxiety. The second is comparing the different recognition performances of the speech and gait models.

In this study, we assumed that the anxiety and depression levels measured by the scales can be identified through the speech and gait data. Based on the speech and gait data, we used machine learning method to establish some supervised learning regression models to predict the scores of the PHQ-9, CES-D, T-AI and GAD-7, trying to develop an objective depression and anxiety assessing method. We also compared the performances of different models which provides more understandings about the reflections of depression and anxiety in speech or gait. In addition, in gait data collection, we used mobile phone for gait video and OpenPose [26] for two-dimensional joint coordinate extraction, which significantly reduces the cost of equipment as a solution in practice.

The study processes included four main stages: data collection, feature extraction, identification model training and evaluation. Firstly, we recruited 88 volunteers and collected: speech voice in the quiet environment, gait video in the natural state by mobile phone, and depression and anxiety questionnaire data. Secondly, we used openSMILE [24, 25] to extract speech features from each volunteer's speech file, and used OpenPose [26] to extract the 2-D coordinate values of the joint points from each volunteer's gait video. Thirdly, we employed Linear Regression, Decision Tree, Random Forest, Naive Bayes, Support Vector Machine and Multi-Layer Perception to build regression models through 10-folds cross-validate. Finally, we used Pearson correlation coefficient and RSME (Root Mean Square Error) to evaluate the accuracy of the regression model. The final results showed that both speech and gait data are effective in identifying depression and anxiety, and the models based on gait data performed better than those on speech data.

2   Method

2.1   Data Collection

We recruited 88 volunteers (male: female = 31:57), with an average age of 23 (max = 39, min = 18, std = 4.4). We used the following criteria for screening: (1) there was no diagnosed depression or anxiety disorder in three generations; (2) there was no history of material dependence, neurological disorder or obvious brain injury in the present and past. Then, besides basic demographic information, each volunteer completed four scales: PHQ-9, CES-D, T-AI and GAD-7. Table 1 shows an overview of the 88 volunteers' scale scores.

Table 1. Summary of 88 volunteers' scale scores.

Scale name	Max score	Min score	Mean of scores	Std of scores
PHQ-9	24	0	5.72	4.99
CES-D	55	16	29.86	8.98
T-AI	58	42	51.17	3.11
GAD-7	16	0	4.65	4.10

Speech Data Collection. Speech data collection experiment was carried out in a quiet room. Each volunteer introduced himself as the guideline displayed by the computer:

(1) Please introduce yourself and your hometown.
(2) Please introduce your major and your research work during your study.
(3) Please describe your plan for the future and what kind of work you want to do.

During the volunteers' speech, there was no interaction. The speech voice was recorded for the following analysis.

Gait Data Collection. The gait data collection experiment was conducted in a rectangular space with 6 m long and 3 m wide. Each participant walked back and forth naturally for 2–3 min. We use a mobile phone on the bracket recorded the video of the volunteers' gaits (the sampling frequency was 25 Hz).

2.2 Data Preprocessing and Feature Extraction

Speech Data Preprocessing and Feature Extraction. We used openSMILE [24] as the tool to preprocess and extract features of speech data. In terms of speech feature

Table 2. 64 low-level descriptors of the baseline feature set [27, 28].

---

4 energy related LLD

sum of auditory spectrum (loudness)

sum of RASTA-style filtered

auditory spectrum RMS energy

zero-crossing rate

54 spectral LLD

RASTA-style auditory spectrum, bands 1–26 (0–8 kHz)

MFCC 1–14

spectral energy 250–650 Hz, 1 k–4 kHz

spectral roll off point 0.25, 0.50, 0.75, 0.90

spectral flux, entropy, variance, skewness, kurtosis

6 voicing related LLD

F0 by SHS + Viterbi smoothing, probability of voicing

logarithmic HNR, jitter (local, delta), shimmer (local)

---

selection, we referred to the feature set from The INTERSPEECH 2012 Speaker Trait Challenge [27, 28]. This feature set is designed to assess speaker's traits, including personality, likability and pathology. Because one of the focuses of this feature set is pathology, such as speech disorders caused by oral and laryngeal cancers, and artic-ulation disorders caused by Parkinson, which is similar to the topic of speech abnor-mality caused by depression and anxiety, we chose this feature set. This feature set includes 64 low-order descriptor baseline features and other features processed by statistical functions [27, 28]. The chosen set of LLDs is shown in Table 2 and the set of applied functions is shown in detail in Table 3.

Table 3. Statistical functions of the baseline feature set [27, 28].

Functions applied to LLD / $\triangle$LLD
quartiles 1–3, 3 inter-quartile ranges
1% percentile ($\approx$min), 99% percentile ($\approx$max)
position of min/max
percentile range 1–99%
arithmetic mean[a], root quadratic mean
contour centroid, flatness
standard deviation, skewness, kurtosis
rel. duration LLD is above/below 25/50/75/90% range
rel. duration LLD is rising/falling
rel. duration LLD has positive/negative curvature[b]
gain of linear prediction (LP), LP coefficients 1–5
mean, max, min, std. dev. of segment length[c]
Functions applied to LLD only
mean value of peaks
mean value of peaks – arithmetic mean
mean/std. dev. of rising/falling slopes
mean/std. dev. of inter maxima distances
amplitude mean of maxima/minima
amplitude range of maxima
linear regression slope, offset, quadratic error
quadratic regression a, b, offset, quadratic error
percentage of non-zero frame[d]

a: Arithmetic mean of LLD/positive $\triangle$LLD.
b: Only applied to voice related LLD.
c: Not applied to voice related LLD except F0.
d: Only applied to F0.

After using openSMILE tool to extract features from each volunteer's speech according to the 2012 Challenge Feature Set, 6125 features were extracted from each volunteer's speech. When all 88 volunteer's speech feature data were combined, we got a 88 * 6125 speech feature matrix.

Gait Data Preprocessing and Feature Extraction. We used the OpenPose [29] tool to identify and extract the information of joint points in gait video. OpenPose is an open source project that can produce 2-D posture estimation of human motion, facial expression, finger movement, etc. It has the advantages of high recognition rate, high robustness and fast running speed [29]. We used OpenPose to capture the 18 main body joints (see Fig. 1) in motion, and output the absolute coordinates (including X and Y directions) of the 18 joints in each frame for the following steps shown in Fig. 2.

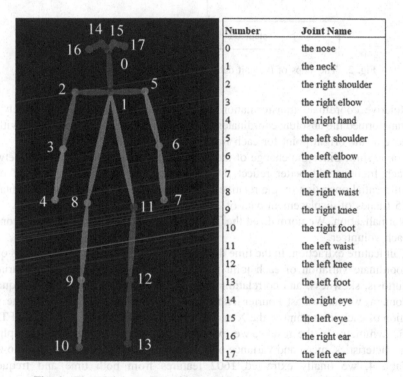

Number	Joint Name
0	the nose
1	the neck
2	the right shoulder
3	the right elbow
4	the right hand
5	the left shoulder
6	the left elbow
7	the left hand
8	the right waist
9	the right knee
10	the right foot
11	the left waist
12	the left knee
13	the left foot
14	the right eye
15	the left eye
16	the right ear
17	the left ear

Fig. 1. The 18 joint points of human body captured from gait video

(1) Continuous frame extraction in which the joint coordinates are all non-zero. Since human gait is relatively periodic, we chose continuous frames with more comprehensive joint information for feature extraction. As OpenPose sets the coordinates of occluded or undetected joints to 0 in a frame, we selected continuous frames with joint absolute coordinate data of each gait video that are all non-zero. Seventy-six consecutive frames (about 3 s) from each volunteer's gait video data were extracted in this step for the following process.

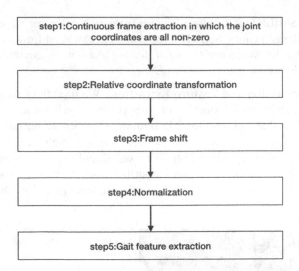

Fig. 2. The steps of the gait data preprocessing and feature extraction

(2) Relative coordinate transformation. In order to calculate more accurately, we transformed the absolute coordinates of all nodes into relative coordinates with the nose as the origin point for each frame of the data.

(3) Frame shift. Since the change of the coordinates of the same joint point between each frame could better reflect the motion during walking, we carried out a differential operation on the relative coordinates of each joint point, and obtained 75 frames of displacement data of each volunteer's gait data.

(4) Normalization. We normalized the 75 frames of displacement data by Z-score for each volunteer.

(5) Gait feature extraction. In the time domain, we chose five statistical features of the coordinate variation of each joint point in the X and Y axes: mean, variance, kurtosis, skewness, and correlation coefficient of the X and Y. In the frequency domain, we make Fast Fourier Transform (FFT, defined as Eq. (1)) of the variation of each joint point in the X and Y axes. According to the results of FFT and the symmetry of the results, we extracted three features: half of the amplitude characteristics, mean and variance of power spectral density (PDD). As shown in Table 4, we finally extracted 1602 features from both time and frequency domains.

$$F_k = \sum_{n=0}^{N-1} x_n e^{-i2\pi k\frac{n}{N}} \quad k = 0, \ldots, N-1 \tag{1}$$

Table 4. Features extracted from gait data.

The features from time domain	Number
Mean and std of x-axis	18 * 2
Mean and std of y-axis	18 * 2
Kurtosis and skewness of x-axis	18 * 2
Kurtosis and skewness of x-axis	18 * 2
PCC for x-axis and y-axis	18
The features from frequency domain	Number
Amplitude of the x-axis	18 * 76/2
Amplitude of the y-axis	18 * 76/2
Mean and std of PSD	18 * 2 * 2
Total	**1602**

3   Result

In the process of data preprocessing and feature extraction, we extracted 6125 features from each volunteer's speech voice and 1602 features from gait video. Based on these features, we used machine learning to train the predictive model. We used the scores of each volunteer in four scales (PHQ-9, CES-D, T-AI and GAD-7) as label data. Then we conducted LDA dimensionality reduction to reduce the dimensionality of speech and gait features. Linear Regression (LR), Decision Tree (DT), Random Forest (RF), Naive Bayes (NB), Support Vector Machine (SVM) and Multi-Layer Perception (MLP) method were used to establish supervised regression model. Finally, we used 10 folds cross-validation to verify the predicted value of the related scale, and evaluate the predictive accuracy of the model by Pearson correlation coefficient (PCC) between predicted value and labeled value, and by RMSE of the model. Table 5 shows the effects of different models based on speech features, and Table 6 shows the results of gait features.

Table 5. Regression results of speech models

	RMSE (PHQ-9)	PCC (PHQ-9)	RMSE (CES-D)	PCC (CES-D)	RMSE (T-AI)	PCC (T-AI)	RMSE (GAD-7)	PCC (GAD-7)
LR	30.71	0.02	154.56	0.41***	10.84	0.13	21.98	0.08
DT	44.77	0.09	142.18	0.01	17.73	0.15	30.35	0.11
RF	25.93	0.18	86.43	0.16	10.13	0	16.29	0.16
NB	25.75	0.25*	81.38	0.35***	9.81	0	16.89	0.35***
SVM	24.87	0.12	88.57	0.43***	9.83	0.01	16.78	0.08
MLP	24.84	0.32**	81.37	0.35***	9.76	0.32**	16.83	0.33**

*p < :05
**p < :01
***p < :001

Table 6. Regression results of gait models

	RMSE (PHQ-9)	PCC (PHQ-9)	RMSE (CES-D)	PCC (CES-D)	RMSE (T-AI)	PCC (T-AI)	RMSE (GAD-7)	PCC (GAD-7)
LR	22.72	0.39***	90.37	0.41***	8.59	0.44***	11.18	0.59***
DT	27.39	0.39***	60.28	0.63***	8.78	0.6***	10.99	0.62***
RF	13.21	0.68***	42.79	0.73***	6	0.62***	8.32	0.73***
NB	23.31	0.3**	68.77	0.41***	8.46	0.39***	11.42	0.55***
SVM	18.3	0.64***	72.79	0.47***	7.62	0.49***	11.06	0.68***
MLP	24.84	0.32**	69.15	0.51***	9.75	0.31**	14.76	0.51***

$*p < :05$
$**p < :01$
$***p < :001$

For the result of the models based on speech data (Table 5), only MLP model's PCC was significant for all the four psychological scales ($p < 0.01$). The highest PCC was 0.43 of CES-D using SVM regression algorithm. For the results of gait (Table 6) models, the PCC obtained by almost all the models were significant ($p < 0.001$). The performances of RF models were slightly better than other models, whose PCC generally reaching more than 0.6 and up to 0.73. In general, the performances of the models based on gaits are better than those on speech data.

4   Discussion

The results of the study support the two hypotheses below: First, based on speech and gait data, machine learning can be used to predict the scores of anxiety and depression scales. The highest PCC (Pearson correlation coefficient) reached the significant level of 0.43 and 0.73 respectively, which corresponds to previous studies [12, 13, 22, 23]. Second, for the same individual, the accuracy of recognition through gait is relatively better than that of speech. However, as this result may be related to many factors such as participants' anxiety and depression level, selection of speech and gait features, machine learning models and so on, this hypothesis needs further verification.

In addition, our research still has some limitations. First of all, the data collection of this study was carried out in the laboratory environment. So to apply this method in practice, more assessing in application scenes is needed. Secondly, most of the volunteers recruited in current study were college students rather than real patients, and their depression and anxiety score distributed in relatively health region, which also affected the ecological validity of the conclusions. Finally, because the main purpose of this study is to verify the validity of speech and gait data in predicting the scores of anxiety and depression, and to compare the different performances of the models based on these two data sources, we were less involved with the psychological significance of those low-level gait and speech characteristics, which is worthy to explore in the future.

From the application point of view, multi-source data fusion could be a direction in future exploration, and some studies have already demonstrated that the recognition accuracy of multi-source data is better than that of single-source data [30–32]. In the current study, we have verified that two single data sources, speech and gait, could be used to predict the degree of depression and anxiety effectively. Whether the different fusion methods with gait and speech data can improve the prediction accurately is a valuable question, and would be focused on in our future research.

In a word, based on 88 subjects' speech and gait data, the experiment demonstrated that both speech and gait data could be used effectively to predict depression and anxiety through machine learning. To some extent, the recognition through gait performed better than the recognition through voice. These results could be a basis of both applications and future studies on the multi-source data fusion in anxiety and depression recognition.

Acknowledgement. This study was partially supported by the Key Research Program of the Chinese Academy of Sciences (No. ZDRW-XH-2019-4).

References

1. Moussavi, S., Chatterji, S., Verdes, E., Tandon, A., Patel, V., Ustun, B.: Depression, chronic diseases, and decrements in health: results from the World Health Surveys. Lancet **370**, 851–858 (2007)
2. Wittchen, H.-U.: Generalized anxiety disorder: prevalence, burden, and cost to society. Depress. Anxiety **16**, 162–171 (2002)
3. Hawton, K., Casañas i Comabella, C., Haw, C., Saunders, K.: Risk factors for suicide in individuals with depression: a systematic review. J. Affect. Disord. **147**, 17–28 (2013)
4. Sareen, J., et al.: Anxiety disorders and risk for suicidal ideation and suicide attempts: a population-based longitudinal study of adults. Arch. Gen. Psychiatry **62**, 1249–1257 (2005)
5. Baxter, A.J., Vos, T., Scott, K.M., Ferrari, A.J., Whiteford, H.A.: The global burden of anxiety disorders in 2010. Psychol. Med. **44**, 2363–2374 (2014)
6. Whiteford, H.A., et al.: Global burden of disease attributable to mental and substance use disorders: findings from the Global Burden of Disease Study. Lancet **2010**(382), 1575–1586 (2013)
7. Kroenke, K., Spitzer, R.L., Williams, J.B.: The PHQ-9, **16**, 606–613 (2001)
8. Radloff, L.S.: The CES-D scale: a self-report depression scale for research in the general population. Appl. Psychol. Meas. **1**, 385–401 (1977)
9. Marteau, T.M., Bekker, H.: The development of a six-item short-form of the state scale of the Spielberger State—Trait Anxiety Inventory (STAI). Br. J. Clin. Psychol. **31**, 301–306 (1992)
10. Löwe, B., et al.: Validation and standardization of the generalized anxiety disorder screener (GAD-7) in the general population. Med. Care **46**, 266–274 (2008)
11. Joshi, J., Goecke, R., Alghowinem, S., Dhall, A., Breakspear, M.: Multimodal assistive technologies for depression diagnosis and monitoring. J. Multimodal User Interfaces **7**, 217–228 (2013)
12. Sobin, C., Sackeim, H.A.: Psychomotor symptoms of depression. Am. J. Psychiatry **154**, 4–17 (1997)
13. Schrijvers, D., Hulstijn, W., Sabbe, B.G.: Psychomotor symptoms in depression: A diagnostic, pathophysiological and therapeutic tool. J. Affect. Disord. **109**, 1–20 (2008)

14. Bennabi, D., Vandel, P., Papaxanthis, C., Pozzo, T., Haffen, E.: Psychomotor retardation in depression: a systematic review of diagnostic, pathophysiologic, and therapeutic implications. Biomed. Res. Int. **2013**, 158746 (2013)
15. Ellgring, H., Scherer, K.R.: Vocal indicators of mood change in depression. J. Nonverbal Behav. **20**, 83–110 (1996)
16. Flint, A.J., Black, S.E., Campbell-Taylor, I., Gailey, G.F., Levinton, C.: Abnormal speech articulation, psychomotor retardation, and subcortical dysfunction in major depression. J. Psychiatr. Res. **27**, 309 (1993)
17. Kuny, S., Stassen, H.H.: Speaking behavior and voice sound characteristics in depressive patients during recovery. J. Psychiatr. Res. **27**, 289–307 (1993)
18. Mundt, J.C., Snyder, P.J., Cannizzaro, M.S., Chappie, K., Geralts, D.S.: Voice acoustic measures of depression severity and treatment response collected via interactive voice response (IVR) technology. J. Neurolinguistics **20**, 50–64 (2007)
19. Lemke, M.R., Wendorff, T., Mieth, B., Buhl, K., Linnemann, M.: Spatiotemporal gait patterns during over ground locomotion in major depression compared with healthy controls. J. Psychiatr. Res. **34**, 277–283 (2000)
20. Giannakakis, G., et al.: Stress and anxiety detection using facial cues from videos. Biomed. Signal Process. Control **31**, 89–101 (2017)
21. Laukka, P., et al.: In a nervous voice: acoustic analysis and perception of anxiety in social phobics' speech. J. Nonverbal Behav. J. Nonverbal Behav. **32**, 195 (2008)
22. Williamson, J.R., Quatieri, T.F., Helfer, B.S., Ciccarelli, G., Mehta, D.D.: Vocal and facial biomarkers of depression based on motor incoordination and timing. In: International Workshop (2014)
23. Zhao, N., et al.: See your mental state from your walk: Recognizing anxiety and depression through Kinect-recorded gait data. PLoS ONE **14**, e0216591 (2019)
24. Eyben, F., Wöllmer, M., Schuller, B.: openSMILE – the Munich versatile and fast open-source audio feature extractor. In: ACM International Conference on Multimedia (2010)
25. Eyben, F., Weninger, F., Gross, F., Schuller, B.: Recent developments in openSMILE, the munich open-source multimedia feature extractor. In: ACM International Conference on Multimedia (2013)
26. Cao, Z., Hidalgo, G., Simon, T., Wei, S.-E., Sheikh, Y.: OpenPose: realtime multi-person 2D pose estimation using part affinity fields (2018)
27. Schuller, B., et al.: The interspeech 2012 speaker trait challenge. In: Thirteenth Annual Conference of the International Speech Communication Association (2012)
28. Schuller, B., et al.: A Survey on perceived speaker traits: personality, likability, pathology, and the first challenge. Comput. Speech Lang. **29**, 100–131 (2015)
29. Cao, Z., Simon, T., Wei, S.E., Sheikh, Y.: Realtime multi-person 2D pose estimation using part affinity fields (2016)
30. Scherer, S., Stratou, G., Morency, L.P.: Audiovisual behavior descriptors for depression assessment. In: ACM on International Conference on Multimodal Interaction (2013)
31. Alghowinem, S., et al.: Multimodal depression detection: fusion analysis of paralinguistic, head pose and eye gaze behaviors, p. 1 (2016)
32. Llinas, J., Hall, D.L.: An introduction to multi-sensor data fusion. In: IEEE International Symposium on Circuits & Systems (1998)

Global Anomaly Detection Based on a Deep Prediction Neural Network

Ang Li[1], Zhenjiang Miao[1], Yigang Cen[1(✉)], Vladimir Mladenovic[2],
Liequan Liang[3], and Xinwei Zheng[3]

[1] Institute of Information Science, Beijing Jiaotong University,
Beijing 100044, China
lianghit@126.com, {zjmiao,ygcen}@bjtu.edu.cn
[2] Department of Information Technologies, University of Kragujevac,
32102 Čačak, Serbia
vladimir.mladenovic@ftn.kg.ac.rs
[3] Information Science School, Guangdong University of Finance Economics,
Guangzhou 510320, Guangdong Province, China
{lianglq,xinweizheng}@gdufe.edu.cn

Abstract. Abnormal event detection in public scenes is very important in recent society. In this paper, a method for global anomaly detection in video surveillance is proposed, which is based on a deep prediction neural network. The deep prediction neural network is built on the Convolutional Neural Network (CNN) and a variant of the Recurrent Neural Network (RNN)-Long Short-Term Memory (LSTM). Especially, the feature of a frame is the output of CNN, which is instead of the hand-crafted feature. First, the feature of a short video clip is obtained through CNN. Second, the predicted feature of the next frame can be gained by LSTM. Finally, the prediction error is introduced to detect that a frame is abnormal or not after the feature of the frame is achieved. Experimental results of global abnormal event detection show the effectiveness of our deep prediction neural network. Comparing with state-of-the-art methods, the model we proposed obtains superior detection results.

Keywords: CNN · LSTM · Prediction error · Global anomaly detection · Video surveillance

1 Introduction

Nowadays, anomaly detection is becoming a hot topic in the research filed of computer vision. With the significant improvement of public safety awareness and the reduction of surveillance equipment cost, it is very common that surveillance cameras are applied in public areas, such as markets, airports, museums, stadiums and train stations. In most surveillance systems, the cameras in public scenes are monitored by human operators, which is inefficient and prone to errors. Thus it is necessary to establish intelligent surveillance systems in modern society.

The established models can be classified into two main categories: hand-crafted feature based models and deep learning based models.

© Springer Nature Switzerland AG 2019
D. Milošević et al. (Eds.): HCC 2019, LNCS 11956, pp. 211–222, 2019.
https://doi.org/10.1007/978-3-030-37429-7_21

1.1   Hand-Crafted Feature Based Models

Established analysis models of a crowd in surveillance videos are divided into three broad classes [1]: (1) Macroscopic models based on the optical flow feature and the spatio-temporal gradient feature. (2) Microscopic models such as the particle filter (PF) framework, improving tracking models using crowded-level cues, the modeling methods to handle occlusions. (3) Abnormal event detection in crowds. During last decade, depending on the developments in the closely related fields, such as mathematical modeling, pattern recognition, computer vision, the research on abnormal event detection in crowds has evolved [2].

By utilizing holistic visual behavior understanding methods, a pixel-level model was described in [3]. By treating the individuals in a crowd as moving particles and utilizing the interaction force between every two particles to analyze the movement, a framework named as social force model was introduced in [4]. A model based on the descriptor of Histogram of Oriented Gradient (HOG) and a Gaussian classifier was presented in [5]. To tackle the limited labeled information of normal events and the unavailable information of abnormal events, a method depending on projection subspace associated with detectors was discovered in [6]. A motion influence map algorithm to describe human activities and detect anomaly was introduced in [7]. Optical flow can reflect the relative distance of moving objects at two different moments in pixel-level, which can provide the information of moving direction and speed. Based on the low-level feature, some models were presented in [8–12].

1.2   Deep Learning Based Models

In the field of computer vision, frameworks based on deep learning have shown their superior performance to tackle various vision tasks, such as image classification, semantic segmentation, object detection, activity recognition [13]. For anomaly detection, the deep learning architectures can be divided in to four main classes [14]: (1) Semi-supervised deep anomaly detection models only using labels of normal data instances. (2) Unsupervised deep anomaly detection models which detect outliers solely based on intrinsic properties of the data instances. (3) Deep hybrid models treating deep neural networks as feature extractors and utilizing traditional anomaly detection methods to detect abnormal events. (4) One-class neural network (OC-NN) methods inspired by kernel-based one-class classification, which combine the ability of deep networks to extract progressively rich representation of data with the one-class objective of creating a tight envelope around normal data.

With extracting motion feature by a convolutional auto-encoder and detecting anomaly by a one-class Support Vector Machine (SVM), a deep hybrid framework was introduced in [13]. A plug-and-play Convolutional Neural Network (CNN) keeping track of the changes in the CNN feature across time was presented in [15]. To approach the problem how to perceive meaningful activities in a long video, an auto-encoder based generative model for regular motion patterns was discovered in [16]. To learn the motion for the ordinary moments and regularity of appearance, an auto-encoder integrating a Convolutional Long Short-Term Memory (ConvLSTM) with CNN was proposed in [17]. Inspired by Generative Adversarial Networks (GAN) for training deep neural networks, an end-to-end deep network to detection anomaly in videos was presented in [18]. A method using Fully Convolutional Neural Networks (FCN) and temporal data to detect abnormal event in scenes was described in [19]. Based on a video prediction framework, a new baseline for anomaly detection was introduced [20]. The Extreme Learning Machine (ELM)-a single layer neural network to detect abnormal events was presented in [21].

In this paper, we propose a deep learning based framework to detect global abnormal events. Our deep prediction neural network contains three main parts corresponding to three main steps in the detection task. First, depending on the superior ability of CNN, we choose a pre-trained convolutional neural network to extract the features of frames instead of the hand-crafted features. Second, utilizing the continuance of surveillance videos, we design a sub-framework to predict the feature of the next frame based on LSTM. Finally, to measure the normalness of a testing frame, we introduce a prediction error detection criterion. In previous deep learning based works, auto-encoders and ConvLSTM frameworks are universal, such as [13, 16, 17], and others are complex to realize, such as [18–20]. Based on a prediction strategy, our model is built on the common neutral networks of CNN and LSTM, which achieves excellent performance in the detection of global abnormal events.

The rest of the paper is organized as follows. In Sect. 2, the model of global abnormal events is presented. In Sect. 3, our experimental results and comparisons with state-of-the-art methods are reported. Finally, some conclusions are described in Sect. 4.

2 Global Anomaly Detection

2.1 Convolutional Neural Network

Many previous works about global anomaly detection are based on hand-crafted feature. In recent years, researchers have shown the strong ability of Convolutional Neural Networks (CNN) to represent objects, such as languages, images, videos. Also, it has been verified that CNN based models can not only obtain superior performance in the tasks of visual recognition, in which CNN frameworks are trained, but also apply to other relevant tasks based on the extracted features [22]. By backpropagation through several layers of convolutional filters, some CNN models have pre-trained on big datasets, e.g., ImageNet including 1.2M images [23]. The pre-trained models contain InceptionNet, AlexNet, ResNet, VGGNet and so on. In our work, we choose one pre-trained CNN framework of VGGNet-VGG16. As shown in Fig. 1, VGG16 contains six sub-blocks: five convolution blocks and one classifier block. In this paper, we only adopt the first five blocks and fine-tune the last convolution block (the blue one). We take the output of convolution block 5 as the extracted feature of a frame. Instead of the softmax classifier block, we employ the Long Short-Term Memory (LSTM) framework and a classifier based on the prediction error.

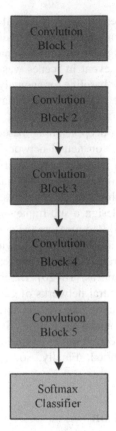

Fig. 1. The architecture of VGG16. Red and blue blocks denote as convolution parts of VGG16, yellow block denotes the pre-trained classifier of VGG16. (Color figure online)

2.2 Long Short-Term Memory Framework

As a special Recurrent Neural Network (RNN) structure, the sequence model Long Short-Term Memory (LSTM) has been proved to be powerful and stable for modeling long-range dependencies [24]. LSTM does not suffer from optimization hurdles such as the gradient vanishing problem and the gradient exploding problem. Depending on the outstanding characters, the LSTM based models have been used to tackle many difficult problems, e.g., language modeling and transaction, speech synthesis, handwriting recognition and generation, analysis of audio and video data [25].

The most important component of LSTM is the memory cell c_t, which is the major innovation of LSTM. The function of c_t is to accumulate the state information. Based on the state of several self-parameterized controlling gates, the memory cell can be written, accessed and cleared. The input gate i_t is to control the new information to be accumulated into the memory cell when a new input appears. The forget gate f_t is to control the past status in the memory cell to be forgotten. The output gate o_t is to control whether c_t, i.e., the latest output of the memory cell, will be propagated to the final state h_t The relevant equations are described as follows:

$$i_t = \sigma(W_{xi}x_t + W_{hi}h_{t-1} + W_{ci} \circ c_{t-1} + b_i)$$
$$f_t = \sigma(W_{xf}x_t + W_{hf}h_{t-1} + W_{cf} \circ c_{t-1} + b_f)$$
$$c_t = f_t \circ c_{t-1} + i_t \circ \tanh(W_{xc}x_t + W_{hc}h_{t-1} + b_c) \tag{1}$$
$$o_t = \sigma(W_{xo}x_t + W_{ho}h_{t-1} + W_{co} \circ c_t + b_o)$$
$$h_t = o_t \circ \tanh(c_t)$$

where '$\circ$' denotes the Hadamard product, x_t denotes the input vector, W is the weight matrix, b is the bias weight, σ and tanh denote activation functions.

2.3 Deep Prediction Neural Network

The feature of a frame can be obtained through a Convolutional Neural Network (CNN). In general, different images have different features and vice versa. Assume that we have T consecutive frames $\{X_1, X_2, \ldots, X_T\}$, and the corresponding feature is $\{F_1, F_2, \ldots, F_T\}$. In our work, we choose VGG16 as the CNN framework to extract features of frames. And a sequence model based on Long Short-Term Memory (LSTM) is integrated with VGG16. The architecture is shown in Fig. 2. $\{y_1, y_2, \ldots, y_T\}$ is the output sequence of LSTM. Especially, when X_T is put into the framework in Fig. 2, the corresponding output y_T is the prediction of the feature of X_{T+1}, i.e., $\hat{F}_{T+1}$.

In the task of global abnormal event detection, only the frames in normal scenes can be obtained in the training stage, which are similar to (a) in Fig. 3. It is easy to be found, the difference between normal and abnormal frames is large. So the neural network in Fig. 2 has a strong ability to predict the feature of normal frames, the performance of which to predict the features of abnormal frames is weak. To distinguish the normal and abnormal frames, we introduce a prediction error (PE) function as follows:

$$PE = \|F_T - \hat{F}_T\|_2 / \|F_T\|_2 \tag{2}$$

The frame X_T, corresponding to F_T and $\hat{F}_T$, is detected as normal if the following criterion is satisfied

$$PE < \tau \tag{3}$$

where τ is a threshold that defined by users to control the sensitivity of the global anomaly detection method. The end-to-end deep prediction neural network based on VGG16 and LSTM is illustrated in Fig. 4.

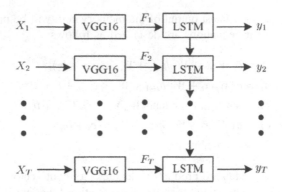

Fig. 2. The architecture of VGG16 combined with LSTM.

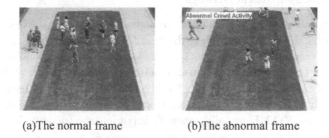

(a)The normal frame   (b)The abnormal frame

Fig. 3. Two different frames in global anomaly detection.

3 Experimental Results

We evaluate our method on the well-known UMN dataset [26]. In UMN dataset, the total frame number is 7739 and the resolution of a frame is 320×240. There are three different crowded scenes in the dataset, which are named as lawn, indoor and plaza respectively. In the dataset, the normal events are people walking randomly in a specific scene, and the abnormal events are people running away in the scene at the same time. In the training stage, we choose the first 400 normal frame in each scene to train and fine-tune the deep prediction neural network. Especially, the value of T in $\{X_1, X_2, \ldots, X_T\}$ is set as 4, and the number of the first frame to be detected is 5 in each scene.

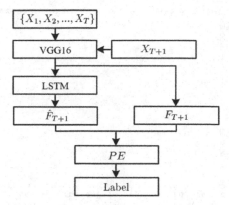

Fig. 4. The end-to-end deep prediction neural network. Label: +1 denotes the detected frame is normal and −1 denotes the detected frame is abnormal.

3.1 Detection in the Lawn Scene

The video sequence of the lawn scene contains 1453 frames in total (i.e., there are 1449 frames to be detected). Two different events in the lawn scene are described in Fig. 5. The detection result in the lawn scene is shown in Fig. 6. The Receiver Operating Characteristic (ROC) curve is provided in Fig. 7, and the Area Under the Curve (AUC) is 99.97%.

(a)The normal event   (b)The abnormal event

Fig. 5. Two different events in the lawn scene.

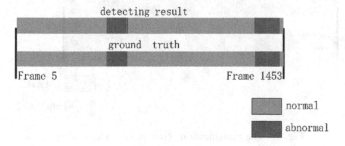

Fig. 6. The classification result in the lawn scene.

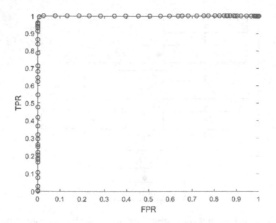

Fig. 7. The ROC curve in the lawn scene.

3.2 Detection in the Indoor Scene

The video sequence of the indoor scene contains 4144 frames in total (i.e., there are 4140 frames to be detected). Two different events in the indoor scene are described in Fig. 8. The detection result in the indoor scene is shown in Fig. 9. The ROC curve is provided in Fig. 10, and the AUC is 99.94%.

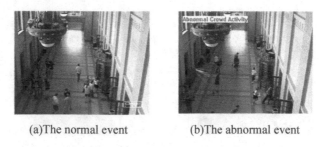

(a)The normal event   (b)The abnormal event

Fig. 8. Two different events in the indoor scene.

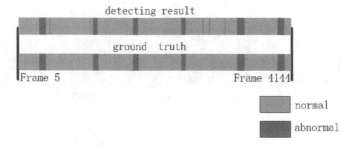

Fig. 9. The classification result in the indoor scene.

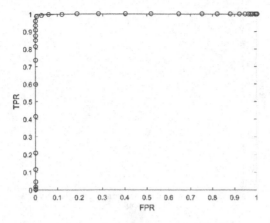

Fig. 10. The ROC curve in the indoor scene.

3.3 Detection in the Plaza Scene

The video sequence of the plaza scene contains 2142 frames in total (i.e., there are 2138 frames to be detected). Two different events in the plaza scene are described in Fig. 11. The detection result of the plaza scene is shown in Fig. 12. The ROC curve is provided in Fig. 13, and the AUC is 99.93%.

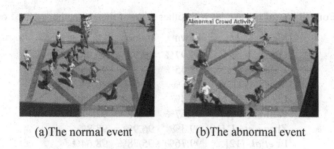

(a)The normal event   (b)The abnormal event

Fig. 11. Two different events in the plaza scene.

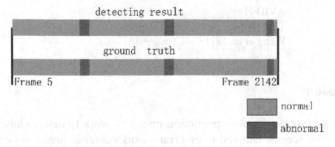

Fig. 12. The classification result in the plaza scene.

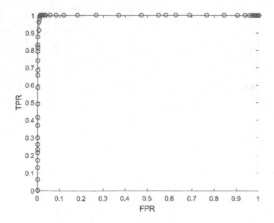

Fig. 13. The ROC curve in the plaza scene

3.4   Comparison

The performances of our proposed deep prediction neural network and the state-of-the-art methods are shown in Table 1. As shown in the table, the AUC of our end-to end global anomaly detection framework outperforms of other methods including hand-crafted feature based models and deep learning based models.

Table 1. Comparison of our model with other state-of-the-art methods based on AUC.

Method	Lawn	Indoor	Plaza
Ours	**99.97%**	**99.94%**	**99.93%**
HOFO [9]	98.45%	90.37%	98.15%
Sparse [11]	99.5%	97.5%	96.4%
Lee et al. [7]	99.4%	90.9%	98.1%
Patil et al. [8]	98.67%	93.68%	97.11%
Zhang et al. [10]	99.3%	96.9%	98.8%
Li et al. [12]	99.76%	95.78%	98.64%
NN [11]		84%	
Optical Flow [4]		93%	
SF [4]		96%	
AVID [18]		99.6%	
TCP [15]		98.8%	
Wang et al. [21]		99.0%	

4   Conclusions

In this work, we proposed a deep prediction neural network to detect global abnormal events in crowded scenes. Our end-to-end framework contains three main components: the Convolutional Neural Network (CNN) VGG16 is to extract the features of frames, the part of Long Short-Term Memory (LSTM) is to predict the feature of the next frame based on a short sequence of frames, and the prediction error function is the key to determine a detected frame being abnormal or not. The proposed model has been tested on UMN dataset with satisfying results about abnormal event detection in global-frame scale.

Acknowledgments. This work is supported by the National Key Technology R&D Program of China (no. 2012BAH01F03), NSFC (nos. 61572067,61872034, 61672089, 61273274, and 61572064), the Science and Technology Program of Guangzhou (201804010271), the Natural Science Foundation of Guizhou Province ([2019]1064).

References

1. Thida, M., Yong, Y.L., Climent-Pérez, P., Eng, H.-l., Remagnino, P.: A literature review on video analytics of crowded scenes. In: Atrey, P.K., Kankanhalli, M.S., Cavallaro, A. (eds.) Intelligent Multimedia Surveillance, pp. 17–36. Springer, Heidelberg (2013). https://doi.org/10.1007/978-3-642-41512-8_2
2. Sjarif, N.N.A., Shamsuddin, S.M., Hashim, S.Z.: Detection of abnormal behaviors in crowd scene: a review. Int. J. Adv. Soft Comput. Appl. **4**(1), 1–33 (2012)
3. Kosmopoulos, D., Chatzis, S.P.: Robust visual behavior recognition. IEEE Signal Process. Mag. **27**(5), 34–45 (2010)
4. Mehran, R., Oyama, A., Shah, M.: Abnormal crowd behavior detection using social force model. In: IEEE Conference on Computer Vision and Pattern Recognition, Miami, pp. 935–942. IEEE (2009)
5. Amraee, S., Vafaei, A., Jamshidi, K., Adibi, P.: Anomaly detection and localization in crowded scenes using connected component analysis. Multimed. Tools Appl. **77**(12), 14767–14782 (2018)
6. Tziakos, I., Cavallaro, A., Xu, L.Q.: Local abnormality detection in video using subspace learning. In: 7th IEEE International Conference on Advanced Video and Signal Based Surveillance, Boston, pp. 519–525. IEEE (2010)
7. Lee, D.G., Suk, H.I., Park, S.K., Lee, S.W.: Motion influence map for unusual human activity detection and localization in crowded scenes. IEEE Trans. Circuits Syst. Video Technol. **25**(10), 1612–1623 (2015)
8. Patil, N., Biswas, P.K.: Global abnormal events detection in surveillance video—a hierarchical approach. In: Sixth International Symposium on Embedded Computing and System Design, Patna, pp. 217–222. IEEE (2016)
9. Wang, T., Snoussi, H.: Detection of abnormal visual events via global optical flow orientation histogram. IEEE Trans. Inf. Forensics Secur. **9**(6), 988–998 (2014)
10. Zhang, Y., Lu, H., Zhang, L., Ruan, X.: Combining motion and appearance cues for anomaly detection. Pattern Recognit. **51**, 443–452 (2016)
11. Cong, Y., Yuan, J., Liu, J.: Sparse reconstruction cost for abnormal event detection. In: IEEE Conference on Computer Vision and Pattern Recognition, Colorado, pp. 3449–3456. IEEE (2011)
12. Li, A., Miao, Z., Cen, Y., Cen, Y.: Anomaly detection using sparse reconstruction in crowded scenes. Multimed. Tools Appl. **76**(24), 26249–26271 (2017)
13. Tran, H.T., Hogg, D.: Anomaly detection using a convolutional winner-take-all autoencoder. In: Proceedings of the British Machine Vision Conference. British Machine Vision Association, London (2017)

14. Chalapathy, R., Chawla, S.: Deep learning for anomaly detection: a survey. Eprint Arxiv (2019)
15. Ravanbakhsh, M., Nabi, M., Mousavi, H., Sangineto, E., Sebe, N.: Plug-and-play cnn for crowd motion analysis: an application in abnormal event detection. In: IEEE Winter Conference on Applications of Computer Vision, Lake Tahoe, pp. 1689–1698. IEEE (2018)
16. Hasan, M., Choi, J., Neumann, J., Roy-Chowdhury, A.K., Davis, L.S.: Learning temporal regularity in video sequences. In: Proceedings of the IEEE Conference on Computer Vision and Pattern Recognition, Las Vegas, pp. 733–742. IEEE (2016)
17. Luo, W., Liu, W., Gao, S.: Remembering history with convolutional LSTM for anomaly detection. In: IEEE International Conference on Multimedia and Expo, Hong Kong, pp. 439–444. IEEE (2017)
18. Sabokrou, M., et al.: AVID: adversarial visual irregularity detection. In: Jawahar, C.V., Li, H., Mori, G., Schindler, K. (eds.) ACCV 2018. LNCS, vol. 11366, pp. 488–505. Springer, Cham (2019). https://doi.org/10.1007/978-3-030-20876-9_31
19. Sabokrou, M., Fayyaz, M., Fathy, M., Moayed, Z., Klette, R.: Deep-Anomaly: fully convolutional neural network for fast anomaly detection in crowded scenes. Comput. Vis. Image Underst. **172**, 88–97 (2018)
20. Liu, W., Luo, W., Lian, D., Gao, S.: Future frame prediction for anomaly detection–a new baseline. In: Proceedings of the IEEE Conference on Computer Vision and Pattern Recognition, Salt Lake City, pp. 6536–6545. IEEE (2018)
21. Wang, S., Zhu, E., Yin, J., Porikli, F.: Video anomaly detection and localization by local motion based joint video representation and OCELM. Neurocomputing **277**, 161–175 (2018)
22. Sharif Razavian, A., Azizpour, H., Sullivan, J., Carlsson, S.: CNN features off-the-shelf: an astounding baseline for recognition. In: Proceedings of the IEEE Conference on Computer Vision and Pattern Recognition Workshops, Columbus, pp. 806–813. IEEE (2014)
23. Russakovsky, O., et al.: ImageNet large scale visual recognition challenge. Int. J. Comput. Vis. **115**(3), 211–252 (2015)
24. Xingjian, S.H.I., Chen, Z., Wang, H., Yeung, D.Y., Wong, W.K., Woo, W.C.: Convolutional LSTM network: a machine learning approach for precipitation nowcasting. In: Advances in Neural Information Processing Systems, Montréal, pp. 802–810 (2015)
25. Greff, K., Srivastava, R.K., Koutník, J., Steunebrink, B.R., Schmidhuber, J.: LSTM: A search space odyssey. IEEE Trans. Neural Netw. Learn. Syst. **28**(10), 2222–2232 (2016)
26. UMN: Unusual crowd activity dataset of University of Minnesota, department of computer science and engineering (2006). http://mha.cs.umn.edu/movies/crowd-activity-all.avi

SCHONA: A Scholar Persona System Based on Academic Social Network

Ronghua Lin, Chengjie Mao$^{(\boxtimes)}$, Chaodan Mao, Rui Zhang, Hai Liu, and Yong Tang

South China Normal University, Guangzhou, Guangdong 510631, China
{rhlin,doris.maocd,liuhai}@m.scnu.edu.cn,
maochj@qq.com, rayzhang@cmbchina.com

Abstract. Recently, social network including academic social network has developed rapidly. It is a challenge to utilize massive academic data including academic social network data and academic achievement data to analyze and mine scholars' important information such as behavioral characteristics and research interests. In this paper, we present a scholar persona system SCHONA which is composed of two parts, data collection and scholar labels generation. It collects three types of data first and finally generates the labels which can accurately represent scholars by extensively using big data analysis methods such as Word2Vec, K-means and TextRank.

Keywords: Scholar persona · Academic social network · SCHOLAT

1 Introduction

With the rapid development of social network [1], the method of interaction and communication between human beings has been transferred from offline to online. People use social media not only to make new friends but also to maintain the friendships of real world in the virtual world. Due to the convenience and practicality of social network, the number of users on social media has increased dramatically. According to the global digital reports from 2016 to 2019 released by Hootsuite and We Are Social, the number of active social media users is increasing year by year (as shown in Fig. 1).

At the same time, academic social media [2] for scholars has also developed rapidly, such as Academia, ResearchGate, SCHOLAT and so on. Due to the frequent interactions between scholars and academic social media, a large amount of academic social network data will be generated every day. In addition, the academic achievement data such as academic paper data have also been exploding with the rise and development of academia. It is crucial for academic social media that how to effectively analyze these academic data including academic social network data and academic achievement data to mine the important information of scholars such as behavioral characteristics and research interests, which can be further used for personalized recommendations [3].

D. Milošević et al. (Eds.): HCC 2019, LNCS 11956, pp. 223–232, 2019.
https://doi.org/10.1007/978-3-030-37429-7_22

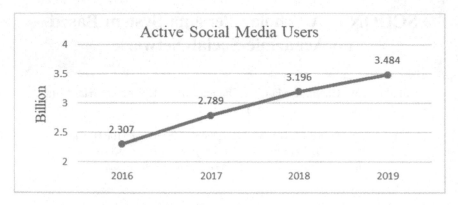

Fig. 1. The number of active social media users in the past 4 years.

Persona [4], i.e. user tagging technique which models the users by using massive real and reliable data can be an available and effective method to address the problem above. There have been several studies on persona already. Zhu et al. [5] collected the user behavior logs and analyzed this data to perform persona process by using LDA topic model [6]. Lin et al. [7] also used LDA to construct a user model of the fusion interest field by analyzing and predicting the user's interests and hobbies.

However, to the best of our knowledge, there are few studies that combine persona with academic social network. In this paper, we propose a scholar persona system named SCHONA based on the academic social media SCHOLAT, which mainly utilizes the user personal profile data, user behavior data from SCHOLAT and academic achievement data (mainly academic paper data) from CNKI (China National Knowledge Infrastructure) to analyze and mine users' important information such as behavioral patterns, attribute characteristics, research interests and so on. We employ three different tools which are JDBC, Flume and Scrapy to collect and manage three types of data above, respectively. Several big data analysis methods which are Word2Vec, K-means and TextRank will then be employed to generate scholar labels for scholars modeling. Our experimental results demonstrate the availability and effectiveness of our system.

In the following, we first review the related work in Sect. 2. We subsequently introduce the detailed design of our system in Sect. 3. In Sect. 4, we describe some implemental details of our methods. We evaluate our system in Sect. 5 and finally conclude our work in Sect. 6.

2   Related Work

Since users of and data generated from social network have increased rapidly, it is vital to analyze and mine these massive data to obtain users' important information such as behavioral characteristics, interests and hobbies and so on for further personalized recommendations.

Recently, there have been several studies on users modeling or user analysis. Zhao et al. [8] employed matrix factorization techniques to model users' behaviors, and further built a topic recommender for predicting users' topical interests. In this paper, we try to model scholars comprehensively by using personal profile data, user behavior data and academic achievement data. Our work can be also used for further user recommendations.

Kazai et al. [9] utilized users' locations and their profiles based on traditional social media such as Facebook and Twitter to automatically infer users' interests, which then are used for news and blog personalized recommendations. It is no doubt that it can be overlapped with our work. Our system is based on academic social network, using more abundant and dimensional academic social network data to analyze and mine users' important information including behavioral characteristics, research interests which can be utilized as bases for academic news or academic papers recommendations.

Word2Vec [10] released by Google in 2013, is a group of related models used for word embeddings [11]. Hu et al. [12] presented a feature extraction method based on the fusion of Word2Vec and TF-IDF [13], and tried to establish a user profile by analyzing the user's search logs in a period of time. In our work, we utilize Word2Vec to build word embedding model for three types of data we collected, which makes our system more accurate and reasonable.

Unsupervised learning such as K-means clustering algorithm has been widely used in big data domain [14]. In 2016, Wang et al. [15] built an unsupervised system to capture dominating user behaviors from clickstream data. They used two large-scale clickstream traces from real social network for evaluation and the results showed that the system they presented could effectively identify previously unknown user behaviors. In this paper, we use K-means for clustering a large number of words of scholars' data and choose the words corresponding to the cluster centers as initial scholars' labels.

Big data analysis methods can often be overlapped for each other to address a real problem. Wen et al. [16] used Word2Vec to calculate the similarity between words and then employed TextRank to build a candidate keywords graph model for improving performance of keyword extraction. Our system SCHONA also relies on several big data analysis methods which are Word2Vec, K-means and TextRank.

Reasonable and effective user analysis can be applied to a wide variety of recommendation systems. Li et al. [17] put forward the music recommendation based on user behavior analysis and user emotion extraction by using Word2Vec and clustering. It demonstrates the importance of user modeling by using massive user's data, which is in line with our main work. We utilize a large amount of scholars' data for scholar persona by extensively employing big data analysis methods.

3   System Design

3.1   Problem Description

Since academic social network has developed rapidly, a large amount of academic data has been generated. However, it is a huge challenge to analyze such a large number of data for mining scholars' important information such as research interests and so on which can be further utilized for various recommendations. To our knowledge, there is no such system used for scholar labeling.

Due to the massive academic data and the difficulty of analyzing them to capture and extract scholars' personal information, behavioral characteristics, research interests and so on, we try to design a simple and convenient scholar persona system SCHONA which can collect scholar personal profile data, user behavior data and academic achievement data regularly, and subsequently analyze them for scholar labeling by using big data analysis methods extensively. We design our system SCHONA in detail in Sect. 3.2.

3.2   A Subsection Sample

As discussed above, we overview our system SCHONA which is shown in Fig. 2. In a nutshell, SCHONA is mainly composed of two components, which are data collection and scholar labels generation.

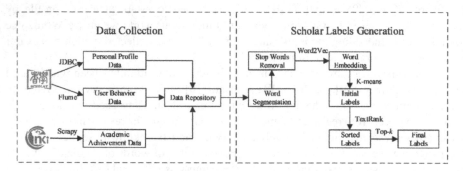

Fig. 2.   Overview of SCHONA.

Data Collection. Data is the core and most important part of our system SCHONA. Scholar labels generation must rely on a large amount of data for analysis. To this end, we utilize scholar personal profile data and user behavior data from academic social networking site SCHOLAT. Besides that, we crawl the academic paper data from CNKI as academic achievement data. All of these three types of data will be stored in our data repository.

For collecting user personal profile data, we connect to SCHOLAT's database and query the data we need by using JDBC which is a Java API for accessing various relational databases. We employ two different collection methods including full collection and incremental collection which will be further introduced in Sect. 4. Although we can collect the complete user personal profile data, however, they are critical to SCHOLAT and are not allowed to be widely publicized. Therefore, we desensitize some user privacy data such as email and address by encryption [18]. Finally, the user personal profile data we obtain mainly includes users' name, gender, degree, work unit, academic title, introduction and so on.

Since user behavior data of SCHOLAT are mainly stored in log files, we utilize Flume to collect and manage them, where Flume is a distributed, reliable and available service for efficiently collecting, aggregating and managing large amounts of log data. SCHOLAT has accumulated a large number and variety of log files since it was officially released in 2008. We mainly collect those for nearly three years which are about academic news posted on SCHOLAT, academic paper data stored in SCHOLAT and keywords that are searched by SCHOLAT users.

For academic achievement data, we mainly use Scrapy which is a Python crawler framework to crawl academic paper data from CNKI. Since CNKI has adopted an anti-crawling strategy [19] (i.e. one IP address cannot request data from CNKI frequently in a short period time), we create an IP pool to automatically switch IP address every once in a while. We mainly crawl the titles, abstracts and keywords of academic papers from CNKI to analyze scholars' research fields and research interests.

Scholar Labels Generation. After collecting three types of data above, we need to clean these data first. For the static attributes such as name and gender in user personal profile data, we do not need to perform additional secondary processing. However, some attributes like work unit, degree and academic title can be changed, so we also try to find out the changes of these dynamic attributes. For long text attributes such as user introduction, we firstly perform word segmentation [20] and stop words removal [21]. Finally, we get the corpus of scholars' introductions $Corpus_{personal}$.

For user behavior data, we also perform word segmentation and stop words removal for long text such as academic news posted on SCHOLAT, the titles and abstracts of academic papers stored in SCHOLAT. However, the keywords searched by SCHOLAT users do not need any other processing since they are concise enough. As the result, we get the corpus of user behavior data $Corpus_{behavior}$.

Our processing on academic paper data crawled from CNKI is relatively simple since most of them are long text. We segment the titles and abstracts and subsequently remove the stop words. It is unnecessary for further processing on keywords in academic papers because they are also concise enough. Finally, we get another corpus of academic paper data $Corpus_{academic}$.

We then utilize these three corpus ($Corpus_{personal}$, $Corpus_{behavior}$, $Corpus_{academic}$) to train for a word embedding model by using Word2Vec. Its main purpose is to convert the words into numerical vectors which can be used to calculate the similarity between these words [22]. We afterward employ K-means algorithm to cluster the word vectors [23] of the same scholar and select the words corresponding to the cluster centers as the initial scholar labels. We choose the parameter k of K-means algorithm by using a heuristic method that when the corresponding data of the scholar are more abundant, we choose a larger k and vice versa.

We finally sort the initial scholar labels by using TextRank [24] which is a graph-based ranking model for text processing. The formula for TextRank is shown in Eq. 1.

$$Rank(w_i) = (1 - d) + d \times \sum_{w_j \in in(w_i)} \frac{weight_{ji}}{\sum_{w_k \in out(w_j)} weight_{jk}} \times Rank(w_j), \qquad (1)$$

where $Rank(w_i)$ is the rank value of word w_i and d is a damping factor which can be set between 0 and 1. In this paper, we set d to 0.85. $weight_{ji}$ is the weight between the nodes w_i and w_j in the graph. $in(w_i)$ represents the set of nodes pointing to node w_i while $out(w_i)$ represents the set of nodes pointed to by w_i. We set default values for each initial labels first and then calculate iteratively by TextRank until convergence.

We get the sorted scholar labels after using TextRank. If the scholar has more labels, we will finally select top-k sorted labels while we will select all sorted labels if the number of labels is small.

4 Implementation

According to the description in Sect. 3, data collection in our system SCHONA is implemented in two different programming languages, Java and Python, while for scholar labels generation, we mainly use Python for development and implementation.

When collecting personal profile data, we use JDBC in Java environment to connect the SCHOLAT database and query the user personal profile data. At the first time, we carry out the full collection, that is, import the personal profile data of all the users in SCHOLAT. When the next collection is performed, we will incrementally collect those have not been collected after last collection.

For the collection of user behavior data, as we mentioned in Sect. 3.2, we use Flume to collect SCHOLAT log files and import them into our data repository. We then employ regular expressions to match and select the data we want.

We use Scrapy framework implemented in Python to crawl academic paper data from CNKI. We create an IP pool to address CNKI's anti-crawling strategy. Our approach is shown in Algorithm 1. We firstly get the first IP address from the IP pool created in advance, and subsequently use it to request academic paper data from CNKI frequently. When this IP address no longer gets the data from CNKI, it indicates that the number of accessing CNKI has reached the upper limit in a short period of time. Therefore, the next IP address is obtained from the IP pool for crawling. When all the IP address in IP pool are used up, we stop crawling and create a new IP pool for the next crawling if necessary.

In scholar labels generation phase, we use jieba word segmentation tool implemented in Python to segment the long text. As described in Sect. 3.2, we utilize Word2Vec to train a word embedding model, and then employ K-means algorithm to initially generate scholar labels and finally sort these labels using TextRank. All of them are developed and implemented based on Python 3.6.

Algorithm 1 Academic paper data crawling from CNKI by using Scrapy.

Input:
  The IP pool we created in advance, Set_{IP};
  The initial URL to be crawled, url_{ini};
1: $ip = getFirstIP(Set_{IP})$; /*get the first IP address from IP pool*/
2: $url = url_{ini}$;
3: **while** $TRUE$ **do**
4:   $data = request(ip, url)$; /*use ip to crawl the paper data corresponding to url*/
5:   **if** $data$ is not null **then**
6:    $saveToRepo(data)$; /*save the data to our data repository*/
7:    $url = getNextURL()$; /*get next URL to be crawled*/
8:   **else**
9:    $ip = getNextIP(Set_{IP})$; /*get next IP address from the IP pool*/
10:    **if** ip is null **then**
11:     $break$;
12:    **end if**
13:   **end if**
14: **end while**

---

5 Evaluation

As we mentioned above, before scholar labels generation, we have collected three types of data. The sizes of these data are roughly shown in Table 1.

Table 1. The sizes of three types of data.

Types	Size
Personal profile data	103,216
User behavior data	244,638
Academic achievement data	>2,000,000

We would like to explain that although the amount of academic achievement data is much larger than the first two types, most of them may be useless because they are not the academic achievements of the 103,216 scholars. Therefore, when we carry out the next step i.e. scholar labels generation, we will take additional processing with these academic achievement data and delete those useless.

We utilize these data we collected to extensively evaluate our scholar persona system SCHONA. However, due to the limited space of this paper, we only select the final labels of three scholars to display. In order to protect the privacy of these scholars, we use different but unique id to represent them respectively, instead of directly displaying their name. The result is shown in Table 2.

Table 2. The final labels of three scholars.

Scholar id	Labels
17	Professor
	South China Normal University
	Temporal Database
	Social Network
1463	Lecturer
	South China Normal University → Guangdong Pharmaceutical University
	WEB System Architecture
	Community Detection
508	Associate Professor
	Sun Yat-sen University
	Recommendation System
	Data Mining

We show the final labels of three scholars generated by our scholar persona system SCHONA. We mainly display their academic titles, work units and some research interests. From the second labels of the scholar whose id is 1463, we see that his (her) work unit has changed from "South China Normal University" to "Guangdong Pharmaceutical University" in the past of 3 years.

With the scholar labels generated by our system SCHONA, they can be further used for personalized recommendations or scientific research collaborations according to work units and research interests, which is important for the development of academic social network.

6   Conclusion

With the rapid development of academic social network, it is crucial for modeling scholars by analyzing massive academic data. In this paper, we propose a scholar persona system named SCHONA based on academic social networking site SCHOLAT. It mainly consists of two parts which are data collection and scholar labels generation. In data collection phase, we mainly collect three types of data including personal profile data and user behavior data from SCHOLAT, and academic achievement data from CNKI by using JDBC, Flume and Scrapy, respectively. In scholar labels generation phase, we firstly carry out additional processing, that is, word segmentation and stop words removal for the raw data if necessary. We then (1) use processed data to train a word embedding model by using Word2Vec, (2) generate initial scholar labels using K-means and (3) finally sort them using TextRank algorithm. Our work can be further utilized for various recommendation systems.

Acknowledgments. Our work is supported by the National Natural Science Foundation of China (No. U1811263), Science and Technology Project of Guangzhou (No. 201807010043), the Research and Reform Project of Higher Education of Guangdong Province, Outcome-based Education on Data Science Talent Cultivation Model Construction and Innovation Practice, and Natural Science Foundation of Guangdong (No. 2016A030313441).

References

1. Van Noorden, R.: Online collaboration: scientists and the social network. Nature **512**(7531), 126–129 (2014)
2. Ovadia, S.: ResearchGate and Academia.edu: academic social networks. Behav. Soc. Sci. Libr. **33**(3), 165–169 (2014)
3. Elkahky, A.M., Song, Y., He, X.: A multi-view deep learning approach for cross domain user modeling in recommendation systems. In: Proceedings of the 24th International Conference on World Wide Web, pp. 278–288. International World Wide Web Conferences Steering Committee (2015)
4. Perlman, H.H.: Persona: Social Role and Personality. University of Chicago Press (2018)
5. Zhu, H., Chen, E., Xiong, H., et al.: Mining mobile user preferences for personalized context-aware recommendation. ACM Trans. Intell. Syst. Technol. (TIST) **5**(4), 58 (2015)
6. Wei, Y., Jia, S., Wang, Q., et al.: The application of LDA model on user profile. In: 2017 3rd International Conference on Economics, Social Science, Arts, Education and Management Engineering (ESSAEME 2017). Atlantis Press (2017)
7. Liu, Q., Niu, K., He, Z., et al.: Microblog user interest modeling based on feature propagation. In: 2013 Sixth International Symposium on Computational Intelligence and Design, pp. 383–386, vol. 1. IEEE (2013)
8. Zhao, Z., Cheng, Z., Hong, L., Chi, Ed.H.: Improving user topic interest profiles by behavior factorization. In: Proceedings of the 24th International Conference on World Wide Web (WWW 2015), pp. 1406–1416. International World Wide Web Conferences Steering Committee, Republic and Canton of Geneva, Switzerland (2015). https://doi.org/10.1145/2736277.2741656
9. Kazai, G., Yusof, I., Clarke, D.: Personalised news and blog recommendations based on user location, Facebook and Twitter user profiling. In: Proceedings of the 39th International ACM SIGIR conference on Research and Development in Information Retrieval (SIGIR 2016), pp. 1129–1132. ACM, New York (2016). https://doi.org/10.1145/2911451.2911464
10. Cui, P., Wang, X., Pei, J., et al.: A survey on network embedding. IEEE Trans. Knowl. Data Eng. **31**, 833–852 (2018)
11. Lilleberg, J., Zhu, Y., Zhang, Y.: Support vector machines and Word2Vec for text classification with semantic features. In: 2015 IEEE 14th International Conference on Cognitive Informatics & Cognitive Computing (ICCI* CC), pp. 136–140. IEEE (2015)
12. Hu, J., Jin, F., Zhang, G., et al.: A user profile modeling method based on Word2Vec. In: 2017 IEEE International Conference on Software Quality, Reliability and Security Companion (QRS-C), pp. 410–414. IEEE (2017)
13. Christian, H., Agus, M.P., Suhartono, D.: Single document automatic text summarization using term frequency-inverse document frequency (TF-IDF). ComTech Comput. Math. Eng. Appl. **7**(4), 285–294 (2016)
14. Dhanachandra, N., Manglem, K., Chanu, Y.J.: Image segmentation using K-means clustering algorithm and subtractive clustering algorithm. Procedia Comput. Sci. **54**, 764–771 (2015)
15. Wang, G., Zhang, X., Tang, S., Zheng, H., Zhao, B.Y.: Unsupervised clickstream clustering for user behavior analysis. In: Proceedings of the 2016 CHI Conference on Human Factors in Computing Systems (CHI 2016), pp. 225–236. ACM, New York. https://doi.org/10.1145/2858036.2858107
16. Wen, Y., Yuan, H., Zhang, P.: Research on keyword extraction based on Word2Vec weighted TextRank. In: 2016 2nd IEEE International Conference on Computer and Communications (ICCC), pp. 2109–2113. IEEE (2016)

17. Li, Q., Liu, D.: Research of music recommendation system based on user behavior analysis and Word2Vec user emotion extraction. In: Xhafa, F., Patnaik, S., Zomaya, A.Y. (eds.) IISA 2017. AISC, vol. 686, pp. 469–475. Springer, Cham (2018). https://doi.org/10.1007/978-3-319-69096-4_65
18. Shuai, Y., Gao, K., Ting, Z., et al.: Data desensitization method of electricity information. DEStech Trans. Eng. Technol. Res. (2017)
19. Liu, Y., Yang, Z., Xiu, J., et al.: Research on an anti-crawling mechanism and key algorithm based on sliding time window. In: 2016 4th International Conference on Cloud Computing and Intelligence Systems (CCIS), pp. 220–223. IEEE (2016)
20. Chen, X., Qiu, X., Zhu, C., et al.: Long short-term memory neural networks for Chinese word segmentation. In: Proceedings of the 2015 Conference on Empirical Methods in Natural Language Processing, pp. 1197–1206 (2015)
21. Jha, V., Manjunath, N., Shenoy, P.D., et al.: HSRA: Hindi stopword removal algorithm. In: 2016 International Conference on Microelectronics, Computing and Communications (MicroCom), pp. 1–5. IEEE (2016)
22. Kenter, T., De Rijke, M.: Short text similarity with word embeddings. In: Proceedings of the 24th ACM International on Conference on Information and Knowledge Management, pp. 1411–1420. ACM (2015)
23. Zhang, X., Zhao, J., LeCun, Y.: Character-level convolutional networks for text classification. In: Advances in Neural Information Processing Systems, pp. 649–657 (2015)
24. Mallick, C., Das, A.K., Dutta, M., Das, A.K., Sarkar, A.: Graph-based text summarization using modified TextRank. In: Nayak, J., Abraham, A., Krishna, B.Murali, Chandra Sekhar, G.T., Das, A.K. (eds.) Soft Computing in Data Analytics. AISC, vol. 758, pp. 137–146. Springer, Singapore (2019). https://doi.org/10.1007/978-981-13-0514-6_14

Customer Evaluation-Based Automobile After-Sale Service Multi-value-Chain Collaborative Mechanism Verification

Dong Liu[1], Wen Bo[2,3], Changmao Wu[2], Hongju Yang[3], and Changyou Zhang[2(✉)]

[1] Southwest Jiaotong University, Chengdu, Sichuan, People's Republic of China
[2] Institute of Software, Chinese Academy of Sciences, Beijing, People's Republic of China
changyou@iscas.ac.cn
[3] Shanxi University, Taiyuan, Shanxi, People's Republic of China

Abstract. In the automotive after-sales service cycle, according to the historical information of the service received by the customer during the period of warranty services of repair, replacement and refund, the customer's willingness to choose the out-of-warranty service is evaluated, and differentiated collaborative services are provided, which can effectively improve the service quality. This paper proposes a collaborative mechanism design and verification method of automobile after-sales service multi-value-chain based on customer evaluation. First, it studies the composition of the service value chain during the warranty period, outside of the warranty period, and the insurance service, explores the source of income and gain factors of each value process to discover the appropriate service business and proving the ability of customer assessment for value creation in the value process. Then, establish a colored Petri Net model for each service value chain, the resources required for after-sales service are mapped to Token in Place and Place and the income is mapped to Transition, which can simulate the gain process of a single service-value-chain. Finally, based on the customer evaluation level, the design provides a differentiated automobile after-sales collaborative service multi-value-chain mechanism, generates the random customer sequence. And the gain effect of multi-value-chain in collaborative status is simulated and operated on the CPN Tools which is a running platform based on colored Petri Net. The results show that the above-mentioned mechanism makes the expenses of important customers relatively reduced, and the income of service providers and insurers increased which is conducive to maintain the ability of auto service providers to extend after-sales service. In addition, it provides a quantitative basis for collaborative optimization of service multi-value-chain.

Keywords: Customer evaluation · After-sales · Collaborative service multi-value-chain · Colored Petri Net · CPN Tools

© Springer Nature Switzerland AG 2019
D. Milošević et al. (Eds.): HCC 2019, LNCS 11956, pp. 233–244, 2019.
https://doi.org/10.1007/978-3-030-37429-7_23

1 Introduction

Since 2009, China has become the world's largest auto market for nine consecutive years. Collaborative value chain can lead the value of reduction and stabilization time [1, 2]. We design a collaboration mechanism for service multi-value-chain help to improve the revenue and efficiency of service providers, it can create corporate value in the relationship between buyers and sellers [3]. Some people proposed a full-chain collaborative model. They believe that the automotive industry should not only pay attention to the punctuality and integrity of services, but also pay attention to the professionalism, integration and coordination of services [4, 5].

In terms of customer assessment, Achim. Thom et al. consider monetary and non-monetary factors, they analyze customer value from the perspective of direct value, indirect value and social value, classify customers into 4 level [6]. Helgesen verifies that loyal customers are profitable and considers profitability based on customer satisfaction, customer (action) loyalty and customer personal level [7]. Haemoon et al. studied customer satisfaction, service quality evaluation and customer value from 2000 to 2015 [8, 9].

Research based on Petri-net, Carolina Gerini et al. propose a discrete event modeling and simulation business management framework using Petri-net [10, 11]. Establishing various CPN (colored Petri Net) models to represent activities and maintenance management, as well as factors that have a major impact, fault logic and fault processes [12–14].

In summary, customer evaluation is an important component in the strategic management of automotive after-sales service. Service multi-value-chain collaboration based on customer evaluation is the basis of service provider strategic management. This paper proposes that based on customer evaluation, we establish a colored Petri-net model for each service chain. According to the historical information of the service during the three guarantees period, then infer the customers' level. Design differentiated synergies for outside the three guarantees period and insurance service value chain and we use the colored Petri net platform CPN Tools to simulate service multi-value-chain collaboration.

2 Related Works

In order to verify the correctness of the customer evaluation-based automobile after-sale service multi-value-chain collaborative mechanism, experiments verify the feasibility and effectiveness of the scheme [11]. This paper studies the concept of value chain, value chain model, Petri Net model.

2.1 Conception of Value Chain

Value chain is a business system that describes in detail the sequence of business operations or functional actions. The value chain is a business system, it describe the order of the business operations or functional behaviors [3]. Every link in the value chain adds value to a product or service and strives to achieve the same goal. Therefore, they do not compete with each other but cooperate with each other.

2.2   Petri Net Model

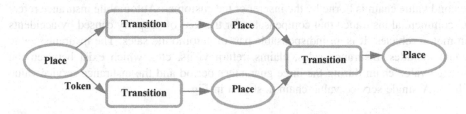

Petri-net model is used to simulate parallel branches with interactions in the system. Dr. C.A. Petri, the United States, first proposed the concept of the network in his dissertation published in his year, this paper is the cornerstone of the development of Petri net theory. Petri net is suitable for describing asynchronous and concurrent computer system models. Petri nets have both strict mathematical expressions and intuitive graphical expressions. They have rich system description methods and system behavior analysis techniques.

3   Petri Net Modeling of Service Value Chain

To research the service process and income process of a single service value chain, According to the resources and gain links needed for after-sales service, we will establish a colored Petri net model for each service value chain in this paper.

Definition 1. Oriented Net
N = (S, T, F) is called oriented net, if and only if:

(1) $S \cup T \neq \emptyset. S \cap U = \emptyset$;
(2) $F \cap (S \times T) \cup (T \times S)$;
(3) $dom(F) \cup cod(F) = S \cup T$, where, $dom(F) = \{x | \exists y : (x, y) \in F\}, cod(F) = \{y | \exists x : (x, y) \in F\}$, Are the domain and value range of F respectively.

S is set of N Place, T is set of N Transition, F is flow relationship.

3.1   Automobile Service Value Chain

At this stage, the automotive after-sales service value chain can be roughly divided into: the service value chain during the three guarantees period, the service value chain outside the three guarantees period and the insurance service value chain.

In our paper, the service value chain during the three guarantees period includes: maintenance, repair, service testing, service return visits and insurance sales. Insert maintenance during regular intervals during each maintenance. Insert repair during regular intervals during twice maintenance. The service value chain outside the three guarantees period refers to the sale of products over the period, the value chain is similar with the service value chain during the three guarantees period, also included: maintenance, repair, service testing, service return visits and insurance sales and so on. The difference is that the fee settlement participants change, the cost in the first value chain of the service is borne by the manufacturer or the insurer, but the cost in the second value chain is borne by the insurer or the customer. Automobile insurance refers to commercial insurance that compensates for the loss of property caused by accidents in motor vehicles. It is an indispensable part of automobile sales. The insurance value chain includes: insurance sales, claims, return visits, etc., which exist between the service value chain during the three guarantees period and the insurance service value chain. A single service value chain is shown in Fig. 1.

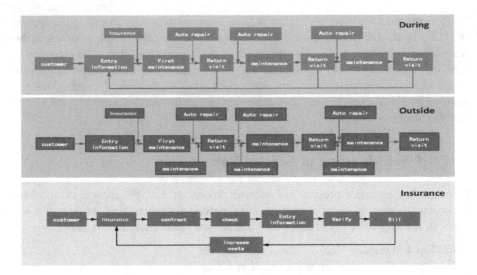

Fig. 1. Single service value chain.

By understanding the gains of a single service value chain, we can design a collaborative solution for service multi-value-chain. The main income of the service provider comes from: maintenance, repair, etc. The main benefit of the insurer comes from insurance sales.

3.2 Service Value Chain Modeling Based on Colored Petri-Net

In the network model of the service value chain, we need to analyze all factors that consider the value chain to increase revenue. For example, the customer's vehicle value and vehicle usage rate should be random, and we should control the customer's initial value distribution. In the service value chain during the three guarantees period, we know cost is not from customers. Maintenance is a random event, we need to introduce random numbers to control the number of repairs. The colored network system is a type of advanced Petri net that defines the token color to distinguish the types of resources and enhances its ability to describe the system. The resources required for after-sales service in the service chain are mapped to tokens in place and the place. The gain link is a Transition. Different resources correspond to token colors. The resource information corresponds to the token value. They are important factors influencing the gains of the value chain. The corresponding relationship of elements is shown in Table 1.

Table 1. The corresponding relationship of elements

Service value chain element	Petri-net element
Gain link	Transition
After-sales resources	Place
Customer information	Token
Service control	Connection
Service resource type	Token color
Resource information	Token value

On the basis of the colored Petri net, the control unit is added to indicate the flow of different customer services, the probability of occurrence of control, resource constraints, etc., and the after-sales service network system is delimited as Definition 2.

Definition 2. Automobile service network system

II = (P, T; F, C, K, D,) sufficient and necessary condition for the automobile after-sales service network system:

(1) $\sum = (P, T; F, C,)$ is colored network system
(2) $t \in C$, D(t) is the control function of Transition t
(3) $p \in P$, K(p) is the resource limit function of the Place p.

3.3   Construct Single Service Value Chain Model

Establish a colored petri net model for each service value chain through CPN-tools. First, model the service value chain during the three guarantees period. In this value chain, the customer's vehicle is maintained three times, and the cost of the first maintenance is paid by the manufacturer to the service provider. Add vehicle detection after vehicle maintenance and randomly insert the repair during the maintenance interval. Its Petri-net model is shown in Fig. 2. The model of the insurance service value chain is relatively simple, and the business has: the customer purchases insurance and the pay for the repair. The service value chain outside the three guarantees period is similar to the service value chain during the three guarantees period, but the fee settlement participant changes, and the cost is borne by the insurer or the customer.

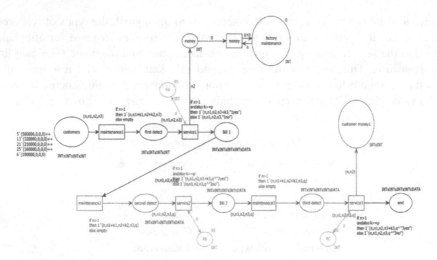

Fig. 2. The model of service value chain during the three guarantees period.

The gain link and the income result of the value chain are represented by different colors. We use random numbers to control the probability of vehicle maintenance, making the model more realistic. According to the model, in the service value chain during the three guarantees period, the service provider benefit comes from the repair and maintenance services funded by the manufacturer, the insurer and the customer. The benefits of the insurance value chain are all derived from the sales of insurance. The benefits of the service value chain outside the three guarantees period come from the repair and maintenance services funded by customers and insurers.

4 Automobile After-Sales Service Multi-value-Chain Collaboration

In multi-value-chain collaboration, by evaluating customer bases of different values, High-level customers can feel the existence of differentiated services, thereby creating more value for service providers and the like. As an important part of service strategy management, customer assessment can be used to make strategic evaluations for the company.

4.1 Collaboration Mechanism of After-Sales Service

Designing differentiated services for service multi-value-chain Business collaboration optimizes multiple service value chains. Therefore, according to the resource information obtained in the service value chain during the three guarantees period, the differentiated service for the automobile insurance and the service value chain outside the three guarantees period, as shown in Fig. 3.

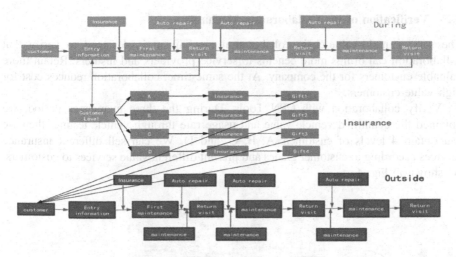

Fig. 3. Service multi-value-chain collaboration.

According to the income link summarized in the previous section, design multi-value-chain collaboration mechanism.

Collaboration point between the service value chain during the three guarantees period and the insurance service value chain:

(1) The service provider can obtain customer or vehicle resource information after the customer's vehicle is maintained. For example: vehicle price, usage rate, average service period, customer base vehicle purchases and service times, etc. Through resource information, we can design a solution to evaluate customers and divide customers into 4 levels.

(2) Different levels of customers receive different value services when purchasing insurance. High-level customers receive a high value of service. Extend the service cycle of high-level customers by providing differentiated services to customers.

Collaboration point between the service value chain outside the three guarantees period and the insurance service value chain:

(1) The cost of the gift to the customer is shared by the insurer and the service provider. Among them, the insurance company's investment ratio is greater than the service provider.

(2) The service provider also provides differentiated services according to the customer evaluation level used through the insurer. High-level customers are willing to continue to receive multiple services to get more revenue for service providers.

(3) Repair costs are derived from: insurance and customers: clear benefit is not only achieving collaboration between the service multi-value-chain, but also facilitating the reasonable distribution of gift services.

4.2   Verification of the Collaboration Mechanism

The service multi-value-chain collaboration mechanism is designed to verify that collaboration can brings more benefits to service providers and insurers. Retain more valuable customers for the company. At the same time, collaboration reduces cost for high-value customers.

Verify collaboration with CPN Tools. During the three guarantees period, we obtained the quantified vehicle value and usage rate through vehicle testing, then we can obtain 4 levels of customers A, B, C, and D. We can sell different insurance services according to customer grades and present different value services to customers. As shown in Fig. 4.

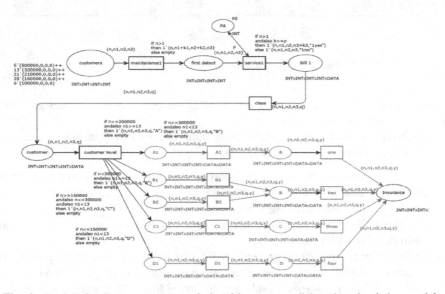

Fig. 4. During the three guarantees period and insurance collaboration simulation model.

Among them, the service fee of during the three guarantees period is from the three parties, 20% of the design comes from the manufacturer, 50% from the insurer and 30% from the customer. In the collaboration between the service value chain outside the three guarantees period and the insurance value chain, the cost of the gift is shared by the insurer and the service provider. The insurer bears 60% of the cost and the service provider assumes 40%. The lower part is the collaborative work content. Finally, the model design of the service multi-value-chain is shown in Fig. 5.

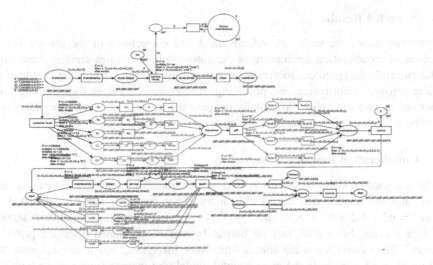

Fig. 5. The service multi-value-chain model.

5   Simulation Experiment

5.1   Experimental Environment and Parameter Settings

Run the system model on CPN Tools 4.0.1 to detect the income of each token and set random number control to verify the effectiveness of multi-service value chain collaboration. CPN Tools supports the standard ML language, which provides basic data type definitions, data manipulation descriptions, etc., to build a simple parametric mathematical model. We define the range of values of the data type and call its random function to generate random numbers within the specified range such as control pass rate and maintenance probability. In this paper, we define the types and abbreviations and descriptions of the main parameters, as shown in Table 2.

Table 2. Description of main parameters

Name	Type	Explain
Vehicle value	int	n
Number of kilometers traveled	int	n1
Maintenance cost	int	n2, N2
Repair fees	int	n3, N3
Insurance fee	int	n4
Gift service fee	int	n5
Control probability	int	p
Number of services	int	time1, time2
Customer evaluation category	string	A, B, C, D

5.2   Expected Results

In the experiment, we verify the effectiveness and correctness of the service multi-value-chain collaboration mechanism of the automobile after-sales service. According to the currently designed collaboration mechanism, by simply quantifying customer or vehicle resource information and providing differentiated services by assessing customer ratings, we can predict to actually increase the revenue of service providers and insurers, as well as reduce some of our customer expenses.

5.3   Collaboration Revenue Observation

In the designed model, the service provider's revenue comes from maintenance and repair, and the service cost observation income per customer can be calculated as income = n2 + N2 + n3 + N3 − n5 * 40%, In the experimental model, the service provider creates the overall income output by the Place named "Service provider income". The insurer's income comes from sales insurance, which can calculate the income n4 − n5 * 60%. In the experimental model, the income from insurer with the name of the Place named "Underwriters income". The high-value customer's reduced car service expense is n5, which can be directly output in the Place.

5.4   Result Analysis

Under the premise of the same initial input and maintenance probability, we have done 10 simulation experiments. The results of the service multi-value-chain collaboration model and the non-collaborative model are shown in Fig. 6.

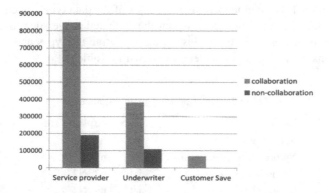

Fig. 6. Comparison of revenues between collaborative model and non-collaborative model.

In the figure, the multi-service value chain collaboration model, the service provider and the insurer's income are: ¥ 849,000 and ¥ 191,000, respectively. The non-collaborative model leads to the loss of high-value customers due to non-differentiation. The service providers and insurers are respectively: ¥ 382,000 and ¥ 109,000. We compared with the non-collaborative model, the service provider's revenue increased by ¥ 467,000 and the insurance income increased by ¥ 82,000. At the same time, we saved the cost from high-level customer about ¥ 70,000.

Compared with the non-collaboration model, the service multi-value-chain collaboration model based on customer evaluation improves the incomes for service providers and insurers. And to retain the high-quality of the customer for the service provider, while relatively reducing the automobile after-sales service costs of high-level customers. High-level customers are more willing to stay in this service provider to purchase insurance, maintain or repair vehicles. The effectiveness of the customer evaluation-based automobile after-sale service multi-value-chain collaborative mechanism is proved by simulation experiments.

6 Conclusion

This paper verifies the effectiveness and practicability of the customer evaluation-based automobile after-sale service multi-value-chain collaborative mechanism, designs a multi-value-chain collaboration scheme which based on customer evaluation, and simulates collaboration process and income. The research mainly content the following aspects.

Firstly, we analyzed the research status of multi-value chain collaboration, customer value and Petri net application. And we also analyzed the composition of single service value chain. Secondly, we determined the mapping rules of parameters, utilized CPN-Tools of colored Petri nets to simulate the model. Finally, due to the gain link of each service chain, we designed customer evaluation-based collaborative mechanism. The results show that the assurer and the service provider have increased the revenue and relatively reduced the cost of the upper-level customers in vehicle service field, which is proved the effectiveness of the mechanism. However, there are still some deficiencies in the paper, the next step of work is to quantify more parameters which related to customer ratings and optimize customer evaluation.

Acknowledgment. The authors would like to thank the anonymous referees for their valuable comments and helpful suggestions. This paper is supported by The National Key Research and Development Program of China (2017YFB1400902).

References

1. Ponte, B., Costas, J., Puche, J., Pino, R., de la Fuente, D.: The value of lead time reduction and stabilization: a comparison between traditional and collaborative supply chains. Transp. Res. Part E Logist. Transp. Rev. **111**, 165–185 (2018)
2. Dania, W.A.P., Xing, K., Amer, Y.: Collaboration behavioural factors for sustainable agrifood supply chains: a systematic review. J. Cleaner Prod. **186**, 851–864 (2018)
3. Gupta, S., Polonsky, M., Lazaravic, V.: Collaborative orientation to advance value co-creation in buyer–seller relationships. J. Strateg. Mark. **27**(3), 191–209 (2019)

4. Yu, Y., Chen, Y., Shi, Q.: Colored petri net model of knowledge flow based on knowledge life cycle. In: Strategy and Performance of Knowledge Flow. International Series in Operations Research & Management Science, vol. 271. Springer, Cham (2018). https://doi.org/10.1007/978-3-319-77926-3_6

5. Ros-Tonen, M.A., Reed, J., Sunderland, T.: From synergy to complexity: the trend toward integrated value chain and landscape governance. Environ. Manag. **62**(1), 1–14 (2018)

6. Walter, A., Ritter, T., Gemunden, H.G.: Value creation in buyer-seller relationships: theoretical considerations and empirical results from a supplier's perspective. Ind. Mark. Manage. **30**(4), 365–377 (2001)

7. Helgesen, Ø.: Are loyal customers profitable? Customer satisfaction, customer (action) loyalty and customer profitability at the individual level. J. Mark. Manage. **22**(3–4), 245–266 (2006)

8. Oh, H., Kim, K.: Customer satisfaction, service quality, and customer value: years 2000–2015. Int. J. Contemp. Hospitality Manag. **29**(1), 2–29 (2017)

9. Balboni, B., Terho, H.: Outward-looking and future-oriented customer value potential management: the sales force value appropriation role. Ind. Mark. Manage. **53**, 181–193 (2016)

10. Gerini, C., Sciomachen, A.: Evaluation of the flow of goods at a warehouse logistic department by petri nets. Flex. Serv. Manuf. J. **31**, 354–380 (2019)

11. Mahato, D.P., Singh, R.S.: Load balanced scheduling and reliability modeling of grid transaction processing system using colored petri nets. ISA Trans. **84**, 225–236 (2019)

12. Chen, H., Wu, N., Li, Z., Qu, T.: On a maximally permissive deadlock prevention policy for automated manufacturing systems by using resource-oriented petri nets. ISA Trans. **89**, 67–76 (2019)

13. Zhao, J., Chen, Z., Liu, Z.: Modeling and analysis of colored petri net based on the semi-tensor product of matrices. Sci. China Inf. Sci. **61**(1), 01–05 (2018)

14. Du, S., Wu, P., Wu, G., Yao, C., Zhang, L.: The collaborative system workflow management of industrial design based on hierarchical colored petri-net. IEEE Access **6**, 27383–27391 (2018)

Research on the Construction of Sharing Service Model in Fresh E-commerce Cold Storage

Xiaofei Liu, Yu Zhang[(⊠)], and Jiahao Xu

Yujiatou Campus, Wuhan University of Technology, Wuhan, Hubei, China
fkevin@whut.edu.cn

Abstract. After analyzing the problem of insufficient utilization of storage location resources in the current fresh e-commerce cold storage, a sharing service model was constructed to apply to the fresh e-commerce cold storage. The idle surplus storage space resources in the cold storage would be shared and rented, thus solving the problem of inadequate utilization of energy resources from the management aspect and optimizing the allocation of resources. The B2C+P2P sharing business model constructed in this paper divides the sharing service into three types according to different users' needs, which provides enterprises with new methods to revenues.

Keywords: Sharing service model · B2C+P2P sharing business model · Fresh e-commerce cold storage

1 Introduction

Under the background of Internet Plus, fresh food e-commerce develops rapidly while cold-chain logistics equipment and system become increasingly perfect. With the increasing demand for fresh food, the continuous expansion of fresh e-commerce market drives the growth of cold storage industry. Cold storage is an indispensable carrier of fresh logistics. The rational allocation of its resources is the important precondition for good development of fresh e-commerce.

However, fresh e-commerce cold storage still has a main problem of insufficient utilization of resources caused by backward operation and management mode and unreasonable internal structure design. At present, the majority of cold storage has low energy efficiency but high costs. The huge power consumption of cold storage directly leads to the increase of its own operating cost Then it indirectly raises the rent of cold storage and cold-chain logistics cost. Since the rent and electricity price of cold storage account for a large proportion of the whole cold-chain cost, many researchers at home and abroad have long been committed to the research of energy-saving technology of cold storage [1]. But few researches pay attention to the reform of cold storage from the aspect of management mode. Sharing mode, regarded as a new mode, has not been applied to the study of fresh e-commerce cold storage yet. Therefore, the construction of sharing service mode in fresh e-commerce cold storage is an exploratory study.

D. Milošević et al. (Eds.): HCC 2019, LNCS 11956, pp. 245–251, 2019.
https://doi.org/10.1007/978-3-030-37429-7_24

2 Construction and Analysis of Sharing Service Model in Fresh E-commerce Cold Storage

2.1 Analysis of Necessity

The sharing economy mode is to obtain the right of use through ownership so that the unused resources are jointly used. Sharing has different meanings under different circumstances, such as Collaborative Consumption [2], Peer Economy or Access-based Consumption [3]. Sharing mode has been already discovered by more and more enterprises and widely used in clothing, eating, housing, transporting, entertainment and other fields at home and abroad, such as foreign Airbnb, Uber, Steam, domestic sharing bikes, sharing dressing rooms, etc. It has a good developing tendency and broad prospects [4]. However, in fresh e-commerce cold storage industry, the application of sharing mode is still in a blank state and there are a lot of unused resources. Fresh trade mostly adopts quick sale strategy because the fresh products have the characteristics of fast updating, frequent transactions, strict storage and distribution conditions [5]. Due to the continuous trade and circulation of commodities, suppliers fail to replenish goods in time. Therefore, it adds to some additional space in the cold storage, resulting in a high empty rate of cold storage [6], thus indirectly leading to the insufficiency and inadequacy of cold storage resources utilization with high energy consumption. The waste of a lot of space resources and energy also violates the sustainable development strategy of green logistics. In order to make rational use of fresh e-commerce cold storage space, the idle surplus storage resources should be shared. In this way, the shelf life of products can be extended to increase the value of products and improve the utilization rate of resources. Therefore, it is of great significance to study the construction of sharing service mode in fresh e-commerce cold storage.

2.2 Analysis of Feasibility

The successful operation of sharing service mode in fresh e-commerce cold storage requires comprehensive consideration from the aspects of credit, user interface, sharing resource, supply demand, infrastructure management and so on [7]. The sharing service environment (shown in Fig. 1) is the basis of the sharing service mode, including the interpretation of the latter four aspects. In terms of credit, no single enterprise or individual user has the ownership of resources, but only the right to manage resources. The management of the cold storage is realized by remote control, and it makes the resources integrated by utilizing the intangible management and the tangible assets. It is driven by value and solves the biggest credit problem in sharing business [8]. At the same time, it establishes a perfect user filtration mechanism: the user who has basic credit scores can obtain certain credit scores in the sharing rental location transaction service after successfully being recorded by cold chain page. The points of users who violate the rules will be deducted, and users whose points do not reach the standard limit will be filtered out, thus improving users' credit as a whole.

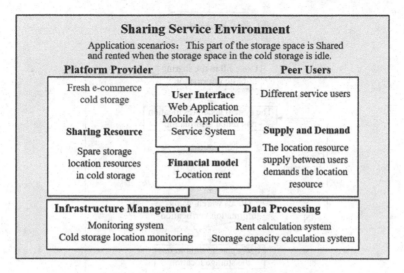

Fig. 1. Sharing service environment

3 The Construction of Sharing Service Mode in Fresh E-commerce Cold Storage

3.1 Basic Service Flow

Fresh e-commerce cold storage platform carries out whole-process monitoring of cold storage goods. And the inventory information including the size and location of the storage location and controllable temperature range will be synchronized to the terminal server as a feedback. The basic flow of sharing service is shown in Fig. 2. The lessee holding the idle surplus storage location resources of the cold storage shall release the storage location sharing information on the service system. The lessee with inventory requirements inputs basic information to the service system, such as the type, quantity, unit size and so on of the items he wants to store (The rented storage space needs to reach a certain amount of space to achieve). The system retrieves the historical storage location information of the cold storage, automatically selects and matches the optimal spare storage location, and calculates the rental price for the lessee to choose. Once the information matching is successful, the management right of the leased part of the cold storage location is transferred to the user. In that case, the free storage location is shared. During the management period, users can achieve the purpose of remote control by intelligent temperature control and real-time monitoring of the cold storage. Intelligent temperature control and adjustment means that the user can control the temperature flexibly. The background system will recommend the temperature control range according to the type and characteristics of the fresh storage so the user can change the temperature intelligently.

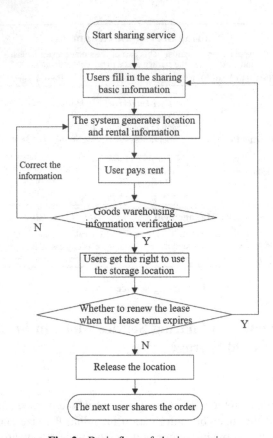

Fig. 2. Basic flow of sharing service

3.2 Multi-user Application Model

There are three main types of sharing service according to the different sharing rental objects:

(1) Sharing between fresh e-commerce cold storage service platform and the first layer users.
(2) Sharing between the first layer users and the second layer users.
(3) Sharing between the second layer users and the second layer users.

The users directly connected with the cold storage service platform are the first layer users, which are mainly large enterprises with large storage capacity that have reached long-term cooperation with cold storage. The connection between the second layer users and the first layer users or the second layer users is not connected with the cold storage service platform, which are mainly small enterprises or individuals with small inventory. The relationship between users at each layer is peer-to-peer.

Since the free storage bits can be rented and shared by other users again after release, the release transfer can be divided into two forms:

(1) The form of full release of storage space with complete transfer of ownership. The transfer shall be in the form of full transfer if all the storage space managed by the

user has been cleared. The management right to the storage location shall be owned by the lessee. The user will no longer have the management right to release the storage location.

(2) The form of partial release of storage space with incomplete transfer of ownership. If there is part of idle space in the storage space managed by the first layer users or the second layer users during the sharing period, the idle storage space can be shared again in the form of consolidation with incomplete transfer of ownership, that is, part of storage space can be released.

The user sets the rental price and term of this part of sharing storage space and publishes the sharing information on the service platform. When the lease term of the transferred user's storage space expires, the right to use the storage space management shall be returned to the original user. In order to prevent high price monopoly, the assignor shall not set the rent price higher than the recommendation of the system. System recommended rent is based on big data processing and uses dynamic algorithm to combine resources in an optimal way. The generation process of rent scheme is shown in Fig. 3. The rent shall be charged according to the users on both sides, the proportion and the number of transactions. For each successful transaction, it adds the service fee equivalent to the homeowner's fee to the basic rent price for the first layer of users, and takes the reverse approach, that is, the proportion which has large transaction amount is low while the proportion which has small transaction amount is high.

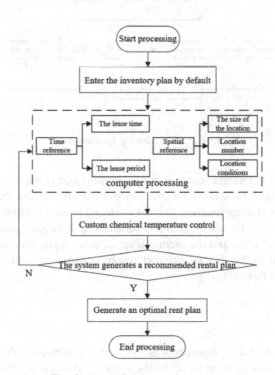

Fig. 3. Rent plan generation process

3.3   Business Model

Sharing models include B2C/C2B, B2B or P2P [9]. A new B2C+P2P sharing business model is constructed in this study (it is shown by Fig. 4). It matches the information according to the needs of different users. The unconfigured storage location resources are provided by the cold storage service platform of the fresh food sharing e-commerce. Information is standardized and presented to users through service system for visualization, thus realizing the use of users' right to use resource management instead of ownership and cooperating with credit point mechanism to ensure a smooth transaction. The spare space resources of cold storage are fully and timely utilized to meet the needs of consumers while the rent of storage space brings additional economic benefits to enterprises.

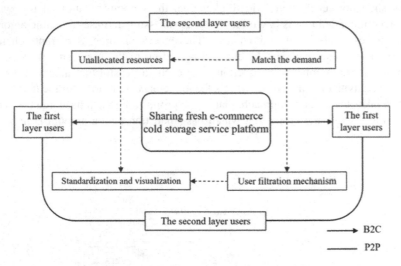

Fig. 4. B2C+P2P sharing business model

Compared with the single transaction relationship and connection mode of other business models, the new sharing business model has advantages that any user with demand can use the sharing function to meet the using needs of different users, and the sharing community can be maximized. The relationship between consumers is divided into the first layer of users and the second layer of users. Each transaction can change the relationship hierarchy due to different resource demands, and the sharing relationship has lots of flexibility.

4   Conclusion

This paper studies the energy resource waste of fresh e-commerce cold storage from the perspective of the management mode rarely involved in the past, pioneering the construction of Shared service mode and B2C+P2P Shared business model in this field.

The construction of Shared service mode solves the major problem that resources in fresh e-commerce cold storage fail to be rationally allocated, which enables the idle surplus resources to be fully utilized and maximizes the utilization of resources. At the same time, different users' needs should be accurately positioned. The construction of a new B2C+P2P business model will bring additional benefits to enterprises and make users achieve optimal demands, setting up the concept of green, open and intelligent development.

References

1. Min, L., Zhan, L., Jiang, X., Ye, B.: Design and performance analysis of the heat pump-based condensing heat of cold storage recovery drying equipment. In: International Conference on Computer Distributed Control & Intelligent Environmental Monitoring (2011)
2. Qin, B., Hou, H.: Research on O2O model of fresh products. In: International Conference on Logistics (2017)
3. Guillen-Royo, M., Wilhite, H.L.: Wellbeing and sustainable consumption. In: Glatzer, W., Camfield, L., Møller, V., Rojas, M. (eds.) Global Handbook of Quality of Life. IHQ, pp. 301–316. Springer, Dordrecht (2015). https://doi.org/10.1007/978-94-017-9178-6_13
4. Botsman, R.: The sharing economy lacks a shared definition, vol. 21. Fast Company (2013)
5. Bardhi, F., Eckhardt, G.M.: Access-based consumption: the case of car sharing. J. Consum. Res. 39(4), 881–898 (2012)
6. Gansky, L.: The Mesh: Why the Future of Business is Sharing. Penguin (2010)
7. Lobbers, J., Hoffen, M.V., Becker, J.: Business development in the sharing economy: a business model generation framework. In: Business Informatics (2017)
8. Johnson, E.A.J., Osterwalder, A., Pigneur, Y.: Business Model Generation: A Handbook for Visionaries, Game Changers, and Challengers. Wiley, Hoboken (2010). 281+iv pages, us34.95, 11 Journal of Product Innovation Management, vol. 29, no. 6, p. 10991100 (2012)
9. Brousseau, E., Penard, T.: The economics of digital business models: a framework for analyzing the economics of platforms. Rev. Netw. Econ. 6(2), 81–114 (2007)

Usability Testing of a Smartphone Telecare Application for Informal Caregivers

Irena Lovrenčič Držanič[1], Vladimir Mladenović[2(✉)],
Matjaž Debevc[1], Vesna Dolničar[3], Andraž Petrovčič[3],
Simona Hvalič Touzery[3], and Ines Kožuh[1]

[1] University of Maribor, Maribor, Slovenia
[2] University of Kragujevac, Cacak, Serbia
vladimir.mladenovic@ftn.kg.ac.rs
[3] University of Ljubljana, Ljubljana, Slovenia

Abstract. Independent living of the elderly has its risks that can be decreased with the use of telecare systems. Smartphone telecare applications such as TeleStiki, are intended primarily for informal carers of independently living elders. The application is aimed at improving quality of life of both the elderly and their caregivers by reducing the caregiver burden and extending the time elders stay at home, while also increasing their safety. The aim of usability testing was to evaluate if users perceive the tested smartphone application as high in usability and user experience. The testing was also aimed at evaluating whether the use of application affects their attitude towards the use of telecare applications for informal care of their elderly relatives in general. In this study Task Completion, User Experience Questionnaire and System Usability Scale, as well as the Think-Aloud protocol has been performed. Results showed the application scores above average in usability and user experience. Findings gained in the qualitative study can be used as guidelines for future re-designs of the application and for improvement of usability in similar applications.

Keywords: Telecare · Usability testing · Independent living · Elderly

1 Introduction

Societal changes, most importantly aging and technological advancements, led to the development of technologies for elders, such as remote monitoring systems [36]. According to research, many elders wish to remain living at home when they age (aging in place), due to a familiar environment and lower costs of living [42]. Aging in place is preferable, even when they struggle with independent living and some decide to apply to institutional care only due to advanced life events occurring, such as hospitalization or frequent falls [27]. Remote monitoring systems such as telecare can help them remain living independently and improve the quality of life of their care-givers [28]. These systems are especially useful for relatives who become informal caregivers, since caring causes them to feel stressed and overworked, especially those who do not live in the same household. Additionally, telecare could counteract the risks of elders living alone, such as falls [7, 39].

D. Milošević et al. (Eds.): HCC 2019, LNCS 11956, pp. 252–265, 2019.
https://doi.org/10.1007/978-3-030-37429-7_25

Telecare systems have been researched in several studies that found them to be beneficial for both elders and caregivers regarding the improvement of their quality of life [4, 20, 44]. Additionally, telecare improves the safety and independence of elders, while reducing caregivers' burdens [22]. A recent study has shown that telecare can also be used for institutional care [6].

When institutionalized care is unavailable or rejected by the elder, telecare can be an opportunity to prolong elders staying at home. Aging in place is favorable in comparison to institutionalized care by many elders, since they wish to remain living in their community independently to some degree, especially those who are homeowners [11, 35, 44]. [43] additionally notes that institutionalized care carries high costs of care, while elders are also unfamiliar with the environment. Furthermore, the use of remote monitoring benefits caregivers, who do not need to check continually on the relative's daily activities, or be worried about accidents happening when they are not present. Applications for smartphones give them the ability to see when certain activities were recorded, or if the elder has been inactive for longer, and be notified about falls and alarming events, such as smoke or flood detection. These applications also encourage family members to remain a part of their relative's life, even if they live in separate households.

Apart from monitoring with sensors, which is limited to indoors, telecare can use other sources to obtain data. Some studies suggest GPS could be used for outdoor tracking, such as smartwatch monitoring [14] in order to minimize the obtrusiveness of telecare sensors. [24] proposed the use of telepresence robots that would improve the quality of life of informal caregivers, while also decreasing the workload of healthcare professionals [21]. For elders with dementia, vision-based monitoring was proposed instead of sensors. The system traces body movement with video cameras and reports dangerous activities. It is suitable for institutionalized care [8]. Other studies proposed other vision-based systems [32], 3D camera-based systems without image recording [5], or infrared motion sensors [2].

Nevertheless, the telecare system that uses sensors to monitor user's' activities still have advantages over other presented solutions. In contrast to wearable sensors, ambient sensors are not as obtrusive as smartwatch monitoring or telepresence robots. For elders with cognitive problems, monitoring with ambient sensors is more reliable, as no wearables need to be worn and, thus, they cannot forget to use it. As for camera monitoring, a major concern could be privacy issues.

Despite the development of telecare systems, current knowledge about needs that the elderly have regarding technology and how such systems should be designed is insufficient [25]. [38] have defined the design requirements for telehealth systems, but usability testing in telecare is not researched thoroughly. A study similar to this was done by [16]. It evaluated system design, but did not include a smartphone application.

Due to this, this study adds to the existing research with findings on how usability testing can improve the design of a telecare smartphone application. The application is primarily intended primarily for informal caregivers who care for independently living elders. In the following sections, we will present how telecare can reduce the caregiver burden and extend the time elders stay at home, while also increasing their safety. In Sect. 2, we present the purpose and objectives of this study. In Sect. 3, related works will be presented regarding the use of telecare by informal caregivers and its benefits and limitations. The methodology will be described in Sect. 4. In Sect. 5, we will present the results of the usability testing, and in Sect. 6 the findings, based on results.

2   Related Work

2.1   Use of Telecare by Informal Caregivers

The telecare application TeleStiki is aimed primarily at informal caregivers who live in a different household than the elder taken care of, and are presumably relatives. The goal of this application is to decrease the caregiver burden and improve the safety of elders. In previous research, informal caregivers reported high levels of stress and health deterioration when caring for family members [34]. This led to them being targeted as the main users of telecare applications, due to convenient and time-efficient use. Over 40% of caregivers of dependent elders across Europe are represented by employed persons [10]. However, this increases to over 50% when we consider only caregivers in the age group 45–64, in which most caregivers of elders belong [33]. Due to this, informal caregiving results in reduced efficiency in the workplace and reduced working hours, which leads to reduced income. Absenteeism causes the caregivers either to not being able to develop their career due to caring, working only occasionally, or being unable to work at all. Due to this, informal caregivers feel stressed, or experience caring as a burden [9].

[19] have found that telecare has improved caregivers' quality of life, especially of those who live in a different household. Those caregivers had decreased stress levels associated with care and reduced absenteeism, which led to better work performance. Telecare mainly reduced caregivers' tasks regarding monitoring and worrying about elders, but also made elders feel more independent, as they could alert caregivers at any time.

Since telecare provides continuous monitoring, it reduces caregivers' perception of task-specific burdens [31] and increases feelings of relief [18]. It also improves the care offered by home care services [22].

2.2   Benefits and Limitations to Telecare

Amongst the most commonly mentioned benefits of telecare is its positive effect on the safety and security of elders, with functions such as fall detectors, alerts and localization (GPS) [4, 22]. These functions lead to the user being more independent and reduce reliance on institutional care [15, 30]. Increased independence also leads to better self-management with increased choice [39]. Elders are able to stay at home for longer, which they value highly [44]. According to [13], 35% of those living in Care Homes could potentially be supported at home with telecare.

Telecare has limitations to its adoption, most importantly with privacy concerns. The authors observe the breach between the independent living of elders that is possible due to telecare and the intrusion of privacy on the other hand. Independent living at their home provides elders with feelings of safety. However, the adoption of telecare can cause feelings, such as loss of control, intrusion into private spheres and disruption

[29, 44]. While some studies have shown no privacy issues [12], others have reported on elders feeling uncomfortable knowing that information on toilet use or time spent outside will be visible to relatives. Participants proposed the possibility to turn off the system when elders do not want their activities to be monitored [40]. In a recent study, elders wanted to see when the camera was active, but otherwise did not have any complaints regarding violation of privacy [22]. In elders with cognitive impairment, where the use of telecare could be the most beneficial, it is difficult to gain informed consent [39].

Telecare has been seen as a tool for promotion of movements "aging in place" and "active aging". However, several authors have noted that both are limited when the technology does not work outside of the person's home [1]. In such instances, location devices (wandering monitors) could be used, which, in turn, leads to questions on liberties and the autonomy of elders [37], but also misinterpretation of data. Apart from privacy issues, elders could also feel stigma or embarrassment, as telecare would remind them of their vulnerabilities [39].

Lastly, some authors have cautioned that elders could feel isolated, due to their relatives being too reliant on telecare applications instead of personal communication [39]. [40] also noted that relatives who have applications for monitoring could put elders at risk by forgetting to monitor their activities. Nevertheless, a recent study has found that most elders felt independence instead of loneliness [22].

3   Purpose and Objectives

The purpose of this study was to test the usability of the telecare application TeleStiki for informal caregivers. Our main objective was to evaluate out if users perceive the tested mobile application as high in usability and user experience. A further objective was to evaluate if the use of the tested application affects their attitude towards the use of telecare applications for informal care of their elderly relatives. More specifically, if the informal caregivers perceive the application as a tool to monitor elder's activities remotely, such as (a) Detect and prevent falls and (b) Detect and notify relatives on an elder's inactivity. Another objective was to receive feedback from users on what functionalities could be changed, added or removed.

In accordance with our objectives, the following research questions were defined:

RQ1. How does interaction with telecare application affect the user's attitude towards telecare:

(a) Familiarity with telecare applications,
(b) Willingness to learn how to use telecare applications,
(c) Recognition of the importance of telecare for understanding elders' needs,
(d) Understanding of telecare applications as a tool for home care,
(e) Recognition of telecare as a compelling new method for home care.

RQ2. How do users evaluate the user experience of the application TeleStiki?
RQ3. How do users evaluate the usability of the application TeleStiki?
RQ4. What is the users' opinion about the application TeleStiki?

4 Methodology

4.1 Participants

The study included a total of 21 participants, out of whom 48% were men and 52% women. They were aged between 19 and 38 years. The majority of participants had a University Degree, Master's Degree or Doctorate (67%), less prevalent were participants with junior college or vocational schools (19%). The least common education level was vocational secondary school.

The participants assessed their ability to save money on the Likert scale (1 = I use savings or borrow money to live, 5 = save more than enough). 57% of participants said they would like to save more, 33% said they save enough and 10% said they do not save anything (M = 3.24, SD = 0.62).

The participants also assessed their use of smartphones, tablets and computers. The evaluation was done on a Likert scale (1 = cannot use, 5 = excellent use). On average, they scored themselves as excellent users of all devices, with the highest evaluation of computers and smartphones (M = 4.9, SD = 0.3) and lowest of tablets (M = 4.7, SD = 0.46). Regarding the use of smartphones, the participants evaluated their daily use on a scale between 1 to 5 (1 = less than one hour, 4 = more than four hours, 5 = do not use a smartphone). The average score was 2.67 (SD = 0.73). More than half of the participants use smartphones between two and four hours (52%), a third between one and two hours (33%) and, less often, more than four hours (10%). The least common use is less than one hour (5%).

4.2 Sampling

The research was approved by the Ethical Committee of the Faculty of Arts at the University of Maribor. We also followed the Declaration of Helsinki by the World Medical Association (WMA).

The participants were invited primarily by researchers involved in this study and by other University staff. Participants were also encouraged to invite others who they thought to be suitable. The testing of the mobile application was done in October and November 2018. All participants were invited to two locations; one at the Faculty of Electrical Engineering and Computer Science in Maribor, and the other at a Community Center in Murska Sobota. Every participant tested the mobile application individually on a given Android phone. Participation in the study was voluntary. The anonymity of participants was assured with the use of random usernames, with which questionnaires and testing data were later matched.

4.3 Materials

All usability testing sessions were done using the Android phone Samsung Galaxy J3 2016 to avoid issues with different operating systems of the participants' phones. The devices were reset after each testing. The participants tested a beta version (v1.0.2.) of the TeleStiki application seen in Fig. 1.

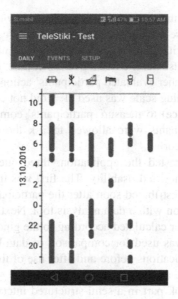

Fig. 1. The user interface in the TeleStiki application.

4.4 Research Design and Procedure

The research process consisted of several steps; first, we collected demographic data and data on the participants' current attitude towards telecare applications. Then, the participants were given tasks that needed to be completed with the mobile application TeleStiki. Afterwards, the participants self-assessed their experience with the said application by completing three questionnaires and conducting an interview.

4.5 Measuring Instruments

Before the testing of the mobile application, the following questionnaires were used to collect demographic data and participants' attitudes towards the use of telecare applications for informal care.

The demographic questionnaire consisted of questions on gender, age, education level, quality of life in regards to savings, use of selected devices (personal computer, smartphone, tablet), smartphone use by functions, relationship and age of the elderly relatives that they care for, distance between their homes, level of dependency, and probability of becoming a caregiver to selected relatives.

The pre-test questionnaire was used to measure the participants' attitudes towards the use of telecare applications before testing the application. The participants were provided with five statements that they evaluated on a 5-point Likert scale (1 = "strongly disagree", 5 = strongly agree"). The statements included (a) Familiarity with telecare applications, (b) Incentives to learn how to use said applications, (c) Belief that the applications are important for understanding the needs of elderly, (d) To be a good addition for informal care and (e) For the application to be more interesting for learning new methods of telecare.

Participants conducted the testing of the application with 12 representative tasks. They were given a use case scenario in which they take care of their independently living mother. The tasks included functionalities that are available in the beta version of the application. Participants were encouraged to follow the think-aloud protocol. During the testing, a researcher evaluated participants' actions with the task assessment questionnaire. A 5-point rating scale was used (1 = did not complete the task, 5 = task completed without assistance) to measure participants' completion of tasks using the mobile application. Participants were allowed to ask the researcher for assistance, which lowered their task scoring.

After the participants tested the application, three questionnaires were used to evaluate their user experience and usability. The first was the User Experience Questionnaire (UEQ), that was distributed soon after the participants finished the tasks. We used the standardized version with a data analysis tool. Next was the System Usability Scale (SUS). Data were later calculated according to the guidelines. Last was the post-test questionnaire, which was used for comparison of data between participants' attitudes towards telecare applications before and after use of the mobile application. The questionnaire was the same as the pre-test questionnaire.

Finally, participants took part in a semi-structured interview that consisted of ten questions not covered by other questionnaires. It was adapted from [23]. The participants were able to express their opinion freely without interruptions. Those who gave plain answers were encouraged to answer more in depth by subquestions.

4.6 Data Analysis

Descriptive statistics were used to describe the participants' use of smartphones and the characteristics of their care for elderly relatives. Data obtained during the testing process were used for the calculation of task completion rate and error rate. Data obtained by the UEQ were analysed with the UEQ data analysis tool. Internal validity was confirmed with Cronbach's alpha. In order for the raw data, obtained by SUS, to be presented as a value, they needed to be calculated according to the Guidelines. Then, the value was presented on a set of three scales, as proposed by [3].

Pretest and posttest questionnaires were used for comparison between participants' attitudes towards the use of telecare applications before and after the use of the application TeleStiki. We used the Wilcoxon signed-rank test due to the use of a small sample.

Qualitative data gathered in the interview were analysed with descriptive statistics. During analysis, data collected with Q5 and Q6 (advantages and additional functionalities) were combined into one question, as participants often answered interchangeably. All answers were coded into categories according to the volume of similar answers.

5   Results

5.1   Participants' Use of Devices

The majority of participants evaluated their device usage knowledge as excellent, meaning they can use devices independently. The highest average scores were seen in the usage of personal computers and smartphones (4.9), lower for tablets (4.7).

Regarding the time spent using smartphones, the majority use smartphones between two and four hours (52%) and 33% use them between an hour and two hours. 10% of participants use them more than four hours a day. In regard to the applications they are using, more than half do not use lifestyle applications such as applications for monitoring physical activity, a third use them less than 30 min per day, 10% between 30 to 60 min, and 5% between one and three hours. Smartphones are most often used for accessing social media and the Internet. Social media are most often used between one and three hours (48%), 38% between 30 and 60 min, and 10% more than three hours. Similarly, the Internet is used by the majority for 30 to 60 min (52%), for more than three hours (19%) or between one and three hours (14%). The next most used application is for messaging (SMS), which is used mostly for less than 30 min (67%), in 19% between 30 and 60 min, and in 10% between one and three hours. The least used functions were calling and games.

5.2   Caring for the Elderly

The majority of respondents were caring for grandparents (70%), less often for parents (15%). 15% of respondents did not take care of any older relative. On average, these elders were 75 years old (age range: 65–89, SD = 9.15). A third of respondents lived in the same household as the cared-for elders, 24% in walking distance, 18% less than 10 min away by car, 18% less than 30 min by car, and 12% one hour or more away. Regarding the elder's independence, more than half were caring for an independent elder who only needs occasional assistance. A quarter of respondents were caring for a partially dependent elder, who needs assistance with supporting activities such as shopping. 18% were caring for a moderately dependent elder, who needs complete assistance with supporting activities, and 6% were caring for a highly dependent elder who is not able to perform basic daily activities.

The participants were also questioned on the probability that they would become their relatives' caregivers. The majority expressed a high probability that they would become their parents' (48%) or grandparents' caregivers (29%). In both instances, the probability to become caregivers was higher for women than men; for mother's caregivers 52% and grandmother's caregivers 38%.

RQ1. How does interaction with telecare application affect the user's attitude towards telecare?

We analyzed the effects on (a) Familiarity and (b) Willingness to learn how to use telecare applications, as well as recognition of (c) Their importance for understanding elders' needs, (d) Telecare applications as a tool for home care and (e) A compelling new method for home care. The Wilcoxon test was used to determine whether the use of the application had a significant effect on the user's attitude towards telecare. As seen in Table 1, statistical significance is seen for familiarity (.01) and understanding of telecare as a tool for home care (.035). Other metrics were not statistically significant.

Table 1. Results of the effects on the user's attitude.

	Familiarity	Willingness to learn	Recognition of importance	Telecare understanding	Recognition of method	Familiarity
Z	−2.580	−1.455	−1.721	−2.111	−1.371	−2.580
Asymp. Sig.	.010	.146	.085	.035	.170	.010

RQ2. *How do users evaluate the user experience of the application TeleStiki?*

User experience was measured with a UEQ. Cronbach's alpha of the scale was .77, which is above the acceptable level. The highest Cronbach's alpha value was recorded for attractiveness (.91) and perspicuity (.88), while the lowest was for dependability (.61). Guttman's lambda-2 coefficient was .77.

As seen in Fig. 2, the application was scored the highest for its efficiency (1.33) and dependability (1.03). The following scales were attractiveness (0.92), perspicuity (0.88), novelty (0.72) and stimulation (0.57). Overall, all scales were evaluated positively. The participants most often described it as "efficient" (M = 2.0, SD = 0.9), "meets expectations" (M = 1.6, SD = 1.2) and "inventive" (M = 1.5, SD = 1.4).

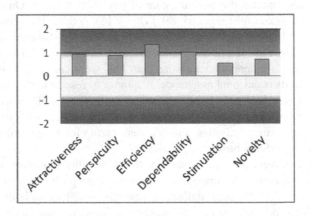

Fig. 2. UEQ score for application TeleStiki.

RQ3. *How do users evaluate the usability of application TeleStiki?*

As seen in Table 2, users evaluated the usability of the mobile application with an average value 68 (SD = 17.8) on SUS, meaning the application has above average usability. The distance between the highest and the lowest ratings was 57.5-points (min. 37.5, max. 95). According to the grading scales proposed by [3], the tested application is graded with a D mark and is marginally acceptable.

Table 2. Usability evaluation results.

SUS value	Minimum value	Maximum value	Grade scale	Acceptability range	Usability
68	37.5	95	D	High marginal acceptance	Above average

RQ4. *What is the users' opinion about the application TeleStiki?*

Users were questioned on their familiarity with and use of telecare. The majority had heard of telecare technologies that use sensors, and mentioned a few of the locally available ones. The tested mobile application had not yet been used by any of them. The most often mentioned advantages of the tested mobile application were activity monitoring function and notifications on inactivity, which they saw as a way to reduce falls. The majority appreciated the graphical display and its ease of use. However, some users found the concept of this display difficult to understand, and preferred the display with timestamps.

Other functions that were occasionally misunderstood were sensor icons, such as those for refrigerator and kitchen. One user also mentioned slow refreshing of data. Users also recognized the importance of privacy, and noted that telecare and similar technologies could be seen as an intrusion into an elder's privacy when used in the toilet, bathroom and bedroom. Nevertheless, all users evaluated mobile applications as useful.

When asked for their recommendations on which functionalities should be changed or added to the application, the majority mentioned adding a legend of icons and change of similar icons (refrigerator-kitchen, toilet-bathroom). A new function that the majority saw the use of was a filter by days or a calendar, and a clear line for separation of activities on consecutive days. In Table 3 is a list of all the proposed solutions.

Table 3. Proposals for improvement.

Usability problems	Proposed solutions
Difficulties with understanding graphical display	• Display of activities that last longer (min. a few minutes) • Clear separation of activities on consecutive days
Misunderstanding of icons	• Change of similar icons (refrigerator-kitchen, toiler-bathroom) • Adding a legend of icons • Adding instructions on how to use application • Filtered display of activities by sensors
Missing functionalities	• Integration with a smartwatch and/or wearable pendant with SOS button • Integration with live camera monitoring • Translation of the application into local language

The application was perceived as the most useful for caregivers who do not live with elders. The group for which they saw the most use were caregivers of elders with dementia, due to the notifications on inactivity and alarm when the person does not return home. Another group were elders with other illnesses that require a family member's care and regular visits when caregivers do not live nearby. Such groups are employed people who do not have time for daily visits and younger caregivers who are more familiar with ICT tools than older caregivers. Participants also noted that retirement homes could use telecare applications for monitoring of their residents, especially those with cognitive illnesses.

6 Discussion

The incentive of this study was to conduct usability testing of a smartphone telecare application for informal caregivers in order to re-design it according to their feedback. Informal caregivers were selected, since previous studies show their high involvement in caring for relatives. Usability testing was done in order to identify issues that could discourage potential users and re-design the application according to their feedback. Through this, the aim was to increase the adoption of telecare applications for informal caregivers. The testing was done amongst relatives who are currently caring for an elder, since this is often the case according to the previous studies [33, 34]. Accordingly, we examined whether the use of a telecare application changes the user's attitude towards the use of telecare applications.

The main positive of this study is the focus on prolonging the independent living of elders by studying telecare as a tool for a better quality of life of both elders and their informal caregivers. As several studies have noted [11, 44], aging in place has many advantages over institutionalized care, and leads to a higher quality of life and lower costs of care. Due to this, telecare could improve caregivers' burdens and improve their quality of life, although, some encounter challenges to its adoption. As [41] found, the involvement of potential users into the application design process can prompt them to use it. For this reason, this usability testing is aimed at improving the application according to the participants' feedback. The next positive of this study is performing testing in an office environment instead of the laboratory, since this application will presumably also be used in such environment.

Results indicate that the attitude of participants towards the use of telecare applications has improved with the testing (RQ1). The highest increase was shown in metrics' familiarity with telecare applications and understanding of telecare as a tool for home care. User experience results (RQ2) show that users evaluate it as an above acceptable level, with the highest values in attractiveness and perspicuity. Regarding usability (RQ3), users also evaluated it as above average according to SUS. They have offered some proposals on which functions should be changed, added and removed. The most often occurring were difficulties with understanding the graphical display that they found filled with unimportant data. The second major remark was misunderstanding of icons that they found either too similar or misrepresenting. Finally, users expressed the need for functionalities that would improve its usability, such as translation into the local language.

This paper presents results of usability testing of a telecare application aimed at informal caregivers. The results show that the application received above average scores in user experience and usability. Use of the application has changed the users' attitude towards telecare, especially regarding familiarity and understanding of telecare as a tool for home care. The participants have proposed several solutions that would improve the application's usability. The participants have noted that the application would be of most use to caregivers who do not live with elders, but could also be beneficial to institutionalized care facilities, such as Retirement Homes. Based on these findings, usability testing in telecare should be researched in future works, in order to broaden the adoption of telecare and promote independent living of the elderly.

Acknowledgements. The authors thank everyone involved in the study. This study was part of the Project ICT Solutions for Active and Healthy Aging: Integrating Informal e-Care Services in Slovenia (Slovenian Research Agency, Project No. L5-7626).

References

1. Aceros, J.C., Pols, J., Domènech, M.: Where is grandma? Home telecare, good aging and the domestication of later life. Technol. Forecast. Soc. Change **93**, 102–111 (2015)
2. Al-Temeemy, A.A.: Multispectral imaging: monitoring vulnerable people. Optik **180**, 469–483 (2019)
3. Bangor, A., Kortum, P., Miller, J.: Determining what individual SUS scores mean: adding an adjective rating scale. J. Usability Stud. **4**(3), 114–123 (2009)
4. Barlow, J., Singh, D., Bayer, S., Curry, R.: A systematic review of the benefits of home telecare for frail elderly people and those with long-term conditions. J. Telemed. Telecare **13** (4), 172–179 (2007)
5. Barriga, A., Conejero, J., Hernández, J., Jurado, E., Moguel, E., Sánchez-Figueroa, F.: A vision-based approach for building telecare and telerehabilitation services. Sensors **16**(10), 1724 (2016)
6. Bjørkquist, C., Forss, M., Samuelsen, F.: Collaborative challenges in the use of telecare. Scand. J. Caring Sci. **33**(1), 93–101 (2019)
7. Brownsell, S., Hawley, M.: Fall detectors: do they work or reduce the fear of falling? Hous. Care Support **7**(1), 18–24 (2004)
8. Cardile, F., Iannizzotto, G., La Rosa, F.: A vision-based system for elderly patients monitoring. In: 3rd International Conference on Human System Interaction, pp. 195–202 (2010)
9. Chiatti, C., et al.: Final report containing case-by-case detailed description and analysis of selected 12 Good practices. The CARICT project: deliverable 4.3 (2011)
10. Czekanowski, P., Mnich, E., McKee, K., Öberg, B., Prouskas, C., Quattrini, S.: Main characteristics of the sample: older care-receivers and their main family carers. In: Family Carers of Older People in Europe, pp. 117–143 (2008)
11. Davey, J.A., de Joux, V., Nana, G., Arcus, M.: Accommodation Options for Older People in Aotearoa/New Zealand. Centre for Housing Research, Christchurch (2004)
12. Demiris, G., Oliver, D.P., Dickey, G., Skubic, M., Rantz, M.: Findings from a participatory evaluation of a smart home application for older adults. Technol. Health Care **16**(2), 111–118 (2008)
13. Department of Health: Building Telecare in England (2005). https://webarchive. nationalarchives.gov.uk/20130124070255/. http://www.dh.gov.uk/prod_consum_dh/groups/ dh_digitalassets/@dh/@en/documents/digitalasset/dh_4115644.pdf. Accessed 8 May 2019
14. Gjoreski, H., Bizjak, J., Gams, M.: Using smartwatch as telecare and fall detection device. In: 12th International Conference on Intelligent Environments, pp. 242–245 (2016)
15. Greenhalgh, T., Wherton, J., Sugarhood, P., Hinder, S., Procter, R., Stones, R.: What matters to older people with assisted living needs? A phenomenological analysis of the use and non-use of telehealth and telecare. Soc. Sci. Med. **93**, 86–94 (2013)

16. Gund, A., Ekman, I., Lindecrantz, K., Sjoqvist, B.A., Staaf, E.L., Thorneskold, N.: Design evaluation of a home-based telecare system for chronic heart failure patients. In: 30th Annual International Conference of the IEEE Engineering in Medicine and Biology Society, pp. 5851–5854 (2008)

17. Holthe, T., Jentoft, R., Arntzen, C., Thorsen, K.: Benefits and burdens: family caregivers' experiences of assistive technology (AT) in everyday life with persons with young-onset dementia (YOD). Disabil. Rehabil. Assistive Technol. **13**(8), 754–762 (2018)

18. Jarrold, K., Yeandle, S.: A weight off my mind. Exploring the impact and potential benefits of telecare for unpaid carers in Scotland. Carers Scotland and the University of Leeds, Glasgow, Scotland (2009)

19. Govind Joshi, S., Woll, A.: A collaborative change experiment: telecare as a means for delivery of home care services. In: Marcus, A. (ed.) DUXU 2014. LNCS, vol. 8519, pp. 141–151. Springer, Cham (2014). https://doi.org/10.1007/978-3-319-07635-5_15

20. Kachouie, R., Sedighadeli, S., Khosla, R., Chu, M.T.: Socially assistive robots in elderly care: a mixed-method systematic literature review. Int. J. Hum.-Comput. Interact. **30**(5), 369–393 (2014)

21. Karlsen, C., Moe, C.E., Haraldstad, K., Thygesen, E.: Caring by telecare? A hermeneutic study of experiences among older adults and their family caregivers. J. Clin. Nurs. **28**(7–8), 1300–1313 (2019)

22. Kim, J.: A qualitative analysis of user experiences with a self-tracker for activity, sleep, and diet. Interact. J. Med. Res. **3**(1), e8 (2014)

23. Koceska, N., Koceski, S., Zobel, P.B., Trajkovik, V., Garcia, N.: A telemedicine robot system for assisted and independent living. Sensors **19**, 834 (2019)

24. Koch, S., Hägglund, M.: Health informatics and the delivery of care to older people. Maturitas **63**(3), 195–199 (2009)

25. Lindquist, L.A., et al.: Advanced life events (ALEs) that impede aging-in-place among seniors. Arch. Gerontol. Geriatr. **64**, 90–95 (2016)

26. Lynch, J.K., Glasby, J., Robinson, S.: If telecare is the answer, what was the question? Storylines, tensions and the unintended consequences of technology-supported care. Critical Soc. Policy **39**(1), 44–65 (2019)

27. McCreadie, C., Tinker, A.: The acceptability of assistive technology to older people. Ageing Soc. **25**(1), 91–110 (2005)

28. Milligan, C., Roberts, C., Mort, M.: Telecare and older people: who cares where? Soc. Sci. Med. **72**(3), 347–354 (2011)

29. Mortenson, W.B., Demers, L., Fuhrer, M.J., Jutai, J.W., Lenker, J., DeRuyter, F.: Effects of an assistive technology intervention on older adults with disabilities and their informal caregivers. Am. J. Phys. Med. Rehabil. **92**(4), 297–306 (2013)

30. Núñez-Marcos, A., Azkune, G., Arganda-Carreras, I.: Vision-based fall detection with convolutional neural networks. Wirel. Commun. Mob. Comput. **2017**(1), 1–16 (2017)

31. Principi, A., et al.: Work restrictions experienced by midlife family caregivers of older people: evidence from six European countries. Ageing Soc. **34**(2), 209–231 (2014)

32. Roth, D.L., Fredman, L., Haley, W.E.: Informal caregiving and its impact on health: a reappraisal from population-based studies. Gerontologist **55**(2), 309–319 (2015)

33. Sarma, S., Simpson, W.: A panel multinomial logit analysis of elderly living arrangements: evidence from aging in Manitoba longitudinal data. Soc. Sci. Med. **65**(12), 2539–2552 (2007)

34. Schulz, R., Wahl, H.W., Matthews, J.T., De Vito Dabbs, A., Beach, S.R., Czaja, S.J.: Advancing the aging and technology agenda in gerontology. Gerontologist **55**(5), 724–734 (2014)

35. Sethi, R., Bagga, G., Carpenter, D., Azzi, D., Khusainov, R.: Telecare: legal, ethical and socioeconomic factors. In: International Conference on Telehealth (2012)
36. Singh, J., Lutteroth, C., Wünsche, B.C.: Taxonomy of usability requirements for home telehealth systems. In: Proceedings of the 11th International Conference of the NZ Chapter of the ACM Special Interest Group on Human-Computer Interaction, pp. 29–32 (2010)
37. Stowe, S., Harding, S.: Telecare, telehealth and telemedicine. Eur. Geriatr. Med. 1(3), 193–197 (2010)
38. Vines, J., et al.: Making family care work: dependence, privacy and remote home monitoring telecare systems. In: Proceedings of the 2013 ACM International Joint Conference on Pervasive and Ubiquitous Computing, pp. 607–616 (2013)
39. Wherton, J., Sugarhood, P., Procter, R., Hinder, S., Greenhalgh, T.: Co-production in practice: how people with assisted living needs can help design and evolve technologies and services. Implementation Sci. 10(1), 75 (2015)
40. Wiles, J.L., Leibing, A., Guberman, N., Reeve, J., Allen, R.E.: The meaning of "aging in place" to older people. Gerontologist 52(3), 357–366 (2012)
41. World Health Organization: Global age-friendly cities: A guide. World Health Organization (2007)
42. Ziefle, M., Himmel, S., Wilkowska, W.: When your living space knows what you do: acceptance of medical home monitoring by different technologies. In: Holzinger, A., Simonic, K.-M. (eds.) USAB 2011. LNCS, vol. 7058, pp. 607–624. Springer, Heidelberg (2011). https://doi.org/10.1007/978-3-642-25364-5_43

High Efficiency Permanent Magnet Synchronous Motor Used in Electric Vehicle

Jinlong Lu[1], Qin Nie[2(✉)], Quanguo Lu[3(✉)], and Zhifang Zhu[3]

[1] 719th Research Institute, China Shipbuilding Industry Corporation,
Wuhan 430200, Hubei, China
[2] High-Efficiency Generator Engineering Technology Center,
Kungfu Sci-Tech Co., Ltd., Nanchang 330096, Jiangxi, China
1134598717@qq.com
[3] Precision Drive and Control Key Laboratory, Nanchang Institute
of Technology, Nanchang 330099, Jiangxi, China
luqg2010@126.com

Abstract. The high efficiency permanent magnet synchronous motor used in electric vehicle was designed, combined with the finite element simulation software Ansys Maxwell, the finite element analyze was carried out. The experimental results showed that the designed motor had excellent electrical performance and overload ability under double torque or double power, combined with speed controller based on weak magnetic effect principle, it can ensure the designed motor for reliable using.

Keywords: Finite element · Weak magnetic effect · Permanent magnet synchronous motor

1 Introduction

With the increasing of car ownership in cities around the world, the excessive carbon-emission from automobile exhaust has aggravated the urban air pollution. At the same time, the mankind is facing the increasing depletion of oil resources, looking for new energy to replace oil resources has became the key to solve the problem [1]. Electric vehicles are powered by electric energy and have the advantages of zero emission, low noise and energy saving [2]. Compared with the traditional internal combustion engine vehicles (ICEV), whether it is the hybrid electric vehicles (HEV) or pure electric vehicles (PEV), or fuel cell vehicles (FCV), all show that the energy sources and structure of the automotive power system have changed [3, 4]. Electric vehicles are mainly composed of driving-motor, controller, energy storage battery, transmission system and other components, among them, the driving-motor is the core component of the pure electric vehicle, and the performance of the driving-motor will affect the pure electric vehicle directly [5]. Permanent Magnet Synchronous Motor (PMSM) has the advantages of small size, light weight, high efficiency and high power density. The electric-driving system based on PMSM has excellent comprehensive performance index, such as high efficiency, high torque-current ratio, strong ability of flux-weakening speed-expansion, low vibration, low noise and fast dynamic response. Thus,

© Springer Nature Switzerland AG 2019
D. Milošević et al. (Eds.): HCC 2019, LNCS 11956, pp. 266–271, 2019.
https://doi.org/10.1007/978-3-030-37429-7_26

PMSM has become the most competitive driving-motor in electric vehicles [6, 7]. China has 80% resource of rare earth in the world, it's also pacing unique advantage for developing high-performance PMSM used in electric vehicle [8].

In this paper, a kind of low-power high-efficiency PMSM used in electric vehicle was designed, the finite element analysis was carried out combined with the Ansys Maxwell, the prototype was manufactured and tested. It has certain theoretical significance and engineering application value.

2 Design of PMSM

2.1 Whole Structure of PMSM

The stator is mainly composed of the frame, core and coils. The stator is externally pressurized structure which superposing the stator flushing into the pendulum tires. And the stator winding is vacuum impregnated. In order to improve the anti-vibration ability and ensure the stator achieves the insulation requirements, the winding are pressed into the frame after soaking (Fig. 1).

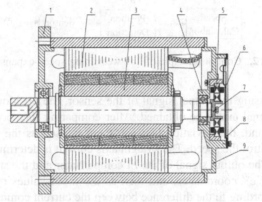

1. Front end cover 2. Stator 3. Rotor 4. Rear end assembly 5. rotary transformer

6. cover plate gasket 7. rotary variable rotor bracket 8. Hall cover plate 9. rear end cover

Fig. 1. The whole structure diagram of PMSM

The rotor is composed of rotating shaft, core, permanent magnet, front bearing and rear bearing. The permanent magnet is installed radially and V-shaped in the rotor core pole slot. Adapting the design can enlarge the reluctance torque and improve the overload capacity of the motor, at the same time, the flux-weakening speed-expansion ability is greatly improved to guarantee the peak speed capability. The rotor adds the permanent magnet cover plate and the locking nut, which can effectively improve the reliability of the motor operation.

2.2   Control Principle of PMSM

The control circuit mainly includes position sensor, control circuit, driving circuit, converter and other parts. The function of the control circuit is to generate different control signals by receiving the electrical signals of the position sensor. The control principle is basically based on the control of the amplitude, or control the phase of stator current, whatever, that is the vector control [9] (Fig. 2).

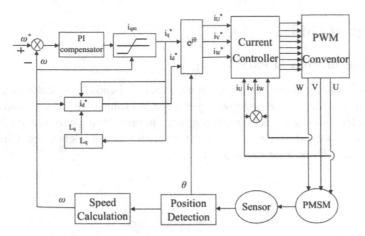

Fig. 2. Control principle of flux-weakening speed-expansion

Through processing the output signal of the sensor, the position signal θ and the rotational speed signal ω can be obtained. After comparing the speed signal with the input speed command, the deviation between them is used as the input signal of PI compensator to control the speed. The straight axis current is determined by the control circuit algorithm. The output signal of the PI compensator and the straight axis current are transformed by $e^{i\theta}$ coordinates, and then as the input values of the current controller. Finally, according to the difference between the current command value and the actual three-phase current of the stator, realizing the turned off or turned on function of converter power element. That is, the stator current of the PMSM is controlled.

3   Finite Element Analysis of PMSM

In order to ensure the performance of PMSM, the electromagnetic design adopts the professional finite element simulation software Ansys Maxwell to carry out the simulation calculation. Through reasonably selecting the coil parameters and the electromagnetic parameters, it can reduce the amount of copper and iron, also, reduce the cost of PMSM. The permanent magnet material adopts Nd-Fe-B, it has excellent magnetism, strong anti-demagnetization ability and unique thermal stability. It can reduce the volume and weight while improving the output power and efficiency of the PMSM.

Through debugging and comparison, the basic simulation parameters are determined as follows. The length of stator and rotor core is 150 mm, the outer diameter of stator is 180 mm, the inner diameter of stator is 110 mm, the number of stator slots is set as 12, the main winding adopt the pattern of fractional slot concentrated winding, the stator wire gauge is 0.71 mm, the number of parallel circuits is set as 4, the air gap is set as 0.7 mm, the outer diameter of rotor is 108.6 mm. The simulation results are shown in the following figures (Fig. 3).

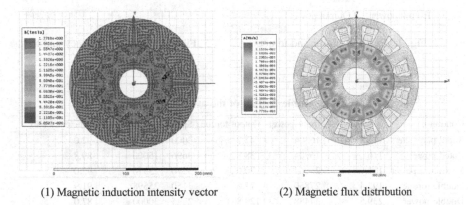

(1) Magnetic induction intensity vector   (2) Magnetic flux distribution

Fig. 3. Finite element analysis result

4 Experiment Test

The front and rear covers of the PMSM are made of high strength cast aluminum alloy ZL102, it can ensure the performance and reduce the overall weight under the premise of structure reliability. The frame is aluminum alloy and the surface is provided with heat dissipation bars. The end of the stator winding is bound with H-class silent glass ribbon. The winding coil is put into the stator core and then dipping, assembled into the base finally. The material of the rotating shaft is 45#steel with the diameter of 50 mm (Table 1 and Figs. 4, 5).

Table 1. Basic parameters of PMSM prototype

Rated voltage	Rated power	Rated speed	Peak speed	Pole-pairs	Rated torque
144 V	15 kW	3000 r/min	6600 r/min	4	47.7 N·m

The actual test shows that the rated speed reaches nearly 3700 r/min when the rated power is 15 kW, which is more than 20% higher than the rated speed, but the torque is only 38.8 N.m, which is lower than the rated torque 47.7 N.m. Part of the test data is shown in the following table (Table 2).

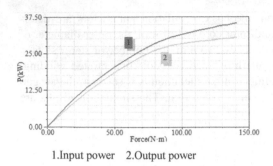

1.Input power   2.Output power

Fig. 4. Curve of power and torque

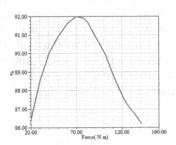

Fig. 5. Curve of efficiency and torque

Table 2. Part of the test data

Condition	Input current	Input power	Output torque	Speed	Output power	Efficiency (%)
Rated parameters	113.2	16304	47.7	3000	15000	92.0
Rated power	116	16704	38.8	3696	15000	89.8
Maximum power	181	26064	70.0	3270	23970	92.0
Double torque	193.4	27850	77.6	3150	25598	91.9
Double power	239.5	34488	128.8	2225	30000	87.0

At the highest efficiency point, the output power of the PMSM reaches nearly 24 kW, it's the 1.6 times higher than the rated power. At this time, the output torque of the motor reaches 70 N·m, and the speed reaches 3270 r/min, which are far higher than the rated value. The input current of the PMSM is 181 A, and the efficiency is 92%.

When the motor reaches the double rated power (30 kW), the torque of the motor reaches 128 N·m, the speed of the motor reaches 2250 r/min, and the input current is close to 240 A. The efficiency still can achieve 87% at a higher level. It can be considered that the motor can meet the requirements of double overload power. The same is true as for the double rated torque. The overload capacity test result shows the PMSM can maintain at a high efficient level under double torque or double power.

5 Conclusion

A kind of low power and high efficiency permanent magnet synchronous motor (PMSM) used for electric vehicles is designed in this paper. Through the prototype experimental test, it is concluded that the designed PMSM has excellent electric-magnetic performance and overload capacity under double torque or double power.

Acknowledgement. This work is supported by the National Natural Science Foundation of China (Grant No. 51165035). Science and Technology Fund of Jiangxi Province of Higher Education (KJLD14094), Youth Science Fund of Jiangxi Province (20133BAB21004), and the Young scientist cultivation plan of Jiangxi Province (20112BCB23025).

References

1. Chen, Y.: Research on direct torque control for low speed and high torque permanent magnetic synchronous motor. Huazhong University of Science & Technology (2008)
2. Fu, Rong-Dou, Dou, M.: D-axis and Q-axis inductance calculation and experimental research on interior permanent magnet synchronous motors for EV. Trans. China Electrotech. Soc. **29**(11), 30–37 (2014)
3. Zhang, B.: Research on electromagnetic-thermal design of permanent magnet synchronous machines for electric vehicle application. Huazhong University of Science & Technology (2015)
4. Xu, J.: Study on permanent magnet synchronous motor drive system in electric vehicle application. Shen yang University of Technology (2003)
5. Li, X., Huang, S., Li, L.: Calculation and analysis of vehicle vibration and noise of permanent magnet synchronous motor applied in electric vehicle. Electr. Mach. Control **17**(8), 37–42 (2013)
6. Yang, Y.: Research on electric brake characteristics of electric vehicle driven by hub motors. Jilin University (2013)
7. Zhao, T.: Design and simulation of power system for a new hybrid electric vehicle and research on electrical drive system. Hefei University of Technology (2006)
8. Feng, G.: Speed adjusting ability with field weakening of permanent magnet motor for electric vehicle. Electr. Mach. Control **18**(8), 55–60 (2014)

The Method of Urban Intelligent Address Coding Based on Spatiotemporal Semantics

Yanling Lu[1,2(✉)], Liliang Huang[1], Caiwei Liu[1], Jingwen Li[1,2], and Jianwu Jiang[1,2]

[1] Guilin University of Technology, Guilin 541004, China
358498163@qq.com
[2] Guangxi Key Laboratory of Spatial Information and Geomatics, Guilin University of Technology, Guilin 541004, China

Abstract. With the development of science and technology and the progress of society, urban construction in China is also developing rapidly. Intelligent address coding plays an important role in urban planning and design. Therefore, the study of smart urban address coding method can make a great contribution to the construction of a smart city. Based on the analysis of the correlation between urban intelligent address coding method and the development of smart city, this paper discusses a scientific and unified standard urban intelligent address coding method based on spatiotemporal semantics. And based on the theory of spatiotemporal semantic model and address model, an intelligent address model based on spatiotemporal semantics is designed according to the address coding standard published in recent years in China. To solve the problems of non-standard city place name and address coding, difficulty in managing address coding data, and imprecision of address coding, thus the unified standardization and refinement of the urban address code, the intelligent management of the urban address code data and the sharing of the urban address code data are realized.

Keywords: Address coding · Spatiotemporal semantics · Smart city

1 Introduction

Since 1980, some provinces and cities in China began to pay attention to the development of urban address coding. At present, some areas of China have basically completed the construction of a geocoding database with high accuracy and a complete address coding system. To unify the address coding standard, the related departments have studied the address coding technology and application in depth, issued a series of address coding standard specifications, and achieved good results in practical application. For example: In December 2007, the release of the "*Classification, description and encoding rules for geographical name and address in the common platform for geospatial information service of digital city*" [1]; Published in May 2009, "*Rule of coding for address in the common platform for geospatial information service of digital city*" and "*Coding rules for urban geographical features-City roads, road intersections, blocks and municipal pipe lines*"; *Technical Specification for urban*

© Springer Nature Switzerland AG 2019
D. Milošević et al. (Eds.): HCC 2019, LNCS 11956, pp. 272–284, 2019.
https://doi.org/10.1007/978-3-030-37429-7_27

geographic coding, published in November 2012; *Information system for digitized supervision and management of city-Part3: Geocoding*, published in August 2016; *Classification and coding of geographic information points of interest* published by the National Basic Geographic Information Center in 2017.

Through the analysis of the existing achievements, it is found that China has formulated and implemented many standard specifications of urban address coding, which has brought positive effects to the development of address coding technology in China. However, the accuracy of address coding in mainland China is low, and the address database contains only provincial and municipal information, which lacks scientific, rational and standard high-precision address coding method and address database. All along, the technologies and standards used in the address coding methods in various industries in China are very different. Variable coding rules cause the difficulty of address matching. However, the address matching and location products launched by GIS software developers in China lack integration with GIS, and these products have a single function and non-standard address models.

There are many problems in the address coding of our country at present [2]: (1) There is no authoritative standard for the classification of geographical information resources published, and it is impossible to share the address coding data by using many address coding methods in a province, city or even district. (2) There are different address codes in different industries, and there is no uniform address coding method among different industries, which result in the phenomenon of multiple address codes in the same address. (3) The address coding elements are complex and various, and there are many artificial elements in the address coding elements. When one of the address elements changes, the entire address coding must be modified, which causes great trouble to the address coding. Therefore, this paper combines the most commonly used address models in our country, finds out the optimal model from these address models, and adds the spatiotemporal semantic model frame on this basis, design a scientific, unified standard spatiotemporal semantic intelligent address model.

2 Spatiotemporal Semantic Intelligent Address Model

2.1 Spatiotemporal Semantic Model

(1) 1D time theory and semantic model of 2D Euclidean space

According to Friedmann's thought, a spatiotemporal semantic model is constructed with the idea of one dimensional time theory. In this model, all kinds of representation methods will be used to represent the model elements, a manifold composed of real number field R will represent the model background elements, and a real number will represent a special moment [3]. There is a coordinate relation between the instantaneous and the mathematical structure of the physical possible world, and a corresponding coordinate system is formed, for example, 1 There is no maximum real number, and there is no last instantaneous, the two correspond to each other; 2 The density of real number field corresponds to the density of time: There must be a real number between two known real numbers, and the time in the physical possible world must be separable [3].

The features in the physical possible world need to be explained when they are expressed in the field of real numbers, for example, (1) The real numbers are different in the direction of increasing or decreasing, and the physical instants are the same continuum in each direction; (2) The real number field is not uniform, and the physical moment is indeed uniform and continuous. In order to solve the above non-correspondence phenomenon, we can carry on the translation transformation and the reflection transformation real number domain such an operation.

Thus, the following model can be constructed:

$$\Delta T = \text{Scale factor} \times |\Delta t| \tag{1}$$

Among them, R real number field, dT scale factor, the coordinate difference of two points in ΔT original coordinate, the coordinate difference of two points in Δt new coordinate.

The scale factor of the standard coordinate system is the same ones, whereas in any other transformation the change in the original coordinate determines the value of the scale factor [12]. Here, the general covariance of the theory is ensured by the dT.

From this, $\langle R\ dT \rangle$ is the semantic model, and the ordinal pairs are expressed by $\langle \rangle$, which is expressed by covariant of linear time theory. With different transformations, the scaling factor dT is also different, so there are an infinite number of semantic models:

$$\langle R\ dT \rangle, \langle R\ dT' \rangle, \langle R\ dT'' \rangle \ldots \tag{2}$$

In model R is a 1D manifold of all real numbers; dT, dT', dT'' is a metric of time [3].

In the same vein, if you extend the spatiotemporal semantic model to a 2D Euclidean space, the semantic model is:

$$\langle R^2 \gamma \rangle \tag{3}$$

In model R^2 is a 2D manifold composed of all real number pairs; γ is a metric tensor of space.

(2) Spatiotemporal semantic model of special relativity and General Relativity

①   The spatiotemporal semantic model of Special Relativity:

$$\langle M\ \eta \rangle \tag{4}$$

M is a 4D manifold; η is a Minkowski spacetime; and the exact representation of a metric η is a symmetric matrix η_{ik} of coefficients $(i, k = 0, 1, 2, 3)$ [3].

②   The spatiotemporal semantic model of General Relativity:

$$\langle M, g, T \rangle \tag{5}$$

In the model, M is a 4D differential manifold, g is a Pseudo-Riemannian manifold and T is a material field.

2.2   Address Model

Before building the address model, we are required to define the address in the address coding system. Different industries and countries have different understanding of the address. To put it simply, an address has the following characteristics: (1) It has a functional symbol that describes the location. (2) It is expressed in natural language according to human's cognitive habit. (3) Commonly used for the exchange of spatial information [4]. So we can define the address as an abstract coding method with life cycle, which describes and organizes the spatial position of an individual through natural language [5].

The address model is the way of word structure [6], and different countries and regions have different ways of address representation, which leads to the diversity of address models [7]. The hierarchical address model and the relational address model are the most commonly used address models in Chinese address coding. The hierarchical relation model classifies the address elements by artificial induction, and the approximate hierarchical relation of the address model is presented in Fig. 1 [8].

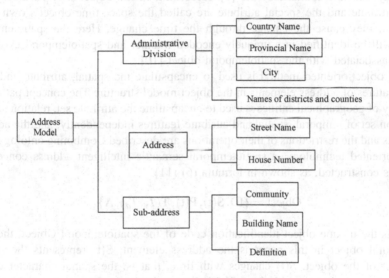

Fig. 1. Address model hierarchy diagram

In the address model hierarchy, the administrative territorial entity is usually used to indicate the administrative region to which the address belongs. The address section is the key part of an address code that displays the main address information by street name and house number. The sub-address section has the function of supplementing the address details.

In the theory of relations and ontology of spatiotemporal semantics, it is based on the structure of the G-metric field and so on. When explaining space-time in the structural realism of space-time theory, we do not pay attention to any causal relationship between object and representation, so the spatiotemporal semantic model is an abstract concept, which is not widely used in daily life. The commonly used address model in China is the hierarchical relation model, which usually constructs the relationship between address elements in the form of a hierarchical model. In previous address model building work, the spatial relation of address elements is often neglected. The complex and confusing phenomenon of address coding in our country brings a lot of inconvenience to address coding management. Therefore, construction of intelligent address model based on spatiotemporal semantics is the basis of designing urban intelligent address coding method.

3   Construction of Intelligent Address Model Based on Spatiotemporal Semantics

3.1   Spatiotemporal Semantic Intelligent Address Conceptual Model

In real life, phenomena are generally ascribed to entities. In the computer application system, the object is the digitized form of the entity, while the spatiotemporal object is the entity with the spatial attribute [9]. According to the object-oriented theory, the space attribute and the special attribute are called the space-time object's own characteristic, they cause the change through the time change. Here the spatiotemporal object will be identified in a uniquely encoded manner and spatiotemporal evolution will be associated with the spatiotemporal object [10].

The object-oriented method is used to encapsulate the spatial, attribute and temporal features of address elements in the object model structure The concept pattern of ontology of geographical entities is used to encapsulate the attribute set, relation set and operation set of temporal, spatial and attribute features independently, and the address elements and the restrictions of their operations are described Combining ontology with object-oriented technology, a spatiotemporal semantic intelligent address conceptual model is constructed, as shown in formula (6) [11].

$$\text{Object} = \{\text{ID}, \text{S(t)}, \text{P(t)}, \text{T}(T_v, T_d), \text{A}\} \tag{6}$$

ID is the unique object identification code of the spatiotemporal Object, the spatiotemporal object in this paper is the address element, S(t) represents the spatial character of the object, S(t) changes with time, that is, the spatial character of the address element changes with time; P(t) indicates the attribute character of the object, P(t) changes with time, that is, the attribute character of the address element changes with time, and T (T_v, T_d) indicates the temporal character of the change of the object's state T_v is the effective time; T_d is the database time; A is the behavior operation of the spatial object, that is, the operation of the time, space and attribute of the address element [10]. The construction of object-oriented spatiotemporal semantic address conceptual model can well represent the change of address elements with time and space.

3.2   Spatiotemporal Semantic Intelligent Address Logic Model

(1)   Abstraction and organization of logical model

In the spatiotemporal semantic intelligent address model, we abstract the address elements, organize the hierarchical relations among address objects, especially the spatial relations among address elements, and construct the intelligent address model by object-oriented method as shown in Fig. 2 [11]:

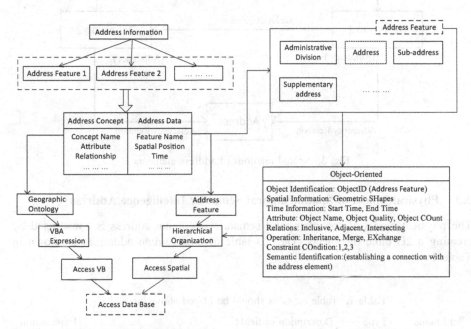

Fig. 2. Object-oriented spatiotemporal semantic intelligent address logic

(2)   Spatial relations of Object-Oriented spatiotemporal semantic address elements

When constructing the logical model of urban intelligent address, the spatial relation of address elements is designed, and the spatial relation of address elements in a complete address code is established. The spatial relation of address elements is shown in Fig. 3 [12]:

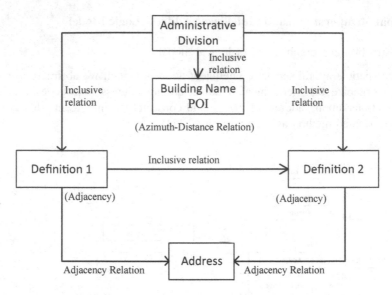

Fig. 3. Spatial relations of address elements

3.3 Physical Model of Spatiotemporal Semantic Intelligence Address

The physical model of spatiotemporal semantic intelligent address is constructed by creating a 2D table of objects. The 2D table of smart urban addresses is shown in Table 1:

Table 1. Table captions should be placed above the tables.

Field name	Type	Description of fields	Explanation
id	String	Primary Key	
address code	Integer	Address Coding	Uniqueness
sjxzdm	Integer	Provincial Administrative Code	
sjxzdm	Integer	Municipal Administrative Code	
qxxzdm	Integer	District and county administrative codes	
jdxzdm	Integer	Street Administration Code	
Sqxzdm	Integer	Community Neighborhood Committee Administrative Code	
zbdm	Integer	Group Code	
dl, jddm	Integer	Road, Street Code	
mpdm	Integer	House numbering	
xqdm	Integer	Cell Code	
ld(z)dm	Integer	Building Code	
dydm	Integer	Unit Code	
fjdm	Integer	Room Code	

(continued)

Table 1. (*continued*)

Field name	Type	Description of fields	Explanation
smartaddress	String	Smart Address	
x	Double	Longitude	
y	Double	Latitude	
gridid	Variant	Grid ID	
addtime	Date	Acquisition Time	
updatetime	Date	Update Time	
disabledtime	Date	Stop time	

The 2D table of provincial coding information is shown in Table 2:

Table 2. 2D table of provincial coding information.

Field name	Type	Null	Description of fields	Explanation
sfmc	String	n	Name of Province	
sfxzdm	Integer	n	Provincial Administrative Code	
qysj	Date	y	Date of launch	
qybs	Integer	y	The sign to start using	
tysj	Date	y	Stop time	

The table of municipal coding information is shown in Table 3:

Table 3. 2D table of municipal coding information

Field name	Type	Null	Description of fields	Explanation
sfmc	String	n	Name of Province	
sfxzdm	Integer	n	Provincial Administrative Code	
qysj	Date	y	Date of launch	
qybs	Integer	y	The sign to start using	
tysj	Date	y	Stop time	

The 2D table of district code information is shown in Table 4:

Table 4. 2D table of district code information

Field name	Type	Null	Description of fields	Explanation
qxmc	String	n	District and county name	
qxxzdm	Integer	n	District and county administrative codes	
qysj	Date	y	Date of launch	
qybs	Integer	y	The sign to start using	
tysj	Date	y	Stop time	

The 2D table of Urban Street coding information is shown in Table 5:

Table 5. 2D table of urban street information coding

Field name	Type	Null	Description of fields	Explanation
csjdmc	String	n	City street name	
csjddm	Integer	n	City street code	
qysj	Date	y	Date of launch	
qybs	Integer	y	The sign to start using	
tysj	Date	y	Stop time	

The 2d table of neighborhood committee coding information is shown in Table 6:

Table 6. 2D table of coding information of community neighborhood committee

Field name	Type	Null	Description of fields	Explanation
csjdmc	String	n	City street name	
csjddm	Integer	n	City street code	
qysj	Date	y	Date of launch	
qybs	Integer	y	The sign to start using	
tysj	Date	y	Stop time	

The 2D table of group coding information is shown in Table 7:

Table 7. 2D table of group coding information

Field name	Type	Null	Description of fields	Explanation
zbmc	String	n	Group name	
zbdm	Integer	n	Group code	
qysj	Date	y	Date of launch	
qybs	Integer	y	The sign to start using	
tysj	Date	y	Stop time	

The 2D table of urban road coding information is shown in Table 8:

Table 8. Table of urban road coding information

Field name	Type	Null	Description of fields	Explanation
csdlmc	String	n	Name of city road	
csdldm	Integer	n	City road code	
qysj	Date	y	Date of launch	
qybs	Integer	y	The sign to start using	
tysj	Date	y	Stop time	

The 2D table of door plate information is shown in Table 9:

Table 9. 2D table of door plate information

Field name	Type	Null	Description of fields	Explanation
mphm	Integer	n	House numbering	
mphmdm	Integer	n	House number code	
qysj	Date	y	Date of launch	
qybs	Integer	y	The sign to start using	
tysj	Date	y	Stop time	

The 2D table of cell coding information is shown in Table 10:

Table 10. 2D table of cell coding information

Field name	Type	Null	Description of fields	Explanation
xqmc	String	n	District name	
xqdm	Integer	n	Cell code	
qysj	Date	y	Date of launch	
qybs	Integer	y	The sign to start using	
tysj	Date	y	Stop time	

The 2D table of building (building) coding information is shown in Table 11:

Table 11. 2D table of building coding information

Field name	Type	Null	Description of fields	Explanation
ld(z)mc	String	n	Building name	
ld(z)dm	Integer	n	Building code	
qysj	Date	y	Date of launch	
qybs	Integer	y	The sign to start using	
tysj	Date	y	Stop time	

The unit coding information is presented in two dimensions as shown in Table 12:

Table 12. 2D table of unit coding information

Field name	Type	Null	Description of fields	Explanation
dymc	String	n	Unit name	
dydm	Integer	n	Unit code	
qysj	Date	y	Date of launch	
qybs	Integer	y	The sign to start using	
tysj	Date	y	Stop time	

The 2D table of household coding information is shown in Table 13:

Table 13. 2D table of household coding information

Field name	Type	Null	Description of fields	Explanation
fjmc	String	n	Room name	
fjdm	Integer	n	Room code	
qysj	Date	y	Date of launch	
qybs	Integer	y	The sign to start using	
tysj	Date	y	Stop Time	

3.4 Design of Urban Intelligent Address Coding Database

At present, the data model of most of the urban address coding databases in our country is confused, there is no unified standard for address coding in modeling, and the spatial relation of address elements in the data model is not strong. In addition, there are still some deficiencies in the current urban address coding database. For example: The database system lacks the rich structure, can't realize the geographic information industry's resources sharing, can't achieve the network interconnection and the address code data intelligent management and so on. In the construction of the frame of urban intelligent address coding database, it is very important to design a scientific and reasonable process of building the database. The process of building the library is as follows [13]:

(1) To determine the key words of the urban address element, and to build the database of the key words of the address element, such as which district the address belongs to, which street and the information of the address's house number, etc.
(2) The effective address information in the existing social resources, such as cadastral management system, Real Estate Registration Management System and the address information from the Internet, is extracted and integrated into an address source database.
(3) Obey the code criterion of urban intelligent address, combine the technology of semantic analysis to analyze the address source database, and from the scientific and standard address data.
(4) The Unified Standard address information is proofread and coordinates are collected to form the address data.

The process of building an urban intelligent address coding database is shown in Fig. 4.

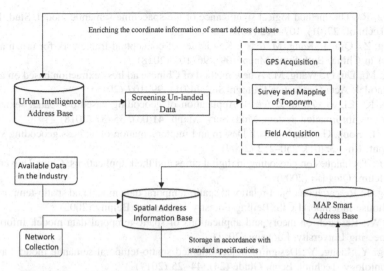

Fig. 4. Process of building intelligent address coding database in city

4 Conclusion

Urban intelligent address coding is one of the basic work of smart city construction, and plays an important role in the frame construction of the smart city. Due to the influence of history, topography and culture, the development of urban address coding technology in China is slow, but overseas urban address coding technology is not in line with China's national conditions. Therefore, this paper studies the encoding method of urban intelligent address based on spatiotemporal semantics, and designs an intelligent address model based on address model theory, smart city framework and spatiotemporal semantic model methodology including spatiotemporal semantic intelligent address conceptual model, spatiotemporal semantic intelligent address logical model and spatiotemporal semantic intelligent address physical model, based on the spatiotemporal semantic intelligent address model, a frame of city intelligent address coding database is designed.

Acknowledgment. The project was funded by Guangxi Key Laboratory of Spatial Information and Geomatics (15-140-07-14,16-380-25-17) and the 'Ba Gui Scholars' program of the provincial government of Guangxi.

References

1. Tong, W., Jiang, Z., Li, X.: Research on the address data content specification. Stand. Sci. (11), 39–42 (2009)
2. Zhang, H., Kong, L., Sun, L. (eds.): Research on development and application of urban GeoCoding. Bull. Surv. Mapp. (07), 58–60 (2008)

3. Cheng, R.: The method logical significance of the spacetime semantic model. Stud. Philos. Sci. Technol. **27**(01), 107–112 (2010)
4. Zhang, Z., Qiu, J., Kang, M. (eds.): Key issues of conceptual framework for urban address model in China. Bull. Surv. Mapp. (09), 96–102 (2018)
5. Kang, M., Du, Q., Wang, M.: A new method of Chinese address extraction based on address tree model. Acta Geodaetica Cartogr. Sin. **44**(01), 99–107 (2015)
6. Kang, K., Li, M., Zhou, Z. (eds.): Application of geocoding based on spatial semantic to smart city information system. Mod. Surv. Mapp. **41**(03), 35–37 (2018)
7. Guo, H., Song, G., Ma, L. (eds.): Design and implementation of address geocoding system. Comput. Eng. **35**(01), 250–252 (2009)
8. Ye, H.: Techniques on geocoding in digital cities and their applications. China University of Petroleum, Qingdao (2009)
9. Zhang, J.: Research on object-relational spatio-temporal data model and spatio-temporal data warehouse in temporal GIS. Beijing Forestry University, Beijing (2002)
10. Cao, W.: Research on theory and applications of spatio-temporal data model. Information Engineering University, Luoyang (2011)
11. Zhang, Y., Liang, Y.: Design of object-oriented spatio-temporal semantic model based on geo-ontology. Technol. Econ. Guide (24), 44+25 (2017)
12. Zang, Y., Wang, B., Qu, X.: Construction method of chinese semantic address model in Chongqing. Geospat. Inf. **13**(03), 122–125+12 (2015)
13. Guo, L.: Application of software project management in standard address library shared service platform project development. Zhejiang University of Technology, Hangzhou (2017)

Advanced Human Centric 5G-IoT in a Smart City: Requirements and Challenges

Dragorad Milovanovic[1(✉)], Vladan Pantovic[2], Natasa Bojkovic[1], and Zoran Bojkovic[1]

[1] University of Belgrade, Belgrade, Serbia
dragoam@gmail.com
[2] Union – Nikola Tesla University, Belgrade, Serbia

Abstract. The concept of a smart city (SC) is developed on advanced information-communication technologies (ICT). The role of new 5G mobile wireless systems and Internet of Things (IoT) is discussed as essential for a successful SC infrastructure for interconnecting citizens in physical and virtual worlds. IoT-based smart cities can provide various kinds of services for citizens, including smart home, smart energy and meter systems, vehicular traffic and transportation, environmental pollution, smart health. The services are transforming cities by improving infrastructure and efficiency of the everyday activities. The huge development of smart devices, novel communication, computing and control technologies have paved the way for 5G-IoT advancement. At the same time the final goal is to provide efficient human-centric communication. In this article smart cities infrastructure evolution is presented. Moreover, an overview of the role of 5G-IoT including up-to-date development is described. Later, we continue with human-centric 5G-IoT in SC. Finally, in order to give future research directions several challenges are highlighted.

Keywords: Smart environment · Smart city · 5G IoT · HCI

1 Introduction

The smart cities (SC) investment in human and social capital, as well as information-communication infrastructure, brings economic growth and a high quality of life [1]. SCs are connecting the physical infrastructure, the ICT and the social infrastructure. Thus, the primary goal is to offer the citizens services which will improve not only the quality, but also efficiency of the everyday activities. SCs understand the information exchange with the large volume of data assuming the range of different application and specific requirements. Today the global amount of urban population is more that 50 percent and is expected to be about 70 percent by 2050 [2]. The continuous need for cost-effectiveness has pushed towards new strategies inspired by ICT research fields.

Smart cities are the product of contemporary development integrating ICT and IoT. Thus, new fifth generation communication technology 5G–IoT will be introduced. There are three fundamental technologies for 5G, massive machine type communication (mMTC), ultra-reliable and low latency communication (uRLLC) and enhanced mobile broadband (eMBB). Service category eMBB is characterized by the human user

© Springer Nature Switzerland AG 2019
D. Milošević et al. (Eds.): HCC 2019, LNCS 11956, pp. 285–296, 2019.
https://doi.org/10.1007/978-3-030-37429-7_28

access to multimedia content and is supported by cloud radio access networks (C-RAN), massive multiple-input-multiple-output (MIMO), small cell, etc. Category mMTC is characterized by a very large number of power-constrained devices, where each device automatically and infrequently transmits a low volume data in bursts. Service category uRLLC has strict requirements concerning availability from one end to another with ultra-high reliability and dead line-based low latency [3]. It should be noted that massive access and low latency networking remain to be discussed and improved beyond 5G [4].

5G networks will have a significant influence in the development of the SC infrastructure. For example, 5G networks provide higher and faster transmission capabilities, larger bandwidth and higher data capacity. To reduce limitation in the SC framework, basis device-to-device (D2D) communication is able to restore large-scale content, while devices are connected directly [5]. Mobile edge computing (MEC) and cloud computing (CC) are complementary and the most promising approach when speaking about 5G content delivery network scenario in SC. Cloud computing introduces processing and storage resources to the lower layers, while MEC reduces pre-processing and gives a corresponding interface between the base station (BS) and the upper layer [6]. With introduction of 5G, we are going to combine network slicing, cloud radio access network, various cell sizes and different kind of supported traffic (machine-to-machine, machine-to-human).

IoT is one of the important ICT technologies to enable smart cities, along with connectivity, data analytics, and artificial intelligence (AI). Many countries have initiated smart city projects with focuses on smart homes, smart transport, and smart health. For homes, IoT enables home automation, enhances home safety, monitoring of children and elderly, and allows efficient use of electric energy. For transportation, IoT has enabled on-demand bus or shared car rides, smart traffic lights, smart streetlights, sensor-based autonomous vehicles. The IoT can be used in conjunction with roads and highways to increase safety, reduce congestion and overall costs of maintenance, improve traffic flow, and provide a more pleasant experience for drivers and passengers. Finally, as for smart health, sensors, artificial intelligence, and data analytics enable health specialists to predict and prescribe medical treatment to patients in an advanced and timely manner, providing better healthcare.

The paper is organized as follows. Smart city evolution, including 5G and IoT technology development, is outlined in the first part. An infrastructure based on 5G-IoT is analyzed in the second part. Finally, human centric communication is pointed out in third part of the paper.

2   Smart City Evolution

The smart city refers to new deployment of information and communication technologies in the combination and integration with urban functions. The term can also be described as the convergence of ICT, energy technologies, and support facilities within urban and residential environments. Smart city evolution includes twenty-year development of traditional applications (traffic, healthcare, power grid, water and waste treatment). SC started to grow as feasible and sustainable concept in the 1990's based on available computing power, introduction of web services and mobile radio networks. It became clear that it was possible to connect and integrate remote devices with processing platform in almost any place and in any time. People and devices became fully connected in the late 2000's based on broadband mobility, initial steps in cloud computing (CC) and sensors in traffic control, smart meters, navigation. Smart applications have became everyone's need (smart home, smart grid, smart healthcare, smart transport, smart government).

Today, almost all cities intend to be more or less smart with regard to a different level of new technologies. Consideration will be completed when monitoring and integration of the critical infrastructure, optimization of resources, planning maintenance activities, monitoring of security as well as maximizing services to the citizens are included. The infrastructure consists of four parts:

- physical (power, water, roads, bridges, rails, airports, seaports)
- ICT infrastructure (telecom networks, data centers, access points, terminal devices, available services)
- social infrastructure (hospitals, media, social clubs, social networks)
- business infrastructure (companies, campuses).

While social and business infrastructure has attracted more attention in the recent years, ICT infrastructure have dominant status in SC. On the other hand, the four technological layers of SC architecture are subject to changes and transformation:

- sensing layer (sensor or terminal devices)
- network layer (telecommunication, media and associated services)
- operation, administration, maintenance and provisioning, security
- application layer.

The evolution of telecommunication technology started from narrowband medium for public use, next more broadband and available access, fixed broadband and mobile, to the current mobile broadband access. The implementation of 5G will enable IoT to make the cities much smarter [3]. Billions of devices will have been connected within a very high speed mobile network by the first phase of 5G to be started in 2020. According to Intellipaat, $6 billion will be spent on IoT technologies from 2016 to 2021, and estimated revenue will be $13 trillion by 2025. According to *Statista*, the size of the worldwide IoT market would hit $2.225 trillion in 2020, and by 2025 there will be over 75 billion IoT connected devices installed worldwide.

IoT is about automation, enabling an intelligent and controllable environment. It is really on the connectivity, data formatting, and handling ends. There are different use cases that will demand different IoT protocols, data formats, data handling, and interfaces. Currently, there is no one universal standard for IoT. Without a single standard, there is a need for interoperability and possibly open standards. The issue is that IoT application and use cases are too diverse, and industry efforts have been focused and segregated.

The 3GPP consortium (3rd Generation Partnership Project) develops conditions and requirements for 5G mobile communications in two phases: the first phase was completed by the end of 2018 (Release 15), the second phase is going to be completed at the end of 2019 (Release 16). 3GPP standardization includes the RAN (Radio

Access Network), network backbone (5G Core), terminal (CT) and service architecture (SA). The 4G LTE-Advanced technical specifications R13 and R14 were enhancements towards 5G. The purpose of the request is to provide cost-effective connections for a large number of IoT/M2M units with very low consumption and excellent coverage of the service zone (Table 1).

Table 1. 3GPP IoT standardization process.

3GPP machine type communication (MTC) development has started with the technical specifications R10 and R11. The R12 specification focuses on the economy and expansion of the service zone. Specification R13 introduces Narrowband-IoT (NB-IoT), eMTC and EC-GSM-IoT (Extended Coverage GSM-IoT) radio access technology that relies on LTE technical components, but also supports standalone operations by introducing individual channel structure. Specification R14 introduces improvements (authorization coverage enhancement, reuse legacy multicast/broadcast system). Specification R15 focuses on completing 5G NR NSA (Non-Standalone Mode) standards for eMBB service, as well as setting the foundation for development SA (Standalone Mode). The standard completed in December 2017 supports the NSA system that is based on the existing LTE infrastructure when added the new 5G radio network. Specification R16 by the end of 2019 is going to support new types of service/devices, new implementation/business models and new spectrum sharing types. The uRLLC communication, the use of the unlicensed spectrum and the new spectrum allocation paradigm, communication with autonomous vehicles (5G NR C-V2X), and continuation of 3GPP LPWA (low-power wide-area) technologies (NB-IoT/eMTC) are being developed.

3   5G-IoT Infrastructure

With the current development of mobile communications technology, the 5G application scenarios defined by the 3GPP provide three communication modes: enhanced mobile broadband (eMBB), massive machine type communication (mMTC), and ultra reliable and low latency communication (uRLLC) [3]. However, the heterogeneity of IoT applications and devices leads to higher requirements for low cost, energy efficiency, wide coverage, low delay, and reliable communication. To satisfy complex and diverse user computing tasks and content requests, operators deploy cloud computing technology to overcome the limitations of storage and computing in IoT devices. Furthermore, the deployment of artificial intelligence (AI) technology in the remote cloud and edge cloud is an effective way of implementing intelligent services. Characteristics of 5G technology for city benefits are as follows:

- 5G technology provides better coverage and performance in open spaces and buildings (densely populated city areas, stadiums, public transport)
- reliable high data flows enable consumers to download high-resolution multimedia, city offices can download high-resolution videos
- 5G represents an alternative to fiber optics, enabling the wireless network to reliably cover areas where today's price is too high for fiber optic applications
- future communications networks are adaptive with programming capability for best support of applications
- high energy efficiency of the network so that the battery life of the IoT devices of low power reaches 10 years, reducing the cost of maintenance and replacement of batteries
- reducing the response time or delay in the 5G network at 1 ms, combination of real-time communication with streaming and capacity for multimedia can become a reality
- combining wireless networks deals with integration gateways, licensed or unlicensed networks to be managed as a single network
- when speaking about quality of experience (QoE), 5G culminates into great reliability, while the overall experience for the person are improved.

5G PPP (Infrastructure Public Private Partnership) is the joint initiative of the EU and the ICT industry which defines the following groups of uses: densely populated urban areas, broadband access at any location, connected autonomous vehicles, smart office of the future, Narrow-band IoT, Tactile Internet/automation. For each individual case, key KPIs (Key Performance Indicators) are defined corresponding to performance requirements (in terms of user service experience) supported by the 5G network. 5G PPP classifies the basic functionality on the following parameters: density of the device in a given space, mobility, infrastructure topology, type of traffic, user data flow, latency, reliability, availability, category 5G communication. The following competencies are suggested: peak data flow 20Gbps, user data flow 100Mbps, latency 1 ms, mobility 500 km/h, density of connection 106/km2, energy efficiency 100x4G, spectral efficiency 3x4G, traffic capacity 10Mbps/m2.

Significant advantages of 5G technology are higher speeds and more connections, enabling wireless connectivity of specific locations; faster and more responsive response time supported by time-sensitive applications, as well as ultra-low power connections. The challenge of sustainable city development and their improvement to become smart is a large amount of effective and efficient investment in infrastructure. There are numerous initiatives focused on the analysis of process conception, implementation methods, and smart city outcomes (Table 2). Understanding the advantages and disadvantages is necessary not only for planners, managers and project directors, but also for people living in smart cities.

Table 2. Sectors and core issues for IoT technologies in a smart city.

Sectors	Key services	Technologies
Automation and transportation	Remote parking management Business fleet management Vehicle telematics	• Autonomous or remote control services • Management systems using IoT technologies • Utilizing individual mobile devices and connected automobiles
Energy and electricity	Automation of transmission and distribution Optimization, management, and reduction of accommodated energy	• Various communication standards • Essential part of smart grid systems • One of the biggest potential markets in IoT technologies
Healthcare and monitoring	Smart healthcare Smart hospital	• Rapidly expanded markets • Tracking medical teams and facilities located in hospitals • Electronic medical records and communication systems
Architecture and building	Building management Building automation Home automation	• Different building preferences • IT solution or telecom service providers with newly built building

3.1   IoT Infrastructure

Smart City is a complex ecosystem characterized by the intensive use of ICT, with the aim of making cities more sustainable as a unique place for innovation and entrepreneurship. The main participants are application developers, service providers, citizens, government and public services, the research community and development platform. The cycle consists of numerous ICT technologies, development platforms, maintenance and sustainability, citizen applications as well as technical, social and economic key performance indicators (KPIs). IoT systems have a fundamental role in the implementation of mass heterogeneous infrastructures. The development of 5G enables G-IoT (Green IoT) and I-IoT (Intelligent IoT) technology for smarter cities.

Intelligent city-based applications based on IoT infrastructure are categorized by the type of network, scalability, coverage, flexibility, heterogeneity, repeatability, and end-user engagement. Applications can be grouped into personal and household, communal, mobile, and business. Utilities include smart energy grid, smart metering/monitoring, water network monitoring, and video surveillance. Similarly, mobile applications include ITS (Intelligent Transport System) and logistics, traffic management, and waste management.

IoT applications in a smart city enforce many demands. For example, IoT is expected to offer low prices, low energy consumption, high quality of service (QoS), broad coverage, greater flexibility, high security and privacy, ultra-dense implementation and interoperability of multiple manufacturers. IoT deployment is based on a number of network topologies to achieve a completely autonomous environment. Capillary IoT networks offer short distance services. Examples include wireless local area network (WLAN), body area network (BAN), and Wireless Personal Area Network (WPAN) wireless networks. Fields of application include e-health services, home automation and street lighting. On the other hand, applications such as intelligent transport system, mobile e-health and waste management use WAN (Wide Area Network) networks, MAN (Metropolitan Area Network) networks and mobile communications networks. These networks have different characteristics in terms of data, size, coverage, latency and capacity requirements. In such an environment, IoT systems have got a huge role in the deployment of large-scale infrastructures. In addition, the integration of IoT and Big data has invoked new challenges like processing and storage for future smart cities [7–9]. The objective is very clear, to promote better management of various sectors [10].

In a networking architecture scenario for IoT-based SC, it is all about integration of technology with sensing networking, information processing and control in real-time in order to efficiently utilize public resources. In that way, smart metering and manufacturing, smart houses, automatic driving, health monitoring are included in offering a broad range of intelligent applications [11–13]. IoT devices are applied to monitor the physical world and multimedia content such as image, video and textual data.

3.2 Smart Grid

The energy infrastructure is the most important feature in any city. Smart cities depend on a smart grid to ensure resilient delivery of energy to supply their many functions, improve efficiencies and enable coordination between infrastructure operators. Advanced metering infrastructures and advanced data analytics gather, assess and formulate essential information to improve operational decision-making. For improving energy efficiency and saving energy cost, intelligent management of energy allocation and consumption (smart metering, smart lighting and smart grid) has lots of benefits. In that way, the application of renewable energy sources (solar and wind energy) is promoted.

The smart grid (SG) is an electrical network that integrate the actions of all users connected to it in order to efficiently deliver sustainable, economic and secure electricity supplies. A smart grid employs innovative products as services together with monitoring, control, communication and self-healing technologies. The main goal when introducing smart grids can be presented as follows:

- accommodate a wide variety of distributed generation and storage systems in order to significantly reduce any impact of the environment, increasing at the same time the efficiency of the electricity supply chain
- provide consumers with relevant information so that they can make their choices and provide services such as integration of electrical vehicles
- improve existing standards of reliability and quality of supply
- allow new business models
- help optimize the utilization and maintenance of assets in a Smart city.

Recently, it has been proved that applying IoT can optimize the management of the renewable energy resources by enabling sensors and data analytics. Indeed, the renewable energy systems based on IoT can facilitate the communication between the renewable energy appliances to optimize the energy efficiency. For example, the energy regeneration will be switched from solar to wind turbine in a particular cloudy day with a lot of wind. On the other side, IoT analytics have a significant effect of renewable energy outperforming. For instance, extracting data from solar panels may help to speed up the maintenance process. Many new ideas and research results in these fields continues to emerge.

With introducing the 5G mobile cellular networks are being developed into SG platform for acquisition, communication, storage and processing. At the same time, 5G will provide on its side, services for real-time applications in smart grids. Mobile edge computing (MEC) and localized processing and management are dramatically decreasing service response latency. The concept is not in collision with mobile cloud computing (MCC), they complement each other in building flexible and configurable 5G network using a network slicing approach, where different services may be easily instantiated using different virtualized architectures on top of the high-performance MEC host nodes. Instrumental to the development of flexible core network are reconfigurable architectures based on SDN (Software Defined Networking) and NFV (Network Function Virtualization).

3.3 Human-Computer Interaction

From the first generation (1G) to the evolving (5G) mobile communication systems, IoT realize the interconnection between things and things as well as people and things based on a huge technology development [14]. Today heterogeneous IoT devices can implement many function that were sometimes ago limited to computers. Next, computing power of the remote cloud is migrating to the edge of the IoT, taking into account that clouds access mode cannot meet the quality of experience (QoE) in human-centric applications.

The deployment of artificial intelligence (AI) technology in the remote cloud and edge cloud (EC) represents a possibility to implement intelligent services [15]. An example of AI-enabled smart edge with heterogeneous IoT architecture is shown in Fig. 1 [16]. In order to monitor and adjust the configuration of network resources and realize real-time data collection of IoT, efficient processing of computation as well as intelligent interaction of smart applications have been carried out. Users are the group with the direct connection to the smart services provided by operators in heterogeneous IoT. They receive feedback on the QoE based on the corresponding effect. The services provided by smart terminals can cover the people's living environments. The role of the smart terminals is to collect the user's multi-modal data. The function of the server is to have the user's audio-visual physiological data resources, performing data mining and data processing using cognitive engine.

Intelligent IoT devices consist of multimedia sensors with the function to collect and transmit user data. There are two types of sensors: human and dynamic. While human sensor is responsible for collecting human physiological information, a dynamic one can sense human motion and audio-visual information. Edge cloud consists of edge

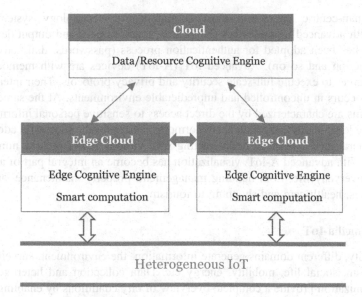

Fig. 1. Heterogeneous IoT architecture with AI-enabled smart edge cloud.

computing nodes (gateways, routers, switches, service servers, wireless access points, micro base stations), whose functions are to migrate the computing power of the remote cloud to the edge of the IoT, to buffer the big data flow generated by smart devices as well as to relieve network congestion during the transmission process. The main part of AI-enabled smart edge with heterogeneous IoT is the cognitive engine. As for computing engines, they consist of data cognitive engines and resource cognitive engines. A data cognition engine includes AI algorithms about data mining, machine learning, deep learning, etc. A resource cognition engine includes software defined networking (SDN), network function virtualization (NFV), self-organizing networks (SONs) and network slicing (NS) technologies. The increase in the number of smart devices and solutions leads to the volume of data communicated across the devices and networks. As for data transfer patterns, they are predicted to degree from man-to-man and machine-to-machine communications models, together with human-computer interaction.

4 Human Centric Communication

The development of wireless communication technologies, the Internet of Things (IoT), and computing science trigger changes in people and machine communication. The IoT technology opens a door for new era into our lives with a wide application of use cases from smart city to smart grid [17]. In order to provide sensitive operation, huge security degree is important [18, 19]. On the other side, by introducing human-centric advanced 5G-IoT infrastructure, multimedia-IoT and smart vehicles are developed.

For human-centric interaction most communication technology systems are equipped with advanced input devices (keyboards, touch screens and output devices). The system has been adopted for authentication process (passwords, daily advanced visual instruction and so on). Advanced IoT (A-IoT) devices are with memory and battery resources to execute full-scale security and privacy protocols. Their interaction with a user occurs in uncontrolled and unpredictable environments. At the same time, A-IoT systems are characterized by the direct access to sensitive personal information. The response level of authentication is of enormous interest for the successful adoption. In future smart cities, users will be interacting with various A-IoT devices numerous time a day. With advanced A-IoT, visualization has become an integral part of a smart city of our everyday life, from building management, property maintenance, autonomous vehicles, health care and shopping to tourism.

4.1 Multimedia-IoT

In a smart city, different domains generate information, the environment, surveillance, traffic, tourism, social life, mobility, energy etc. Data collection and heterogeneous sensors organization provide a complete overview of city conditions by enabling early warning and development of predictive models, limiting adverse effects to citizens who derive from unpredictable events. In this scenario, environmental monitoring and visualization are specific applications. A smart video surveillance system is capable of automatically analyzing video streams of a large number of distributed high resolution 4 K cameras. Scenarios of 5G eMBB communications are the monitoring of visual descriptors and management of different areas of the city, recognition and management of traffic incidents, recognition of dangerous and undesirable events.

In a visual Internet view, sensors such as cameras and the distribution of multimedia content are key components of smart cities. However, the high bandwidth requirements for visual data and the gap between computer capabilities and communication are challenging the development of visual Multimedia-IoT (M-IoT). Key components of the system are intelligent perception of sensitive environment, smart video analysis for compression of visual data, software-defined video for generating elastic visual streams, flexible control of optimal adaptation, and efficient transmission for resource utilization enhancement.

4.2 Vehicle-to-P Communication

While some small systems and proofs of concept have already been demonstrated, no comprehensive IoT-based transportation systems have been fully deployed. Connected vehicles involve wireless technologies that enable interactions between specially equipped cars. Connected vehicles can also V2V communicate with wireless access points on or near the roadside (vehicle-to-vehicle communications). These special access points can be incorporated into the roadway, installed in smart traffic lights, or utilized as mobile communication points such as cell towers that can connect with various cloud-based services. The potential benefits are significant. IoT-enabled services can aggregate information shared by connected vehicles and provide analytics that enable advanced information and navigation services. IoT-based traffic monitoring systems can use a combination of V2V, vehicle-to-infrastructure (V2I), and infrastructure-to-infrastructure communications and analytics to manage traffic situations. Recently, there is also vehicle-to-everything (V2X), effectively talking to any object or device within proximity. Wireless Access in Vehicular Environments (WAVE) IEEE 802.11p standardizes V2V and V2I communications [20]. The standard addresses at the PHY and MAC layers, focusing on operating frequency, higher rate of mobility, real-time requirements, use of alert messages. It is envisaged that all future smart cars (including autonomous vehicles) will have some sort of V2X capability. The IEEE 1609 family of standards for WAVE define the architecture and standardized set of services and interfaces for secure V2V and V2I wireless communications.

Cellular Vehicle-to-everything (C-V2X) communication is the passing of information from a vehicle to any entity, and vice versa. It is a vehicular system that incorporates V2P (vehicle-to-pedestrian), V2I (vehicle-to-infrastructure), V2V (vehicle-to-vehicle), V2D (vehicle-to-device) and V2G (vehicle-to-grid) communication. 3GPP started standardization work of cellular V2X in Release 14 in 2014, based on LTE technology and specifications that were published in 2016. Standardization was continued based on 5G technology, and Release 15 specifications were published in 2018. In Release 16, the work is currently in progress. Direct V2X communication technology leads to the possibility to communicate with road users (pedestrian, cyclist) by having PC5 interface to be integrated into smartphones. Vehicle-to-person (V2P) includes vulnerable road user scenarios to detect pedestrians and cyclists to avoid accident and injuries involving those road users [21].

5 Concluding Remarks

This work reviews recent advances in smart cities development including the role of 5G-IoT networking, their requirements and challenges. The important features of 5G network connectivity are high data rate transmissions and very low latencies. SC systems have been applied in different regions for examining the potential of new smart systems along with identifying the infrastructure needs for supporting the future services for a large number of devices. There are still much more to be done in order to develop integrated solutions and create novel application to build smart cities around the world with a sustainable society. The final objective is to improve human and citizen social life including economic growth.

IoT is one of the important ICT technologies to enable smart cities, along with connectivity, data analytics, and artificial intelligence. Many countries have initiated smart city projects and invested billions into the development of smart cities. For example, in China alone, more than 300 pilot smart city projects have been initiated. In Singapore, the country has introduced the *Smart Nation* program that focuses on smart homes, smart transport, and smart health. Understanding the advantages and challenges of 5G-IoT technology is crucial to the development of future smart cities.

References

1. Qian, Y., et al.: The Internet of Things for smart cities: technologies and applications. IEEE Netw. **33**(2), 4–5 (2019)
2. Urban Population Growth: World Health Organization, GHO Data 2019
3. Milovanovic, D., Bojkovic, Z., Pantovic, V.: Evolution of 5G mobile broadband technology and multimedia services framework. In: Fleming, P., Lacquet, B.M., Sanei, S., Deb, K., Jakobsson, A. (eds.) ELECOM 2018. LNEE, vol. 561, pp. 351–361. Springer, Cham (2019). https://doi.org/10.1007/978-3-030-18240-3_32
4. Ford, R., et al.: Achieving ultra-low latency in 5G Milimeter Wave cellular networks. IEEE Commun. Mag. **55**(3), 196–203 (2017)
5. Wu, D., Zhan, L., Cai, Y.: Social-aware rate based content sharing mode selection for D2D content sharing scenarios. IEEE Trans. Multimedia **19**(11), 2571–2582 (2017)
6. Hu, H., et al.: Joint content replication and request routing for social video distribution over cloud CDN: a community clustering method. IEEE Trans. CSVT **26**(7), 1320–1333 (2016)
7. Chaurabi, H., et al.: Understanding smart cities: an integrative framework. In: Proceedings of HICSS, pp. 2289–2297 (2012)
8. Fan, W., Bifet, A.: Mining Big Data: current status and forecast to the future. ACM SIGKDD Explor. Newsl. **14**(2), 1–5 (2013)
9. Yaqoob, L., et al.: Enabling communication technologies for smart cities. IEEE Commun. Mag. **55**(1), 112–120 (2017)
10. Mehmood, Y., et al.: Internet-of-Things-based smart cities: Recent advances and challenges. IEEE Commun. Mag. **55**(9), 16–24 (2017)
11. Zhou, L., et al.: Greening for smart cities: energy-efficient massive content delivery via D2D communications. IEEE Trans. Ind. Inform. **14**(4), 1626–1634 (2018)
12. Zhao, L., et al.: Optimal edge resource allocation in IoT-based Smart Cities. IEEE Netw. **33**(2), 30–35 (2019)
13. Wu, S., et al.: Survey on prediction algorithms in smart homes. IEEE J. Internet Things **4**(3), 636–644 (2017)
14. Wu, D., et al.: Optimal content sharing mode selection for social-aware D2D communications. IEEE Wirel. Commun. Lett. **7**(6), 910–913 (2018)
15. Zhou, L., et al.: When collaboration hugs intelligence content delivery over ultra-dense networks. IEEE Commun. Mag. **55**(12), 91–95 (2017)
16. Hao, Y., et al.: Smart-Edge-CoCaCo: AI-enabled smart edge with joint computation, caching and communication in heterogeneous IoT. IEEE Netw. **33**(2), 58–64 (2019)
17. Yan, Y., et al.: An efficient security protocol for advanced metering infrastructure in smart grid. IEEE Netw. **27**(4), 64–71 (2013)
18. Lin, Y., et al.: A survey on Internet of Things: Architecture, enabling technology, security and application. IEEE J. Internet Things **4**(5), 1125–1142 (2017)
19. Ometov, A., et al.: Challenges of multi-factor authentication for securing advances IoT applications. IEEE Netw. **33**(2), 82–88 (2019)
20. Cano, J.C., et al.: Evolution of IoT: an industry perspective. IEEE IoT Mag. **1**(2), 82–88 (2018)
21. Merdrignac, P., et al.: Fusion of perception and V2P communication systems for the safety of vulnerable road users. IEEE Trans. Intell. Transp. Syst. **18**(7), 1740–1751 (2017)

Design of Electrical Equipment Integration System for LNG Single-Fuel Bulk Carrier

Qin Nie[1(✉)], Jinlong Lu[2], Quanguo Lu[3(✉)], and Zhifang Zhu[3]

[1] High-Efficiency Generator Engineering Technology Center,
Kungfu Sci-Tech Co., Ltd., Nanchang 330096, Jiangxi, China
1134598717@qq.com
[2] 719th Research Institute, China Shipbuilding Industry Corporation,
Wuhan 430200, Hubei, China
[3] Precision Drive & Control Key Laboratory, Nanchang Institute of Technology,
Nanchang 330099, Jiangxi, China
luqg2010@126.com

Abstract. This paper mainly designed the electrical equipment integrated system for LNG (Liquefied Natural Gas) single fuel bulk carrier, aiming at providing a solution scheme for realizing real-time, reliable and expandable communication between the ship and the shore. The core communication layer of the system adopts star-structure to communicate with the monitored equipment. According to the actual communication ability of electrical equipment, different communication configurations are made. Data access is accomplished by centralized communication protocol conversion through NX control components, which can solve the problem of communication matching under different communication protocols and transmission rates. In addition, for ship-shore communication, the 4G mobile communication links is configured. Under the special sections without mobile signal coverage, the network communication monitoring and data replenishment functions are designed to ensure the integrity and reliability of data remote transmission.

Keywords: Integrated system · Communication protocol · Remote transmission

1 Introduction

Energy saving and emission reduction is the theme of the development of green shipping industry. LNG fuel-powered ships have good environmental protection benefits. With the increasing demand for energy saving and emission reduction in the shipping industry, countries and regions around the world are increasingly green in developing LNG fuel-powered ships [1]. For example, the EU has formulated the layout of alternative fuel infrastructure in Europe, the strategic exchange mechanism of alternative energy in Europe, and the action plan of applying LNG integrated framework in EU. Norway took the lead in formulating LNG ship inspection rules as early as 2001, which has become the standard for LNG ship construction and inspection for

D. Milošević et al. (Eds.): HCC 2019, LNCS 11956, pp. 297–304, 2019.
https://doi.org/10.1007/978-3-030-37429-7_29

many years, at present, one hundred and ten LNG fuel vessels are in operation, and fifteen LNG filling stations have been built along their coasts [2, 3].

Domestic LNG fuel power application started in 2009, and its development started late. However, breakthroughs have been made in the construction of LNG fuel-powered ships and filling equipment in China. According to the statistics, by March 2018, two hundred and seventy-nine LNG fuel-powered ships had been built in China (19.4% of the total number of completed plans), including two hundred and seventy-six inland ships and three seagoing ships, one hundred and sixty-two new ships and one hundred and seventeen renovated ones [4, 5]. The main types of ships were general dry cargo ships, container ships and port tugs. seventeen inland LNG filling docks (73.9% of the total number of completed plans) were built, mainly distributed along the Yangtze River, the Beijing-Hangzhou Canal and the Xijiang River.

In the early pilot demonstration construction, the demonstration zones combined with local reality and took the demonstration construction as an opportunity to promote the establishment of a new inland waterway transport model. Foe example, in the application demonstration zone of Guangxi section of Xijiang trunk line, the undertaking units of Guangxi Xijiang Investment Group Co., Ltd. jointly promoted the relevant logistics, finance, energy and shipyard units to build a symbiotic water transport ecological industry chain [6–9]. With the development of the whole industry chain, the voice of "shipping big data" for LNG ship operation is becoming higher and higher. The establishment of shipping big data depends on real-time, reliable and expandable communication between the ship and the shore [10, 11]. Based on this, this paper presents the electrical equipment integrated system for LNG single fuel bulk carrier, aiming at providing a solution scheme for realizing real-time, reliable and expandable communication between the ship and the shore.

2   Design of General Scheme

2.1   Setting Boundary Conditions

Intelligence of electrical equipment is the most important basis for realizing shipping big data. However, there are many kinds of electrical equipment on ships, for example, large equipment include generators, main distribution boards, small equipment include radar, GPS navigation, and so on. Generally, there are many kinds of mechanical and electrical equipment, and the different manufacturers of the corresponding equipment will lead to the coordination difficulty among the electrical equipment. Based on this, this paper presents a new design scheme of integrated system for electrical equipment, taking the 1000-tons LNG single-fuel bulk carrier as an example.

The ship navigated in inland river, such as Class A, B, C and Beijing-Hangzhou Canal. It is driven by two engines and two propellers, and has steel welded structure. Bulk yellow sand, stones and municipal construction waste (municipal dregs) are the major transport goods. The head and tail of the ship are single deck, single bottom and double hull transverse skeleton. The cargo hold area of the ship is single bottom, transverse skeleton and double side transverse skeleton. There is a 5 m^3 LNG tank at the tail, and a sinking deck chamber at the head and tail.

According to the actual ship type of 1000-tons LNG single fuel power bulk carrier, monitoring boundary conditions are set as follow. There are many kinds of monitoring equipment, and the situation of equipment is relatively complex. The specific monitoring boundary conditions are shown in the table below (Table 1).

Table 1. Boundary conditions

Equipment		Signal type	Signal	Data acquisition
Loading		Analog	Index Information of Weight	NX
Draught		Analog	Water level	NX
steering		Analog	Angle, Hydraulic Pressure	NX
Engine	#1 ECU	Digital	throttle position/speed/Oxygen content/temperature	485
	#2 ECU	Digital	throttle position/speed/Oxygen content/temperature	485
Power Station	Branch 1	Digital	Current/Voltage/Frequency/Alarm Information	485
	Branch 2			485
	Branch 3			485
	Branch N			485
	CFCVG*	Digital	Speed/power/voltage/current/power factor/frequency/alarm signal	485
	Generator		Speed/power/voltage/current/temperature/frequency	485
GPS		Digital	Longitude/Latitude	485
Depth		Digital	depth/width of Channel	485
Video surveillance		Digital	image information	Ethernet
Anemoclinograph		Digital	speed and direction of Wind	485

* CFCVG is short for Constant Frequency Constant Voltage Generator.

2.2 Design of Function Modules

The whole ship-electric integrated system can be divided into the shore-terminal management monitoring platform and the ship-terminal management monitoring platform.

The Shore-terminal Management Monitoring Platform. The shore-terminal management monitoring platform can manage the daily operation and maintenance of the whole ship, and optimize the operation and maintenance management scheme, so as to improve the utilization efficiency of equipment, save energy consumption and capital costs continuously. The six main functional modules of the shore-terminal management monitoring platform are shown as follows.

(1) Data acquisition module. Quasi-real-time ship information is acquired through 4G network.
(2) Real-time monitoring module. Real-time information of each ship is displayed in graphics, tables and other ways.
(3) Efficiency analysis module. The energy efficiency of ships can be analyzed by analyzing the information of load, LNG flow, output power and speed, and the

functions comparison can be carried out to provide data basis for improving the working efficiency of ships.

(4) Maintenance module of ship equipment. According to the operation data and maintenance information issued by the ship, the shore-terminal management monitoring platform can reasonable dispatch of maintenance personnel arrived at the ship site in time for equipment maintenance.

(5) Stowage and transportation management module. Through real-time monitoring of the ship's transport situation, combined with the state of cargo to be transported, it can provide decision-making basis for reasonable and efficient ship-dispatch.

(6) Information publishing module. Through mobile phone APP and other mobile terminal modes, ship information can be released to facilitate relevant users to grasp ship dynamics in time.

The Ship-Terminal Management Monitoring Platform. The ship-terminal management monitoring platform can realize reliable and reasonable monitoring of key electrical equipment. At the same time, uploading the real-time dynamic data to the shore-terminal management monitoring platform. The six functional modules of the ship-terminal management monitoring platform are shown as follows.

(1) Security module. Prevent all kinds of viruses and hackers from intruding, with firewall functions.

(2) Equipment parameter management module. It can realize operation parameters management of various electrical equipment on board.

(3) Communication management module. It can communicate with NX controller, also can communication with the equipment which has the communication interface. The module also has the functions of communication alarm and dynamic change.

(4) Database management module. The design of database is strictly in accordance with relational database specifications, and incorporates the object-oriented database design method. The database has the functions of automatic backup and automatic fault recovery.

(5) Alarm-processing module. This module can classify the received electrical equipment event alarms, also including the alarm event status changing. According to the user's needs, different event levels can be set differently.

(6) Client module. This module shows equipment operation management, operation monitoring, chart management, report management, etc.

2.3   Implementation Principle

As it mentioned above, the whole ship-electric integrated system has been divided into the shore-terminal management monitoring platform and the ship-terminal management monitoring platform. Also, there are many kinds of monitoring equipment in the ship-terminal management monitoring platform, and the situation of equipment is relatively complex, so the ship-terminal management monitoring platform has been divided into the monitoring object layer and the application service layer (Fig. 1).

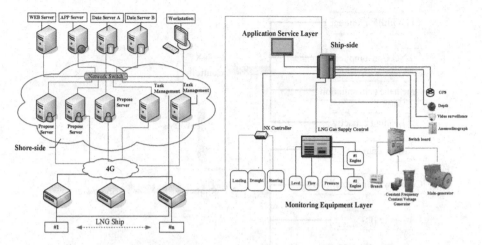

Fig. 1 Diagram of implementation principle

Monitoring Equipment Layer. This layer mainly divides the monitoring object into two parts according to the communication ability of the monitoring object equipment.

The first part, this part mainly consists of the monitoring equipment which has its own communication capabilities and communication interface. For example, the equipment with 485 communication interface can realize communication by configuring network serial port conversion device. Through analyzing the communication protocol of the monitoring equipment, the data access can be directly completed. The ship-terminal management monitoring platform can meet the requirements of equipment access and information acquisition under different communication protocols.

The second part, this part mainly consists of the monitoring equipment without its own communication function. So for this part, the NX controller is adopted to complete the collection and integration of key data. After completing the data integration, the ship-terminal management monitoring platform interacts with NX controller through analyzing the communication protocol of NX controller.

Considering the scattered location of monitoring experiment, the anti-interference ability of electrical signals is poor. Therefore, the network/485 line is the main transmission medium in communication.

Application Service Layer. It is equipped with a vessel terminal main device on each ship, and the monitoring service procedures are deployed in the main device. The ship-terminal management monitoring platform is composed of main device, local/remote monitoring terminal, profibus, local network, remote network, etc. When the main device needs to be maintained, it can visually display and configure the data of the main device through the maintenance module.

The main device uploads real-time equipment information to the shore-terminal management monitoring platform through the deployed 4G router (Fig. 2).

According to the Information transmission process above, the NX controller is used to monitor the real-condition of the monitoring equipment without communication protocol. After the information of the NX controller is collected, the communication

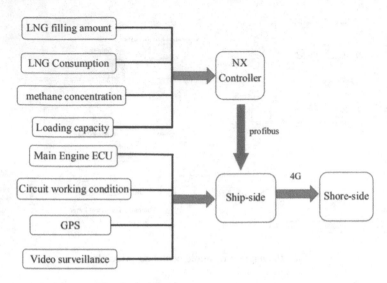

Fig. 2. Information transmission process

protocol is opened and integrated into the ship-terminal management monitoring platform. Based on this principle, the ship-electrical integration system can collect and monitor the key operation data of ship-electrical equipment, at the same time, solving the problems of port closure and protocol differences in data communication due to different levels and types of manufacturers.

3 NX Controller

The ship-terminal management monitoring platform uses NX controller to logically control and monitor the key equipment on board. After information is collected, the communication protocol is opened and transmitted to the shore-terminal management monitoring platform. NX controller is the data center of the monitoring equipment layer, also, it's the data center of the whole ship. The structure diagram is shown as below (Fig. 3).

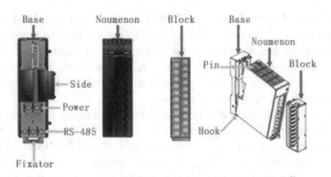

Fig. 3. Structure diagram of NX controller

The communication modes of NX electrical interface are divided into RS485 mode and Ethernet mode. Ethernet mode means the Ethernet communication, RS485 mode means the serial communication. The serial communication adopts master-slave communication mode. One master station can connect up to 31 slave stations. The connection mode is divided into parallel connection and decentralized connection (Fig. 4).

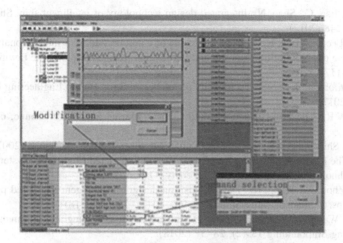

Fig. 4. Engineering software package

The NX controller module mainly involve five categories: digital input module, digital output module, AI module, management module and communication module. Digital input and output module realize the acquisition of data signals. AI module is also the regulator module to adjust the collected data signals (PID, etc.). When it comes to more complex logic control, we can adopt management module. The management module is accompanied by a special engineering software package for programming. And ultimately, through the Ethernet communication module, the integrates data and information get access to the ship-terminal management monitoring platform.

4 Conclusion

Under the background of informationization and digital upgrade, it is necessary to build large shipping data. Integrating the state information of key electrical equipment into a large data system,it is not only a simple listing and stacking of software and hardware products, but also closely related to software development, hardware upgrade and deep integration of software and hardware. Someday in future, by formulating a unified ship data standard, we can integrated the multi-source heterogeneous system into a single integrated system, and gradually realize the platform management of ship-electric equipment.

Acknowledgement. This work is supported by the National Natural Science Foundation of China (Grant No.51165035). Science and Technology Fund of Jiangxi Province of Higher Education (KJLD14094), Youth Science Fund of Jiangxi Province (20133BAB21004), and the Young scientist cultivation plan of Jiangxi Province (20112BCB23025).

References

1. Liu, X., Miao, C., Shao, N.: Integrated design technology of intelligent ship. Ship Eng. **39** (1), 245–248 (2017)
2. Kim, S., Lee, J., Yoon, H.: Conceptual design of a portal system for international shipping's greenhouse gas monitoring reporting and verification. Korean Soc. Marine Environ. Saf. **22** (1), 108–117 (2016)
3. Liu, P.: Information Service Technology of Intelligent Ship. Harbin Engineering University, Harbin (2017)
4. Cairns, W.R.: The E-navigation revolution. Proc. Marine Saf. Secur. Counc. **64**(2), 8–15 (2007)
5. Yang, J.: Ship Integrated Automation. Dalian Maritime University, Dalian (2007)
6. Xia, J.: The Research of Tianjin Port Small Vessels Integrated Monitoring System based on GPS Technology. Dalian Maritime University, Dalian (2014)
7. Lin, C., Zhou, X., Wang, M.: Date acquisition design of intelligent integrated platform for smart ships. J. Shanghai Ship Ship. Res. Inst. **41**(2), 59–64 (2018)
8. Jiang, T., Wang, Y., Zheng, C., Li, S.: Design and sea trial of integrated automation system. Guangdong Shipbuilding **159**(2), 28–31 (2018)
9. Sun, L.: Research of Integrated Platform Management System and Implementation of Embedded Can/Ethernet Gateway. Dalian Maritime University, Dalian (2004)
10. Fagin, R.: Data exchange getting to the core. ACM Trans. Database Syst. 30(l), 174–182 (2005)
11. Chen, K.-Z., Gao, Y., Shen, X.-B.: An analysis of single point positioning with real-time internet-based precise GPS data. Wuhan Univ. J. Nat. Sci. **82**(1), 64–68 (2003)

Protocol Analysis Method Based on State Machine

Bin Li[1], Jun Peng[2(✉)], and XueFeng Han[2]

[1] Changchun Institute of Optics, Fine Mechanics and Physics,
Chinese Academy of Science, Changchun, China
mykey008@126.com
[2] College of Computer Science and Technology, Jilin University,
No.2699 Qianjin Street, Changchun, China
pengjun@jlu.edu.cn

Abstract. Based network protocol analysis method. By capturing the data flow through the network interface, using data flow management and data stream reorganization technology, multiple data packets are identified as different data streams. Calling Protocol Characteristic Library to Identify Protocol Types Used by Different Data Streams. Complete the functions of network packet content extraction and private protocol identification in high-speed and high-traffic environment.

Keywords: State machine · Network protocol analysis · P2P · Private protocol

1 Introduction

With the rapid development of TCP/IP-based Internet and the popularity of P2P (Peer to Peer) design concept [1], the network architecture mode has shifted to the P2P network architecture mode, and a large number of proprietary protocols use random ports for communication. Identification technology has not achieved the desired results, making network protocol analysis work difficult.

In 2005, Karajannis et al. proposed that P2P traffic has two characteristics different from other traffic on the transport layer: a. Most P2P protocols use both TCP and UDP for data transmission; b. Each in P2P networks The nodes use different port numbers when transmitting data, while the number of peers of the communicating peers is the same as the number of ports, and the ports used by P2P traffic are usually high port numbers. The protocol identification method for these two characteristics can effectively detect P2P traffic.

In 2012, scholars from Beijing Jiaotong University analyzed the behavioral characteristics of P2P traffic in the application layer and proposed an improved DFI (Deep Flow Inspection) deep flow detection technology [2]. The technology is based on P2P protocol high port connection rate characteristics, P2P protocol traffic occupancy characteristics, P2P node ingress and outbound characteristics, P2P protocol packet length characteristics and other attributes to complete the identification of specific types of P2P traffic.

© Springer Nature Switzerland AG 2019
D. Milošević et al. (Eds.): HCC 2019, LNCS 11956, pp. 305–314, 2019.
https://doi.org/10.1007/978-3-030-37429-7_30

In this paper, the protocol analysis work is divided into two parts: feature extraction of protocol messages and network data flow analysis based on protocol features: firstly, a large number of network data streams are collected, and the application layer load is analyzed to find out the feature strings different from all other protocols. Renormalized to a regular expression that identifies the current protocol service type. Then, the pre-separated feature string is loaded, and then according to a certain algorithm, the data packets are aggregated into a "stream", and the protocol type used in the stream is detected. In the implementation process, the characteristics of the TCP protocol are comprehensively considered, and the concept optimization system analysis capability of the state machine [3] is introduced, and an anti-attack TCP flow state management [4] strategy is applied to improve the system security.

2   TCP Flow State Machine Model

The purpose of the TCP protocol design is to provide reliable, connection-oriented data transport services over the network layer. There are 6 commonly used flag bits in the TCP header. Each flag bit occupies 1 bit. The default bit is 0. When it is set to 1, it is valid. ACK, SYN, FIN are the most important flag bits, SYN indicates that the datagram is the first packet to initiate the connection, and ACK indicates that the acknowledgment sequence number of the datagram is valid, usually after receiving the SYN datagram at the receiving end. The text will set the ACK flag to be valid. FIN indicates that the data of the current Socket interface has been sent and the request to close the connection is initiated.

According to the process of TCP stream connection establishment and connection release, the concept of state machine is introduced when implementing TCP protocol. The 11 states involved in the state machine are defined as follows:

CLOSED: no connection status;

LISTEN: The listening status, usually set manually by the receiving end;

SYN-SENT: Requests the connection waiting state, and the sender sends a SYN to establish the connection request message status;

SYN-RCVD: Connection establishment confirmation status, the status of the receiving end after receiving the SYN message;

ESTABLISHED: the status of the connection, after the sender receives the ACK acknowledgement message, in this state, the sender can send data to the other party of the communication;

FIN-WAIT-1: Release the connection waiting state, and the sender sends a FIN to release the connection request message status;

FIN-WAIT-2: Connection release confirmation status, the status after the sender receives the ACK acknowledgement message. In this state, the sender can no longer send data to the other party of the communication;

CLOSE-WAIT: Releases the connection waiting state, and the receiving end receives the status after the FIN message;

LAST-ACK: Connection release confirmation status, the status of the receiver after receiving the ACK message in the CLOSE-WAIT state;

TIME-WAIT: Waiting state, after receiving the FIN packet, the sender usually waits for 2MSL (the maximum lifetime of the packet) and then enters the CLOSED state.

CLOSING: Two-way connection close confirmation status,

Figure 1 shows the operation of the TCP state machine. It can be seen that the transmitting end of the FIN-WAIT-2 state still has the ability to receive data, so FIN-WAIT-2 is also called a semi-closed state. As for the purpose of designing the TIME-WAIT state, it is because a packet loss event may occur in a specific network data transmission [5]. If the ACK packet of the sender does not reach the opposite point, then the state machine of the point is after a certain period of time. The FIN packet is resent. At this time, the port has released the port resource, but the FIN packet of the receiver cannot be processed. Waiting for the 2MSL time is to resend the ACK packet that may be lost.

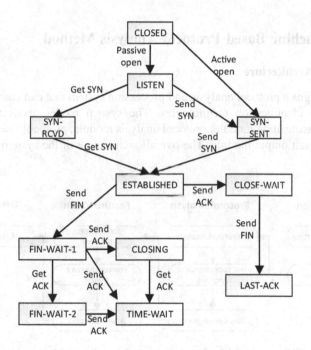

Fig. 1. TCP flow state machine

The formalization of a finite state automaton is defined as: $M = (S, \sum, \delta, S_0, F)$.

(1) S represents a finite state set, which is a set of all defined states in the state machine;

(2) Σ represents a finite input set, which is a collection of all input parameters that the state machine may receive;

(3) δ represents a state transition function. When a running state machine receives an input parameter in Σ , the state machine changes its current state according to the state transition function;

(4) S_0 indicates a unique initial state, $S_0 \subseteq S$;

(5) F indicates the only final state $F \subseteq S$

The finite state automaton can be represented by state transition diagrams and regular expressions [6]. In the case of a known regular expression, we can arbitrarily represent it as a state transition diagram and a finite state automaton. A typical algorithm for constructing a finite state machine using regular expressions is the Thompson ε-NFA construction algorithm [7]. The Thompson ε-NFA algorithm first constructs an NFA state machine for each field of the regular expression, and then glues each field together through the transformation to obtain the whole state transition diagram of the NFA state machine.

3 State Machine Based Protocol Analysis Method

3.1 System Architecture

This paper designs a protocol analysis and processing system that can comprehensively apply multiple identification technologies. The system model needs to have four modules: data acquisition module, protocol analysis module, protocol feature matching module, and result output module. The overall architecture of the system is as follows (Fig. 2).

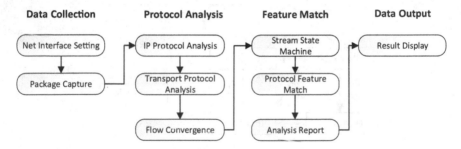

Fig. 2. State machine based network protocol analysis system architecture

The data acquisition module captures the data packets flowing through the network interface and submits them to the protocol analysis module. After receiving the data packet, the protocol analysis module mainly extracts the lower layer protocol information, that is, the physical interface layer, the network layer, and the part of the transport layer. The main fields are the source IP address, the destination IP address, the source port, the destination port, and the TCP protocol. The identification field, by organizing the field information, can divide the network data packet into different data streams, and then submit the data stream entry to the protocol feature matching module.

After receiving the data stream, the protocol feature matching module starts to analyze the static load of the protocol, and obtains the protocol feature field from the feature library for pattern matching, and generates an analysis report of each data stream, and all the analyzed protocol connection information and application layer. The protocol data content and protocol identification results are submitted to the data output module. The data output module manages all source data packets flowing through the network interface, and all the analysis data, presents key information of the protocol analysis to the user by providing a simple window interface, and provides partial menu functions and file configuration functions for system setting, The preservation of network data and analysis reports, etc.

The data acquisition module uses the libpcap data packet to capture the underlying interface, captures the network data packet from the host's network interface in real time, provides buffering for the system by means of channel transmission, and the protocol analysis module extracts the network packet data from the channel.

The protocol analysis module, the gopacket toolkit interface, parses the format of the lower layer protocol, and offloads the network data packet by extracting the information of the quaternary group (source IP address, destination IP address, source port, destination port). When the flow belonging to a certain quaternion has just arrived, the protocol feature matching module is started for this data flow, and the subsequent data packets belonging to the same quaternary group directly enter the protocol feature matching module for the next analysis work.

The protocol feature matching module distinguishes P2P and non-P2P traffic by the behavior characteristics of the statistical flow, and performs different analysis work. Non-P2P traffic can use the port mapping method to identify the type of application layer protocol used by the data stream. The P2P traffic needs to be loaded in the signature field of the signature database to match. If the matching fails, it indicates that the system has no way to identify the protocol type used by the current data stream.

The data output module is designed to implement a simple window interface using the interface provided by the third-party toolkit walk. The window uses commonly used window components such as LineText, LableText, and TableView to display the protocol analysis information generated by the protocol analysis system, and uses Menu, Action, and other components to provide the user with settings related to system parameters.

3.2 Protocol Analysis

The TCP/IP reference model abstracts the communication process into a physical interface layer, a network layer, a transport layer, and an application layer, and implements different protocols at each layer. On the host that is the sender, each layer encapsulates the upper layer data packet and its own protocol header as a new data packet and sends it to the lower layer protocol. On the host as the receiving end, each layer decapsulates the lower layer data packets into its own protocol header and data packet, and sends the data packet to the upper layer. Each level only performs related calculation processing on its own protocol header, and does not care about the content of the data packet, and realizes the decoupling of each layer protocol.

Protocol analysis is a behavior that decodes packets in a strictly defined format. This behavior parses the incoming packet according to the reverse order of packet encapsulation. The key operation of this behavior is to check the type field in the protocol header of each layer, that is, the upper protocol type. Common protocol type numbers are shown in Table 1.

Table 1. Common protocol and field.

Protocol	ARP	RARP	IP	ICMP	TCP	UDP
Field	0×0806	0×0835	0×0800	1	6	17

Sometimes protocol analysis requires only a certain level of information, without having to know the protocol header information of the previous level. In this way, a more efficient protocol decoding can be realized. By extracting a small number of bytes in a specified position in the data packet, the analysis and judgment are performed, and only when a certain layer of protocol information is needed, the byte area is decoded.

In this paper, the data stream-based protocol analysis method is used to aggregate the captured network packets into a "flow" for analysis. Generally, we consider network packets that belong to the same quad (source IP address, destination IP address, source port, and destination port) collected over a period of time as data on a data stream, generally the same data stream. Each message in the same belongs to an application protocol type. The flow of data aggregation is shown in the following Fig. 3:

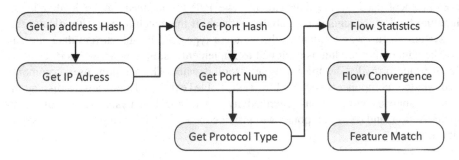

Fig. 3. Protocol analysis process

(1) Extract the quaternion information, call the packet.NetworkLayer() function, and return the network layer interface networkLayer of the data packet. This interface provides NetworkFlow() flow management, which can decode the network layer byte to obtain the source IP address and destination IP address. Hash value calculation can be performed on two IP addresses. Call the packet.TransportLayer() function to return the transport layer interface transportLayer of the packet. This interface provides TransportFlow() stream management, which can decode the transport layer byte to obtain the source port and destination port. In addition, TransportFlow() and NetworkFlow() has the same method.

(2) Calculate the stream hash value, call the FastHash() function provided by the NetworkFlow() and TransportFlow() interfaces, and return the stream hash value of uint 64. The hash value is symmetric, that is, the hash value is still the same when the source IP address and the destination IP address are swapped.

(3) Add or supplement the stream information in the Map, call the custom Map Add() function, use the ipHash and portHash of the packet as the Key value, and the packet interface as the Value value to save the network packet. In this way, multiple network data packets are divided into different streams through the Map table, so as to facilitate the next processing.

3.3 Protocol Feature Matching

The protocol feature matching module identifies the data stream as a unit. In the analysis, only the first few packets of one data stream are considered, or the first few packets whose total payload is less than a certain value, if the feature matching method can be adopted. If the type of the protocol used by the packet is identified, the subsequent packet on the data stream should be the same protocol type. If the packet cannot be identified, the subsequent packets on the data stream are unrecognizable.

This module defines a data flow with two protocol ready states: flow preparation, flow confirmation. The flow preparation state refers to the requirement that the packet size does not meet the protocol feature matching, such as the SYN packet, the ACK packet, and the RST packet for port redirection in the TCP handshake mechanism. The flow confirmation status means that the packet size can be matched by protocol features, such as the first few packets of the TCP stream and almost all packets of UDP. The module also defines that the data stream should have an end state identified by three protocols: unrecognized, unrecognized, and identified. Unrecognized status means that the captured data stream is an abnormal stream, such as a TCP three-way handshake. The identified state means that when detecting the first few messages in the stream, the payload matches a certain pattern of the protocol signature library. An unrecognized state means that the match failed.

(1) Initialize the stream state machine. When the protocol analysis module calls the pflow.Add() function, a packet is added to a stream. If the packet is the first arriving packet, the stream preparation state is generated for the stream, and the recognition status is initialized., window data, timeout time, number of messages, and other information. If this packet.

It is a subsequent message on the stream, and it is judged according to the window data size of the stream whether the data load matching requirement is met to determine whether to perform feature matching.

(2) Flow status detection. For the data stream using the UDP protocol at the transport layer, there is no obvious end of flow flag, so the timeout mechanism is used for UDP stream management. For the data stream using the TCP protocol at the transport layer, the status of the stream can be confirmed by the RST and FIN messages, and the timeout mechanism is used in parallel for TCP flow management.

(3) Protocol feature matching based on data payload. There are two feature matching methods, single message matching mode and multiple message matching mode. The single-message matching method is to match the packets in the stream with the protocol features. The multi-message matching method is to sort and reassemble each packet in the stream, and then match the protocol features.

(4) Generate an analysis report. After the protocol feature matching is completed, the flow state machine should be in one of three end states. The module shall generate a data flow analysis report for saving and clear the current stream data.

There are two kinds of data packets in the network, one is a packet for communication between two communication ports, and the packet contains a data request or its control information, which can be called a signaling packet, and the packet length is between 40B and 100B, and the other is It is a package that actually transfers file streams. The packets are mostly binary files, videos, pictures, and other stream data. They can be called file packages, and the packet length is 1 KB or more. In general, developers use the UDP protocol to transmit command information and the TCP protocol to transfer file data, but sometimes use the TCP protocol to transmit command information in JSON format. For the protocol analysis system, the useful data packet is the signaling packet, so for the data packet larger than 1 KB, the discarding strategy is adopted, and the remaining data packets are subjected to protocol feature matching.

4 Case Study

A protocol identification system has multiple design indicators, and often each indicator will affect each other. Therefore, comprehensive consideration should be made from many aspects, including the following aspects: packet throughput (based on network speed), accurate of protocol identification, flexibility (maintainable, scalable), etc.

Considering the above several indicators comprehensively, this paper uses the characteristics of channel to establish a buffer between packet collection and protocol analysis. If the protocol analysis takes a long time, multiple goroutine concurrent methods are used to analyze multiple streams simultaneously. The protocol analysis and processing system based on state machine is designed by means of bufio, ini and regexp. Bufio is an IO programming interface supported by the golang language. It uses multiplexing and buffering to make real io events asynchronous. Ini is a package of configuration files, which is used by major systems because it can load and generate various types of files. Regexp is a package of regular expressions used to hold protocol features.

Figure 4 is a screenshot of the Fiddler 4 system. As can be seen from the data table in the figure, the main job of Fiddler 4 is to capture and detect HTTP and HTTPS data packets to provide users with package content resolution.

#	Result	Protocol	Host	URL	Body	Caching	Content-Type	Process	Comments	Custom
1	200	HTTPS	www.fiddler2.com	/UpdateCheck.aspx?isBeta=False	514	private	text/plain; charset=utf-8			
2	502	HTTP	config.pinyin.sogou.com	/api/toolbox/geturl.php?h=5570481963440088C...	582	no-cac...	text/html; charset=UTF-8	sgtool:9552		
3	200	HTTP	update.helper.2345.cc	/e/index.php	200		text/html; charset=UTF-8	helper_haozip:9180		
4	200	HTTP	ts.helper.2345.cc	/helper/index.php	3,092		text/html; charset=UTF-8	helper_haozip:9180		
5	200	HTTP	t.helper.2345.cc	/helper/index.php	0		text/html	helper_haozip:11164		
6	200	HTTP	update.khd.2345.cc	/dmdt/dmdt_data.php	690		text/html; charset=UTF-8	helper_haozip:7628		
7	200	HTTP	t.helper.2345.cc	/helper/index.php	0		text/html	helper_haozip:9180		

Fig. 4. Screenshot of the Fiddler4 system

Figure 5 is a screenshot of the case study. The content of the figure is the partial result of the protocol analysis report. The attributes such as the source IP address and the destination IP address are all analyzed based on a data stream. The system can also display records of data transmitted by two communication ports of a data stream, or even byte data therein. Of course, not all byte data is useful. If you use it as a tool like TCPDUMP, it will provide you with a file of all the original data packets captured by the packet, such as test.pcap format. Files can be used as a data source for this or other systems.

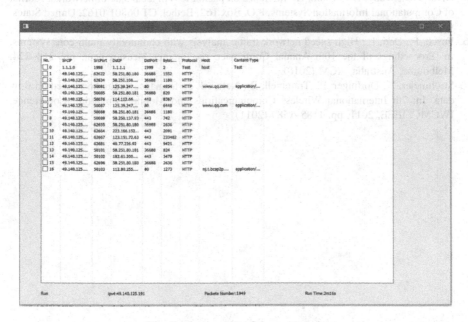

Fig. 5. Screenshot of the case study

5 Conclusion

This paper analyzes the common protocol analysis software architecture, and designs and implements a state machine based protocol analysis method for diversified and privatized high-capacity high-speed network environments. Integrate and optimize technologies such as data stream management, TCP stream reassembly and stream state machine to realize feature-based network packet content extraction and private protocol identification.

References

1. Ni, J., Tatikonda, S.: Network tomography based on additive metrics. IEEE Trans. Inf. Theory **57**, 7798–7809 (2011)
2. Nechaev, B., Allman, M., Paxson, V., et al.: A preliminary analysis of TCP performance in an enterprise network. In: Proceedings of the 2010 Internet Network Management Conference on Research on Enterprise Networking. USENIX Association, p. 7 (2010)
3. Bando, M., Artan, N.S., Wei, R.: Range hash for regular expression pre-filterin. In: IEEE Symposium on ANCS 2010, La Jolla, CA, USA, vol. 10, pp. 1–12 (2010)
4. Bing, X., Xiaosu, C., Ning, C.: An efficient TCP flow state management algorithm in high-speed network. In: Muhin, V.E., Ye, Z. (eds.) Proceedings of 2009 International Symposium on Information Engineering and Electronic Commerce. Ternopil, Ukraine. May 2009. Los Alamitos, California: IEEE Computer Society, 106110 (2009)
5. Xiao, Y.D., Wang, H.J., Lium, K.: Inference on packet loss in local side of networks. Journal of Computational Information Systems, P.O. Box 162, Bethel, CT 06801-0162, United States, vol. 8, pp. 4695–4703 (2012)
6. Fusco, F., Deri, L.: High speed network traffic analysis with commodity multi-core systems. In: Proceedings of the 10th Annual Conference on Internet Measurement, pp. 218–224. Melbourne, Australia, ACM (2010)
7. Strohmeier, F., Dorfinger, P., Trammell, B.: Network performance evaluation based on flow data. In: 7th International Wireless Communications and Mobile Computing Conference, IWCMC, IEEE, 2011, pp. 1585–1589 (2011)

News Recommendation Model Based on Improved Label Propagation Algorithm

Zelin Peng[1], Ronghua Lin[1], Yi He[1], Liu Wang[1], Hanlu Chu[1],
Chengzhou Fu[1,2(✉)], and Yong Tang[1]

[1] School of Computer Science, South China Normal University,
Guangzhou, China
{zelin,rhlin,yihe,lwang,hlchu,ytang}@m.scnu.edu.cn,
fuchzhou@hotmail.com
[2] College of Medical Information Engineering,
Guangdong Pharmaceutical University, Guangzhou, China

Abstract. Facing the ever-expanding scale of news information, how to filter and redundant information in complex and diverse data to accurately recommend information to users has become an important challenge. News recommendations have become a powerful tool for dealing with information overload. Scholars have done a lot of research on personalized news recommendation. However, due to the sparseness of user data, the traditional collaborative filtering algorithm is not only too time-consuming but also less accurate. Therefore, aiming at the characteristics of academic social networks, we propose a news recommendation model based on community detection, which builds a friend relationship community by using the users friend relationship and coherent neighborhood propinquity algorithm, and then integrates the collaborative filtering algorithm to recommend news to users. Through verification on the dataset of the academic social network SCHOLAT, we can prove that the recommendation model can achieve good accuracy while improving recommendation efficiency.

Keywords: Label propagation · Community detection · Collaborative filtering · News recommendation

1 Introduction

With the rise and development of academic social network, the news information of social network has attracted more and more attention. However, the massive amount of news has also led to the problem of information overload. People often get lost in excessive information and can't see the news they are really interested in [1], which not only wastes the users' time and energy, but also reduce users' Satisfaction with social networking sites. More and more users are looking for an excellent smart technology to provide them with personalized recommendations. Therefore, how to personalize recommend news to users on academic social networks has become an important task for social networking sites to improve subscribers' loyalty. Facing this challenge, collaborative filtering recommendation algorithm is a powerful tool to meet the needs.

© Springer Nature Switzerland AG 2019
D. Milošević et al. (Eds.): HCC 2019, LNCS 11956, pp. 315–324, 2019.
https://doi.org/10.1007/978-3-030-37429-7_31

With extensive research and application, the collaborative filtering recommendation algorithm is considered to be one of the most popular and successful personalized recommendation algorithms. In the field of news recommendation, traditional content-based recommendation algorithms are typically based on the users' browsing history or by extracting the users' personal information to discover user interest characteristics for recommendation. For example, [2] proposed a keyword extraction algorithm based on rapid automatic key word extraction, which performs news recommendation by keyword extraction and scoring of user's personal information. Such an approach can indeed be very user-friendly, but extracting users' interest characteristics and news features is not a simple task. More importantly, this recommendation ignores the social relationships of users in academic social networks. We believe that the interests of users with a close social relationship are mutually influential. For example, on an academic social network, if the friends of a user reprints a high-quality news report, even if the news is not related to the user's interest characteristics, the user will read this news because of the influence of the friend. Content-based recommendation algorithms often ignore this situation. However, the collaborative filtering algorithm also has many shortcomings. The most obvious is that when the number of users is large and the user data is sparse, the performance of collaborative filtering will decrease. Therefore, it is necessary to improve the collaborative filtering algorithm.

Based on the characteristics and complexity of academic social networking sites [3–5], this paper proposes a collaborative filtering news recommendation model based on improved label propagation algorithm. In this model, we propose a label propagation algorithm based on the coherent neighborhood propinquity to perform community detection which consider the user's friend relationship. Then, based on the results of community segmentation, we can apply collaborative filtering algorithms to personalize recommend news for users in the community. Finally, we conducted experimental tests on the dataset of the large academic social networking site Scholar Network to prove the validity of this model.

2 Related Work

The news recommendation system has been widely studied at home and abroad. There are two basic methods for the most commonly used news recommendations. One is content-based recommendation and the other is collaborative filtering recommendation.

Content-based recommendation [6, 7] is to construct the user interest feature vector by exploring the characteristics of the user's past favorite items, and then extract the content features of the item to be recommended, calculate the similarity between the two, and generate recommendations for the user. There are many content-based recommendation methods, such as keyword-based methods, named entity-based methods, and news tag-based methods. [8] proposed a method using TF-IDF to calculate the relevance of news tags, and used this method to design a personalized news recommendation method based on the probability map between tags. [9] improve the quality of news recommendations by using a user-reviewed ranking model to process noisy comments and eliminate redundant comments in a diverse pattern and leverage the valuable information provided by user reviews. [10] conducting news recommendations by integrating user personal information, popularity, trends, and location.

Collaborative filtering recommends news to users based on people with similar interest and preferences. The advantage of collaborative filtering recommendation is that it does not need to know the relevant corpus information of the news. [11] using a linear function to integrate MinHash clustering, probabilistic latent semantic indexing (PLSI), and Covisitation counts to recommend. [12] proposed a collaborative filtering method based on rough set, which is used to predict the news category scores that have missed, and proposes a new novelty detection method to improve the ranking of new products.

Although collaborative filtering has many advantages, the traditional collaborative filtering algorithm is very time consuming and the accuracy is reduced when the number of users is large and the user data is sparse. Therefore, users' data need to be pre-processed before using collaborative filtering algorithm to generate recommendation results. Community discovery is now also widely used in recommendation systems. [13] proposed an improved community-based microblog recommendation collaborative filtering algorithm, which improved the recommendation results of the collaborative filtering recommendation system through community discovery. [14] proposed a multi-labels community detection technology based on LPA, mapping the binary network composed of users and articles into the integrated user network, and then detecting the overlapping communities of the integrated user network, thereby alleviating the influences of large number of sparse data in collaborative filtering algorithms. [15] proposed a personalized recommendation method for fusion label propagation and trust diffusion. According to the trust network between users in the large community of each user, the trust preprocessing algorithm is given to predict the new trust relationship of the user, thereby expanding the user's trust network. And use the hybrid trust diffusion algorithm to make the trust between users and other users in the big community more differentiated.

However, the above work ignores the efficiency issues in community detect and does not have an accurate measure of the distance among nodes. Therefore, this paper proposes a label propagation algorithm based on coherent neighborhood propinquity, considering the relationship between users' friends, and improving the speed of community discovery without reducing its quality. Based on the community structure obtained by the label propagation algorithm, we combine the collaborative filtering algorithm to generate news recommendations for users.

3 Recommend Approach

3.1 Coherent Neighborhood Propinquity

The coherent neighborhood propinquity(CNP) is a parameter proposed by [16] reflects the probability that a pair of nodes participate in the same community structure. According to the proof of [17], the diameter of the correlation diagram is not more than 2. Therefore, assuming that the obtained community structure is consistent, the topological distance of each node should be within 2. With this in mind, the correlation between two nodes can be calculated by directly connected edge of two nodes, the shared neighbors of the two nodes, and the edges between the shared neighbors of the two nodes. The calculation formula is as shown in Eq. 1:

$$P(v_1, v_2) = |E(v_1, v_2)| + |N(v_1) \cap N(v_2)| + |E(G[N(v_1) \cap N(v_2)])| \tag{1}$$

Where $P(v1, v2)$ represents the degree of closeness of nodes $v1$ and $v2$, $|E(v1, v2)|$ represents the number of edges directly connected by nodes $v1$ and $v2$, and $|N(v1) \cap N(v2)|$ represents the number of shared neighbor nodes of $v1$ and $v2$, $|E(G[N(v1) \cap N(v2)])|$ represents the number of edges of the connection between the intersection nodes of the set of neighbor nodes shared by v_1 and v_2. Figure 1 shows an example of CNP calculation. Figure shows a complete graph with 5 nodes. According to Eq. 1, there is a directly connected edge between nodes v_0 and v_1. There are 3 shared neighbor nodes v_2, v_3, v_4, and 2 directly connected edges between v_2, v_3 and v_4. Therefore, the CNP of nodes v_0 and v_1 tend to $P(v_1, v_2) = 1 + 3 + 2 = 6$.

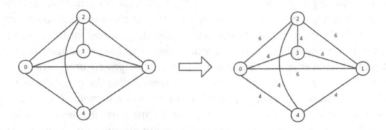

Fig. 1. Calculation of coherent neighborhood propinquity

3.2   Coherent Neighborhood Propinquity Based Label Propagation

In order to obtain the user's social relationship, we use the community discovery algorithm to analyze the relationship diagram of users' friends. However, the iterative process of the traditional label propagation algorithm is unstable and accuracy is not high with too many iterations. Therefore, we use the coherent neighborhood propinquity to calculate the weight between nodes instead of selecting the most occupied tags in the neighbor nodes to propagate in the traditional tag propagation, we improves the stability and accuracy of the algorithm and greatly reduces the number of iterations required by the algorithm. According to formula 1, the implementation process of the label propagation algorithm based on the CNP is as follows:

Obtain the relationship among users, and create a social network relationship diagram $G = \{V, E\}$ according to the friend relationship of users, where V represents the set of user nodes in the social network relationship diagram, and E is the edge set of the graph, which also represents the collection of user friend relationships.

All user nodes are randomly assigned a unique label, and the CNP between each user node and its neighbor nodes is calculated according to formula 1.

Randomly select a node v_i, obtain the label of the neighbor node v_j whose relationship with the node v_i is most related, and update the label of the node v_i according to the label of the v_j.

Repeat step 3 until the tag of the nodes no longer changes or the maximum times of iterations is reached, and the set of nodes S which tags has been assigned is obtained.

Aggregate the nodes with the same tags in the node set S to obtain a community of multiple user friendships.

According to the above rules, the implementation process of the label propagation algorithm can be represented by Fig. 2. Suppose that there is a complete graph with 8 nodes at the beginning, each node randomly assigns a unique label, and then calculates the intimacy between each node according to Eq. 1. For example, node 2 has a directly connected edge with node 1, three shared neighbor nodes, and one edge between shared neighbor nodes. Therefore, the coherent neighborhood propinquity of node 2 and node 1 tends to be 5. Calculate each node and its neighbors. After the calculation, the label can be propagated according to the intimacy of the nodes. For example, select the node with label 2, which has 4 adjacent nodes with labels 0, 1, 3, 4, and the intimacy is 2, 5, 3, 4 respectively. Therefore, label 1 with a density of 5 is selected to update its own label. This process is repeated, and finally when the node no longer changes or reaches the maximum number of iterations, the eight nodes are divided into two communities.

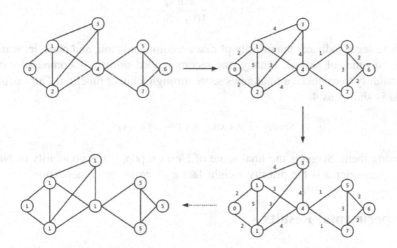

Fig. 2. Implementation process of CNP-LPA

3.3 News Recommendation Based on Community Detection

In social networks, the traditional recommendation algorithm usually needs to calculate the similarity of users. In the case of a large number of users with sparse data, this is a very time consuming process, and only a small percentage of users can really contribute to the final recommended results. However, by introducing a community discovery algorithm, we can reduce the number of users who need to be calculated, and users in the same community often have similar interests or academic areas. After using CNP-LPA for community segmentation, we can get the community of the target user and get the news history that other users in the community have viewed. Then, the user's similarity with other users and the priority of the news are calculated, and a final score is obtained based on these two scores. Finally, the user is recommended for news according to the final score.

We calculate the user similarity by taking the structured data in the user's personal information. The structured data includes the user's work unit, scholar title, degree, and research field, and then converts the structured data into a one-dimensional feature vector for calculation. The news similarity calculation formula is as follows:

$$sim_i = \frac{1}{n}\sum_{j=1}^{n} \cos(s_{user}, s_{userj}) \qquad (2)$$

Where simi represents the similarity score of news i, Suser represents the personal information of the target user, Suserj represents the personal information of user j who has visited news i, and n represents the number of users who have browsed news i.

Then, the recommendation priority of the news is calculated by the number of clicks of the news, and the priority calculation formula is as shown in 3:

$$level_i = \frac{click_i}{10 + click_i} \qquad (3)$$

Where $level_i$ indicates the priority of news recommendation, and $click_i$ indicates the number of times of news i. Finally, the recommended priority is compared with the user similarity to calculate a final news score through a linear function. The calculation formula is shown as 4:

$$Score_i = \alpha * sim_i + (1 - \alpha) * level_i \qquad (4)$$

Among them, $Score_i$ is the final score of News i, pop_i is the popularity of News i, and the parameter α is the priority weight factor of news recommendation.

4   Experimental Results

In this section, we first introduce the dataset used in this experiment, then introduce the evaluation method used in this experiment, and finally analyze the result of the experiment and the comparison experiment.

4.1   Dataset

The dataset used in the experiment is a dataset of users' news and friend relationships of the academic social network (SCHOLAT) from March 2018 to November 2018. The dataset recorded news published by scholars and scholars' friendship. After the noise reduction process, there were 2,308 valid news and 20446 users with 39867 edges.

4.2   Evaluation Method

In the experiment, we used three general indicators [18] to evaluate our experimental results: Precision, Recall, F-Measure.

The precision refers to the ratio of the total length of the recommended list of the items that the user likes to the user's recommendation result. Its calculation formula is as shown in 5:

$$\Pr ecision = \frac{1}{T} \sum_u \frac{N_t}{L} \tag{5}$$

Where T represents the number of experimental samples, Nt represents the number of recommended items in the recommendation result, and L represents the length of the recommendation list.

The recall expresses the possibility that the user's favorite object is recommended by the recommended system. Its calculation formula is as shown in 6:

$$Recall = \frac{1}{T} \sum_u \frac{N_t}{A_u} \tag{6}$$

Where Au indicates the number of items that should be recommended in the test dataset.

Under ideal conditions, the higher the accuracy and recall rate, the better the recommended effect of the model, but in the actual calculation, the accuracy and recall values are often negatively correlated. In order to better measure the model's recommendation, we introduce the F1-Measure, and F1-Measure is the harmonic mean of Precision and Recall, as shown in 7:

$$F1 = \frac{2 * \Pr ecision * Recall}{\Pr ecision + Recall} \tag{7}$$

F1-Measure combines the results of Precision and Recall. The higher the value of F1-Measure, the better the performance of the recommended system.

4.3   Result Analysis

First, we use the CNP-LPA to divide the community of 20446 users into the scholar network dataset. Compared with the traditional label propagation algorithm, the traditional LPA needs to iterate 10 to 15 times to converge and its module degree is 0.21552, while CNP-LPA only need to iterate 5 times, the time is almost one-third of the original algorithm, and its module degree is 0.285918.

After the user is divided into communities, the collaborative filtering algorithm is used to personalize recommend news to user. Through experiments, we found out that we can get the best recommendation result when we set the threshold α to 0.35. We selected three recommended algorithms to compare with our experiment. The first algorithm is content based recommendation using LDA topic model. We use the LDA topic model to extract the topic of the news and the interest characteristics of the scholars, and then recommend the news according to the scholar's interest characteristics. The second algorithm is traditional collaborative filtering algorithm for news recommendation, hoping to highlight our advantage. The last algorithm is a hybrid recommendation algorithm combines the above two. Figs. 3 and 4 show the accuracy and recall of the three experiments. We can see that the accuracy and recall of CL-CF are higher than LDA and traditional collaborative filtering algorithm in most cases. When CL-CF are recommended between 10 and 50, the accuracy rate is above 0.4. The result of hybrid algorithm is better than LDA and collaborative filtering algorithm but they are all worse than CL-CF.

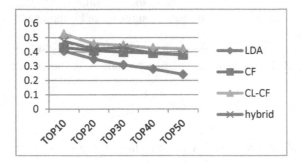

Fig. 3. Precision

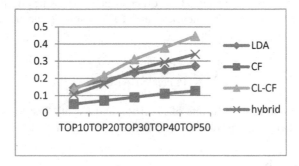

Fig. 4. Recall

Figure 5 shows the results of the F1-Measure result for three experiments. We can see that the F1-Measure of CL-CF is slightly lower than LDA and hybrid algorithm when the recommended number is 10 to 20, but higher than the traditional collaborative filtering. While the recommended number is greater than 30, the F1-Measure of CL-CF is significantly higher than the other two algorithm, and when the recommended number is 40, the F1-Measure value of CL-CF reaches its highest, so it is concluded that CL-CF can obtain the best recommendation results with a recommended number of 40.

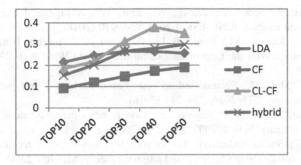

Fig. 5. F1-Measure

5 Conclusion

Aiming at the characteristics of academic social networks, this paper proposes a label propagation algorithm based on related neighborhood propensity. By calculating the CNP of each user node and its trusted nodes, the correlation degree of each node is obtained, and then according to the CNP of each node, we finally get multiple communities. Then we use the CL-CF algorithm to recommend news in the community for each user. After comparing the datasets on the academic social network SCHOLAR, the recommendation model is proved to be more accuracy and faster.

Acknowledgements. Our works were supported by the National Natural Science Foundation of China (No. U1811263, No. 61772211) and Innovation Team in Guangdong Provincial Department of Education (No. 2018-64/8S0177).

References

1. Jannach, D., Zanker, M., Felfernig, A., Friedrich, G.: Recommender Systems-An Introduction. Cambridge University Press 2010, vol. 1–335, pp. I–XV (2010). ISBN 978-0-521-49336-9
2. Wang, Z., Hahn, K., Kim, Y., Song, S., Seo, J.-M.: A news-topic recommender system based on keywords extraction. Multimedia Tools Appl. **77**(4), 4339–4353 (2018)
3. Gruzd, A., Staves, K., Wilk, A.: Connected scholars: examining the role of social media in research practices of faculty using the UTAUT model. Comput. Hum. Behav. **28**, 2340–2350 (2012)
4. Nández, G., Borrego, Á.: Use of social networks for academic purposes: a case study. Electron. Libr. **31**, 781–791 (2013)
5. Gu, F., Widén-Wulff, G.: Scholarly communication and possible changes in the context of social media: a finnish case study. Electron. Libr. **29**, 762–776 (2011)
6. Lu, Z., Dou, Z., Lian, J., et al.: Content-based collaborative filtering for news topic recommendation. In: Proceedings of the Twenty-Ninth AAAI Conference on Artificial Intelligence, pp. 217–223. Austin Texas, USA (2015)
7. Son, J., Kim, S.B.: Content-based filtering for recommendation systems using multiattribute networks. Expert Syst. Appl. **89**, 404–412 (2017)

8. Shen, Y., Ai, P., Xiao, Y., Zheng, W., Zhu, W.: A tag-based personalized news recommendation method. ICNC-FSKD **2018**, 964–970 (2018)
9. Meguebli, Y., Kacimi, M., Doan, B.-L., Popineau, F.: Towards better news article recommendation - With the help of user comments. World Wide Web **20**(6), 1293–1312 (2017)
10. Natarajan, S., Moh, M.: Recommending news based on hybrid user profile, popularity, trends, and location. CTS **2016**, 204–211 (2016)
11. Das, A., Datar, M., Garg, A., Rajaram, S.: Google news personalization: scalable online collaborative filtering. WWW **2007**, 271–280 (2007)
12. Saranya, K.G., Sudha Sadasivam, G.: Personalized news article recommendation with novelty using collaborative filtering based rough set theory. MONET **22**(4), 719–729 (2017)
13. Qiu, H.-H., Liu, Y., Zhang, Z.-J., Luo, G.X.: An improved collaborative filtering recommendation algorithm for microblog based on community detection. IIH-MSP, pp. 876–879 (2014)
14. Qiang, H., Yan, G.: A method of personalized recommendation based on multi-label propagation for overlapping community detection. In: International Conference on System Science, pp. 360–364 (2012)
15. Bowen, C., GongShen, L., Haolin, Z., et al.: Personalised recommendation method fused with label propagation and trust diffusion. Comput. Eng. **40**(12), 33–38 (2014)
16. Zhang, Y., Wang, J., Wang, Y., Zhou, L.: KDD. In: Iv, J.F.E., Fogelman-Soulié, F., Flach, P.A., Zaki, M.J. (eds.) Parallel Community Detection on Large Networks with Propinquity Dynamics, pp. 997–1006. ACM, New York (2009)
17. Pei, J., Jiang, D., Zhang, A.: On mining cross-graph quasi-cliques. In Proceedings of the Eleventh ACM SIGKDD International Conference on Knowledge Discovery in Data Mining, pp. 228–238, New York, NY, USA (2005)
18. Lichtenwalter, R.N., Lussier, J.T., Chawla, N.V.: New perspectives and methods in link prediction. In: ACM SIGKDD International Conference on Knowledge Discovery and Data Mining, pp. 243–252. Washington, DC, USA (2010)

Route Optimization of Robot Groups in Community Environment

Qiaohong Zu[✉] and Shuwen Yang[✉]

School of Logistics Engineering, Wuhan University of Technology,
Wuhan 430063, People's Republic of China
zuqiaohong@foxmail.com, 873498686@qq.com

Abstract. The paper studies the path planning and task assignment of robots in the low-efficiency distribution of express delivery in the community. The grid method is used to model the environment and a community is analyzed as an example. In the ant colony optimization (ACO), the heuristic function is reconstructed by the valuation function of the A* algorithm to improve the convergence speed of the ACO. The algorithm has enhanced global search ability in the early stage, and the convergence speed is fast in the later stage with the improvement of pheromone volatilization coefficient, and the experimental parameters simulation analysis is done in MATLAB software. The experimental results show that the improved ACO has faster convergence and higher efficiency than the basic ACO. The rationality of the path planning model and the effectiveness of the optimized ACO are verified.

Keywords: Multi-destination robot · Path planning · ACO

1 Introduction

With the popularity of online shopping, the increase of express delivery business has been gradually unable to meet the needs of the community. Recently, many experts and institutions have conducted research on express robots, and express robots take advantage of group advantages at home and abroad. The task of courier delivery is efficiently completed in the community. In order to solve the problem of unreasonable task assignment and path planning in the current community express delivery, Jingdong Express robot (car) is used as an example to study the path planning and task assignment of multi-delivery robots. Thus this study of express robots is of great significance.

2 Modeling of Path Planning and Task Assignment

2.1 Environmental Modeling

We don't discuss the avoidance since the courier robot in this topic can have automatic obstacle avoidance function. Taking a community as the experimental environment, the robot takes 4 entrances as the starting point, 9 buildings as the mission points. The task

© Springer Nature Switzerland AG 2019
D. Milošević et al. (Eds.): HCC 2019, LNCS 11956, pp. 325–332, 2019.
https://doi.org/10.1007/978-3-030-37429-7_32

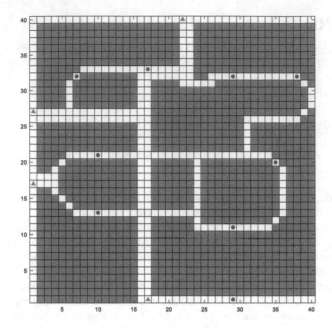

Fig. 1. Multi-delivery robot working environment model (Color figure online)

of the robot is to complete the delivery of the courier between the designated locations. A two-dimensional plan view of the working environment of the express robot is obtained based on the grid method, as shown in Fig. 1.

In the formula, n is the grid dimension, and mod is the calculation to remove the remainder, and fix is used to calculate the remainder.

Among them, the white grids indicate the area that the multi-destination robot can move. (i.e. the path), while the black grids represents the obstacles (the building and the lawn, etc.), and the blue circles (297, 327, 349, 358, 770, 835, 1090, 1189, 1589) represents the mission point, and the red triangle (22, 521, 921, 1577) represents the starting point of the multi-destination robot.

2.2 Multi-destination Robot Task Assignment Model Establishment

The total cost of a multi-delivery robotic system is expressed as the cost of all individual courier robots. The task assignment problem of multi-delivery robots in the smart community can be considered as a scientific allocation method to ensure that the total cost of the system is minimized.

When dealing with the task assignment problem of multi-delivery robots in the near-principle, the first step is to calculate the distance from the departure point of each courier robot to the arrival task point. Second step is to select the shortest path to perform the task. Therefore, the farther the distance is, the higher the cost pays. The closer the distance the courier robot starts, the less cost it will pay to reach the mission point. Taking the actual situation into account, the formula (2) for finding the distance in the principle of proximity is as follows:

$$D = \sum_{k=1}^{n} d_k \tag{1}$$

In the principle of proximity, the distance from the starting point to the point of arrival of the courier robot can be calculated using the improved ACO. Therefore, the flow chart of the principle of multiple courier robots in the community task allocation is as follows (Fig. 2):

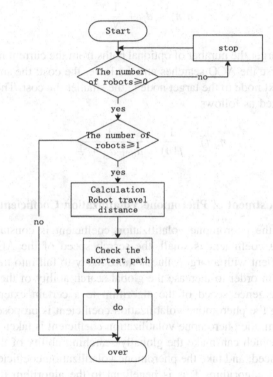

Fig. 2. Near principle flow chart

3   Improvement of ACO

3.1   Multi-node Based Heuristic Function Design

In the basic ACO, the heuristic function is only related to the length of the path between the current node and the next node. If the length of the path is shorter, the probability that the next node is selected is larger, so the ACO searches for the path in the early stage. It has a lot of blindness and reduces the search efficiency of the algorithm. In order to improve the search speed and accuracy of the ACO and prevent it from falling into the local optimal solution, the ACO is reconstructed considering the idea of the valuation function of the A* algorithm, so as to solve the problem of slow convergence of the algorithm. The valuation function is shown in equation:

$$f(n) = g(n) + h(n) \tag{2}$$

In the formula, the valuation function representing the node, representing the cost from the node to the optional node, consists of two parts: one is the cost of the optional node to the target node g, and the other is the number of optional nodes. The function expression is as follows:

$$h(n) = d_{jg} + \frac{1}{N_j} \tag{3}$$

In the formula, the larger the number of optional paths from the current node to the next node, the more diverse the ACO searches and the lower the cost; the smaller the Euler distance from the next node to the target node g, the smaller the cost. The new heuristic function is constructed as follows:

$$\eta_{ij}(t) = \frac{1}{f(n)} = \frac{1}{d_{ij} + d_{jg} + \frac{1}{N_j}} \tag{4}$$

3.2 Dynamic Adjustment of Pheromone Volatilization Coefficient

In the basic ACO, the pheromone volatilization coefficient is constant at a certain threshold. When the coefficient is small, the search speed of the ACO is lowered. However, the coefficient with a large value is more likely to fall into the local optimal solution. Therefore, in order to increase the global search ability of the algorithm and accelerate the convergence speed of the algorithm to a certain extent, a two-stage method for adjusting the pheromone volatilization coefficient is proposed. In the early stage of the algorithm, the pheromone volatilization coefficient is taken as any number between (0.1, 0.3), which can make the globally-searching ability of the algorithm in the early stage enhanced, and take the pheromone volatilization coefficient (0.4, 0.6) in the later stage of the algorithm. This is beneficial to the algorithm to speed up the convergence at a later stage. The pheromone volatilization coefficient is improved as follows:

$$\begin{cases} \rho = unifrnd\,(0.1, 0.3) & k < 0.25K \\ \rho = unifrnd\,(0.4, 0.6) & k \geq 0.25K \end{cases} \tag{5}$$

Where, K represents the maximum number of iterations of the algorithm, and k represents the current number of iterations.

3.3   Improved ACO

In summary, the specific steps of the improved ACO for multi-destination robots in community path planning are as follows:

1. Multi-deployment robot environment modeling and parameter initialization. The grid method is used to divide the running environment of the multi-destination robot into a 40*40 grid map. Each rectangle grid is the same size, wherein the grid marked with 1 is an obstacle, and the grid marked with 0 indicates a movable space;
2. Setting various experimental parameters of the improved ACO;
3. Placing only the ants at the initial point;
4. The ant k selects the next node from which the starting node can move, calculates the probability value of the node from the pheromone of each optional node, and determines the next starting point by the formula (5);
5. update the path;
6. Repeat steps 4 and 5 until the ant reaches the destination or has no way to go;
7. Repeat steps 4, 5, and 6 until all ants have finished iterating over a generation;
8. Update the pheromone matrix according to formula (6), excluding ants that ave not arrived;
9. Repeat steps 4–8 until the end of the nth generation ant iteration.

4   Multi-express Robot Path Planning Experiment and Analysis

4.1   Simulation Results and Analysis

In order to verify the feasibility and effectiveness of the improved ACO multipath robot proposed in this paper in community path planning. Through the simulation platform, the basic ACO and the improved ACO algorithms were tested 20 times so that the average could be finally obtained. We select task points 3, 5, 8 and starting points 9, 2 and 1, and the simulation results of the two algorithms obtained are shown in Table 1:

Table 1. Comparison of simulation results

	Basic ACO	Improved ACO
Optimal path length	50.97	48.73
Worst path length	54.73	51.56
Average path length	51.862	50.523
Minimum number of iterations	87	57
Maximum number of iterations	159	147
Average number of iterations	95	76
Average time/s	325.3532	291.6748

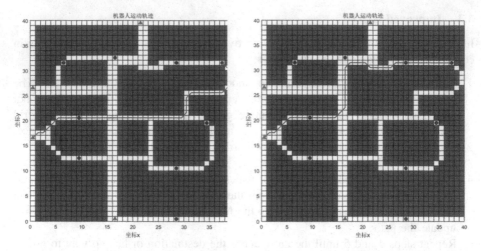

Fig. 3. Optimal path comparison (left side is traditional, right side is optimization)

The comparison of the optimal path comparison chart and the convergence curve of the two algorithms is shown in Figs. 3 and 4.

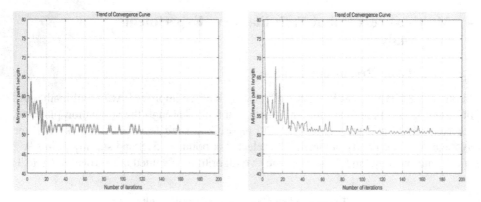

Fig. 4. Convergence curve comparison (left side is traditional, right side is optimization)

It can be seen from the simulation results that the improved ACO is superior to the basic ACO in terms of path length, number of iterations and running time. It can be seen from Fig. 7 that the optimal path obtained by the improved ACO is shorter than the path length of the basic ACO, and the improved ACO has a faster average convergence speed and a shorter average running time of the algorithm.

4.2   Task Assignment Implementation

The optimal path length from each starting point to each task point obtained by the improved ACO is shown in Table 2.

Table 2. Length from the starting point to the task point

Task point	Starting point			
	A (22)	B (521)	C (921)	D (1577)
1 (297)	11.41	20.24	28.49	32
2 (327)	24.07	9.83	19.49	41.73
3 (349)	15.24	32.9	39.73	47.38
4 (358)	28.49	42.49	48.73	50.73
5 (770)	45.14	25.24	10.66	26.66
6 (835)	47.14	38.07	37.49	38.07
7 (1090)	36.9	20.31	10.66	17.83
8 (1189)	43.31	42.14	30.49	25.66
9 (1589)	52.31	50.14	41.31	12

Considering the proximity principle strategy, it can be seen that task point 1 (297), task point 3 (349) and task point 4 (358) are assigned to the multi-destination robot starting point A (22); task point 5 (770), task point 6 (835) and task point 7 (1090) are assigned to the multi-destination robot starting point C (921); task point 8 (1189), and task point 9 (1589) are assigned to the multi-destination robot starting point D (1577) from the results of Table 3. It can make the total path of the multi-destination robot to complete the task the shortest, the specific allocation results are shown in Table 4.

Table 3. Assignment result

Starting point	A (22)	B (521)	C (921)	D (1577)
Task point	1, 3, 4	2	5, 6, 7	8, 9

5 Conclusion

This paper studies the basic theory of path planning and task allocation for multi-robot systems based on the community. By reconstructing the heuristic function of the ant colony algorithm and modifying the pheromone volatilization coefficient, the convergence speed and search speed of the algorithm are improved. In the MATLAB software environment, the path planning results under different algorithms are compared. The rationality of the path planning model and the effectiveness of optimizing the ACO are verified.

References

1. de Almeida, J.P.L.S., Nakashima, R.T., Neves-Jr, F., et al.: Bio-inspired on-line path planner for cooperative exploration of unknown environment by a multi-robot system. Robot. Auto. Syst. **112**, 32–48 (2018)

2. Purcaru, C., Precup, R., Iercan, D., et al.: Multi-robot GSA- and PSO-based optimal path planning in static environments. In: 2013 IEEE 9th International Workshop on Robot Motion and Control, Kuslin, pp. 197–202 (2013)
3. Hao, W., Xu, X.: Immune ant colony optimization network algorithm for multi-robot path planning. In: 2014 IEEE 5th International Conference on Software Engineering and Service Science, Beijing, pp. 1118–1121 (2014)
4. Das, P.K., Behera, H.S., Jena, P.K., et al.: Multi-robot path planning in a dynamic environment using improved gravitational search algorithm. J. Electr. Syst. Inf. Technol. 3(2), 253–313 (2016)

Matching User Accounts Across Social Networks Based on LDA Model

Shuting Zhang and Hong Qiao

Shandong Normal University, Jinan, China
shutingzhang114@163.com, qiaohongsd@126.com

Abstract. Identifying users across social networks has received more and more attention in recent years. In this paper we propose a model that combines ATM and DTM. The model uses the ATM model to study the user's potential behavior information and extract user's topic preferences. Then it analyzes the trend of user topic preferences changing with time through the DTM. The purpose of this paper is to improve the accuracy of account identification across social networks. The experimental results show that the model performs well on Chinese text or mixed text in Chinese and English because it obtains higher accuracy and F1 scores.

Keywords: LDA · User identification · Account matching

1 Introduction

In the past decade, as online social networks have flourished, an increasing number of people have accounts on social networks. According to the data released by Facebook in 2017 [1], the number of active users per month has exceeded 2 billion, which is about a quarter of the world's population. The quarterly report of WeChat, the primary social platform in China, shows that its monthly active users reached 980 million, and about 38 billion messages were sent daily. An increasing number of netizens use social networks to share information and communicate. As different social networking sites offer different services, people have begun to use different social networks to meet different needs, such as using WeChat to keep in touch with friends and share pictures, using Weibo, Twitter and other software to focus on hot events and spread information, and using Zhihu to answer questions and share knowledge. According to statistics [2], 42% of adults have accounts on multiple social platforms, and people use different social networks to record their lives. However, at present, user information is scattered among various social networks, and it is very difficult to obtain complete user information. User identification across social networks attempts to link users' accounts in multiple social networks, which has many advancements for practical applications such as personage search, cross platform recommendation and user portraits.

At present, the related work of account identification is mainly based on structure, user attribute information and user behavior information. There is also some scholars' research based on comprehensive information. In terms of behavioral information, Nie et al. [3] identified each user by relying on the status of the user's published status, and

D. Milošević et al. (Eds.): IICC 2019, LNCS 11956, pp. 333–340, 2019.
https://doi.org/10.1007/978-3-030-37429-7_33

proposed a method to analyze the dynamics of user interests. Almishari et al. [4] put forward a method of using writing style for identification. Based on the status and comments published by users on the network, and Sha et al. [5] adopted Doc2Vec to convert user-generated text into vectors for matching user accounts. Liu [6] built the behaviour similarity model and the structure information model, and used multi-objective optimization with missing information to identify linkage across social networks.

This paper uses the ATM model [7] to study the user's potential behavior information, extract user's topic preferences, and analyze the trend of user topic preferences changing with time through the DTM. By calculating the similarity of the two features between two authors from Zhihu and Jianshu, we successfully completed accounts matching. The experimental results verify the effectiveness and rationality of the model, which is good at handling text in different languages and is not limited by the length of the user generated content.

2   Data Sources

Our research is based on China's two largest Q&A communities, Zhihu and Jianshu as objects. First, we manually searched for a number of accounts belonging to the same user on both platforms. If some users provided important information such as the same email address, they were considered to be the same user. According to the content characteristics of most users, we selected the text released before June 2016 as the research data, which not only ensures the complete user's data, but also facilitated the segmentation of time segments. Then, we collected those users' names, the articles published before June 2016, and the release time. Based on these users, we collected the same information for all followers. Finally, all users who have not published content are eliminated, the rest of the data constitutes the data source for this study, in which 203 users have been selected from various areas of interest. The original data includes the user name, article content, and article release time.

3   Topic Preferences

3.1   The ATM Model

The same author in different social networks has the same interest, and even if you focus on or post information on different social networking sites, you should have similar topic preferences. To extract user topic preference features, the Author-Topic-Model [7] (ATM) is adopted. The ATM is a text-implicit topic-generation model that fuses the LDA implicit topic model and author information. In the ATM, document collection is divided into two levels: the collection of documents and authors, documents and words. Its working process is: first, select an author from the author's concentration by equal probability, and then randomly sample a topic according to the author's polynomial distribution on the topic, after that, sample a word with the same method; finally, the process is iterated until all the words are obtained. The LDA model fails to consider the author's topic preference while synthesizing the document content, the ATM model solves this problem.

In order to generate the best ATM, some parameters need to be adjusted. In the ATM, the number of topics K is a very important parameter to be determined. Being too large or too small may seriously affect the validity of the model. In order to obtain the best number of topics, we use the consistency c_v to measure the model performance under each K, and search for the optimal value of K. The c_v ranges in $[-1, 1]$. The larger the c_v is, the better the model effect is. Therefore, when the number of topics gets higher consistency, it means that the number of topics will give the ATM a better result. We set the search range of the K to an integer of $[2, 50]$ because there are at least 2 topics, and 50 is enough to satisfy the maximum number of topics. As shown in Fig. 1, the best number of topics found is 21.

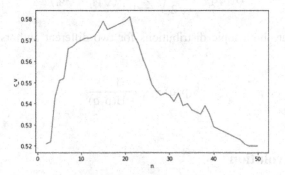

Fig. 1. Distribution of coherence coefficient.

In this paper, the data was trained by the ATM to obtain the topic distribution of each author as shown in the following Fig. 2:

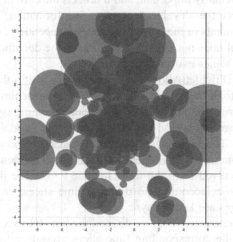

Fig. 2. Topic distribution of each author.

Each circle in the figure represents an author. The size of the circle indicates the number of articles written by the author. The overlap of the circles indicates the overlap of the authors' topic distribution. As we can see from the author-topic map, the topic preferences of different authors are quite different. In this case, the overlapping part can be used to identify similar authors to achieve the accounts link.

3.2 Similarity of Topic Preference

To calculate the similarity of topic preference of two authors, we use Hellinger distance that is usually adopted to measure the similarity between two probability distributions, which is defined as follows:

$$H(p,q) = \frac{1}{\sqrt{2}} \sqrt{\sum_{i=1}^{K} \left(\sqrt{p_i} - \sqrt{q_i} \right)^2} \tag{1}$$

where p and q are both topic distributions for two different authors. We define the similarity as:

$$S(p,q) = \frac{1}{1 + H(p,q)} \tag{2}$$

4 Interest Evolution

4.1 The DTM Model

With the increasing demands for information on text data, and that demand has not been satisfied with the extraction of data information topics under static time. "Time" as a label is extremely important for text data should not be ignored in the model. The time of the text is particularly important, and it reflects the current information situation and profoundly affects the development of post-information. The original model assumes that the documents in the corpus are order-independent, which means that the traditional LDA model does not consider the time of the document as a label, but the reality is that the text topics evolve under time series.

The LDA (Latent Dirichlet Allocation) model consists of a three-tier structure: the document layer, the topic layer and the word layer. To achieve dynamic keyword extraction, we can increase the time layer. Blei added the time layer to the LDA topic model, and he proposed the DTM [8] (Dynamic Topic Model), which can extract dynamic topic words. The hierarchical allocation takes into account the influence of discrete time. The topic model is adopted to model the documents of each time slice. The topic in slice t evolved from the topic in slice t − 1. The document distribution generates different topics according to different time slices, and the acquired topic words and time changes are kept in sync.

Based on the data characteristics, we divided the time into six slices, and put all the author's articles into the corresponding time slices according to the release time. In order to improve identification accuracy, we selected five candidates with the highest similarity of topic-preference (calculated in the previous step) for each author. Then, we used the DTM to analyze the topic-interest evolution. For the author and each candidates, we selected the same topics, and drew the curve of the importance degree of the topic, as shown in Fig. 3. There is the time-varying curve of the author and one of the candidates under topic 1. It can be seen that the trends of topic 1 of these two authors are roughly the same.

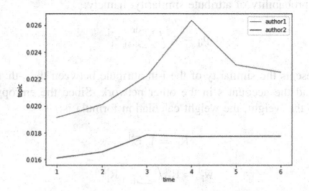

Fig. 3. Topic interest evolution of two authors

4.2   Similarity of Topic Evolution

To calculate the similarity of the two authors' interest evolutions over time for the same topic, we used the Residual Sum of Squares (RSS). The RSS calculation formula is as follows:

$$\text{RSS} = \sum\nolimits_{i=1}^{m} (y_1 - y_2)^2 \tag{3}$$

For multiple similar topics for two authors, we calculated the average of the RSS as the final similarity of interest evolution, as shown in Sect. 5.

5   Account Matching

To calculate the final similarity of the users of the two platforms, we first normalized the data for reducing the error. Then, an Information Entropy based Multiple Social Networks User Identification Algorithm (IE-MSNUIA) proposed by Wu [9] was adopted to weight the two similarities of topic preference and topic evolution. According to the definition of information entropy, when the system may be in many different states, the probability of each state j is p_{ij}, and then the information entropy of the system is:

$$E_i = - \sum_{j=1}^{n} (p_{ij} * lb \, p_{ij}) \tag{4}$$

In this paper, the weight of the user is determined according to the similarity of each attribute. The greater the difference between the similarity of the matching account and the unmatched account is, the more orderly the information is and the smaller the entropy is. In other words, this attribute that carries more information is more valuable. In this case, the weight should be larger. Based on the above analysis, p_{ij} in Eq. (4) is defined as the probability of attribute similarity, namely:

$$p_{ij} = v_i^{sj} / \sum_{j=1}^{n} v_i^{sj} \tag{5}$$

where v_i^{sj} represents the similarity of the i-th attribute between the j-th account in the one network and the account s in the other network. Since the entropy is inversely proportional to the weight, the weight calculation formula is:

$$R_i^{sj} = 1/E_i^{sj} \tag{6}$$

$$w_i^{sj} = R_i^{sj} / \sum_{i=1}^{n} R_i^{sj} \tag{7}$$

The final similarity for measuring the similarity of two users is:

$$Sim(s,j) = \sum_{i=1}^{n} (w_i^{sj} * v_i^{sj}) \tag{8}$$

Finally, we calculated the similarity matrix of two social network users.

For user matching, first selected five candidates with the highest similarity to the source account as the account to be matched. If both accounts are the most similar to each other, and the total similarity is greater than 0.5, it is considered to be a match; otherwise we did not consider the two accounts to be a match.

6 Model Evaluation

In this paper, accuracy, recall, and comprehensive evaluation index F1 are used as evaluation criteria for measuring model performance. The specific definition is as follows:

$$precision = tp/(tp + fp) \tag{9}$$

$$recall = tp/(tp + fn) \tag{10}$$

$$F1 = 2 * precision * recall/(precision + recall) \tag{11}$$

Precision refers to the ratio of correct matching, where tp indicates the number of accounts that match correctly using the model; fp indicates the number of accounts that match incorrectly; recall refers to the correctly identified ratio, where fn represents the accounts that belong to the same user but are not matched by the model. F1 is the harmonic mean of the first two, and it is a comprehensive evaluation index of the performance of the model. The conclusion is as follows (Table 1):

Table 1. Evaluation of the model.

Pre.	Rec.	F1
0.618	0.807	69.9

It can be seen that our model has obtained higher accuracy and F1 scores in Chinese, as well as in the Chinese-English mixed texts, indicating that the model is effective and reasonable.

7 Conclusion

In this paper, a model for matching user accounts across social networks is proposed based on user behavior information. By collecting the public user behavior information of Zhihu and Jianshu, the ATM model was adopted to extract the topic preferences of different authors. The author's interest was fully extracted as the basic information for account identification. Then, the DTM model was used to study the evolution of users' interest over the past three years, and extract the topic evolution curve. Finally, by combining the two models, constructing the similarity vector of the account to be matched, and assigning weights to different attributes with the entropy weight method, the overall similarity of the authors for account matching was calculated. The experimental results verify the effectiveness and rationality of the model. The model used in this article is featured with several advantages. For example, the length of the text has little impact on the results, and it can handle text or mixed text in different languages. In addition, the amount of data in this paper is limited, and the validity of the model will be verified on a larger data set in the future.

References

1. Tang, J., Chen, W.: In-depth analysis and mining for large social data. Chin. Sci. Bull. 509–519 (2015)
2. Kai, S., Wang, S., Tang, J., Zafarani, R., Liu, H.: User identity linkage across online social networks: a review. ACM SIGKDD Explor. Newsl. **18**, 5–17 (2017)
3. Nie, Y., Yan, J., Li, S., Xiang, Z., Li, A., Zhou, B.: Identifying users across social networks based on dynamic core interests. Neurocomputing **210**, 107–115 (2016)
4. Almishari, M., Tsudik, G.: Exploring linkability of user reviews. In: Foresti, S., Yung, M., Martinelli, F. (eds.) ESORICS 2012. LNCS, vol. 7459, pp. 307–324. Springer, Heidelberg (2012). https://doi.org/10.1007/978-3-642-33167-1_18
5. Sha, Y., Liang, Q., Zheng, K.: Matching user accounts across social networks based on users message. Procedia Comput. Sci. **80**, 2423–2427 (2016)

6. Liu, S., Wang, S., Zhu, F., Zhang, J., Krishnan, R.: HYDRA: large-scale social identity linkage via heterogeneous behavior modeling (2014)
7. Rosen-Zvi, M., Griffiths, T.L., Steyvers, M., Smyth, P.: The author-topic model for authors and documents (2012)
8. Blei, D.M., Lafferty, J.D.: Dynamic topic models. In: Proceedings of International Conference on Machine Learning, pp. 113–120 (2006)
9. Wu, Z., Yu, H., Liu, S., Zhu, Y.: User identification across multiple social networks based on information entropy. J. Comput. Appl. 37, 2374–2380 (2017)

Sales-Forecast-Based Auto Parts Multiple-Value Chain Collaboration Mechanism and Verification

Yufang Sun[1,2], Changmao Wu[3], Wen Bo[3,4], Lei Duan[3,5], and Changyou Zhang[3(✉)]

[1] School of Computer Engineering, Chengdu Technological University, Chengdu, Sichuan, People's Republic of China
[2] Chengdu Guolong Information Engineering Co., Ltd., Chengdu, Sichuan, People's Republic of China
[3] Laboratory of Parallel Software and Computational Science, Institute of Software, Chinese Academy of Sciences, Beijing, People's Republic of China
changyou@iscas.ac.cn
[4] School of Computer and Information Technology, Shanxi University, Taiyuan, Shanxi, People's Republic of China
[5] School of Computer and Communication Engineering, University of Science and Technology Beijing, Beijing, People's Republic of China

Abstract. Auto parts in the after-sales service flow across parts supplier, manufacturer, and transit store as well as service provider. The accurate and timely demand for auto parts plays a key role in the value-added process of the parts multiple-value chains. This paper presents a collaboration mechanism and evaluation method for auto parts multiple-value chains based on sales prediction. First, we study the composition of inner value chain of independent accounting unit, and discuss factors of value increment of the inner value chain. Moreover, dominated elements are selected as main factors for future value. Second, a colored petri nets model is created for each independent accounting unit all over the multiple-value-chains, of which value-added process maps to Transition, repository to Place and auto parts to Place's Token. Such model is utilized to simulate the value-added process over the multiple-value chains. Finally, a collaboration mechanism for multiple-value chains is designed based on the prediction amount for auto parts. By means of randomly generating a sequence of custom's arrival and Monte Carlo method, we use CPN Tools to simulate the value-added process over the multiple-value chains. Experiments show that the offered auto parts collaboration mechanism can fully utilize existed resources in the multiple-value chains, gets maximized rise in value, and further seek the fundamental path to value increment. The model implemented in this paper is able to provide some quantification references for the multiple-value chains.

Keywords: Auto parts · Sales prediction · Parts value chain · Multiple value chain collaboration · Colored petri network

© Springer Nature Switzerland AG 2019
D. Milošević et al. (Eds.): HCC 2019, LNCS 11956, pp. 341–350, 2019.
https://doi.org/10.1007/978-3-030-37429-7_34

1   Introduction

Since China's economy and auto industry has been booming in the last decade, China has gradually become the world's largest automobile production and consumption country, and then makes large contribution to the scale of automotive service. The auto parts play a paramount role in after-sales service of the entire automotive industry. Consequently, with the gradual maturity of the after-sales service industries, more and more auto companies focus on the demand of auto parts. The accurate forecast of the demand for auto parts and its timely supply both have played a more and more critical role in terms of improving the quality of service [1–4].

At present, on one hand, the traditional automobile manufacturers have adopted the business units to submit the parts plan, and then using these plans to predict the demand for the parts during the three guarantees period. However, the demand of amount of parts in each business unit is often determined by experience, which usually yields error. On the other hand, auto factory generally makes a relatively simple "addition" operation on the submission of parts plan, which results in more and more errors as time flies [3, 4].

Auto parts are the material basis for vehicle maintenance and repair [5]. We define each units and products for auto as auto parts. The auto parts mainly consist of car body, engine, braking device and maintenance tools, and so on. Delivering auto parts in time is the basic condition of ensuring the quality of vehicle maintenance and repair time. Auto service provider and maintenance station are the main places for selling auto parts. For the purpose of earning profit in the strong competition with others, they should offer excellent maintenance service. In other words, the auto service provider and maintenance station have ability to timely and accurately provide auto parts service [6, 7].

Auto parts' forecasting is one kind of important technique for improving the quality of after-sales service of the entire automotive industry, in the past had many scholars to conduct the auto parts prediction in this aspect. Paper [8] presents a combination forecasting method to auto parts prediction, which utilizes multi-factor Gray Model [9] and neural network to model the process of auto parts forecasting. In paper [10], many techniques of auto parts prediction have been analyzed, and on the basis of Gray Model and BP neural network, auto parts prediction model has been offered and optimized.

In general, for current auto parts selling management process, auto parts prediction is inaccurate as most of auto parts plan is submitted by experience. Moreover, only upper and lower limit for the auto parts are taken into consideration. Weak auto parts prediction usually leads to postpone parts supply, and lower service quality.

In this paper, we implement a cooperation mechanism and evaluation method for auto parts multiple-value chains based on sales prediction (in Sect. 2), and by means of randomly generating a sequence of custom's arrival and Monte Carlo method, we use CPN Tools to simulate the value-adding process over the multiple-value chains (in Sect. 3).

2 Collaborative Process of Multiple Auto Parts

In this section, we will offer our collaborative method for auto parts multiple-value chains based on sales prediction. In addition, for accurately modeling auto part's circulation process across parts supplier, manufacturer, and transit store as well as service provider, we introduce Poisson distribution and Normal distribution to simulate customer's arrival process and auto parts replacement, respectively.

2.1 Storage Forecast of Multiple Auto Parts

Generally speaking, auto parts are produced when parts supplier receiving parts purchase order from auto manufacturer. After batch production, auto parts are either sent to auto manufacturer's center parts store or transit store for future use. It clearly shows in upside of Fig. 1, just as light blue and green portion depict. Auto manufacturer will suffer financial pressure or lower quality of service as the amount of auto parts order is too big or small.

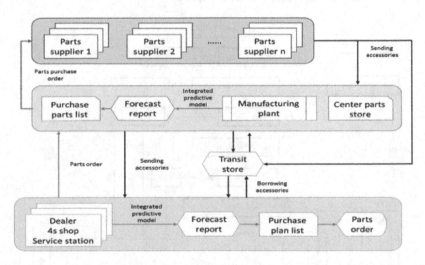

Fig. 1. Application of predictive model in multi-value-chain collaboration (Color figure online)

As auto manufacturers, they usually build their own central store for auto parts, and some of transit stores all over the country. When one kind of auto parts is unavailable in service provider, they firstly contact their local counterparts for borrowing. If still unavailable, they ask transit store for help till auto manufacturers. Borrowing auto parts from the local counterparts is the time-money-saving way for service provider. While borrowing auto parts from transit store or auto manufacturer will gradually increase the cost of time and money. What's more, the last two borrowing way reduces customer's experience and may lead to loss of customer.

In order to maximize profits, two factors should carefully are taken into consideration: one is the customer's arrival time; the other is probability of auto part's replacement. If we can accurately model the two parameters, we are able to predict the demands for auto parts per month or season. This is the sub-section's subject.

2.2 Random Model of Multiple-Value-Chain Collaboration

Auto parts selling mainly come from customer's buying an auto and replacing parts when auto maintenance in 4S store, maintenance station and retailer. When something is wrong with an auto, the auto needs to have maintenance and repair in 4S store, maintenance station and retailer. The auto parts are provided to 4S store, maintenance station and retailer directly by auto manufacturer or through transit store.

When there is a shortage of auto parts in 4S store or maintenance station or retailer, they may obtain the parts through the following paths: (1) through local 4S store or maintenance station or retailer. It is the most convenient and fast way to get the parts; (2) through the transit store. It is the second convenient and fast way to get the parts, compared to local 4S store or maintenance station or retailer; (3) through the auto manufacturer. It is the most time-consuming way to get the parts. In addition, this often needs more money and decrease the service quality.

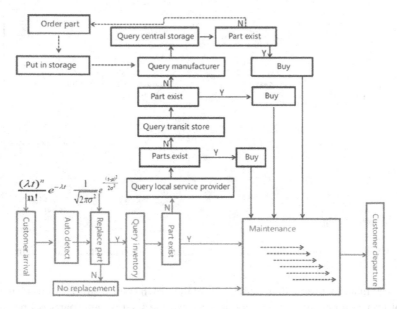

Fig. 2. Implementation of our cooperation mechanism for auto parts multiple-value chains based on sales prediction. We simulate customer's arrival process by Poisson distribution, the auto parts replacement abides by Normal Distribution.

Such an auto parts circulation process just as Fig. 1 depicts. In Fig. 1, we utilize Poisson distribution to randomly simulate customer's arrival process. For a costumer arrives at 4S store or maintenance station or retailer in an interval, we call costumer's arrival as an event. An event can occur 0, 1, 2, …, times in an interval. The average number of events in an interval is designated λ (lambda). λ is the event rate, also called the rate parameter. If we let $N(t)$ denote the number of events that occur between now and time t, then the probability that n events occur

$$P(N(t) = n) = \frac{(\lambda t)^n e^{-\lambda t}}{n!} \tag{1}$$

Where:

- λ is the average number of arrival events per interval
- t is time unit, which can be minute, hour or day or year, determined by your event
- e is the number 2.71828... (Euler's number) the base of the natural logarithms
- n takes values 0, 1, 2, ...
- $n! = n \times (n-1) \times (n-2) \ldots \times 2 \times 1$ is the factorial of n

For example, a 4S store receives 4 auto maintenances per hour on average. Assuming the Poisson model is appropriate. Thus, the time-unit t is one hour and the rate λ is 4. Knowing these, we can answer questions such as:

- What is the probability that the 4S store receives 6 customer's maintenance in the next 2 h?

$$P(N(2) = 6) = \frac{(4 * 2)^6 e^{-4*2}}{6!} = 0.12213821545677231$$

- What is the probability that the 4S store receives at least 2 auto maintenances in the next 3 h?

$$P(N(3) \geq 2) = 1 - P(N(3) = 1) - P(N(3) = 0)$$
$$= 1 - \frac{(4 * 3)^1 e^{-4*3}}{1!} - \frac{(4 * 3)^0 e^{-4*3}}{0!} = 0.9999201252394068$$

In addition, in Fig. 1, we using Normal Distribution to judge whether an auto parts is replaced or not. The probability density of the normal distribution is

$$P(t|\mu, \sigma^2) = \frac{1}{\sqrt{2\pi\sigma^2}} e^{-\frac{(t-\mu)^2}{2\sigma^2}} \tag{2}$$

Where:

- t is use time of an auto parts
- μ is mean or expectation of the distribution. In our case, μ is an auto parts' average probability of replacement
- σ is the standard deviation, and
- σ^2 is the variance.

3 Petri-Net-Based Model and Analysis

3.1 Basic Petri-Nets

To research cooperation mechanism and evaluation method for auto parts multiple value chains based on sales prediction. According to the resources and gain links needed during the three guarantees period, we will establish a colored Petri net model for multiple auto parts value chains in this paper.

Definition 1. Oriented Net.

N = (S, T, F) is called oriented net, if and only if:

(1) $S \cup T \neq \emptyset \cdot S \cap U = \emptyset$;
(2) $F \cap (S \times T) \cup (T \times S)$;
(3) $dom(F) \cup cod(F) = S \cup T$, where $dom(F) = \{x | \exists y : (x, y) \in F\}$, $cod(F) = \{y | \exists x : (x, y) \in F\}$, Are the domain and value range of F respectively.

3.2 Colored Petri-Nets

Colored Petri-nets (CP-nets or CPNs) is a graphical language for constructing models of concurrent systems and analyzing their properties [11]. CP-nets is a discrete-event modelling language combining Petri nets and the functional programming language CPN ML which is based on Standard ML [12]. A CPN model of a system describes the states of the system and the events (transitions) that can cause the system to change state. By making simulations of the CPN model, it is possible to investigate different scenarios and explore the behaviors of the system. Different resources correspond to token colors. The resource information corresponds to the token value. They are important factors influencing the gains of the value chain. The corresponding relationship of elements is shown in Table 1.

Table 1. The corresponding relationship of elements

Service value chain element	Petri-net element
Gain link	Transition
After-sales resources	Place
Customer information	Token
Service control	Connection
Service resource type	Token color
Resource information	Token value

On the basis of the colored Petri net, the control unit is added to indicate the flow of different customer services, the probability of occurrence of control, resource constraints, etc., and the after-sales service network system is delimited as Definition 2.

Definition 2. Automobile service network system.

Π = (P, T; F, C, K, D), sufficient and necessary condition for the automobile after-sales service network system:

(1) $\sum = (P, T; F, C)$, is colored network system
(2) $t \in C$, D(t) is the control function of Transition t
(3) $p \in P$, K(p) is the resource limit function of the Place p.

3.3 CPN Tools

In this paper, the experimental environment is CPN Tools 4.0.1. CPN Tools is a colored Petri-Net tool for editing, design, simulation and analysis. It has strong mathematical modeling capabilities and it is closely related to mathematics. It is often used for modeling analysis of complex, concurrent systems. The user of CPN Tools works directly on the graphical representation of the CPN model. The graphical user interface (GUI) of CPN Tools has no conventional menu bars and pull-down menus, but is based on interaction techniques such as tool palettes and marking menus. Users take file operations (create, save, load, print, etc.), graphic elements editing (create, modify, delete, etc.), model simulation, performance analysis and so on. Part of the graphical interface is shown in Fig. 2. A license for CPN Tools can be obtained free of charge via the CPN Tools web pages [13]. CPN Tools is currently licensed to more than 4,000 users in more than 115 different countries and is available for MS Windows and Linux. CPN Tools combined the Colored Petri nets semantics and CPN ML language to construct a description of discrete event model. CPN ML provides basic data type definition. It operates on the data to build a compact parametric model. ML language data type suit to describe the color set of Petri nets. It defines the Token color to distinguish the types of resources and enhances its ability to describe the system. By defining the range of data types, the random function (ran ()) generate network delay, processing rate and other random numbers in a specified range (Fig. 3).

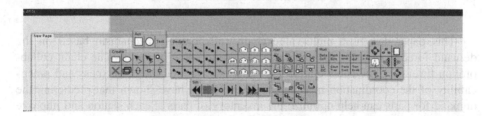

Fig. 3. Interface of CPN Tools.

In addition to modeling, CPN Tools also provides a wealth of model analysis tools such as monitors, state space analysis, and more. They enhance its model analysis capabilities and model validation capabilities. In the process of simulating the CPN model, we can generate some quantitative information about system performance, such as queue length, response time, throughput and resource utilization. The CPN test method used in this paper is a simulation execution test method. The simulation execution test method models the application system. Then it verifies the state transitions at each step of the system and verifies that the system model meets expectations.

3.4   Experimental Environment and Parameters

Very often, the goal of simulation is to debug and investigate the system design. CP-nets can be simulated interactively or automatically. An interactive simulation is similar to single-step debugging. It provides a way to "walk through" a CPN model, investigating different scenarios in detail and checking whether the model works as expected. During an interactive simulation, the modeler is in charge and determines the next step by selecting between the enabled events in the current state. It is possible to observe the effects of the individual steps directly on the graphical representation of the CPN model. Automatic simulation is similar to program execution. The purpose is to simulate the model as fast as possible and it is typically used for testing and performance analysis. For testing purposes, the modeler typically sets up appropriate breakpoints and stop criteria. For performance analysis the model is instrumented with data collectors to collect data concerning the performance of the system.

We define the range of values of the data type and call its random function to generate random numbers within the specified range such as control pass rate and maintenance probability. In this paper, we define the types and abbreviations and descriptions of the main parameters, as shown in Table 2 (Fig. 4).

Table 2.   Description of main parameters

Name	Type	Explain
Vehicle value	int	n
Maintenance cost	int	n2, N2
Control probability	int	p
Number of services	int	time1, time2

This paper designs a multi-value chain collaboration mechanism based on the demand forecast of parts and simulates the service flow of multi-value chain collaboration through CPN Tools platform. Finally, we verified the correctness and practicability of the coordination mechanism. Experiments have shown that predicting the demand for parts can help optimize the inventory of parts at each station and improve collaborative efficiency.

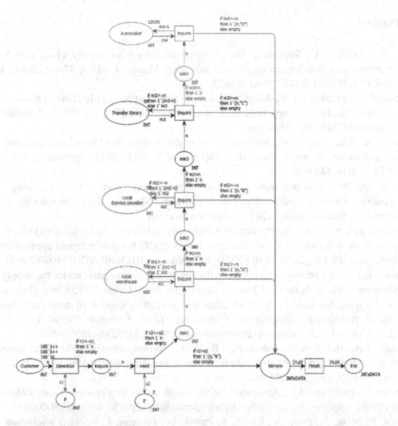

Fig. 4. Simulation process for sales-forecast-based auto parts multiple-value chain collaboration mechanism on CPN Tools platform.

4 Conclusion

In this paper, we implement a collaboration mechanism and evaluation method for auto parts multiple-value chains based on sales prediction. A colored petri nets model is created for an independent accounting unit all over the value chain, of which value-adding process maps to Transition, repository to Place, as well auto parts to Place's Token, such model is utilized to simulate the value-adding process over the value chain. Experiments show that the offered auto parts cooperation mechanism can fully utilize existed resources in the chains, maximum the value-adding, and further seek the fundamental path for increment of value. The model implemented in this paper is able to provide some quantification references for the multiple-value chains.

Acknowledgment. The authors would like to thank the anonymous referees for their valuable comments and helpful suggestions. This paper is supported by The National Key Research and Development Program of China (2017YFB1400902).

References

1. Ho, D., Kumar, A., Shiwakoti, N.: A literature review of supply chain collaboration mechanisms and their impact on performance. Eng. Manag. J. **31**(1), 47–68 (2019). https://doi.org/10.1080/10429247.2019.1565625
2. Howard, M., Hopkinson, P., Miemczyk, J.: The regenerative supply chain: a framework for developing circular economy indicators. Int. J. Prod. Res. **57**, 1–19 (2018). https://doi.org/10.1080/00207543.2018.1524166
3. Gupta, M., Andersen, S.: Throughput/inventory dollar-days: ToC-based measures for supply chain collaboration. Int. J. Prod. Res. **56**(13), 4659–4675 (2018). https://doi.org/10.1080/00207543.2018.1444805
4. Wang, Q.: Research and implementation of automobile parts sales forecasting system supporting industrial chain collaboration platform. Master's thesis, Southwest Jiaotong University, Chengdu, China (2017). http://kns.cnki.net
5. Gunasekaran, A., Kobu, B.: Performance measures and metrics in logistics and supply chain management: a review of recent literature (1995–2004) for research and applications. Int. J. Prod. Res. **45**(12), 2819–2840 (2007). https://doi.org/10.1080/00207540600806513
6. Zhang, H.: On the strategies to optimize the inventory of auto parts under the supply chain environment. Master's thesis, Changan University, Xian, China (2013). http://kns.cnki.net
7. Shin, J., Kim, S., Lee, J.-M.: Production and inventory control of auto parts based on predicted probabilistic distribution of inventory. Digit. Commun. Netw. **1**(4), 292–301 (2015). http://www.sciencedirect.com/science/article/pii/S2352864815000577
8. Torabi, S.A., Hatefi, S.M., Saleck Pay, B.: ABC inventory classification in the presence of both quantitative and qualitative criteria. Comput. Ind. Eng. **63**(2), 530–537 (2012). https://doi.org/10.1016/j.cie.2012.04.011
9. Verma, R., Pullman, M.E.: An analysis of the supplier selection process. Omega **26**(6), 739–750 (1998). http://www.sciencedirect.com/science/article/pii/S0305048398000231
10. Avila, P., Mota, A., Pires, A., Bastos, J., Putnik, G., Teixeira, J.: Supplier's selection model based on an empirical study. In: Procedia Technology 4th Conference of Enterprise Information Systems-Aligning Technology, Organizations and People (CENTERIS 2012), vol. 5, pp. 625–634 (2012). http://www.sciencedirect.com/science/article/pii/S2212017312005002
11. Yang, T.: Study and application of demand forecast in auto parts company. Master's thesis, Shanghai Jiaotong University, Shanghai, China (2014). http://kns.cnki.net
12. Gray control system. J. Huazhong Univ. Sci. Technol. (Nat. Sci. Ed.) **3**, 11–20 (1982)
13. Shi, H.: Demand forecasting system based on auto parts industry chain collaboration platform research and implementation. Master's thesis, Southwest Jiaotong University, Chengdu, China (2015). http://kns.cnki.net
14. Jensen, K., Kristensen, L.M., Wells, L.: Coloured Petri Nets and CPN tools for modelling and validation of concurrent systems. Int. J. Softw. Tools Technol. Transf. **9**, 213–254 (2007)
15. Standard ML of New Jersey (2019). http://www.smlnj.org
16. Ratzer, A.V., et al.: CPN tools for editing, simulating, and analysing Coloured Petri Nets. In: van der Aalst, W.M.P., Best, E. (eds.) ICATPN 2003. LNCS, vol. 2679, pp. 450–462. Springer, Heidelberg (2003). https://doi.org/10.1007/3-540-44919-1_28

Multi-scale Attentive Residual Network for Single Image Deraining

Jing Tan, Yu Zhang, Huiyuan Fu$^{(\boxtimes)}$, Huadong Ma, and Ning Gao

Beijing Key Lab of Intelligent Telecommunications, Software and Multimedia,
Beijing University of Posts and Telecommunications, Beijing 100876, China
{tan_jing, zhangyu2015, fhy, mhd}@bupt.edu.cn,
ceen@21cn.com

Abstract. Removing rain streaks from a single image is extremely challenging since the appearance of rain streaks in shapes, scales and densities is ever changing. Therefore, we propose a novel end-to-end two- stage multi-scale attentive residual network that is both location-aware and density-aware, in order to preferably remove various rain streaks. Specifically, in the first stage, a multi-scale progressive attention sub- network is designed to automatically locate the distribution of diverse rain streaks and further to guide the following deraining. Then the second stage with the guidance of the attention map generated in the former stage aims to efficiently remove various rain streaks. To aggregate the characteristics of rain streaks with different scales and densities, we construct a multi-scale residual sub-network in which dilated convolution and residual learning are used to combine these features. As a result, these two sub-networks make up the whole network, and accomplish the process of joint detection and removal of diverse rain streaks fairly well. Extensive experiments on both synthetic and real-world rainy images demonstrate that our proposed method significantly outperforms several recent state-of-the-art approaches.

Keywords: Single image deraining · Multi-scale · Visual attention mechanism · Residual learning

1 Introduction

Single image deraining which aims to remove rain streaks and restore the original clean image from a rainy image is highly essential for many outdoor computer vision applications. Images captured under rainy conditions suffer from a series of visibility degradations and detail loss. Rain streaks that have various shapes and directions can hinder, deform and blur the background scenes. This will have a negative impact on the performance of many computer vision algorithms, such as detection, tracking and recognition, since these algorithms mostly rely on images taken under well weather conditions as their input. Thus image deraining is highly significant.

In the past decades, single image deraining has drawn considerable research attention, and a series of methods have been proposed to address this problem [1, 2, 5, 6, 10, 13–15, 18, 20–23]. Some treat this as a decomposition problem since a rainy image can be formed as the consist of a rain layer and a clean background layer

© Springer Nature Switzerland AG 2019
D. Milošević et al. (Eds.): HCC 2019, LNCS 11956, pp. 351–362, 2019.
https://doi.org/10.1007/978-3-030-37429-7_35

[1, 2, 10, 14, 21]. In addition, driven by the success of deep learning in computer vision tasks, a few methods based on convolution neural networks have been proposed for image deraining [4–6, 13, 15, 20, 22]. Despite the continuous progress, there are still several deficiencies, which can be observed from Fig. 1(b) or Fig. 1(c):

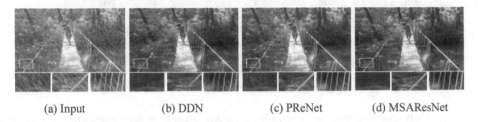

    (a) Input   (b) DDN   (c) PReNet   (d) MSAResNet

Fig. 1. Single image deraining results. It can be observed that rain streaks tend to be left over in the result of DDN [6] and PReNet [15], and some details are often lost by DDN [6] method, while our method can address these issues better.

- The shape, direction, density and scale of rain are varied and complicated. Most of the existing approaches do not take these factors fully into account, thus fail in dealing with images in various rain density levels.
- Because rain streaks overlap with the background texture, most methods tend to remove texture details in regions without rain, resulting in over-smoothing of these regions.

To address these issues, we propose a novel end-to-end multi-scale attentive residual network (MSAResNet) for single image deraining, which is both density-aware and location-aware. Specifically, our proposed method is composed of two stages including both rain streaks detection and removal. In the first stage, a multi-scale progressive attention network that recursively unfold a multi-scale recurrent module is proposed to automatically detect the location of various rain streaks and produce the corresponding attention maps, which aims to guide the following deraining process. Then we construct a multi-scale residual network to remove different rain streaks. It is designed to aggregate convolutional features of different scales to fully extract the rain streak components with different scales and density. And it takes both a rainy image and the last attention map generated by the multi-scale progressive attention network as input to make use of the location information. Extensive experiments have been conducted to evaluate the performance of our proposed MSAResNet, and Fig. 1(d) shows a derained example from our network.

In summary, the contributions of this paper are as follows:

- A novel end-to-end two-stage MSAResNet method is proposed, which first automatically detects the location of various rain streaks and generates the attention maps indicating where we should pay attention to due to the cover of rain streaks, and then efficiently removes various rain streaks guided by the attention map.
- We introduce visual attention mechanism by unfolding a novel multi-scale recurrent module into multiple stages iteratively to detect the distribution of rain streaks in

different density levels. In order to aggregate features of different scales to learn a better representation of the characteristics of various rain streaks, we propose two different structures in the detection stage and the removal stage, respectively.

- Comprehensive experiments have been conducted on two challenging datasets including both synthetic and real-world datasets. Our proposed method achieves superior performance compared against several state-of-the-art approaches.

2 Related Work

2.1 Prior-Based Single Image Deraining

A common mathematical model for single image deraining is defined as follows:

$$O = B + R \qquad (1)$$

where O, B and R are the rainy image, the background image and the rain streak image, respectively. That is, a derained image can be obtained by separating a rain streak component from the rainy image. Hence, some methods regard single image deraining as a layer decomposition problem on basis of some prior information due to its ill-posed nature. Among these methods, sparse coding representation [10], low rank representation [2] and Gaussian mixture model (GMM) [14] have been applied to model the background image or the rain streak image. However, they tend to produce over-smoothing results.

2.2 Deep Learning-Based Single Image Deraining

Recently, deep learning has been widely applied to single image deraining because of its success in many computer vision tasks. Fu et al. [6] propose a deep detail network which takes the detail layer separated from the rainy image as input and outputs the negative residual. Yang et al. [20] apply a multi-task recurrent dilated network for joint rain detection and removal. In [22], a density-aware multi-stream densely connected convolutional neural network (DID-MDN) is proposed to remove rain streaks of different densities and scales with the guidance of the estimated rain density information. Ren et al. [15] incorporate ResNet, recurrent layer, and multi-stage recursion to construct a simpler and more efficient deraining network. In comparison to these approaches, we propose an end-to-end two-stage multi-scale attentive residual network to aggregate both the distribution and density information of diverse rain streaks to facilitate deraining.

3 Multi-scale Attentive Residual Network

Given that rain streaks can have quite diverse appearance occurring in practice, we propose the MSAResNet to detect and remove rain streaks by aggregating the characteristics in various scales with the use of dilated convolutions of different dilated factors and residual learning. As is shown in Fig. 2(a), the proposed MSAResNet architecture consists of two sub-networks: a multi-scale progressive attention network and a multi-scale residual network. The multi-scale progressive attention network is applied to identify areas of concern due to the cover of rain streaks and generate the corresponding attention maps, so that it can guide better rain removal. Then the multi-scale residual network along with the final attention map from the multi-scale progressive attention network as one of the input performs removing rain streaks and restoring the original clean image from a rainy image.

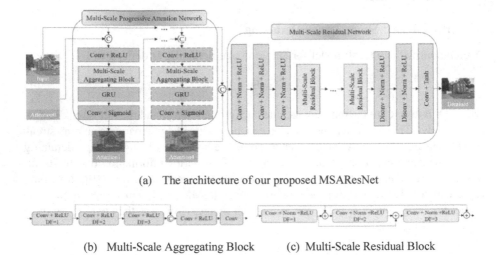

(a)   The architecture of our proposed MSAResNet

(b)   Multi-Scale Aggregating Block   (c)   Multi-Scale Residual Block

Fig. 2. The architecture of our proposed MSAResNet. (a) It comprises a multi-scale progressive attention sub-network and a multi-scale residual sub-network, in which the former plays the role to generate the attention map of the distribution of diverse rain streaks to guide the following deraining, and the later aims to remove various rain streaks. (b) It shows the details of the multi-scale aggregating block. (c) It shows the details of the multi-scale residual block. Note that DF represents the dilated factor, and © represents the concatenate operation.

3.1   Multi-scale Progressive Attention Network

Considering the process of image deraining as mapping from a rainy image to a derained image, it is obvious that it requires different mapping strategies between regions covered with and without rain streaks. It tends to loss texture details in non-rainy regions or fail to restore texture contents in rainy regions if the distribution of rain streaks is ignored. Researches on human perception process [17] have confirmed that visual attention mechanism can automatically focus on regions of interest so that it is beneficial to extract local features more significant for the task. Inspired by this, we propose a multi-scale progressive attention network to locate the distribution of rain streaks to further guide the process of deraining, as illustrated in the left of Fig. 2(a).

The proposed network is organized by unfolding only one multi-scale recurrent module into multiple stages recursively, which is on the basis of inter-stage recursive computation [11, 15, 16]. In this way, it shares the same parameters of the module without introducing extra parameters or bringing the decline of the performance.

(a)   Various rain streaks among images   (b) various in single image

Fig. 3. Rain streaks vary tremendously both among images and in a single image.

Multi-scale Recurrent Module. It is shown in the Fig. 3 that rain streaks have diverse appearance in shapes, densities and scales among images and even in a single image, so it is a better choice to aggregate features in different scales favourable to extract diverse rain streaks components. Therefore, we consider a multi-scale recurrent module to learn the distribution of various rain streaks. There are four parts of it: a convolution-ReLU layer fin, a multi-scale aggregating block $f_{msaggre}$, a recurrent block f_{GRU} and a convolution-Sigmoid layer f_{out}, as demonstrated in the left of Fig. 2(a). In particular, the fin layer aims to capture shallow features with the concatenation of the rainy image and the attention map generated by the former stage as input. Then in the $f_{msaggre}$ block, to better extract the characteristics of different rain streaks, we use three dilated convolution layers with the dilated factors set to 1, 2 and 3, respectively. Besides, we design several skip connections to further aggregate the information flow of rain streaks in various scales, in which the outputs of the three dilated convolution layers are concatenated as input of the later convolution layer. Figure 2(b) shows the details of the architecture of the $f_{msaggre}$ block. We further introduce the convolutional Gated Recurrent Unit (GRU) [3] as the recurrent block f_{GRU} in order to propagate the feature dependencies across stages. Finally, the f_{out} layer takes the output of the f_{GRU} as input and produce an attention map ranging from 0 to 1. A larger value of the element in the attention map suggests that the region is so more likely to be covered with rain streaks that more attention should be paid to.

Loss for the Multi-scale Progressive Attention Network. As mentioned above, we execute the process of various rain streaks detection by unfolding the multi-scale recurrent module recursively into multiple stages. There is an attention map A_t generated at stage t, and the loss function at stage t is defined as the mean squared error

(MSE) between A_t and the binary mask M in which value 1 indicates an area covered with rain while value 0 means a background region. To get the binary mask M, we subtract the corresponding clean image B from the rainy image R according to the widely used formulation of rain model [7, 14, 21]. Then we use an experience threshold to determine whether an area is covered with rain or not. As a result, the loss function of the multi-scale progressive attention network is expressed as:

$$\mathcal{L}_A = \sum_{t=1}^{N_1} t\theta \mathcal{L}_{MSE}(A_t, M) \tag{2}$$

where N_1 is the number of stages. θ controls the proportion of the attention map of different stages.

3.2   Multi-scale Residual Network

Taken the fact that rain streaks vary in shapes, scales and densities, we present a multi-scale residual network to more efficiently combine characteristics presented by diverse rain streaks and further to facilitate rain streaks removal. The architecture of the multi-scale residual network is illustrated in the right of Fig. 2(a). It includes three convolution-normalization-ReLU layers, several multi-scale residual blocks, two disconvolution-normalization-ReLU layers and a convolution-Tanh layer. Besides, to leverage the distribution information of various rain streaks, it takes the concatenation of the rainy image and the attention map generated at final stage by the multi-scale progressive attention network as input, and finally outputs a derained image.

Multi-scale Residual Block. The multi-scale residual block is the backbone of the multi-scale residual network, whose detail framework is illustrated in Fig. 2(c). Since the appearance of rain streaks is extremely irregular, the purpose of the multi-scale residual block is to better capture the characteristics of various rain streaks. In each multi-scale residual block, there are three dilatedConvolution- normalization-ReLU layers with dilated factors set to 1, 2, and 3 respectively. Then a skip connection is designed to combine features before and after a dilated Convolution-normalization-ReLU layer to aggregate multi-scale characteristics. More specifically, the input of the later layer is the combination output of the former two layers. Finally, the input of the multiscale residual block is concatenated with the output of the last layer as the overall output of this block to further aggregate different features and reduce the information loss.

Loss for the Multi-scale Residual Network. We apply three loss functions to the multi-scale residual network: L1 distance, perceptual loss [9] and SSIM loss [24]. The L1 distance is expressed as:

$$\mathcal{L}_{L_1} = \mathbb{E}\big[\|Y - B\|_1\big] \tag{3}$$

where Y and B are the derained image and the corresponding clean image, respectively. The L1 distance is a loss at the pixel level and mainly used to promote the same low-level characteristics between the derained image and the ground truth, while perceptual loss captures more high-level features to guarantee the visual quality. We consider a pre-trained VGG-19 network as the feature extractor to capture features in both the derained image and the ground truth. These features are from different ReLU layers of the VGG19 network and together used to determine the differences between the two images. The perceptual loss is calculated as follows:

$$\mathcal{L}_P = \sum_{i=1}^{m} \theta_i \mathbb{E}\left[\|\phi_i(Y) - \phi_i(B)\|_1\right] \tag{4}$$

where m is the number of ReLU layers. ϕ_i represents features from the i-th ReLU layer. θ_i represents the corresponding weight. In addition, to further ensure the structural authenticity of the derained image, we apply SSIM loss, which is defined as:

$$\mathcal{L}_{SSIM} = 1 - SSIM(Y, B) \tag{5}$$

$$SSIM(Y, B) = \frac{2\mu_y\mu_b + C_1}{\mu_y^2 + \mu_b^2 + C_1} \cdot \frac{2\sigma_{yb} + C_2}{\sigma_y^2 + \sigma_b^2 + C_2} \tag{6}$$

where μ_y and μ_b are the mean value of the derained image Y and the clean image B, respectively. σ_y^2 and σ_b^2 b are the variance value. σ_{yb} is the covariance value. Meanwhile, C1 and C2 are two small constants to avoid division by 0.

3.3 Loss for the Multi-scale Attentive Residual Network

The full objective during training is as follows:

$$\mathcal{L} = \lambda_A \mathcal{L}_A + \lambda_{L_1} \mathcal{L}_{L_1} + \lambda_P \mathcal{L}_P + \lambda_{SSIM} \mathcal{L}_{SSIM} \tag{7}$$

where each λ indicates a relative importance of each term.

4 Experiments

In this section, we evaluate the performance of the proposed MSAResNet on both synthetic and real-world rainy datasets. Qualitative and quantitative evaluations of deraining performance on the synthetic dataset are conducted since each rainy image has a corresponding clean image as the ground truth. We apply two full-reference image quality assessment metrics as the quantitative comparison criteria including PSNR [8] and SSIM [19]. For comparison on real-world rainy dataset, we only present the qualitative results due to the lack of the ground truth. Besides, we compare the proposed MSAResNet with several recent state-of-the-art methods.

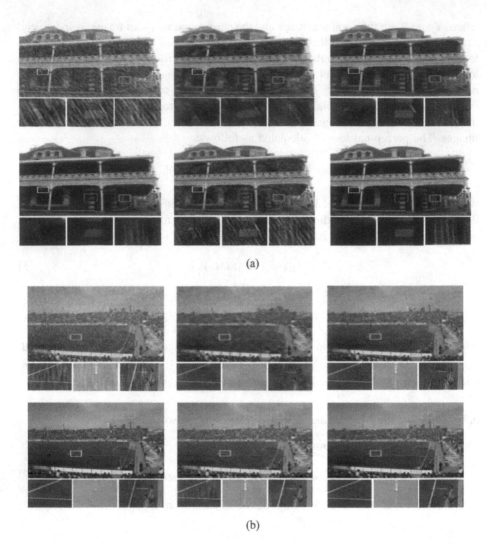

Fig. 4. Visual quality comparisons on both the whole image and some local details. The deraining results of two scenes are displayed here, from left to right and from top to down are: Input, JORDER [20], DDN [6], DID-MDN [22], PReNet [15] and our proposed MSAResNet.

Table 1. Quantitative results among different methods on the synthetic dataset.

Methods/Metrics	PSNR	SSIM
Rainy images	22.2957	0.6928
JORDER [20]	25.4771	0.7968
DDN [6]	28.7371	0.8503
DID-MDN [22]	28.3183	0.8726
PReNet [15]	29.1923	0.8508
Ours MSAResNet	**29.6649**	**0.8902**

4.1 Datasets and Training Details

Synthetic Datasets. Synthetic images are used for both training and testing. We collect a dataset containing 19200 synthetic rainy images that are selected from [20, 22] as training dataset, named as Train. The shapes, scales and densities of rain streaks vary from image to image in Train, and there are approximately 3600 to 4000 images per rain streaks density level. Moreover, we perform the evaluation of the existing methods on the synthetic test set from [22], which is composed of 1200 images with different scales and densities of rain streaks and named as Test.

Real-World Rainy Dataset. To evaluate the effectiveness of our proposed method on real-world rainy data, we randomly select 42 real rainy images downloaded from the Internet as the real-world rainy test dataset. The images collected of different scenes also vary in directions, scales and densities of rain streaks.

Training Details. During training, we apply Adam [12] solver with momentum parameters $\beta_1 = 0.5$ and $\beta_2 = 0.999$, weight decay $= 5e-5$ and batch size $= 2$. The learning rate is initialized to $2e-4$, and then divided by 10 after epoch 19. In addition, we set N1 in Eq. 2 to 4, as well as the number of the multi-scale residual block in the multi-scale residual network N2. The parameter θ in Eq. 2 is set to 0.2. And we set the threshold which is used to obtain the binary attention mask M to 30. In Eq. 4, we set m to 5, and $\theta1$, $\theta2$, $\theta3$, $\theta4$, $\theta5$ to 1/32, 1/16, 1/8, 1/4 and 1, respectively. The parameters $\lambda_A, \lambda_{L1}, \lambda_P$ and λ_{SSIM} in Eq. 7 are set to 0.1, 8, 8 and 10, respectively. As for training samples, two image patches with the size of 256×256 are randomly cropped from each image in one batch as input. The training process ends with 24 epochs.

4.2 Baselines

Our proposed MSAResNet is compared with the following recent state-of-the-art image deraining methods: JORDER [20], DDN [6], DID-MDN [22] and PReNet [15]. For fair comparison, we conduct all the comparative experiments in the same testing environment using the released pre-trained models of the existing methods.

4.3 Comparison with Baselines on Test

Qualitative Evaluation. The visual comparison results of different methods on the synthetic dataset Test are presented in Figs. 4 and 5. It can be clearly seen from Fig. 4 that some of the existing methods frequently remain some rain streaks, such as [6, 15, 20], and some tend to over de-rain the image resulting in over-smoothing and loss of details, like [6, 20, 22]. However, our proposed method can commendably remove various rain streaks as well as preserve details well. In addition, the performance on removing rain streaks of different directions, scales and densities between our method and three quite good methods [6, 15, 22] are illustrated in Fig. 5. It demonstrates that our proposed method can not only efficiently deal with various rain streaks but also well retain the image details.

Figure 6 shows an example of attention maps during testing. As can be seen, the attention map focuses on rain streaks more general and accurate as the stage increases. As a consequence, guided by the last attention map that indicates the distribution of rain streaks, our method achieves superior performance.

Quantitative Evaluation. Table 1 shows the average quantitative results of different methods on Test. As can be observed, our approach outperforms the peer methods in terms of both PSNR and SSIM. It is because the proposed method can locate the distribution of different rain streaks, and further remove them accurately with details well preserved.

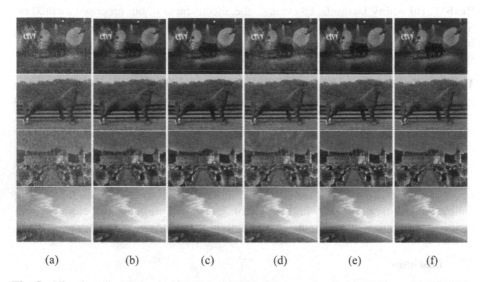

(a)   (b)   (c)   (d)   (e)   (f)

Fig. 5. Visual quality comparisons on images with different rain streaks. (a) Input. (b) DDN [6]. (c) DID-MDN [22]. (d) PReNet [15]. (e) MSAResNet. (f) Ground truth.

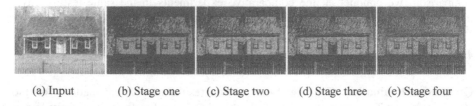

(a) Input   (b) Stage one   (c) Stage two   (d) Stage three   (e) Stage four

Fig. 6. Visualization of attention maps generated by the multi-scale progressive attention subnetwork over multi stages.

4.4   Comparison with Baselines on Real-World Rainy Dataset

We also verify the effectiveness of the proposed method on real-world rainy dataset. Figure 7 shows two deraining results of images with different rain streaks. The state-of-the-art methods either tend to over de-rain causing image blurring or under de-rain resulting in rain streaks being left over. In contrast, our proposed method removes diverse rain streaks better while restoring the image details fairly well.

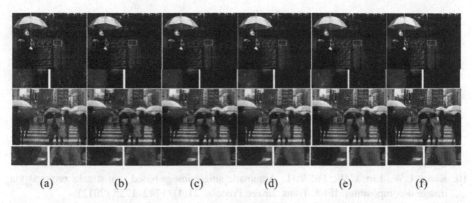

| (a) | (b) | (c) | (d) | (e) | (f) |

Fig. 7. Visual quality comparisons on real-world rainy dataset. (a) Input. (b) JORDER [20] (c) DDN [6]. (d) DID-MDN [22]. (e) PReNet [15]. (f) MSAResNet.

5 Conclusion

In this paper, we propose an end-to-end two-stage method to efficiently remove various rain streaks from a single image. The proposed method is both location-aware and density-aware. Introducing visual attention mechanism in the form of progressively unfolding a multi-scale recurrent module makes it possible to locate the distribution of various rain streaks. Besides we aggregate characteristics of various rain streaks by the use of dilated convolutions with different dilated factors, skip connections and residual learning. Comprehensive experimental evaluations show that our method is superior to several state-of-the-art approaches.

Acknowledgment. The research reported in this paper is supported by the Natural Science Foundation of China under Grant No. 61872047, 61732017, the NSFC-Guangdong Joint Found under No. U1501254, and the National Key R&D Program of China 2017YFB1003000.

References

1. Chen, D.Y., Chen, C.C., Kang, L.W.: Visual depth guided color image rain streaks removal using sparse coding. IEEE Trans. Circ. Syst. Video Technol. **24**(24), 1430–1455 (2014)
2. Chen, Y.L., Hsu, C.T.: A generalized low-rank appearance model for spatiotemporally correlated rain streaks. In: IEEE International Conference on Computer Vision (2013)
3. Cho, K., Merrienboer, B.V., Gulcehre, C., Bougares, F., Bengio, Y.: Learning phrase representations using RNN encoder-decoder for statistical machine translation. Computer Science (2014)
4. Eigen, D., Krishnan, D., Fergus, R.: Restoring an image taken through a window covered with dirt or rain. In: IEEE International Conference on Computer Vision (2014)
5. Fu, X., Huang, J., Ding, X., Liao, Y., Paisley, J.: Clearing the skies: a deep network architecture for single-image rain removal. IEEE Trans. Image Process. **PP**(99), 2944–2956 (2016)

6. Fu, X., Huang, J., Zeng, D., Yue, H., Ding, X., Paisley, J.: Removing rain from single images via a deep detail network. In: IEEE Conference on Computer Vision and Pattern Recognition (2017)
7. Huang, D.A., Kang, L.W., Yang, M.C., Lin, C.W., Wang, Y.C.F.: Context-aware single image rain removal. In: IEEE International Conference on Multimedia and Expo (2012)
8. Huynh-Thu, Q., Ghanbari, M.: Scope of validity of PSNR in image/video quality assessment. Electron. Lett. **44**(13), 800–801 (2008)
9. Johnson, J., Alahi, A., Fei-Fei, L.: Perceptual losses for real-time style transfer and super-resolution. In: Leibe, B., Matas, J., Sebe, N., Welling, M. (eds.) ECCV 2016, Part II. LNCS, vol. 9906, pp. 694–711. Springer, Cham (2016). https://doi.org/10.1007/978-3-319-46475-6_43
10. Kang, L.W., Lin, C.W., Fu, Y.H.: Automatic single-image-based rain streaks removal via image decomposition. IEEE Trans. Image Process. **21**(4), 1742–1755 (2012)
11. Kim, J., Kwon Lee, J., Mu Lee, K.: Deeply-recursive convolutional network for image super-resolution. In: IEEE Conference on Computer Vision and Pattern Recognition (2016)
12. Kingma, D.P., Ba, J.: Adam: a method for stochastic optimization. CoRR abs/1412.6980 (2015)
13. Li, X., Wu, J., Lin, Z., Liu, H., Zha, H.: Recurrent squeeze-and-excitation context aggregation net for single image deraining. In: Ferrari, V., Hebert, M., Sminchisescu, C., Weiss, Y. (eds.) ECCV 2018, Part VII. LNCS, vol. 11211, pp. 262–277. Springer, Cham (2018). https://doi.org/10.1007/978-3-030-01234-2_16
14. Li, Y., Tan, R.T., Guo, X., Lu, J., Brown, M.S.: Rain streak removal using layer priors. In: IEEE Conference on Computer Vision and Pattern Recognition (2016)
15. Ren, D., Zuo, W., Hu, Q., Zhu, P., Meng, D.: Progressive image deraining networks: a better and simpler baseline. In: IEEE Conference on Computer Vision and Pattern Recognition (2019)
16. Tai, Y., Yang, J., Liu, X.: Image super-resolution via deep recursive residual network. In: IEEE Conference on Computer Vision and Pattern Recognition (2017)
17. Wang, F., et al.: Residual attention network for image classification. In: IEEE Conference on Computer Vision and Pattern Recognition, pp. 3156–3164 (2017)
18. Wang, T., Yang, X., Xu, K., Chen, S., Zhang, Q., Lau, R.W.: Spatial attentive single-image deraining with a high quality real rain dataset. In: IEEE Conference on Computer Vision and Pattern Recognition (2019)
19. Wang, Z., Bovik, A.C., Sheikh, H.R., Simoncelli, E.P., et al.: Image quality assessment: from error visibility to structural similarity. IEEE Trans. Image Process. **13**(4), 600–612 (2004)
20. Yang, W., Tan, R.T., Feng, J., Liu, J., Guo, Z., Yan, S.: Deep joint rain detection and removal from a single image. In: IEEE Conference on Computer Vision and Pattern Recognition, pp. 1357–1366 (2017)
21. Yu, L., Yong, X., Hui, J.: Removing rain from a single image via discriminative sparse coding. In: IEEE International Conference on Computer Vision (2015)
22. Zhang, H., Patel, V.M.: Density-aware single image de-raining using a multi-stream dense network. In: IEEE Conference on Computer Vision and Pattern Recognition, pp. 695–704 (2018)
23. Zhang, H., Sindagi, V., Patel, V.M.: Image de-raining using a conditional generative adversarial network. arXiv preprint: arXiv:1701.05957 (2017)
24. Zhao, H., Gallo, O., Frosio, I., Kautz, J.: Loss functions for neural networks for image processing. Computer Science (2015)

Intention Classification Based on Transfer Learning: A Case Study on Insurance Data

Shan Tang[1,2(✉)], Qiang Liu[2], and Wen-an Tan[1]

[1] Shanghai Polytechnic University, Shanghai, China
{tangshan,watan}@sspu.edu.cn
[2] Shanghai Zhipan Intelligent Technology Co., Ltd., Shanghai, China
liuqiang@foxmail.com

Abstract. With the rapid development of Artificial Intelligence and Big Data technology, intelligent chatbot in insurance industry has become the major technical means to reduce labor costs and improve the quality of service. The core technology of this application is to understand and classify the users' intentions accurately. However, insurance as a product with complex knowledge system and long service cycle, users' intentions and the corresponding corpus is rather scattered. The initial corpus is especially scarce at the early stage of new business. So it is very important to classify the customers' intentions accurately based on the rare corpus. This paper offers an empirical case study on intention classification of insurance data by using transfer learning model BERT. The experimental comparative analysis result shows that method based on BERT model can better reduce the error rate than other existing model methods (TextCNN, HAN, ELMo).

Keywords: Intention classification · Insurance data · Chatbot · Transfer learning · BERT model

1 Introduction

Different from the agent marketing mode of traditional insurance, Internet insurance provides service of products and sales through network channels directly. So online customer service becomes the most important communication form instead of insurance agents. In recent years, along with the breakthrough and increasingly mature of AI technology in the field of Natural Language Processing (NLP) and Deep Learning technology, Intelligent chatbot has become the major technical means to reduce repetitive human work and improve the quality of customer service. The core technology of this application is to understand and classify the users' intentions accurately.

Intention classification is to categorize the inquire sentences into the corresponding type of intention. When customers input a sentence or a text, intention recognition system can identify which class it belongs to, and then assign the corresponding domain chatbot to deal with the post-processing. This can improve the accuracy of question matching obviously.

However, insurance as a product with complex knowledge system and long service cycle, there are a large number of queries and communication requirements in the

© Springer Nature Switzerland AG 2019
D. Milošević et al. (Eds.): HCC 2019, LNCS 11956, pp. 363–370, 2019.
https://doi.org/10.1007/978-3-030-37429-7_36

stages of pre-sale, sale, and after-sale. Users' intentions and the corresponding corpus is rather scattered. The initial corpus is especially scarce at the early stage of new business. Therefore, it is a very complicated work to develop intelligent chatbot for insurance industry. Transfer learning [1] is a new technology, which has the ability to recognize and apply knowledge and skills learned in previous tasks to novel tasks, in recent years it has emerged as a new learning framework to address the problem mentioned above.

In this paper, we conduct a case study on insurance data classification by using BERT (the most prominent transfer learning model) and offer an experimental comparative analysis result between BERT and the others typical text classification approaches (TextCNN, HAN, ELMo).

2 Related Works

Our research work is related to many other text classification and language representation approaches developed by other researchers. Here we briefly discuss the most widely-used approaches in this direction.

TextCNN is a classic model published in 2014. Kim and his team apply the convolutional neural network (CNN) to the task of text classification, and use multiple kernels of different sizes to extract the key information in sentences. They show that a simple CNN with little hyper-parameter tuning and static vectors achieves excellent results on multiple benchmarks [2].

In [3] Zhang et al. extend the application of CNN for text classification to the character-level. They treat text as a kind of raw signal at character level and compare with a large number of traditional and deep learning models using several large-scale datasets. The empirical study shows that character-level CNN is an effective method.

Yang et al. of Carnegie Mellon University propose a hierarchical attention network (HAN) for document classification in 2016. This model has two levels of attention mechanisms applied at word-level and sentence-level. They conduct experiments on 6 large-scale text classification tasks and demonstrate that their model performs significantly better than previous methods [4].

Bojanowski et al. of Facebook argue that popular models to learn continuous word representations ignore the morphology of words, by assigning a distinct vector to each word. And they propose an approach based on the skipgram model to solve this problem, which can train models on large corpora quickly and allow to compute word representations for words that did not appear in the training data [5]. And their vectors obtain state-of-the-art results in 2017.

Peters et al. propose ELMo model, which is a new type of deep contextualized word representation, can be easily integrated into existing models, and significantly improves the state of the art in every considered case across a range of challenging language understanding problems [6]. And in [7], Kant et al. apply ELMo model to the emotion classification tasks and achieve high quality results on real-world sentiment classification.

Devlin et al. of Google propose BERT model, which is designed to pre-train deep bidirectional representations from unlabeled text by jointly conditioning on both left and right context in all layers. This language representation model is conceptually simple and empirically powerful. It obtains new state-of-the-art results on eleven natural language processing tasks [8]. Sun et al. of Fudan University apply BERT model to the intention classification and obtain new state-of-the-art results on eight widely-studied text classification datasets [9].

3 Case Study

3.1 Datasets

We evaluate our approach on a real customer question sample dataset in insurance domain. The statistics for insurance data is presented in Table 1. In order to facilitate supervised training and learning, all of the samples were labeled with intention categories by experts of this domain. It is important to note that we conduct the case study based on **Chinese corpus**. So we retain Chinese words in this paper to better illustrate the experiment.

Table 1. Statistics of insurance data classification datasets

Dataset	Classes	Average lengths	Max lengths	Train samples	Test samples
Insurance data	10	32	117	20000	5000

This dataset has 25000 samples covering 10 common intention categories: Claim Settlement (核赔), Check Insurance Liability (查保险责任), Conservation (保全), Check Premiums (查保费), Underwriting (核保), Insurable Diseases (查可投保疾病), Insurable Occupations (查可投保职业), Check Policies (查保单), Check Insurance Amount (查保额), Check insurance clauses (查保险条款). The average length of each question is 32. The longest question contains 117 characters. Table 2 shows five sample sentences and their corresponding intention categories.

Table 2. Sample sentences and the corresponding intention categories

Sample	Intention Category
可以报销门诊费用吗？	Claim Settlement(核赔)
主要赔付什么疾病？	Check Insurance Liability(查保险责任)
保单还能修改吗？	Conservation(保全)
保费有优惠么？	Check Premiums(查保费)
如何获得投保资格？	Underwriting (核保)

3.2   Transfer Learning

In our work, we adopt BERT$_{BASE}$ model to do transfer learning in insurance field. BERT enables the representation to fuse the left and the right context, which allows us to pretrain a deep bidirectional Transformer. As a ready-to-use component, it can help us to save the time, effort, knowledge, and resources required to train the model from scratch.

There are two procedures in BERT: pre-training and fine-tuning. Apart from output layers, these two procedures can share the same network architecture illustrated in Fig. 1. The official BERT$_{BASE}$ model consists of an encoder with 12 Transformer blocks, 12 self-attention heads, and a hidden size of 768. And it takes an input of a sequence of no more than 512 tokens and outputs the representation of the sequence. The sequence has one or two segments that the first token of every sequence is always a special classification token ([CLS]), and another special token [SEP] is a separator (e.g. used for separating questions/answers). That is to say, the sequence may be a single sentence or two sentences packed together. As shown in Fig. 1, we use a single question sentence as a running example and denote input embedding as E, the final hidden state corresponding to the first token [CLS] token as $C \in R^H$ (which is used as the aggregate sequence representation for classification tasks), and the final hidden state for the ith input token as $T_i \in R^H$.

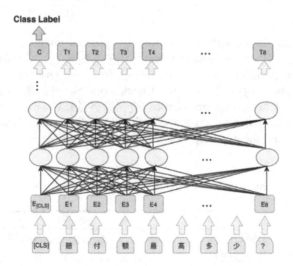

Fig. 1. Network architecture based on BERT model

We first initialize the BERT model by using the pre-trained parameters, and then fine tuning all of the parameters by using the labeled data from the downstream tasks. Algorithm 1 describe the fine-tuning process. In this algorithm, the self-attention function can be defined as mapping a query and a set of key-value pairs to an output, where the query, keys, values, and output are all vectors. The output is computed as a weighted sum of the values, where the weight assigned to each value is computed by a compatibility function of the query with the corresponding key [10].

Algorithm 1. Fine-tuning based on BERT
Input: sentences X, labels Y
1. **for** words in X and y in Y **do**
2.   tokens ← concat([CLS],words,[SEP]);
3.   embedding ← embed(tokens)
4: **for** layer i in Transformer layers **do**
5:     hidden_state ← self-attention(query(hidden_state), key(hidden_state), value(hidden_state))
6: **end for**
7: predicted label y' ← softmax(final hidden state);
8: update the parameters by minimizing the cross entropy loss between y' and y;
9: **end for**

After we fine-tune the BERT model, we can do the intention classification task by using this model as the Algorithm 2 shows.

Algorithm 2. Intention Classification
Input: Contain Sentences = $\{s_1, s_2, s_3, ..., s_n\}$
Output: ClassLabels
1. **initialize:** Contain ClassLabels = {}
2. **for** t = 1,2,3...n
3.   Embedding = Concat [CLS] and s_t
4.   Feed Emdbedding to BERT
5.   Insert to ClassLabels
6. **End for**

3.3 Experiment Results and Evaluation

We perform the evaluation on four typical text classification approaches based on the insurance data. Figure 2 presents the test error rates comparison result on a variety of training samples of different size. And Table 3 shows the classification error rates comparison result among TextCNN, HAN, ELMo and BERT on **20000** training samples.

Obviously, method based on BERT model can better reduce the error rate than other existing model methods (TextCNN, HAN, ELMo). When the experiment performed on all the 20000 samples, the error rate of classification based on BERT is only 5.29%. Such score has enough qualification to be applied in reality. In comparison, the error rates of TextCNN and HAN are greater than 10%. The high error rate would seriously reduce the customers' satisfaction. Especially, if the training sample is insufficient (e.g., 5000), the error rates of TextCNN and HAN reaches 16.24% and

Table 3. The classification error rates comparison result on 20000 training samples

Model	Error rate (%)
TextCNN	13.25
HAN	10.32
ELMo	6.27
BERT	5.29

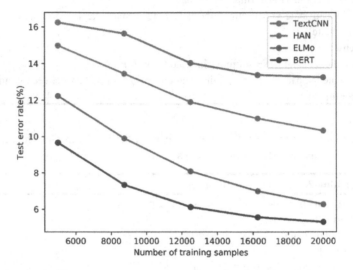

Fig. 2. The test error rates comparison result among TextCNN, HAN, ELMo and BERT

14.98% respectively. As a result, a large number of artificial customer service will need to support the customer service business of the new product at the early stage of new business, which seriously hamper the application of intelligent chatbot in insurance industry. When comparing the error rates based on BERT model and ELMo model, under different quantity of training samples, BERT achieves a reduction of around 2% in error rate.

In order to further compare the characteristics and operation mechanism of these four models, we use t-SNE to visualize the transformed features generated by them respectively. The results as Fig. 3 show that the method based on BERT maps the samples of the same classification to a very close space, and the distance between different classifications is extended sufficiently. Whereas the other models (TextCNN, HAN, ELMo) don't work as well as BERT model. The visualization of transformed features based on them show that the intra-class distances of the same category samples are loose, and the inter-class distances of the different category samples are not pulled apart enough. There are even mixture areas of samples from different categories. Thus it can be seen that BERT based model has higher precision.

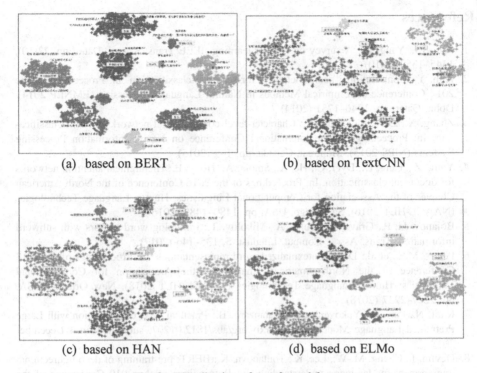

(a) based on BERT

(b) based on TextCNN

(c) based on HAN

(d) based on ELMo

Fig. 3. Visualization of transformed features

4   Conclusions and Future Works

Although BERT has achieved amazing results in many natural language understanding tasks, its potential has yet to be fully explored. In this paper, we perform an empirical case study on intention classification of insurance industry based on BERT model and offer the experimental comparison result between BERT and the others typical models (TextCNN, HAN, ELMo). The experimental results show that the method based on BERT can obtain the best results in reducing the classification error rate. In the future, we will establish a multi-modal emotional classification model by combining customer voice information to implement a more comprehensive intelligent chatbot.

Acknowledgment. This work is sponsored by Shanghai Pujiang Program under Grant No. 18PJ1433400, Key Disciplines of Computer Science and Technology of Shanghai Polytechnic University under Grant No. XXKZD1604, and Leap Funding of SSPU Scientific Research under Grant No. EGD19XQD09.

References

1. Pan, S.J., Yang, Q.: A survey on transfer learning. IEEE Trans. Knowl. Data Eng. **22**(10), 1345–1359 (2010)
2. Kim, Y.: Convolutional neural networks for sentence classification. In: Proceedings of the 2014 Conference on Empirical Methods in Natural Language Processing (EMNLP 2014), Doha, Qatar, pp. 1746–1751 (2014)
3. Zhang, X., Zhao, J., Lecun, Y.: Character-level convolutional networks for text classification. In: Proceedings of the International Conference on Neural Information Processing Systems (NIPS 2015), Montreal, Canada, pp. 1–9 (2015)
4. Yang, Z., Yang, D., Dyer, C., He, X., Smola, A., Hovy, E.: Hierarchical attention networks for document classification. In: Proceedings of the 2016 Conference of the North American Chapter of the Association for Computational Linguistics: Human Language Technologies (NAACL-HLT 2016), San Diego, USA, pp. 1480–1489 (2016)
5. Bojanowski, P., Grave, E., Joulin, A., Mikolov, T.: Enriching word vectors with subword information. Trans. Assoc. Comput. Linguist. **5**, 135–146 (2017)
6. Peters, M.E., et al.: Deep contextualized word representations. In: Proceedings of the 2018 Conference of the North American Chapter of the Association for Computational Linguistics: Human Language Technologies (NAACL-HLT 2018), New Orleans, USA, pp. 2227–2237 (2018)
7. Kant, N., Puri, R., Yakovenko, N., Catanzaro, B.: Practical Text Classification with Large Pretrained Language Models. https://arxiv.org/abs/1812.01207, submitted on 4 December 2018
8. Devlin, J., Chang, M.-W., Lee, K., Toutanova, K.: BERT: pre-training of deep bidirectional transformers for language understanding. In: Proceedings of the 2019 Conference of the North American Chapter of the Association for Computational Linguistics: Human Language Technologies (NAACL-HLT 2019), Minneapolis, USA, pp. 4171–4186 (2019)
9. Sun, C., Qiu, X., Xu, Y., Huang, X.: How to Fine-Tune BERT for Text Classification? https://arxiv.org/abs/1905.05583, submitted on 14 May 2019
10. Vaswani, A., Shazeer, N., Parmar, N., Uszkoreit, J., et al.: Attention Is All You Need. https://arxiv.org/abs/1706.03762, submitted on 12 June 2017

Network Optimization Under Traffic Uncertainties Based on SDN

Junjie Teng[1(✉)], Yuxue Hu[2], Yong Zhang[3], and Mo Chen[3]

[1] China Financial Certification Authority, Beijing, China
junjieteng@foxmail.com
[2] Bank of China Software Center, Beijing, China
huyuxue@bupt.edu.cn
[3] Beijing University of Posts and Telecommunications, Beijing, China
yongzhang@bupt.edu.cn

Abstract. Software Defined Networking (SDN) is an emerging network architecture that separates the control plane from the data plane to simplify and improve network management with a high degree of flexibility. Network optimization under traffic uncertainties is one of the most challenging topics in communication networks to optimize network performance and traffic delivery. Although the traffic optimization technology has been extensively studied in the industry, a traffic optimization solution different from the traditional network is needed in the SDN network, which can utilize global network information and traffic characteristics to control and manage traffic in a better way. In this paper, a Mixed Linear Geometric Programming Traffic Optimization Algorithm (MLGP-TOA) is proposed for the problem of traffic uncertainties in SDN. Aiming at minimizing the maximum link utilization (MLU), the initial problem is transformed into a convex optimization problem by monomial approximation and variable substitution. Then, the inner point method is used to find the global optimal solution, and the optimal split ratio at each node is obtained. Finally, the configuration information is sent to the data plane. The simulation results show that the algorithm can reduce MLU, so that the traffic can fully utilize network resources and avoid congestion.

Keywords: Software Defined Networking · Traffic optimization · Mixed Linear Geometric Programming

1 Introduction

Traffic engineering (TE) has always been one of the most challenging topics in communication networks. An important issue in TE is how to handle the uncertainty of traffic information. Under the traffic uncertainties and the rapid growth of network size and traffic requests, the uneven distribution of network traffic is becoming more and more obvious. When some links in the network are congested, other links are often in a light load state, and link utilization is not balanced, affecting end-to-end network quality [1]. A straightforward way to find an optimal scheduling scheme within the network is to measure the utilization of all links in the network and to move traffic from

© Springer Nature Switzerland AG 2019
D. Milošević et al. (Eds.): HCC 2019, LNCS 11956, pp. 371–382, 2019.
https://doi.org/10.1007/978-3-030-37429-7_37

high-utility links to lower-utilization links. The key challenge is how to handle the uncertainties to reduce MLU.

As a new paradigm for future networks, SDN has attracted the attention of academia and industry with its flexibility, programmability, simplified network management and other important features [2]. SDN is divided into application layer, control layer and data layer. The interaction between the control layer and the data layer is through the south-bound interface, which is well defined in protocols such as Open-Flow [3]. Since the northbound interface is open, SDN can implement custom development quickly and efficiently.

At present, researchers have made some research on the issue of link load balancing under the uncertainty of traffic. The QoS routing algorithm is a mechanism to determine the best path based on network performance such as latency and bandwidth. OpenQoS [4] classifies traffic into data streams and multimedia streams, and the routing of multimedia streams is optimized using QoS routing algorithms. HiQoS [5] utilizes multiple paths between source and destination and queuing mechanisms to optimize QoS. Equal-cost multi-path routing (ECMP) [6] is a solution to multiple optimal routes and is considered to be more efficient than single-path QoS routing. The main feature of ECMP is to separate and forward streams to multiple paths for load balancing. However, ECMP has obvious shortcomings in both routing calculation and offloading: in the case of a given link weight configuration, traffic from the source node to the destination node in the network can only be forwarded along the shortest path. Only equal divisions can be made between paths. Weighted Cost Multipath (WCMP) [7] distributes traffic by path weight, which is a special case of ECMP. However, this method is more complicated and costly in the traditional switch. In a software-driven wide area network (SWAN) architecture, SWAN enables software-driven data center-to-area WANs to carry more traffic. The SWAN is updated with a small amount of temporary capacity on the link to ensure that transient congestion does not occur. SWAN improves bandwidth utilization while providing a preference for higher priority services and fairness between similar services. In the Cisco CONGA [8] architecture, TCP flows are divided into multiple streams and distributed to multiple paths, enabling CONGA to effectively balance the load of the data center, but with poor scalability in current network architectures. In Hedera [9], traffic exceeding 10% of the link capacity is considered to be a large stream. The SDN controller sends a traffic statistics request to the edge switch to detect the large stream within a certain period of time. Whenever large traffic is detected, it is rescheduled to light-load links that meet their bandwidth needs. However, Internet video streams typically account for most network traffic, and each traffic is well below 10% of the link capacity. Aggregated streams (such as video streams) from the same video sharing site may exceed 10% of the link capacity and introduce congestion. In a sense, they may be considered an elephant stream for traffic engineering to improve QoS. In [10], the authors propose a distributed load balancer that uses Open Flow to measure the state of the network and achieve dynamic, flexible and scalable load balancing by directly controlling the forwarding path. Reference [11] uses network information (collected by centralized SDN controllers) to improve network utilization and reduce packet loss and latency. In addition, they have developed optimization issues for SDN controllers for partially deployed traffic engineering. In [12], a generalized framework for constructing multi-stream update sequence scheduling and a general framework for optimal traffic migration sequencing solution is proposed, which minimizes the MLU and significantly reduces the possibility of congestion.

Based on the above analysis, this paper focuses on how to perform network optimization under traffic uncertainties to achieve load balancing of links. With the goal of minimizing MLU, based on the determined traffic path, the split ratio at each node is optimized to achieve link load balancing. How to efficiently and optimally solve the problem of calculating the traffic split ratio in a given set of traffic paths is still a more complicated problem. There is no direct solution to the common set of TE math tools (e.g., integer programming and linear programming). In this paper, the initial problem is transformed into Mixed Linear Geometric Programming (MLGP) by means of monomial approximation and variable substitution. Then the optimal solution is obtained by the convex optimization method, and the final configuration information is formed and sent to the data plane. The simulation results show that the traffic optimization algorithm proposed in this paper can achieve smaller MLU and load balancing.

The remainder of this paper is organized as follows: Sect. 2 presents the system model and Sect. 3 gives detail algorithms. The experiment results are presented in Sect. 4 and Sect. 5 concludes the paper.

2 System Model

The system framework is as shown in Fig. 1. The task of the control plane is to optimize the traffic distribution based on the collected traffic information to form configuration information, which is sent to the forwarding device. The task of the data plane is to record the traffic information, and then perform the traffic distribution configuration to implement the load balanced.

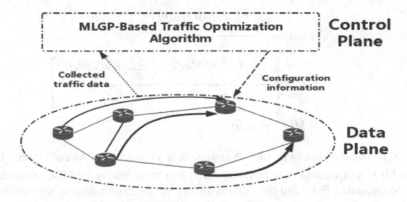

Fig. 1. System model

Consider the SDN network $G = (V, E)$, $V = \{v_i\}_{i=1,2,...,N}$ represents the set of all SDN nodes in the network, and $E = \{e_i\}_{i=1,2,...,L}$ represents the set of directed links. N represents the number of nodes, L represents the number of links and c_e represents the capacity of link e. Each node in the network can be either a source node or a destination node, so the number of possible OD pairs in the network is $N(N-1)$. For any two nodes s and d in the network, the value of t_{sd} represents the traffic size of OD pair (s, d). Due to the uncertainty of network traffic and the limitations of traffic measurement and estimation techniques, only actual estimates of traffic size $\hat{t}_{sd}$ can be obtained. We assume that ε is the estimation error. That is, $\varepsilon = \hat{t} - t$. We also assume that ε is subject to a certain distribution, and $E(\varepsilon) = \mu$.

We assume that the path of each traffic from the source node to the destination node has been determined. For any node $d \in V$, E_d all directed links representing the flow to the node d. For link $e = (i,j) \in E$, the value of $\phi_d(e)$ represents. The proportion of traffic destined for node d to node j at node i. The set of $\phi_d(e)$ decides how to split the traffic between two nodes. For OD pair (s, d) and node $i \in V$, the value of $f_{sd}(i)$ represents the proportion of the flow into the node i in the flow of s to d. For example, when the traffic between (s, d) is t_{sd}, the load of link (i, j) is $t_{sd} f_{sd}(i)\phi_d(i,j)$. Consider all OD pairs in the network, traffic matrix $T = \{t_{s_1 d_1}, \ldots, t_{s_k d_k}\}$ represents the set of traffic sizes between each pair of nodes, so the total load of link (i, j) is $\sum_{(s,d)\in T} t_{sd} f_{sd}(i)\phi_d(i,j)$. The definition of MLU is below:

$$MLU = \max_{e \in E} \frac{\sum_{(s,d)\in T} t_{sd} f_{sd}(i)\phi_d(i,j)}{c_e} \tag{1}$$

The expectation of MLU is:

$$E(MLU) = E\left[\max_{e \in E} \frac{\sum_{(s,d)\in T} t_{sd} f_{sd}(i)\phi_d(e)}{c_e}\right]$$

$$= \max_{e \in E} \frac{\sum_{(s,d)\in T} E(t_{sd}) f_{sd}(i)\phi_d(e)}{c_e} \tag{2}$$

$$= \max_{e \in E}\left[\frac{\sum_{(s,d)\in T} \hat{t}_{sd} f_{sd}(i)\phi_d(e)}{c_e} - \frac{\sum_{(s,d)\in T} \mu f_{sd}(i)\phi_d(e)}{c_e}\right]$$

$$= \hat{MLU} + X(\mu)$$

In (2), $X(\mu)$ is a function of μ, and we assume it is a fixed value. As can be seen from (2), the MLU's expectation can be divided into two parts, one is the $\hat{MLU}$ obtained by using the estimated flow, and the other is related to the expectation of the estimation error. Therefore, the expectation of the MLU depends on $\hat{MLU}$. The goal of traffic optimization is to minimize $\hat{MLU}$:

$$\min \hat{MLU} \tag{3}$$

s.t.

$$\sum_{(i,j)\in E_d} f_{sd}(i)\phi_d(i,j) \leq f_{sd}(j), \ \forall d \tag{4}$$

$$\sum_{(i,j)\in E_d} \phi_d(i,j) = 1, \quad \forall d \tag{5}$$

$$f_{sd}(i)\phi_d(i,j) \leq c_e p_e(s,d), \qquad \forall (s,d), \forall i,j \tag{6}$$

The traffic in the network needs to meet the following constraints: (4) indicates that the traffic flowing into node j through link (i, j) does not exceed the total traffic flowing into node j in the network; (5) indicates that the sum of the split rates is 1; (6) represents the split ratio constraint according to the shortest path, where $p_e(s,d)$ denotes the shortest path length from s to d.

3 Solution

In Sect. 2, the optimization problem represented by (3)–(7) is a non-linear (NLP) problem, and it is difficult to find the global optimal solution. Therefore, this section uses the monomial approximation method to solve this NLP. The NLP is transformed into the MLGP problem by the monomial approximation, and then the variable substitution is used to transform into the convex optimization problem. Finally, the inner point method is used to obtain the global optimal solution. The process of the traffic optimization algorithm is shown in Fig. 2.

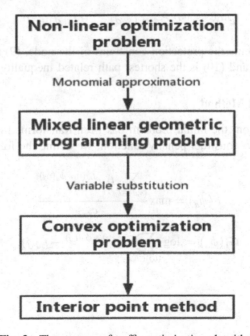

Fig. 2. The process of traffic optimization algorithm

3.1 Monomial Approximation and Variable Substitution

In the above NLP problem, since the split flow rate constraint does not satisfy the form of geometric programming, we adopt the method of single item approximation to solve the problem. We define $S_{id}(\phi) = \sum\limits_{(i,j)\in E_d} \phi_d(i,j)$, where ϕ is an array of all ϕ_d, $\phi(k)$ represents the k-th variable of ϕ. From literature [13], we know that given a point ϕ_0, we can approximate $S_{id}(\phi)$ to $m_d \prod\limits_{k=1}^{n} (\phi(k))^{a_d(k)}$, where $a_d(k) = \dfrac{\phi_0(k)}{\sum\limits_{k}\phi_0(k)}$, $m_d = \dfrac{\sum\limits_{k}\phi_0(k)}{\prod\limits_{k=1}^{n}(\phi(k))^{a_d(k)}}$.

Therefore, the split ratio constraint can be written as:

$$m_d \prod_{k=1}^{n} (\phi(k))^{a_d(k)} = 1 \tag{7}$$

We define $\tilde{f}_{sd} = \log f_{sd}$, $\tilde{\phi}_d = \log \phi_d$, the whole optimization problem can be written as:

$$\min \hat{MLU}$$

s.t.

$$\tilde{f}_{sd}(j) \geq \log \sum_{(i,j)\in E_d} e^{\tilde{f}_{sd}(i) + \tilde{\phi}_d(i,j)}, \quad \forall (s,d) \in T,\ j \in V \tag{8}$$

$$\log m_d + \sum_{e=(i,j)\in E_d} a_t(e)\tilde{\phi}_d(e) = 0, \quad \forall d \tag{9}$$

$$e^{\tilde{f}_{sd}(i) + \tilde{\phi}_d(i,j)} \leq c_e p_e(s,d), \quad \forall (s,d) \in T \tag{10}$$

Inequality (8) represents the protection constraint of the node, (9) represents the split flow rate constraint, and (10) is the shortest path related inequality constraint.

3.2 Interior Point Method

In the previous section, the initial problem has been transformed into a convex optimization problem. To construct penalty function, we define the following functions:

$$F(\tilde{\phi}_d) = \max \frac{\sum\limits_{(s,d)\in T} t_{sd} e^{\tilde{f}_{sd}(i) + \tilde{\phi}_d(i,j)}}{c_e} \tag{11}$$

$$G_1(\tilde{\phi}_d) = \log \sum_{(i,j)\in E_d} e^{\tilde{f}_{sd}(i) + \tilde{\phi}_d(i,j)} - \tilde{f}_{sd}(j) \tag{12}$$

$$G_2(\tilde{\phi}_d) = e^{\tilde{f}_{sd}(i) + \tilde{\phi}_d(i,j)} - c_e p_e(s,d) \tag{13}$$

Then the penalty function is:

$$\varphi(\tilde{\phi}_d, r^{(k)}) = F(\tilde{\phi}_d) + r^{(k)} \left| \sum_{u=1}^{2} \ln |G_u(\tilde{\phi}_d)| \right| = F(\tilde{\phi}_d) - r^{(k)} \sum_{u=1}^{2} \ln \left[-G_u(\tilde{\phi}_d) \right] \tag{14}$$

In the above equation, $r^{(k)}$ is penalty factor, and is a decreasing positive sequence:

$$r^{(0)} > r^{(1)} > r^{(2)} > \cdots > r^{(k)} > r^{(k+1)} > \cdots > 0 \tag{15}$$

$$\lim_{k \to \infty} r^{(k)} = 0 \tag{16}$$

Usually, $r^{(k)} = 1.0, 0.1, 0.01, 0.001, \ldots$. The second term on the right side of the above penalty function expression is the penalty term. Through iteration, the optimal solution in the feasible domain can be obtained. The iteration steps are as follows:

(1) Define the penalty factor $r^{(0)} > 0$, we allow error $\tilde{\varepsilon} > 0$ and let $\tilde{\varepsilon} = 0.001$;
(2) Adopt the initial point $\tilde{\phi}_d^{(0)}$ in the feasible domain D, let $k = 1$;
(3) Construct penalty function $\varphi(\tilde{\phi}_d, r^{(k)})$, at the point $\tilde{\phi}_d^{(k-1)}$, use unconstrained optimization method to solve the extreme point $\tilde{\phi}_d^*(r^{(k)})$ of the penalty function;
(4) Judge the iteration termination condition, if the following condition is satisfied, then stop the iteration computation, and $\tilde{\phi}_d^*(r^{(k)})$ is the optimal solution. Otherwise, step into (5);
(5) Let $r^{(k+1)} = Cr^{(k)}, \tilde{\phi}_d^{(0)} = \tilde{\phi}_d^*(r^{(k)}), k = k+1$, go back to step (3). C is decreasing coefficient, usually is 0.1.

4 Simulation Results

As shown in Fig. 3, the experimental simulation topology consists of 20 nodes and 60 links, and the link capacity is 100 Mbps. The traffic matrix is randomly generated, and the number of OD pairs generated at each moment and the amount of traffic between each OD pair are randomly distributed. The computer environment used in this experiment was Intel(R) Core(TM) i5-4210U CPU @1.70 GHz 2.39 GHz, 4G RAM.

The directed graph formed by the traffic demand forwarding path for each destination is called a Directed Acyclic Graph (DAG). Studies have shown that generating DAGs using the shortest path can produce good routing results [14]. The path generation strategy assigns weights to each link and then generates a shortest path DAG based on a single destination based on ECMP or other traditional routing protocols.

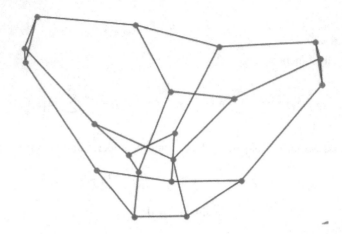

Fig. 3. Simulation topology

Two heuristics are evaluated in [14] for setting link weights: reverse link capacity and local search algorithms. The reverse link capacity approach is to set the link weight to the reciprocal of the link capacity, which is compatible with Cisco's recommendations for ECMP link weights. The heuristic method based on local search uses the technique in [15] to optimize the load-independent ECMP routing configuration. The research results show that the heuristic algorithm based on local search can achieve better performance. Therefore, we use the local search-based heuristic method to set the link weight, and then generates a DAG based on a single destination according to the shortest path principle.

More specifically, the link weight is initially set to the reciprocal of the link capacity, and then the worst case traffic matrix relative to the current link weight is calculated, adding this matrix to a set of traffic matrices T (initially set to null). Then, the weight of the individual link is changed to improve the worst case link utilization of the matrix in the set {TM}, and the above process is iterated until the maximum link utilization is less than the threshold. The search technique in [16] is used to iteratively try to reduce the utilization by locally increasing the path diversity at the most congested nodes. In the simulation environment, MLGP-TOA is compared with the most traditional algorithm (ECMP).

Figure 4 shows the variation of the MLU corresponding to ECMP and MLGP-TOA as the number of OD pairs increases. It can be seen that as the number of OD pairs increases, the total traffic in the network gradually increases, and the MLUs corresponding to the two algorithms increase. When the traffic load in the network is light, the MLU of the two algorithms is not much different, but when the number of OD pairs exceeds 5, the MLU of MLGP-TOA is significantly lower than that of ECMP. When the number of OD pairs is 10, ECMP's MLU is over 80%, while MLGP-TOA's MLU is only 70%. In an actual network, an increase in the number of OD pairs means an increase in the number of users. Traffic based on different sources and destinations needs to share link resources. Since MLGP-TOA optimizes the split rate at each node according to the available resources of the entire network. It can coordinate the split ratio of different flows at the node, and ECMP simply divides all traffic flowing into the node in a proportional manner. Therefore, MLGP-TOA can better plan the traffic in the network, make the traffic distribution more balanced, and reduce congestion. Improve the overall utilization of network resources.

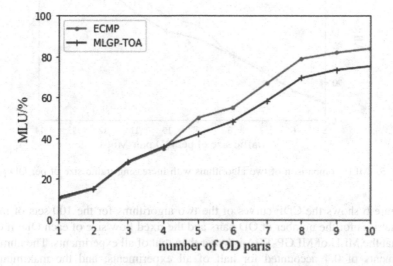

Fig. 4. MLU comparison of two algorithms with the change of OD pairs

Figure 5 shows the change of the MLU corresponding to the two algorithms as the number of OD pairs increases. It can be seen that as the OD increases the traffic volume, the total traffic in the network gradually increases, and the MLUs corresponding to the two algorithms increase. When the traffic load in the network is light, the MLU of the two algorithms is not much different. However, when the OD traffic exceeds 9 Mbps, the MLU of the MLGP-TOA is significantly lower than that of the ECMP. The traffic volume of the OD is 14 Mbps. At the time, ECMP's MLU exceeded 80%, while the MLGP-TOA's MLU was only 70%. In the actual network, the number of users is fixed. As each user's traffic request increases, the network load also increases. MLGP-TOA can take advantage of the SDN network and dynamically adjust the nodes according to the resource utilization of the entire network. The split ratio allocates traffic on the heavily loaded link to the lightly loaded link, while ECMP only randomly divides all traffic flowing into the node, so MLGP-TOA can better schedule the network. Traffic makes the traffic distribution more balanced and improves the overall utilization of network resources.

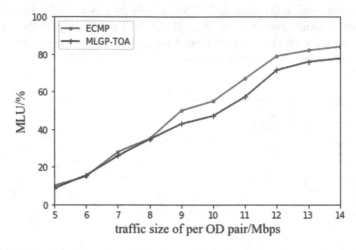

Fig. 5. MLU comparison of two algorithms with increasing traffic size of per OD pair

Figure 6 shows the CDF curves of the two algorithms for the 100 sets of random flow matrices for the number of OD pairs and the fixed flow size of each OD. It can be seen that the MLU of MLGP-TOA is lower than that of all experiments. The number of experiments of 0.4 accounted for half of all experiments, and the maximum link utilization rate of all experiments was less than 0.8, while the number of experiments with ECMP of MLU below 0.4 was only 40% of the total number of experiments, and the largest of all experiments. Link utilization is close to 0.9. In the actual network, the number of users and user traffic requests are fixed, but the location of the users in the network is different, and the distribution of traffic in the network is also different. MLGP-TOA can adjust the split ratio at the node according to the resource utilization of the entire network under different distribution conditions, and distribute the traffic on the heavily loaded link to the lightly loaded link, while ECMP only fixedly flows to the ingress node. All traffic is equally distributed, so MLGP-TOA can better plan traffic in the network, make traffic distribution more balanced, and improve the overall utilization of network resources.

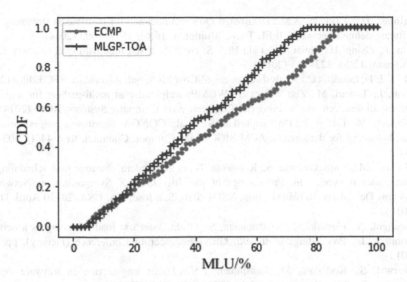

Fig. 6. Cumulative distribution function curve

5 Conclusion

Traffic optimization is one of the most challenging topics in communication networks to optimize network performance and traffic delivery. In this paper, MLGP-TOA is proposed for the problem of traffic uncertainties in SDN. Aiming at minimizing MLU, the initial problem is transformed into a convex optimization problem by monomial approximation and variable substitution. Then, the inner point method is used to find the global optimal solution, and the optimal split ratio at each node is obtained. Finally, the configuration information is sent to the data plane. The simulation results show that, when the traffic load is heavy, compared with the traditional method ECMP, MLGP-TOA can reduce MLU obviously, so that the traffic can fully utilize network resources and avoid congestion.

Acknowledgement. This work is supported by National Key R&D Program of China No. 2018YFB1201500, National Natural Science Foundation of China under Grant No. 61871046, and Beijing Natural Science Foundation under granted No. L171011.

References

1. Netze, H.: Overview and principles of internet traffic engineering. RFC **121**(6), 239–242 (2002)
2. Nunes, B.A.A., Mendonca, M., Nguyen, X.N., et al.: A survey of software-defined networking: past, present, and future of programmable networks. IEEE Commun. Surv. Tutor. **16**(3), 1617–1634 (2014)
3. Mckeown, N., Anderson, T., Balakrishnan, H., et al.: OpenFlow: enabling innovation in campus networks. ACM SIGCOMM Comput. Commun. Rev. **38**(2), 69–74 (2008)

4. Egilmez, H.E., Tekalp, A.M.: Distributed QoS architectures for multimedia streaming over software defined networks. IEEE Trans. Multimed. **16**(6), 1597–1609 (2014)
5. Yan, J., Zhang, H., Shuai, Q., et al.: HiQoS: an SDN-based multipath QoS solution. China Commun. **12**(5), 123–133 (2015)
6. &Lt C E H, Edu&Gt C.: Analysis of an Equal-Cost Multipath Algorithm. RFC Editor (2000)
7. Zhou, J., Tewari, M., Zhu, M., et al.: WCMP: weighted cost multi-pathing for improved fairness in data centers. In: European Conference on Computer Systems. ACM (2014)
8. Alizadeh, M., Edsall, T., Dharmapurikar, S., et al.: CONGA: distributed congestion-aware load balancing for datacenters. ACM SIGCOMM Comput. Commun. Rev. **44**(4), 503–514 (2015)
9. Al-Fares, M., Radhakrishnan, S., Raghavan, B., et al.: Hedera: dynamic flow scheduling for data center networks. In: Proceedings of the 7th USENIX Symposium on Networked Systems Design and Implementation, NSDI 2010, San Jose, CA, USA, 28–30 April. DBLP (2010)
10. Handigol, N., Flajslik, M., Seetharaman, S., et al.: Aster* x: load-balancing as a network primitive. In: Proceedings of the 9th GENI Engineering Conference (Plenary), pp. 1–2 (2013)
11. Agarwal, S., Kodialam, M., Lakshman, T.V.: Traffic engineering in software defined networks. In: 2013 Proceedings of the IEEE INFOCOM (2013)
12. Bai, W., Chen, L., Chen, K., et al.: PIAS: practical information-agnostic flow scheduling for commodity data centers. IEEE/ACM Trans. Netw. **25**(4), 1954–1967 (2017)
13. Awduche, D., Berger, L., Gan, D., et al.: RSVP-TE: Extensions to RSVP for LSP Tunnels. Ietf RFC (2001)
14. Chiesa, M., Rétvári, G., Schapira, M.: Oblivious routing in IP networks. IEEE/ACM Trans. Netw. **26**(3), 1292–1305 (2018)
15. Altin, A., Fortz, B., Ümit, H.: Oblivious OSPF routing with weight optimization under polyhedral demand uncertainty. Networks **60**(2), 132–139 (2012)
16. Vissicchio, S., Vanbever, L., Rexford, J.: Sweet little lies: fake topologies for flexible routing. In: ACM Workshop on Hot Topics in Networks (2014)

Design of Crane Virtual Simulation Teaching Assessment System

Aoxiang Tian[1], Taotao Li[2(✉)], Jiquan Hu[1,2], and Bin Zhao[1,2]

[1] Yujiatou Campus, Wuhan University of Technology, Wuhan, Hubei, China
Taxll05@163.com
[2] China Ship Development and Design Center, Wuhan, Hubei, China
taotaoliwhut@163.com

Abstract. With the rapid development of cranes, crane teaching aids cannot follow up with new products in real time, and the teaching cost is high and the space utilization rate is low. In order to solve these problems, we used Solid-Works and 3ds Max to carry out mechanical modeling in the virtual environment. Based on the Unity3D development engine, we designed a virtual simulation teaching evaluation system for cranes, which completed the design of modules for cognitive teaching, maintenance, fault repairing and teaching assessment. The system takes advantage of virtual technology's features of immersion, interaction and imagination. Therefore, it solves the problem of the slow update of teaching aids in the teaching process of cranes effectively, saves teaching costs greatly, overcomes the traditional teaching boring shortcomings, and improves the efficiency of port machinery education. And It plays an important role in promoting the application of virtual reality technology in education and training.

Keywords: Unity3D · Virtual simulation · Interaction

1 Introduction

With the development of cranes [1, 2], the corresponding crane teaching aids lag behind emerging products. This article relies on the Unity3D development engine to virtualize the crane [3], and finally import it into the Unity3D development engine, moving the traditional teaching to the computer screen and mobile terminal. Virtual reality technology's features of immersion, interaction and imagination can be used to realize teachers' teaching assisting and it enables students to obtain good practical operation learning effects [4].

2 System and Function Design

In order to ensure the systemic and integrity of system development, the system architecture design is completed first. According to the development requirements, the system architecture is divided into three levels: interaction layer, logical layer and data layer. Each layer is divided into multiple sub-modules, and each sub-module is independent and related to each other. The implementation of the function of interaction layer interface depends on the design of the operation logic of logical layer, and the call of logical layer to data of data layer. The system architecture diagram is shown in Fig. 1.

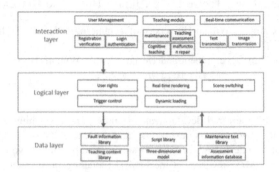

Fig. 1. Diagram of system architecture

Secondly, the interaction layer design is completed [5]. The interaction layer operation process flow is designed as the overall logic idea and layout of the project, and all functions are connected in series according to the operation process to form an organic whole. The main interface consists of four branches, including cognitive teaching, maintenance, fault repairing, and teaching assessment.

2.1   Cognitive Teaching

Cognitive teaching has a total of five sub-menus, which are Main components, Hoisting mechanism, Running mechanism, Slewing mechanism, and Luffing mechanism [6]. The interface of main component lists seven main component selection buttons. After selection, the icon and text will pop up in the middle and right (see Fig. 2).

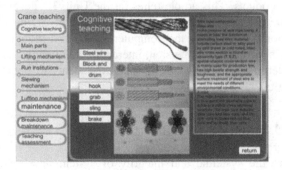

Fig. 2. Interface of wire rope teaching

The illustration shows the cognitive teaching module of the lifting mechanism. You can select the left view, top view, front view, and axial view of the model for demonstration teaching. There are two slider in the lower left corner, and drag the slider to achieve left and right rotation and up and down rotation of the lifting mechanism model, to view the details of the model at any angle. The text on the right side is the introduction for the lifting mechanism (see Fig. 3).

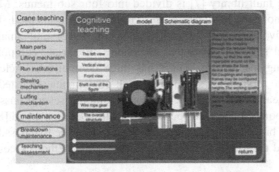

Fig. 3. Schematic diagram of the lifting mechanism

2.2 Maintenance

The maintenance interface is mainly divided into three modules, Inspection and maintenance system, Inspection and maintenance items, and Precautions. The Inspection and maintenance system explains the cycle of maintenance from a time perspective, including Daily inspection and maintenance, Weekly inspection and maintenance, Monthly inspection and maintenance, Half-year inspection and maintenance, and Annual inspection and maintenance. The Inspection and maintenance project explains the maintenance items of the main modules, including the main beam

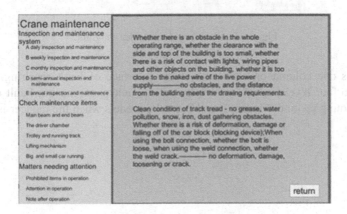

Fig. 4. Interface of crane maintenance

and end beam, the driver's cab, the trolley and the running track, the lifting mechanism, the large and small car operating mechanism, and other projects. A comprehensive introduction to maintenance work through the cycle and project dimensions has a good effect on maintenance work. The interface of crane maintenance is shown in Fig. 4.

2.3   Fault Repairing

The design of the fault library [7] is divided into four sub-menus: Module and component, Fault phenomena, Fault cause, and Troubleshooting method. The Fault library is divided into five modules: Lifting mechanism, Large and small car operating mechanism, Metal structure, Various components, Electrical equipment failure. For example, the lifting mechanism mainly has three components, including a brake, a reducer, and a wire rope pulley system. All fault phenomena are listed under the three components, and each fault phenomenon corresponds to several possible causes, and each cause corresponds to the corresponding troubleshooting method [8]. The four fault phenomena of the brake and reducer are listed, as well as the five possible causes of the brake's failure and their corresponding troubleshooting methods (see Fig. 5).

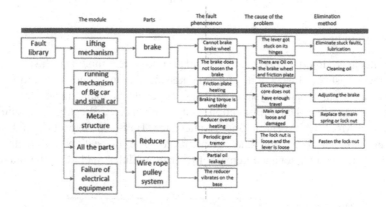

Fig. 5. Fault library design idea diagram

There is one column on each side of the interface of Fault repairing, which is divided into four levels, which are Faulty parts, Fault phenomenon, Fault cause and Troubleshooting method. Figures 6 and 7 are the models when the reducer is selected.

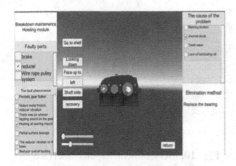 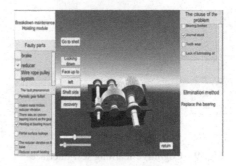

Fig. 6. Interface of lifting mechanism maintenance teaching module

Fig. 7. Remove the shell of the lifting module and rotate angle

2.4 Teaching Assessment

There are four modules in the teaching assessment, which are Assessment project, Error question bank, Question bank practice, and Assessment results. The Assessment project module is divided into four projects, followed by Cognitive teaching, Maintenance, Fault repairing, and Comprehensive testing. After completing the practice of a certain project, you can open the Assessment results to view the test scores. And it will show whether each question is correct. The teaching assessment operation flow chart is shown in Fig. 8. After the test is completed, the Test results interface will pop up the results of this test., and the annotation about whether each question is correct or not. As shown in Figs. 9 and 10.

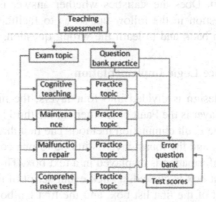

Fig. 8. Flow chart of teaching assessment operation

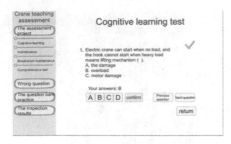   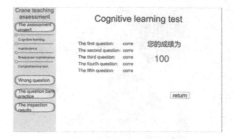

Fig. 9. Cognitive learning test interface   **Fig. 10.** Assessment Results Interface

3 System Development and Implementation

The development and implementation of the system is mainly for the realization of the functions in the interface operation process, corresponding to the logic layer and the data layer in the overall architecture. The development and implementation of the logic layer is divided into interactive layer logic implementation and fault library logic design. The implementation of the data layer is to satisfy the logic layer's call to the data.

3.1 Interaction Layer Logic Implementation

The assessment items in the teaching interface are divided into cognitive learning, maintenance, fault repairing, and comprehensive testing. There are Error question bank and Question bank practice alongside the assessment items. The logic implementation of this part is consistent. Does the statistics whether answer is correct to check the correctness of each question in the follow-up, so as to facilitate the collation of the following error question bank and to learn the wrong question.

3.2 Fault Maintenance Logic Implementation

The fault repair logic design is divided into four layers, the first layer is the Faulty component, the second layer is the Fault phenomenon, the third layer is the Fault cause, and the fourth layer is the Fault elimination method. The first three layers are displayed in the form of a list box, and the fourth layer is in one-to-one correspondence with the third layer, so the text information is displayed in a text box. The design form is: when the option in the first list box is clicked, a click event script of the option is triggered, the script calls the script of the first list box, and the first list box script completes the update of the options in the second list box, and so on. You need to click three times to select the options corresponding to the three list boxes, and finally get the solving method for the fault. The logical relationship is shown in Fig. 11.

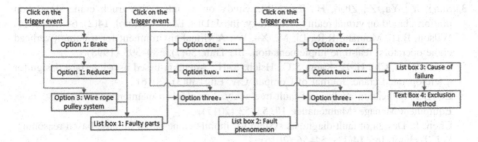

Fig. 11. Fault maintenance logic relationship

3.3 Data Layer Implementation

The header part data is bound with the code of the logical relationship implementation, and the subtitle part is stored as an array in the logical relationship code. The rest of the data is stored in the mySQL database. The result of the logical layer processing is to call the corresponding data in the database and transfer the data to the interactive layer to display.

4 Conclusion

At present, crane teaching is still based on physical or model teaching, and the price of manufacturing physical objects or models is very expensive. A small gantry crane needs hundreds of thousands of yuan, and the teaching efficiency and the utilization rate is low. However, the use of virtual simulation teaching assessment system can greatly reduce the budget. According to the developer's evaluation, the development of a relatively complete system requires only about 50,000 yuan, and the utilization rate is high. A complete system can be used by almost all schools and enterprises across the country. It has strong reproducibility and great economic value.

Based on the virtual reality technology and relying on the Unity3D development engine, we designed a virtual simulation teaching evaluation system for cranes, and carries out specific analysis and integration of the cognitive teaching, maintenance, fault repairing and teaching assessment of the crane [9], and a relatively complete fault library has been established. This system is not only important for crane teaching [10], but also plays a good role in promoting the application of virtual reality technology in teaching and maintenance.

References

1. Zhu, C., Yan, C., Li, Y.: Analysis of intelligent research foundation and future development trend of mobile crane. Constr. Mach. Maintenance **11**, 76–78 (2015)
2. Han, C., Liu, Z.: On the development of crane automation technology. Commun. World **22**, 296–297 (2017)

3. Sang, Y., Yu, Z., Zhao, H., Tang, M.: Study on an interactive truck crane simulation platform based on virtual reality technology. Int. J. Dist. Educ. Technol. **14**(2), 64–78 (2016)
4. Wilson, B.H., Mourant, R.R., Li, M., Xu, W.: A virtual environment for training overhead crane operators: real-time implementation. IIE Tran. **30**(7), 589–595 (1998)
5. Chen, Y.-C., Chi, H.-L., Kang, S.C., Hsieh, S.-H.: Attention-based user interface design for a tele-operated crane. Journal of Comput. Civ. Eng. **30**(3) (2016)
6. Lu, D.: Application analysis of fault tree analysis method in maintenance of gantry crane. Equipment Manage. Maintenance **10**, 52–53 (2017)
7. Chen, J.: Design of fault diagnosis system for mobile crane based on case-based reasoning. Sci. Technol. Inf. **14**(32), 54–56 (2016)
8. Li, W., Lai, L., You, Z., Liu, J., Liu, M.: Design of intelligent fault diagnosis database and management system for truck cranes. Chin. J. Constr. Mach. **8**(03), 348–352 (2010)
9. An, N.M., Ying, T.Y.: Portal crane virtual assembly system based on virtools. Adv. Mater. Res. **694–697**, 2423–2427 (2013)
10. Jennifer, B., Mario, L., Marc, B., Niels, H., Manfred, T., Philipp, K.: Being there again – presence in real and virtual environments and its relation to usability and user experience using a mobile navigation task. Int. J. Hum. Comput. Stud. **101**, 76–87 (2017)

Fog Concentration Grade Judgment for Meter Reading Based on SVM

Zhiren Tian[1], Guifeng Zhang[1], Kaixin Cao[2(✉)], Yongli Liao[1],
and Ruihai Li[1]

[1] Electric Power Research Institute, CSG, Guangzhou 510663, China
[2] School of Electrical and Information Engineering, Tianjin University,
Tianjin 300072, China
ckxin@tju.edu.cn

Abstract. Robot reading energy meter devices automatically is a basic task for smart grid. Unfortunately, the predicted accuracy is influenced by environment dramatically. In order to solve the problem, we propose a method which can metric fog concentration in environment before meter reading. Moreover, based on the characteristics of image and fog influence, joint features and SVM are used to discriminate the concentration level of foggy image. Specially, Contrast feature and DFT feature are extracted as joint features to describe the foggy image and SVM is used to study and train the joint feature. Finally the fog concentration level is discriminated naturally. The experimental results demonstrate the effectiveness of proposed method in term of judging the meter image with five level.

Keywords: Contrast feature and DFT feature · SVM · Fog grade judgment

1 Introduction

In recent years, the number of transformer substation increases rapidly. With the development of smart grid, the research on meter reading algorithm become more and more popular. Such as the automatic interpretation method for pointer-type meter reading from complex environment based on visual saliency was proposed in [1, 2] introduced a reading method based on improved ORB and Hough algorithm for pointer meters, in [3] a reading recognition method based on pointer region was presented and for the inspection robot in intelligent substation, [4] proposed a robust method of pointer meter reading recognition and etc. The meter reading method requires a clear dial, however it is easy to fog in meters or air due to weather and other factors. Therefore, those problems cause a great inconvenience to the meter reading. That means, in order to meet the needs of reading, it is necessary to find an effective fog removal algorithm to deal with fog meters. Fog reduces the visibility of the environment [5], this is caused by the absorption or scattering of light by the atmosphere [6]. Therefore, the atmospheric scattering model or the modified atmosphere scattering model is often used to study the defog algorithm [7–11].

However, the effects of different fog concentrations on meter readings are also not alike. For example, When the fog concentration is low, the meter reading is almost

D. Milošević et al. (Eds.): HCC 2019, LNCS 11956, pp. 391–401, 2019.
https://doi.org/10.1007/978-3-030-37429-7_39

no different from that without fog, and it can be read accurately. If the fog concentration is moderate, it can still be read, but sometimes the reading is inaccurate. And the meter reading is almost impossible when the fog concentration is too high. But the aforementioned meter-to-fog algorithm or meter reading algorithm does not propose an effective solution to this problem. Therefore, it is necessary to find an algorithm to prejudge the fog concentration of the meter before reading or defogging.

At present, there are few studies on the classification of fog concentration levels at home and abroad. Visibility (visible distance) [6] is often used as an indicator of fog concentration. For instance, [12] introduced a method base on psychovisual model and on contrast estimation with wavelet transform. In order to solve the problem of traffic safety in the fog environment, [13, 14] proposed a method that measure visibility distances under foggy weather conditions using a camera mounted onboard a moving vehicle. Aiming at the problem detection of fog concentration at night, a methodology to detect the presence of night fog and characterize its density in images grabbed by an in-vehicle camera is then presented in [15]. However, all visibility methods can not classify fog density.

In addition to aforementioned methods, there are also some algorithms that estimate fog density directly. For example, in order to estimate fog density correctly and to remove fog from foggy images appropriately, [16] proposed a novel feature based on the hue, saturation, and value color space, which correlate well with the perception of fog density. Concerning low efficiency detection for foggy density in traffic image, a novel algorithm [17] was presented to check foggy density based on distribution characteristics of natural statistics. The daytime fog is characterized by its extinction coefficient, which is equivalent to the visibility distance. In this situation, [18] introduced an original algorithm, based on entropy minimization, to detect the fog and estimate its extinction coefficient by the processing of stereo pairs. In [19] a referenceless perceptual fog density prediction model based on natural scene statistics (NSS) and fog aware statistical features was proposed, which can predict the visibility of a foggy scene from a single image without reference to a corresponding fog-free image etc.

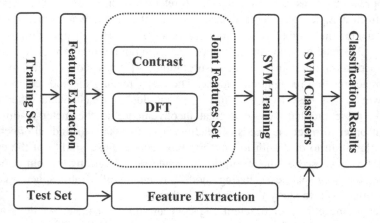

Fig. 1. Flow chart of fog grade classification algorithm

The above-mentioned fog density estimation model can be used for rough estimation of the fog concentration, but the accurate classification is not performed. And the meter reading is more focused on the edge information, but the aforementioned fog density estimation model is more directed to the overall image information. Based on this research background, in this paper, a classification algorithm of fog concentration level for meter is proposed. The algorithm is based on the combined feature and support vector machine, the implementation process is shown as Fig. 1.

2 Methodology

2.1 Contrast Feature

An image contains different grayscale levels, and its gray level distribution contains rich image information, which is an important feature of the image. Image histogram is an image processing tool to describe the gray level distribution of images. It is simple to calculate and has many advantages, such as image translation, rotation, scaling invariance and so on, so it is widely used in various image processing algorithms.

The image histogram is a function of the number of occurrence of each gray level in the image. If the gray level of the image f(x, y) is L, the image gradation value is distributed between [0-L-1]. The histogram of the image is a row vector with dimension L, wherein the coordinates of the vector are the gray level of the image, and the value corresponding to the coordinates is the total number of pixels of the gray level.

When the size of an image is very large, the number of pixels in each gray level of the statistical histogram will also be too many. In this situation, the extracted features will be not conducive to training, so it is necessary to normalize them to the (0, 1) interval. The principle of histogram normalization is first to calculate the histogram of the image, then count the total number of pixels of the image, and at last calculate the probability of the appearance of each gray level pixel, which can be expressed as the following functions:

$$p(x_i) = \frac{n_i}{n} \quad i = 0, 1, \ldots, L - 1 \tag{1}$$

Among them, n is the total number of pixels in the image, and n_i is the number of pixels with pixel value i in the image, $p(x_i)$ reflects the probability of each gray level distribution, and is a statistic that reflects the overall characteristic of the image.

Image contrast refers to the measurement of different gray levels in the light and dark regions of the image, that is, the magnitude of the gray contrast of an image. Fog is mainly composed of water vapor or fine particles in the air, so when light passes through the fog in the air, will cause scattering, refraction, absorption and other phenomena must reduce the contrast of the collected image. Therefore, the contrast of the image is an effective feature to reflect the fog concentration. In order to extract the contrast features of the image, the image histogram is first required. Then the image histogram needs to be normalized. And then, the normalized image histogram vector is divided into multiple equal length intervals, and the gray level pixels contained in the interval are divided. At last, the contrast eigenvector is obtained by adding the probability of the value.

The normalized histogram vector of the image can be obtained by the above method, and the dimension is 256. A 32-dimensional contrast eigenvector can be obtained by evenly dividing it into 32 cells and adding the probability of pixel values of each gray level on each interval. The formula is as follows (Fig. 2):

$$C_i = \sum_{i*8}^{i*8+7} p(x_i) \quad i = 0, 1, \ldots, 31 \tag{2}$$

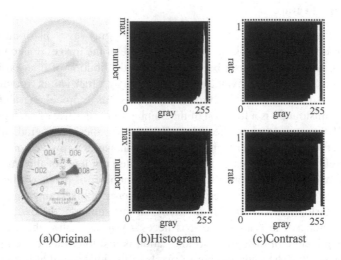

(a)Original   (b)Histogram   (c)Contrast

Fig. 2. Image histogram and contrast feature

As shown above, (a) is the original foggy and non-foggy image (the upper is the foggy image, and the lower is the non-foggy image), (b) is the histogram image, max is the number of pixels with maximum gray value, (c) is the contrast image of the original image. From the above image, it can be seen that the image histogram reflects the difference between foggy image and non-foggy image to a certain extent. The image contrast reflects the difference between foggy image and non-foggy image more obviously on the basis of image histogram, so the contrast feature is an effective feature of fog grade discrimination.

2.2   DFT Feature

All signals in time domain can be expressed as the sum of infinitely many sinusoidal functions and cosine functions. Fourier transform is based on this idea, which can decompose an image into sinusoidal and cosine components, that is, from spatial domain to frequency domain. The two-dimensional Fourier transform formula of the image is as follows:

$$F(u,v) = \frac{1}{MN} \sum_{x=0}^{M-1} \sum_{y=0}^{N-1} f(x,y) e^{-j2\pi(\frac{ux}{M} + \frac{vy}{N})} \tag{3}$$

Where u = 0, 1, ..., M−1, v = 0, 1, ..., N−1, f(x, y) represents the M × N size image matrix, F(u, v) is the Fourier transform of the image. Because $e^{jx} = \cos(x) + j\sin(x)$, the Fourier transform of the image can also be represented by trigonometric function.

According to the above Fourier transform principle, when the input is a two-dimensional real matrix, the Fourier transform result is a two-dimensional complex matrix, that is, the Fourier transform result of the image is composed of a real image and a imaginary image. The spectrum of Fourier transform is needed in the actual image processing algorithm, because the Fourier spectrum contains all the structural information of the image. The Fourier spectrum can be obtained from real and imaginary images through the following formulas:

$$|F(u,v)| = [R^2(u,v) + I^2(u,v)]^{1/2} \tag{4}$$

Where, R(u, v) is a real number image, I(u, v) is an imaginary image, and F(u, v) is the spectrum of Fourier transform. The Fourier transform spectrum calculated by formulas (2–3) and (2–4) has not been centralized, that is, it is surrounded by low frequency components and the center is high frequency components. In order to facilitate frequency domain analysis and frequency domain filtering, we need to translate the frequency domain center, that is, the four weeks of low frequency components are concentrated to the center position, and the high frequency components are moved to four weeks. There are two ways to centralize the general spectrum, one is to multiply the input image $(-1)^{x+y}$, because $e^{j\pi} = -1$, so $(-1)^{x+y} = e^{j\pi(x+y)}$, plug it into (3):

$$F(u,v) = \frac{1}{MN} \sum_{x=0}^{M-1} \sum_{x=0}^{N-1} f(x,y) e^{-j2\pi(\frac{(u-\frac{M}{2})x}{M} + \frac{(v-\frac{N}{2})y}{N})} \tag{5}$$

From the above formula, the original coordinate (0, 0) is moved to the (M/2, N/2) position, that is to say, the spectrum centralization is completed. The other way is to use formulas (3) and (4) to get the Fourier spectrum map, then according to M/2, N/2 divides the spectrum map into four regions, and then exchanges the left upper right and lower left and lower right upper and lower left regions respectively to complete the frequency domain center translation.

As shown in Fig. 3, (a) and (c) are the original images, and (b) and (d) are the corresponding DFT spectrum diagrams. The center of the frequency spectrum is the low-frequency component, the periphery is the high-frequency component, and the visible image energy is almost concentrated in the low-frequency component, where the F(0, 0) is the direct-current component, that is, the average brightness of the image. Effective DFT eigenvector can be obtained by removing DC component and extracting fourth-order low-frequency component.

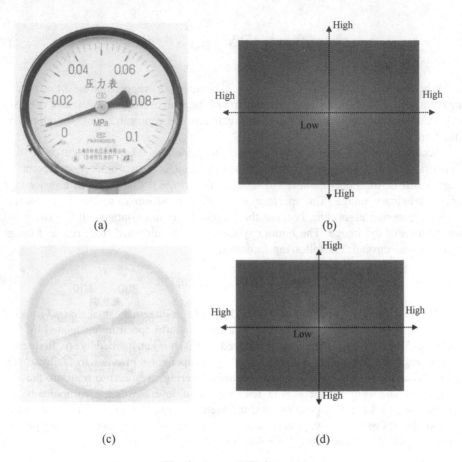

Fig. 3. Image DFT feature

2.3   Support Vector Machine(SVM)

Support Vector Machine (SVM) is a very important classification algorithm for tra-ditional machine learning, which is a general type of feed forward network. SVM is a kind of generalized linear classifier for binary classification of data in a supervised learning way. The classification of SVM is generally two kinds of thought:

For the simple linear separable case, the original problem is converted into the optimization problem of the secondary convex function plus the constraint condition, then the Lagrange algorithm is used for optimization, and the existing algorithm is used to solve the problem.

In a complex case, that is non-linear separable, the kernel function is used to project the sample into the high dimensional space, so that it becomes linearly detachable, and uses kernel functions to reduce the amount of high latitude calculations.

Suppose there is a set of training samples, and the expected response is that we are represented by classes +1 and −1 to show that the samples are linearly divisible.

$$w^T x_i + b \geq 0 \quad d_i = +1$$
$$w^T x_i + b < 0 \quad d_i = -1 \tag{6}$$

Where x is the input vector, that is, the vector in the sample set; w is the adjustable weight vector, each vector can be adjusted; b represents the offset of the hyperplane from the origin. All samples above the upper interval boundary belong to $d = \frac{2}{||w||}$ the positive class and the samples below the lower interval boundary belong to the negative class. The distance between the two spaced boundaries is $d = \frac{2}{||w||}$.

Maximizing the separation edge between two classes is equivalent to minimizing the Euclidean norm of the weight vector w, which is equivalent to the optimization $||w||^2$, so the following conditional extremum problem can be constructed:

$$\min_{\gamma,\omega,b} \frac{1}{2} ||w||^2 \tag{7}$$
$$s.t. y^{(i)}(w^T x^{(i)} + b) \geq 1, i = 1, \ldots, m$$

The Lagrangian equation is constructed:

$$L(w, b, \partial) = \frac{1}{2} ||w||^2 - \sum_{i=1}^{m} a_i [y^{(i)}(w^T x^{(i)} + b) - 1] \tag{8}$$

The partial derivation of Lagrangian equation can be obtained:

$$w = \sum_{i=1}^{m} a_i y^{(i)} x^{(i)} \tag{9}$$

Obtaining a_i by SMO algorithm, w^* is obtained from formula (2–9), b^* is obtained by the following formula, and finally the solution of SVM can be obtained.

$$b^* = -\frac{\max_{i:y^{(i)}=-1} w^{*^T} x^{(i)} + \min_{i:y^{(i)}=1} w^{*^T} x^{(i)}}{2} \tag{10}$$

3 Experiments

3.1 Sample Set

It is necessary to collect the sample set for training and testing by constructing the fog concentration level model with SVM. The following points shall be paid attention to the image sample set:

The sample set should be rich enough, and the sample set should contain a variety of meters.

The number of samples should be sufficient. SVM training model needs to be cross-verified, which requires the sample set to be divided into multiple drawing libraries.

(iii) The sample set should contain images of all stages of fog concentration from dense to light.

Finally, 1000 representative images are selected from the collected images to form a sample set, and they are divided into five image databases, each of which contains 200 images. Each images library is divided into five grades according to fog concentration, which the first grade fog concentration is the largest and the fifth grade fog concentration is the smallest.

3.2   Discriminant Standard

For the fog grade discriminant model proposed in this paper, the error rate and accuracy are mainly used to evaluate the performance of the system. The error rate, as the name implies, is the proportion of the number of samples that are classified wrongly in the classification result in the total number of samples. In contrast, the accuracy rate is the proportion of the number of samples that are correctly classified as the total number of samples in the classification result. If the sample set is set to D, the classification error rate and accuracy rate are defined as:

$$E(f;D) = \frac{1}{m}\sum_{i=1}^{m}I(f(x_i = y_i)) \tag{11}$$

$$acc(f;D) = \frac{1}{m}\sum_{i=1}^{m}I(f(x_i) = y_i) = 1 - E(f;D) \tag{12}$$

Where $I(\cdot)$ is an indicator function, it takes 1 when $\cdot$ is true and takes 0 when $\cdot$ is false. m is the total number of samples, $f(x_i)$ is the predicted value, y_i is the real sample value.

3.3   Results Comparison with Different Features

In this paper, the performance of the fog grade discriminant model is tested on the collected sample set, a total of 1000 images were collected from the sample set, which were divided into five image libraries. Each library is divided into five grades according to the concentration of fog from dense to light, marked as 1–5 respectively. In the test, each image library is taken as the test set, and the other four image libraries are used as the training set for cross-verification, the classification results are shown in Table 1, and finally, the average error rate and accuracy are counted as the results of performance evaluation. In this paper, the recognition accuracy of extracting contrast features, extracting DFT features separately and extracting contrast and DFT joint features is tested respectively, and the results of the comparison are as follows (Table 2):

Table 1. Classification results with joint features

Images			
Classification results	1	2	3
Real results	1	2	3
Images			
Classification results	4	5	1
Real results	4	5	2

Table 2. Results comparison with different features

Extracted features	Average error rate	Average accuracy
Contrast feature	3.3%	96.7%
DFT feature	2.4%	97.6%
Contrast and DFT joint features	0.5%	99.5%

As can be seen from the above tables, the overall classification results of each fog concentration grade are relatively accurate, the classification accuracy of a single feature is lower than the classification accuracy of the joint features. This is because a single contrast feature does not remove the effects of light and noise, while a separate DFT feature lacks a foggy and non-foggy feature comparison. The combined features synthesize their advantages. On the basis of extracting the effective contrast features, the DFT low-order coefficients which remove the DC components are used to remove the effects of light and noise, so the accuracy of classification is effectively improved.

4 Conclusion

In this paper, a fog concentration grade discrimination algorithm based on SVM and joint characteristics is proposed, which provides a reference for solving the problem of fog concentration pre-judgment before meter reading. Firstly, the contrast features and DFT features of the image are extracted as the joint feature vector. Then the joint feature vector set is obtained based on the sample set, and the SVM classifier is trained by using the joint feature vector set. Finally, the trained SVM classifier is used to classify and distinguish the fog concentration level of the test image. The experimental results show that the algorithm can effectively classify five levels of fog concentration images, and the accuracy reaches 99.5%. The next step is to continue to expand the sample set, further refine the fog concentration level, and enhance the classification accuracy of the algorithm, add distance variables to consider the influence of fog concentration and distance synthesis factors on meter readings.

References

1. Zhang, L., Fang, B., Zhao, X., et al.: Pointer-type meter automatic reading from complex environment based on visual saliency. In: International Conference on Wavelet Analysis & Pattern Recognition. IEEE (2016)
2. Jianlong, G., Liang, G., Yaoyu, L.V., et al.: Pointer meter reading method based on improved ORB and Hough algorithm. Comput. Eng. Appl. (2018)
3. Wei, S., Wenjie, Z., Jiaqi, Z., et al.: Meter reading recognition method via the pointer region feature. Chin. J. Sci. Instrum. (2014)
4. Gao, J.W., Xie, H.T., Zuo, L., et al.: A robust pointer meter reading recognition method for substation inspection robot. In: 2017 International Conference on Robotics and Automation Sciences (ICRAS). IEEE (2017)
5. Narasimhan, S.G., Nayar, S.K.: Contrast restoration of weather degraded images. IEEE Trans. Pattern Anal. Mach. Intell. 25(6), 0–724 (2003)
6. Tan, R.T.: Visibility in bad weather from a single image. In: 2008 IEEE Computer Society Conference on Computer Vision and Pattern Recognition (CVPR 2008), 24–26 June 2008, Anchorage, Alaska, USA. IEEE (2008)
7. Zhang, T., Shao, C., Wang, X.: Atmospheric scattering-based multiple images fog removal. In: International Congress on Image & Signal Processing. IEEE (2011)
8. Chuangbai, X., Hongyu, Z., Jing, Y., et al.: Traffic image defogging method based on WLS. Infrared Laser Eng. 44(3), 1080–1084 (2015)
9. Dong, H.Y., Bai, H., Wang, X.W.: A method of restoring the fog degraded image based on the monochromatic atmospheric scattering model. Adv. Mater. Res. 461, 849–853 (2012)
10. Sun, W., Wang, H., Sun, C., et al.: Fast single image haze removal via local atmospheric light veil estimation. Comput. Electr. Eng. 2015:S0045790615000348 (2015)
11. Ju, M., Gu, Z., Zhang, D.: Single image haze removal based on the improved atmospheric scattering model. Neurocomputing, 2017:S0925231217307051 (2017)
12. Busch, C., Debes, E.: Wavelet transform for analyzing fog visibility. IEEE Intell. Syst. 13 (6), 66–71 (1998)
13. Hautiére, N., Tarel, J.P., Lavenant, J., et al.: Automatic fog detection and estimation of visibility distance through use of an onboard camera. Mach. Vis. Appl. 17(1), 8–20 (2006)
14. Hautiere, N., Aubert, D., Dumont, E.: Hautière, Mobilized and mobilizable visibility distances mobilized and mobilizable visibility distances for road visibility in fog. In: Session of the Cie (2007)
15. Gallen, R., Cord, A., Hautiere, N., et al.: Nighttime visibility analysis and estimation method in the presence of dense fog. IEEE Trans. Intell. Transp. Syst. 16(1), 310–320 (2015)
16. Jiang, Y., Sun, C., Zhao, Y., et al.: Fog density estimation and image defogging based on surrogate modeling for optical depth. IEEE Trans. Image Process. 26(7), 3397–3409 (2017)

17. Limin, W., Yongfeng, J.U., Maode, Y.: Inspection of fog density for traffic image based on distribution characteristics of natural statistics. Acta Electronica Sinica **45**(8), 1888–1895 (2017)
18. Caraffa, L., Tarel, J.P.: Daytime fog detection and density estimation with entropy minimisation (2014)
19. Choi, L.K., You, J., Bovik, A.C.: Referenceless prediction of perceptual fog density and perceptual image defogging. IEEE Trans. Image Process. **24**(11), 3888–3901 (2015)

A Robust Optimization Model for Closed-Loop Supply Chain Network Under Uncertain Returns

Changqiong Wang[1](✉), Hui Jiang[1], Qi Luo[1], and Shuncai Li[2]

[1] School of Logistics Engineering, Wuhan University of Technology,
Wuhan 430063, China
zqwang@whut.edu.cn
[2] School of Management, Huazhong University of Science and Technology,
Wuhan 430074, China

Abstract. An effective closed-loop supply chain (CLSC) is increasingly important for corporate sustainable development, where used products are returned, remanufactured and/or recycled. Optimized planning of CLSC network is required and is influenced by uncertainties of recovery. So, this study seeks to establish a robust optimal model for CLSC network, considering both uncertainties in quantity and quality of returned products. After the general framework of CLSC is discussed, the measurements for uncertain quantity and quality of returned products are formulated mathematically, respectively as a number of discrete scenarios and Quality Index. A mixed integral linear programming (MILP) model for a CLSC network design is established and then translated into a robust optimization model based on regret value, to determine facilities' locations and quantity of flows between facilities in the network. A numerical example is given, and the simulation results show that the operation strategies of the CLSC are relatively stable under different recycling scenarios. Therefore, the optimization model for CLSC network has good robustness.

Keywords: Closed-loop supply chain (CLSC) · Network design · Robust optimization · Robustness · Uncertainty

1 Introduction

A Closed Loop Supply Chain (CLSC), where used products are returned, remanufactured and/or recycled, can improve the ecological benefits of products in the whole life cycle [1]. There are much more uncertainties in CLSC than in general supply chain such as uncertain demands, returned products and recovery ways. Therefore, designing and operating effectively a CLSC system under uncertain circumstance have been a hot research area over decade years [2]. Uncertainties and stochastic variations in CLSC make the network decisions complicated, and moreover, difficult to alter in future [3]. Khalilzadeh and Derikvand [4] proposed a multi-objective model to identify optimal suppliers for a green supply chain network under uncertainty, considering the probability of defective finished products and using stochastic programming. Karimi et al. [5]

© Springer Nature Switzerland AG 2019
D. Milošević et al. (Eds.): HCC 2019, LNCS 11956, pp. 402–412, 2019.
https://doi.org/10.1007/978-3-030-37429-7_40

presented a multi-objective multi-facility network model of green CLSC under uncertain environment, where the recovery possibility of products was considered.

Robustness is a characteristic describing a system's ability to effectively perform while its variables are altered. Supply chain robustness refers to the ability of a supply chain to resist or avoid change [6]. A supply chain network design is robust, if it is capable of providing sustainable value creation under all possible future scenarios [7]. Therefore, robust optimization, as a modeling methodology under uncertain circumstance, is increasingly used in optimization of supply chain network [8]. Hosseini et al. [9] extended a mixed integer linear programming mode (MILP) to a robust optimization framework to determine the optimal parameters of a three-tiered supply chain under uncertainty. Pishvaee et al. [10] built a linear robust optimization model for CLSC network considering uncertainties in returned products, demands for recovered products and transportation costs. Vahdani et al. [2] set up a bi-objective robust optimization model for a CLSC network, and a self-adaptive imperialist competitive algorithm was put forward to solve the model.

Most of relevant literatures considered the uncertainties in demand and/or the quantity of recycle, but few considered the uncertainty in quality of returned products. However, the quality of returned product has decisive effects on choosing the recovery way, and then on facilities location and CLSC network. So, this paper will focus on uncertainties both in quantity and quality of recycling products, and establish a robust optimization model for CLSC network.

The remainder of this paper is as follows: the framework of a CLSC network is described in Sect. 2. The model assumptions and variables definition are proposed in Sect. 3. The mathematical methods for characterizing uncertainties in quantity and quality of returned products are provided in Sect. 4. A robust optimization model for CLSC network design is formulated in Sect. 5. Numerical experimentation and results are given in Sect. 6. Finally, conclusions are discussed in Sect. 7.

2 The Framework of a CLSC

A CLSC for household appliances is studied in this paper. Generally, a CLSC system consists of facilities and firms, such as material suppliers, manufacturing centers (MCs), distribution centers (DCs), markets/recycling market, recycling centers (RCs) and disposal centers. Flows of materials/products between facilities in CLSC are illustrated in Fig. 1, including reverse and forward flows.

The reverse flows in a CLSC are driven by returned products, while the forward flows are pulled by customer demands. The quality of returned products will determine the decision of recovery ways, i.e. reusing, repairing/remanufacturing, recycling or disposal, which will further affect the function and location of facilities. That is to say, the quantity and quality of the returned products have a great impact on design and operation of CLSC. Therefore, the uncertainties in quantity and quality of recycling must be taken account into the CLSC network design.

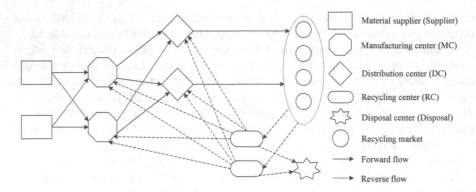

Fig. 1. Framework of a closed-loop supply chain

3   Model Assumptions and Variables Description

3.1   Model Assumptions

Six assumptions are postulated to simplify modelling a CLSC network design.

(1) Only a single-cycle and single-product will be considered in modelling.
(2) Fixed set-up costs, product-related costs and transportation costs are considered, but time costs are ignored.
(3) The transportation rate per unit between facilities is constant for new and used products; the rate difference in different transport modes is ignored also.
(4) Refurbished products/parts in recycling centers are transferred into distributors, manufacturers/remanufacturers, and only transportation costs are considered.
(5) The remanufactured products have the same quality with the original new products, meeting the same market demands.
(6) Capacities of facilities and routes are known, but the capacities of recycling centers are not limited.

3.2   Index

I = {1, 2, ..., i} Set of manufacturing centers;
J = {1, 2, ..., j} Set of DCs;
L = {1, 2, ..., l} Set of supplier;
M = {1, 2, ..., m} Set of recycling centers (RC); $m = 1$;
N = {1, 2, ..., n} Set of disposal centers.

3.3   Decision Variables

$Xlinew$: Quantity of material shipped from the supplier - l to the MC - i;
Xij: Quantity of product shipped from the MC - i to the DC - j;
Xjd: Quantity of product shipped from the DC - j to the market;
Xmi: Quantity of product shipped from the RC - m to the MC - i;

Xmj: Quantity of product shipped from the RC - m to the DC - j;
Xmn: Quantity of products shipped from the RC - m to the disposal center - n;
Yl: If the supplier l is selected, $Yl = 1$; otherwise, $Yl = 0$;
Yi: If a manufacturing center is opened at i, $Yi = 1$; otherwise $Yi = 0$;
Yj: If a DC is opened at j, $Yj = 1$; otherwise $Yj = 0$;
Yn: If a disposal center is opened at n, $Yn = 1$; otherwise $Yn = 0$.

3.4   Parameters

D: Demand of certain market;
R_m: The number of recycled products in the RC - m;
ch_i: Capacity of the MC - i;
ch_j: Capacity of the DC - j;
ch_n: Capacity of the disposal center - n;
θ_{mj}: Refurbishment ratio of used products;
θ_{mi}: Ratio of remanufacturing of used parts;
θ_{mn}: Ratio of disposal products;
f_i: Fixed set-up cost of the MC - i;
f_j: Fixed set-up cost of the distributing center - j;
f_n: Fixed set-up cost of the disposed center - n;
c_{mi}: Unit remanufacturing cost of used products from the RC - m to the MC - i (including recycling and manufacturing costs);
c_{mj}: Unit rebuilding cost of used products from the RC - m to the DC - j (including recycling and rebuilding cost);
c_{mn}: Unit disposal cost of used products from the RC - m to the disposal center - n (including recycling and disposal cost);
c_{linew}: Unit purchase cost of the MC - i paid for the supplier - l;
c_{inew}: Unit manufacturing cost of new product in the MC - i;
v_g: Unit transporting rate per products/materials ($/kg km);
g_{ij}: Shipping distance from the MC - i to the DC - j;
g_{mn}: Shipping distance from the RC - m to the disposal center - n;
g_{mi}: Shipping distance from the RC - m to the MC -i;
g_{mj}: Shipping distance from the RC - m to the DC - j;
g_{li}: Shipping distance from the material supplier - l to the MC - i.

4   Mathematical Methods to Characterize Uncertainties in Quantity and Quality of Returned Products

The possibility of various quality and quantity of returned products is called as a recycling scenario. The following methods are used to formulate the uncertainty.

4.1 Uncertainty in Quality of Returned Products

The quality of a used product is difficult to record in digital, but its service life can be recorded. Normally, the longer a product was used, the shorter the residual life, so the lower the remaining quality level. Therefore, the quality of returned products can be expressed as the ratio of the remaining useful life to the expected life, named as quality index, q.

Assuming that the expected life of a product is T, in year(s), its quality index q(t) at a specific stage - t is defined as Eq. (1):

$$q(t) = \frac{t}{T}, \quad t1 < t < t2 \tag{1}$$

Here, t1 and t2 are respectively the lower and the upper limits of the residual life of products, and (t1, t2) $\in$ [0, T]. Let a and b stand for the lower and upper limits of quality level of used products. Obviously, $\frac{t1}{T} = a$, $\frac{t2}{T} = b$, and q(t) $\in$ (a, b).

The reprocessing way of a returned product is different with the quality level q(t). Given two thresholds w1 and w2, and a $\leq$ w1 $\leq$ w2 $\leq$ b, then used products can be sorted into the following three levels: A, B and C.

Used products belong to the level A, if the quality index is q(t) A $\in$ (w2, b). They have the highest quality and may be probably refurbished to recovery the products' value. The probability is $\theta_{mj} = \frac{b - W_2}{b - a}$.

Used products belong to the level C, if the quality index is q(t) C $\in$ (a, w1). And they have the lowest quality and are probably disposed, such as incineration or landfill. The probability is $\theta_{mn} = 1 - \frac{b - w_2}{b - a} - \frac{w_2 - w_1}{b - a}$.

Otherwise, used products belong to the level B, and they are probably remanufactured to reuse components or materials, and the probability is $\theta_{mi} = \frac{w_2 - w_1}{b - a}$.

4.2 Uncertainty in Quantity of Returned Products

Assume that the number of returned products, R_m is uncertain. The uncertainty can be described by S known recycling scenarios. The probability of the scenarios is ps, and $\sum_{s=1}^{S} p_s = 1$. When the scenarios happens, the flows between facilities in the CLSC can be adjusted according to the relations of supply and demand. This means all the continuous decision variables can effectively achieve the goals in all scenarios.

Therefore, for the scenarios, ds is the market demand, R_m^s is the number of returned products, and X_{linew}^s, X_{ij}^s, X_{jd}^s, X_{mi}^s, X_{mj}^s and X_{mn}^s are flows between facilities, and (a^s, b^s) is the quality level interval of returned products, and θ_{mj}^s, θ_{mi}^s and θ_{mn}^s are respectively the ratios of refurbishment, remanufacturing and disposal under the scenarios.

5 Robust Optimization Model for CLSC Network Design

A robust optimization model focuses on the expected objective function values and also evaluates the difference among the function values when uncertain parameters change. According to the optimal objective function in the model, there are three modelling methods: the difference value model, the preference value model and the regret value model [8]. The regret value model will be adopted in this study.

5.1 Robust Optimization Model Based on the Regret Value

In the regret value model, the regret value of a scenario is measured by the difference between the optimal objective value and the feasible objective value under the scenario.

Given a constant $\pi \geq 0$, a scenario set S, a specific scenarios, $\forall s \in S$, and let Z_s^* and $Zs(X)$ respectively be the optimal and the feasible objective value under the scenarios.

If exists s, $\forall s \in S$, and $\frac{Zs(X)-Z_s^*}{Z_s^*} \leq \pi$, then X is called as the robust solution of the original question. Here, $Zs(X) - Z_s^*$ is called as the absolute regret value, and $\frac{Zs(X)-Z_s^*}{Z_s^*}$ is the relative regret value, and π is the regret value limit coefficient.

For a certain problem, there may be a number of robust solutions. The optimal robust solution will be chosen to help making design. The model is as follows.

$$min \sum_s qsZs(x) \qquad (2)$$

$$s.t. Z_z(x) \leq (1 + \pi)Z_s^*$$

$$x \in X \qquad (3)$$

The robust solution of the above model is the best solution for the original question. X is the set of feasible solutions in the certain circumstance, and qs is the probability of the scenarios.

5.2 Robust Optimization Model for CLSC Network Based on the Regret Value

Define the optimization criteria that the objective function is to minimize the maximum deviation rate W, i.e., the relative regret value, which measures the deviation between a feasible cost and the optimal target cost of the network, under all recovery scenarios.

For the recycling scenarios, $\forall s \in S$ and a given constant $\pi \geq 0$, the objective function is minimization of the average total cost under all scenarios, which is the same with Eq. (2).

$$min \ W_\pi = \sum_{s \in S} q_s W_s$$

The constraints are as follows.

$$
\begin{aligned}
s.t.\ W_s = & \sum_{i \in I} f_i Y_i + \sum_{i \in J} f_j Y_j + \sum_{n \in N} f_n Y_n + \sum_{j \in J} c_{mj} X_{mj}^s + \sum_{i \in I} c_{mi} X_{mi}^s \\
& + \sum_{n \in N} c_{mn} X_{mn}^s + \sum_{l \in L} \sum_{i \in I} c_{linew} X_{linew}^s + \sum_{j \in J} \sum_{i \in I} c_{linew} X_{linew}^s \\
& + \sum_{l \in L} \sum_{i \in I} X_{li}^s g_{li} v_g + \sum_{i \in I} \sum_{i \in L} X_{ij}^s g_{ij} v_g + \sum_{m \in M} \sum_{i \in I} X_{mi}^s g_{mi} v_g \\
& + \sum_{m \in M} \sum_{j \in J} X_{mj}^s g_{mj} v_g + \sum_{m \in M} \sum_{n \in N} X_{mn}^s g_{mn} v_g
\end{aligned}
\tag{4}
$$

$$
\frac{W_S(X) - W_s^*}{W_s^*} \leq \pi, \forall s
\tag{5}
$$

$$
W_s^* = \min W_s
\tag{6}
$$

$$
\sum_{j \in J} X_{jk}^s \geq ds, \forall s
\tag{7}
$$

$$
\sum_{l \in L} X_{li}^s + \sum_{m \in M} X_{mi}^s = \sum_{j \in J} X_{ij}^s, \forall i, \forall s
\tag{8}
$$

$$
\sum_{i \in I} X_{ij}^s + \sum_{m \in M} X_{mj}^s = \sum_{k \in K} X_{jd}^s, \forall j, \forall s
\tag{9}
$$

$$
\sum_{j \in J} X_{mj}^s = \theta_{mj}^s \cdot R_m^s, \forall s
\tag{10}
$$

$$
\sum_{i \in I} X_{mi}^s = \theta_{mi}^s \cdot R_m^s, \forall s
\tag{11}
$$

$$
\sum_{n \in N} X_{mn}^s = \theta_{mn}^s \cdot R_m^s, \forall s
\tag{12}
$$

$$
\sum_{l \in L} X_{li}^s + \sum_{m \in M} X_{mi}^s \leq Y_i \cdot chi, \forall i, \forall s
\tag{13}
$$

$$
\sum_{i \in I} X_{ij}^s + \sum_{m \in M} X_{mj}^s \leq Y_j \cdot chj, \forall j, \forall s
\tag{14}
$$

$$
\sum_{m \in M} X_{mn}^s \leq Y_n \cdot chn, \forall n, \forall s
\tag{15}
$$

$$
X_{linew}^s, X_{ij}^s, X_{jd}^s, X_{mi}^s, X_{mj}^s, X_{mn}^s \geq 0, \forall l, i, j, n, s
\tag{16}
$$

$$
Yl, Yi, Yj, Yn \in \{0, 1\}, \forall l, i, j, n
\tag{17}
$$

Equation (4) shows the total cost under the scenarios. Equation (5) restrains the deviation between the feasible and the optimal function values within a given range. Equation (6) describes the optimal solution. Equation (7) states that the demand does not exceed the distribution capacity. Equations (8)–(12) show that the sum of the flow entering each facility equal to the flow leaving the facility. Equations (13)–(15) ensure that the total flows exiting a plant do not exceed the plant's capacity. Equations (16)–(17) define ranges of values of variables.

6 A Numerical Experiment

6.1 Data Exhibition

A small-size household appliance company is planning to establish a CLSC network to recycle used products. The product's expected life $T = 12$ years. All the returned products are within the service life cycle. Four recycling scenarios may happen according to investigation, and the relative information and data are listed in Table 1.

Table 1. Data in recycling centers in the numerical experiment

Scenarios	Probability ps	Recovery R_m^s	Service life t	Demand ds
1	0.25	4000	(2, 10)	5500
2	0.25	3800	(2, 9)	5000
3	0.25	2900	(4, 9)	5700
4	0.25	3200	(3, 11)	5900

The quality index and reprocessing probability can be calculated according to given parameters $\pi = 0.1$, w1 = 0.4, w2 = 0.6. The results are shown in Table 2.

Table 2. Value of parameters related to the returned products

Scenarios	Quality level (a^s, b^s)	Ratio of revision θ_{mj}^s	Ratio of remanufacturing θ_{mi}^s	Ratio of disposal θ_{mn}^s
1	(0.17, 0.83)	0.348	0.303	0.349
2	(0.17, 0.75)	0.258	0.345	0.397
3	(0.33, 0.75)	0.357	0.476	0.167
4	(0.25, 0.92)	0.478	0.299	0.223

The alternative facilities have been chosen after investigation, which are two potential suppliers, two potential manufacturing/remanufacturing centers (MCs), two potential DCs and two potential disposal centers. The unit transportation rate vg = 2 ($/kg km). Information and data of the potential facilities are shown from the Tables 3, 4, 5 and 6.

Table 3. Capacity of facilities (kg)

Capacity	1	2
MC chi	5000	4500
DC chj	6000	7500
DC chn	8000	7500

Table 4. Fixed cost of facilities ($)

Fixed cost	1	2
MC fi	300000	280000
DC fj	400000	420000
Disposal center fn	400000	530000

Table 5. Variable cost per unit ($/kg)

Variable cost per unit	1	2
Material supplier clinew	260	250
MC cinew	13	15
MC cmi	10	9
DC cmj	8	7
Disposal center cmn	23	30

Table 6. Transportation distance between facilities (km)

l/i	Supplier-MC, gli	i/j	MC-DC, gij	m/i	DC-RC, gmi	m/j	RC-DC, gmj	m/n	RC-disposal, gmn
1/1	424	1/1	1106	1/1	787	1/1	220	1/1	22
1/2	457	1/2	1031	1/2	684	1/2	183	1/2	31
2/1	341	2/1	1380	–	–	–	–	–	–
2/2	361	2/2	1006	–	–	–	–	–	–

6.2   Results

The optimal value is $W_s = 12429620$ by using the software Lingo 10.0 based on data above. Then sites selected and flows between facilities in the four scenarios are inferred, shown in Tables 7 and 8 respectively.

Table 7. Results of sites selected

Scenarios	Supplier Yl		Manufacturer Yi		DC Yj		Disposal center Yn	
	l = 1	l = 2	i = 1	i = 2	j = 1	j = 2	n = 1	n = 2
1	0	1	0	1	0	1	1	0
2	0	1	0	1	0	1	1	0
3	0	1	1	1	1	1	1	0
4	0	1	0	1	0	1	1	0

Table 8. Flows between nodes (Kg)

Item		Scenario 1	Scenario 2	Scenario 3		Scenario 4
Supplier-MC	Route (l, i)	(2, 2)	(2, 2)	(2, 1)	(2, 2)	(2, 2)
	Flow Xli	3158.4	2708.6	164.7	3119.6	2413.6
MC-DC	Route (i, j)	(2, 2)	(2, 2)	(1, 1)	(2, 2)	(2, 2)
	Flow Xij	4370.4	4019.6	164.7	4500	4370.4
RC-MC	Route (m, i)	(1, 2)	(1, 2)	(1, 2)		(1, 2)
	Flow Xmi	1212	1212	1380.4		956.8
RC-DC	Route (m, j)	(1, 2)	(1, 2)	(1, 2)		(1, 2)
	Flow Xmj	1392	980.4	1035.3		1529.6
RC-disposal	Route (m, n)	(1, 1)	(1, 1)	(1, 1)		(1, 1)
	Flow Xmn	1396	1508.6	484.3		713.6

Table 7 shows that the facilities selected change very little in different scenarios, i.e. different quantity and quality of returned products. Comparing the results, the optimal network configuration is that from the supplier 2 to the manufacturing 2, to DC 2, to the recycling center, and finally to the disposal center 1. Consequently, the flows between facilities and the relative differences are shown in Table 9.

Table 9. Flows between facilities and the relative difference under scenarios

Item	Scenario 1	Scenario 2	Scenario 4	Average
Flows from supplier to MC;	3158.4	2708.6	2413.6	2760.2
Relative difference (%)	14.4	−1.9	−12.6	
Flows from MC to DC;	4370.4	4019.6	4370.4	4253.5
Relative difference (%)	2.7	−5.5	2.7	
Flows from RC to MC;	1212	1212	956.8	1126.9
Relative difference (%)	7.6	7.6	−15.1	
Flow from RC to DC;	1392	980.4	1529.6	1300.7
Relative difference (%)	7.0	−24.6	17.6	
Flows from RC to disposal center;	1396	1508.6	713.6	1206.1
Relative difference (%)	15.7	25.1	−40.8	

Table 9 shows that the relative difference in flows from an upstream facility to the downstream facility is mostly within +10%. This means the operational strategy of the CLSC is stable, and therefore the model is robust.

7   Conclusions

The network design of a CLSC aims at locating facilities and determining the flows between facilities. Uncertainties in recycling have a strong influence on CLSC network design. So, the uncertainties in quantity and quality of returns products were characterized mathematically, and then, a MILP model of CLSC network design was transformed into a robust optimization model based on the regret value. The numerical experiment results have verified that the model is robust and the operational strategy of CLSC is stable in certain stage. However, the work has limitations. In fact, there are many other uncertainties in CLSC, such as the trade-off between economic and ecological benefits, uncertainty in return time and so on. And the study can be further expanded to other uncertainties in CLSC system.

References

1. Vahdani, B., Mohammadi, M.: A bi-objective interval-stochastic robust optimization model for designing closed loop supply chain network with multi-priority queuing system. Int. J. Prod. Econ. **170**(A), 67–87 (2015)
2. Inderfurth, K.: Impact of uncertainties on recovery behavior in a remanufacturing environment: a numerical analysis. Int. J. Phys. Distrib. Logist. Manag. **35**(5), 318–336 (2005)
3. Prakash, S., Soni, G., Rathore, A.P.S.: Embedding risk in closed-loop supply chain network design: case of a hospital furniture manufacturer. J. Model. Manag. **12**(3), 551–574 (2017)
4. Khalilzadeh, M., Derikvand, H.: A multi-objective supplier selection model for green supply chain network under uncertainty. J. Model. Manag. **13**(3), 605–625 (2018)
5. Karimi, B., Niknamfar, A.H., Gavyar, B.H., Barzegar, M., Motashami, A.: Multi-objective multi-facility green manufacturing closed-loop supply chain under uncertain environment. Assem. Autom. (2019). https://doi.org/10.1108/AA-09-2018-0138
6. Durach, C.F., Wieland, A., Machuca, J.A.D.: Antecedents and dimensions of supply chain robustness: a systematic literature review. Int. J. Phys. Distrib. Logist. Manag. **45**(1/2), 118–137 (2015)
7. Klibi, W., Martel, A., Guitouni, A.: The design of robust value-creating supply chain networks: a critical review. Eur. J. Oper. Res. **203**(2), 283–293 (2010)
8. Ben-Tal, A., Nemirovski, A.: Robust optimization – methodology and applications. Math. Program. **92**, 453–480 (2002)
9. Hosseini, S., Farahani, R.Z., Dullaert, W., Raa, B., Rajabi, M., Bolhari, A.: A robust optimization model for a supply chain under uncertainty. J. Manag. Math. **25**(4), 387–402 (2014)
10. Pishvaee, M.S., Rabbani, M., Torabi, S.A.: A robust optimization approach to closed-loop supply chain network design under uncertainty. Appl. Math. Model. **35**(2), 637–649 (2011)

Design and Implementation of Energy Saving Monitoring and Management System on Campus Based on Internet of Things Technology

Hongrui Wang[1], Ran Wei[2(✉)], Peng Gong[1], and Zhimin Yang[1]

[1] Shandong University at Weihai, Weihai 264200, Shandong, China
hong.rui88@163.com
[2] Harbin Institute of Technology, Weihai, Weihai 264200, Shandong, China
weiran@hitwh.edu.cn

Abstract. Campus energy consumption has many kinds with large total amount, which leads to great energy-saving potential. With the rapid development of Internet of Things (IoT) technology and sensor technology, it is possible to achieve large-scale campus energy management. Taking the energy saving monitoring and management platform of Shandong University at Weihai as an example, this paper designs and realizes an energy management system based on IoT. Using distributed acquisition and centralized management technology, an IoT system is established by installing data acquisition terminals in the whole campus, which can transmit real-time data through network and realize data collection, data storage, data organization, data analysis, and data publishing in real time. This paper introduces the architecture design of the platform and its software implementation, describes the key technical segments in detail, and displays some functions of the system. The establishment of the platform improved the efficiency of campus energy use, which provides technical means for upgrading the modernization level of campus energy management.

Keywords: Internet of Things · Energy saving monitoring and management platform · Energy consumption acquisition · Architecture design · Intelligent remote transmission · Wireless public network

1 Introduction

University buildings have always been a major energy consumer in society for its high level of population density. With large amount of people studying, teaching, doing scientific research and living in campus, the proportion of energy consumption on campus reaches 8% of China [1]. Therefore, it will make significant impact if we could explore the great potential of campus energy-saving. However, the traditional way of energy management in colleges and universities is extensive. In most cases, it needs to check energy consumption meter manually for data collecting and analyzing. Plenty of labor power is consumed without getting comprehensive and real-time data in this way,

© Springer Nature Switzerland AG 2019
D. Milošević et al. (Eds.): HCC 2019, LNCS 11956, pp. 413–418, 2019.
https://doi.org/10.1007/978-3-030-37429-7_41

which means managers could neither get efficient conclusion of energy consumption nor accurate data for policy decision.

With the rapid development of computer technology, the emergence of new methods and technologies provides a way for energy management in colleges and universities. Internet of Things (IoT) technology is a computer technology which takes the internet as the core and foundation to exchange and communicate information about things [2]. Energy monitoring and management in colleges and universities requires large number of sensor points, wide distribution and real-time transmission. It is very suitable to adopt IoT technology. The IoT system is established by installing data acquisition terminals in the whole campus, which saves a lot of manpower. Efficient and accurate data aggregation and analysis are achieved through the transmission of real-time data, which changes the energy management mode in colleges and universities fundamentally. It also provides a reliable decision-making basis for energy conservation in colleges and universities.

2 Overview of Energy Saving Monitoring and Management Platform of Shandong University at Weihai

The platform is composed of terminal equipment, network transmission system, data center and so on. It cost half a year to construct the platform, with a total investment of 3 million Yuan. More than 1200 terminals are laid on the platform, collecting and transmitting real-time data of electricity and water consumption in the whole campus per hour (the highest frequency is four times per hour). It can monitor the total energy consumption, energy consumption of buildings and energy consumption of some rooms. The system application interface provides a flexible data query method, which can display the queried data through hydroelectric analysis reports, charts and other forms to form trend contrast. The data center sends the data to the provincial data center regularly. The system has been running for three years. It runs steadily and reliably, and can complete data acquisition and transmission of energy consumption.

3 Architecture Design of the Platform

The platform system is designed with five layers: field equipment layer, network communication layer, core data layer, application services layer and customer terminal layer. The architecture diagram of the system is shown in Fig. 1.

Big sensing data is prevalent in both industry and scientific research applications where the data is generated with high volume and velocity [3]. Modern sensing devices play a pivotal role in achieving data acquisition, communication, and dissemination for the IoT [4]. The adjacent meters are connected in parallel by RS485 bus and communicate with the data collectors by DL/T645 protocol. The RS485 bus has the characteristics of simple structure, low cost and convenient docking with field instruments [5].

On the basis of the original campus network, the network transmission layer divides the relevant switches into virtual local area network. Network addresses are distributed

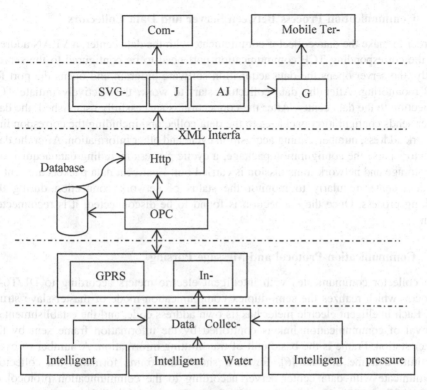

Fig. 1. Architecture diagram of energy saving monitoring and management platform system

to the data collectors to access the network. The data collectors transmit the collected data to the data acquisition server through network.

The core data layer uses Oracle database to store and manage all data. On this basis, the system carries out various analyses of data, forms various reports such as summary and statistics, and deeply excavates the problems behind the data. As the system is designed with B/S architecture, users can use computers, notebooks, mobile phones and other devices to view the system web pages at any time at the mission sites where they can access the campus network, which is extremely convenient for staff to work.

4 Implementation of Key Software Technologies of the Platform

The development of system software is accomplished on JDK 6.0 and Tomcat 8.0 platform using JAVA language. Based on J2EE three-tier architecture, the presentation logic, business logic and data logic are separated. Oracle 11g database is used in the database. The data storage types include basic information data, original value data, energy consumption data and analysis data. The following is the implementation process of key technologies of communication and data acquisition.

4.1   Communication Process Between Server and Data Collectors

In order to make the data collector communicate with the data center, a VLAN address and the corresponding TCP communication port need to be configured to the server. Firstly, the server opens the data acquisition software program and opens the port for TCP monitoring. After the data collectors start to work, they actively initiate TCP connection to the data center. After the connection is successfully established, the data center sends configuration packages to the data collectors, including the corresponding network address, number, timing acquisition cycle and other information. After the data collectors parse the configuration package, a cyclic process including data acquisition, data storage and network transmission is carried out. Heartbeat data packets are sent to the data center regularly to monitor the status of network connection during the working process. Once the connection is found to be disconnected, it is reconnected again.

4.2   Communication Protocol and Message Parsing

Data collector communicates with intelligent electric meters according to DL/T645 protocol, which realizes the semi-duplex communication mode of master-slave structure. Each intelligent electric meter has its own address code, and the establishment or removal of communication link is controlled by the information frame sent by the master station. Frame is the basic unit of transmitting information. A number of bytes constitute a frame of data [6]. Figure 2 shows the frame format. Data collectors communicate with data center server according to the communication protocol of Q/GDW 376.1-2009. Figure 3 shows the frame format of data link layer.

0x68	A0A1A2 A3A4A5	0x68	C	L	DATA	CS	0x16
Frame start character	Address field	Frame start character	Control code	Data Domain Length	Data Domain	Check code	Terminat or

Fig. 2. Format of the frame

The data collector maps and reorganizes the collected intelligent meter data, converts them into data link layer data, and transmits them to the data center server through the network. Energy usage data are stored in application layer.

0x 68	Length L	Length L	0x 68	C	A	Link user data	CS	0x16
Fixed length header				Control code	Address field	Application layer	Frame check sum	Terminator

Fig. 3. The frame format of data link layer

5   System Display

5.1   Energy Consumption Map

The platform is connected with three-dimensional GIS spatial data engine, which realizes the display of campus three-dimensional scene map. By using the 3D geographic information system, the data acquisition terminal can be positioned intuitively to form energy consumption map. Figure 4 shows the use of energy in the energy consumption map of the academy of arts.

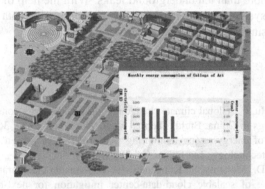

Fig. 4. Interface of energy consumption map

5.2   Data Aggregation and Analysis

For the huge data collected, the system application layer design achieves the comparison of energy consumption in different time periods of the same building, and also the comparison of energy consumption in different buildings in the same time period. It provides an intuitive comparison of horizontal and vertical data, which provides a basis for decision-making. Figure 5 shows the hourly electricity consumption of the library on May 27, 2019.

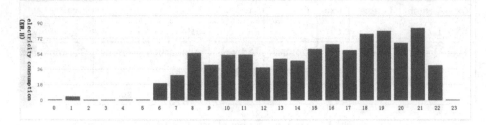

Fig. 5. Hourly electricity consumption of the library on May 27, 2019

6 Conclusion

Energy saving monitoring and management platform of Shandong University at Weihai, using the IoT technology, has realized the energy monitoring covering the whole campus. The system provides a scientific and visual energy management tool for university energy management. Through the operation of the platform system, managers can find the long-standing energy consumption crux of the campus, and targeted improvement in energy use can be carried out. Through this system, managers have found and treated more than ten underground leaks. With the help of the data collected by the system, the system can provide scientific and reasonable data support in power distribution, increasing power load and other occasions.

References

1. Tan, H., Xu, Y., Hu, C.: Global climate change response and energy conservation regulation of campus buildings in China. Build. Thermal Vent. Air Cond. **29**(1), 36–40 (2010)
2. Kang, J.: Design of energy-saving monitoring platform for tea garden irrigation based on internet of things technology. Tea Fujian (6), 482 (2018)
3. Yang, C., Puthal, D., Mohanty, S., Kougianos, E.: Big-sensing-data curation for the cloud is coming: a promise of scalable cloud-data-center mitigation for next-generation IoT and wireless sensor networks. IEEE Consum. Electron. Mag. **6**(4), 48–56 (2017)
4. Yang, C., Chen, J.: A scalable data chunk similarity based compression approach for efficient big sensing data processing on cloud. IEEE Trans. Knowl. Data Eng. (TKDE) **29**(6), 1144–1157 (2017)
5. Wang, F., Wu, M.: Design of intelligent automatic reclosing based on RS485. Power Syst. Prot. Control **38**(1), 100–106 (2011)
6. Sun, Y., Chen, R.: Wireless sensors network and the ethernet frame format transformation. In: Qian, Z., Cao, L., Su, W., Wang, T., Yang, H. (eds.) Recent Advances in CSIE 2011. LNEE, vol. 127, pp. 479–485. Springer, Heidelberg (2012). https://doi.org/10.1007/978-3-642-25769-8_68

A Deep Reinforcement Learning Approach Towards Computation Offloading for Mobile Edge Computing

Qing Wang[1], Wenan Tan[1,2(✉)], and Xiaofan Qin[1]

[1] College of Computer Science and Technology,
Nanjing University of Aeronautics and Astronautics, Nanjing 211100, China
qwangchn@foxmail.com, wtan@foxmail.com
[2] School of Computer and Information Engineering,
Shanghai Polytechnic University, Shanghai 201209, China

Abstract. In order to improve the quality of service for users and reduce the energy consumption of the cloud computing environment, Mobile Edge Computing (MEC) is a promising paradigm by providing computing resources which is close to the end device in physical distance. Nevertheless, the computation offloading policy to satisfy the requirements of the service provider and consumer at the same time within a MEC system still remains challenging. In this paper, we propose an offloading decision policy with three-level structure for MEC system different from the traditional two-level architecture to formulate the offloading decision optimization problem by minimizing the total cost of energy consumption and delay time. Because the traditional optimization methods could not solve this dynamic system problem efficiently, Reinforcement Learning (RL) has been used in complex control systems in recent years. We design a deep reinforcement learning (DRL) approach to minimize the total cost by applying deep Q-learning algorithm to address the issues of too large system state dimension. The simulation results show that the proposed algorithm has nearly optimal performance than traditional methods.

Keywords: Mobile edge computing · Deep reinforcement learning ·
Computation offloading · Deep Q-learning · Cost minimization

1 Introduction

Cloud computing is the most successful computing model after decades of development due to its scalable resource allocation and low economic cost [1]. The cloud computing architecture is a two-layer structure in which the end device and the cloud computing center (End-Cloud) are interconnected. The end devices need to offload a part of the computation tasks to the cloud computing center because of its limited computing power and low battery energy, and then the cloud computing center allocates corresponding resources (computation and bandwidth resource) to meet the user requirements. However, with the development of the Internet of Things (IoT), more and more mobile devices will access to the cloud environment, Cisco expects to have 50 billion mobile devices accessing the network by 2020 [2]. If the huge amounts of

D. Milošević et al. (Eds.): HCC 2019, LNCS 11956, pp. 419–430, 2019.
https://doi.org/10.1007/978-3-030-37429-7_42

tasks and data are offloaded to the cloud computing center for analysis and processing, it will bring high bandwidth cost and be difficult to meet delay-time sensitive service (i.e. virtual reality, auto-driving image processing). In order to satisfy service requests while decrease system energy consumption, European Telecommunications Standards Institute (ETSI) has proposed a promising paradigm, Mobile Edge Computing (MEC).

Instead of the traditional cloud computing such as Amazon Web Services and Alibaba Cloud, MEC enhances the radio access networks (RANs) with computing capability, which is in close proximity to end devices [3, 4]. It enables end devices to offload computation workloads to the MEC server rather than cloud center which improves the QoS of applications with considerably reduced latency and power consumption. However, compared with cloud computing center, the MEC server also has limited computation capability considering the economic and scalable deployment constraints. Therefore, the emerging of MEC is not to replace the cloud computing. Only by fully considering end devices, MEC server and cloud computing center, which form a collaborative network structure, can meet the enormous challenges brought by the era of Internet of Everything.

Motivated by the difference between Edge Computing and Cloud Computing, we dedicate to design a computation offloading mechanism for MEC. A dynamic computation offloading policy is formulated as a cost minimization problem, which is subject to specified maximum tolerable delay and computation capacity. Reinforcement Learning (RL) [5] has been used to solve this problem in recent years, which takes the future reward feedback from the environment into consideration, the RL agent can adjust its policy to achieve a best long-term goal. Specifically, we propose a Deep Reinforcement Learning (DRL) based theme to solve the offloading optimization problem and evaluate the performance.

The remainder of this paper is organized as follows. The related works are introduced in Sect. 2. We present the architecture of the MEC system and formulate the problem in Sect. 3. In Sect. 4, we describe the DRL solution in detail. The experiment results are presented in Sect. 5. Section 6 concludes the paper.

2 Related Work

In order to solve the problem of insufficient computing power and low battery capacity of the end device, the computation offloading has been attracting significant attention in recent years. The early research direction was to offload computing tasks to the cloud computing center, Mobile cloud computing (MCC) was widely studied, which executed mobile applications on resource providers external to the mobile device [6, 7]. The literature [8] proposed a game theoretic approach for achieving efficient computation offloading for mobile cloud computing. The literature [9] fully considered the mobility of end devices and instability of the network, proposed a offloading decision which subjects to the dependency relations among component services and aimed to optimize execution time and energy consumption of executing mobile services. The literature [10] provided an energy-efficient dynamic offloading and resource scheduling policy to reduce energy consumption and shorten application completion time. However, due to the long physical distance between the cloud computing center and the end device, the inevitable excessive energy consumption and delay time are caused. The literature [11] formulated the task offloading problem as a mixed integer non-linear program to investigate the task offloading problem in ultra-dense network, which is based on the mobile edge computing. The literature [12] researched energy-efficient computation offloading (EECO) mechanisms for MEC in 5G heterogeneous networks. The literature [13] proposed an iterative search algorithm combining interior penalty function with D.C. (the difference of two convex functions/sets) programming to find the optimal offloading policy. In addition to using traditional optimization methods to solve the offloading problem, more and more scholars use RL algorithm to solve this problem. The literature [14] formally formulated the resource allocation problem (in each decision epoch) as a convex optimization problem, and presented a novel RL-based framework for power-efficient resource allocation. The literature [15] proposed a Deep Reinforcement learning-based Online Offloading (DROO) framework that implements a deep neural network as a scalable solution that learns the binary offloading decisions from the experience.

In the existing works, strategies are focus on computation offloading to the MEC server or remote cloud, so the current research is basically a two-level offloading strategy. In summary, we propose a computation offloading policy based on the three-level architecture of "End-Edge-Cloud", which making better use of resource in different places. Meanwhile, we design different algorithm of deep reinforcement learning to solve this problem.

3   System Model and Problem Formulation

3.1   Architecture of MEC System

At present, there is no uniform standard for the architecture of edge computing, the reason is that definition of devices for the edge level and the end level is not clear [16]. This paper defines the three-level architecture of "End-Edge-Cloud" as shown in Fig. 1. In this structure, there are N independent end devices (EDs), an MEC server and a cloud computing center. The end devices and the MEC server are connected through a wireless local Area Network (LAN), meanwhile the MEC server and the cloud computing center are connected through a wired Wide Area Network (WAN). Each end device i has a computation request $R_i \triangleq (D_i, C_i, \delta_i)$, where D_i stands for the size of input data, including program codes and input parameters, while C_i stands for the total number of CPU cycles required to complete the request R_i. δ_i is the maximum delay time for completing task R_i, and it's an important limiting indicator because it affects user's QoS. This paper defines that each task cannot be split, so the request task may be processed locally at the end device, at the MEC server, or at the cloud computing center by offloading.

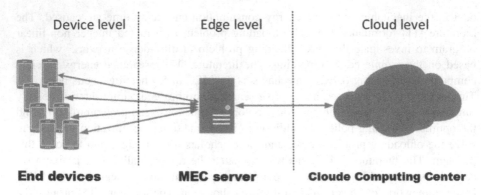

Fig. 1. The architecture of the MEC system

3.2   Cost of Local Processing

When the user task R_i chooses not to perform the offloading operation, the task is executed directly on the end device. T_{l_i} indicates the total delay time, since there is no transmission delay in the processing performed by the end device, the total delay time includes only the processing time of the end device's CPU. Using p_{l_i} to represent the CPU processing capability of the ED i, so the delay time is:

$$T_{l_i} = \frac{D_i}{p_{l_i}} \tag{1}$$

E_{l_i} represents the current energy consumption of task R_i:

$$E_{l_i} = \kappa(p_{l_i})^2 D_i \tag{2}$$

where κ is the effective switched capacitance depending on the chip architecture [17]. We set $\kappa = 10^{-26}$ according to the practical measurement in [18]. Considering the delay time (1) and the energy consumption (2), the total cost C_{l_i} performed by the task R_i at the end device is expressed as:

$$C_{l_i} = \omega T_{l_i} + (1 - \omega)E_{l_i} \tag{3}$$

where $\omega(0 \leq \omega \leq 1)$ represents the weight of delay time and energy consumption. When $\omega = 0$, it means only considering energy consumption. When $\omega = 1$, it only considers the delay time, so it can meet the demand focus of different task type.

3.3   Cost of MEC Processing

When the user task R_i decides to offload to the edge level, the task R_i uploads the execution code and data D_i to the MEC server. In this mode, the delay time and energy consumption need to be considered in the three stages which is uploading, processing, and downloading. T_{o_i} indicates the total delay time of the task R_i offloading to the MEC server, which is composed of the upload time T_{u_i}, the MEC server processing time T_{p_i}, and the download time T_{d_i}. The upload time T_{u_i} is determined by the data amount D_i, and the wireless transmission rate λ_{up} between the end device and the MEC server:

$$T_{u_i} = \frac{D_i}{\lambda_{up}} \qquad (4)$$

In the processing task phase of the MEC server, We define P_e as the total CPU processing capability of the MEC server, and define p_{e_i} as the assigned CPU-cycle frequency to complete user task R_i, then the delay time of the processing phase:

$$T_{p_e^i} = \frac{C_i}{p_{e_i}} \qquad (5)$$

In the download task processing result stage, according to the research results in [19], the data download rate is much higher than the upload rate, and the amount of downloaded data is much smaller than the original data amount to be processed, so the delay time and energy consumption in the download phase can be ignored. According to (4) and (5), the total delay time T_{e_i} of the offloading to MEC server processing mode is:

$$T_{e_i} = T_{u_i} + T_{p_e^i} \qquad (6)$$

The corresponding energy consumption is similar to the delay time. It is described as in three stages: the energy consumption E_{u_i} generated during the upload phase and the energy consumption E_{p_i} generated by the MEC server, the energy consumption in the download phase is ignored. In the data upload phase, the energy consumption per unit time is defined as e_{u_i}, then the total energy consumption E_{u_i} at this stage is:

$$E_{u_i} = e_{u_i} T_{u_i} \qquad (7)$$

In the processing phase, the energy consumption per unit time of the MEC server is defined as e_{p_i}, then the energy consumption E_{p_i} at this stage is:

$$E_{p_e^i} = e_{p_i} T_{p_i} \qquad (8)$$

The total energy consumption E_{e_i} of the offloading to MEC server processing mode is:

$$E_{e_i} = E_{u_i} + E_{p_i} \qquad (9)$$

According to the above formulas (6) and (9), the total cost C_{e_i} executed by the user task R_i on the MEC server is expressed as:

$$C_{e_i} = \omega T_{e_i} + (1 - \omega) E_{e_i} \qquad (10)$$

3.4 Cost of Cloud Processing

The cloud computing center has almost unlimited computing resources compared to the MEC server. Therefore, this paper considers that the user task can further offload to the cloud computing center based on the offloading to the MEC server. The delay time at this time needs to consider the time T_{u_i} of user data uploaded to the MEC server, the transmission time T_{t_i} of the MEC server and the cloud computing center, and the time $T_{p_c^i}$ of the cloud computing center processing task, the download time is ignored. The delay time uploaded to the MEC server is obtained by the formula (4). During the transmission phase of the MEC server and the cloud computing center, the transmission time is determined by the data amount D_i and the MEC server and the WAN transmission rate λ_t of the cloud computing center:

$$T_{t_i} = \frac{D_i}{\lambda_t} \tag{11}$$

In the processing task phase of the cloud computing center, the delay time is determined by the computing resource p_{c_i} assigned to the task by the cloud computing center and the total computing resource C_i required for the task:

$$T_{p_c^i} = \frac{C_i}{p_{c_i}} \tag{12}$$

According to formulas (3), (11), (12), the total delay time T_{c_i} for the task to be offloaded to the cloud computing center is:

$$T_{c_i} = T_{u_i} + T_{t_i} + T_{p_c^i} \tag{13}$$

In the data transmission phase, the energy consumption per unit time of this phase is defined as e_{t_i}, then the energy consumption E_{t_i} of this phase is:

$$E_{t_i} = e_{t_i} T_{t_i} + E_{u_i} \tag{14}$$

In the processing phase of the cloud computing center, the energy consumption per unit time of this phase is defined as $e_{p_c^i}$, then the energy consumption $E_{p_c^i}$ of this phase is:

$$E_{p_c^i} = e_{p_c^i} T_{p_c^i} \tag{15}$$

Then the total energy consumption E_{c_i} of the offloading to cloud computing center processing is:

$$E_{c_i} = E_{t_i} + E_{p_c^i} \tag{16}$$

According to the above formulas (13) and (16), the total cost C_{c_i} executed by the user task R_i on the MEC server is expressed as:

$$C_{c_i} = \omega T_{c_i} + (1 - \omega)E_{c_i} \tag{17}$$

3.5   Problem Formulation

The user first decides whether its task R_i should be processed at the end device locally or offloaded to the MEC server, using α_i to represent the offloading decision:

$$\alpha_i = \begin{cases} 0, & \text{task } R_i \text{ is processed at the end device,} \\ 1, & \text{task } R_i \text{ is offloaded to the MEC server.} \end{cases}$$

When user task R_i is offloaded to the MEC server, system need to determine whether the task should be processed by the MEC server or further offloaded to the cloud computing center, using β_i to represent the offloading decision:

$$\beta_i = \begin{cases} 0, & \text{task } R_i \text{ is processed at the MEC server,} \\ 1, & \text{task } R_i \text{ is offloaded to the Cloud.} \end{cases}$$

At any moment, the MEC system total cost is the weighted sum of the delay time and energy consumption of all user tasks, and the total cost is minimized by optimizing the offloading decision variables α_i and β_i. Regardless of where the user task is executed, the system must be serviced for less than the maximum delay time, and the MEC server's computing power must be considered when offloading to the edge layer. The computing resources allocated to the current time task cannot exceed its total computing power, then optimization problem is formulated as follows:

$$\min_{\alpha_i, \beta_i} \sum_{i=0}^{N} (1 - \alpha_i)C_{l_i} + \alpha_i(1 - \beta_i)C_{e_i} + (\alpha_i\beta_i)C_{c_i} \tag{18}$$

$$s.t. \quad (1 - \alpha_i)T_{l_i} + \alpha_i(1 - \beta_i)T_{e_i} + (\alpha_i\beta_i)T_{c_i} \le \delta_i$$

$$\sum_{i=0}^{N} p_{e_i} \le P_e$$

$$\alpha_i, \beta_i \in \{0, 1\}$$

Formula (18) describes the minimum goal of this paper, taking into account the energy consumption and delay time of different offloading locations. At the same time, the problem is subject to the maximum tolerance time of the user task and the computing power of the MEC server. At last, the values of α_i and β_i determine the action strategy of the system.

4   Problem Solution

4.1   Reinforcement Learning

Reinforcement learning is a kind of algorithm in machine learning. This kind of algorithm simulates the human-defined agent learning strategy from the specific environment. The basic composition is agent and environment, basic element is the state s_t at each moment, the selected action a_t, and the instant reward r_t at any moment. The agent selects the most appropriate action a_t from the action set in the current state s_t, then the environment jumps to the next state s_{t+1} and obtains the instant reward r_t. Therefore, the agent adjusts the strategy of selecting actions according to the instant reward to minimize the cost of the entire system. In order to get the optimal offloading decision and resource allocation, we need to define the system state, action and reward in detail, which are described below:

System State. The state needs to accurately describe the environment at the moment, and the current request volume of the user task and the remaining system resources are the focus of our attention. We define the state as $s_t = \{C_t, L_t\}$, and C_t is the total cost of the system described previously. Compared with edge server, cloud computing does not need to consider whether the computation resources are sufficient or not, so L_t is the remaining computing resources of edge server.

Action Space. Each end equipment offloading decisions can be expressed as (α_i, β_i), then the system of optional action expressed as: $a_t = \{(\alpha_1, \beta_1), (\alpha_2, \beta_2), \ldots, (\alpha_i, \beta_i)\}$.

Reward Function. The immediate reward of the system represents the direction we want to optimize the target. The reward can guide the agent to choose the action that is more and more suitable for the current environment. Therefore, we define the immediate reward as $\frac{Cost_{local} - Cost_{curr}}{Cost_{local}}$, which describe the difference between the current cost and local computing cost.

4.2   Deep Q Network

Given the basic elements of learning: state, action, reward, and it wants to find optimal strategy, which generally referred to as model-based reinforcement learning. Q-learning [20, 21] is an important method in model-based reinforcement learning problems that predicts the expected value (Q-value) of all actions in a certain state. The agent generates a new status and instant reward by selecting the action with the highest Q-value, and the current instant reward and the desired reward that is subsequently available to update the Q-value. This step continues until the value function converges. The goal of Q-learning is to solve the function $Q(s_t, a_t)$, which is to estimate the expected value of the action based on the current system state, expressed as:

$$Q_{new}(s_t, a_t) \leftarrow (1 - \alpha) \cdot Q_{old}(s_t, a_t) + \alpha \cdot (r_{t+1} + \beta \cdot max_a Q(s_{t+1}, a))$$

s_t is the current state, a is the action executed in the current state, r_{t+1} is the reward after executing the action, and s_{t+1} is the other state. $r_{t+1} + \beta \cdot max_a Q(s_{t+1}, a)$ is the reward obtained by the current action plus the maximum expected value that can be obtained in the next state.

Q-Learning is very flexible and effective for small reinforcement learning problems. It needs to maintain a state-action Q-value table in memory, and agent can look up the table directly when selecting actions. However, in this paper, the possible state may be very large, can even lead the Q-value table far beyond the memory, which limits the using scenario of Q-Learning in this problem. After the rise of deep learning, reinforcement learning based on deep learning begins to dominate, so that a Deep Neural Network to estimate $Q(s_t, a_t)$ instead of computing Q value for every state-action pair, this is the original idea of Deep Q Network (DQN). Algorithm 1 shows the detail of DQN algorithm.

Algorithm 1: A DQN algorithm to computation offloading problem.

Begin
Randomly initialize all parameters w of the Q network.
Initialize replay memory $\mathcal{R}$.
Repeat
Randomly generate an initial state s, use feature vector $\varphi(s)$ an input in the Q network
Generate a random probability p
if $p < \varepsilon$ **then**
    select $a = max_a Q^*(\varphi(s), a; w)$
else
    select a randomly
Execute action a, get new stat s', feature vector $\varphi(s')$, reward r and end state flag is_end
Store tuple $(\varphi(s), a, r, \varphi(s'), is\_end)$ in replay memory $\mathcal{R}$
Set $s = s'$
Take out the $m(j = 0, 1, ..., m)$ sample from $\mathcal{R}$, calculate the current Q-value y_j:

$$y_j = \begin{cases} r_j, & is\_end_j \ is \ true, \\ r_j + \eta max_{a'} Q^*(\varphi(s'_j), a'_j; w), & is\_end_j \ is \ false. \end{cases}$$

Use $\frac{1}{m}\sum_{j=0}^m (y_j - Q^*(\varphi(s), a; \theta))^2$ as loss function, update parameters w of Q network through back propagation
until s is the end state
End

---

5 Experiment and Results Analysis

The parameters of the end device are derived from the literature [18]. We set the number of user tasks in this MEC system to be $\mathcal{N} = 10$, the CPU frequency of any end device is 0.5 GHz/sec and the processing energy consumption is 7.3×10^{-8} J/cycle. We define upload rate is 72.2 Mbit/s (e.g., IEEE 802.11n) between the end user and the MEC server, and the transmission energy consumptions of the mobile user is 1.42×10^{-7} J/bit according to the literature [18]. We assume that the input data size D_i of the task R_i is uniformly distributed from 10 to 30 MB, the number of CPU cycles C_i (in Mega-cycles) obeys uniform distribution between (900, 1100). The CPU rates of the MEC server and the Cloud center are 5 GHz/sec and 10 GHz/sec, respectively. When tasks are offloaded to the cloud, the transmission rate λ_t is 15 Mbit/s. This article treats the delay time and energy consumption equally, so the weight parameter is set to $\omega = 0.5$.

For comparison, we consider review the following methods: (1) the local processing only method where all tasks are processed by the end device, (2) the cloud processing only method where all tasks are offloaded to the cloud, (3) the Local-MEC (LM) offloading method where the two-layer offloading structure considered by most MEC system, and (4) the random mapping method where all α_i and β_i are chosen with equal probability.

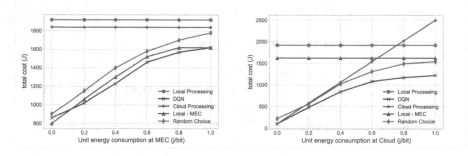

Fig. 2. The total cost vs unit energy consumption

In Fig. 2, it shows the change of total cost of the system when the unit energy consumption increases. As can be seen from left of Fig. 2, local processing and cloud processing do not change because these algorithms do not interact with the MEC server. It is also proved that the DQN three-layer offloading algorithm proposed in this paper is superior to the Local-MEC two-layer offloading algorithm. In the right of Fig. 2, when the unit energy consumption changes, cloud processing presents a linear growth trend, which is exactly the problem of excessive energy consumption when large data volume is faced with in cloud computing. Meanwhile, when the energy consumption of cloud computing center increases, user tasks are more willing to perform tasks in MEC server, so the algorithm in this paper can get a lower system cost.

6 Conclusion

In this paper, we design an integrated framework for multi user tasks computation offloading and resource allocation with mobile edge computing. We consider a MEC system consisting of multiple end devices, a MEC server, a Cloud computing center and formulate the computation offloading decision problem. Then, minimizing a weighted total cost of energy computation and delay through optimal tasks offloading that is our objective. We use the DRL-based solutions to solve the question, which considering the long-term benefits of the system. Simulation results suggest that the proposed algorithm gives nearly optimal performance than some baseline solutions. Future work is in progress to consider a more complex framework which takes interference and radio resource into consideration.

Acknowledgments. The paper is supported in part by the National Natural Science Foundation of China under Grant No. 61672022, and Key Disciplines of Computer Science and Technology of Shanghai Polytechnic University under Grant No. XXKZD1604.

References

1. Varghese, B., Buyya, R.: Next generation cloud computing: new trends and research directions. Future Gener. Comput. Syst. **79**, 849–861 (2018)
2. Dave, E.: How the next evolution of the Internet is changing everything. The Internet of Things (2011)
3. Abbas, N., Zhang, Y., Taherkordi, A., Skeie, T.: Mobile edge computing: a survey. IEEE Internet Things J. **5**(1), 450–465 (2017)
4. Shi, W., Cao, J., Zhang, Q., Li, Y., Xu, L.: Edge computing: vision and challenges. IEEE Internet Things J. **3**(5), 637–646 (2016)
5. Kiumarsi, B., Vamvoudakis, K.G., Modares, H., Lewis, F.L.: Optimal and autonomous control using reinforcement learning: a survey. IEEE Trans. Neural Netw. Learn. Syst. **29**(6), 2042–2062 (2018)
6. Guan, L., Ke, X., Song, M., Song, J.: A survey of research on mobile cloud computing. In: 2011 10th IEEE/ACIS International Conference on Computer and Information Science, pp. 387–392. IEEE (May 2011)
7. Fernando, N., Loke, S.W., Rahayu, W.: Mobile cloud computing: a survey. Future Gener. Comput. Syst. **29**(1), 84–106 (2013)
8. Chen, X.: Decentralized computation offloading game for mobile cloud computing. IEEE Trans. Parallel Distrib. Syst. **26**(4), 974–983 (2014)
9. Deng, S., Huang, L., Taheri, J., Zomaya, A.Y.: Computation offloading for service workflow in mobile cloud computing. IEEE Trans. Parallel Distrib. Syst. **26**(12), 3317–3329 (2014)
10. Guo, S., Xiao, B., Yang, Y., Yang, Y.: Energy-efficient dynamic offloading and resource scheduling in mobile cloud computing. In IEEE INFOCOM 2016-The 35th Annual IEEE International Conference on Computer Communications, pp. 1–9. IEEE (April 2016)
11. Chen, M., Hao, Y.: Task offloading for mobile edge computing in software defined ultra-dense network. IEEE J. Sel. Areas Commun. **36**(3), 587–597 (2018)
12. Zhang, K., et al.: Energy-efficient offloading for mobile edge computing in 5G heterogeneous networks. IEEE Access **4**, 5896–5907 (2016)

13. Zhang, J., et al.: Energy-latency tradeoff for energy-aware offloading in mobile edge computing networks. IEEE Internet Things J. **5**(4), 2633–2645 (2018)
14. Xu, Z., Wang, Y., Tang, J., et al.: A deep reinforcement learning based framework for power-efficient resource allocation in cloud RANs. In: 2017 IEEE International Conference on Communications (ICC), pp. 1–6 (2017)
15. Huang, L., Bi, S., Zhang, Y.J.A.: Deep reinforcement learning for online offload in wireless powered mobile-edge computing networks. arXiv preprint. arXiv:1808.01977 (2018)
16. Mach, P., Becvar, Z.: Mobile edge computing: a survey on architecture and computation offloading. IEEE Commun. Surv. Tutor. **19**(3), 1628–1656 (2017)
17. Zhang, W., Wen, Y., Guan, K., Kilper, D., Luo, H., Wu, D.O.: Energy-optimal mobile cloud computing under stochastic wireless channel. IEEE Trans. Wirel. Commun. **12**(9), 4569–4581 (2013)
18. Miettinen, A.P., Nurminen, J.K.: Energy efficiency of mobile clients in cloud computing. HotCloud **10**, 4 (2010)
19. Chen, X., Jiao, L., Li, W., Fu, X.: Efficient multi-user computation offloading for mobile-edge cloud computing. IEEE/ACM Trans. Netw. **24**(5), 2795–2808 (2015)
20. Watkins, C.J., Dayan, P.: Q-learning. Mach. Learn. **8**(3–4), 279–292 (1992)
21. Van Hasselt, H., Guez, A., Silver, D.: Deep reinforcement learning with double Q-learning. In: Thirtieth AAAI Conference on Artificial Intelligence (March 2016)

Joint Neural Collaborative Filtering with Basic Side Information

Shuo Wang$^{(\boxtimes)}$, Hui Tian$^{(\boxtimes)}$, Shaoshuai Fan$^{(\boxtimes)}$, Boyang Hu$^{(\boxtimes)}$, and Baoling Liu$^{(\boxtimes)}$

State Key Laboratory of Networking and Switching Technology,
Beijing University of Posts and Telecommunications, Beijing 100876, China
{wangshuo16, tianhui, fanss,
goodmark37, blliu}@bupt.edu.cn

Abstract. Nowadays, deep neural network has been greatly developed and widely used in many areas. However, the research of the deep neural network on recommendation system is inadequate. Most research focuses on analyzing the textual descriptions of items and comments of users, making use of the neural network to get feature vectors from texts or pictures. In this paper, we directly adopt the deep neural network to better fit the non-linear relationship of users and items and effectively integrate some side information (basic information and statistical information) into the neural network. Utilizing deep neural network, we explore the impact of some basic information on neural collaborative filtering. To the best of our knowledge, it is the first time to combine the basic information, statistical information and rating matrix by the deep neural network. Finally, we use the benchmark data set (MovieLens) to demonstrate the effectiveness of the proposed deep neural network model with side information.

Keywords: The deep neural network · Recommendation system · Matrix factorization · Neural collaborative filtering · Side information

1 Introduction

In recent years, Information overload has become a serious obstacle in many online services. Recommendation systems provide a promising way to alleviate the dilemma. There are many prominent approaches employed in recommender systems. Many of them are based on Collaborative Filtering (CF) techniques [1]. The CF recommendation algorithm models users' tastes on items, which could recommend a list of items to users based on the interaction of users and items. CF is widely used in many fields. One application of CF, the Matrix Factorization (MF), becomes the most popular CF technique with the rise of the Netflix competition in 2007 [13]. MF maps users and items to low-dimensional feature vectors in the latent space to exploit the interaction between users and items [5, 8] by rating matrix. It combines the latent vectors linearly by the inner product operation and the rating matrix is usually sparse matrix, which lead to a weak exploitation of the complex relationship between users and items. To overcome these shortcomings, deep neural network is applied in the recommendation.

© Springer Nature Switzerland AG 2019
D. Milošević et al. (Eds.): HCC 2019, LNCS 11956, pp. 431–442, 2019.
https://doi.org/10.1007/978-3-030-37429-7_43

There are many neural network techniques including Autoencoder [9], Restricted Boltzmann Machine [16], Multilayer Perceptron and Neural Collaborative Filtering [4] etc. Besides, some researchers use the deep neural network to model item properties and user behaviors from review text [19]. It can address this lack of data to use the information in review text [10]. They use natural language processing methods including convolutional neural networks [6], recurrent neural networks [12] and so on to get meaningful word embedding vectors. In addition, word vectors are added to rating matrix to improve the performance of recommender systems [11]. These approaches need get the word vectors first. Then integrate them into the recommendation model. This information and the ratings of users are separated in the recommendation systems [18]. Besides, it is also hard to obtain this information sometimes.

In this paper, we mainly consider some basic information, like users' age, sex, occupation and items' genres. Further, consider some deep statistical information of users and items, like click-through rate of specific users, proportion of specific users, proportion of specific items and so on. We directly integrate the side information and the click-behaviors of users in the deep neural network. Through combining the two parts, we jointly build the recommendation model to fully characterize the interaction between users and items.

Our contributions are summarized as follows: (1) We propose a model which integrates this side information into the neural collaborative filtering; (2) It is the first time that an interaction network which directly employs the basic information and statistical information to get better performance is constructed; (3) We perform extensive experiments on real-world data sets to demonstrate the effectiveness of our model.

2   Preliminary

In this section, we primarily state the recommendation problem, and discuss the implicit feedback and the key side information.

2.1   Implicit Data

In the real world, people do not like to give a rate sometimes, which is inconvenient and unnecessary. Therefore, implicit feedback becomes more important for people to exploit. We can get the implicit data in different recommendation systems, such as in the interaction where users buy items online, browse the web pages and read news, etc.

We get the data sets that include some interactions between users and items. In the data sets, we use U and I to denote users and items respectively, and the user $u \in U$ and item $i \in I$. We define the user-item interaction matrix Y [4] as:

$$y_{u,i} = \begin{cases} 1 & \text{if the user } u \text{ has interaction with the item } i \\ 0 & \text{otherwise} \end{cases} \tag{1}$$

If the value of $y_{u,i}$ equals 1, this only indicates that the user u has interaction with the item i. However, it does not mean that the user u actually likes item i. Similarly if $y_{u,i}$ equals 0, it does not mean that the user u does not like the item i. The interaction is the noisy signal, but we can get users' interest by the overall interactive information. Usually, the items which have an interaction (the label $y_{u,i}$ is 1) with the user should be ranked in a higher position in the recommendation list than the items that do not have an interaction (the label $y_{u,i}$ is 0) with the user.

2.2   Side Information

In the recommendation problem, we add side information to the recommendation system to improve the performance. Define the side information as T. T_u denotes users' side information and T_i denotes items' side information. We call a user-item-side information triplet u, i, t an interaction triplet. For the convenience of data processing, we decompose the triplet to two tuples, u, t and i, t. They denote the corresponding user behavior and item behavior respectively. Meanwhile u, i denote the rating behavior of user u on item i. The three kinds of relationships are showed in Fig. 1.

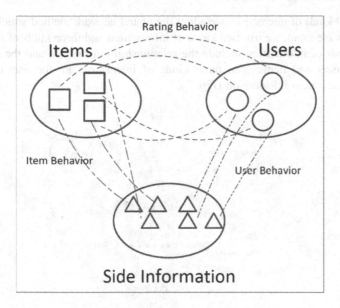

Fig. 1. Network relationship

The rating behavior is comprehensible. It is denoted by $y_{u,i}$ which indicates whether user u has interaction with the item i. The user behavior is some basic information like age, sex and occupation and the item behavior is the genres. Besides, we get some deep information from the interaction between users and items to better exploit the internal connections between users and items.

The side information plays an important role in the recommendation system. It can increase the prediction accuracy, reduce the training time and increase the robustness of the model [15]. Furthermore, the side information can ensure a better initialization of the recommendation system to alleviate the cold start problem. Thus we utilize all this relationship information to jointly improve the performance of recommendation system with deep neural network.

Based on the description of the two sections above, we aim to get the score $y_{u,i}$ of non-interactive information in the recommendation problem. We use the recommendation algorithm with all relationship information to get the prediction scores, and then rank them to recommend the high score items to users.

3   Proposed Methods

In this section, we propose a novel deep model with side information. We present the global framework picture of our model. Then we elaborate the learn process with the binary property implicit data in detail.

3.1   General Framework Picture

To model all kinds of interaction, we adopt the neural network method which is shown in Fig. 1. As the input, we use two kinds of information and three kinds of interaction. The two kinds of information include the relationship information and the side information of users and items. The three kinds of interaction are the user-item, user-information and item-information (Fig. 2).

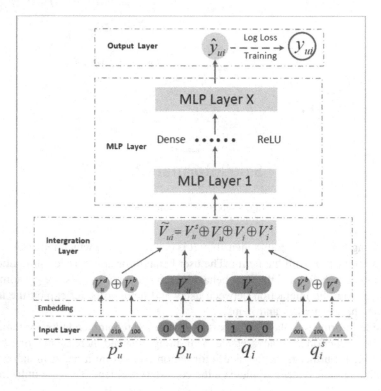

Fig. 2. HNCF framework

In the process of integration, we encode the user-item pair, transforming the user u and the item i to a binarized sparse vector p_u and q_i respectively with the one-hot encoding. p_u equals 1 only in the u-th entry while in the other entries, p_u equals 0. The q_i is similar. Then, we add the side information p_u^s and q_i^s as the input. It includes the basic information and deep statistical information. We define the basic information of user and item as p_u^b and q_i^b respectively. The q_u^b includes the users' age, sex and occupation and q_i^b includes items' genres. Because the basic information is discretized, we also encode it by the one-hot. Through the embedding layer, the one-hot encodings are transformed to dense vectors V_u, V_i, V_u^b and V_i^b. Besides we add some deep statistical information, like click-through rate of specific users, proportion of specific users, proportion of specific items and so on. We define them as V_u^d and V_i^d respectively. They are continuous values. Therefore, we can directly input them into the neural network without passing through the embedding layer. We take the V_u^d for example. We define C_u as the click-through rate of user u, P_{sex} as the click proportion that users' sex is female or male and R_{sex} as the average rating that users' sex is female or male. They are given as follows

$$C_u = \frac{Number\ of\ items\ clicked\ by\ user\ u}{Total\ number\ of\ items}. \tag{2}$$

$$P_{male} = \frac{Number\ of\ items\ clicked\ by\ users\ whose\ sex\ is\ the\ male}{Total\ number\ of\ items\ click\ by\ users}. \tag{3}$$

$$R_{male} = \frac{Sum\ of\ users'\ rate\ whose\ sex\ is\ the\ male}{Number\ of\ users\ whose\ sex\ is\ the\ male}. \tag{4}$$

These two parts make up the side information V_u^s and V_i^s. Then we combine these vectors into a unified representation as:

$$\tilde{V}_{u,i} = V_u^s \oplus V_u \oplus V_i \oplus V_i^s \tag{5}$$

where "$\oplus$" denotes the vectors concatenation operation, V_u^s and V_i^s denote the vectors of the side information of the user and the item respectively, and V_u and V_i denote the vector of the user and the item respectively.

These side information is not separated from the representations of users and items. We add the side information which is unified with the vectors of users and items in the framework. In the neural network framework, the relationship of users and items can be fully captured.

3.2 Deep Model with Side Information

The relationship between users and items is complex. It is difficult to clearly describe the inner relationship. Previous studies mainly focus on the linear relationship of users and items like matrix factorization. Besides, researchers mainly make use of deep neural network to study some textual descriptions to better exploit the relationship between users and items. In this paper, we don't use deep neural network to analyze textual feature like comments of users, descriptions of items and so on. We mainly focus on some basic information and the interaction of users and items. We use deep neural network to integrate them together. The model has a higher level of flexibility and nonlinearity which can better fit the complex relationship to enhance the performance of recommendation.

We combine this information by concatenating them, which is very intuitive in the deep learning network. Generally, the deep neural network is the forward structure neural network, and it is one layer after layer. For our method, the model is defined as

$$Z_1 = \phi_1\left(p_u^s, p_u, q_i, q_i^s\right) = \left[p_u^s, p_u, q_i, q_i^s\right], \tag{6}$$

$$\phi_2(Z_1) = \alpha_2\left(W_2^T Z_1 + b_2\right),$$

$$\ldots\ldots$$

$$\phi_L(Z_L - 1) = \alpha_L\left(W_L^T Z_L - 1 + b_L\right),$$

$$\tilde{y}_{u,i} = \sigma\left(H^T \phi_L(Z_L - 1)\right)$$

where W_j, b_j and a_j denote the weight matrix, bias vector, and the activation function for the j-th layer's perceptron respectively. For the activation function, we choose the ReLU, which can effectively prevent gradient disappearance and reduce over-fitting. Besides, it is more suitable for the sparse data. For the output layer, we endow the $\tilde{y}_{u,i}$ with a probabilistic explanation. By constraining the output value in the range of [0, 1], we adopt the sigmoid function which is defined as

$$\sigma(x) = \frac{1}{1 + \exp(-x)} \tag{7}$$

whose result represents the conditional probability of the label being 1. These parameters could be obtained by the back-propagation method.

For the implicit data, the label $\tilde{y}_{u,i}$ is a binarized 0 or 1. We assume this recommendation problem as a binary classification problem, whose rationality is demonstrated in [4]. We adopt the binary cross-entropy loss (log loss) as the objective function

$$L = -\sum_{(u,i)\in t\cup t^-} y_{u,i} log\tilde{y}_{u,i} + \left(1 - y_{u,i}\right)\log\left(1 - \tilde{y}_{u,i}\right) \tag{8}$$

where t and t^- represent the positive set and negative set respectively. Due to the implicit data, we cannot get the explicit negative cases as training set. We randomly choose the negative cases from the unobserved interactions. Following [4], we sample four items which do not have interaction with the user as the negative cases. We minimize objective function with the stochastic gradient descent.

In this paper, we propose two kinds of methods. First we only add the basic information to the neural network. We call it the basic information deep neural network (BNCF). We add the users' age, sex, occupation and items' genres and directly encode it by the one-hot, then pass through the embedding layer. The initialization of embedding is random. In the second method, we add the deep statistic information which is the continuous value without the one-hot. So it doesn't pass through the embedding layer. We use the embedding vectors of users, items and basic information which are train by the BNCF. We call it the side information deep neural network (SNCF).

4 Experiments

4.1 Database

In our experiment, we use the public data set-1 M MovieLens provided by GroupLens [2], which has been widely used to evaluate recommendation methods. It contains 10^6 ratings from 6040 users on 3706 movies and it's sparsity is 95.53%. Each user has at least 20 ratings on scale of 1 to 5 which are the explicit ratings. We can transform them into implicit data. If the user rates the item, we mark the rating as 1, indicating that user has interaction with the item. If not, we mark the rating as 0. We split data into two parts. The latest interaction of each user is chosen as the test set and the remaining data is chosen as the training set. We use the leave-one-out method which has been widely used in other research [4, 15].

4.2 Baselines

We compare our proposed method SNCF and BNCF with the following methods:

ItemKNN [17]. It is the classic recommendation algorithm which is an item-based collaborative filtering method.

BPR [14]. It optimizes the classic MF model with a pairwise ranking loss. It is a highly competitive baseline for item recommendation.

NCF [4]. It is neural collaborative filtering which use the deep neural networks to exploit the key factor in collaborative filtering the interaction between user and item features. It is a state-of-the-art method. Our methods are mainly based on it.

For those baseline methods' setting, we followed [4].

Our proposed methods aim to model the relationship between users and items. We only consider how to make better use of the basic information to better exploit the relationship between users and items. We don't consider the textual reviews information, picture information and so on.

4.3 Evaluation Metrics

In this paper, we adopted the leave-one-out evaluation [7] and adopt the Hit Ratio (HR) and Normalized Discounted Cumulative Gain (NDCG) [3] to evaluate the performance which have been widely used in the top-n recommendation. For each user, we randomly choose 100 items which do not have interaction with the active user and rank them with the test item of to get a ranking list. Then we can calculate these evaluation metrics by the ranking list.

Hit Ratio (HR): HR measures the ratio of the number of the test set contained in the top-n recommendation list for each user$_{u,i}$. If a test item appears in the top-n recommendation list, it is considered as a hit. It can be given by

$$HR@K = \frac{Number\ of\ hits@k}{Number\ of\ tests} \tag{9}$$

The HR only evenly measures whether the test item appears in the recommendation list. It does not reflect the order of the list. So we introduce the Normalized Discounted Cumulative Gain (NDCG).

Normalized Discounted Cumulative Gain (NDCG): NDCG can be gotten by the order relationship which is defined as:

$$NDCG@K = Z_K \sum\nolimits_{i=1}^{K} \frac{2^{r_i} - 1}{log_2(i+1)} \tag{10}$$

where Z_K is a normalizer. For the r_i in the formula, if the item is in the test set, the r_i equals 1, otherwise it equals 0.

For the both metrics, we truncated the recommendation list at 10. That is to say the K equals 10. We use the average score of each user to get the final results of HR@k and NDCG@K.

4.4   Parameter Settings

Our experiment is based on the deep learning framework keras in python and use the theano as the backend. Keras is easy and convenient to build a deep learning model. It is very popular in the simulation of research work.

We adopt the cross-validation to choose the parameters of the model optimize the objective function. In the initialization of model, we randomly initialize the model with standard gaussian distribution. We use the mini-batch Adam to optimize the model parameters [7]. The first and second momentum coefficients of Adam are the default values which are suggested in the paper [7]. The first momentum is 0.9 and the second momentum is 0.999. We test the learning rate of [0.0001, 0.0005, 0.001, 0.005] to get the optimal value. We choose the 0.001 as the learning r atc. We also test the batch size of [128, 256, 512, 1024] and the batch size is set as 256. For the architecture of model, we adopt the tower type. The last hidden layer determines the model capability. We term it as predictive factors and test the last hidden layer size of [8, 16, 32, 64].

In the embedding layer, we use concatenation operation. So the embedding size of the user and item are the half of whole embedding layer size in the baseline method NCF. In our method, we add the basic information into the embedding layer. For the basic information. We can only get the sex, age, occupation of users and the genres of items in the 1M MovieLens. We transform them into the number with the one-hot encoding. In order to keep the number of neuron consistent in the embedding layer, so we set the user and item size as the one-quarter of embedding layer size, and set the size of the basic information as embedding layer size/16. The statistical information is directly added to the embedding.

4.5   Simulation and Discussion

In the simulation, we compare the baseline methods with our proposed methods. The Fig. 3 shows the performance HR@10 and NDCG@10 with respect to the number of predictive factors.

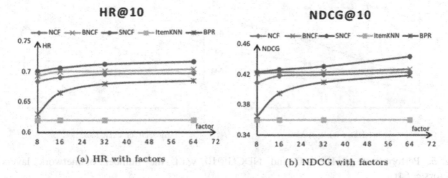

Fig. 3. Performance of HR@10 and NDCG@10 w.r.t. the number of predictive factors (layers = 4)

We can see that the SNCF achieves the best performance and significantly out-performing the classic methods ItemKNN and BPR, and the state-of-the-art NCF. Besides the BNCF that only adds the basic information also show quite strong per-formance. For the NDCG, The SNCF is more better than the other method in big predictive factors. It shows the stronger fitting ability.

The Fig. 4 shows the change of train loss, HR@10 and NDCG@10 with the number of iteration on MovieLens.

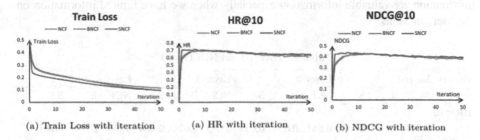

Fig. 4. Training loss and performance metric of our methods w.r.t. the number of iterations (factors = 64)

We change the number of hidden layers to investigate the influence. Comparing the NCF with BNCF and SNCF, we can get the influence of side information on the recommendation result. The simulation is as follow:

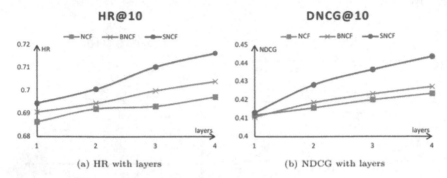

(a) HR with layers   (b) NDCG with layers

Fig. 5. Performance of HR@10 and NDCG@10 w.r.t. the number of network layers (factors = 64)

From the Fig. 5, It can also be seen that as the number of layers increases, the recommendation performance improves. It verified the effectiveness of deep neural network. It can improve the recommendation performance by increasing the number of hidden layers. Besides, compared with the baseline method, our methods are better. It shows that some basic information is helpful to exploit the relationship of users and items. It can alleviate the sparsity problem in a way. It make the relationship of users and items clearer. In the Fig. 5, The SNCF is the best, no matter the number of layers. It shows that the deep statistic information is very important. The relationship information is hardly captured by the deep neural network. Through some statistics, we can exploit deeper relationship. The result shows that the basic information and statistic information are valuable information especially when we have limited information on the users or items.

Table 1. HR@10 and NDCG@10

Factors	Layer-1			Layer-2			Layer-3			Layer-4		
	NCF	B	S	NCF	B	S	NCF	B	S	NCF	B	S
HR@10												
8	**0.616**	0.577	0.581	**0.661**	0.634	0.639	0.675	0.676	**0.684**	0.684	0.693	**0.701**
16	**0.659**	0.637	0.643	0.677	0.674	**0.680**	0.685	0.685	**0.692**	0.691	0.700	**0.706**
32	**0.684**	0.670	0.674	0.684	0.690	**0.693**	0.690	0.701	**0.705**	0.696	0.701	**0.712**
64	0.686	0.691	**0.694**	0.692	0.695	**0.701**	0.693	0.700	**0.710**	0.697	0.704	**0.716**
NDCG@10												
8	**0.346**	0.323	0.324	**0.386**	0.364	0.369	0.403	0.399	**0.409**	0.409	0.422	**0.423**
16	**0.382**	0.366	0.370	**0.403**	0.397	0.401	0.408	0.408	**0.419**	0.418	0.423	**0.427**
32	**0.406**	0.394	0.396	0.410	0.413	**0.417**	0.419	0.424	**0.429**	0.420	0.424	**0.431**
64	0.412	0.411	**0.413**	0.415	0.419	**0.428**	0.420	0.423	**0.437**	0.423	0.427	**0.444**

In the table, the layer-4 indicates that those methods have four hidden layers, not including the embedding layer. The NCF is the baseline \cite{article10}. The B represents the BNCF. The S represents the SNCF.

Observing Table 1 in major terms, no matter HR@10 or NDCG@10, we can find that the proposed method SNCF is more better than other methods in most cases, especially when the number of predictive factors and the number of layers are large. However when the number of predictive factors and the number of layers are small (in the upper left corner of the table), the NCF is the best. This indicates that when the number of predictive factors and the number of layers are large, the model has the strong fitting ability. The model can better capture the side information. The side information play a more important role. But when the number of predictive factors and the number of layers are small, the model has the weak fitting ability. It doesn't fully exploit the relationship of users and items. The collaborative filtering relationship of user and item play a more important role.

In detail, the SNCF is always better than BNCF. It can be seen that the deep statistic information is very important. Only explicitly adding those basic information can't completely express the relationship. The deep neural network hardly capture this layer of statistic information from the basic information.

5 Conclusion

In this paper, we propose a novel neural network model SNCF. From the experimental result, it can be seen that we can recommend more accurate items to a user by adding the side information. The experimental results show that our method outperforms the baseline methods in most cases, demonstrating the effectiveness of adding the side information in the neural network. In the future, we hope to extend our method to more scenarios and add the side information more effectively to fully exploit the relationship of users and items.

Acknowledgement. This work was supported by the Funds for Creative Research Groups of China (NSFC [61421061]).

References

1. Ekstrand, M.D., Riedl, J.T., Konstan, J.A.: Collaborative filtering recommender systems. Found. Trends Hum.-Comput. Interact. **4**(2), 81–173 (2011)
2. Harper, F.M., Konstan, J.A.: The MovieLens datasets: history and context. ACM **5**(4), 19 (2016)
3. He, X., Chen, T., Kan, M.Y., Chen, X.: Trirank: review-aware explainable recommendation by modeling aspects, pp. 1661–1670 (2015)
4. He, X., Liao, L., Zhang, H., Nie, L., Hu, X., Chua, T.S.: Neural collaborative filtering, pp. 173–182 (2017)
5. He, X., Zhang, H., Kan, M.Y., Chua, T.S.: Fast matrix factorization for online recommendation with implicit feedback. In: International ACM SIGIR Conference on Research Development in Information Retrieval, pp. 549–558 (2016)

6. Kalchbrenner, N., Grefenstette, E., Blunsom, P.: A convolutional neural network for modelling sentences. Arxiv 1 (2014)
7. Kingma, D., Ba, J.: Adam: a method for stochastic optimization. Computer Science (2014)
8. Koren, Y., Bell, R., Volinsky, C.: Matrix factorization techniques for recommender systems. Computer **42**(8), 30–37 (2009)
9. Li, X., She, J.: Collaborative variational autoencoder for recommender systems. In: The ACM SIGKDD International Conference, pp. 305–314 (2017)
10. Ling, G., Lyu, M.R., King, I.: Ratings meet reviews, a combined approach to recommend, pp. 105–112 (2014)
11. Lu, Y., Dong, R., Smyth, B.: Coevolutionary recommendation model: mutual learning between ratings and reviews. In: World Wide Web Conference, pp. 773–782 (2018)
12. Mikolov, T., Karafiát, M., Burget, L., Černocký, J., Khudanpur, S.: Recurrent neural network based language model. In: Eleventh Annual Conference of the International Speech Communication Association (2010)
13. Program, W.: Proceedings of KDD cup and workshop 2007 (2007)
14. Rendle, S., Freudenthaler, C., Gantner, Z., Schmidt-Thieme, L.: BPR: Bayesian personalized ranking from implicit feedback. In: Proceedings of the Twenty-Fifth Conference on Uncertainty in Artificial Intelligence, pp. 452–461. AUAI Press (2009)
15. Rendle, S., Freudenthaler, C., Gantner, Z., Schmidt-Thieme, L.: BPR: Bayesian personalized ranking from implicit feedback, pp. 452–461 (2012)
16. Salakhutdinov, R., Mnih, A., Hinton, G.: Restricted Boltzmann machines for collaborative filtering. In: International Conference on Machine Learning, pp. 791–798 (2007)
17. Sarwar, B., Karypis, G., Konstan, J., Riedl, J.: Item-based collaborative filtering recommendation algorithms. In: Proceedings of the 10th international conference on World Wide Web, pp. 285–295. ACM (2001)
18. Strub, F., Mary, J., Gaudel, R.: Hybrid collaborative filtering with autoencoders (2016)
19. Zheng, L., Noroozi, V., Yu, P.S.: Joint deep modeling of users and items using reviews for recommendation. In: Tenth ACM International Conference on Web Search and Data Mining, pp. 425–434 (2017)

Reaching Consensus for SDN Multi-controller Networks

Yun Wang[✉] and Liuhe Tian

School of Computer Science and Engineering, Key Laboratory of Computer Network and Information Integration, Ministry of Education, Southeast University, Nanjing, China
ywang_cse@seu.edu.cn

Abstract. Multiple SDN controllers architecture has been proposed to improve scalability and to avoid single point of failure. In order to resolve consistent network state among SDN multi-controllers, an efficient consensus mechanism to synchronize the control state of each controller is required. Raft is a consensus algorithm used in the OpenDayLight (ODL) Clustering. However, Raft algorithm may suffer from the leader election timeout and the system loading may fall on a certain controller. Thus, we propose a High Performance Paxos-based Consensus algorithm (HPPC) and implement it in the ODL Clustering to maintain a consistent global network state. The proposed HPPC simplifies the original Paxos protocols and guarantees that execution among interference commands when committing the client requests simultaneously. Since HPPC has no leader and every controller can commit requests concurrently, it has lower average consensus time than Raft. Meanwhile, experiment results show that HPPC is 25.2% faster at retrieving data and 66.3% faster at storing data via REST API comparing with Raft.

Keywords: SDN · Consensus algorithm · Multiple controllers

1 Introduction

Software-defined networking (SDN) technology is a novel approach to cloud computing that facilitates network management and enables programmatically efficient network configuration in order to improve network performance and monitoring [1]. The static architecture of traditional networks is decentralized and complex while current networks require more flexibility and easy troubleshooting. SDN architectures decouple network control and forwarding functions, enabling network control to become directly programmable and the underlying infrastructure to be abstracted from applications and network services [2]. The control plane consists of one or more controllers which are considered as the brain of SDN network where the whole intelligence is incorporated. However, the intelligence centralization has its own drawbacks when it comes to security [3], scalability and elasticity.

© Springer Nature Switzerland AG 2019
D. Milošević et al. (Eds.): HCC 2019, LNCS 11956, pp. 443–457, 2019.
https://doi.org/10.1007/978-3-030-37429-7_44

To overcome these limitations, several approaches have been proposed in the literature that fall into two categories, hierarchical and fully distributed approaches. However, both of these two methods need to synchronize the network state information among controllers, known as the controller consensus problem. Many consensus algorithms have been proposed to solve these consensus problems, such as Paxos [4], Raft [5], etc. In SDNs, most applications need to synchronize network state information among controllers as quickly and efficiently as possible. For example, the SDN routing application needs the latest topology information to generate a correct routing path. Thus, it is necessary to design and implement a consensus algorithm to achieve low latency, high throughput and high availability among multiple controllers.

The main challenge of this paper is how to design a High-Performance Paxos-Based Consensus algorithm (HPPC). The key idea of HPPC is to simplify the original Paxos protocols. The client requests can be committed concurrently in HPPC. Eventually, HPPC still guarantees that execution among interference commands.

The paper is organized as follows. Section 2 introduces some consistency mechanisms used in major multi-controller SDN networks. Section 3 presents the SDN controller architecture and details the design of the proposed HPPC algorithm. Simulations and experimental results are conducted in Sect. 4. Finally, we summarize this paper and demonstrate future work in Sect. 5.

2   Related Work

In this section, some relate work on distributed consensus protocols used in SDN multi-controller networks is summarized. In distributed architectures, every controller operates on their local view or they may exchange related topology messages to enhance their knowledge. However, while synchronizing network state information to other controllers, each of these distributed architectures has different solutions to deal with consistency issues. Multi-controller SDN networks generally apply two different consistency models, strong consistency and eventual consistency.

With strong consistency, controllers have higher synchronization delay but can get the latest updated network state information relative to eventual consistency. DISCO [6] provides a distributed control plane for WAN and constrained networks based on a message-oriented communication bus. A DISCO controller manages its own network domain and communicates with other controllers based on the AMQP (advanced message queuing protocol) to discover neighboring controllers. GRACE [7] supplies two key features, reconfigurability and reusability. Their implementations are used to deal with the asynchrony of data plane states and the complexity of service control states respectively. The OpenDayLight [8] utilizes a multi-controller SDN architecture named OpenDayLight Clustering [9]. OpenDayLight clustering applies multiple controller processes to form a powerful logical entity. Each controller has its own data slice. ODL uses Raft algorithm to achieve controller consistency.

With eventual consistency, controllers have lower synchronization delay but may not get the latest updated network state information relative to strong consistency. HyperFlow [10] implements a universal control plane based on distributed events, logically centralized but physically distributed, maintaining good scalability. Multiple controllers in HyperFlow exploit an event propagation system to maintain a consistent, shared network view, and each switch uses the OpenFlow protocol for communication with its neighboring controllers. Onix [11] operates on one or more different servers and generates a common network interface to allow applications on other servers to read or modify all devices in the network, enabling synchronization and consensus between devices and applications. A data structure called NIB is used to save all network status, the controllers exploit NIB to share the global network view. A summarization of the controller consensus mechanism is shown in Table 1.

Table 1. Comparison of multi-controller consensus mechanisms

Design	Property		Consistency
	Leader	Controller consensus	
DISCO [6]	Leader free	Publish/subscribe	Strong
GRACE [7]	Leader free	Byzantine Paxos	Strong
ODL clustering [9]	Leader free	Raft	Strong
HyperFlow [10]	Leader free	Publish/subscribe	Eventual
Onix [11]	Leader free	Onix API	Eventual
MCN-clustering (proposed)	Leader free	HPPC (proposed) by expanding Egalitarian Paxos	Strong

3   Design of HPPC

We propose the HPPC based Paxos for SDN multi-controller networks to achieve consistency among multiple controllers. With the strong consistency model, it is guaranteed that all network updates can be synchronized to all controllers. Each peer controller can get a global consistency state to make optimal decisions for the local network.

The proposed HPPC simplifies the original Paxos protocols and expands the Egalitarian Paxos [12]. As long as concurrent proposals do not interfere, they will be committed after only one round of communication between the command leader and a fast-path quorum of peers. When commands interfere, they acquire dependencies on each other attributes that commands are committed with, used by controllers to determine the correct order in which to execute the commands. Figure 1 represents a simplified example that how HPPC works.

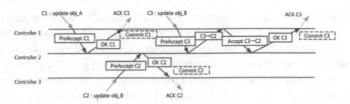

Fig. 1. Message flow in HPPC algorithm. Commands C2 and C3 interfere (they update the same object) so C3 requires an additional round of communication. C3 → C2 signifies that C3 has acquired a dependency on C2. For clarity, we did not illustrate the asynchronous commit messages.

Assumption III. 1: *The controllers are connected by a network and the network may fail to deliver request messages, delay them, duplicate them, or deliver them out of order.*

The model we established is described as follows: Each controller C has a unique identifier, and all identifiers can be totally ordered. For every controller C there is an unbounded sequence of numbered requests (C.id, 1), (C.id, 2), (C.id, 3),..., (C.id, i), where C.id is the identifier of controller C, and i is the incremental instance number of controller C. Table 2 presents the key objects and the notations are used in the proposed HPPC.

Table 2. Key object and notation used in the proposed HPPC

Object	Notation	Definition
Controller	id	The identifier of the controller
	i	The incremental instance number of the controller
	commands	All request messages received by the controller
Command	proposerid	The identifier of the original proposer
	deps	The list of all instances that contain commands (not necessarily committed) that interfere with the command
	seq	A sequence number used to break dependency cycles during the Reply Protocol
	operation	The operation of the command. The update topology or link capacity message and the read message
	state	The state of the request, can be one of pre-accepted, accepted, committed, replaying and replayed
	ballotNum	Current round number of the command

The proposed HPPC algorithm is mainly composed of three Protocols:

- the commit protocol for controllers is to submit commands concurrently without concern for the order between the commands.
- the execution protocol for controllers is to assign an order to the commands according to their dependencies and execute them following that order concurrently.
- the failure recovery protocol for controllers is to actively learn the submission decision of an instance.

The relationship among these three protocols is explained as follows: When a controller receives a Request message from a switch, it will apply the commit protocol to commit the command in the request. Then the controller invokes the execution protocol to execute the commands in the assigned order. Finally, the controller obtains the execution result and replies to the switch. Once a controller times out while waiting for the relevant command to commit, it will start failure recovery protocol to learn this command resolution.

3.1 The Commit Protocol

The commit protocol for multi-controller is to commit commands concurrently without concern for the order, and then use the seq and deps attributes of that command to track their dependencies. Among the following eight algorithms in this subsection, the first three algorithms correspond to the propose phase of the classical paxos algorithm, the intermediate algorithm 4, 5 and 6 correspond to the accept phase, and the last two algorithms correspond to the commit phase. These algorithms are running in controllers in the ODL clustering and are driven by message events.

Once a controller receives a Request message including command γ from an OpenFlow switch, it will become the leader of a new instance, just as Algorithm 1 shows. First, the controller who has received the Request increases its own instance number. Then, it attaches what it believes are the correct attributes for that command in the instance. Finally, the controller sends PreAccept messages to all other controllers.

Any controller, upon receiving a PreAccept message, Algorithm 2 is executed. First, update the seq and the deps attributes of the command γ according to the contents of its commands log. Then, the controller records the command γ and updates its own commands log with the attributes of the received command. Finally, the controller replies a PreAcceptOK message to the leader of the command.

Algorithm 1 Controller L on receiving $Request(\gamma)$ from a switch

Input: L: the controller who has received the $Request$; γ: the command in $Request$;
Output: Send $PreAccept(\gamma, L.i)$ to all other controllers in Ω, where Ω is the fast quorum that includes L

1: **function** ONRECEIVEREQUEST($L.i, \gamma$)

2:     $L.i \leftarrow L.i + 1$

3:     { $Interf_{L,\gamma}$ is the set of instances $Q.j$ such that the command recorded in $L.commands[Q.id][j]$ interferes γ}

4:     $\gamma.seq \leftarrow 1 + \max(\{L.commands[Q.id][j].sep|Q.j \in Interf_{L,\gamma}\{0\})$

5:     $\gamma.deps \leftarrow Interf_{L,\gamma}$

6:     $\gamma.state \leftarrow pre - accepted$

7:     $L.commands[L.id][L.i] \leftarrow \gamma$

8: **end function**

Algorithm 2 Controller R on receiving $PreAccept(\gamma, L.i)$

Input: R: controller who has received the $PreAccept(\gamma, L.i)$;

      γ: the *command* in the *PreAccept* message;

Output: Reply $PreAcceptOK(\gamma, L.i)$ to Controller L

1: **function** ONRECEIVEPREACCEPT(R, γ)

2:    update

      $\gamma.seq \leftarrow \max(\{1 + R.commands[Q.id][j].sep | Q.j \in Inter\ f_{R,\gamma}\} \cup \{\gamma.seq\})$

3:    update $\gamma.deps \leftarrow \gamma.deps \cup Inter f_{R,\gamma}$

4:    $R.commands[L.id][L.i].sep \leftarrow \gamma$

5: **end function**

Algorithm 3 Controller L (command leader for γ) on receiving at least $\lfloor \frac{N}{2} \rfloor$ $PreAcceptOK$ responses

---

Input: L: the controller who has received the majority $PreAcceptOK$ message;

1: **function** ONRECEIVEPREACCEPTOK(L)

2:    **if** received $PreAcceptOK's$ from all controllers in $\Omega\backslash\{L\}$, with $\gamma.seq$ and $\gamma.deqs$ the same in all controllers

      then

3:      run Commit phase for (γ) at $L.i$

4:    **else**

5:      update $\gamma.seq \leftarrow \max(\{\gamma.seq$ of all controllers$\})$

6:      update $\gamma.deqs \leftarrow Union(\gamma.deqs$ from all controllers$)$

7:      run Paxos-Accept phase for (γ) at $L.i$

8:    **end if**

9: **end function**

Algorithm 4 Command leader L, for γ at instance $L.i$

Input: L: the command leader controller; γ: the command;

1: **function** PROPOSE(L, γ)

2:    $\gamma.state \leftarrow accepted$

3:    $L.commands[L.id][L.i] \leftarrow \gamma$

4:    Send $Accept(\gamma, L.i)$ to at least $\lfloor \frac{N}{2} \rfloor$ other controllers

5: **end function**

Algorithm 5 Any Controller R on receiving $Accpet(\gamma, L.i)$ message

---

Input: R: the controller who has received the $Accept$ message; γ: the command;

1: **function** ONRECEIVEACCEPT(C,$command$)

2:     $R.commands[L.id][L.i] \leftarrow \gamma$

3:     reply $AcceptOK(\gamma, L.i)$ to L

4: **end function**

Algorithm 6 Command leader L on receiving at least $\lfloor \frac{N}{2} \rfloor$ $AcceptOk$ responses

---

Input: L: the controller who has received the majority $AcceptOK$ message;

    γ: the command;

1: **function** ONRECEIVEACCEPTOK(L,γ)

2:     run Commit phase for (γ) at $L.i$

3: **end function**

When the command leader controller receives replies from majority controllers, Algorithm 3 is invoked. If the majority controllers can constitute a fast-path quorum and all the updated attributes are the same, it commits the command. If the command leader controller does not receive enough response messages, or the attributes in some replies are different from others, then the command leader updates the attributes based upon a simple majority controllers (taking the union of all deps, and the highest seq), and tells at least a majority controllers to accept these attributes.

In the Paxos-Accept phase, Algorithms 4, 5 and 6 is executed. First, in the Algorithm 4, the leader controller updates the state of the received command and saves the received command. Then, the controller sends the Accept message to majority controllers. In the Algorithm 5, the controller who receives the Accept Message will just store the command then replay AcceptOK Message to the leader. In the Algorithm 6, once upon the leader controller has received the majority replies, it will enter the commit phase and commit the command.

In the commit phase, Algorithms 7 and 8 is executed. First, the leader controller updates the state of the received command and saves the received command. Then, it notifies the switch while can be seen as the client the executed result. Finally, it sends the Commit message asynchronously to the other controllers.

Algorithm 7 Command leader L, for γ at instance $L.i$

Input: L: the command leader controller; γ: the command;

1: **function** COMMIT(L, γ)

2:     $\gamma.state \leftarrow commited$ $L.commands[L.id][L.i] \leftarrow \gamma$

3   send commit notification for γ to the switch

4:     send $Commit(\gamma, L.i)$ to all other controllers

5: **end function**

Algorithm 8 Any Controller R on receiving $Commit(\gamma, L.i)$ message

---

Input: R: the controller who has received the $Commit$ message; γ: the command;

1: **function** ONRECEIVECOMMIT($C, command$)

2:     $R.commands[L.id][L.i] \leftarrow \gamma$

3: **end function**

3.2   The Execution Protocol

The execution protocol for controllers is to assign an order to the commands according to their dependencies and execute them following that order concurrently. For commands that do not conflict with each other, the commands can be directly committed and executed. Hence, only the conflicting commands are discussed. For simplicity, we define the commands which are interference among them.

Command Interference: $\exists \Sigma$ such that

$$SE(\Sigma, \gamma, \delta) \neq SE(\Sigma, \delta, \gamma) \tag{1}$$

SE function represents the result of the serial execution of these commands, Σ is a sequence of commands. γ and δ are the two interference commands.

Every controller must assume that any unexecuted command in self dependency or concurrent dependency instances are possible dependencies and independently check them at execute time. The execution process of command is shown in Fig. 2. The execution protocol is a topological sorting of a DCG. When a cycle in DCG is detected, it is simply broken by the ascending order of the controllers.

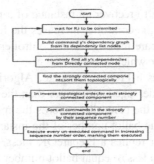

Fig. 2. The execution process of command γ in instance R.i

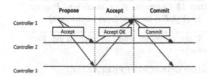

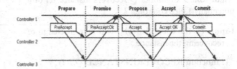

Fig. 3. The messages of the failure recovery protocol when there is a command that has been committed or been accepted by at least a majority of controllers in the instance to be recovered, like Paxos without the accept phase.

Fig. 4. The message of the failure recovery protocol when there is a command that has been accepted by less than half controllers or no controller has seen a command in the instance to be recovered, like Paxos.

3.3 The Failure Recovery Protocol

The controllers may fail to deliver request messages, delay them, duplicate them and response some messages timeout, etc. The failure recovery protocol tolerates maximum F controller errors with N = 2F + 1 controllers in a network.

The failure recovery protocol for controllers is to actively learn the submission decision of an instance. Figure 3 shows the messages of the failure recovery protocol when there is a command that has been committed or been accepted by at least a majority of controllers in the instance to be recovered, it likes the messages in Paxos without the accept phase. Figure 4 shows the messages of the failure recovery protocol when there is a command that has been accepted by less than half controllers or no other controller has seen a command in the instance to be recovered, it likes the complete messages in Paxos.

When a controller is going to recover an instance of a potentially failed controller, Algorithm 9 is invoked. First the controller increases the highest ballot number of the instance. Then, it sends a Prepare message to all controllers (including itself) with the new the highest ballot number which it believes.

When a controller receives a Prepare message, Algorithm 10 is executed. If the ballot number in the Prepare message is not larger than the previously received largest ballot number of the instance, the controller ignores the message and takes no action. Then, the controller updates the ballot number of the instance with the received ballot number. Finally, the controller replies a PrepareOK message to promise to ignore all future Prepare messages with a less or equal ballot number and Propose messages with a less ballot number in the instance.

Algorithm 9 Controller Q recovers instance $L.i$ of a potentially failed controller L

Input: Q: the controller who will take in charge of recovering the instance ;

1: **function** RECOVERINSTANCE($Q,L.i$)

2:      increment ballot number to $epoch.(b+1).Q$,

   (where $epoch.b.R$ is the highest ballot number Q is aware of in instance $L.i$)

3:      send $Prepare(epoch.(b+1).Q, L.i)$ to all controllers (including self)

4: **end function**

Algorithm 10 Controller R on receiving $Prepare$ $(epoch.(b+1).Q, L.i)$ message from Controller Q

Input: R: the controller who has received the $Prepare$ message;

1: **function** ONRECEIVEPREPARE(R)

2:      **if** $epoch.b.Q$ is larger than the most recent ballot number $epoch.x.Y$ accepted for instance $L.i$ **then**

3:          reply $PrepareOK(R.commands[L.id][L.i], epoch.x.Y, L.i)$

4:      **else**

5:          NACK

6:      **end if**

7: **end function**

Algorithm 11 Controller Q received at least $\lfloor \frac{N}{2} \rfloor$

PrepareOk responses

Input: Q: the controller who has received the majority *PrepareOK* message;

1: **function** ONRECEIVEPREPAREOK(Q)

2:    let ϕ be the set of controllers with the highest ballot number

3:    **if** ϕ contains a $(\gamma, \gamma.seq, \gamma.deps, committed)$ **then**

4:        run Commit phase for (γ) at $L.i$

5:        **else if** ϕ contains a $(\gamma, \gamma.seq, \gamma.deps, accepted)$ **then**

6:        run Paxos-Accept phase for (γ) at $L.i$

7:        **else if** ϕ contains at least $\lfloor \frac{N}{2} \rfloor$ identical controllers $(\gamma, \gamma.seq, \gamma.deps, pre-accepted)$ for the default ballot $epoch.0.L$ of instance $L.i$, and none of those controllers is from L **then**

8:        run Paxos-Accept phase for (γ) at $L.i$

9:        **else if** ϕ contains a $(\gamma, \gamma.seq, \gamma.deps, pre-accepted)$ **then**

10:        run Algorithm 1 (at line 2) for (γ) at $L.i$

11:    **else**

12:        run Algorithm 1 (at line 2) for $no-op$ at $L.i$

13:    **end if**

14: **end function**

When a controller receives a $\lfloor N/2 \rfloor + 1$ PrepareOK message, Algorithm 11 is invoked. If the controller receives PrepareOK messages of the instance, there are five situations according to the set of replies with the highest ballot number: (1) if the set of controllers contains a command whose state is committed, which indicates the instance is committed, then the controller just enters the commit phase; (2) if the set of controllers contains a command whose state has been accepted, which indicates the command has not yet been replicated to a majority of controllers, then the controller enters the Paxos-Accept phase; (3) if the set of controllers excluding the proposer controller contains at least $\lfloor N/2 \rfloor + 1$ commands whose states are pre-accepted for the default ballot number 0, which indicates the instance could be accepted, then the controller enters the Paxos-Accept phase; (4) if the set of controllers contains a command whose state is pre-accepted, which indicates the command just has been proposed, the controller need run the commit protocol from the origin; (5) if the set of controllers is not in any situation above, the controller initializes a no-op command and sends Propose messages to all other controllers to start the commit protocol.

4 Simulation

Table 3 shows our simulation parameters. In our simulation experiment, the Implementation of control plane is OpenDayLight OpenFlow controller [11], and we use mininet to implement the data plane. The operation system of the VM we used is Ubuntu 14.04 LTS. Intel core i7 processor is used to compute.

The 3-node OpenDayLight clustering is shown in Fig. 5. The 5-node OpenDayLight clustering architecture is the same as the 3-node, only two controller nodes have been added. HPPC is implemented on OpenDayLight Clustering project.

Most SDN multi-controllers architectures apply only three OpenFlow controllers. We evaluate the proposed HPPC with 3-node and 5-node OpenDayLight clustering. Toaster [14] is a test model for testing OpenDayLight MD-SAL (Model-Driven SAL) functionality. It provides a sample consumer application that uses MD-SAL APIs and a unit test suite. We used toaster model and modified the data value store in it as a request. We executed these requests (write toaster data) on each controller.

We will compare the proposed HPPC with the Raft consensus algorithm used in OpenDayLight clustering, in terms of latency, throughput. The latency is the time elapsed from a transaction created to the transaction end. The throughput is the number of requests the system can handle within a certain period of time.

Table 3. Key object and notation used in the proposed HPPC

Parameter	Value
Control plane	OpenDayLight OpenFlow controller [8]
Data plane	Mininet OpenFlow network emulator [13]
Environment	Ubuntu 14.04 LTS
Hardware	Intel(R) Core(TM) i7-6700 CPU@3.40 GHz

Table 4. Comparison of multi-controller consensus mechanisms

Scheme	Read data	Write data
Raft (request from leader)	5.415 ms	11.691 ms
Raft (request from follower)	10.323 ms	60.513 ms
Raft (average)	8.687 ms	44.239 ms
HPPC	6.495 ms	14.910 ms

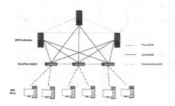

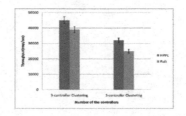

Fig. 5. 3-node OpenDayLight SDN Clustering architecture

Fig. 6. Throughput in SDN Clustering for 3 controllers and 5 controllers (with 95% CI). -

4.1 The Consensus Time

The consensus time of the HPPC is more stable and the overall consensus time is lower, compared to that of the Raft. Having a low and stable consensus time can improve the performance of some REST APIs which need to access the data store and can also decrease the flow setup time. Table 4 shows the REST API data access time after executing 1000 requests on each controller with 3-node clustering. The proposed HPPC has lower data access time than the Raft.

4.2 The Throughput in Multi-controller Network

In the throughput experiment. A client on a separate server sends requests in an open loop, and measures the rate at which it receives replies. For HPPC and Raft, the client sends each request to a replica chosen uniformly at random. Figure 6 shows the throughput for HPPC and Raft in LAN using three controllers and five controllers. HPPC has higher throughput than Raft. Raft cannot achieve load balance among controllers since the leader controller must process more messages than other controllers.

4.3 Latency vs. Throughput

Latency will increase with throughput within a certain range. It has been validated that the latency increases more slowly with the throughput in the proposed HPPC than that in Raft when used in the 3-controller SDN network. Each command can be committed after one at most two rounds of communication in the proposed HPPC. High throughput will increase the probability of command interference. Figure 7 shows the relationship between latency and throughput for the proposed HPPC and Raft.

4.4 Service Availability Under Failures

Figure 8 shows the evolution of the throughput in a OpenDayLight clustering using three controllers in LAN that experiences the failure of one controller. For HPPC, the failure controller is a random one in the system, and for the origin ODL clustering used Raft, the failure replica is the leader. A client sends requests at the same appropriate rate of approximately 10,000 requests per second for every system.

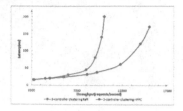 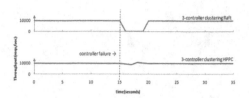

Fig. 7. Latency with throughput in the 3-controller SDN network

Fig. 8. Throughput when one of three controllers in the OpenDayLight clustering fails. For the origin ODL Clustering, the Raft leader fails.

5   Conclusion

In this paper, we present a High Performance Paxos-based Consensus algorithm (HPPC) to handle the consensus issue of multiple SDN controllers. The proposed HPPC simplifies the original Paxos protocols and can alleviate the complexity in developing and implementing Paxos. Simulation results show that the proposed HPPC has lower average consensus time than the Raft. With a low consensus time, the proposed HPPC is 25.2% faster at retrieving data and 66.3% faster at storing data via REST API comparing with Raft in the OpenDayLight clustering.

References

1. Benzekki, K., El Fergougui, A., Elbelrhiti Elalaoui, A.: Software-defined networking (SDN): a survey. Secur. Commun. Netw. **9**(18), 5803–5833 (2016)
2. Software-Defined Networking (SDN) definition (2014). Opennetworking.org
3. Benzekki, K., El Fergougui, A., Abdelbaki, E.B.E.A.: Devolving IEEE 802.1X authentication capability to data plane in software-defined networking (SDN) architecture. Secur. Commun. Netw. **9**(17), 4369–4377 (2016)
4. Lamport, L.: Paxos made simple. ACM Sigact News **32**(44), 18–25 (2016)
5. Ongaro, F.P.D., Ousterhout, J.: In search of an understandable consensus algorithm. In: Proceedings of the USENIX Annual Technical Conference, pp. 305–320 (2014)
6. Phemius, K., Bouet, M., Leguay, J.: Disco: distributed multi-domain SDN controllers. In: Network Operations and Management Symposium, pp. 1–4 (2014)
7. Zhou, B., Wu, C., Hong, X., Jiang, M.: Programming network via distributed control in software-defined networks. In: IEEE International Conference on Communications (2014)
8. Opendaylight Linux Foundation Collaborative Projects. http://www.OpenDayLight.org/
9. OpenDaylight SDN Controller Platform (OSCP): Clustering. https://wiki.OpenDayLight. org/view/OpenDayLight. SDN Controller Platform (OSCP): Clustering
10. Tootoonchian, A., Ganjali, Y.: Hyperflow: a distributed control plane for openflow. In: Internet Network Management Conference on Research on Enterprise Networking, p. 3 (2011)
11. Koponen, T., et al.: Onix: a distributed control platform for large-scale production networks. In: Usenix Conference on Operating Systems Design and Implementation, pp. 351–364 (2010)

12. Moraru, I., Andersen, D.G., Kaminsky, M.: There is more consensus in egalitarian parliaments. In: Twenty-Fourth ACM Symposium on Operating Systems Principles, pp. 358–372 (2013)
13. Lantz, B., Heller, B., Mckeown, N.: A network in a laptop: rapid prototyping for software-defined networks. In: ACM Workshop on Hot Topics in Networks, HOTNETS, pp. 1–6 (2010)
14. Opendaylight Toaster. https://wiki.OpenDayLight.org/view/OpenDayLight. Controller: MD-SAL: Toaster Step-By-Step

A Model Based on a Fuzzy Petri Net for Scenario Evolution of Unconventional Emergencies

Wei-dong Huang[1(✉)], Qian Wang[1], Bang-lan Ding[1], and Jie Cao[2]

[1] Nanjing University of Posts and Telecommunications, Nanjing, Jiangsu, China
huangwd@njupt.edu.cn
[2] Nanjing University of Information Science and Technology,
Nanjing, Jiangsu, China

Abstract. Improving the response precision of unconventional-emergency management has been difficult. As a new type of decision-making model, "scenario–response" can help solve this problem. From the perspective of scenario evolution, this research designed the system structure of event scenario evolution as a hierarchical network structure of hazard factors, key hazard-affected bodies, and derivative events. In addition, a model for scenario evolution based on a fuzzy Petri net was constructed. Taking as an example an earthquake scenario evolution based on a fuzzy Petri net. A population (P) from the key hazard-affected objects was selected as the sample object, and the scenario evolution path of P by constructing a reasoning tree based on the fuzzy reasoning algorithm can be obtained to solve the response sequence of P. The feasibility of this model was verified by the example of an earthquake.

Keywords: Emergency management · Fuzzy Petri net · Scenario evolution · Unconventional emergency

1 Introduction

An Unconventional Emergency (UE) is an emergency with no substantial warning, complex variability, strong destructiveness, and which can easily trigger many kinds of derivative hazardous events. The effects of UEs and their derivative disasters are serious; these emergencies have no precursor and are difficult to predict. The process of variations and the inherent mechanisms of the event are complex [1]. The decision-making model for "prediction–response," which is just to take measures by constructing case scenarios based on empirical rules and historical events, can't meet the demands of real-time decision making in handling emergencies [2]; an analysis of the evolutionary process of unconventional emergencies is needed. The results show that the event environments of UEs are characterized by complicated variety and fuzzy uncertainty. In addition, those factors have complicated correlations, such as derivative events, the state of the key events, and hazard factors. Affected by external environmental factors, UEs have a chain response behavior [4]. Handling the emergencies requires sequential decisions for multiple scenario reconstruction and matching, and needs to be aimed at the decision-making demands of multiple stages and multiple scenarios.

© Springer Nature Switzerland AG 2019
D. Milošević et al. (Eds.): HCC 2019, LNCS 11956, pp. 458–467, 2019.
https://doi.org/10.1007/978-3-030-37429-7_45

2   Analysis of the Scenario Evolutionary Mechanism of a UE

The evolutionary process of events was divided into 4 stages from the perspective of evolution: occurrence, development, evolution, and termination [5]. The evolution of a UE can be of 2 types; one is a short development process, such as that of an earthquake, and the other is a long one, such as that of a forest fire. For the former, the scenario evolution analysis should focus on the changes between the event itself and derivative events. We can control the occurrence and development of derivative events by decision making so that the evolutionary scope of the event can be controlled. For a long development process, it should be pay attention to the state scenarios of key events and the changes of scenario and hazard factors and try our best to control the evolution of the events within the scope of their own development to avoid a wider range of evolutionary influences [15].

Because the scenarios of derivative, key, and hazard events change frequently, the scenario evolution of UEs was divided into 3 types: the chain scenario evolution, the diffusion scenario evolution, and the cyclic scenario evolution. The main characteristic of the chain scenario evolution is that there is a single causal relation between the first scenario and its induced subsequent scenarios. The diffusion scenario evolution shows that the destructive effects of the event are diffused, and that there is a complex causal relation among the scenarios. In the cyclic scenario evolution, the initial scenario is renewed after triggering other disaster scenarios. The scenario structure is a closed-loop sequence.

The introduction of variable structure Petri net modeling can make the main structure of the system modular storage and object classification. Pertaining to the evolutionary mechanism of UE, a 3-layer scenario structure network diagram is shown in Fig. 1. The 3 layers, from bottom to top, are the hazard factors layer (F), the key event state layer (S), and the derivative event layer (E). The hazard factors layer consists of a network of scenario relations that are caused by many factors such as initiation, aggravation, or mitigation. The changes of scenarios in this layer mark the changes of the hazard factors' states. The key event state layer consists of a network of scenario relations that are the evolution of UEs between important event state scenarios. The changes of scenarios in this layer mark the changes to the internal states of the events. The derivative event consists of a network of scenario relations between the event itself and derivative events. The changes of scenarios in this layer mark the evolutionary changes between the events. There are also some evolutionary characteristics between the 3 layers; in the first-level evolution from layers F to S and in the second-level evolution from layers S to E. That is to say, the 3-layer network has a certain causal relation vertically. In Fig. 1, solid lines depict associations between the scenario factors within a layer, and dotted lines depict associations of factors between layers.

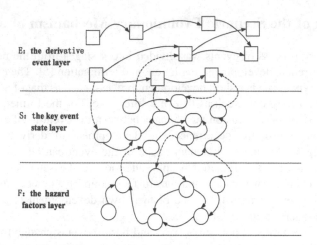

Fig. 1. Three-layer scenario structure network diagram. Solid lines depict associations between the scenario factors within a layer; dotted lines depict associations of factors between layers.

3 Construction of a Scenario Evolution Fuzzy Petri Net Model Based on Fuzzy Reasoning

3.1 Introduction and Rules of FPN

A Fuzzy Petri Net (FPN) is a combination of a Petri net and knowledge representation; it is used to describe the fuzzy generative rule [11]. The fuzzy generative rule describes the fuzzy relations between 2 propositions [12]. Assuming that R is a combination of fuzzy generative rules, $R = \{R_1, R_2, \cdots, R_n\}$. The ith fuzzy rule can be expressed as

$$R_i : \text{IF } d_j \text{ THEN } d_k(\text{CF} = \mu_i) \quad (i = 1, 2, \cdots, n) \tag{1}$$

Where d_j and d_k are the propositions that usually contain the fuzzy variables, such as "high," "low," and "heat" μ_i is the value of the certainty factor and shows the confidence level of the fuzzy rule; the higher the value of μ_i, the higher the confidence level. Combining the basic definition of the Petri net, defined the structure of the FPN as 8 critical elements:

$$\text{FPN} = (P, T, D, I, O, f, \alpha, \beta) \tag{2}$$

Where, $P = \{p_1, p_2, \cdots, p_n\}$ is the finite set of place; $T = \{t_1, t_2, \cdots, t_n\}$ is the finite set of transition; $P \cap T = \Phi$; $D = \{d_1, d_2, \cdots, d_n\}$ is the finite set of propositions; and $|P| = |T|$. I and O are the inputting and outputting functions respectively, f is the mapping function from T to real numbers from 0 to 1, α is the mapping function from P to real number from 0 to 1, and β is the bidirectional mapping function from P to the set of propositions.

3.2 Construction of a UE Scenario Evolution Model

The elements of the scenario evolution system of a UE are not independent, but interact with each other [6]. Considering all kinds of scenario elements as a series of objects. The relations between the objects can be expressed by defining a special kind of gate change. Based on the structure of the scenario evolution model and the theory of the object-oriented Petri net model [7, 8], we created an object-oriented Petri net model. A hierarchical network structure of scenario evolution was maintained as shown in Fig. 2, and the scenario elements correspond corresponded to 3 kinds of objects; they are were O_f, O_s, and O_e. Additionally, $O_f = \{O_{f1}, O_{f2}, \cdots O_{fm}\}$, $O_s = \{O_{s1}, O_{s2}, \cdots O_{sn}\}$, $O_e = \{O_{e1}, O_{e2}, \cdots O_{ez}\}$, m, n and z were the total number of objects of O_f, O_s, and O_e in the system. Taking the information transmission relation between O_{fm} and O_{sn} for example, it could be expressed as $R_{fmsn} = \{OM_{fmsn}, g_{fmsn}, IM_{sn}, C(OM_{fm}), C(IM_{sn}), C(g_{fmsn}), I_{fmsn}, O_{fmsn}\}$, among this, $g_{fmsn} = G_6$ and when the place and the change of color were the same, the value of I_{fmsn} and O_{fmsn} was 1; otherwise, 0.

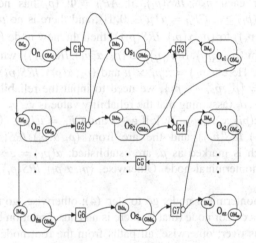

Fig. 2. Sketch map of object-oriented scenario evolution model

3.3 Fuzzy Inference Algorithm

Due to the variability and uncertainty of the decision factors in the emergency response, Expert decision making of the prognosis and evolution of environmental factors needs to be introduced into the fuzzy system description method. Based on this, we needed to design an iterative algorithm based on fuzzy reasoning that could automatically generate all the reasoning paths from the starting place to the target place. In an FPN, the fuzzy reasoning algorithm can be represented as a tree structure, where each node of the structure is represented as 3 main elements: $(p_k, \alpha(p_k), IRS(p_k))$. Here, $p_k \in P$ and $IRS(p_k)$ is the immediately reachable set of p_k. Assume that λ is a threshold value, and CF_{xy} represents the certainty factors that are associated with the transition of p_x and p_y. Supposing that there are some places (including p_x), and all of them take p_y

as the immediately reachable place, AP_{xy} is used to represent a set that takes p_y as the immediately reachable place (next to the p_x).

Fuzzy reasoning algorithm:

Inputting: y_s is the reliability of proposition d_s, $y_s \in [0, 1]$.
Outputting: y_j is the reliability of proposition dj, $y_j \in [0, 1]$.

Reasoning is divided into 4 steps:

(1) It begins at the root node $(p_s, \alpha(p_s), IRS(p_s))$. The root node is a nonterminal node and p_s is the starting node, $\beta(p_s) = d_s$, $\alpha(p_s) = y_s$.

(2) A nonterminal node is chosen $(p_i, \alpha(p_i), IRS(p_i))$, and then if $IRS(p_i) = \Phi$, or all $p_k \in IRS(p_i)$, and if $p_j \notin RS(p_k)$, then this node is labeled as a terminal node. If $p_j \in IRS(p_i)$, $\alpha(p_i) \geq \lambda$ and $CF_{ij} = \mu$ ($\mu \in [0, 1]$), then a new node $(p_j, \alpha(p_j), IRS(p_j))$ and an arc from $(p_i, \alpha(p_i), IRS(p_i))$ to this new node are built into the tree and marked as μ. Here, $\alpha(p_j) = \mu \times \alpha(p_j)$ and $(p_j, \alpha(p_j), IRS(p_j))$ is a successor node.

Otherwise, for each $p_k \in IRS(p_i)$, if $AP_{ik} = \Phi$ (p_i has no adjacent place), $p_j \in RS(p_k)$, $\alpha(p_i) \geq \lambda$, $CF_{ij} = \mu$ ($\mu \in [0, 1]$), and there is no p_k in the path from $(p_s, \alpha(p_s), IRS(p_s))$ to $(p_j, \alpha(p_j), IRS(p_j))$, then the new node $(p_k, \alpha(p_k), IRS(p_k))$ and the arc from $(p_j, \alpha(p_j), IRS(p_j))$ to $(p_k, \alpha(p_k), IRS(p_k))$, which is marked as μ, are established. Here $\alpha(p_k) = \alpha[p_i] \times \mu$ and $(p_k, \alpha(p_k), IRS(p_k))$ is a nonterminal node. If $AP_{ik} = \{p_a, p_b, \cdots, p_z\}$, we need to input the reliability of the propositions $d_a, d_b, \cdots, d_z$ (assuming that the reliability value is $y_a, y_b, \cdots, y_z$). Supposing that $g = Min(\alpha(pi), y_a, y_b, \cdots, y_z)$: if $g \geq \lambda$ and $CF_{ij} = \mu$ ($\mu \in [0, 1]$), then the new node $(p_k, \alpha(p_k), IRS(p_k))$ and the arc from $(p_i, \alpha(p_i), IRS(p_i))$ to $(p_k, \alpha(p_k), IRS(p_k))$, which is marked as μ, are established; $\alpha(p_k) = g \times \mu$ and $(p_k, \alpha(p_k), IRS(p_k))$ is a nonterminal node. Otherwise, $(p_i, \alpha(p_i), IRS(p_i))$ is marked as the terminal node.

(3) If there is no nonterminal node, go to step (4); otherwise, go to step (2).

(4) If there is no successor node (that is, there is no causal relation between d_s and d_j), the reasoning is over; otherwise, all paths from the root node to each successor node compose a reasoning path, and we mark the set of successor nodes as Q. Additionally,

$$Q = \left\{ (p_j, s_1, IRS(p_j)), (p_j, s_2, IRS(p_j)), \cdots, (p_j, s_m, IRS(p_j)) \right\} \qquad (3)$$

Here, $s_i \in [0, 1]$ and $0 \leq i \leq m$. Assuming that $z = Max\{s_1, s_2, \cdots, s_m\}$, it can be obtain that the reliability of the d_j is z.

4   Applied Analysis

We used cases to analyze the application of the above algorithm. First, constructing a model for the scenario evolution of UE and using an earthquake as an example to verify and analyze this model.

4.1 Structural Design of a Scenario Evolution Model for an Earthquake

According to the above structure of the UE scenario evolution model, we listed 3 levels of factors in an earthquake scenario combining with the specific characteristics of an earthquake:

(1) The object layer of hazard factors: earthquake, weather, and time. The state of the earthquake object includes its magnitude and intensity, the state of the weather object includes precipitation and air temperature, and the state of the time object includes day and night. The information source of this layer is mainly by computer or common knowledge, and the change of the internal state of the objects is neither controllable nor predictable, and is obtained directly. Using fuzzy statistical analysis of the relevant parameters of these objects, we can obtain 3 kinds of damage factors and assign the initial value of the model to carry on scenario evolution reasoning.

(2) The object layer of key hazard-affected bodies: buildings, communications, transportation, water supply facilities, power supply facilities, and population. The buildings state includes collapsed, completely accessible, and completely inaccessible. The communications state includes interruption, maintenance, and recovery. The transportation state includes disrupted roads, congested roads, and smooth roads. The water supply facilities state includes faults, repair, and recovery. The population state includes buried, injured, dead, and safe transfer.

(3) The object layer of the derivative event chain: landslides, explosions, fire, infectious diseases, and dangerous goods leakage. The objects in this layer can be understood as the interface of the earthquake and other UEs, and we defined the states of the object as occurrence and non-occurrence. When the state of any object in this layer is occurrence, managers would execute the scenario evolution process of the corresponding UE object, which can also be understood as a process of nested execution.

The initial position of the fuzzy reasoning was the information input place of the hazard factor. In the process of the scenario evolution, we mainly studied the state change trend of the key hazard-affected body to make specific and effective responses to the implementation activities.

For the hazard factors object—Earthquake (E), Weather (W), and Time (T)—the degree of damage to key hazard-affected bodies and the probability value can be determined by the method of fuzzy statistics mentioned in this paper. The state change of the key hazard-affected bodies is affected by the combined effects of the 3 objects, and the relation between the 3 is "or." The most complete response possible is made, considering that the situation of the maximum extent of damage can simulate the greatest extent of the development of the situation. Assuming that the degrees of damage to key hazard-affected bodies are z_1, z_2 and z_3 for the 3 objects, then the initial token value of the FPN of the key hazard-affected body object is $Max(z_1, z_2, z_3)$.

The key hazard-affected body layer contains the objects of Population (P), Transportation (T), Communication (C), Building (B), Electricity Supply (SE), and Water Supply (SW). The comprehensive effect of the object of the hazard factors layer is abstracted as a place of FPN (P_0).

$$\beta(P_0) = D_0 = \{d_0\} = \left\{ \begin{array}{c} earthquake \\ disaster \end{array} \right\} \qquad (4)$$

$$\alpha(p_0) = y_0 = Max(z_1, z_2, z_3). \qquad (5)$$

The analysis of the scenario evolution of the key hazard-affected body layer is simplified in Fig. 3. It can be see that P_0 is the initial place and its token has 6 flow directions. Its compound rule is the relation of single inputting and multiple outputting, and the relation between the 6 output places is independent; namely, the FPN model diagram can be decomposed into the 6 branches of FPN. Then the development trend of the scenario evolution of the key hazard-affected body of earthquake can be obtained by doing research on the sub-FPN model.

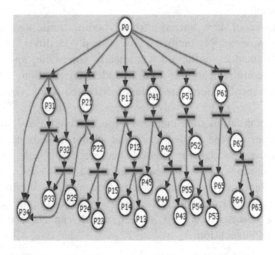

Fig. 3. FPN model for scenario evolution of earthquake disaster

Selecting the Population (P) in the left column as the research object. There were 3 transitions in the object's internal system, and any transition involved multiple inputting and multiple outputting. The T_{31} was decomposed into T_{311}, T_{312}, and T_{313} for different output places; T_{32} was decomposed into T_{321}, T_{322}, and T_{323}; T_{33} was decomposed into T_{331} and T_{332}. Assuming that the threshold value is λ, the fuzzy reasoning rules of the scenario evolution can be represented as shown in Fig. 4. The initial value and the CF parameters of the model are given:

$$y_0 = Max(z_1, z_2, z_3) = Max(0.9, 0.8, 0.7) = 0.9, \mu_{31} = \{\mu_{311}, \mu_{312}, \mu_{313}\} = \{0.3, 0.9, 0.8\},$$

$$\mu_{32} = \{\mu_{321}, \mu_{322}, \mu_{323}\} = \{0.4, 0.8, 0.7\}, \mu_{33} = \{\mu_{331}, \mu_{332}\} = \{0.8, 0.4\}.$$

According to the fuzzy reasoning algorithm, the growth of the reasoning tree is shown in Fig. 5. P_{33} and P_{34} represent the safe transfer and death of the population respectively. Both of them were the target place, and their evolutionary paths were not

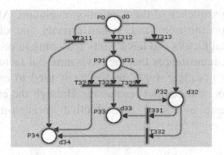

Fig. 4. FPN of population object P

only one. For P_{33}, there were 3 paths of evolution: (1) $P_0 \rightarrow P_{32} \rightarrow P_{33}$, (2) $P_0 \rightarrow P_{31} \rightarrow P_{33}$, and (3) $P_0 \rightarrow P_{31} \rightarrow P_{32} \rightarrow P_{33}$. The numerical values of the possibility of evolution were 0.504, 0.648, and 0.5185 respectively. For P_{34}, there were 4 paths of evolution: (1) $P_0 \rightarrow P_{34}$, (2) $P_0 \rightarrow P_{32} \rightarrow P_{34}$, (3) $P_0 \rightarrow P_{31} \rightarrow P_{34}$, and (4) $P_0 \rightarrow P_{31} \rightarrow P_{32} \rightarrow P_{34}$. The numerical values of the possibility of evolution were 0.27, 0.254, 0.324, and 0.2268 respectively.

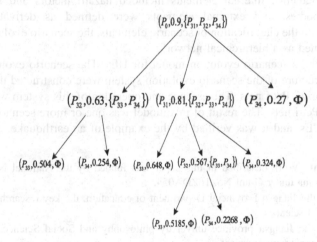

Fig. 5. Growth of the tree of reasoning based on the FPN

The above evolutionary paths scenario can be divided into 2 categories; one is the deteriorating deduction of the event, which is called a reverse evolution, and the other is the better deduction of the event, which is called a positive evolution. For a positive evolution the path of the highest evolutionary possibility is selected to respond, and for a reverse evolution the path of the lowest possibility is selected. According to this principle, we selected path (2), which took P_{33} as the destination place, and path (3), which took P_{34} as the destination place, as the final scenarios for the above 7 evolutionary paths. Considering the generation of implementation activities based on the 2 final scenarios; finally, the union of the corresponding set was taken as the final response implementing sequence of the population evolution. After introducing fuzzy Petri nets reasoning, on the one hand, it can directly and effectively express the inaccurate environmental factors and uncertainty reasoning in production rules. On the other hand, under the circumstances that the environmental factors are not completely clear or the solution is not clear, fuzzy reasoning is used to carry out effective rule matching, rule activation and conflict resolution. Through the effective integration of the positive and reverse evolution, the effective decision-making of "scenario-response" is realized.

5   Conclusion

This paper is based on the National Natural Science Foundation of China; we did a preliminary exploration and analysis of the law of UEs and found that the analysis of scenario evolutions plays an important part in emergency management. The main results of this paper are summarized as follows:

(1) Identifying elements of the scenario evolution system of UE and designing a system structure. The elements of the system were divided into internal and external elements. Internal elements included hazard factors and key hazard-affected bodies, and external elements were defined as derivative events. According to the classification of scenario elements, the scenario evolution system was designed as a hierarchical network.

(2) Constructing a scenario evolution model for UEs. The scenario evolution model and the structure of the scenario evolution system were constructed using object-oriented fuzzy Petri nets. The essence and operation of this system were realized by fuzzy Petri nets. The result of this model was one or more scenario evolution paths of UEs, and it was verified by the example of an earthquake.

Acknowledgment. Work described in this paper was funded by the National Natural Social Foundation of China under Grant No. 16ZDA054.

Supported by the Jiangsu Provincial Department of education, the key research base of philosophy and social sciences.

Supported by the Jiangsu province university philosophy and Social Sciences outstanding innovation team construction project.

References

1. Yun, J., Liu, Y., Wang, D.: Research on emergency management of unconventional emergencies in ethnic areas based on complex system. China Saf. Sci. J. **20**(3), 172–175 (2010)
2. Luo, A.C., Chen, S.W., Fang, C.Y.: Gaussian successive fuzzy integral for sequential multi-decision making. Int. J. Fuzzy Syst. **2**(17), 321–336 (2015)
3. Couso, I., Sánchez, L.: Upper and lower probabilities induced by a fuzzy random variable. Fuzzy Sets Syst. **165**, 1–23 (2011)

4. Nefti-Meziani, S., Oussalah, M., Soufian, M.: On the use of inclusion structure in fuzzy clustering algorithm in case of Gaussian membership functions. J. Intell. Fuzzy Syst. **28**(4), 1477–1493 (2015)
5. Bodwell, W., Chermack, T.: Organizational ambidexterity: integrating deliberate and emergent strategy with situation planning. Technol. Forecast. Soc. Chang. **77**(2), 193–202 (2010)
6. Feng, N., Zheng, C.: A cooperative model for IS security risk management in distributed environment. Sci. World J. **2014**, 1–10 (2014)
7. Kang, Q., Zheng, R.: Design and implementation of non-conventional emergency accident command platform. China Saf. Sci. J. **20**(3), 161–165 (2010)
8. Gao, X.X.: Establishment and application of intrinsic coal-workman evaluation model based on wavelet neural network. Procedia Eng. **45**, 225–230 (2012)
9. Wang, W.M., Lee, A.H.I., Peng, L.P., Wu, Z.L.: An integrated decision making model for district revitalization and regeneration project selection. Decis. Support Syst. **54**(2), 1092–1103 (2013)
10. Bian, R., He, J., Zhuang, Y.: Propagation and evolution model and Simulation of unconventional emergencies based on complex networks. Stat. Decis. **04**, 22–24 (2011)
11. Luo, X., Le, X., Zhou, L.: An efficient algorithm for parameter optimization of fuzzy Petri nets. Comput. Eng. Appl. **25**, 40–43 (2010)
12. Gu, Y., Wang, Z., Wu, Q.: Application of object-oriented Petri net technology in system modeling. J. Tongji Univ. (Nat. Sci.) **03**, 437–441 (2010)
13. Zhang, D.Y., Chen, H.S.: Security analysis of disaster recovery system using Stochastic Petri Nets. J. Tsinghua Univ. **51**(10), 1281–1286 (2011)
14. Pan, Y., Meng, J.E.: Enhanced adaptive fuzzy control with optimal approximation error convergence. IEEE Trans. Fuzzy Syst. **21**(6), 1123–1132 (2013)
15. Pohl, D., Bouchachia, A., Hellwagner, H.: Online indexing and clustering of social media data for emergency management. Neurocomputing **172**, 168–179 (2016)

Semi-physical Simulation of Fuel Loss for Unmanned Aerial Vehicle

Wenhao Xiang[1], Shufen Liu[2], Tong Yu[2], and Shuqiu Li[2(✉)]

[1] Systems Engineering Research Institute, Beijing, China
xiangwh2018@163.com
[2] Jilin University, Changchun, China
{liusf,shuqiu}@jlu.edu.cn, yutong2109@yeah.net

Abstract. There are errors in the simulation of UAV (Unmanned Aerial Vehicle) fuel gauge. To solve this problem, by analyzing the fuel consumption factors of UAV, a flight fuel consumption calculation model is established, which makes the data generated by the simulator closer to the real environment. At the same time, the semi-physical simulation of computer simulation is realized, which effectively compensates for the disadvantage of the great difference between computer simulation and real environment.

Keywords: UAV fuel consumption · Computational model · Semi-physical simulation

1 Introduction

At present, UAV is widely used and plays its role in various fields. The fuel consumption of UAV has also been raised to the research agenda. The fuel gauge of UAV is located in the instrument monitoring area, which is the second important display area in the flight control part [1]. The instrument monitoring area presents the fuselage status of UAV in real time during the mission, including some data: Oil meter propeller temperature, waste air temperature, rotor speed, link communication status, cooling water temperature and so on [2–4]. This series of data is an important reference for the normal flight of UAV. In the current UAV simulation control system, instrument monitoring area usually only realizes visual simulation, according to the actual flight situation [5], this flight control part establishes a calculation model of fuel loss in the instrument monitoring area, and uses the data in the calculation model to carry out semi-physical simulation of the fuel loss area [6–8].

2 Analysis of Flight Fuel Consumption Factors

2.1 Important Parameters

(1) Basic aerodynamic parameters
The aerodynamic parameters of UAV include lift coefficient, drag coefficient, lift-drag ratio and so on. The basic aerodynamic parameters of aircraft with different

© Springer Nature Switzerland AG 2019
D. Milošević et al. (Eds.): HCC 2019, LNCS 11956, pp. 468–475, 2019.
https://doi.org/10.1007/978-3-030-37429-7_46

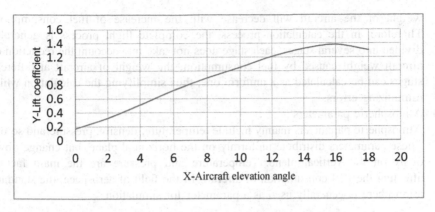

Fig. 1. Elevation and lift coefficient curves of aircraft

mod[p] between different aircraft individuals is different, so it can not be expressed by functions, as shown in Figs. 1 and 2. The basic aerodynamic parameters of different types of aircraft are measured by wind tunnel test or flight test, and then stored in the database for use.

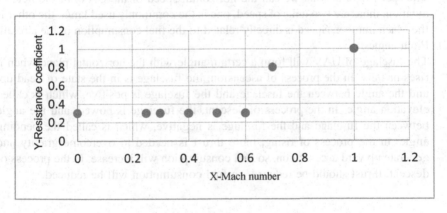

Fig. 2. Scatter plot of aircraft mach number and drag coefficient

(2) Start Engine Performance Parameters

The performance of the engine has a linear relationship with the fuel consumption of the UAV. The fuel consumption of the engine is determined by the type, speed and height. An important reference index of engine is fuel consumption rate TSFC. Unlike the basic aerodynamic parameters, the fuel consumption rate of the engine is a function of speed, height and thrust.

(3) Aircraft weight

The weight of UAV refers to the sum of the weight of the fuselage and the carrying load. The larger the weight of UAV, the greater thrust is needed when it rises, thus increasing fuel consumption. In the course of UAV's mission, the weight of the aircraft will decrease with the increase of fuel consumption. Therefore, in the calculation process, the complete flight process is generally divided into several stages, each stage does not take into account the reduction of aircraft weight caused by fuel consumption, the weight of aircraft at different stages can be calculated as a uniform one, thus simplifying the calculation while minimizing errors.

(4) Atmospheric parameters

Atmospheric parameters mainly include temperature, density, pressure and so on. These parameters distribute uniformly on the horizontal plane, but change obviously on the vertical plane. Temperature and pressure are the main factors affecting the fuel consumption of aircraft. In the field of aerospace, the standard atmosphere is generally used as a parameter for simulation.

2.2   Aircraft Trajectory

The basic motion state of UAV is basically displayed in HUD area, including speed, altitude, pitch angle, roll angle, deflection angle, acceleration, heading and so on. These are the factors that affect the fuel consumption of UAV.

(1) Speed

The speed here is what we call the horizontal speed of the UAV. In the field of aviation, there are two units of speed, one is the commonly used km/h, the other is the Mach number. Speed is directly related to the fuel consumption of the aircraft.

(2) Pitch angle

The fuselage of UAV will form a certain angle with the horizontal plane when it rises or falls. In the process of ascension, the fuselage is in the state of head-up, and the angle between the fuselage and the fuselage is positive, which is called elevation angle. In the process of descent, the fuselage is bowed, and the angle between the fuselage and the fuselage is negative, which is called the bending angle. In the process of rising, more thrust is needed to overcome gravity and generate upward acceleration, so fuel consumption will increase. In the process of descent, thrust should be reduced, so fuel consumption will be reduced.

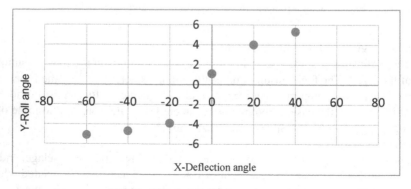

Fig. 3. Roll and deflection scatter plots

(3) Roll angle and deflection angle

When UAV is turning, it is affected by centrifugal force, gravity and resistance. The fuselage will form an inclined angle, which is called roll angle (also known as roll angle). The turning radius and turning rate are determined by roll angle. Deflection angle is the heading change of UAV when turning. When the aircraft is turning, it needs more thrust to maintain the hover without changing its altitude. Figure 3 shows the angles of a series of roll angles and deflection angles intercepted during UAV turning.

(4) Acceleration

The acceleration of aircraft is divided into horizontal acceleration and vertical acceleration. Figure 4 is a scatter plot of the vertical acceleration and fuel volume intercepted by the UAV during its mission.

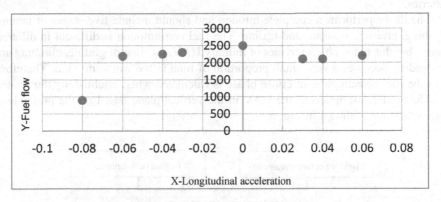

Fig. 4. Longitudinal acceleration and fuel volume scatter plot

3 Flight Fuel Consumption Calculation Model

Among the factors affecting flight fuel consumption summarized in the previous section, aerodynamic parameters are a complex variable: Firstly, the aerodynamic parameters of different types of aircraft are different. Secondly, the relationship between different aerodynamic parameters can not be expressed in the form of functions. Therefore, in this paper, the aerodynamic parameters of aircraft are not considered when establishing flight fuel consumption model.

There is a linear relationship between engine performance parameters and fuel consumption (with velocity v, height h, thrust F), therefore, the engine performance parameters can be used to establish the calculation model.

The most direct relationship between aircraft trajectory parameters and fuel consumption. Shear speed, altitude, pitch angle, roll angle and so on can be used to establish the calculation model, so the parameters of aircraft trajectory are also included in the flight fuel consumption model. Because the weight of UAV fuselage and load can be accurately obtained, and the calculation is not complicated, the weight of UAV can be used to build flight fuel consumption model.

Normal atmospheric parameters have a certain impact on fuel consumption in flight, but in actual flight, the critical difference of altitude range is not big in plane flight, so atmospheric parameters have little influence on fuel consumption model. In the process of establishing the flight fuel consumption calculation model in this paper, the atmospheric factors are not considered.

In this paper, based on the principle of energy balance, the following conditions are assumed before the flight fuel consumption model is established:

The weight change of UAV caused by the decrease of fuel consumption does not affect the calculation.

When the UAV is accelerating, it is regarded as uniformly accelerating motion, that is, the change of speed and height is linear.

Fuel consumption is used for the motion of UAV, neglecting the energy provided for other devices. In fact, in UAV, most of the energy of other devices is supplied by batteries.

The drone performs a complete mission and should include five stages of taxiing, climbing, cruising, descent, and taxiing. The fuel consumption is different in different stages, but due to the characteristics of military UAV, the time of gliding, climbing and descending occupies a very small proportion of time in the whole mission. Therefore, only the fuel consumption in cruise phase is calculated when establishing the model.

The model description of the UAV in the vertical plane based on the principle of energy balance is shown in Fig. 5:

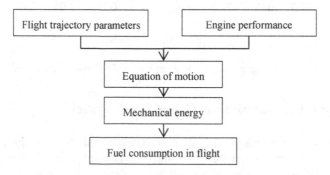

Fig. 5. Energy principle to calculate aircraft fuel consumption

Suppose: the height is h, the speed is v, the acceleration of gravity is g, the mass of UAV is m1, load mass on board is m2, the thrust produced by the engine is T, air resistance f, lift is L, the pitch angle is α, roll angle is γ, the forward and longitudinal coordinates of UAV are x and y, respectively. Thus, the following equations can be obtained:

$$\mathrm{m} \times \frac{dv}{dt} = T \times \cos\alpha - f - (m1 + m2)g \times \sin\gamma \tag{1}$$

$$\mathrm{m} \times \mathrm{v} \times \frac{d\gamma}{dt} = T \times \sin\alpha + L - (m1 + m2)g \times \cos\gamma \tag{2}$$

$$\frac{dy}{dt} = v \times \sin\gamma \tag{3}$$

$$\frac{dx}{dt} = v \times \cos\gamma \tag{4}$$

The energy height of UAV is:

$$\mathrm{E} = \mathrm{h} + \frac{1}{2g} \times v^2 \tag{5}$$

The product of the energy height and the mass of the UAV is the total energy of the UAV. According to the principle of energy balance, the following state models can be obtained:

$$\begin{cases} E = h + \frac{1}{2g}v^2 \\ E = \frac{v \times (T \times \cos\alpha - f)}{mg} \\ x = v \times \cos\gamma \end{cases} \tag{6}$$

The Mf formula of fuel consumption is as follows:

$$\mathrm{dMf} = C_e P dt \tag{7}$$

In this formula, Ce is the fuel consumption rate and P is the rated thrust of the engine. So the formula of fuel and energy consumption height for UAV is:

$$\frac{dE}{dMf} = \frac{dE/dt}{dMf/dt} = \frac{dE/dt}{CeP} \tag{8}$$

The fuel consumption rate of engine Ce varies with the type and type of aircraft. The fuel consumption rate of passenger aircraft is generally 0.1–0.5. The fuel consumption rate of fighter aircraft is generally 0.7–1.5. The fuel consumption rate of UAV in this paper is about 0.1/kg thrust per hour. The fuel consumption of UAV in cruising phase can be deduced from the Eqs. 7 and 8.

4   Oil Meter Simulation and Verification

The following verification methods are used:

(1) Write the aerodynamic parameters, engine performance parameters and weight of a certain type of UAV into the program;
(2) Estimate the endurance capability of the UAV in horizontal flight according to the actual situation;

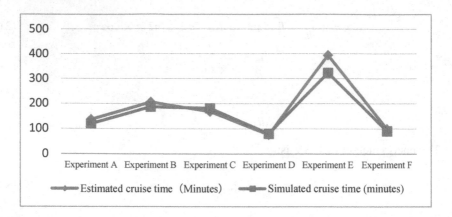

Fig. 6. Schematic simulation of oil meter

(3) In the flight simulation control system, the data are simulated manually according to the performance of the UAV.

(4) Observing the endurance of the UAV in the flight control section;

(5) To verify the validity of the oil meter simulation, 6 experiments were carried out on the UAV with different fuel loads and different parameters.

Figure 6 shows the intention of simulating the fuel consumption of a UAV. The trend of simulated fuel consumption is consistent with that of real fuel consumption. This shows that the semi-physical simulation is closer to the real fuel consumption than the pure computer simulation.

5   Conclusion

In the traditional flight control system, due to the deficiency of computer simulation in real environment simulation, in order to make computer simulation closer to physical simulation, the flight control part of this paper introduces the UAV flight model, and takes the fuel consumption of UAV as an example to simulate the flight data. According to the fuel consumption factors of UAV, the fuel consumption model is established, and the semi-physical simulation of UAV fuel consumption is designed and realized.

This paper only completes the semi-physical simulation of fuel consumption. In the process of UAV's mission, there are more data in computer simulation which are quite different from the real environment, such as rotor speed, cooling temperature and so on. All of these can be used for semi-physical simulation based on the analysis of flight model, so that the computer simulation can be closer to the real environment.

References

1. Nouri, J., Lotfi, F.H., Borgheipour, H., Atabi, F., Sadeghzadeh, S.M., Moghaddas, Z.: An analysis of the implementation of energy efficiency measures in the vegetable oil industry of Iran: a data envelopment analysis approach. J. Cleaner Prod. **52**, 84–93 (2013)
2. Wu, Z., Wang, L., Lin, J., Ai, J.: Investigation of lateral-directional aerodynamic parameters identification method for fly-by-wire passenger airliners. Chin. J. Aeronaut. **27**(4), 781–790 (2014)
3. Jeong, S.H., Jung, S.: A quad-rotor system for driving and flying missions by tilting mechanism of rotors: From design to control. Mechatronics **24**(8), 1178–1188 (2014)
4. Zhao, Q., Yu, L., Song, G.: Comparison of microscopic on-road fuel consumption model based on VSP and Ln(TAD). J. Syst. Simul. **27**(11), 2662–2669 (2015)
5. Xie, Y., Savvaris, A., Tsourdos, A., Laycock, J., Farmer, A.: Modelling and control of a hybrid electric propulsion system for unmanned aerial vehicles. In: IEEE Aerospace Conference (2018)
6. Yao, Z., Wu, S.: Intermittent gliding flight control design and verification of a morphing unmanned aerial vehicle. IEEE Access **7**, 40991–41005 (2019)
7. Aksaray, D., Griendling, K., Mavris, D.: UAVs for law enforcement: a case study for connectivity and fuel management. In: AIAA Aviation Technology, Integration, and Operations Conference (2014)
8. Wang, Y., Shi, Y., Cai, M., Xu, W., Yu, Q.: Efficiency optimized fuel supply strategy of aircraft engine based on air-fuel ratio control. Chin. J. Aeronaut. **32**(2), 489–498 (2019)

An Early Warning System of Tram Safety Protection Based on Multi-information Detection

Binjie Xiao[✉]

Shanghai Second Polytechnic University, Shanghai, China
Binjiexiao@163.com

Abstract. Aiming at the existing collision accidents with motor vehicles and pedestrians in the operation of tram, a tram safety protection early warning system is proposed, which can automatically analyze and interpret dangerous objects in the direction of moving forward. The system uses ranging sensor (Ultrasonic, Camera, Ranging radar) and infrared video information to obtain the accurate distance between the obstacle and the locomotive. The obstacle type and potential running speed can be obtained by infrared image recognition and classification. According to the direction of the tram, the distance, type and speed of the obstacles, the safety level of the obstacles can be determined. The above analysis and interpretation results are displayed intuitively on the on-board display and reminder unit. Relevant technologies can effectively improve the ability of the tram driver to obtain information, automatically detect obstacles and judge the degree of danger, and improve the safety of the tram.

Keywords: Trams · Safety protection · Obstacle detection · Safety level · Sensor detection

1 The Introduction

With the construction and operation of trams in China, there are many traffic accidents of trams and motor vehicles collision. Even for tram systems operated for many years in European, the annual statistical safety accidents also include the collisions between trams and motor vehicles and non-motor vehicles, and accidents such as trams hitting pedestrians [1–3]. How to ensure the safety of tram operation under the condition of mixed road right is the key technology of tram safety operation. In view of the complicated traffic condition and large traffic volume in China, the driver adopts emergency braking according to the driver visual, which causes the braking distance to be big. The above factors cannot guarantee the smooth operation of trams under a mix of modes of transportation, nor can they avoid the safety threats of trams to motor vehicles and pedestrians.

© Springer Nature Switzerland AG 2019
D. Milošević et al. (Eds.): HCC 2019, LNCS 11956, pp. 476–482, 2019.
https://doi.org/10.1007/978-3-030-37429-7_47

2 Demand Analysis of Tram Active Safety Protection System

Tram is an effective way of public transport, which uses a hybrid way under the condition of driver visual, operating control for central control mode. The driver take the functions of the main driving, which need to face more uncertainty and risk factors, such as motor vehicles and pedestrians into, locomotive fault, road traffic accident etc. In the running of trams, drivers should pay close attention to the traffic conditions at the front and side, and take timely and effective measures to ensure traffic safety [4–6].

If the tram can take the initiative to obtain the dangerous information in front of the driving and display the warning in the driver control room; emergency braking can be carried out through the vehicle controller to prevent risks when an emergency is confirmed, reducing accident hazards and ensure driving and traffic safety along the way. In the construction and design stage of trams, video monitoring and on-board cameras along the way are mostly adopted to deal with emergencies combined with the operation and dispatch personnel of the control center. Video monitoring can only provide visual images and post-forensics information, but cannot issue early warning and perceive traffic safety risks.

According to the operation characteristics of trams and road traffic conditions, the sensors, intelligent analysis and other technical means are used to achieve active safety monitoring and protection early warning, to overcome the physical and climatic constraints on lookout driving. Under various conditions, the system can automatically acquire and identify the potential obstacle information in the direction of the tram. The driver can judge the danger level of the obstacle according to the type, distance, speed and other information of the obstacle, and acquire the danger information in the on-board display equipment and voice equipment. The system effectively improves the ability of the tram driver to obtain information and improves the safety of the tram.

3 Tram Active Safety Protection System Scheme

3.1 System Composition

The tram safety protection warning system is composed of several ranging sensors, infrared video camera and safety protection control mechanism (see Fig. 1). The ranging sensor and infrared video camera are electrically connected with the safety protection control mechanism respectively. There are 3 infrared video cameras. One infrared video camera is set in the middle of the upper edge of the front part of the tram, and the other two infrared video cameras are set at the junction of the front part and the left side wall of the tram front part and the right side, 0.2–0.4 m above the ground. There are three ranging sensors, one of which is located in the middle of the upper edge of the front wall of the tram head, and the other two are located 0.2–0.4 m above the ground on the left side wall and right side wall of the tram head respectively. The fully protected control mechanism includes a safety protection controller, an on-board display device and a voice playback device. The safety protection controller is electrically connected with the ranging sensor, infrared video camera, vehicle-mounted display device and voice playback device respectively.

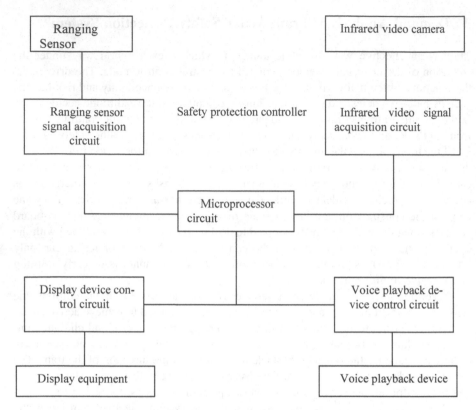

Fig. 1. System composition diagram

The safety protection controller includes the ranging sensor signal acquisition circuit, infrared video signal acquisition circuit, display equipment control circuit, speech playback control circuit and microprocessor circuit. Distance sensor signal acquisition circuit and distance sensor electrical connection, infrared video signal sampling circuit and the infrared video camera electrical connection, the display equipment control circuit and the on-board equipment electrical connection, voice broadcast control circuit and voice broadcast equipment electrical connection, distance sensor signal acquisition circuit, infrared video signal sampling circuit, display device, control circuit and voice broadcast control circuit with microprocessor circuit electrical connection. The microprocessor circuit is electrically connected to the braking control system of the tram. The ranging sensor is an ultrasonic ranging sensor. The distance sensor of the tram safety protection warning system is the laser radar distance sensor, which is composed of the distance sensor (ultrasonic wave, camera, Ranging radar), infrared video camera, vehicle host, vehicle display and reminder unit.

3.2 System Solutions

The system uses ranging sensors (ultrasonic, camera, Ranging radar) to detect approaching obstacles in the direction of the tram. The infrared video camera collects the image of the tram's forward direction, which is not limited by light, weather and time. It can accurately collect the image of the tram's forward direction in rain, snow or at night. The infrared video analysis module of the vehicle host of the system will combine with the ranging sensor to detect signals and analyze obstacles in the direction of the tram.

The obstacles detected by the on-board host of the system are classified into motor vehicles, two-wheeled non-motor vehicles and pedestrians according to the pattern features. The on-board host of the system can automatically divide the early warning safety level according to the classification result of obstacle detection and the distance from the trolley head, and display and remind in the on-board display and remind unit. At the same time, the system takes the track of the tram into consideration and lights dangerous obstacles in the corresponding predicted area.

The sensor are installed at the front and rear locomotives of a tram. After the detected obstacle is identified and classified, it is displayed on the vehicle display and alarm unit with the detection distance of the ranging sensor. The vehicle-mounted host collects 2–1 to 2–3 information of the locomotive and interprets the distance of obstacles.

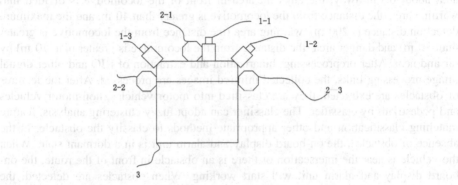

Fig. 2. System schematic diagram of sensor installation

The installation diagram of the sensor is shown in Fig. 2, where 1 is the infrared camera, 1–1 is installed in the central position directly in front of the locomotive, 1–2 and 1–3 are installed in the left front and right front of the locomotive at an Angle of 15 degrees; 0.3 m above the ground; 2 is the ranging sensor (ultrasonic, camera, or radar detection unit), 2–1 installed in the front of the central position, 2–3 installed on the left and right side of the locomotive; 0.3 m above the ground; Fig. 3 is the schematic diagram of the locomotive. In addition, the on-board host and on-board display and alarm unit are installed in the cab to facilitate direct vision of the appropriate area.

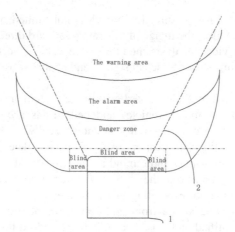

Fig. 3. System display and warning ranges

As shown in Fig. 3, in addition to the blind area in front of the locomotive and the area about 0.5 m away, the adjacent area in front of the locomotive is divided into warning area (the distance from the locomotive is greater than 40 m, and the maximum detection distance is 200 m), warning area (the distance from the locomotive is greater than 30 m) and danger area (the distance from the locomotive is greater than 10 m) by far and near. After preprocessing, binarization and extraction of RIO and other digital image processing links, the collected infrared images are processed. After the features of obstacles are extracted, they are classified into motor vehicles, non-motor vehicles and pedestrians by classifier. The classifier can adopt fuzzy clustering analysis, feature matching classification and other appropriate methods to classify the obstacle. In the absence of obstacles, the on-board display and alarm unit is in a dormant state. When the vehicle is near the intersection or there is an obstacle in front of the route, the on-board display and alarm unit will start working. When obstacles are detected, the images displayed after classification and recognition are displayed.

The system need detect whether there are any other obstacles within 200 m ahead of the train that threaten the safety of the tram. According to the distance of obstacles in front of the car, left front and right front, it is divided into three levels of display and alarm. When the ranging sensor detects an obstacle that has not yet affected the trolley, the warning area is light green. The warning area is shown to be orange and yellow, which may be harmful to the tram. Obstacles requiring proper braking are required. The danger zone is shown in red with obstacles that are extremely hazardous to the trolley and require immediate braking.

According to the degree of danger, the following alarm prompt will be given when the obstacle appears in the warning area. The vehicle display and the alarm unit will display the obstacle flashing and voice prompt when the obstacle appears in the dangerous area, the on-board display and alarm unit will display the obstacle flashing and voice prompt, and link the automatic braking unit for the tram driver to take emergency measures. The Angle between the two infrared video cameras and the extension cord of the front wall is 15°. It has a larger setting angle, so that the monitoring range is larger.

The on-board display device is set in a position that is convenient for the driver to look straight at.

The ranging sensor is an ultrasonic ranging sensor and other types of ranging sensors may be used in other embodiments. Infrared video cameras gain video of the direction of the train left to right in front of a range of infrared image. Infrared video signal is sent to microprocessor circuit, in order to extract obstacles. The types of obstacles are classified according to the characteristics of obstacles by using the pattern recognition classification algorithm, such as motor vehicles, non-motor vehicles or pedestrians.

4   The Early Warning Function of Tram Active Safety Protection

Ultrasonic ranging sensors accept the reflection of obstacles of ultrasonic measurement distance by a barrier of launch ultrasonic wave. The sensor signal acquisition circuit will collect and transmit obstacle distance information from ultrasonic distance sensor to the microprocessor circuit. The different risk levels are confirmed combined with the kinds of obstacles and the obstacles information. Obstacle information, danger level and braking measures are transmitted to the on-board display equipment and voice playback equipment.

The specific criteria for the classification of the danger level are determined according to the actual situation of the project and the specific traffic of the running area of the train, as well as the running speed and braking performance of the train itself. When the pedestrian appears 100 m ahead of the train, the on-board unit will display the pedestrian warning information and flash, and the voice of emergency brake warning sound. If the driver fails to take braking measures within five seconds after sending the above warning information and voice prompt, the microprocessor circuit will give instructions to the braking control system of the train and take automatic braking measures. The early warning of active safety protection of tram can remind the driver to attach importance to the current safety hidden danger by the way of sound and light alarm according to the danger level. In case of emergency or when the driver fails to take measures, the system realizes emergency braking through linkage with the vehicle controller to avoid or reduce accident losses.

5   Conclusion

Under various conditions, the system can automatically acquire and identify the potential obstacle information of the tram's direction, and overcome the influence of physical and climatic factors. The system judges the danger level of the obstacle according to the type, distance, speed and other information of the obstacles, and at the same time provides warning information to the driver through the on-board display equipment and voice playback equipment. The system effectively improves the ability of the tram driver to obtain information and improves the safety of the tram.

References

1. Ming, R.: Safety analysis of tram system. Urban Traffic (4), 14–20 (2016)
2. Wang, Y.: A brief analysis of safety assistance system for modern tram operation. Commun. World (11), 68–70 (2016)
3. Yang, H., Ye, M., Wu, A., et al.: Hazard identification and risk assessment for the safe operation of modern trams. Mod. Transp. Technol. (02), 79–84 (2017)
4. Shen, H., Hu, C., Ye, Y., et al.: Urban tramcar safety assurance system based on RFID, WIFI, GPS and infrared monitoring technology. Railway Energy Conserv. Environ. Prot. Saf. Health 4(3), 23–26 (2014)
5. Wen, Z.: Analysis of main factors affecting the operation safety of modern trams and countermeasures. Mod. Urban Rail Transit (4), 69–71 (2018)
6. Tao, S., Liang, Y., Wang, Z.: Safety improvement measures for modern tram operation in Hunnan new district, Shenyang city. Urban Rail Transit Res. 19(7), 116–121 (2016)

A Low-Frequency Broadband Triboelectric Energy Harvester Based on Cantilever Beam with a Groove

Xin Hu, Fang Cheng, Gang Tang$^{(\boxtimes)}$, Bin Xu, Zhibiao Li, Xiaoxiao Yan, and Dandan Yuan

Jiangxi Province Key Laboratory of Precision Drive & Control, Nanchang Institute of Technology, Nanchang 330099, China
tanggangnit@163.com

Abstract. This paper introduces a contact separation triboelectric energy harvester (TEH) using a cantilever beam with a groove to work at low-frequency. The designed TEH exhibits broadband behavior which is induced in the cantilever motion due to contact between two triboelectric surfaces. The open-circuit peak output voltage and output power of this fabricated prototype are 64 V and 5.4 µW at low resonant frequency of 13 Hz, respectively, when it matches an optimal loading resistance of 1.2 MΩ under the excitation of 0.9 g acceleration. Moreover, an operating frequency bandwidth of 9.2 Hz for the TEH device can be obtained at an acceleration level of 1.1 g.

Keywords: Triboelectric · Energy havesting · Broadband · Low-frequency

1 Introduction

With the rapid development of science and technology, many low-power electronic devices and sensors are gradually emerging [1–5]. The power consumption of these devices is becoming more and more low and even with the level of microwatts. Thus, it is feasible to use energy harvesters instead of limited lifetime batteries to power these devices. Due to the abundance in our daily life, mechanical vibration energy has been considered an effective energy source and can be transformed into useful electrical power by any kind of electromechanical transduction, including electrostatic [6, 7], electromagnetic [8, 9], piezoelectric [10, 11] and triboelectric [12, 13]. Compared with other types of energy harvesters, triboelectric energy harvester (TEH) has several advantages including frequency independency which make it broadband, high energy conversion efficiency, outstanding output power density and low cost [14, 15].

In this work, we proposed a triboelectric vibrational energy harvester based on a cantilever-structure with a groove to realize low-frequency and broadband operation. Moreover, Ecoflex was used as triboelectric materials and Ecoflex micropollars to enhance the triboelectric generation. The fabrication process of the designed TEH was discussed and the formed prototype was characterized.

© Springer Nature Switzerland AG 2019
D. Milošević et al. (Eds.): HCC 2019, LNCS 11956, pp. 483–492, 2019.
https://doi.org/10.1007/978-3-030-37429-7_48

2   Device Design and Fabrication

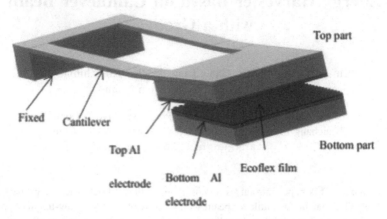

Fig. 1. The structural diagram of the triboelectric energy harvester

The structure of the proposed triboelectric energy harvester is shown in Fig. 1. The structure mainly includes a cantilever beam with a groove, the top triboelectric part and the bottom triboelectric part. The top triboelectric part is fixed on the cantilever beam, consist of a glass substrate coated with Aluminum tape, the size of the glass substrate is 2 cm × 2 cm. The bottom triboelectric part constitutes of an array of Ecoflex micropillars on a glass substrate coated with an aluminum film, which is kept fixed when the device is in operation.

The cantilever beam with a groove is fabricated by 3D printing using PLA material. The fabrication steps of top and bottom triboelectric part are shown in Fig. 2. Both the top and bottom triboelectric part are fabricated starting with preparation of glass substrate. As shown in Fig. 2(a), the top triboelectric part of TEH, a glass substrate is laid with a 100 μm thick laminated aluminum layer. The aluminum film is used as the top electrode and the first triboelectric layer. The fabricated top triboelectric part is then attached to the printed cantilever beam using epoxy adhesive. For the fabrication of bottom triboelectric part in Fig. 2(b), the key technique is the preparation of Ecoflex micropad structure. Ecoflex can be more negative than PDMS and other common materials, making it easier to obtain electrons. The fabrication processes begin with bare silicon wafer which was cleaned by acetone and IPA followed by oxygen plasma. The micro round holes are then prepared on silicon wafer by laster cutting, which are used as a mold to fabricate Ecoflex micropad structures. The micro structure can increase the contact area and improve the output performance [16]. Thereafter, a mixture is prepared according to a volume ratio of 1:1 for A and B of the EcoFlex 00-30 and poured into the mold for thin film casting followed by a 20 min baking at 70 °C for curing. After curing EcoFlex mixture, the EcoFlex film can be peeled off from silicone mold and micropillars can be seen on the EcoFlex film. Finally, the EcoFlex film with micropillars is attached to a glass substrate coated with aluminum tape which is used as the bottom electrode of TEH.

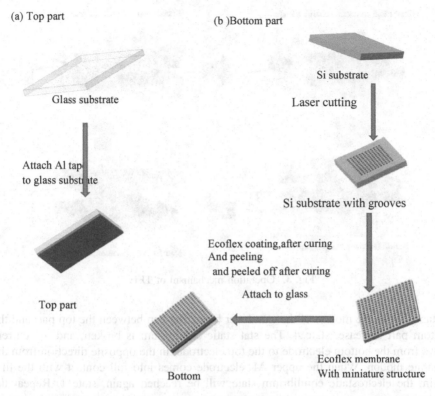

(a) Top part

Glass substrate

Attach Al tape
to glass substrate

Top part

Bottom

(b)Bottom part

Si substrate

Laser cutting

Si substrate with grooves

Ecoflex coating,after curing
And peeling
and peeled off after curing

Attach to glass

Ecoflex membrane
With miniature structure

Fig. 2. (a) Fabrication steps for the top parts of TEH. (b) Fabrication steps for the bottom parts.

3 Operating Mechanism and Analysis

3.1 Operating Mechanism

The working process of TEH is illustrated in Fig. 3. The periodic contacting and separating motion between top and bottom parts is realized by the vibrating motion of cantilever when excited by mechanical vibrations. In the initial state, there is a gap between the two parts, at this time, no electric potential difference between the two plates. Compare with aluminum, ecoflex00-30 has stronger electron acquisition ability, and it is easy to form negative charge. Therefore, in state 1, when the cantilever beam starts to vibrate under stress. the top aluminum electrode contacts the bottom ecoflex00-30 film, and equal amount of different charges are evenly distributed on the material surface. Top aluminum electrode has a positive charge, and bottom ecoflex00-30 film has a negative charge. At this moment, there is no charge transfer. The cantilever continus to vibrate and the plates begin to separate, the potential of top aluminum electrode being higher than that of bottom aluminum electrode. Now, electrons are transferred from the bottom electrode to the top electrode, and the current flows from the top electrode to the bottom electrode, state 2. The current continues until state 3 disappears. At this point, the two parts reach electrostatic balance, the maximum

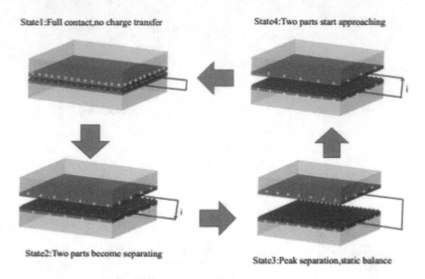

State1:Full contact,no charge transfer

State4:Two parts start approaching

State2:Two parts become separating

State3:Peak separation,static balance

Fig. 3. Operation mechanism of TEH

distance. With the motion of the cantilever beam, the gap between the top part and the bottom part decrease, state 4. The stat static equilibrium is broken, and the current flows from the bottom electrode to the top electrode, in the opposite direction from the separate motion. When the upper AL electrode comes into full contact with the film again, the electrostatic equilibrium state will be reached again, state 1. Repeat the process.

3.2   Frequency Reduction

In order to reduce the working frequency of TEH and get closer to practical application, the cantilever beam structure is designed as a slotted structure on the beam. Figure 4 is the finite element analysis of slotted cantilever beam and unslotted cantilever beam structure. From the figure, we can see the total deformation of the cantilever beam and the first order resonace frequency. The first order resonace frequency of slotted can-tilever beam structure is 21.979 Hz, and the another cantiliver is 36.656 Hz. The reason is the mechanical damping and bending stiffness of the slotted cantilever beam are lower [17]. In general, the working frequency is near the resonance frequency. It can be seen that the working frequency decreases, achieving the purpose of reducing the working frequency.

(a)

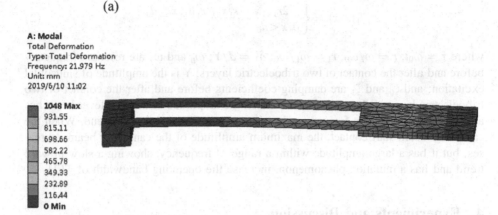

A: Modal
Total Deformation
Type: Total Deformation
Frequency: 21.979 Hz
Unit: mm
2019/6/10 11:02

1048 Max
931.55
815.11
698.66
582.22
465.78
349.33
232.89
116.44
0 Min

(b)

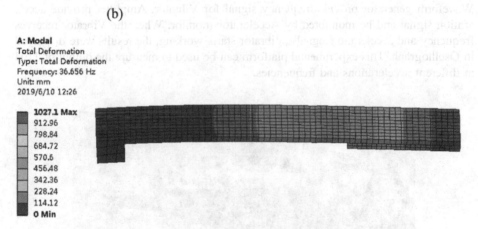

A: Modal
Total Deformation
Type: Total Deformation
Frequency: 36.656 Hz
Unit: mm
2019/6/10 12:26

1027.1 Max
912.96
798.84
684.72
570.6
456.48
342.36
228.24
114.12
0 Min

Fig. 4. (a) Finite element analysis and first order frequency of cantilever beam with special structure. (b) Finite element analysis and first order frequency of complete cantilever beam.

3.3   Analysis of Broadband Behavior

According to the present research, the modeling of broadband behavior can be simulated by the piecewise linear oscillator model. The equation can be written as in Eq. (1) [18, 19]

$$\ddot{x} + 2\zeta_0\dot{x} + x = r^2 \sin(r\tau) + f_1(x, \dot{x}) \tag{1}$$

The expression for $f_1(x, \dot{x})$ is given by Eq (2):

$$f_1(x,\dot{x}) = \begin{cases} -2r_1\zeta_1\dot{x} - r_1^2x + r_1^2\delta_1, x \geq d_0 \\ 0, x < d_0 \end{cases} \tag{2}$$

where $\tau = \omega_0 t$; $r = \omega/\omega_0$; $r_1 = \omega_1/\omega_0$; $\delta_1 = d/Y$; ω_0 and ω_1 are resonant frequency before and after the contact of two triboelectric layers; Y is the amplitude of sinusoidal excitation; and ζ_0 and ζ_1 are damping coefficients before and after the contact of two triboelectric layers. When there is no contact, the amplitude of the cantilever beam has a peak value at a certain frequency, showing a sharp rise and decline trend. When the cantilever beam start contact, the maximum amplitude of the cantilever beam decreases, but it has a large amplitude within a range of frequency, showing a slow upward trend and has a mutation phenomenon, increase the operating bandwidth of TEH.

4 Experiments and Discussion

4.1 Experimental Setup

Figure 5 is the test platform for the experiment, mainly by the Vibrator, Amplifier, Waveform generator, Oscillograph and Accelerator monitor composition and so on. Waveform generator provide frequency signal for Vibrator, Amplifier provide acceleration signal and be monitored by Accelerator monitor. When the Vibrator receives frequency and acceleration signals, Vibrator starts working, the results were displayed in Oscillograph. This experimental platform can be used to measure the output of TEH at different accelerations and frequencies.

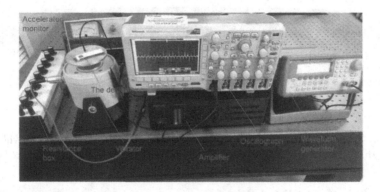

Fig. 5. Test platform for TEH

4.2 Voltage and Power Characteristics

In order to explore the output performance of TEH, the above experimental platform is used to measure the open-circuit TEH output under the acceleration of 0.5 g to 1.1 g shown in Fig. 6. Theoretically, with the increase of acceleration, the output of TEH also increases. As the acceleration increases, the contact force between the top and bottom contact surfaces increases, and the chemical bond formed between ecoflex00-30

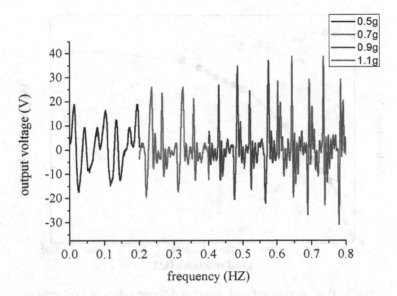

Fig. 6. Voltage response of TEH in time domain at different acceleration levels

film and metal Al becomes stronger, leading to the increase of surface charge and the amount of charge transfer, thus making the output performance better. The test data also verified this point. With the increase of acceleration, the open circuit voltage of TEH increased from 0.5 g and 36 V to 1.1 g and 64 V.

To further investigate the output performance of TEH, the load resistor was connected to the test circuit, the output power under different load resistances is show in Fig. 7. The resistance of TEH is $Z_s = R_s + jx_s$, the external load is $Z_L = R_L + jX_L$, the output power can be obtained by calculation Eq (3) [20]:

$$P = \frac{R_S V_S^2}{(R_S + R_L)^2 + (X_L + X_S)^2} \tag{3}$$

According to Eq (3), when the resistance value of TEH is the same as the load, the maximum power can be reached. As shown in Fig. 7, the output power increases with the increase of the load. When the peak value reaches 5.4 μw, the output power starts to decrease with the increase of the load. At this time, the load is 1.2 MΩ, so the resistance value of TEH is about 1.2 MΩ.

In practical application, since the energy generated by TEH is AC pulses and unstable, it cannot directly supply power to the load, so it is necessary to convert the AC power into DC power, and charge the capacitor or rechargeable battery. Use a full-wave bridge rectifier for rectification, shown in Fig. 8(b). When the TEH charges different figure capacitors, Fig. 8(a), the rate of charging decreases as the capacitance value increases. The capacitor charges at maximum speed, then slowly decreases until it reaches saturation.

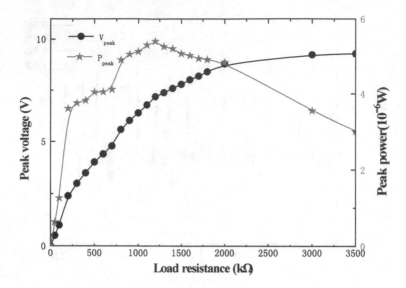

Fig. 7. Peak voltage and peak power at different values of load resistor

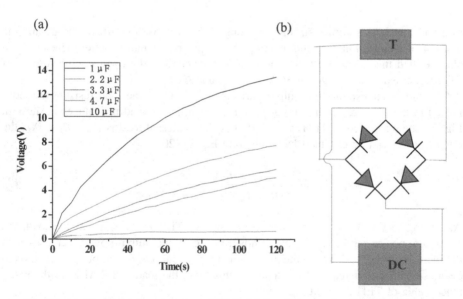

Fig. 8. (a) Charging characteristics of TEH. (b) Rectifier circuit schematic diagram.

4.3   Bandwidth Widening

According to theoretical analysis, when cantilever beam start to contact, the working bandwidth of TEH will be increased. In Fig. 9, when the frequency rises from 5 Hz to 45 Hz, the working frequency width is greatly increased under the same output voltage in different acceleration conditions.

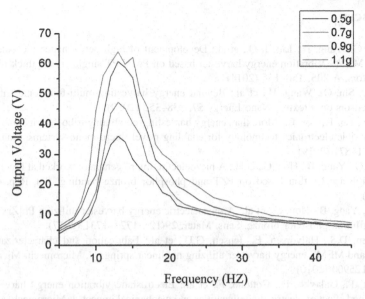

Fig. 9. The output voltage at different frequency and acceleration

It can be seen from Fig. 9 that the first-order resonance frequency of TEH is about 14 Hz, which is different from the theoretical analysis. The reason for this phenomenon may be that the cantilever beam made by 3D printing technology has different density and young's modulus to the theoretical value. Meanwhile, the impact of cantilever beam may also affect the resonance frequency.

5 Conclusion

This paper presents a cantilever based TEH with slot structure on the beam. Using ecoflex00-30 material as friction material, the output performance is better than before.3D printing process to manufacture cantilever beam, convenient and fast. The cantilever beam is designed as a slotted structure to reduce the working frequency of TEH. At the same time, collision is added to the cantilever beam to increase the working bandwidth of TEH. Through experimental research, the maximum output voltage of TEH is 64 V under the acceleration of 1.1 g. The maximum output power is 5.4 μw at load of 1.2 MΩ.

Acknowledgments. This work was supported by the National Natural Science Foundation of China (51565038) and the Science and Technology Project of Jiangxi Provincial Education Department (GJJ170986, GJJ180938).

References

1. Tang, G., Yang, B., Liu, J.-Q., et al.: Development of high performance piezoelectric d_{33} mode MEMS Vibration energy harvester based on PMN-PT single crystal thick film. Sens. Actuators, A **205**, 150–155 (2014)

2. He, T., Shi, Q., Wang, H., et al.: Beyond energy harvesting-multi-functional triboelectric nanosensors on a textile. Nano Energy **57**, 338–352 (2019)

3. Shi, Q., He, T., Lee, C.: More than energy harvesting-combining triboelectric nanogenerator and flexible electronics technology for enabling novel micro-/nano-systems. Nano Energy **57**, 851–871 (2019)

4. Tang, G., Yang, B., Hou, C., et al.: A piezoelectric micro generator worked at low frequency and high acceleration based on PZT and phosphor bronze bonding. Sci. Rep. **6**, 38798 (2016)

5. Li, G., Yang, B., Hou, C., et al.: A piezoelectric energy harvester utilizing Pb[ZrxTi1-x]O3 thick film on phosphor bronze. Sens. Mater. **29**(12), 1723–1731 (2017)

6. Nguyen, D.S., Halvorsen, E., Jensen, G.U., et al.: Fabrication and characterization of a wideband MEMS energy harvester utilizing nonlinear springs. J. Micromech. Microeng. **20**(20), 125900 (2010)

7. Basset, P., Galayko, D., Cottone, F., et al.: Electrostatic vibration energy harvester with combined effect of electrical nonlinearities and mechanical impact. J. Micromech. Microeng. **24**(3), 501–522 (2017)

8. Saha, C.R., Odonnell, T., Wang, N., et al.: Electromagnetic generator for harvesting energy from human motion. Sens. Actuators, A **147**(1), 248–253 (2008)

9. Rome, L.C., Flynn, L., Goldman, E.M., et al.: Generating electricity while walking with loads. Science **309**, 1725–1728 (2005)

10. Yi, Z., Yang, B., et al.: High performance bimorph piezoelectric MEMS harvester via bulk PZT thick films on thin beryllium-bronze substrate. Appl. Phys. Lett. **111**, 013902 (2017)

11. Tang, G., Liu, J.-q., Yang, B., et al.: Fabrication and analysis of high-performance piezoelectric MEMS generators. J. Micromech. Microeng. **22**, 065017 (2012)

12. Dhakar, L., Tay, F.E.H., Lee, C.: Development of a broadband triboelectric energy harvester with SU-8 micropillars. J. Microelectromechan. Syst. **24**(1), 91–99 (2014)

13. Dhakar, L., Tay, F.E.H., Lee, C.: Investigation of contact electrification based broadband energy harvesting mechanism using elastic PDMS microstructures. J. Micromech. Microeng. **24**, 104002 (2014)

14. Li, H., Li, R., Fang, X., et al.: 3D printed flexible triboelectric nanogenerator with viscoelastic inks for mechanical energy harvesting. Nano energy **58**, 447–454 (2019)

15. Xie, Y., Wang, S., Niu, S., et al.: Grating-structured freestanding triboelectric-layer nanogenerator for harvesting mechanical energy at 85% total conversion efficiency. Adv. Mater. **26**, 6599–6607 (2014)

16. Zhu, Y., Yang, B., Liu, J., et al.: A flexible and biocompatible triboelectric nanogenerator with tunable internal resistance for powering wearable devices. Sci. Rep. **6**, 22233 (2016)

17. Tian, Y., Li, G., Yi, Z., et al.: A low-frequency MEMS piezoelectric energy harvester with a rectangular hole based on bulk PZT film. J. Phys. Chem. Solids **117**, 21–27 (2018)

18. Dhakar, L., Liu, H., Tay, F., Lee, C.: A new energy harvester design for high power output at low frequencies. Sens. Actuators, A **199**, 344–352 (2013)

19. Liu, H., Lee, C., Kobayashi, T., Tay, C.J., Quan, C.: Investigation of a MEMS piezoelectric energy harvester system with a frequency-widened-bandwidth mechanism introduced by mechanical stoppers. Smart Mater. Struct. **21**, 035005 (2012)

20. Hu, Y., Yue, Q., Yu, H., et al.: An adaptable interface conditioning circuit based on triboelectric nanogenerators for self powered sensors. Micromachines **9**, 105 (2018)

Micro Heater with Low Temperature Coefficient of Resistance for ICF Target

Bin Xu[✉], Zhibiao Li, Gang Tang, Yulong Bao, and Huang Wang

Jiangxi Province Key Laboratory of Precision Drive and Control,
Department of Mechanical and Electrical Engineering,
Nanchang Institute of Technology, Nanchang 330099, China
xubin84115@163.com

Abstract. A micro heater with low temperature coefficient of resistance (TCR) at liquid hydrogen temperature was designed and fabricated by micro fabrication technology. The NiCr heater annealed in N_2 at 250 °C for 9 min achieves a smallest TCR of 9.36 ppm/K at 20 K. The crystal structures of NiCr film annealed in nitrogen were analyzed by scanning electron microscope (SEM) and X-ray diffraction (XRD). The crystallization of NiCr film improved with the annealing temperature increasing. The fabricated micro heater applied in the temperature control test achieves the accuracy of ±0.5 mK, which is qualified for the temperature control accuracy requirement of the ignition target of inertial confinement fusion (ICF).

Keywords: Heater · ICF · Cryogenic target · MEMS · TCR

1 Introduction

As the energy problems become more urgent, fusion and solar-energy technology, as sub-critical clear energy, are solutions to the energy problem, and many researches are committed to this field. In the ICF experiment, ignition target as an essential part to induce a fusion reaction is supported at the ends of the cooling arms, as shown in Fig. 1. The support cooling arm grips the hohlraum to maintain the temperature for heat conduction. In the fusion experiment, the capsule must be precisely kept in liquid hydrogen temperature, and the temperature control accuracy must be less than 0.5 mK. In order to precisely control the temperature of cooling arm, a micro heater and a temperature sensor are employed on the cooling arm [1–3]. The temperature is controlled by adjusting the input power of the heater according to the difference of reference temperate and the measured value by the temperature sensor. The resistance variation of the micro heater affects the temperature control accuracy. So, the micro heater applied in the fusion ignition target should have a nearly zero temperature coefficient of resistance (TCR).

© Springer Nature Switzerland AG 2019
D. Milošević et al. (Eds.): HCC 2019, LNCS 11956, pp. 493–503, 2019.
https://doi.org/10.1007/978-3-030-37429-7_49

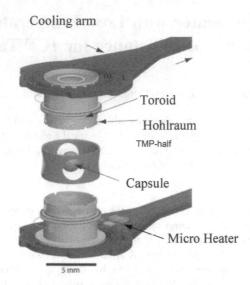

Fig. 1. The model of fusion energy ignition target (exploded) [2]

NiCr alloy is normally used as the material of the micro heater due to its low TCR, good stability and adhesion force with many kinds of substrates, such as silicon, Sapphire, and polymers [4]. Many research have found that the film deposition condition and post deposition treatment can influence the NiCr thin films on its crystalline [5–13]. Nguyen *et al.* [6] investigated the chromium concentration has relationship with the crystalline nature of the NiCr films. The resistance of NiCr alloys increases and the TCR decreases along with the chromium concentration increasing. And the films would change from crystalline to amorphous phase when the chromium concentration increase to 40% or above. As the deposited NiCr films with chromium concentration above 40% have a negative TCR and annealing treating can increase the TCR of the film into a positive value. Therefore, a micro heater with nearly zero TCR is possible by annealing at a certain temperature in nitrogen ambient. Iida and Nakamura [7] found that the NiCr film with a Ti layer underneath could improve the crystalline. And Nachrodt et al. [8] acquired a NiCr film with a TCR below 10 ppm/K by adding an exact 5 nm Ti layer and an optimized annealing condition. However, the thickness of the Ti layer is hard to control in nanometer level by normal sputtering. Most of the literatures study the TCR under ambient temperature (about 300 K), but the micro heater for fusion energy application works at the temperature around liquid hydrogen temperature. So, this paper put emphasis on the investigation of the TCR at cryogenic temperature (around 20 K).

2   The Design of Micro Heater

Micro heater in fusion energy application is mounted on the cooling arm. During the heating process, the Joule heat of the NiCr film transfers though the substrate to the cooling arm by conduction mechanism. In order to improve the heat-transfer efficiency, the substrate should have high thermal conductivity at cryogenic temperature. Silicon as the most commonly used semiconductor material with high thermal conductivity of 2400 W/m.K at liquid hydrogen temperature is selected as the substrate of micro heater.

The NiCr thin film (60/40 at. %) is deposited on SiO_2/Si. There are five different patterns of NiCr film designed in this paper, which are shown in Fig. 2. As the heater is mounted on the cooling arm to conduct heat, the temperature on the bottom surface should distribute as uniformly as possible, so that the temperature control of the cooling arm can be more precise. In order to calculate the temperature distribution of the micro heater, the simulation software COMSOL was applied, and the temperature distributions on the bottom surface of the heaters with different patterns are illustrated in Fig. 3. As can be seen from the legend of each simulation, the temperature range of the heater with pattern (c) is the smallest, so the pattern (c) of the NiCr film is established as the final design.

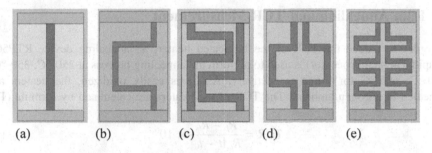

(a)　　　　　　(b)　　　　　　(c)　　　　　　(d)　　　　　　(e)

Fig. 2. The pattern of NiCr film on top of micro heater

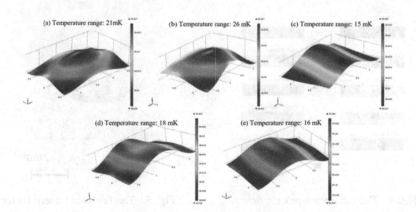

Fig. 3. The temperature distribution on the bottom of the micro heater

3 Fabrication Process of Cooling Arm

The micro heater is fabricated by micro fabrication technology. The fabrication process is described in Fig. 4. In step (1), the oxide silicon wafer is ultrasonic cleaned by acetone, isopropanol and deionized water separately. In step (2), approximately 200 nm-thickness NiCr film is deposited on the silicon wafer by magnetron sputtering from a 60%Ni-40%Cr target. In step (3), 5 μm photoresist (AZ4620) is spin-coated on the NiCr thin film. In step (4), the photo resist is exposed and developed. In step (5), the NiCr thin film is etched by plasma etching method for 20 min. In step (6), the photo resist is removed by acetone, isopropanol and deionized water. In step (7), 10 μm photo resist (AZ4620) is spin-coated on the pattern surface. In step (8), the photo resist is exposed and developed. In step(9), 200 nm Cr/Au electrode is deposited on the wafer. In step (10), the photo resist was lifted off from the surface of the NiCr thin film in acetone. The fabricated micro heater is shown in Fig. 5.

4 Film Annealing and TCR Measurement

In order to reduce the TCR of the fabricated heater, an annealing device RTP500 (Rapid Thermal Processor) is used to perform the annealing process at 250 °C, 350 °C, 450 °C, 600 °C for 9 min respectively. As Cr is easily oxidized, the heaters are annealed in nitrogen ambient. The TCR of the heaters are calculated by formula (1).

$$TCR = \frac{R_t - R_{t_0}}{R_{t_0}|t - t_0|} \times 10^6 \tag{1}$$

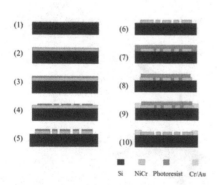

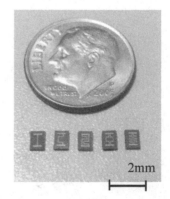

Fig. 4. The fabrication-process flow   **Fig. 5.** The fabricated micro heater

Where R_t and R_0 are the resistance of thin-film heaters at the temperature of t and 300 K respectively. In order to figure out the optimal annealing temperature for the smallest TCR of the micro heater, the Physical Property Measurement System (PPMS) equipment, as shown in Fig. 6(a), is used to measure the resistances of the samples at

        (a) The measurement system   (b) The sample holder

Fig. 6. The Physical Property Measurement System equipment and the sample test holder

the temperature range from 10 K to 300 K,and the micro heaters are fixed on a sample test holder during the measurement, as shown in Fig. 6(b).

As can been seen from the R-T curve of the testing samples in the Fig. 7, the as-deposited heater has a negative TCR as the resistance decreasing along with the temperature increasing, and the TCR becomes positive after annealing. Also, the resistance of the heaters decreases as the annealing temperature increase. The TCR of each sample at 20 K is calculated by formula (1) and listed in Table 1. The sample annealed at 250 °C has the lowest TCR at 20 K, which is 9.36 ppm/K.

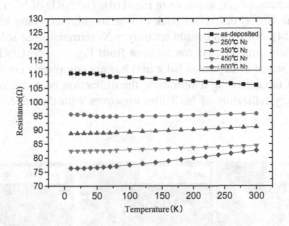

Fig. 7. The R-T curve of the micro-heaters

Table 1. TCR of the heaters under different annealing conditions

Annealing temperature	Before annealing	250 °C	350 °C	450 °C	600 °C
TCR (ppm/K) at 20 K	−145	9.36	87.22	119.11	293.16

5   Discussion

The resistance of most metal is generally due to the electron scattering caused by the thermal vibrations of the atoms. As the scattering probability increases with temperature increasing, the resistance of metal will increase correspondingly. As for the amorphous alloys, the electrical transport properties research is still at the stage of establishing theoretical model. Following the theory of Mooij and Tsuei, the amorphous alloys could exhibit a negative TCR. On the basis of the literature [11], it is known that the sputtered NiCr film with chromium concentration of 40% exhibits an amorphous nature, so the TCR of as-deposited NiCr film is negative. Along with the annealing temperature increase, the crystallization of NiCr film will improve, and the TCR value will become positive or even higher.

In order to verify the above theoretical analysis, the crystal structures of NiCr film annealed in nitrogen were analyzed by scanning electron microscope (SEM) and X-ray diffraction (XRD). Figure 8 shows the film surface morphology of the NiCr films. The surface morphology of films annealed under 350 °C are smooth, and the crystal grains can hardly be found, because the film still exhibits amorphous nature. When the annealing temperature increased to 450 °C, the crystals grown above 20 nm is observed. When the annealing temperature is 600 °C, the formed crystals combined each other and became larger, as shown in Fig. 8(d). The XRD of NiCr films are shown in Fig. 9, and the Ni(Cr) diffraction peak can be found. According to the NiCr alloy phase diagram [14], NiCr film constitutes only γ-Ni terminal solid solution with face-centered cubic (fcc) structure. As can be seen from Fig. 9, γ-Ni(Cr)(111) has a wide diffraction peak, so the as-deposited NiCr film has not completely crystallization. With the increasing of the annealing temperature, the diffraction peak intensity increases. It proves that the crystallization of NiCr film improves with the annealing temperature increasing.

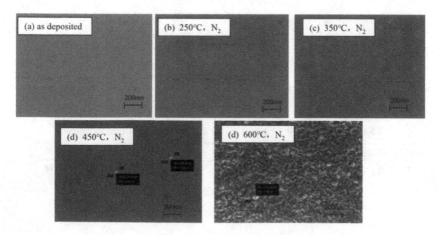

Fig. 8. The SEM of NiCr films

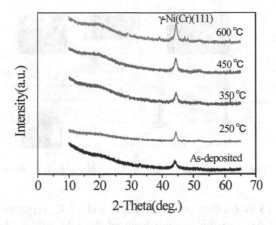

Fig. 9. The XRD patterns of NiCr films

6   Performance Test of Micro Heater

6.1   Response Time Measurement

In order to test the performance of the micro heater, a experiment system, which is illustrated in Fig. 10, is built to measure and calculate the response time of the micro heater. The clamping apparatus is design to bind the micro heater and the thermocouple measuring end. The micro heater and the thermocouple measuring end are attached to the cushions separately. By adjusting the nuts in the clamping apparatus, the micro heater will contact well with the thermocouple measuring end on its bottom face. The compensating end of the thermocouple is placed in the ice point cell where the temperature is constant at zero degree Celsius. The universal analog input module (NI 9219, National Instruments) is used to transfer the voltage data of the thermocouple into data on temperature, and input the data to the PC. While a constant voltage of 6 V is inputted to the micro heater by the DC power supply, the temperature will rise from room temperature to steady-state temperature, and the PC will store all the temperature data of NI 9219 collected from the thermocouple. As can be seen from the temperature-rise period of the micro heater which is illustrated in Fig. 11, the temperature rises from 292.87 K to 346.42 K. Normally, 63.2% of the temperature incensement is defined as the respond time of the micro heater. Thus, the response time of the fabricated heater is 11.3 s, which has a quick response own to the chosen substrate material silicon has a large coefficient of thermal conductivity.

6.2   Temperature Control Accuracy Test

The assembly of the cooling arm is shown in Fig. 12. The micro heater and a micro temperature sensor are mounted on the cooling arm by epoxy (Sytcast Epoxy 2850-FT), and connect to the golden lead by wire bonding respectively. The setup test system for evaluate the performance of the micro heater is shown in Fig. 13. A GM cryocooler (Sumitomo SRDK408) is used as the cold source, whose second stage cold head can

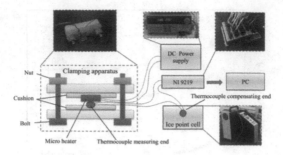

Fig. 10. The measurement system of the response time of micro heater

provide 1 W and 13 W cooling power at 4.2 K and 18 K, respectively. The cooling arm assembly was fixed to the cooling head of the GM cooler. And a temperature controller (Cryocon 24C) is used to monitor the temperature of the temperature sensor and automatically adjust the power loaded on the heater by PID control method.

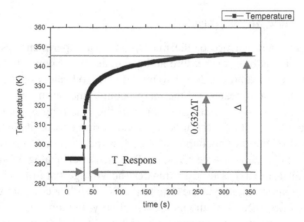

Fig. 11. The temperature-rise period of micro heater

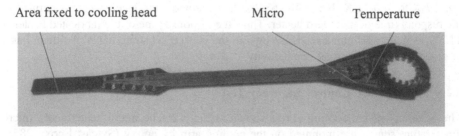

Fig. 12. The cooling arm assembly

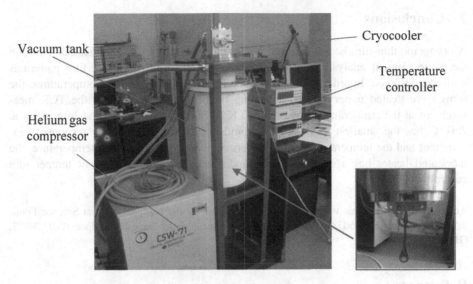

Fig. 13. The test system

In the temperature control experiment, the temperature of cooling arm is controlled in PID mode. Before the temperature controller works, the temperature of the cooling arm is cooled down to 18.18 K. As can be seen from Fig. 14(a), the temperature reaches the steady state in 220 s by the temperature controller adjusting the power of the micro heater. As shown in Fig. 14(b), the temperature fluctuation of the cooling arm in steady state is ±0.5 mK, which meets the accuracy requirement of the temperature control.

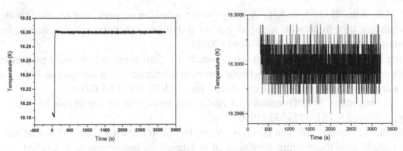

(a) Temperature control process   (b) Temperature fluctuation in steady
state

Fig. 14. The temperature of cooling arm in control process

7 Conclusions

A cryogenic thin-film heater made of NiCr alloy was design and fabricated. Through the finite element analysis of the temperature distribution, the NiCr film pattern is optimize designed. In order to obtain a small TCR at liquid hydrogen temperatures, the films were treated under various annealing treatments. According to the TCR measurement at the temperature range from10 K to 300 K, the heater annealing in N_2 at 250 °C has the smallest TCR of 9.36 ppm/K at 20 K. Through the response measurement and the temperature control experiments at liquid hydrogen temperature, the fabricated heater has a quick response and meets the requirements of temperature control accuracy.

Acknowledgments. This work was supported in part by the National Natural Science Foundation of China (51565038), the department of education of Jiangxi province (GJJ170987, GJJ180936).

References

1. Miles, R., Hamilton, J., Crawford, J., Ratti, S., Trevino, J., Graff, T., Stockton, C., Harvey, C.: Microfabricated deep-etched structures for ICF and equation-of-state targets. Fusion Sci. Technol. **55**(3), 308–312 (2009)
2. Miles, R., et al.: Micro-fabrication techniques for target components. Fusion Sci. Technol. **55**(3), 308–312 (2009)
3. Khater, H., Brereton, S.: Radiological assessment of target debris in the National Ignition Facility. In: 23rd IEEE/NPSS Symposium, Fusion Engineering, SOFE, pp. 1–4 (2009)
4. Kazi, I.H., Wild, P.M., Moore, T.N., Sayer, M.: Characterization of sputtered nichrome (Ni–Cr 80/20 wt.%) films for strain gauge applications. Thin Solid Films **515**(4), 2602–2606 (2006)
5. Kim, H.S., Kim, B.O., Seo, J.H.: A study on the adhesion properties of reactive sputtered molybdenum thin films with nitrogen gas on polyimide substrate as a Cu barrier layer. J. Nanosci. Nanotechnol. **15**, 8743–8748 (2015)
6. Phuong, N.M., Kim, D.-J., Kang, B.-D., Yoon, S.-G.: Structural and electrical properties of NiCr thin films annealed at various temperatures in a vacuum and a nitrogen ambient for π-type attenuator applications. J. Electrochem. Soc. **153**(7), 660–663 (2006)
7. Iida, A., Nakamura, S.-I.: Orientation of Ni-Cr thin films with an underlying Ti layer. Jpn. J. Appl. Phys. **35**(3A), 335–337 (1996)
8. Nachrodt, D., Paschen, U., Ten Have, A., Vogt, H.: Ti/Ni (80%)Cr (20%) Thin-film resistor with a nearly zero temperature coefficient of resistance for integration in a standard CMOS process. IEEE Electron Device Lett. **29**(3), 212–214 (2008)
9. Belič, L.I., Požun, K., Remškar, M.: AES, AFM and TEM studies of NiCr thin films for capacitive humidity sensors. Thin Solid Films **317**(1–2), 173–177 (1998)
10. Au, C.L., Jackson, M.A., Anderson, W.A.: Structural and electrical properties of stable Ni/Cr thin films. J. Electron. Mater. **16**(4), 301–306 (1987)

11. Phuong, N.M., Kim, D.-J., Kang, B.-D., Kim, C.S., Yoon, S.-G.: Effect of chromium concentration on the electrical properties of NiCr thin films resistor deposited at room temperature by magnetron cosputtering technique. J. Electrochem. Soc. **153**(1), 27–29 (2006)
12. Lai, L., Zeng, W., Fu, X., Sun, R., Du, R.: Anneling effect on the electrical properties and microstructure of embedded Ni-Cr thin film resistor. J. Alloys Compd. **538**, 125–130 (2012)
13. Kwon, Y., Kim, N.-H., Choi, G.-P., Lee, W.-S., Seo, Y.-J., Park, J.: Structural and surface properties of NiCr thin films prepared by DC magnetron sputtering under variation of annealing conditions. Microelectron. Eng. **82**(3–4), 314–320 (2005)
14. Rolke, J.: Nichrome thin film technology and its application. Electrocompon. Sci. Technol. **9**(1), 51–57 (1981)

Random Convolutional Neural Network Based on Distributed Computing with Decentralized Architecture

Yige Xu[1], Huijuan Lu[1(✉)], Minchao Ye[1], Ke Yan[1], Zhigang Gao[2], and Qun Jin[3]

[1] College of Information Engineering, China Jiliang University,
258 Xueyuan Street, Hangzhou, China
hjlu@cjlu.edu.cn
[2] College of Computer Science and Technology, Hangzhou Dianzi University,
1158 Baiyang Street, Hangzhou, China
[3] Faculty of Human Sciences, Waseda University, Tokorozawa 359-1192, Japan

Abstract. In recent years, deep learning has made great progress in image classification and detection. Popular deep learning algorithms rely on deep networks and multiple rounds of back-propagations. In this paper, we propose two approaches to accelerate deep networks. One is expanding the width of every layer. We reference to the Extreme Learning Machine, setting big number of convolution kernels to extract features in parallel. It can obtain multiscale features and improve network efficiency. The other is freezing part of layers. It can reduce back-propagations and speed up the training procedure. From the above, it is a random convolution architecture that network is proposed for image classification. In our architecture, every combination of random convolutions extracts distinct features. Apparently, we need a lot of experiments to choose the best combination. However, centralized computing may limit the number of combinations. Therefore, a decentralized architecture is used to enable the use of multiple combinations.

Keywords: Distributed computing · Random convolution · Decentralized architecture

1 Introduction

Deep CNN have been proved to have powerful capabilities in various aspects [1], such as image classification, detection, segmentation. However, the development of deep learning is hampered by the extremely long training time and the huge demand for computing power. So, we need a network structure that can train quickly.

Obviously, in the deep neural networks, a lot of training time is generated by back-propagations. So, feedforward neural networks are faster than back-propagation neural networks. The feedforward neural network is time-saving, however the accuracy may be reduced. Extreme learning machine (ELM) [2] is a representative feedforward neural network, which balances the time and accuracy. The randomly assigning the hidden layer weights is the structural characteristic of ELM. This paper refers to the structure

© Springer Nature Switzerland AG 2019
D. Milošević et al. (Eds.): HCC 2019, LNCS 11956, pp. 504–510, 2019.
https://doi.org/10.1007/978-3-030-37429-7_50

of ELM. We design a network which contains one random convolution layer. The random convolutional network will inherit the characteristics of ELM which reduces the time cost of back-propagation network. The network expects better accuracy in small sample scenarios.

In this paper, we draw on the GoogLeNet structure. The first layer of the network consists of three convolution kernels and one Max Pooling. These outputs are spliced on the feature layer. The second layer contains several convolution kernels (greater than three) and one Max Pooling. Several outputs are integrated using the Boosting method. The parameters of the network are tuned through a decentralized distributed architecture.

The rest of this paper is organized as follows. Section 2 introduces the basic concept of CNN and blockchain application. Section 3 discusses the proposed approach of random convolution architecture. Section 4 shows the experimental results and analysis. Finally, conclusions are drawn in Sect. 5.

2 Related Works

AlexNet [3] is a classical image classify networks in the area of image classification. It leads the development of convolutional neural network to deep networks. But, GoogLeNet [4] works better by increasing the width of the network. We reference the design of GoogLeNet that build the first layer of our network.

ELM was proposed by Huang because of its fast and powerful generalization [5]. In our research, the architecture is different from the popular convolutional neural network that the first layer is a similar feedforward neural network.

John et al. proposed decentralized computing is a way beyond centralized computing [6]. Bitcoin Blockchain [7] and Ethereum [8] are representative decentralized distributed systems. Hassain et al. discuss the use of blockchain for computing [9]. Therefore, we experimented on training convolutional neural network on the blockchain architecture.

3 Basic Definitions

3.1 Random Convolution Architecture

The design of random convolution architecture is to reduce the computational cost of back-propagation in traditional CNN. The deep networks extract features base on multi-layer features, which can touch deep information in pictures. Apparently, different convolution kernels can extract multiscale features. We consider to extent the width of CNN.

Our network takes advantages of GoogLeNet that the first layer of our network contains multiple convolution kernel. These outputs will be concatenated by features, and it will be input to the next layer. This layer will be frozen which avoid back-propagations. This structure would be shown like Fig. 1:

The 1 × 1 convolution followed by 2 × 2 Max Pooling will extent the thickness of Pooling layer that match the output of the convolution kernels. The next layer will obtain multiscale features that contain more information than one output.

In the next, multiple convolution kernels parallelly distribute in this layer. We integrate multiple convolution kernels into one output. There are two kernels that 1 × 1 convolution and 2 × 2 Max Pooling will not be replaced. In addition, there are at least four convolution outputs in this layer. We call this layer as random convolution layer. This layer could be demonstrated as Fig. 2.

This architecture will be lighter than traditional convolution neural network. In our experience, this whole train process about 30 min. And 800 images will be trained on GeForce GTX 1060 per second. Decentralized distributed computing can be built on such speeds.

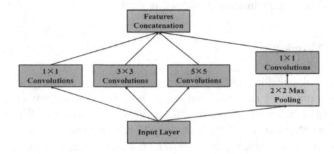

Fig. 1. The first layer of random convolution network

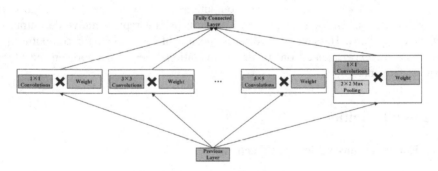

Fig. 2. The random convolution layer

3.2 Decentralized Distribute Computing Architecture

To record the parameter of random convolution, we define some new attributes to meet the demands. There are two key parameters that the number of features and cores are recorded in blocks. Furthermore, precision, loss, steps and time should be recorded in every block. So, we list all attributes in Table 1:

Table 1. The attributes in blocks

Attribute	Remark
Index	The index of blocks
Previous_Hash	Hash of the previous block
Cores	Amount of the cores in random convolution layer
Features	The features in the first layer.
Step	The steps of the previous train
Loss	The loss of the previous train
Precision	The loss of the previous train, it replaces the proof to validate the new block

In conclusion, we build new models by mining new blocks. The features and cores will be depended on train time and train steps. When the time longer, the model needs more features and cores to be more efficient.

4 Experiments

4.1 Dataset Introduction

From the CIFAR-10 benchmark classification dataset. We choose 15000 images from datasets for our searching and further test. The CIFAR-10 dataset consists of 60000 32×32 color images in 10 classes, with 6000 images per class. There are 50000 training images and 10000 test images. For creating the tiny dataset condition, we choose one tenth train data in CIFAR-10 to simulate tiny dataset and keep 10000 test images. The example of CIFAR-10 dataset (Fig. 3):

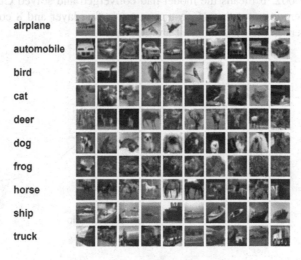

Fig. 3. The Example of CIFAR-10

These parameters will define number of cores and features that C and F. The parameter function is (1):

$$C = {}^S\!/_{StepFactor} + 2$$

(1)

$$F = 2^{AimTime/TimeFactor + 1}$$

C is the number of cores. *StepFactor* is an artificially defined parameter which determine the number of cores in random convolution layer. F is the number of features, Averagely, a new block will be created by every *AimTime*. The *TimeFactor* keeps features grow at a constant rate.

There is another function in our algorithm. The reduce function control the deceleration of loss value. It can be described as (2):

$$L = L_0 \times (1 - precision)$$

(2)

Apparently, the deceleration of loss value will be extremely high when the **mine** begins. Besides, L_0 is the average loss value over the latest mines. When the accuracy is low, L_0 is averaged over last 10 losses, while the accuracy is closer to 100%, L_0 is averaged over last 100 losses to ensure the convergency.

4.2   Decentralized Training

We make serious experiments on decentralized network by four computers. The results are shown as Figs. 4 and 5.

Figure 4 shows that after 9 valid blocks the precision reach 0.9796, and the value of loss reach to 0.0002. It means the model had converged and solved CIFAR-10 problem. Figure 5 demonstrates that 16 feature maps in first layer and 8 cores in random convolution layers is the most efficient architecture.

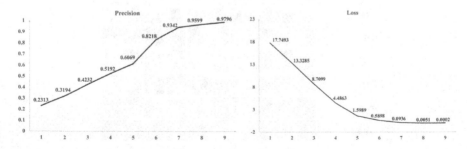

Fig. 4. The value of precision and loss

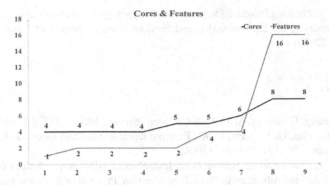

Fig. 5. The number of cores and features

We save these architectures and retrain it in local. The results are shown as Table 2.

Table 2. Compare between two architectures

Cores	Features	Precision	Time
4	2	0.88	207.308629
8	16	0.978	254.5754378

It shows that more core and features will use more time, and the precision will better than the cores and features in low level.

5 Conclusion

In this paper, we proposed an architecture to solve the image classification problems. We used decentralized network to train this classification models. Generally, the random convolution architecture is faster than traditional convolution neural networks. And it can fit better than traditional convolution neural networks. The experimental results shown that 6 cores and 16 features is a reasonable configuration for solving CIFAR-10 problem. We expect develop this architecture to solve more complex dataset, such as ImageNet.

Moreover, algorithm competition is a new type of decentralized network application. Model parameter will be adjusted by every competition. But the data transmission problem should be considered cautiously. Higher bandwidth could make this design more efficient and more useable.

Acknowledgments. This study is supported by National Natural Science Foundation of China (Nos. 61272315, 61602431, 61701468, 61572164, 61877015 and 61850410531), International Cooperation Project of Zhejiang Provincial Science and Technology Department (Nos. 2017C34003), the Project of Zhejiang Provincial Natural Science Foundation (LY19F020016), and the Project of Zhejiang Provincial Science and Technology Innovation Activities for College Students University (Nos. 2019R409030) and Student research project of China Jiliang university (2019X22030).

References

1. Xing, H., Zhang, G., Shang, M.J.I.J.o.S.C.: Deep learning **10**(03), 417–439 (2016)
2. Huang, G.-B., Zhu, Q.-Y., Siew, C.-K.: Extreme learning machine: theory and applications. Neurocomputing **70**(1–3), 489–501 (2006)
3. Krizhevsky, A., Sutskever, I., Hinton, G.E.: ImageNet classification with deep convolutional neural networks. In: Advances in Neural Information Processing Systems, pp. 1097–1105 (2012)
4. Szegedy, C.: Going deeper with convolutions. In: Proceedings of the IEEE Conference on Computer Vision and Pattern Recognition, pp. 1–9 (2015)
5. Zhang, M., Wen, Y., Chen, J., Yang, X., Gao, R., Zhao, H.: Pedestrian dead-reckoning indoor localization based on OS-ELM. IEEE Access **6**, 6116–6129 (2018)
6. King, J.L.: Centralized versus decentralized computing: organizational considerations and management options. ACM Comput. Surv. (CSUR) **15**(4), 319–349 (1983)
7. Nakamoto, S.: Bitcoin: a peer-to-peer electronic cash system, pp. 1–9 (2008)
8. Buterin, V.: A next-generation smart contract and decentralized application platform, pp. 1–36 (2014)
9. Hossain, S.A.: Blockchain computing: prospects and challenges for digital transformation. In: 2017 6th International Conference on Reliability, Infocom Technologies and Optimization (Trends and Future Directions) (ICRITO), pp. 61–65. IEEE (2017)

Emotion Recognition from Human Gait Features Based on DCT Transform

Penghui Xue[1], Baobin Li[1(✉)], Ning Wang[2], and Tingshao Zhu[3]

[1] School of Computer Science and Technology,
University of Chinese Academy of Sciences, Beijing 100190, China
libb@ucas.ac.cn
[2] Beijing Institute of Electronics Technology and Application,
Beijing 100871, China
[3] Institute of Psychology Chinese Academy of Sciences, Beijing 100101, China
tszhu@psych.ac.cn

Abstract. Emotion recognition is of great value in human-computer interaction, psychology, etc. Gait is an important pattern of emotion recognition. In this paper, 59 volunteer's gait data with angry or happy emotion, have been collected by the aid of Microsoft Kinect. The gait data are treated as discrete time signals, and we extract a series of frequency features based on the discrete cosine transform. Simultaneously, we have established emotion recognizing models with SVM, the K-nearest neighbors, and decision tree. The best recognition rate can exceed 80%, which indicates that our proposed features are useful for recognizing emotions.

Keywords: Emotion recognition · Discrete cosine transform · Gait · Kinect

1 Introduction

Emotion is a complex behavioral phenomenon involving many levels of neural and chemical integration [16]. Recent scientific findings suggest that emotion is a ubiquitous element of any human-computer interaction [3]. How to recognize human emotions and respond accordingly becomes an important aspect of improving the human-computer interaction experience of intelligent electronic devices [8]. There are many application scenarios based on emotion recognition such as intelligent voice assistant [9], smart robot [13], and interactive games [4]. People with different emotions are also different in gait to some extent. Montepare et al. have found that angry gaits are relatively more heavy-footed than the others, and sad gaits have less arm swing than the others [14]. When one walks, he will do some repetitive actions. At the same time, all joints in the body will move in a unique pattern. In our study, we utilize 59 volunteers' gait data with angry or happy emotion collected by Microsoft Kinect. The gait information is a series of discrete time signals that are suitable to obtain frequency features with discrete cosine transform.

© Springer Nature Switzerland AG 2019
D. Milošević et al. (Eds.): HCC 2019, LNCS 11956, pp. 511–517, 2019.
https://doi.org/10.1007/978-3-030-37429-7_51

2 Related Work

In recent years, affective computing has gained increasingly intensive attention [15]. Although human emotion is manifested in many aspects, current computer-based research mainly recognizes emotions by speech signals [6], facial patterns [18], or multimodal information including mentioned above [7]. Atkinson et al. proved that the body movement information is also an important pattern that can be used to identify emotions [2]. Gait studies usually focus on motion features such as stance phase, frequency, footstep length, and inclination of the body. Some of these researches are shown in [5, 12, 17]. Li et al. utilized Microsoft Kinect to get non-contact gait information containing 3-dimensional coordinates of 25 joints. By the discrete Fourier transform and statistic methods, some time-frequency features related to neutral, happy and angry emotion were extracted and used to establish the classification model to identify these three emotions. The accuracy of identifying anger and happiness could reach up to 80% [11]. In this paper, we also take gait frequency features into consideration. The main contributions of this paper are as follows:

1. The gait data is processed with DCT and effective frequency features for emotion classification are extracted.
2. We use machine learning to build an automatic emotion recognition system with the abovementioned features.

3 Discrete Cosine Transform

Discrete cosine transform (DCT) was proposed by Ahmed et al. in 1974 and has a history of 45 years [1]. DCT approximates a discrete time signal with a set of sinusoidal functions of different magnitudes and frequencies. Although DCT is proposed later than the discrete Fourier transform, its performance is closer to the ideal Karhunen-Loève transform. Ahmed et al. showed that DCT can be used in the area of digital processing for the purposes of pattern recognition [1]. DCT has a strong characteristic of energy concentration, and most of the natural signal energy is concentrated in the low-frequency part of the DCT. Thus, we utilize DCT to process gait data and extract effective frequency features.

In this paper, we will track the change of the 3-D coordinates of 25 human joints during one's walking. For each of the axes of a human joint, we treat the change in the axis over time as a one-dimensional discrete time signal. Therefore, in our study, we use one-dimensional DCT to process gait data. One-dimensional DCT has eight forms, and the second form has energy-concentration properties, which is often used for signal processing and image processing for lossy data compression of signals and images. Therefore, we use the second form of DCT (DCT-II). DCT-II is defined as:

$$\begin{cases} F(u) = a(u) \sum_{x=0}^{N-1} f(x) \cdot \cos\left[\frac{\pi(2x+1)u}{2N}\right], \\ a(0) = \frac{1}{\sqrt{N}}, a(u) = \sqrt{\frac{2}{N}} (u = 1, \ldots, N-1), \end{cases}$$

where $\{f(x)\}(x = 0, 1, \ldots, N - 1)$ represents a one-dimensional real signal sequence with length of N. a(u) is the compensation coefficient of one-dimensional DCT. $\{F(u)\}$ is the result of the one-dimensional DCT of $\{f(u)\}$, which is a one-dimensional real signal sequence with length of N as well.

4 Data Set and Gait Features

The data set we use is the walking data of 59 volunteers collected by Kinect under happy and angry emotion respectively, which is the same as in [11]. It provides three-dimensional coordinate sequences of 25 joints of the human body. The 25 joints of the human body are shown in Fig. 1. Raw gait data is noisy, complicated and inconsistent, so we utilize the preprocessing method as in [11]. For each sample in the data set, we intercept 64 frames to maintain data consistency. Thus, for each volunteer's angry or happy gait data, we get 75 (25 joints × 3-D coordinates) discrete time signals, each of which is 64 in length.

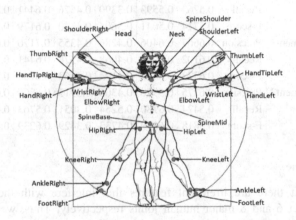

Fig. 1. The 25 joints on human body recorded by Kinect.

For each of the 75 signals in anger or happy of a volunteer, we calculate its DCT and obtain 64 frequency features. In the end, we extract 4800 features (75 × 64) for each sample. We perform the same frequency feature extraction on gait data of all volunteers. Because the range of values for different features is very different, in order to prevent features with small values from being ignored during model training, we utilize Z-score normalization for all 4800 features.

5 Models and Results

We utilize three classifiers including Linear Support Vector Machine (LSVM), the K-Nearest Neighbors (KNN) and Decision Tree (DTree). In our study, we utilize leave-one-out cross-validation to evaluate model performance which is suitable for cases with a small number of samples [10]. Each training set is created by taking all the samples except one, the test set being the sample left out. Repeat until all samples have been chosen as the test set only once. For the evaluation of the model, we choose precision, recall, and F-score (harmonic average of precision and recall). In our study, we use the average of the experimental results as the result of the evaluation indicators.

Table 1. The precision, recall and F-score of the classification results of 3 experiments utilizing frequency features.

		LSVM		KNN		DTree	
		Angry	Happy	Angry	Happy	Angry	Happy
experiment1	Precsion	0.5667	0.5690	0.3846	0.4091	0.5938	0.6111
	Recall	0.5763	0.5593	0.3390	0.4576	0.6441	0.5593
	F-score	0.5714	0.5641	0.3604	0.4320	0.6179	0.5841
experiment2	Precsion	0.6667	0.6406	0.4286	0.4355	0.7170	0.6770
	Recall	0.6102	0.6949	0.4068	0.4576	0.6441	0.7458
	F-score	0.6372	0.6667	0.4174	0.4463	0.6786	0.7097
experiment3	Precsion	0.6275	0.5970	0.4306	0.3913	0.6800	0.6324
	Recall	0.5424	0.6780	0.5254	0.3051	0.5763	0.7288
	F-score	0.5818	0.6349	0.4733	0.3429	0.6239	0.6772

5.1 Result One

Considering that there are irrelevant features that interfere with the classification results, we select 6 and 8 major human joints respectively. Thus, we conduct three groups of experiments. In experiment1, we adopt all frequency features of 25 human joints. In experiment2, we adopt frequency features of 8 human major joints, including ElbowLeft, ElbowRight, FootLeft, FootRight, HandLeft, HandRight, KneeLeft, and KneeRight. In experiment3, we adopt frequency features of 6 human major joints, including WristLeft, WristRight, KneeLeft, KneeRight, AnkleLeft, and AnkleRight. The results of the three experiments with frequency features are shown in Table 1.

In Table 1, we can see that the highest classification accuracy for anger and happiness appears in experiment2 when the classifier is DTree. The best classification accuracy can reach approximately 70%, which proves that the frequency features extracted by DCT are effective for emotion recognition of anger and happiness. Experiment2 and experiment3 perform better than experiment1, indicating that the human joints we select are effective for recognizing emotions. In all three experiments, the classification accuracy of KNN is even lower than 0.5, so the frequency features we extract are not suitable for classification by KNN. In experiment2 and experiment3, the accuracy of LSVM is lower than that of DTree, but it is an acceptable result.

Table 2. The precision, recall and F-score of the classification results of 3 experiments utilizing motion features and frequency features.

		LSVM		KNN		DTree	
		Angry	Happy	Angry	Happy	Angry	Happy
experiment1	Precsion	0.6200	0.5882	0.4225	0.4507	0.7797	0.8182
	Recall	0.5254	0.6780	0.3390	0.5424	0.8305	0.7627
	F-score	0.5688	0.6299	0.3774	0.4923	0.8033	0.7895
experiment2	Precsion	0.7778	0.7344	0.5490	0.5373	0.8596	0.8361
	Recall	0.7119	0.7966	0.4746	0.6102	0.8305	0.8644
	F-score	0.7434	0.7642	0.5091	0.5714	0.8448	0.8500
experiment3	Precsion	0.7586	0.7500	0.4923	0.4906	0.8571	0.8226
	Recall	0.7458	0.7627	0.5424	0.4407	0.8136	0.8644
	F-score	0.7521	0.7563	0.5161	0.4643	0.8348	0.8430

5.2 Result Two

In order to improve the classification accuracy, for the three groups of experiments in the Sect. 5.1, we add some motion features of gaits including speed, stride distance, and the time one takes a step. The results of the three experiments with the combination of motion features and frequency features are shown in Table 2. We can see that the best result is in experiment2 when the classifier is DTree, and the precision for anger and happiness reaches 85.96% and 83.61% respectively. In most cases, the F-score for happiness is higher than F-score for anger. Generally speaking, DTree has the best classification result, LSVM is the second, and KNN is the worst of the three classifiers. KNN's performance is still much worse than the other two classifiers. In almost all experiments, F-score is higher than the F-score in Table 1.

In [11], Li et al. performed discrete Fourier transform(DFT) on every coordinate series of joints and selected the amplitudes and phases of the first 20 components as features. For comparison, we select 64 amplitudes of every coordinate series of joints after DFT as frequency features. Just like in Sect. 5.1, we also conduct three groups of experiments. The results are showed in Table 3. We can see that the highest precision appears in experiment2 when using DTree as a classifier and utilizing our frequency features. When the classifier is KNN, the classification results are poor in all experiments. In each experiment, when using DTree as a classifier and utilizing our frequency features, the classification effect is the best. When using the LSVM classifier, we can get an acceptable classification accuracy, but the accuracy utilizing the frequency features of Li is better. The experimental results show that the best classification result is obtained when using DTree as the classifier and adopting our proposed frequency features.

Table 3. The accuracy of the classification results of 3 comparative experiments. CF represents the frequency features we utilize in this paper. FF represents the frequency features in [11].

	LSVM		KNN		DTree	
	CF	FF	CF	FF	CF	FF
experiment1	0.5678	0.5932	0.3983	0.4229	0.6017	0.5593
experiment2	0.6525	0.6780	0.4322	0.4661	0.6949	0.6610
experiment3	0.6102	0.6356	0.4153	0.3814	0.6525	0.6186

6   Conclusion

Human gait is an important clue to reveal human emotion. In this paper, we focus on how to automatically recognize emotions through gait information. We utilize the gait data of volunteers in anger and happiness collected by Kinect, extract the frequency features with DCT, and then establish the emotion recognition model. The experimental results show that the recognition model works well and has a good classification accuracy. This shows that our proposed feature extraction method is effective for emotion recognition. However, we do not get the physical meaning of these frequency characteristics to explain why they are valid. At the same time, the experimental results also show that the main joints of the human body we selected are effective for emotion recognition. Although we now have more quantities of subjects than many other emotion recognition research based on human gait information, the quantity of experimental samples is still small. We only identify anger and happy from each other through gait information while human emotions are very rich. In the future, we will collect gait data of other emotions and explore whether our proposed feature extraction method is effective in identifying other emotions.

References

1. Ahmed, N., Natarajan, T., Rao, K.R.: Discrete cosine transform. IEEE Trans. Comput. **100** (1), 90–93 (1974)
2. Atkinson, A.P., Tunstall, M.L., Dittrich, W.H.: Evidence for distinct contributions of form and motion information to the recognition of emotions from body gestures. Cognition **104** (1), 59–72 (2007)
3. Brave, S., Nass, C., Hutchinson, K.: Computers that care: investigating the effects of orientation of emotion exhibited by an embodied computer agent. Int. J. Hum.-Comput. Stud. **62**(2), 161–178 (2005)
4. Brooks, A.G., Gray, J., Hoffman, G., Lockerd, A., Lee, H., Breazeal, C.: Robot's play: interactive games with sociable machines. Comput. Entertain. (CIE) **2**(3), 10 (2004)
5. Chiu, M., Shu, J., Hui, P.: Emotion recognition through gait on mobile devices. In: 2018 IEEE International Conference on Pervasive Computing and Communications Workshops (PerCom Workshops), pp. 800–805. IEEE (2018)
6. El Ayadi, M., Kamel, M.S., Karray, F.: Survey on speech emotion recognition: features, classification schemes, and databases. Pattern Recognit. **44**(3), 572–587 (2011)

7. Go, H.J., Kwak, K.C., Lee, D.J., Chun, M.G.: Emotion recognition from the facial image and speech signal. In: SICE 2003 Annual Conference (IEEE Cat. No. 03TH8734), vol. 3, pp. 2890–2895. IEEE (2003)
8. Hibbeln, M.T., Jenkins, J.L., Schneider, C., Valacich, J., Weinmann, M.: How is your user feeling? Inferring emotion through human-computer interaction devices. MIS Q. **41**(1), 1–21 (2017)
9. Hoy, M.B.: Alexa, siri, cortana, and more: an introduction to voice assistants. Med. Ref. Serv. Q. **37**(1), 81–88 (2018)
10. Kohavi, R., et al.: A study of cross-validation and bootstrap for accuracy estimation and model selection. In: IJCAI, vol. 14, pp. 1137–1145, Montreal, Canada (1995)
11. Li, B., Zhu, C., Li, S., Zhu, T.: Identifying emotions from non-contact gaits information based on microsoft kinects. IEEE Trans. Affect. Comput. **9**, 585–591 (2016)
12. Lim, A., Okuno, Hiroshi G.: Using speech data to recognize emotion in human gait. In: Salah, A.A., Ruiz-del-Solar, J., Meriçli, Ç., Oudeyer, P.-Y. (eds.) HBU 2012. LNCS, vol. 7559, pp. 52–64. Springer, Heidelberg (2012). https://doi.org/10.1007/978-3-642-34014-7_5
13. Manohar, V., Crandall, J.W.: Programming robots to express emotions: interaction paradigms, communication modalities, and context. IEEE Trans. Hum.-Mach. Syst. **44**(3), 362–373 (2014)
14. Montepare, J.M., Goldstein, S.B., Clausen, A.: The identification of emotions from gait information. J. Nonverbal Behav. **11**(1), 33–42 (1987)
15. Poria, S., Cambria, E., Bajpai, R., Hussain, A.: A review of affective computing: from unimodal analysis to multimodal fusion. Inf. Fusion **37**, 98–125 (2017)
16. Stevens, S.S.: Handbook of Experimental Psychology (1951)
17. Venture, G.: Human characterization and emotion characterization from gait. In: 2010 Annual International Conference of the IEEE Engineering in Medicine and Biology, pp. 1292–1295. IEEE (2010)
18. Zhang, Y.D., et al.: Facial emotion recognition based on biorthogonal wavelet entropy, fuzzy support vector machine, and stratified cross validation. IEEE Access **4**, 8375–8385 (2016)

Person Search with Joint Detection, Segmentation and Re-identification

Rui Xue, Huadong Ma, Huiyuan Fu$^{(\boxtimes)}$, and Wenbin Yao

Beijing Key Lab of Intelligent Telecommunications, Software and Multimedia,
Beijing University of Posts and Telecommunications, Beijing 100876, China
{xrl995,mhd,fhy,yaowenbin}@bupt.edu.cn

Abstract. Person search is a new and challenging task proposed in recent years. It aims to jointly handle person detection and person re-identification in an end-to-end deep learning neural network. In this paper, we propose a new multi-task framework, which jointly learn person detection, person instance segmentation and person re-identification. In this framework, a segmentation branch is added into the person search pipeline to generate a high-quality segmentation mask for each person instance. Then, the segmentation feature maps are concatenated with corresponding convolution feature maps in the re-identification branch, which results as a self-attention mechanism, provides more discriminative feature for person re-identification. The experimental results on the public dataset PRW demonstrate the effectiveness of the framework.

Keywords: Person search · Person detection · Instance segmentation · Person re-identification · Deep learning

1 Introduction

Person re-identification (re-ID) is a fast-growing research area. It aims at matching query image in large-scale gallery images, which is to confirm whether the person targets recorded by different cameras at different time and locations are the same person, so as to achieve the purpose of tracking or searching. Because of the difference of illumination, angle of view and person posture, the same person may present different appearances in different pictures. Although person re-ID has developed rapidly in recent years due to the proposal of deep learning, there is still a big gap between the problem and the actual application needs. At present, most of the methods are carried out under the prior condition of person detection, so a very robust person detection model is needed. If person detection and person re-ID can be combined in an end-to-end framework, it will be more in line with the actual application requirements. So, some end-to-end methods based on person detection and person re-ID, which we call person search, have been proposed in recent years. In 2014, Xu et al. [23] proposed a method that combining person detection and person re-ID based on sliding windows, which combines these two independent tasks for the first time. In 2017, Xiao et al. [22] designed a person search framework by using deep learning, which jointly learns person detection and person re-ID in an end-to-end way. More recently, many researchers have made further improvements base on [22]. [16] proposed an improved

© Springer Nature Switzerland AG 2019
D. Milošević et al. (Eds.): HCC 2019, LNCS 11956, pp. 518–529, 2019.
https://doi.org/10.1007/978-3-030-37429-7_52

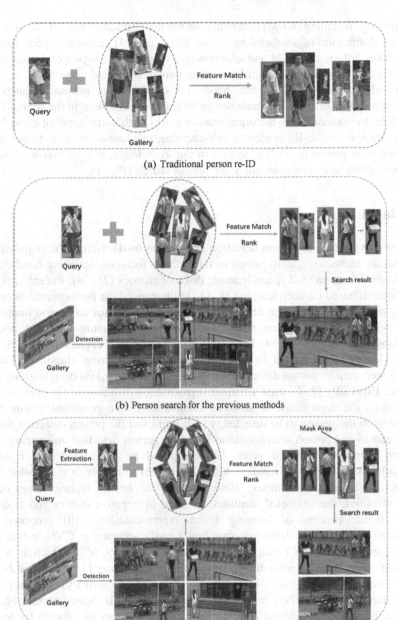

(a) Traditional person re-ID

(b) Person search for the previous methods

(c) Person search in this paper

Fig. 1. Comparison between traditional person re-ID, person search of previous methods and the method proposed in this paper. In the first part, traditional person re-ID requires manually cropped person images. In the second part, we can see that the previous methods returned the wrong ranking result of the query image, which resulting in incomplete person matching. In the third part, the use of semantic information provides more discriminative feature for person re-ID. The red border around the person corresponds to the semantic information for each identity. (Color figure online)

online instance matching (IOIM) loss to harden the distribution of labeled identities and soften the distribution of unlabeled identities. Liu et al. [14] proposed NPSM module, which makes full use of LSTM and attention idea to shrink the region of person search recursively in the whole image until the target person is located accurately.

Inspired by [4, 14, 22], in this paper, person detection, person instance segmentation and person re-ID are combined to be learned end-to-end in the frame- work. Meanwhile, by concatenating the segmentation masks with corresponding convolution features maps in the re-ID branch, a self-attention mechanism is used to get better performance on person re-ID. To the best of our knowledge, it's the first work which learn these three tasks simultaneously for person search (Fig. 1).

2 Related Work

Person search consists of person detection and person re-ID. Before the popularity of deep learning technology, most person re-ID methods focus on designing hand-crafted discriminative features [8, 12] and learning distance metrics [21, 24]. Recently, person re-ID methods based on deep learning [1, 11, 15] improved the performance of person re-ID by designing deep neural network to extract better feature of person images.

For person detection, at first, people used hand-crafted feature extractors, such as Deformable Part Model (DPM) [5], Aggregated Channel Features (ACF) [3] and Locally Decorrelated Channel Features (LDCF) [18], and linear classifier to detect person. Later, various person detection methods [2, 4, 17] based on deep learning were proposed. Especially [2, 4], used semantic segmentation information for person detection, through the supervision of semantic segmentation, the generated feature maps focus more on the person to be detected, which improved the person detection results.

The end-to-end person search method based on person detection and person re-ID has been proposed in recent years. [22] proposed an end-to-end person search framework, and used Online Instance Matching (OIM) to deal with the labeled identities and the unlabeled identities. While learning the feature representation of the labeled identities, the unlabeled identities are used as negative classes and also participate in the process of learning feature representation. [10] proposed ID-discriminative embedding (IDE) to obtain better re-ID accuracy in CNN feature classification, and proposed Confidence Weighted Similarity (CWS), which is a new evaluation criterion, to improve the person search performance. [14] proposed NPSM module, which makes full use of LSTM and attention idea to shrink the area of person search recursively in the whole image until the target person is located accurately. [16] improved the OIM loss function [22], proposed the IOIM loss function to harden the distribution of labeled identities while softening the unlabeled identities distribution.

Inspired by [2, 4], we add a segmentation branch to the original person search framework to generate a high-quality segmentation mask for each detected person instance. In this way, each pedestrian instance has its own unique segmentation mask based on its own feature. By concatenating the segmentation mask with the corresponding feature maps in the person re-ID branch, the framework forms a self-attention mechanism to provide more discriminative feature for person re-ID, thereby improving the performance of person search.

3 The Proposed Approach

In this section, we will introduce the design details of the whole neural network architecture, the training mechanism of the network, as well as the overall inference process for person search.

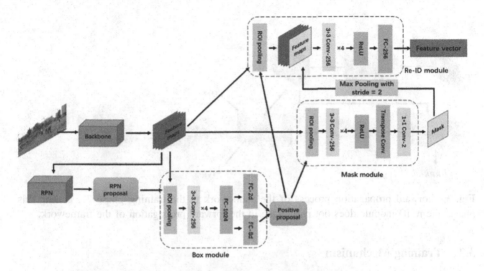

Fig. 2. The overall architecture of the framework. On the basis of Faster RCNN, mask module and re-ID module are added into the whole framework.

3.1 Architecture Overview

The overall architecture of the network is shown in the Fig. 2. It can be seen that the whole network is based on the Faster RCNN architecture [20], on which a mask module and a re-ID module are added to output the segmentation mask and the person re-ID feature vector respectively. The network mainly includes three parts: the Backbone network, the Region Proposal Network (RPN) and the ROI modules. And there are three branches included in the ROI modules: the Box module, the Mask module and the Re-ID module.

In inference phase, given a full-scene picture, we first transform the original pixel information into convolution feature maps $F_{backbone}$ through backbone network, and then send $F_{backbone}$ to RPN to predict the possible person proposals P_{rpn}. $F_{backbone}$ and P_{rpn} are then fed into the box module, after ROI pooling and another bounding box regression and classification, the final proposals P_{box}, which containing the positive sample, are output by the box module. Then $F_{backbone}$ and P_{box} are fed into the mask module to generate the semantic segmentation mask M for the corresponding image area of P_{box}. $F_{backbone}$ and P_{box} are also fed into the re-ID module. After ROI pooling in the re-ID module, we get the feature maps F_{reid}, then concatenate F_{reid} with M_p, which is obtained by down-sampling M, to provide more discriminative feature for person re-ID. Then after some feature extraction operations, the re-ID module output the feature vectors V_{reid} for feature matching with the query image.

That is to say, the output of the whole framework is the bounding box for person detection, the segmentation mask and the corresponding feature vector for each person instance, which is used to match the feature vector of query image to determine whether it is the same person.

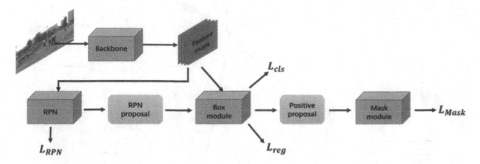

Fig. 3. Forward propagation process of the framework in the training phase T_c. During this phase, the re-ID module does not participate in the forward propagation of the framework.

3.2   Training Mechanism

Since there is no public dataset that can satisfy these three tasks of person detection, instance segmentation, person re-ID at the same time, we use two datasets to training the whole network. We choose COCO dataset [13] to train person detection and segmentation, we call this training phase T_c. And PRW dataset [10] is used to train person detection and person re-ID, we call this training phase T_p.

In phase T_c, as shown in Fig. 3, the re-ID module does not participate in the forward propagation process. The final loss function of the network in phase T_c is defined as:

$$L_{T_c} = L_{RPN} + L_{Box} + L_{Mask} \tag{1}$$

L_{RPN} stands for the loss function of RPN. In [20], it is defined as:

$$L_{RPN} = \frac{1}{N_{cls}} \sum_i L_{cls}(p_i, p_i^*) + \lambda \frac{1}{N_{reg}} \sum_i p_i^* L_{reg}(t_i, t_i^*) \tag{2}$$

Here, i is the index of the proposal in a mini-batch, p_i is the predicted probability of the proposal containing an object, p_i^* is the ground-truth label for the proposal, p_i^* is 1 if the proposal contains an object, and is 0 if the proposal does not contain an object. t_i is the predicted 4 coordinates for the proposal relative to the corresponding anchor, t_i^* is the ground truth relative to the corresponding anchor. L_{cls} is log loss over two classes (foreground or background). $L_{reg}(t_i, t_i^*) = R(t_i - t_i^*)$ where R is the robust loss function (smooth L1) defined in [6]. $p_i^* L_{reg}$ means only positive proposal is used to calculate the regression loss. More detailed parameter setting can be referred to [20].

L_{Box} stands for the loss function of the Box module, that is Fast CNN loss [6]. It has the same definitions of L_{RPN}, except for L_{Box} is aimed at multi-class proposals, not just foreground and background. L_{Box} can be referred to [6].

L_{Mask} stands for the loss function of the Mask module. The mask module generates a k × m × m dimension output for each ROI region, m × m represents the resolution of the feature map, and each pixel corresponds to the prediction results of k classes. To calculate mask loss, we apply a per-pixel sigmoid function on the output of the mask module. For a pixel in the proposal region whose ground truth is class K, the loss value on this pixel is a binary cross-entropy corresponding to the predicted value of class K. The loss for the whole ROI is the sum average of loss for each pixel.

In phase T_p, as shown in Fig. 4, although the mask module participates in the forward propagation process, it does not participate in loss computation. The final loss function of the network in phase T_p is defined as:

$$L_{T_p} = L_{RPN} + L_{Box} + L_{Re-ID} \tag{3}$$

L_{RPN} and L_{Box} have been explained above. For L_{Re-ID}, in this paper, to minimize the features discrepancy among the instances of the same person, while maximize the discrepancy among different person, we use Online Instance Matching (OIM) loss proposed in [22]. In the re-ID module, the normalized feature vector of a labeled identity is denoted as $x \in R^D$, where D is the feature dimension. x is fed into the labeled matching layer. A lookup table (LUT) $V \in R^{L \times D}$ is used to store the features of all the labeled instances, where L stands for the number of all the person ID. In the forward propagation phase T_p, we calculate the cosine similarity between x and all the labeled identities by $x V^T$. During the backward propagation phase, if the target class-id is t, the t-th row of the lookup table is updated by

$$v_t \leftarrow \gamma v_t = (1 - \gamma)x \tag{4}$$

where $\gamma \in [0, 1]$, then the t-th row should be l_2-normalized. For the unlabeled identities, a circular queue (CQ) $U \in R^{Q \times D}$ in the unlabeled matching layer is used to store the features of unlabeled identities, Q is the queue size. In the forward propagation phase T_p, we calculate the cosine similarities between x and all the unlabeled identities by xU^T. And in the backward propagation phase, the new feature vectors of the unlabeled identities are pushed into the queue and the out-of-date ones are popped out. The probability of x being recognized as the identity with id i is computed by a new SoftMax function bellow:

$$p_i = \frac{\exp(xV_i^T/\gamma)}{\sum_{j=1}^{L} \exp\left(xV_j^T/\gamma\right) + \sum_{j=1}^{Q} \exp\left(xU_j^T/\gamma\right)} \tag{5}$$

and the probability of x being recognized as the i-th unlabeled identity can be computed in a similar way. And the objective of OIM loss is to maximize the expected log-likelihood $L_{Re-ID} = E_x[p_t]$. For more detail, it can be referred to [22].

To make the overall training steps clear, we show the training steps in Algorithm 1. And the inference process of the overall framework can be seen in Algorithm 2.

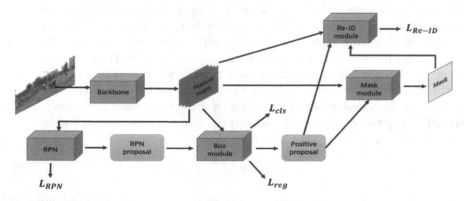

Fig. 4. Forward propagation process of the framework in the training phase T_p. During this phase, the mask module participates in the forward propagation process, it does not participate in calculating the loss.

Algorithm 1 Training steps of the framework

Input:
  Training data $\{G_c, G_p\}$, which representing the dataset used in phase T_c and phase T_p;
  Initialized parameters $W \leftarrow W_0$ of the network;
  Initialized (LUT) V and (CQ) U;
  Iteration number $I \leftarrow 0$ and learning rate, max_iteration.
Output: The parameters W.
1:    **while** not converge and $I <$ max_iteration **do**:
2:      $I \leftarrow I + 1$.
3:      Forward propagation of phase T_c using G_c.
4:      Compute the loss L_{T_c} by Formula 1.
5:      Update the parameters W except the re-ID module by BP algorithm.
6:      Forward propagation of phase T_p using G_p.
7:      Compute the loss L_{T_p} by Formula 3.
8:      Update the W by BP algorithm.
9:      Update U by pushing the new feature vectors of the unlabeled identities into the queue, and popping out the out-of-date ones.
10:    Update V by Formula 4.
11: **end while**
12: **return** W;

Algorithm 2 Person Search Inference Process

Input:

  The query person image q;

  The gallery images for person search;

  Load W_t into the network, the network parameters W_t is trained by Algorithm 1.

Output: Gallery images containing the most matching person with the query person.

1:    Input the query image into the network, the RPN and the box module do not partici-
       pate in the forward propagation phase, output the re-ID feature vector $V_q \in R^D$ of the
       query image, where D is the feature dimension.

2:    **for** image I in gallery images **do**:

3:        Input I into the network, output the detection results, instance segmentation re-
       sults

           and the re-ID feature vectors $V_g \in R^{N \times D}$ for image I, where N is the number of
           bounding boxes for the detection results of image I.

4:        By calculating the inner product of V_q and V_g^i, where V_g^i is the i-th vector in V_g,
           find out the i which not only makes the inner product value smallest and but also
           makes the value smaller than a certain threshold.

5:        **if** such i exists **then**:

6:           Draw the bounding box of the i-th detection result on image I, indicating that
           this person instance has the same person id with the query image on image I.

7:           Save the search result image of image I.

8:        **end if**

9:    **end for**

---

4 Experiments

4.1 Dataset

Since there is no public dataset that can satisfy these three tasks for person detection, segmentation and re-ID at the same time, we use two datasets to train the network. COCO dataset [13] is used to train person detection and segmentation. PRW dataset [10] is used to train person detection and re-ID. In the evaluation phase, we use PRW dataset to evaluate the performance of the framework.

The PRW dataset [10] is a large-scale public person search dataset. It contains 11816 video frames captured by 6 cameras and provides 43110 manually annotated person bounding boxes. Among them, 34304 persons are annotated with 932 IDs and the overage are annotated as id −2. It provides 5134 frames with 482 persons for training. And the testing set contains 2057 query person with the gallery size is 6112.

4.2 Evaluation Protocol

We use mean average precision (mAP) and the top-1 matching rate as performance metrics. The mAP metric, reflecting the accuracy of detecting the query person from the gallery images. To calculate the mAP, we should calculate the average precision (AP) for each query, and then calculate the average of the AP values for different queries to get the final mAP. The top-1 matching rate makes a matching counted if a bounding box among the top-1 predicted boxes overlaps with the ground truth larger than the threshold 0.5.

4.3   Implementation Details

In this paper, Faster RCNN [20] is used as the basic framework for person detection, and we use Pytorch [19] deep learning framework to complete our work. We use ResNet-50 [9] as our backbone and initialize the network using the pre-trained ResNet-50 on ImageNet during the training phase. During training, we set the mini-batch size to 4, and the base learning rate to 0.0025 with the SGD solver, max iteration number to 120000, and when the number of iterations reaches 50000 and 90000, the learning rate is reduced by 10 times, respectively. The training lasted 3 days on a Tesla P100 GPU.

4.4   Person Search Result

At first, we do several analysis experiments on PRW dataset to explore the contribution of mask module and self-attention mechanism in the whole framework. We use OIM [22] as our baseline, as it does not contain these two parts mentioned above. As shown in Table 1, our method has improved OIM by 3.05% and 3.83% on mAP and top-1 respectively by gradually adding mask module and self-attention mechanism, which proves the effectiveness of our framework.

Table 1. Results of ablation study on PRW dataset.

Method	Mask module	Self-attention	mAP (%)	top-1 (%)
OIM (baseline)	×	×	21.3	49.9
Ours (without self-attention)	√	×	22.28	52.67
Ours	√	√	24.35	53.73

Figure 5 shows the comparison between our method and OIM method. In the figure, the red rectangles are the result of person detection, and the green rectangles are the result of person search. We can find that many person instances are not detected by OIM. At the same time, there are some person matching error in the re-ID stage. Our method performs better in both detection and re-ID.

On PRW dataset, we compare our method with some state-of-the-art methods that combine person detection (respective R-CNN [7] detectors of ACF [3], LDCF [18], DPM [5]) and person re-identification (LOMO + XQDA [12], IDEdet [10], CWS [10]). At the same time, we compare our algorithm with the end-to-end algorithms proposed in recent years. The baseline result and the NPSM result is obtained from [14]. As shown in Table 2, our method has shown better performance on both mAP value and top-1 accuracy.

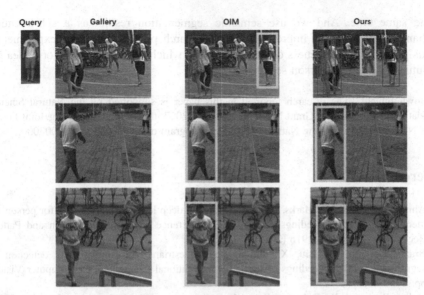

Fig. 5. Comparison of person search results between our method and OIM. The red rectangles are the results of person detection, and the green rectangles are the results of person search. For all these three gallery images, it is clear that some person identities have not been detected by OIM. And for the first gallery image, the person search result for OIM is also wrong. (Color figure online)

Table 2. Performance comparison on the PRW dataset

Method	mAP (%)	top-1 (%)
LDCF + LOMO + XQDA	11.0	31.1
LDCF + IDEdet	18.3	44.6
LDCF + IDEdet + CWS	18.3	45.5
ACF-Alex + LOMO + XQDA	10.3	30.6
ACF-Alex + IDEdet	17.5	43.6
ACF-Alex + IDEdet + CWS	17.8	45.2
DPM-Alex +LOMO + XQDA [12]	13.0	34.1
DPM-Alex + IDEdet [10]	20.3	47.4
DPM-Alex + IDEdet + CWS [10]	20.5	48.3
OIM (Baseline) [22]	21.3	49.9
NPSM [14]	24.2	53.1
Ours	**24.35**	**53.73**

5 Conclusion

In this paper, we propose a multi-task learning framework based on person detection, instance segmentation and person re-identification for person search. As far as we know, this is the first time that these three tasks are implemented in a single framework at the same time. And we use semantic segmentation results as a self-attention mechanism to effectively improve the person search performance. The experimental results prove the effectiveness of our framework, which also provides a good idea for the future research on person search.

Acknowledgment. The research reported in this paper is supported by the Natural Science Foundation of China under Grant No. 61872047,61732017, the NSFC-Guangdong Joint Found under No. U1501254, and the National Key R&D Program of China 2017YFB1003000.

References

1. Ahmed, E., Jones, M., Marks, T.K.: An improved deep learning architecture for person re-identification. In: Proceedings of the IEEE Conference on Computer Vision and Pattern Recognition, pp. 3908–3916 (2015)
2. Brazil, G., Yin, X., Liu, X.: Illuminating pedestrians via simultaneous detection & segmentation. In: Proceedings of the IEEE International Conference on Computer Vision, pp. 4950–4959 (2017)
3. Dollár, P., Appel, R., Belongie, S., et al.: Fast feature pyramids for object detection. IEEE Trans. Pattern Anal. Mach. Intell. **36**(8), 1532–1545 (2014)
4. Zhou, C., Wu, M., Lam, S.K.: SSA-CNN: semantic self-attention CNN for pedestrian detection. arXiv preprint arXiv:1902.09080 (2019)
5. Felzenszwalb, P.F., Girshick, R.B., McAllester, D., et al.: Object detection with discriminatively trained part-based models. IEEE Trans. Pattern Anal. Mach. Intell. **32**(9), 1627–1645 (2009)
6. Girshick, R.: Fast R-CNN. In: Proceedings of the IEEE International Conference on Computer Vision, pp. 1440–1448 (2015)
7. Girshick, R., Donahue, J., Darrell, T., et al.: Region-based convolutional networks for accurate object detection and segmentation. IEEE Trans. Pattern Anal. Mach. Intell. **38**(1), 142–158 (2015)
8. Gray, D., Tao, H.: Viewpoint invariant pedestrian recognition with an ensemble of localized features. In: Forsyth, D., Torr, P., Zisserman, A. (eds.) ECCV 2008. LNCS, vol. 5302, pp. 262–275. Springer, Heidelberg (2008). https://doi.org/10.1007/978-3-540-88682-2_21
9. He, K., Zhang, X., Ren, S., et al.: Deep residual learning for image recognition. In: Proceedings of the IEEE Conference on Computer Vision and Pattern Recognition, pp. 770–778 (2016)
10. Zheng, L., Zhang, H., Sun, S., et al.: Person re-identification in the wild. In: Proceedings of the IEEE Conference on Computer Vision and Pattern Recognition, pp. 1367–1376 (2017)
11. Li, W., Zhao, R., Xiao, T., et al.: DeepReID: deep filter pairing neural network for person re-identification. In: Proceedings of the IEEE Conference on Computer Vision and Pattern Recognition, pp. 152–159 (2014)
12. Liao, S., Hu, Y., Zhu, X., et al.: Person re-identification by local maximal occurrence representation and metric learning. In: Proceedings of the IEEE Conference on Computer Vision and Pattern Recognition, pp. 2197–2206 (2015)
13. Lin, T.-Y., Maire, M., Belongie, S.: Microsoft COCO: common objects in context. In: Fleet, D., Pajdla, T., Schiele, B., Tuytelaars, T. (eds.) ECCV 2014. LNCS, vol. 8693, pp. 740–755. Springer, Cham (2014). https://doi.org/10.1007/978-3-319-10602-1_48
14. Liu, H., Feng, J., Jie, Z., et al.: Neural person search machines. In: Proceedings of the IEEE International Conference on Computer Vision, pp. 493–501 (2017)

15. Liu, H., Feng, J., Qi, M., et al.: End-to-end comparative attention networks for person re-identification. IEEE Trans. Image Process. **26**(7), 3492–3506 (2017)
16. Liu, H., Shi, W., Huang, W., et al.: A discriminatively learned feature embedding based on multi-loss fusion for person search. In: 2018 IEEE International Conference on Acoustics, Speech and Signal Processing (ICASSP). IEEE, pp. 1668–1672 (2018)
17. Liu, W., Liao, S., Ren, W., et al.: High-level semantic feature detection: a new perspective for pedestrian detection. In: Proceedings of the IEEE Conference on Computer Vision and Pattern Recognition, pp. 5187–5196 (2019)
18. Nam, W., Dollár, P., Han, J.H.: Local decorrelation for improved pedestrian detection. In: Advances in Neural Information Processing Systems, pp. 424–432 (2014)
19. Paszke, A., Gross, S., Chintala, S., et al.: Automatic differentiation in Pytorch (2017)
20. Ren, S., He, K., Girshick, R., et al.: Faster R-CNN: towards real-time object detection with region proposal networks. In: Advances in Neural Information Processing Systems, pp. 91–99 (2015)
21. Tao, D., Guo, Y., Song, M., et al.: Person re-identification by dual-regularized kiss metric learning. IEEE Trans. Image Process. **25**(6), 2726–2738 (2016)
22. Xiao, T., Li, S., Wang, B., et al.: Joint detection and identification feature learning for person search. In: Proceedings of the IEEE Conference on Computer Vision and Pattern Recognition, pp. 3415–3424 (2017)
23. Xu, Y., Ma, B., Huang, R., et al.: Person search in a scene by jointly modeling people commonness and person uniqueness. In: Proceedings of the 22nd ACM International Conference on Multimedia, pp. 937–940. ACM (2014)
24. Zhang, L., Xiang, T., Gong, S.: Learning a discriminative null space for person re-identification. In: Proceedings of the IEEE Conference on Computer Vision and Pattern Recognition, pp. 1239–1248 (2016)

Analysis of Influencing Factors of PV Based Ensemble Modeling for PV Power and Application in Prediction

Lingfan Yang, Qian Liu, Zhihao Zhou, Yujin Zhang,
and Hangxia Zhou[✉]

China Jiliang University, XueYuan Street 258, Hangzhou, China
fanlingyang00030@163.com, zhx@cjlu.edu.cn

Abstract. According to the volatility and intermittent characteristics of photovoltaic power generation. Integrating PV power to the grid have an impact on the stability and safety. To address this challenge, the work learns the effect of support vector machine (SVM) and several algorithms on forecast. An algorithm model for improving the prediction accuracy of training data for multiple groups of factors has been proposed. The model consists of gradient boosting decision tree (GBDT), Particle Swarm Optimization (PSO) and SVM. Finally, according to the integrated algorithm, assigning weak learners' weights and integrating become strong learners. The GBDT algorithm is able to find the factors with high correlation coefficient in the data to establish the model, avoiding of using the empirical method to select the factors. The PSO algorithm finds the optimal parameters of the SVM algorithm and the optimal weight of the weak learner. Compared with BP and traditional SVM, the model established by the data without determining the weather type can obtain better prediction accuracy.

Keywords: PV power generation · SVM · GBDT algorithm · Prediction model

1 Introduction

In recent years, the world's energy consumption is increasing drastically. The supply of fossil fuels has been insufficient. Solar energy as an important renewable and clean energy source, the benefit of converting into electricity has developed rapidly, and is gradually developing from an independent system to a large-scale grid connection [1]. In order to reduce the adverse influence of photovoltaic power generation access on the grid-connected system, it is necessary to predict the PV output power and obtain the development curve of its future output power [2–5].

As reported in [6], according to the time-varying and non-property characteristics of network traffic time series, the prediction accuracy of traditional time series prediction methods is relatively low. A paper [7] which published recently proposed an algorithm to search step size and search direction for the problem that least squares support vector machines (LS-SVM) is difficult to obtain the best model for processing large-scale data. These two Improved seeker optimization algorithm (ISOA) optimizes the model parameters in the LS-SVM modeling process. The improved crowd search algorithm is a hybrid Iterative optimization algorithm. Malvoni *et al.* [8] studies the accuracy of PV

D. Milošević et al. (Eds.): HCC 2019, LNCS 11956, pp. 530–536, 2019.
https://doi.org/10.1007/978-3-030-37429-7_53

power prediction in different data processing techniques applied to input data set. First, wavelet decomposition and principal component analysis are used to decompose meteorological data as predictive inputs. The LS-SVM and Group method of data handing are combined, and the time series prediction method of data processing group least squares support vector machine (G-LS-SVM) is applied to the predict the photovoltaic power generation of the day. Literature [9], in order to make the infinite sensor not be captured and manipulated by attackers, proposed an improved support vector machine multi-classification algorithm, which combines the ideas of Hadamard algorithm and sparse random coding algorithm with support vector machine. In literature [10], the air temperature and total solar radiance are used as input data. The two sets of data are predicted by the dynamic artificial neural network model in this paper to obtain data of the air temperature and total solar radiance for the next hour. A research [11] uses BP networks to predict irradiance, temperature, and cloud cover to indirectly predict PV generation.

2   Model Construction

2.1   Analysis Affecting PV Power

At present, domestic and foreign research considers the external environmental meteorological factors as the primary factor training prediction model. It is generally believed that the light intensity received by photovoltaic panel components is the direct cause of photovoltaic power generation and an important factor affecting the power generation output [12]. Many experimental results show that the power output of the power generation increases with the increase of the light intensity in a certain range. According to the literature [13], the solar elevation angle, solar illumination intensity and sky cloud volume are the main meteorological factors affecting photovoltaic power generation. Many other literatures think that temperature, wind, humidity, etc., are also factors influencing photovoltaic power generation.

The influencing factors of the photovoltaic power generation from the above-mentioned many literature experiments have brought a fixed research thinking to the later scholars. If the experimental data of the research have above-mentioned factors of the photovoltaic power generation, the scholars always rely on the literature of the predecessors to classify this kind of data into the training set, but the data of these dimensions together as a training set is easy to make problems. With such multiple dimensions of data together as a training set, it is not sure to be a good training model, and one or more dimensions data have outlier that can make an impact on the final accuracy. More likely, the combination of these dimensions and other PV generation factors will result in more accurate predictions.

The integration of photovoltaic power generation involves the use of three-phase inverters [14], and the photovoltaic power plants that obtained the experimental data use three-phase controllers. The experimental data studied in this paper includes temperature, input of two DC currents and DC voltage, current, voltage, frequency and phase of three-phase AC. This paper excludes the reference of the empirical method and uses the GBDT algorithm [15] to find out the importance relationship between the factor data and the target value in multi-dimensional data. The most important factor is 100%, and the results of the remaining factors are displayed as a percentage (Fig. 1).

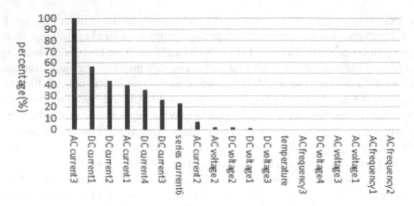

Fig. 1. Importance of PV power influence factor

From the above analysis, it can be seen that in the experimental data of this group, the week correlation between the temperature data and the power generation. According to the empirical method, the temperature is directly selected as the training set, or combine with another data as the training set, to train the model. It is proved by fact that the temperature in the experimental data is not an important factor affecting the power generation. The prediction error obtained by using the temperature data as the training set is very large, and it is not suitable as the training set to train model. Moreover, it is concluded that the training set model generated by the combination of the factors with different importance values above 70% will have better precision.

2.2 Construction of Prediction Model

This prediction model is constructed using the GBDT algorithm and the PSO algorithm [16] multiple times. The model construction method of this paper is based on the K-fold cross-validation [17–20].

After obtaining the historical data set of photovoltaic power generation, and find out the factors affecting the power generation of each day through the GBDT algorithm, and then take the first 90% of the daily data as the training set, the data. The last 10% is used as the test set. The SVM algorithm is used to build the model. After the model is built, the improved PSO algorithm which uses the MAPE error index as the fitness is used to search optimal parameters of each SVM model, the data of that day with the smallest MAPE error is extracted again and is used as the test set, the remaining PV data of each day is used as the training set. The SVM algorithm is used to build the model. After the model is completed, the improved PSO algorithm is used to use the MAPE error index as the fitness to find optimal parameters of each SVM model, this way can get the MAPE error results of each day, with MAPE error value 0 to 1 combining, 1 to 2 combining, 2 to 3 combining, and so on, this classification combination method increases the overall dimensions of the data and the amount of data that can find a better hyperplane which is a cutting surface of high-dimensional data [21]. This experimental data obtained six PSO-SVM models using this combined method. Then, because the PSO algorithm is looking for the weights of the six weak learners, the weight sum after each iteration cannot be made to 1, so the PSO algorithm is improved, and finally the prediction model of this study is obtained. And because when the PSO algorithm is looking for the weights of the six weak learners, the weight sum after each iteration cannot be made to 1, the PSO algorithm has improved, and finally the prediction model of this study is obtained.

In the above construction model, the PV data with the smallest MAPE error value is used as the test set, and the data of remaining days are used as the training set to establish the SVM model. The idea of obtaining the MAPE error value of the remaining days is that the photovoltaic power generation of the day which is stable when the MAPE error value is the smallest, and the external factors are the least interference. It is considered that the day is relatively sunny. The relatively sunny data is used as the test set, the MAPE error of the remaining days can be judged by the size of the MAPE, which days of the remaining days are seriously interfered by external factors, and which days are less affected by external interference.

3 Experiment Analysis

The data in this paper comes from the measured monitoring data of a small photovoltaic power station in Shaoxing, Zhejiang. The power station records the temperature every 6 min, the input of two DC currents and DC voltage, the current, voltage, frequency and phase of the three-phase AC. For using the improved PSO algorithm, the experimental running time is too long, and the experimental data needs to be reduced. This paper selects data of December 2017 to train predict model, and examine the advantages and disadvantages of the model and other models.

Taking the PV data of December 2017 as the experimental data, the BP algorithm is used to construct the model, and the PSO algorithm is used to optimize the weight of the BP model. The relatively sunny data is the test set. In Fig. 2, the ordinate is power, and the abscissa is the number of times of 6 min per day. The model prediction results are shown in Fig. 2.

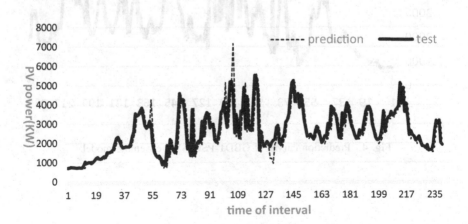

Fig. 2. Prediction curves of GBDT-PSO-BP

GBDT was used to find out the importance of power generation factors, and established the PSO-SVM model. As shown in Fig. 1, the experimental results show that when the training set is two-dimensional data of AC current 1 and DC current 4, the prediction model obtained has the best prediction effect, and the test set is relatively sunny. In Fig. 3, the ordinate is power, and the abscissa is the number of times of 6 min per day. The model prediction results are shown in Fig. 3.

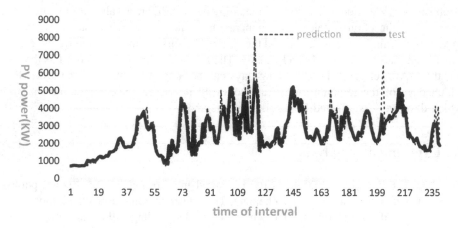

Fig. 3. Prediction curves of GBDT-PSO-SVR model

The GBDT-PSO-SVR integrated model prediction results are shown in Fig. 4.

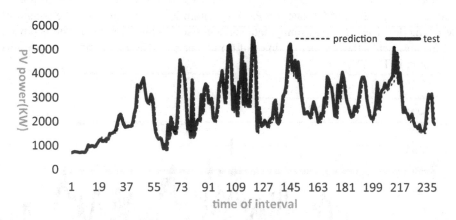

Fig. 4. Prediction curves of GBDT-PSO-SVR ensemble model

Table 1. MAPE and R2 of three models

	MAPE (%)	MAE	MSE	RMSE	R2
GBDT-PSO-BP	18.01	455.21	474 168.68	688.60	0.80
GBDT-PSO-SVR	18.32	486.97	602 120.50	775.96	0.49
GBDT-PSO-SVR ensemble	12.65	414.99	386 537.71	621.72	0.91

The mean absolute percentage error MAPE values and the goodness of fit R2 values predicted by the three models are shown in Table 1.

It can be seen from Table 1, Figs. 2, 3 and 4 that the proposed GBDT-PSO-SVR integration model has a significant improvement in prediction accuracy compared to the two other models, especially the R2 coefficient, GBDT-PSO-SVR integrated model predicted curve extremely fit test curve. It can be seen from the two error indicators of RMSE and MSE that the abnormal value of the GBDT-PSO-SVR integrated model prediction.

Compared with the target value in the test set is much less than that of the other two models, and the model is closer to the prediction curve. Therefore, by comparing the prediction effects of GBDT-PSO-SVR model, GBDT-PSO-BP model and GBDT-PSO-SVR integration model, it can be concluded that the method of constructing the model in this experiment can help to improve the prediction effect.

4   Conclusion

The photovoltaic power generation prediction helps to solve the problem of real-time balance between the two independent variables of random power generation and random power consumption, which is beneficial for the grid staff to complete the coordinated dispatching work, and to maintain the safety and stability better of the long-term operation of the power grid. Based on the measured data of PV, this paper proposes a prediction model which is GBDT-PSO-SVR integration through analyzing the influence of various factors in the data on PV system power generation. The simulation results analysis leads to the following conclusions.

When establishing a predictive model, the training set does not need to consider the weather category. The data was classified by the data combination method in the above construction model method. Once the predictive model is built, it is not necessary to consider the weather categories in the test set.

Acknowledgment. This work was supported by Basic Public Welfare Research Project of Zhejiang Province, China (No. LGF18F020017).

References

1. Zhou, L., Ren, W., Liao, B., et al.: Reactive power and voltage control for grid-connected PV power plants. Trans. China Electrotech. Soc. J. **30**(20), 168–175 (2015)
2. Bao, G., Yang, G., Yang, Y., et al.: Improved genetic algorithm-based optimal planning for grid connected distributed generation. Comput. Eng. Appl. J. **52**(16), 251–256 (2016)

3. Scolari, E., Reyes-Chamorro, L., Sossan, F., et al.: A comprehensive assessment of the short-term uncertainty of grid-connected PV systems. IEEE Trans. Sustain. Energy **9**, 1458–1467 (2018)

4. Chen, W., Ai, X., Wu, T., et al.: Influence of grid-connected photovoltaic system on power network. Electr. Power Autom. Equip. J. **33**(02), 26–32, 39 (2013)

5. Wang, S., Gao, R., Du, J.: With super parent node bayesian network ensemble regression model for time series. Chin. J. Comput. J. **40**(12), 2748–2761 (2017)

6. Tian, Z., Gao, X., Li, S.J., et al.: Prediction method for network traffic based on genetic algorithm optimized echo state network. J. Comput. Res. Dev. J. **52**(05), 1137–1145 (2015)

7. Gao, X.J., Fu, L.X., Wu, C.X., et al.: Energy consumption prediction model of air-conditioning system in subway station based on ISOA-LS-SVM (10), 36–43 (2018)

8. Malvoni, M., De Giorgi, M.G., Congedo, P.M.: Forecasting of PV power generation using weather input data preprocessing techniques. Energy Procedia J. **126**, 651–658 (2017)

9. Yan, K., Du, Y., Ren, Z.: MPPT perturbation optimization of photovoltaic power systems based on solar irradiance data classification. IEEE Trans. Sustain. Energy (Early Assess) **10**, 514–521 (2018)

10. Almonacid, F., Perez-Higueras, P., Eduardo, J., Fernandez, F., et al.: A methodology based on dynamic artificial neural network for short-term forecasting of the power output of a PV generator. Energy Convers. Manag. J. **85**(09), 389–398 (2014)

11. Cervone, G., Clemente-Harding, L., Alessandrini, S., et al.: Short-term photovoltaic power forecasting using artificial neural networks and an analog ensemble. Renew. Energy J. **108**, 274–286 (2017)

12. Saez-de-Ibarra, A., Milo, A., Gaztanaga, H., et al.: Co-optimization of storage system sizing and control strategy for intelligent photovoltaic power plants market integration. IEEE Trans. Sustain. Energy **7**, 1749–1761 (2016)

13. He, L., Ding, L.: Solar thermal physics-based architecture, pp. 42–81 (2011)

14. Zhang, C.: Research on key technology based on current source inverter photovoltaic grid-connected generation system. China University of Mining and Technology (2016)

15. Zhang, Y., Wang, D.: A speech model cluster method based on GBDT algorithm. Inf. Res. J. **39**(03), 23–27 (2013)

16. Cheng, Z., Lu, K., Qian, L., et al.: New method of MAP fixed-point selection based on improved PSO algorithm. Appl. Res. Comput. J. **10**, 1–6 (2019)

17. Jiang, G., Wang, W.: Error estimation based on variance analysis of k-fold cross-validation. Pattern Recogn. **69**, 94–106 (2017)

18. Wong, T.-T.: Parametric methods for comparing the performance of two classification algorithms evaluated by k-fold cross validation on multiple data sets. Pattern Recogn. **65**, 97–107 (2017)

19. Ling, H., Qian, C., Kang, W., et al.: Combination of support vector machine and k-fold cross validation to predict compressive strength of concrete in marine environment. Constr. Build. Mater. **206**, 355–363 (2019)

20. Wiens, T.S., Dale, B.C., Boyce, M.S., et al.: Three way k-fold cross-validation of resource selection functions. Ecol. Model. **212**(3–4), 244–255 (2008)

21. Xu, J.: Data set separability discriminant algorithm based on linear discriminant analysis. Bull. Sci. Technol. J. **29**(04), 31–32, 35 (2013)

Design of Searchable Algorithm for Biological Databased on Homomorphic Encryption

Minglang Yang[1(✉)], Yi Man[1(✉)], Ningning Liu[2(✉)],
Yixin Zhang[3(✉)], and Xiao Xing[3(✉)]

[1] Beijing University of Posts and Telecommunications, Beijing, China
yangml_95@foxmail.com, manyi@bupt.edu.cn
[2] Neusoft Corporation, Beijing, China
15575953840@163.com
[3] National Computer Network Emergency Response Technical
Team/Coordination Center of China, Beijing, China
1952902819@qq.com, 1521840143@qq.com

Abstract. With the rapid development of biotechnology, researchers are able to obtain large number of genome data sets. However, biological data often involves high privacy and data security issues. Thus when storing, transferring or analyzing these data, a safe and effective method is highly needed. This paper aims to propose a practical scheme using searchable homomorphic encryption. We combined the inverted index mechanism with the interactive operation on the homomorphic encrypted ciphertext data files, so as to realize the management and protection of biological data.

Keywords: Homomorphic encryption · Searchable encryption · Inverted index · Biological data · Data privacy

1 Introduction

In recent years, with the rapid development of biological science and technology, biological data resources have expanded rapidly. Since these data contain much private information, it can be very dangerous to reveal any part of them.

Currently, mainstream biological websites store data by anonymous. However, studies [1–3] have shown that identifiable personal information contained in genes cannot be completely eliminated. Attackers can even extract unique characteristics from small gene fragments. Study [4] shows that individual genomes can recover the last name of the person; Research [5] proposed a REIDIT algorithm that could link genomic data with specific individuals in public records. Therefore, the protection of genomic data cannot simply be solved by anonymous. Security issues is particularly important.

Based on the background above, homomorphic encryption algorithms aroused researchers interest. Through homomorphic encryption method, operations can be carried out on ciphertext, and has same effect to operate the plaintext directly.

In this paper, we present a method to establish inverted index on homomorphic encrypted data to obtain encrypted search result. In our approach, the genomic data owner provides only the encrypted data, and the sever can perform the sequence search

© Springer Nature Switzerland AG 2019
D. Milošević et al. (Eds.): HCC 2019, LNCS 11956, pp. 537–545, 2019.
https://doi.org/10.1007/978-3-030-37429-7_54

without decryption. The result can be decrypted only by the data owner or authenticated user holding the decryption key.

2   Theoretical Basis of Homomorphic Encryption and Search Algorithm

Many privacy protection computing protocols have been introduced in fully homomorphic cryptographic genome testing. Study [6] presents a practical solution, which uses LNV scheme [7] to calculate multiple hamming distance values on encrypted data. Study [8] describes how to calculate edit distance on homomorphic encrypted data. Study [9] discussed how to use homomorphic encryption to conduct private genome test on encrypted genome data, and focused on BGV scheme and YASHE scheme.

2.1   Introduction to Homomorphic Encryption Algorithm

Definition 2.1 (homomorphic encryption): Let E be an asymmetric encryption scheme, the public key is pk, the private key is sk. If an algebraic operation on the ciphertext is converted to the same operation on the corresponding plaintext, then E is said to be a homomorphic encryption scheme.

Definition 2.2 (Fully homomorphic encryption): Let E be an asymmetric encryption scheme, the public key is pk, the private key is sk. If E can achieve any homomorphic addition and multiplication at the same time without the need for decryption of the private key sk, then E is called the all-homomorphic encryption scheme.

Definition 2.3 (Somewhat homomorphic encryption): Let E be an asymmetric fully homomorphic encryption scheme, public key pk and private key sk. If E can only perform homomorphic evaluation on low-degree polynomial circuits, E is called SHE homomorphic encryption scheme.

Early homomorphic encryption schemes only support addition or multiplication operation. The FHE (Fully homomorphic encryption) concept was put forward by Rivest et al. [10] in 1978 and achieved breakthroughs by Gentry [11] until 2009. Gentry et al. [12] also constructed an efficient partial homogenous encryption scheme of the type BGV, based on the RLWE sampling and using the Chinese remainder theorem (CRT). In December 2018, Halevi-Polyakov-Shoup built BFV homomorphic encryption algorithm using CRT.

2.2   Introduction to Search and Searchable Encryption

Traditional search methods are usually based on the plaintext search technology. That is, both the keywords submitted by users and the data information in the server database are given in plaintext form. In such scenario, any malicious server can obtain these informations. The security of data is seriously lack of protection.

Based on this difficulty, searchable encryption is proposed. The research on searchable encryption starts from the search of a single keyword. This concept was first proposed by Song et al. [13]. Lu et al. [14] proposed a genome-wide association study security outsourcing scheme. Ayday et al. [15] constructed a privacy protection DNA retrieval scheme by order preserving encryption. Confounding circuit [16] and differential privacy technology [17] can also be used to protect genetic data, but there are noise problems. Research [18] used attribute encryption to solve the sequence matching problem, but could not obtain specific target gene location information.

In addition, regarding the scene of searching ciphertext documents in the cloud server on mobile devices, Chang et al. proposed the PPSED scheme [19], which requires the storage of preset dictionaries in the cloud server and requires multiple interactions between users and the server. Curtmola et al. [20] proposed two symmetric searchable encryption schemes based on the idea of inverted index of "index-files": sse-1 and sse-2.

3 BFV Algorithm and Inverted Index

In this section, we describe the two primary algorithm we used.

3.1 Brakerski/Fan-Vercauteren (BFV)

We focus here on describing the basic version of the BFV encryption scheme (Table 1).

Table 1. Symbol description

Symbol	Description
λ	Denotes the desired security level of the scheme. For instance, 128-bit security ($\lambda = 128$) or 256-bit security
PT	Denotes the underlying plaintext space
K	Denotes the dimension of the vectors to be encrypted
B	Denotes an auxiliary parameter that is used to control the complexity of the programs/circuits
M	Message from the message space
C	Ciphertext

The BFV schema mainly has six parts.

(1) $ParamGen(\lambda, PT, K, B) \rightarrow Params$

Set and generate the necessary parameters, including error-distributions $D1$, $D2$, a ring R and its corresponding integer modulus q. Integer modulus p for the plaintext, etc.

(2) $SecKeygen(Params) \rightarrow SK, EK$

The secret key SK of the encryption scheme is a random element s. The evaluation key consists of L LWE samples encoding the secret s in a specific fashion. In particular, for $i = 1, \ldots, L$, sample a random a_i from R/qR and error e_i from $D2$, compute

$$EK_i = \left(-(a_i s + e_i) + T^i s^2, a_i\right). \tag{1}$$

(3) $PubKeygen(Params) \rightarrow SK, PK, EK$

The public key is a random LWE sample with the secret s. Particularly, it is computed by sampling a random element a from R/qR and an error e and setting:

$$PK = (-(as + e), a). \tag{2}$$

(4) $PubEncrypt(PK, M) \rightarrow C$

To encrypt a message M, parse the public key as a pair $(pk0, pk1)$. Encryption consists of two LWE samples using a secret u where $(pk0, pk1)$ is treated as public randomness. In particular,

$$C = (pk0u + e1 + WM, \ pk1u + e2) \tag{3}$$

where u is a sampled from $D1$ and $e1, e2$ are sampled from $D2$.

(5) $SecEncrypt(SK, M) \rightarrow C$

The algorithm outputs a ciphertext C.

(6) $Decrypt(SK, C) \rightarrow M$

The message M can be recovered by dividing the polynomial $C(s)$ by W, rounding each coefficient to the nearest integer, and reducing each coefficient modulo p.

3.2   Inverted Index

Inverted indexes concept come from the need to find records based on the value of the property in a practical application. Each entry in the index table contains an attribute value and the address of each record that has that attribute value. Since the attribute value is determined not by the record, but by the attribute value, it is called inverted index.

The pseudo-code for establishing inverted index is as follow:

Algorithm : inverted index

```
Input: D (data file)
Output: res (search results)
1   while (read a document D) {
2       while (read a term T in D) {
3           if (Find (Dictionary, T ) == false )
4               Insert (Dictionary, T );
5           Get T's posting list;
6           Insert a node to T's posting list;
7       }
8   }
9   Write the inverted index to disk;
```

---

4   Searchable Homomorphic Encryption Based on Inverted Indexes

Traditional searchable encryption can achieve ciphertext search by constructing index of preset keywords [22, 23]. The common ciphertext retrieval algorithm doesn't encrypt the index directly, but encrypts the index through more complex operation to ensure security. In this paper, we take a different approach. The ciphertext retrieval algorithm is based on fully homomorphic encryption. It directly submits the index encrypted by the fully homomorphic encryption algorithm. Because of the feature of fully homomorphic encryption, the encrypted index can be homomorphic in the process of ciphertext retrieval.

4.1   Scheme Description

The whole scheme of the system can be described as follows:

Firstly, generate the public key and private key using homomorphic encryption algorithm proposed in the previous section. Use the public key to encrypt the data file. Keep the private key for decrypting the search results. Secondly, establish inverted index and the data to the sever. Thirdly, when the user initiates the retrieval request, the user side gives the retrieval term, uses the ciphertext retrieval algorithm to encrypt the index information in the query statement, generates the new retrieval statement and sends it to the database server. After that, the server computes and returns the retrieved result (ciphertext). Finally, the user obtains the encrypted data and decrypts the ciphertext with the private key to get the plaintext.

The ciphertext retrieval algorithm presented in this paper can be simply expressed in Fig. 1.

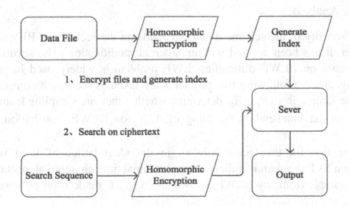

Fig. 1. Overall process of ciphertext retrieval scheme

4.2   Homomorphic Encryption of Biodata Based on Inverted Index

Inverted index was established by ElasticSearch (ES). ES provides an inverted index on each field. When retrieving, ES uses string matching. While homomorphic encryption of the same plaintext may have indifferent outputs. Thus, we rewrote the function which is responsible for the retrieval. The logical structure of the index file can be described as Fig. 2.

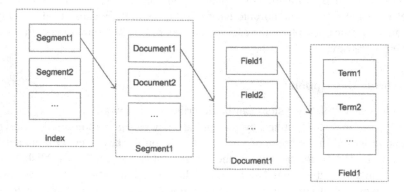

Fig. 2. Logical structure of index file

4.3   Theoretical Analysis and Experimental Results

In this part, we demonstrate the security of the algorithm from the theoretical level, and verify the feasibility through actual data experiments.

4.4   Safety Analysis

The overall security of the scheme mainly depends on the security of BFV algorithm. In this paper, it has been applied without internal modification. The security of this encryption relies on RLWE difficulties. LWE problem is widely used in public key cryptography, and the following two problems are taken as security assumptions [21]:

Given the sample from $R_q * R_q$, determine whether they are sampling from $A_{s(x),\chi}$ or $R_q * R_q$. Given the independent sampling of $A_{s(x),\chi}$ on RLWE distribution, to solve $s(x) \in R_q$.

Assuming that for the factor $poly(n)$, in the ideal lattice of R at worst, the approximation SVP problem is difficult for polynomial time quantum algorithms. Then, any $poly(n)$ sample from any RLWE distribution $A_{s(x),\chi}$ is random for polynomial time attackers.

In terms of data security, the data owner has processed the data into ciphertext before uploading. The pretreatment process is also completed locally. When user do a retrieve, the search sequence and index are also in the form of ciphertext. Thus data servers and visitors can not know the real information of the data, nor do they know the frequency details.

4.5   Experimental Results

Combined with ElasticSearch's open source search engine, 10 gene files were selected from the experimental data set, and the search sequences contained 5, 10, 15, 20, 25 characters, respectively. The encrypted search was conducted according to the designed scheme.

The search hit ratio is shown in Fig. 3.

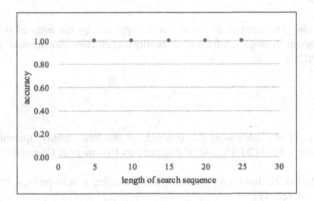

Fig. 3. Hit ratio of the search

It can be seen that in the form of ciphertext, the search accuracy can also be guaranteed.

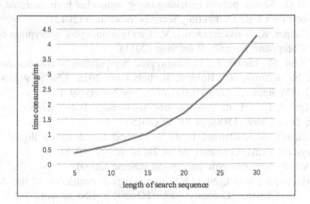

Fig. 4. Effect of sequence length on search time

Figure 4 shows the effect of sequence length on search time.

5   Conclusion

In this paper, we discussed how to privately perform sequence search on encrypted data using homomorphic encryption and inverted index strategy. We proposed an algorithm to encrypt sequence and file by BFV scheme and establish inverted index on encrypted data. Our algorithm returns the ciphertext of retrieval results. The proposed method enables us to perform any sequence search over encrypted data without worrying about privacy leakage.

Acknowledgments. The research of this article is supported by the national key research and development program "biological information security and efficient transmission" project, project No.2017YFC1201204.

References

1. Humbert, M., et al.: Addressing the concerns of the lacks family: quantification of kin genomic privacy. In: ACM SIGSAC Conference on Computer & Communications Security (2013)
2. Yaniv, E., Arvind, N.: Routes for breaching and protecting genetic privacy. Nat. Rev. Genet. **15**(6), 409–421 (2014)
3. Naveed, M., et al.: Privacy in the Genomic Era. ACM Comput. Surv. **48**(1), 1–44 (2015)
4. Melissa, G., et al.: Identifying personal genomes by surname inference. Science **339**(6117), 321–324 (2013)
5. Malin, B., Sweeney, L.: How (not) to protect genomic data privacy in a distributed network: using trail re-identification to evaluate and design anonymity protection systems. J. Biomed. Inform. **37**(3), 179–192 (2004)
6. Yasuda, M., et al.: Secure pattern matching using somewhat homomorphic encryption. In: ACM Workshop on Cloud Computing Security Workshop (2013)
7. Naehrig, M., Lauter, K., Vaikuntanathan, V.: Can homomorphic encryption be practical? In: ACM Cloud Computing Security Workshop (2011)
8. Cheon, J.H., Kim, M., Lauter, K.: Homomorphic computation of edit distance. In: Brenner, M., Christin, N., Johnson, B., Rohloff, K. (eds.) FC 2015. LNCS, vol. 8976. Springer, Heidelberg (2015). https://doi.org/10.1007/978-3-662-48051-9_15
9. Miran, K., Kristin, L.: Private genome analysis through homomorphic encryption. BMC Med. Inform. Decis. Mak. **15**(Suppl 5), S3 (2015)
10. Rivest, R., Shamir, A., Adleman, L.M.: A method for obtaining digital signatures and public-key cryptosystems. Commun. ACM **26**(2), 96–99 (1978)
11. Gentry, C.: A fully homomorphic encryption scheme (2009)
12. Gentry, C., Halevi, S., Smart, Nigel P.: Homomorphic evaluation of the AES circuit. In: Safavi-Naini, R., Canetti, R. (eds.) CRYPTO 2012. LNCS, vol. 7417, pp. 850–867. Springer, Heidelberg (2012). https://doi.org/10.1007/978-3-642-32009-5_49
13. Song, D.X., Wagner, D., Perrig, A.: Practical techniques for searches on encrypted data. In: IEEE Symposium on Security & Privacy (2002)
14. Lu, W., Yamada, Y., Sakuma, J.: Efficient secure outsourcing of genome-wide association studies. In: Security & Privacy Workshops (2015)
15. Ayday, E., Raisaro, J.L., Hengartner, U., Molyneaux, A., Hubaux, J.P.: Privacy-preserving processing of raw genomic data. In: Garcia-Alfaro, J., Lioudakis, G., Cuppens-Boulahia, N., Foley, S., Fitzgerald, W. (eds.) DPM 2013, SETOP 2013. LNCS, vol. 8247. Springer, Heidelberg (2014). https://doi.org/10.1007/978-3-642-54568-9_9

16. Riazi, M.S., et al.: GenMatch: secure DNA compatibility testing. In: IEEE International Symposium on Hardware Oriented Security & Trust (2016)
17. Simmons, S., Berger, B.: Realizing privacy preserving genome-wide association studies. Bioinformatics **32**(9), 1293–1300 (2016)
18. Bing, W., et al.: Privacy-preserving pattern matching over encrypted genetic data in cloud computing. In: IEEE INFOCOM - IEEE Conference on Computer Communications (2017)
19. Chang, Y.-C., Mitzenmacher, M.: Privacy preserving keyword searches on remote encrypted data. In: Ioannidis, J., Keromytis, A., Yung, M. (eds.) ACNS 2005. LNCS, vol. 3531, pp. 442–455. Springer, Heidelberg (2005). https://doi.org/10.1007/11496137_30
20. Curtmola, R., et al.: Searchable symmetric encryption: improved definitions and efficient constructions. In: ACM Conference on Computer & Communications Security (2006)
21. Lyubashevsky, V., Peikert, C., Regev, O.: On ideal lattices and learning with errors over rings. In: Gilbert, H. (ed.) EUROCRYPT 2010. LNCS, vol. 6110. Springer, Heidelberg (2010). https://doi.org/10.1007/978-3-642-13190-5_1
22. Ning, C., et al.: Privacy-preserving multi-keyword ranked search over encrypted cloud data. In: INFOCOM. IEEE (2011)
23. Bing, W., et al.: Privacy-preserving multi-keyword fuzzy search over encrypted data in the cloud. In: INFOCOM. IEEE (2014)

Speech-Based Automatic Recognition Technology for Major Depression Disorder

Zhixin Yang[✉], Hualiang Li, Li Li, Kai Zhang, Chaolin Xiong,
and Yuzhong Liu

Electric Power Research Institute of Guangdong Power Grid Corporation,
No. 8 Shuijungang, Dongfengdong Road, Yuexiu District,
Guangzhou 510080, China
67677483@qq.com

Abstract. Depression is one of the common mental illnesses nowadays. It can greatly harm the physical and mental health of patients and cause huge losses to individuals, families and society. Because of the lack of hardware and social prejudice against depression, there are a large number of misdiagnosis and missed diagnosis in hospitals. It is necessary to find an objective and efficient way to help the identification of depression. Previous studies have demonstrated the potential value of speech in this area. The model based on speech can distinguish patients from normal people to a great extent. On this basis, we hope to further predict the severity of depression through speech. In this paper, a total of 240 subjects were recruited to participate in the experiment. Their depression scores were measured using the PHQ9 scale, and their corresponding speech data were recorded under the self-introduction situation. Then, the effective voice features were extracted and the PCA was conducted for feature dimensionality reduction. Finally, utilizing several classical machine learning method, the depression degree classification models were constructed. This study is an attempt of the interdisciplinary study of psychology and computer science. It is hoped that it will provide new ideas for the related work of mental health monitoring.

Keywords: Depression · Speech · Machine learning · Classification · Prediction

1 Introduction

With the development of science and technology and the social system, material life is gradually guaranteed. People begin to pay attention to mental health, and the common mental illness such as depression has also received more and more attention. Depression affects people's normal study, work and life to varying degrees, and even severely leads to suicide. Although there are reliable medical treatments, many factors make it impossible for people with depression to receive effective treatment at the first time. These factors include lack of resources, lack of well-trained medical staff and discrimination against mental patients in society. At the same time, the inability to accurately diagnose depression also affects whether patients are effectively treated. If we can develop efficient, accurate and low-cost diagnostic methods, it can greatly help patients with depression to receive effective treatment.

Depression is a serious mental illness centered on depressive emotions and accompanied by symptoms such as low self-esteem, low vitality, and low interest in daily activities. Depression in the narrow sense refers to Major Depression Disorder (MDD), and the depression mentioned in this article, unless otherwise specified, refers to major depression. According to data from the World Health Organization (WHO) in 2017, more than 300 million people worldwide suffer from depression. Depression can lead to a very high rate of death. Reports indicate that th e suicide rate of depressed patients is more than 20 times that of the general population. At the same time, studies have indicated that depression increases the risk of cardiovascular disease and stroke. In addition, depression can badly affect the life, work and study of patients. Patients may cause work efficiency and economic loss due to work-related problems. There are reports that the economic loss in the United States due to depression in a year is as high as 36.6 billion dollars. The World Health Organization (WHO) believes that depression will become the world's major disease in 2030. Depression is a heavy burden for individuals, families, and society.

Although depression is a serious hazard, it can be effectively improved after proper medical treatment, psychotherapy or physical therapy. However, less than one-half of the patients receive effective treatment (according to the country's economic and cultural level, many regions are even less than one tenth). There are many factors that influence effective treatment, such as lack of resources and medical conditions, and lack of trained medical staff. Despite lack of hardware condition, in order to effectively treat patients with depression, we also need to ensure that the symptoms are correctly diagnosed. However, the misdiagnosis of depression is very serious. Due to the influence of traditional culture in China, the society has discrimination and misunderstanding of mental illness such as depression. People are afraid of medical treatment for mental illness, and they are not willing to go to the psychiatric department for treatment. Because of people's lack of sufficient knowledge in mental illness, they visit different wrong departments such as neurology, gastroenterology, cardiology. Besides, patients may not be able to express their mental discomfort through self-report, resulting in a high rate of misdiagnosis and missed diagnosis. Lilan [1] found that 54.2% of misdiagnosis and missed diagnosis was caused by the misdiagnosis of the first-time doctor. The doctors in general hospital had insufficient understanding of depression, only paid attention to the patient's physical symptoms and ignored their mental symptoms, therefore failed to detect those mental symptoms which patients could not express.

Most of the current diagnosis of depression is judged by a specialized psychiatrist with the diagnostic and statistical manual of mental disorders (DSM) [2], and the Hamilton Depression Scale (HAMD), Beck Depression Inventory (BDI) and other scales are also used to assist judgment. The DSM is published by the American Psychiatric Association and is the most commonly used instruction manual for diagnosing mental illness. Since its promulgation, it has been widely concerned. The American Psychiatric Association has continued to collect and sort out opinions to improve its clinical utility. However, because DSM can only be measured in a psychiatric clinic, its actual use is very limited. HAMD, which is commonly used in clinical practice, requires a professional assessor to communicate with the patient or to observe the patient's condition. It takes about half an hour for a single assessment. The self-assessment scale has the advantage of being able to use at any time and place without the need of professional medical staff, such as BDI, PHQ (Patient Health Questionnaire) and other self-rating scales. As a more simplified version of the BDI scale, PHQ-9 is often used in patients' self-assessment to judge the recovery status and the severity of depression. It is a widely recognized self-rating depression scale.

If we can automatically identify depression through the patient's ecological behaviors such as voice, gait, facial expressions, etc., we can give the doctors some advice without the patient's self-report or psychiatric expert's diagnosis, suggesting whether they should pay attention to the patient's mental state. In order to establish a model to automatically detect and recognize depression, people choose speech as the material for feature extraction. Speech is an easily accessible, non-invasive message, and slight psychological or physiological changes can result in significant changes in speech characteristics. Depression is thought to cause changes in neurophysiological and neurocognitive effects that affect human behavior, language, and cognitive function, and thus affect human speech. The voice of patients with depression is described as low-pitched, slow, long pauses, lack of stagnation, and dull. The speech characteristics of patients with depression have changed, which may be related to cognition and memory. The cognitive impairment caused by depression affects the working memory of individuals, and working memory is closely related to language planning. The effect of depression on speech indicates that the feature extraction of speech can be used to predict depression. The speech feature itself has nothing to do with the content of the speech, and cannot be concealed by the speaker. It can reflect the true psychological state of the speaker and is more objective and reliable.

Use of speech to predict depression has yielded preliminary results [3–8]. Cohn et al. [9] used the cross-validation to distinguish between depression and healthy people, the prediction accuracy rate reached 79%, suggesting the correlation between speech characteristics and depression. Cummins et al. [10] used the readings of 47 subjects under neutral and depressed emotions as the original input data and extracted the features for classification training. They found that the detailed features of the speech can be well adapted to the classification task, and the classification accuracy could reach 80%. Williamson et al. [11] used the data of the 6th International Audio/Video Emotion Challenge (AVEC) to classify depression and healthy people and score of Patient Health Questionnaire (PHQ). It is predicted that the average F1 score is 0.70.

The current research on speech and depression makes people notice the change of speech in patients with depression. Self-rating depression scale can reflect the degree of depression of the person filling the form with numerical value, and it is also recognized by the society for its convenience. Suppose we can combine the two, extract the characteristics of speech, prove the correlation between speech and depression scores through machine learning and mathematical statistics, and establish a model to predict the depression scale. The score is used to diagnose whether or not the participant is suffering from depression and predict the degree of depression on the basis of the score. This can provide a supplementary means for the diagnosis of depression in reality.

In this paper, we use the PHQ-9 scale as a basis for quantifying the severity of depression. PHQ-9 is often used as a self-assessment of patients' depression status to judge the therapeutic effect and degree of depression, and the measures that should be taken afterward. The score of the scale explains the severity of depression, and whether it can be combined with speech to predict if a participant has depression is also a problem we want to explore.

2 Materials and Methods

2.1 PHQ-9 Self-rating Depression Scale

PHQ is called Patient Health Questionnaire and is often used as a diagnosis of mental health diseases, such as depression, anxiety, alcoholism, binge eating disorder and other common mental illnesses. The PHQ scale initially contained 11 multiple-choice questions, which were supplemented with continuous improvement. PHQ-9 specifically refers to 9 scales for depressive symptoms, which are measured by the past two weeks, from loss of pleasure, depression, sleep disorders, lack of energy, eating disorders, low self-evaluation, difficulty in concentration, slowness and negative attitudes. The scores of the options are: 0 points not at all, 1 point for several days, 2 points for more than half of the day, 3 points for almost every day, and a total score of 0–27 points. The specific evaluation indicators are listed in Table 1. In the experiment, each participant was asked to fill in the questionnaire, and the score was used as a criterion for judging whether or not depression occurred.

Table 1. PHQ-9 scoring rules

Score	The degree of depression
0–4	No depression
5–9	May have mild depression and suggest to consult a psychiatrist
10–14	May have moderate depression and suggest to consult a psychiatrist
15–19	May have moderate to severe depression and suggest to consult a psychiatrist
20–27	May have severe depression and should consult a psychiatrist

2.2 Speech Collection

First, brief the participants on the purpose and steps of the experiment, let participants sign the informed consent form, register the information of the participants, give the participants an outline of the speech, and leave a preparation time of 5 min. The outline of the speech is: Please introduce yourself and introduce your hometown in detail; please introduce your major in detail, your research work during your study; please tell us about your future planning and what kind of work you want. Participants were recorded in a state where they were not intentionally induced to have positive or negative emotions. During the process, we would not interfere or remind. The duration of the recording was about 1 min each. The participants then used the pad to fill in the questionnaire.

Data collection was carried out in two batches. In August 2018, 88 speech and questionnaire data were collected, of which 88 were valid data; in January 2019, 152 speech and questionnaire data were collected, one of which appeared error, and the actual effective data was 151 copies. A total of 239 data were collected, in which, there are 130 women and 109 men.

2.3   Data Processing

OpenSMILE is an open source toolkit for feature extraction in signal processing and machine learning. It mainly extracts the characteristics of the audio signal. In this study, we used the emotion recognition feature extraction configuration file—emobase.conf that came with the software to extracted 989 acoustic features, including the following low-level descriptors (LLD): intensity, loudness, 12 MFCC, pitch (F0), voiced probability, F0 envelope, 8 LSF (line spectral frequency), zero-crossing rate, and Delta regression coefficients for these LLDs.

The openSMILE used in the experiment is version 2.3.0. The extracted features are processed differently, for example, directly using raw data, using normalized data, and using PCA (Principal Component Analysis) to reduce the data.

2.4   Depression Classification Model

In the experiment, the WEKA software was used to train the extracted features using different machine learning methods, and the model was extracted. Then we used 10-fold cross-validation method to test the accuracy of the model, doing the regression and classification analysis respectively.

In the regression analysis, differently processed features were used to predict the scores of PHQ-9. During each training process, we respectively used Random Forest, Linear Regression, Sequential Minimal Optimization (SMO), and Random Tree, Simple Linear Regression and other algorithms to train the data.

According to the score of the PHQ-9, the person filling the form can be divided into five categories according to the severity of depression (Table 1). According to the research needs, we focus on the classification and identification of major depression disorder, so we divide the sample into two groups with 14. The score is greater than 14 means able to have severe depression, it is recommended to consult a doctor, and less than 14 means depression tendency is not serious or no depression. The processed data is trained using different algorithms, such as Bayes Net, Support Vector Machine (SVM), Support Vector Machine Minimum Optimization Algorithm (SMO), Random Tree (Random Tree), decision tree J48, etc.

In addition, because the speaker's speech may be affected by the content, the extracted features may be more related to the content. We try to use the deep learning method to learn the feature to reduce the influence of the speech content on the prediction result. Here, the BP (Back Propagation) algorithm in the neural network algorithm is used, that is, a multi-layer feedforward network. BP network for classification and regression analysis can be implemented in weka. After selecting the MultilayerPerceptron algorithm in the classifier in weka explorer, we can set the hidden layer, learning rate, regularization and iteration number and other parameters in the graphical interface.

3 Results

Different classification algorithms are used for binary classification prediction. The ten-fold cross-validation is used to test the performance of classification model. The extracted data are normalized and standardized separately, and different algorithms are used for classification prediction. The results of depression classification on speech features without feature dimensionality reduction are shown in Table 2.

Table 2. The results of depression classification on speech features without feature dimensionality reduction

Algorithm	Correctly classified rate
BayesNet	72.38%
NaiveBayes	61.51%
Logistic	60.67%
SMO	62.34%
J48	69.46%
RandomTree	64.44%

PCA is performed on the normalized data, and then different algorithms are used for prediction. The results are shown in Table 3.

Table 3. The results of depression classification on speech features with feature dimensionality reduction

Algorithm	Correctly classified rate
BayesNet	77.82%
NaiveBayes	66.53%
Logistic	65.69%
J48	71.13%
RandomTree	68.20%

The BP network is used to learn the normalized data. The learning rate is set to 0.3.

The default regularization attribute is used to improve the utility of the network. The batch size is 100, and the number of training iterations is based on the actual number. The speed and effect of the operation are determined, and the initial setting is 500 to observe the actual learning effect. Set different hidden layers, where 0 means no hidden layer, and numbers indicate the number of nodes in each layer. The results are shown in Table 4.

Table 4. The results of depression classification on speech features using BP network

Hiden layers	Correctly classified rate
0	71.97%
2	70.71%
20	66.53%
50,4	64.02%

4   Discussion

The emobase.conf configuration file can extract 989 features related to emotion recognition. These features may be affected by the speaker's content and cannot reflect the acoustic characteristics completely. Too many features may cause a lot of noise without discrimination, which increases the difficulty of calculation and affects the actual training process. When using PCA to reduce dimension, the redundant features were removed and the prediction accuracy were improved. And the Bayesian network is better with 77.82% correctly classified rate.

There are still many shortcomings in this study. The first problems encountered when classifying were the sampling bias and uneven distribution. Due to the limitation of experimental conditions, it is difficult to recruit too many subjects with severe depressive symptoms. Most of the 239 subjects are normal (scores less than 5), while those who may have severe depression (with scores greater than 14) are less than one-twentieth.

5   Conclusion

In this study, we use a variety of machine learning methods to analyze the voice data acquisition in self-introduction context, and explored the association pattern between the voice and the severity of depression. The classification model of depression tendency were constructed and the classification accuracy is higher than 77%. It further validates the validity of the depression recognition method based on speech analysis.

Acknowledgements. The work was supported financially by the China Southern Power Grind (Grant No. GDKJXM20180673).

References

1. Huang, L.: Analysis of misdiagnosis of anxiety and depression in grass-roots hospitals Asia-pacific traditional medicine **8**(04), 207–208 (2012)
2. Kocsis, R.N.: Diagnostic and statistical manual of mental disorders: fifth edition (DSM-5). Int. J. Offender Ther. Comp. Criminol. **57**(12), 1546–15468 (2013)
3. Hashim, N.W., Wilkes, M., Salomon, R., et al.: Evaluation of voice acoustics as predictors of clinical depression scores. J. Voice **31**(2), 6 (2017)

4. Helfer, B.S., Quatieri, T.F., Williamson, J.R., et al.: Classification of depression state based on articulatory precision. In: Bimbot, F., Cerisara, C., Fougeron, C., et al. (eds.) 14th Annual Conference of the International Speech Communication Association, pp. 2171–2175 (2013)
5. Mundt, J.C., Vogel, A.P., Feltner, D.E., et al.: Vocal acoustic biomarkers of depression severity and treatment response. Biol. Psychiatry **72**(7), 580–587 (2012)
6. Pan, W., Wang, J., Liu, T., et al.: Depression recognition based on speech analysis. Chin. Sci. Bull. **63**(20), 2081–2092 (2018)
7. Scherer, S., Stratou, G., Gratch, J., et al.: Investigating voice quality as a speaker-independent indicator of depression and PTSD. In: Bimbot, F., Cerisara, C., Fougeron, C., et al. (eds.) 14th Annual Conference of the International Speech Communication Association, pp. 847–851 (2013)
8. Scherer, S., Stratou, G., Lucas, G., et al.: Automatic audiovisual behavior descriptors for psychological disorder analysis. Image Vis. Comput. **32**(10), 648–658 (2014)
9. Cohn, J.F., Kruez, T.S., Matthews, I., et al.: Detecting depression from facial actions and vocal prosody (2009)
10. Cummins, N., Epps, J., Breakspear, M., et al.: An Investigation of Depressed Speech Detection: Features and Normalization (2011)
11. Williamson, J.R., Godoy, E., Cha, M., et al.: Detecting Depression using Vocal, Facial and Semantic Communication Cues. Assoc Computing Machinery, New York (2016)

Multi-layer Filtering Webpage Classification Method Based on SVM

Yiwen Chen[✉] and Zhilin Yao[✉]

Jilin University, Changchun, China
snailbusterl@163.com, yaozl@jlu.edu.cn

Abstract. This paper presents a classification method based on web structure strategy and support vector machine, which can be applied to the classification of a large number of web pages. The algorithm designs the initial filter layer to recall the web pages quickly according to the structure of the web pages, and then trains the SVM classifier to do two classifications to improve the accuracy. Experiments show that the classification algorithm is feasible.

Keywords: Web page classification · DOM structure · Support vector machine

1 Introduction

The web pages in the Internet is numerous, varied and of varying quality. How to efficiently find out the pages that people need to browse is an important issue. Filtering technology can effectively improve the efficiency of obtaining information. For classification tasks with large data volume and high dimensionality, it is difficult to achieve a certain accuracy rate and recall rate while guarantee the classification efficiency at the same time. This paper proposes a svm-based multi-layer filtering web page classification method. Experiments show that this method has certain feasibility and efficiency while with high accuracy and recall rate.

2 Data Preprocessing

2.1 Document Object Model DOM

DOM tree connects each node into a tree structure in a hierarchical and sequential way through the parent-child relationship between nodes [1]. Some web information extraction methods based on DOM tree path are a simplification and expansion of Web information extraction methods based on keyword group clustering and node distance. DOM is very effective in processing web information. After the content in HTML is encapsulated as objects, the content in these documents can be easily manipulated or accessed. Through DOM, the document object model, we can effectively search and analyze various areas of the web page for denoising. Denoising method based on DOM tree text statistics has good effect.

© Springer Nature Switzerland AG 2019
D. Milošević et al. (Eds.): HCC 2019, LNCS 11956, pp. 554–559, 2019.
https://doi.org/10.1007/978-3-030-37429-7_56

2.2   Webpage Denoising

After obtaining the DOM tree of a web page, we can denoise the web page through the structure of the web page, such as advertisements, menu bars and so on. These noises are unnecessary for us to use, but they will appear frequently in various web pages and need to be eliminated. Single algorithm can be used to generate the training data we need, and then the training set is used to train the classification model. Finally, the trained classifier is used on DOM nodes to remove the noise [2]. This method has good scalability and is more efficient because it does not need to label a large number of data manually.

3   Web Page Classification Method

3.1   Semantic Analysis Participle

Semantic analysis and natural language understanding need word segmentation first. By setting a dictionary, dividing symbols and spaces we can get related words or groups of words, then we can filter out the stop words, such as "is, are, then", etc. Because such words do not contain important semantics, so these meaningless high-frequency words should be filtered out. Then we should extract stems. The purpose of extracting stem is to merge the same semantics of a word [3]. The text to be classified should match the words in the dictionary, and if the word is found in the dictionary, it is output according to the word. The effect of this mechanical will be determined by the correctness and completeness of the dictionary.

3.2   Feature Selection

Feature words of different importance have different degrees of discrimination. The effect of feature selection is based on the function of appropriate feature evaluation. The words in the features set should better express the content of the text topic, and distinguish the target text from other texts [4]. Chi-square is a widely used hypothesis test on counting data. It mainly compares two or more sample rates (composed ratios) and the correlation analysis of two categorical variables. The fundamental idea is to compare the degree of agreement between the theoretical frequency and the actual frequency or the goodness of fit. When the deviation is small, it is considered as the result of normal sample error, and it can be determined that the two independent hypotheses are established; otherwise, the two are actually related, and the alternative hypothesis is accepted, which is a non-parametric test of free distribution [5]. The formula for calculating the chi-square value is:

$$\chi^2 = \sum \frac{(A - E)^2}{E} = \sum_{i=1}^{k} \frac{(A_i - E_i)^2}{E_i} = \sum_{i=1}^{k} \frac{(A_i - np_i)^2}{np_i} (i = 1, 2, 3, \ldots, k) \quad (1)$$

Where Ai is the actual observed frequency of i, that is, the number of times the target situation actually occurs, Ei is the expected frequency of i, that is, the result of theoretical calculation, n is the total frequency, and pi is the desired frequency of i level. When n is relatively large, the χ^2 statistic approximates the chi-square distribution of k − 1 degrees of freedom. The most common use of chi-square test is to investigate whether the distribution of each level of an unordered categorical variable is consistent between two or more groups.

3.3   Multi-layer Filtering Web Page Classification Method Initial Filtration

The multi-layer filtering web page classification can be summarized as a three-layer classification structure: the first layer is the pre-processing layer, and the source webpage is correspondingly pre-processed; the second layer is the initial filtering layer, it uses high-recall rate and low-accuracy standard to collect and classify the web pages. The third layer is the svm filter layer. The support vector machine principle is used to judge the web pages that have passed the first two layers to obtain the category of the web pages.

For the initial filtering layer, the manual classification strategy is used to make the classifier have a good recall rate, and there is no need to excessively require accuracy in this layer. The recall rate needs to reach more than 90% to ensure that the classifier can effectively select the web pages that are needed for each category from a batch of quantitative web pages, thus ensures the classification accuracy of the entire classifier.

First, we classify feature words as strong, medium, and weak. Different levels of words have different scores. For example, "football" has a higher weight and should have a higher score. The appearance of that word often is the key to the sports category. There are two factors in the semi-structured data of web pages that can affect the intensity of the feature. One is the frequency of occurrence of feature words, and the other is the location where feature words appear. Considering the different impact of each type of node on topic information, it is appropriate to set the corresponding impact factors for different types of nodes [6]. The content in the url and <title> tags can often summarize the content of the web page, so the system gives a high weight of 0.8; the words in the location of the name = "keyword" in the <meta> tag are keywords [7], so the system gives a weight of 0.7 to them; In <meta> tag, name = "description" is a description of the web page, but this description is just introduction to the overall business of the website, so it is not accurate enough, and the system gives a low weight of 0.5 to it.

The classification rule for the initial filtering classification system is:

$$J = \sum_{i=1}^{m} L_i * \log_2(\sum_{j=1}^{n} V_j) - T \tag{2}$$

Where Li is a different weight corresponding to the occurrence of different parts of the web page structure, and Vj represents a different weight value corresponding to the weight of different vocabulary categories. T is the judgment threshold, and the T value of the system should not be set too high on the premise of requiring the recall rate. This category will recall the page when the J value is greater than zero. In order to ensure that the initial filtering has sufficient recall rate, the recall algorithm is optimized by the unrecalled negative sample iterative algorithm logic, which is beneficial to the overall performance of the classification system.

In practical engineering applications, the effect of the initial filtering determines the performance of the entire classifier. On the other hand, the web pages recalled by the initial filtering can be used for people to mark up to get the training web pages faster.

3.4 Support Vector Machine (SVM) Filter Layer

After the initial filtering, the classifier has ensured the recall of a large number of theme-related pages. When training web classifier, it is possible to improve the performance of Web classifier by effectively screening the web sample set. SVM can find support vectors with good classification ability after training. The constructed classifier can maximize the interval between classes and has high classification accuracy [8]. SVM can train a classifier with good performance in the case of fewer samples.

To construct and solve the optimization problem, we need to select the appropriate kernel function K (x, x') and the appropriate parameter C.

$$\min_a \frac{1}{2} \sum_{i=1}^{l} \sum_{j}^{l} y_i y_j a_i a_j (K(x_i, y_j) + \frac{1}{C} \delta_{ij}) - \sum_{j=1}^{l} a_j \tag{3}$$

Satisfy: $\sum_{i=1}^{l} y_i a_i = 0$ 且 $a_i \geq 0$, i = 0, 1, 2, ..., l.

Among them: $\delta = \{^{0(i=j)}_{1(i \neq j)}$

Optimal solution: $\alpha^* = (\alpha_1^*, \alpha_2^*, \ldots, \alpha_t^*)^T$

The polynomial kernel function with wide convergence region and easy parameters control is better than the radial basis function. It can be applied to low-dimensional, high-dimensional, large and small sample data. Balanced constrained programming model can be used to optimize the selection of parameters [9]. The support vector machine corresponding to the polynomial kernel function is a p-order polynomial classifier, and the classification decision function is:

$$f(x) = sign(\sum_{i=1}^{N_i} a_i^* y_i (x_i \cdot x + 1)^p + b^*) \tag{4}$$

After the initial filtering, SVM has good generalization performance by annotating all the web pages classified by the topic and training the classifier. However, the requirement for training set is relatively high, so it is very important to ensure the reliability of training data.

4 Experiment and Result Analysis

4.1 Experimental Result

The classification indexes of the webpage classification system are the recall rate and the accuracy rate. The recall rate and the accuracy rate have certain relationship. The comprehensive consideration of the recall rate and the accuracy rate can be the evaluation index of the classification effect. The experimental data is the web pages captured from the Internet. They are the Top data set, the Schedule data set and the History data set. The Top data set is a set of high-quality web pages that each website pushes to the user, and the Schedule data set is a set of randomly crawled web pages. The History dataset is a set of previously stored web pages. There are 1000 pages in each data set, which are manually labeled as "sports" or "criminal" content. The trained classification system sets parameters to classify sports and crimes. The initial filtering results and final results are as follows (Tables 1, 2, 3 and 4):

Table 1. Initial filtration effect of criminal categories

Index / Category	Top	Schedule	History	Average
Accuracy	58.6%	55.3%	51.1%	55.0%
Recall	98.3%	96.5%	95.7%	96.8%

Table 2. Initial filtration effect of sports categories

Index / Category	Top	Schedule	History	Average
Accuracy	56.1%	53.9%	48.2%	52.7%
Recall	96.5%	95.2%	97.7%	96.4%

Table 3. The final classification effect of criminal categories

Index / Category	Top	Schedule	History	Average
Accuracy	94.6%	93.8%	93.1%	93.8%
Recall	98.3%	96.5%	95.7%	96.8%

Table 4. The final classification effect of sports categories

Index / Category	Top	Schedule	History	Average
Accuracy	93.2%	92.1%	92.5%	92.6%
Recall	96.5%	95.2%	97.7%	96.4%

The experimental data has a higher recall rate because of the results of the iterative strategy. It can be seen that the improvement of the accuracy is based on the svm layer. Therefore, the training data of the svm layer needs better quality to ensure the training effect of the classifier, and the accuracy of the preprocessing layer is low, but the classifier can ensure the classification efficiency of large-scale web pages. The overall classification is good.

4.2 Conclusion

This paper proposed a svm-based multi-layer filtering web page classification method, which has certain feasibility. The performance of the original classification system has the efficiency of processing large-scale web pages and guarantees the recall rate. After passing the svm layer, the accuracy can be greatly improved, so the entire classification system has a reliable classification ability, and can meet the requirements of numerous web pages classification in engineering.

References

1. Ma, Y., Zheng, X., Xianmin, Li, Y.: XML web page classification method based on web mining and document object model tree. Microcomput. Appl. **32**(07), 47–49+52 (2016)
2. Chakrabarti, D., Kumar, R., Punera, K.: Page-level template detection via isotonic smoothing. In: Proceedings of the 16th International Conference on World Wide Web, pp. 61–70. ACM, New York (2007)
3. Zhang, D.: Research and Implementation of Content-Oriented Web Page Classification Method. Nanjing University of Posts and Telecommunications (2017)
4. Ye, L.: Design and Research of Web Classification Scheme Based on SVM. Beijing University of Posts and Telecommunications (2014)
5. Guo, X.: Application of information gain rate and chi-square test (IGRAC) in manufacturing industry. J. Comput. Sci. Inst. Inf. Technol. Appl. Southwest Univ. Financ. Econ. **3** (2009)
6. Wang, W., Guo, X.: The method of selecting kernel function. J. Liaoning Normal Univ. (Nat. Sci. Ed.) **01**, 1–4 (2008)
7. Gu, M., et al.: Research on web page classification technology based on structure and text features. J China Univ. Sci. Technol. **47**(04), 290–296 (2017)
8. He, N., Wang, J., Zhou, Q., Cao, Y.: Chinese web page classifier based on decision support vector machine. Comput. Eng. **02**, 47–48 (2003)
9. Ding, S., Qi, P., Tan, H.: Review of support vector machine theory and algorithms. J. Univ. Electron. Sci. Technol. **40**(01), 2–10 (2011)

Character-Level Attention Convolutional Neural Networks for Short-Text Classification

Feiyang Yin[1(✉)], Zhilin Yao[1(✉)], and Jia Liu[2(✉)]

[1] Jilin University, Changchun, China
yinfy17@mails.jlu.edu.cn, yaozl@jlu.edu.cn
[2] Systems Engineering Research Institute, Beijing, China
liujia02416@163.com

Abstract. This paper proposes a character-level attention convolutional neural networks model (ACNN) for short-text classification task. The model is implemented on the deep learning framework which named tensorflow. The model can achieve better short-text classification result. Experimental datasets are from three different categories and scales. ACNN model are compared with traditional model such as LSTM and CNN. The experimental results show that ACNN model significantly improves the short-text classification results.

Keywords: ACNN · Attention mechanism · LSTM · Short-text classification

1 Introduction

Text classification is a very important issue in the field of natural language processing. Joachims used support vector machines which is traditional machine learning method to classify text [1]. In 2012, Wang used linear model and kernel methods of machine learning [2]. In recent years, deep learning methods have widely used. In 2014, Kim used the word2vec to train to obtain word vectors, and then used convolutional neural networks to classify sentences [3]. In 2015, Zhang used convolutional neural networks which achieve significant results in character-level text compared with word bag models, n-gram models, and TFIDF [4]. A lot of research shows that RNN can model in time series greatly compared with traditional model such as DNN and CNN [5]. In 2017, Yogatama compared the advantages and disadvantages of RNN generative model and RNN discriminative model in text classification in detail [6]. Furtherly, In order to solve the problem that RNN does not support long-term dependence. Researchers have put forward Long short-term memory neural network (LSTM) [7]. In 2018, K Gao further proposed a BiLSTM neural network based on the attention mechanism [8].

This paper puts forward a character-level attention convolution neural networks model. The attention mechanism is used to the model focus on text location information, and different weights are assigned to different characters. Moreover, the dropout mechanism is added to the model to weaken the joint adaptability of neural nodes and enhances the generalization ability of the model. In these ways, we can improve the short-text classification results.

D. Milošević et al. (Eds.): HCC 2019, LNCS 11956, pp. 560–567, 2019.
https://doi.org/10.1007/978-3-030-37429-7_57

2 The Research of Model Structure

Convolution Neural network (CNN) that is a kind of feedforward neural network structure with deep convolution operation. It is one of the representative algorithms in deep learning. The traditional CNN structure contains the data Input layer (Input layer), Convolution layer (Conv layer), Pooling layer (Pooling layer) and Full connection layer (FC layer) [9, 10]. The structure of ACNN model which adds the attention mechanism and Dropout layer to the traditional CNN structure is shown in Fig. 1. The input layer preprocesses the text to obtain the embedded representation of the text, then undergoes the six convolutional layers and pooling layers crossover operation. Finally the output is obtained through the three full connection layers, in which two Dropout layers is included.

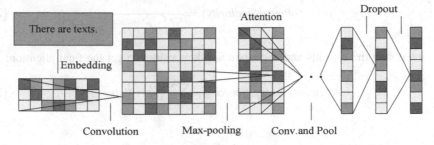

Fig. 1. The structure of ACNN model

2.1 Dropout

The left of Fig. 2 shows the hidden layer network structure of simple neurons. In the forward propagation, all neurons are delivered to the neurons of next layer after calculation. The right of Fig. 2 shows the Dropout structure. Red neurons are discarded with a certain probability, then forces a neuron work with the neurons of next layer selected randomly. This operation weakens the joint flexibility of the neurons, but increases the diversity of neuronal connections, thereby increases the generalization ability of the model.

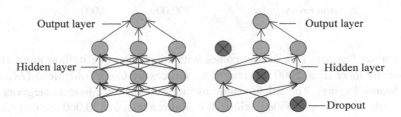

Fig. 2. The principle of Dropout (Color figure online)

2.2   Attention Mechanism

Attention mechanism first appeared in the field of computer vision, human vision captured the focus of attention through rapid scanning, paying more attention on the area, improving the efficiency and accuracy of visual information processing. And text processing is similar to that, the more weight represents the more importance of information in the text [11]. The main principle is as follows by three steps:

In this paper, perceptron is used to calculate the similarity between Query and each Key to obtain the weight:

$$Similarity(Query, Key_i) = MLP(Query, Key_i) \qquad (1)$$

Use Softmax function to normalize these weights and convert them to 0–1;

$$a_i = Softmax(Similarity) = \frac{e^{Similarity}}{\sum_{j=1}^{L_x} e^{Similarity}} \qquad (2)$$

Weighted sum the weights and corresponding key values to get the final attention;

$$Attention(Query, Source) = \sum_{i=1}^{L_x} a_i \cdot Value_i \qquad (3)$$

3   Experiment

3.1   Experiment Datasets

The following three datasets were mainly used in this experiment, and the specific information was referred to Table 1.

Table 1. Experiment datasets

Dataset	Classes	Train samples	Test samples
Ag's news	4	120,000	7,600
Amazon review full	5	3,000,000	650,000
Amazon review polarity	2	200,000	80,000

AG news Dataset: This dataset comes with more than one million news articles collected by more than 2,000 different news sources by ComeToMyHead (Academic News Search Engine). The article mainly includes these four different categories such as the world, sports, business and sci/tec. It contains a total of 120,000 training samples and 7600 test samples. The sample contains three columns of information which are category index, title and description information.

Amazon Review Full Dataset: The dataset is based on Amazon reviews and covers five levels of score from 1 to 5, with a total of 3,000,000 training samples and 650,000 test samples. The sample contains three columns of information which are category, review titles and comment.

Amazon Review Polarity Dataset: This dataset is used for sentiment analysis of texts. The dataset uses negative category with rating levels 1 and 2, and positive category with rating levels 4 and 5. The article contains 200,000 training samples and 80,000 test samples.

3.2 Parameter Setting

The convolutional layer and pooling layer parameters setting are shown in Table 2. In addition, the first and second fully connected layers are all set to 1024, and the third layer fully connected layer is different depending on the training situation.

Table 2. Parameter setting of convolution layer and pooling layer

Layer	Feature	Kernel	Pool
1	256	7	3
2	256	7	3
3	256	3	N/A
4	256	3	N/A
5	256	3	N/A
6	256	3	3

3.3 Comparison Model

LSTM model: LSTM (Long Short Term Memory) model is a special RNN network structure, which mainly includes three structures such as forgotten gate, input gate and output gate [12]. It is shown in Fig. 3.

Step 1: The first step determines what information we will discard from the cell state. The forgotten gate reads h_{t-1} and x_t, and outputs a value between 0 and 1 for each cell state C_{t-1}. 1 means "completely reserved" and 0 means "completely discarded". Where h_{t-1} represents the output of the previous cell and x_t represents the input of the current cell. σ represents the sigmoid function.

$$f_t = \sigma\left(W_f \cdot [h_{t-1}, x_t] + b_f\right) \tag{4}$$

Step 2: The second step of the input gate determines how much new information is added to the cell state. The need to implement this involves two steps, first, the sigmoid layer of the input gate determines which information needs to be updated. A tanh layer generates a vector, which is the alternative content $\tilde{C}_t$ for updating. Next we combine the two parts and update the state of the cell. Then it is time to update the old cell state, and C_{t-1} is updated to C_t. We multiply the old state by f_t and discard the information we determined to be discarded. Finally add $i_t * \tilde{C}_t$. This is the new candidate value.

$$i_t = \sigma(W_i \cdot [h_{t-1}, x_t] + b_i) \tag{5}$$

$$\tilde{C}_t = \tanh\left(W_C \cdot [h_{t-1}, x_t] + b_C\right) \tag{6}$$

$$C_t = f_t * C_{t-1} + i_t * \tilde{C}_t \tag{7}$$

Step 3: Finally, we need to determine what value to output. First, we run a sigmoid layer to determine which part of the cell state will be output. Next, we process the cell state through tanh (get a value between −1 and 1) and multiply it by the output of the sigmoid gate. Finally we will only output the part of we determined.

$$o_t = \sigma(W_o \cdot [h_{t-1}, x_t] + b_o) \tag{8}$$

$$h_t = o_t * \tanh(C_t) \tag{9}$$

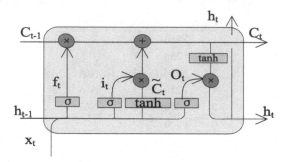

Fig. 3. The structure of LSTM model

3.4 Results and Analysis

In this paper, nine groups of comparison experiments perform on the three different data sets which shown in Fig. 4.

The LSTM, CNN, and ACNN model validation models are used to verify the model in the AG's news dataset. The error rates are 13.94 (LSTM), 12.82 (CNN), and 13.28 (ACNN). The scale of the AG's news dataset which is compared to the other two datasets is small, and the CNN model achieves the best results. It can be seen that the Dropout layer is more advantageous in the large-scale dataset or deep network structure.

The LSTM, CNN, and ACNN models are used to verify the model in the Amazon review full dataset. The error rates are 40.57 (LSTM), 41.31 (CNN), and 34.37 (ACNN). The ACNN model which is compared to the other two models achieves the best results. The scale and category of dataset is the largest. The Dropout layer has achieved good results in a large number of data sets, and the attention mechanism better assigns weight information to make the model perform better.

Using the LSTM, CNN, and ACNN models to verify in the Amazon review polarity dataset, the error rates are 6.1 (LSTM), 5.51 (CNN), and 5.49 (ACNN). The comparison also shows that the ACNN model has achieved better results.

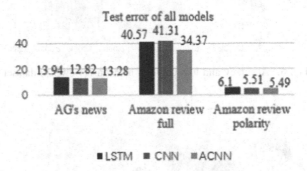

Fig. 4. Test error of LSTM/CNN/ACNN

In this paper, the accuracy and loss curve of the ACNN model on different datasets are given to verify the validity of the model during training:

In the AG's new dataset, use the first 10,000 rounds of training results, the data curves around six thousand rounds are shown in Fig. 5, we can see that the model has achieved nearly 95% accuracy. In the Amazon review full dataset, take the first 20,000 rounds of training result curve is shown in Fig. 6. Around four thousand rounds, the accuracy rate is close to 53%. In the Amazon review polarity dataset, the first 10,000 rounds of training result curves are shown in Fig. 7. Around the two thousand rounds, it is close to 90% accuracy, but there is a short-term convergence of the loss curve in the training process, and the Loss curve is on the rise. It may be that the model structure and parameter settings lead to the non-convergence of Loss. This part of the work needs to be improved in the future work.

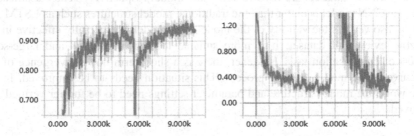

Fig. 5. Accuracy and loss curve of AG news dataset

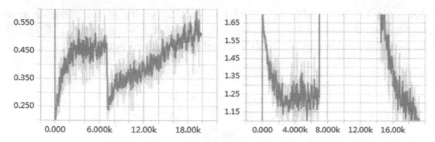

Fig. 6. Accuracy and loss curve of amazon review full dataset

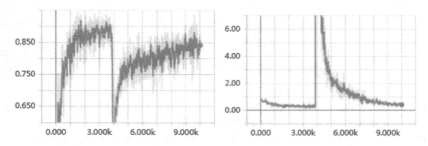

Fig. 7. Accuracy and loss curve of amazon review polarity dataset

4 Conclusion

This paper proposes character-level attention convolutional neural networks model which use the dropout mechanism to reduce the risk of overfitting in multi-layer model structure, increasing the connection diversity between neurons, improving generalization ability, and use the attention mechanism to measure the importance of text information, so that the model can be more specific for classifying the semantic information. In the nine groups of comparison experiments, the different magnitudes and categories of datasets are used to verify the model respectively. The classification effect of ACNN model is better than the traditional model structures such as LSTM and CNN. It also verifies the validity of the model. The model is especially effective in the Amazon review full dataset. It is more suitable for the multi-class and large-scale datasets of classification task. However, there is a short-term non-convergence of the loss curve in the model training process. This situation needs to be improved. In the future work, the model structure and parameter settings need to be further verified and improved.

References

1. Joachims, T.: Text categorization with support vector machines: learning with many relevant features. In: Nédellec, C., Rouveirol, C. (eds.) ECML 1998. LNCS, vol. 1398, pp. 137–142. Springer, Heidelberg (1998). https://doi.org/10.1007/BFb0026683
2. Wang, S., Manning, C.D.: Baselines and bigrams: simple, good sentiment and topic classification. In: Proceedings of the 50th Annual Meeting of the Association for Computational Linguistics (2012)
3. Kim, Y.: Convolutional neural networks for sentence classification. arXiv preprint arXiv: 1408.5882 (2014)
4. Zhang, X., Zhao, J., LeCun, Y.: Character-level convolutional networks for text classification. In: Advances in Neural Information Processing Systems, vol. 28 (2015)
5. Lai, S., Xu, L., Liu, K., Zhao, J.: Recurrent convolutional neural networks for text classification. In: AAAI (2015)
6. Yogatama, D., Dyer, C., Ling, W., Blunsom, P.: Generative and discriminative text classification with recurrent neural networks. arXiv preprint arXiv:1703.01898 (2017)
7. Hochreiter, S., Schmidhuber, J.: Long short-term memory. Neural Comput. 9(8), 1735–1780 (1997)
8. Gao, K., Xu, H., Gao, C., et al.: Attention-based BiLSTM network with lexical feature for emotion classification. In: International Joint Conference on Neural Networks (2018)
9. Zhang, Y., Wallace, B.: A sensitivity analysis of (and practitioners' guide to) convolutional neural networks for sentence classification. arXiv preprint arXiv:1510.03820 (2015)
10. Ke, C., Bin, L., Wende, K., et al.: Chinese micro-blog sentiment analysis based on multi-channels convolutional neural networks. J. Comput. Res. Dev. 55(5), 945–957 (2018)
11. Bin, L., Liu, Q., Jin, X., et al.: Aspect-based sentiment analysis based on multi-attention CNN. J. Comput. Res. Dev. Chin. 54(8), 1724–1735 (2017)
12. Graves, A., Jaitly, N.: Towards end-to-end speech recognition with recurrent neural networks. In: International Conference on Machine Learning (2014)

Machine Learning in Short Video APP User Activity Prediction

Fuwei Zeng[1(✉)], Tie Bao[1(✉)], and Wenhao Xiang[2(✉)]

[1] Jilin University, Changchun, China
zengfw17@mails.jlu.edu.cn, baotie@jlu.edu.cn
[2] Systems Engineering Research Institute, Beijing, China
xiangwh2018@163.com

Abstract. In order to improve the accuracy and reduce the cost of forecasting, this paper uses machine learning related technology to solve this problem in the user activity prediction model of short video industry. Continuous use of short video APP by active users is a sufficient and necessary condition for its success. The prediction of user activity has a direct guiding effect on the subsequent user loss warning. Based on the analysis of the impact on user activity, this paper extracts the characteristics according to registration log, startup log, shooting log and behavior log, and proposes a prediction algorithm based on model fusion for user activity. Based on the experimental data, the results show that the predicted AUC value reached 0.9514.

Keywords: Short video APP · Activity · Machine learning · LightGBM · AUC

1 Introduction

Judging whether the user is active is a two-category problem [1], and many classification algorithms have been proposed. These algorithms have been applied to the user loss warning model. The current research status shows that machine learning performs better than deep learning in terms of structured data, while deep learning is superior to machine learning in unstructured data (such as image, audio, etc.). Because machine learning is a data-based autonomous learning method, it is applied to user activity prediction. Compared with traditional methods, machine learning can improve work efficiency, improve prediction accuracy, and reduce costs.

The user activity prediction model established in this paper is established by using machine learning algorithms and model fusion methods [2], combining a large amount of data, conducting in-depth analysis, and mining features related to user activity. That is, through the user's login, shooting and other information, using the data within one month for training, using machine learning technology, using SVM and lightgbm algorithm, the results of the two models are evaluated and integrated to improve the accuracy of the prediction.

© Springer Nature Switzerland AG 2019
D. Milošević et al. (Eds.): HCC 2019, LNCS 11956, pp. 568–575, 2019.
https://doi.org/10.1007/978-3-030-37429-7_58

2 Theory of Machine Learning

Machine learning can be broadly divided into Supervised Learning, Semi-Supervised Learning, Unsupervised Learning, and Reinforcement Learning [3].

2.1 Supervised Learning

The task of supervised learning is to learn a model that can make a good prediction of the corresponding output for any given input. The data trained by the supervised learning algorithm consists of two parts, one is the input variable with many features, and the other is the label. The algorithm derives a model by training the training data, and then uses the model to predict the test data without the label and to obtain the label of the test data. The algorithms that belong to supervised learning are: decision tree, K-nearest neighbor algorithm, logistic regression, and so on.

User activity prediction is essentially a two-category problem in supervised learning. We can extract features and set tags based on existing data. Train the training data until the results are optimal in the performance metrics we specify. Therefore, we need to use classification algorithms. This article selects two models, LightGBM [4] and Support Vector Machine (SVM) [5]. The theories of the two algorithms used in this article are introduced below.

2.2 LightGBM

LightGBM is released by the Microsoft DMTK team and is an optimized version of the Gradient Boosting algorithm [4]. LightGBM is a lightweight GB framework. It is a learning algorithm based on decision tree. The core idea of the Gradient Boosting algorithm is to promote the weak classification algorithm to a strong classification algorithm, so as to improve the classification accuracy. The main process of the algorithm is as follows. Initialize n decision trees, the weights of the training example is 1/n. Training to obtain a weak classifier f(x). Determine the weights of the weak classifier. Update weights. Get the final strong classifier.

$$F(\mathrm{x}) = \partial_0 \mathrm{f}_0(x) + \partial_1 \mathrm{f}_1(x) + \partial_2 \mathrm{f}_2(x) + \ldots + \partial_n \mathrm{f}_n(x) \tag{1}$$

LightGBM (Light Gradient Boosting Machine) supports efficient parallel training and has the following advantages: faster training speed, lower memory consumption, better accuracy and support distributed, which can process massive data quickly.

2.3 Support Vector Machine (SVM)

Support Vector Machine (SVM) is a two-category model. Its basic model is an interval-maximizing linear classifier defined in the feature space. The learning strategy of support vector machine is to maximize the interval, which can be formalized as a problem of solving convex quadratic programming. It is also equivalent to the problem of minimizing the regularized hinge loss function. The learning algorithm of support vector machine is the optimal algorithm to solve the convex quadratic programming. The main process of the algorithm is as follows:

(1) Chose the appropriate kernel function K(x, z) and the appropriate parameter C to construct and solve the optimization problem.

$$\min_{\alpha} \frac{1}{2} \sum_{i=1}^{N} \sum_{j=1}^{N} \alpha_i \alpha_j y_i y_j K\left(x_i, x_j\right) - \sum_{i=1}^{N} \alpha_i \tag{2}$$

$$s.t. \ \sum_{i=1}^{N} \alpha_i y_i = 0 \tag{3}$$

$$0 \leq \alpha_i \leq C, \ i = 1, 2, \ldots N \tag{4}$$

Find the optimal solution $\alpha^* = (\alpha_1^*, \alpha_2^*, \ldots, \alpha_N^*)^T$.

(2) Select a positive component $0 < \alpha_j^* < C$ of α^*, calculate

$$b^* = y_j - \sum_{i=1}^{N} \alpha_i^* y_i K(x_i \bullet x_j) \tag{5}$$

(3) Construct the decision function:

$$f(x) = sign\left(\sum_{i=1}^{N} \alpha_i^* y_i K(x \bullet x_i) + b^*\right) \tag{6}$$

SVM has the following advantages and disadvantages:

Advantages: Good adaptability to high-dimensional space, even if the data dimension is larger than the number of samples; only need to support vector (a subset of the training set), so it can save memory; different kernel functions can be used according to different targets.

Disadvantages: SVM has a poor effect when the number of features of the data is larger than the number of samples, SVM can not provide a quantitative possibility prediction, which needs to be implemented by a costly cross-validation algorithm.

3   Experimental Modeling

Introduce the data source of the model and the interpretation of the attributes, the process of designing the model, the pre-processing of the data, and the application effect of the model.

3.1 Data Source and Interpretation of the Field

The experimental data set is derived from the data within one month of more than 50,000 users published by a short video APP company, and the data is desensitized. The data is provided in the form of logs. It contains four logs, which are registration log, start up log, shooting log, and behavior log. Tables 1, 2, 3 and 4 show samples of the four logs.

Table 1. Registration log

user_id	register_day	register_type	device_type
571220	1	1	2

Table 2. Start up log

user_id	launch_day
14808	4

Table 3. Shooting log

user_id	video_day
1139496	9

Table 4. Behavior log

user_id	activity_day	page	video_id	author_id	action_type
191890	14	1	90887	29437	0

3.2 Main Steps of Modeling

A user who has used APP (appears in any of the above types of logs) "in the next 7 days" is defined as "active user". The purpose of preparing the data is to form a wide table [6]. The wide table integrates characteristics related to whether the user is active in the future into a table. These characteristics are extracted by the original data, which facilitates the prediction of user activity in the later stage.

In the data exploratory analysis, when analyzing the number of registered users per day, it is found that there are obvious abnormalities in the amount of registration on some dates, and the bar chart of the amount of daily registration in 30 days is shown in Fig. 1.

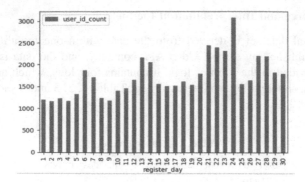

Fig. 1. Bar chart of the amount of daily registration

As can be seen from the above figure, overall, the number of users registered daily is increasing. The number of users registered will increase significantly in two days after five days. This article concludes that these two days should be weekends, 21, 22 and 23 these three days may be a small holiday, and the registration amount on the 24th day is obviously abnormal, which is supposed to be caused by abnormal users. To verify this conjecture, a bar chart of the number of daily activities of the users for 30 days is drawn, as shown in Fig. 2.

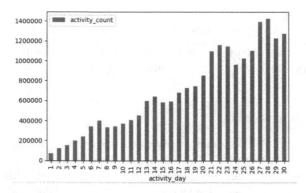

Fig. 2. Bar chart of the number of daily activities of users

It can be seen from Fig. 2 that overall the number of daily activities is rising within 30 days, and the date when the local peak appears is basically consistent with the chart of daily registration. Only on the 24th day, the registration amount of users is the maximum value within 30 days, but the amount of daily activities is the minimum value within three days before and after the 24th day. Therefore, it can be concluded that a large number of abnormal users appeared on the 24th day. It is necessary to identify and remove users who are only registered but not active on the 24th day.

Since predicting whether the user is active in the future is based on timing, this paper uses the sliding window method to increase the number of training samples when constructing the training set. The specific content of the sliding window method can be found in the literature [7]. The characteristics are extracted according to the behavior of the user in the past 16 days, and the tag is extracted according to whether the user is active in the 7 days later. Therefore, from the 30-day data, eight windows of length 16 for extracting characteristics and eight windows of length 7 for extracting labels can be obtained. The final required wide table is obtained by characteristic engineering [8] and setting labels.

The data is divided into training, validation, and test sets in a ratio of 3:1:1. Considering that the sample labels are balanced, and the performance metrics such as accuracy, recall rate and F1 score are sensitive to the classification threshold setting, this paper selects the AUC score as the evaluation index [9].

3.3 Experimental Results of the Model

Adjusting the parameters of LightGBM to obtain the current optimal parameters, the algorithm single-mode AUC score is 0.9066, and the ROC curve is shown in Fig. 3. The Gaussian kernel function is chosen as the kernel function of the SVM algorithm, the single-mode AUC score is 0.9491, and the ROC curve is shown in Fig. 4.

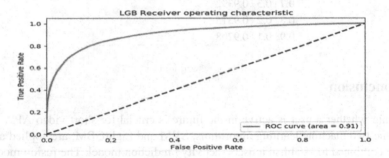

Fig. 3. ROC curve of the LightGBM algorithm

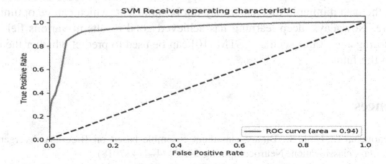

Fig. 4. ROC curve of the SVM algorithm

In the process of model fusion [2], the weighted average fusion method is used for model fusion, and the formula is shown in (7).

$$Model\_pred\_prob = \alpha * LGB\_pred\_prob + \beta * SVM\_pred\_prob(\alpha + \beta = 1) \quad (7)$$

Using the active probability values and tags of the obtained test set samples, the AUC scores after the model fusion can be calculated.

The values of α and β range from 0.1 to 0.9 (step size is 0.1), and the scores of model fusion are shown in Table 5.

Table 5. Table of model fusion scores

α	β	AUC score after model fusion
0.1	0.9	0.9512
0.2	0.8	0.9514
0.3	0.7	0.9504
0.4	0.6	0.9492
0.5	0.5	0.9475
0.6	0.4	0.9444
0.7	0.3	0.9392
0.8	0.2	0.9315
0.9	0.1	0.9208

4   Conclusion

Predicting whether a user is active in the future is crucial for short video APP. In this paper, the two machine learning algorithms, SVM and LightGBM, are applied and the two are combined to establish a users' activity prediction model. The fusion model has a good application effect, and the experimental optimal AUC value reaches 0.9514. The characteristics of the data determine the upper limit of machine learning, and the models and algorithms approximate this upper limit. In the process of feature engineering, the data mining in this paper may not be sufficient, which can be optimized in the future. Nowadays, deep learning has achieved good results in various fields, and deep learning algorithms such as LSTM [10] can be used to predict whether the user is active in the future.

References

1. Takenouchi, T., Ishii, S.: Binary classifiers ensemble based on Bregman divergence for multi-class classification. Neurocomputing **273**, 424–434 (2018)
2. Hoegh, A., Leman, S.: Correlated model fusion. Appl. Stoch. Model Bus. **34**, 31–43 (2018)
3. Li, H.: Statistical learning method. Tsinghua University Press (2012)
4. Xiong, S.S.: Identifying transportation mode based on improved LightGBM algorithm. Comput. Mod. 68–73 + 126 (2018)

5. Thai, V., Cheng, J., Nguyen, V., Daothi, P.: Optimizing SVM's parameters based on backtracking search optimization algorithm for gear fault diagnosis. J. VibroEng. **21**, 66–81 (2019)
6. Lian, J.Y.: The loss of telecommunications customers prediction model base on data mining. Zhongshan University (2008)
7. Chang, J.H., Lee, W.S.: estWin: online data stream mining of recent frequent itemsets by sliding window method. J. Inf. Sci. **31**, 76–90 (2005)
8. Xin, H.L., Wang, S.W., Ma, X.F., Li, Z., Achenie, L.: Feature engineering of machine-learning chemisorption models for catalyst design. Abstr Pap Am Chem S 252 (2016)
9. Huang, J., Ling, C.X.: Using AUC and accuracy in evaluating learning algorithms. IEEE Trans. Knowl. Data Eng. **17**, 299–310 (2005)
10. Tao, F., Liu, G.: Advanced LSTM: a study about better time dependency modeling in emotion recognition. In: 2018 IEEE International Conference on Acoustics, Speech and Signal Processing (ICASSP), pp. 2906–2910 (2018)

SF-KCCA: Sample Factoring Induced Kernel Canonical Correlation Analysis

Bisheng Zhan[✉], Ernest Domanaanmwi Ganaa, Na Qiang,
and Xiaozhen Luo

School of Computer Science and Communication Engineering,
Jiangsu University, Zhenjiang, China
309413860@qq.com, ennyganze@gmail.com,
1365031872@qq.com, 837564072@qq.com

Abstract. The Canonical Correlation analysis (CCA), such as linear CCA and Kernel Canonical Correlation Analysis (KCCA) are efficient methods for dimensionality reduction (DR). In this paper, a method of sample factoring induced KCCA is proposed. Different from traditional KCCA method, sample factors are introduced to impose penalties on the sample spaces to suppress the effect of corrupt data samples. By using a sample factoring strategies: cosine similarity metrics, the relationships between data samples and the principal projections are iteratively learned in order to obtain better correlation projections. By this way, the authentic and corrupt data samples can be discriminated and the impact of the corrupt data samples can be suppressed. Extensive experiments conducted on face image datasets, such as Yale, AR, show our approach has better classification and DR performance than that of linear CCA and KCCA, especially in noisy datasets.

Keywords: Sample factoring · Cosine similarity metrics · Correlation projection · CCA · KCCA

1 Introduction

Most machine learning (ML) tasks involve high dimensional datasets which lead to curse of dimensionality [1]. Besides the curse of dimensionality problem these datasets pose, they are usually incomplete, and made of missing and corrupt instances due to several reasons, such as recent upsurge in social media. Due to these challenges posed by datasets to ML tasks, dimensionality reduction (DR) techniques which transform the usually high dimensional data to low dimensional spaces [2] are indispensable for efficient and effective ML tasks. Because of this, several DR techniques have been presented [3–5]. These include Principal Component Analysis (PCA), Linear Discriminant Analysis (LDA), Locality Preserving Projection (LPP), Canonical Correlation Analysis (CCA), and Neighborhood Preserving Embedding (NPE).

Among these classical DR techniques, CCA is a well-known technique that characterizes the relationship between two sets of multidimensional variables by finding linear combinations of variables with maximal correlation. It is however sensitive to

© Springer Nature Switzerland AG 2019
D. Milošević et al. (Eds.): HCC 2019, LNCS 11956, pp. 576–587, 2019.
https://doi.org/10.1007/978-3-030-37429-7_59

outlying observations since its standard computation is based on sample covariance matrices [6]. Existing canonical correlation analysis (CCA) techniques are generally divided into two categories: linear CCA [7] and kernel CCA. While linear CCA finds the largest possible correlation between a linear combination of variables; KCCA [8] tries to overcome the limitation of linear CCA if there is nonlinear relationship between two variables by mapping the original sample space to a feature space through non-linear mapping.. Therefore KCCA often has better performance than linear CCA. Despite this advantage of KCCA over its linear counterpart, it is sensitive to outliers [9]. In solving a linear identification problem, an iterative procedure that alternates between canonical correlation analysis (CCA) to estimate the linear parts and kernel canonical correlation (AKCCA) to estimate the nonlinearities was proposed [10]. Regularized generalized CCA (RGCCA) [11] models the linear relationships between several blocks of variables observed on the same set of individuals. A two-dimensional CCA (2D-CCA) [12] which directly sought relationships between two different image data without reshaping the images into vectors is presented. Despite the existence of several adaptations of CCA, most of these existing methods directly accept all data samples, regardless of the effect of corrupt or noisy data samples which lead to per-formance degradation of models in terms of classification and DR.

Based on the above discussions, we propose the methods of sample factoring Induced relative Canonical Correlation analyses, such as Canonical Correlation Analysis (CCA), Kernel Canonical Correlation Analysis (KCCA). Those are called SF-KCCAs. To adequately address the sensitivities of the traditional CCAs to corrupt data samples or instances, we introduce a sample factor into their models which impose a penalty on each instance to suppress the impact of corrupt instances in pursuing pro-jections. We further use cosine similarity metric to iteratively learn the relationships between each data sample and the principal projections in the feature space. This leads to discrimination between authentic and corrupt instances by the proposed methods.

The main contributions of this paper are as follows:

(1) A novel framework by introducing sample factors into kernel CCA model is proposed to impose penalties on samples to suppress the effect of corrupt samples.
(2) The cosine similarity as sample factoring strategy is used to discriminate samples if they are authentic and corrupt data by iteratively learning the coupled rela-tionships among the principal correlation projections and data samples.
(3) Extensive experiments on ORL and AR datasets demonstrate the superiority of our method over state-of-the-art methods.

The rest of this paper is organized as follows: we briefly present related work in Sect. 2. Section 3 presents the proposed methods. Experiments and results analysis is given in Sect. 4, and conclusions and future works are described in Sect. 5.

2   Related Work

This section briefly reviews related works on linear CCA and KCCA.

2.1 Linear Canonical Correlation Analysis

Different dimensionality reduction techniques have different results due to their mathematical theoretical basis and different application ranges. But they have a common goal. For example, Foster et al. employed CCA method for dimensionality reduction in multi-view learning [13].

However, as CCA can only address the problem of fusing two sets of features, the application of this approach is limited. A novel correlation analysis algorithm based on CCA, called ApproxCCA [14] was proposed to explore the correlations between two multidimensional data streams in the environment with limited resources. Recently, the Fractional-order embedding CCA (FECCA) [15] proposed by Sun et al. can accurately estimate the correlation matrices in CCA, even if there are noises in the dataset. Following similar idea, Sun and Chen proposed a novel locality preserving method for CCA, named LPCCA [16] for multi-view data. A new Locality Preserving CCA (ALPCCA) [17] have been proposed to discover the inherent data structure.

2.2 Kernel Canonical Correlation Analysis

The major drawback of CCA is that it cannot capture nonlinear relations among data variables. A kind of commonly used technique is the kernel trick for the nonlinear extension of CCA, resulting in kernel CCA. The aim of such kernel canonical correlation analysis (KCCA) is to find canonical vectors in terms of expansion coefficients $\alpha, \beta \in \Re^p$.

Apart from KCCA, sparse kernel CCA algorithm (SKCCA) [18] introduced sparsity of the dual transformations which is based on a relationship between kernel CCA and least squares. Alternating kernel canonical correlation analysis (AKCCA) [10] was proposed to obtain linear identification through an iterative procedure to estimate the linear parts, while KCCA was applied to estimate the memoryless nonlinearities. Cost sensitive transfer kernel canonical correlation analysis (CTKCCA) [19] was proposed for cross-project defect prediction in the scenario where source and target projects have different metric sets.

Kernel methods usually require users to design a particular kernel, which critically affects the performance of the algorithm. To make the design more flexible, multiple kernel learning (MKL) method have shown its necessity to consider multiple kernels or the combination of kernels rather than a single fixed kernel. Generally, MKL has proven that it can offer some needed flexibility and well manipulate the case that involves multiple, heterogeneous data sources [20]. Since MKL considers multiple kernels, it can be effectively employed for the heterogeneous data sources under the common framework of kernel learning. To a certain extent, MKL also relaxes the model selection of kernels. For example, kernel multi-view CCA (KMCCA) [21] is described as a method that could be treated as a multi-view extension of KCCA. Localized multi-kernel discriminative canonical correlation analysis (LMKDCCA) [22] was proposed to identify persons captured in videos from non-overlapping cameras.

3   The Proposed Method

A detail formulation of the proposed method is presented in this section. To be able to suppress the impact of corrupt data samples in the training set, we discriminate between authentic and corrupt data samples using the cosine sample factoring strategy.

3.1   CCA with Sample Factoring

We implement our idea of suppressing the effect of corrupt data samples by observing the objective function of the traditional CCA:

$$\arg\ \min_{s.t. v^T X X^T v = w^T Y Y^T w = I} J(v, w) = \|X^T v - Y^T w\|_2^2 \tag{1}$$

It can be seen that CCA uses a least square framework to minimize the sum of distances between the reconstructed datasets $v^T X$ and $w^T Y$, and the original datasets X and Y respectively. This geometrically will force the projection vectors v and w to pass through the densest data points to minimize this sum distance. So this gives us the motivation to consider the relationships between the projection vectors and the data samples. How close a data sample is to the principal projections v or w determines how important it is in pursuing projections.

Motivated by this relationship between data samples and projections, we reformulate formula (1) by introducing sample-factors to impose penalties on the sample spaces to suppress the effect of corrupt data samples as follows:

$$\arg\ \min_{s.t.\ v^T X D_x^2 X^T v = w^T Y D_Y^2 Y^T w = I} \sum_{i,j} \|v^T x_i d_{x_i} - w^T y_j d_{y_j}\|_2^2 \tag{2}$$

where $D_X = \mathrm{diag}(d_{x_i})$ and $D_Y = \mathrm{diag}(d_{y_j})$ are diagonal sample factoring matrices. Based on reasonable sample factoring, the method can improve the classification performance.

3.2   Strategy of Building Matrix D_X and D_Y

We model the relationship between sample-factors D_X, D_Y and the principal projections v and w using the cosine similarity metric.

By using cosine similarity metric, we build the sample factors by iteratively learning the angle relationship between data samples x_i, y_j and the principal projections v and w respectively. The coordinates of data samples x_i and y_j which are projected onto the principal projections v and w respectively can be computed respectively as follows:

$$a_i = v^T x_i,\ b_i = w^T y_j \tag{3}$$

By normalizing Eq. (3) respectively, the angles between x_i and the principal projection v, and that between y_j and the principal projection w are computed respectively as follows:

$$\cos \theta_i = \frac{v^T x_i}{\|v\|\|x_i\|}, \ \cos \eta_i = \frac{w^T y_i}{\|w\|\|y_i\|} \tag{4}$$

From Eq. (4), if $\cos \theta_i$ or $\cos \eta_i$ is smaller, that will translate to a bigger angle between x_i and v, angle between y_i and w and vice versa. As shown in Fig. 1. it can be seen that, angle φ of sample x_j is relatively smaller than angle θ of sample x_i. Similarly, angle β of y_j is relatively smaller than angle α of y_i. Thus, x_j and y_j will be considered probably more important in finding best projections than x_i and y_i which might be noisy. Recall that d_{x_i} and d_{y_j} are negative factors, we compute d_{x_i} and d_{y_j} through the similarity metrics as follows:

$$d_{x_i} = \frac{1}{abs(\cos \theta_i) + \varepsilon}, \ d_{y_i} = \frac{1}{abs(\cos \alpha_i) + \varepsilon} \tag{5}$$

where $\varepsilon = 0.0001$ is a parameter to avoid d_{x_i} or d_{y_j} approaching infinity.

Therefore the impact of corrupt or noisy samples can greatly be minimized through iteratively scaling the data using total distance and cosine similarity metrics which will lead to better low-rank projections.

3.3   Sample Weighting Induced Kernel Cannonical Correlation Analysis

We further extend the proposed model to kernel space. The major drawback of CCA is that it cannot capture nonlinear relations among data variables. So we map the samples X into Kernel space and reconsider CCA method to capture nonlinear relations among data variables. The compact representation of the samples in kernel feature space are given as $\varphi(x) = [\varphi(x_1), \varphi(x_2), \ldots, \varphi(x_n)]$ and likewise $\varphi(y) = [\varphi(y_1), \varphi(y_2), \ldots, \varphi(y_n)]$. Accordingly, the canonical variates are $v = \varphi(x)\alpha$ and $w = \varphi(y)\beta$. Then the formula (1) can be reformulated by extending CCA to kernel space, resulting in sample weighting induced kernel CCA. The proposed formula is expressed as follows:

$$\operatorname{argmin} J(\alpha, \beta) = \|\alpha^T \varphi(x)^T \varphi(x) D_x' - \beta^T \varphi(y)^T \varphi(y) D_y'\|_2^2$$
$$\text{s.t } \alpha^T K_x D_x'^2 K_x \alpha = \beta^T K_y D_y'^2 K_y \beta = I \tag{6}$$

where K_x and K_y is the kernel Gram matrices of dataset x, y, respectively. $K_x = \varphi(x)^T \varphi(x), K_y = \varphi(y)^T \varphi(y), \ D_x' = \operatorname{diag}(d_{x_i}')$ and $D_y' = \operatorname{diag}(d_{y_j}')$ are diagonal sample factoring matrices for the sample weighting induced kernel CCA.

When the proposed model is extended to the kernel space, we also use cosine similarity metric to build the sample factor. The cosine similarity metric for the sample factor can be expressed as follows:

$$\cos \theta_i = \frac{\alpha^T \varphi(x)^T \varphi(x_i)}{\left\| \alpha^T \varphi(x)^T \right\| \left\| \varphi(x_i) \right\|} = \frac{\alpha^T K_{x,i,:}}{\sqrt{\alpha^T K_x \alpha} \sqrt{K_{x,ii}}} \tag{7}$$

where $K_{x,i,:}$ is the i-th row of K_x.

The factor of sample x_i can be expressed as follows:

$$d'_{x_i} = \frac{1}{abs(\cos \theta_i) + \varepsilon} \tag{8}$$

Similarly, the factor of label of sample y_j can be expressed as follows:

$$d'_{y_j} = \frac{1}{abs(\cos \eta_j) + \varepsilon} \tag{9}$$

4 Experiments Analysis

To validate the effectiveness of the proposed SF-KCCA framework, we conduct experiments on ORL and AR datasets against four state-of-the-art DR methods such as CCA, KCCA, AKCCA and LMKDCCA. To further demonstrate the efficacy of the proposed SF-KCCA on corrupt datasets, we corrupt the ORL and AR datasets with corruption densities of 5% and 10% each. We then run experiments on these noisy scenarios using the proposed SF-KCCA in comparison with all the above comparative methods.

4.1 Parameter Settings

For each dataset, we randomly sampled 80% of the samples for training and 20% for testing in our experiments. We set the k-nearest-neighbors parameter K to 5 in SF-KCCA models and all other comparative methods, in order to make a very fair comparison. We finally make use of the K-nearest neighbor (KNN) classifier for classifications. We record results for our framework as SF-KCCA. The experiments are repeated 20 times and we record the average classification accuracies and standard deviations for the various methods.

4.2 Dataset Description

We select four benchmark publicly available datasets for the evaluation of the performance of our algorithm: the ORL and AR datasets.

The ORL face dataset has 40 subjects, each with 10 faces at different poses making a total of 400 images of size 112×92. However, the images were resized to 32×32

for our experiments. These images were taken at different times, lighting and facial expressions. The faces are in an upright position in frontal view with a slight left-right rotation.

The AR face dataset was created by Aleix Martinez and Robert Benavente in the Computer Vision Center (CVC) at the U.A.B. It contains over 4,000 color images corresponding to 126 people's faces (70 men and 56 women). Images feature frontal view faces with different facial expressions, illumination conditions, and occlusions (sun glasses and scarf). The pictures were taken at the CVC under strictly controlled conditions. No restrictions on wear (clothes, glasses, etc.), make-up, hair style, etc. were imposed to participants. Each person participated in two sessions, separated by two weeks (14 days) time. The same pictures were taken in both sessions.

4.3   Experiment Comparisons

To demonstrate the proposed SF-KCCA can suppress the impact of noise in corrupt datasets, we run extensive experiments on two noisy scenarios by corrupting each of the ORL and AR face datasets with 5% and 10% corruption densities. Sample images of the original and corrupt images from the ORL and AR datasets are shown in Fig. 1(a) and (b) respectively.

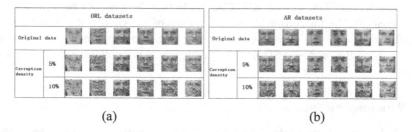

(a)   (b)

Fig. 1. Examples of original and corrupt images under different corruption densities from (a) the ORL and (b) the AR datasets

To validate effect of dimensionality reduction for relative methods, the experiments have been conducted by varying dimension for reduction. Figure 2 shows the trend of classification accuracies of the various methods in different dimensions on the ORL dataset with the 0% and 5% corruption densities. It can be seen from both Fig. 2(a) and (b) that as dimensions increase, the performance of each method increases, but SF-KCCA is in the lead. When dimension reaches 225, the best performance of SF-KCCA can be obtained. However, the performances of CCA and KCCA begin to decline in dimensions beyond 200, and the best performance of AKCCA and LMKDCCA were gotten when dimension reaches 250. From these, we can see SF-KCCA is more effective for dimensionality reduction.

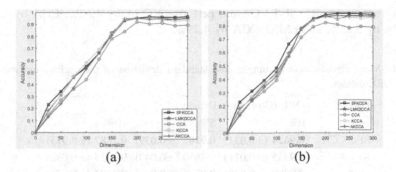

(a)   (b)

Fig. 2. Classification accuracies for the various methods against the variation of dimensions on the ORL dataset with the (a) 0% and (b) 5% corruption densities

Figure 3 illustrates variations in accuracies for each method with best dimension for reduction on ORL dataset with the (a) 0% and (b) 5% corruption densities across 20 random runs. The performance of the proposed SF-KCCA is stable as indicated in both Fig. 3(a) and (b). It is evident from the results that our proposed method has superior performance in all cases because it is able to detect and suppress the impact of noisy data points more than all the comparative methods.

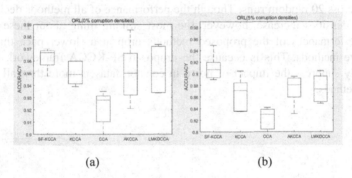

(a)   (b)

Fig. 3. Classification accuracies for the various methods on ORL dataset with the (a) 0% and (b) 5% corruption densities across 20 runs

We present results for each method for 0%, 5% and 10% corruption densities for the ORL dataset in Tables 1 respectively with best results in bold in each case. Table 1 shows that, in all cases, SF-KCCA obtain superior face recognition accuracies as compared to all the comparative methods. For the 0% corruption density, SF-KCCA outperforms KCCA by 1.89%, CCA by 4.24%, AKCCA by 0.99% and LMKDCCA by 0.69%. The difference in recognition accuracies between the 0% and 5% corruption densities for SF-KCCA is 5.23%, KCCA is 8.5%, CCA is 9.95%, AKCCA is 7.7% and LMKDCCA is 7.8%. It is very clear that SF-KCCA has the lowest decrease in recognition accuracy. As corruption increase, the performance difference between the proposed method and the comparative methods become very significant. Also, for the

10% corruption density, SF-KCCA out performs KCCA by 4.62%, CCA by 4.83%, AKCCA by 3.82%, and LMKDCCA by 4.29%.

Table 1. Mean classification accuracies and standard deviation of the various methods on a corrupt ORL dataset.

Dataset	ORL (Corruption Density)		
	0%	5%	10%
SF-KCCA	**0.9624 (±0.011)**	**0.9102 (±0.039)**	**0.7887 (±0.073)**
KCCA	0.9435 (±0.011)	0.8650 (±0.040)	0.7425 (±0.082)
CCA	0.9200 (±0.030)	0.8205 (±0.040)	0.7404 (±0.075)
AKCCA	0.9525 (±0.035)	0.8735 (±0.042)	0.7505 (±0.078)
LMKDCCA	0.9555 (±0.021)	0.8746 (±0.031)	0.7458 (±0.070)

Figure 4 shows the trend of classification accuracies of the various methods in different dimensions for the AR dataset with the 0% and 10% corruption densities. From both Fig. 4(a) and (b), we can also see SF-KCCA is more effective for dimensionality reduction.

Figure 5 shows varying classification accuracies for the various methods with best dimension for reduction on the AR dataset with the (a) 0% and (b) 10% corruption densities across 20 random runs. Though the performance of all methods decline as the corruption densities increase, however, the most striking thing about these results is that the performances of the proposed methods drop at a slower pace than all the comparative methods. This is because the proposed SF-KCCA framework is able to significantly suppress the impact of noise in corrupt datasets more than all the comparative methods.

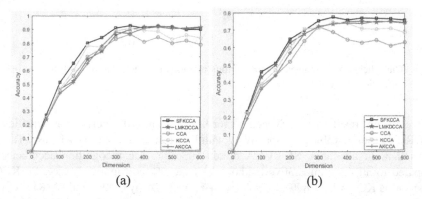

(a)   (b)

Fig. 4. Classification accuracies for the various methods against the variation of dimensions on the AR dataset with the (a) 0% and (b) 10% corruption densities

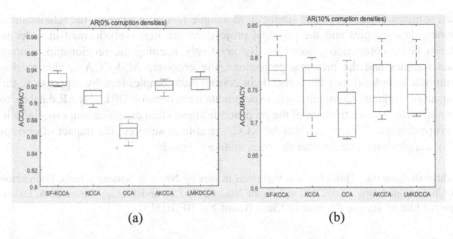

(a)   (b)

Fig. 5. Classification accuracies for the various methods on the AR dataset with the (a) 0% and (b) 10% corruption densities across 20 runs

As Table 2 shows, the proposed SF-KCCA framework again proves its superiority over all other comparative methods in terms of face recognition accuracy notwithstanding the presence of corrupt data points. For the 0% corruption density, with a superior recognition accuracy of 92.82%, SF-KCCA performs 2.17% more than KCCA, 6.16% more than CCA, 0.68% more than AKCCA and 0.20% more than LMKDCCA. For the 5% corruption density, SF-KCCA out performs KCCA by 2.80%, CCA by 6.54%, AKCCA by 2.54% and LMKDCCA by 2.80%. SF-KCCA obtains the best accuracy 78.65% for the 10% corruption density. Thus, SF-KCCA has a superior performance of 3.9% over KCCA, 6.59% over CCA, 3.4% over AKCCA and 3.35% over LMKDCCA.

Table 2. Classification accuracies and standard deviation of the various methods on a corrupt AR dataset.

Dataset	AR (Corruption Density)		
	0%	5%	10%
SF-KCCA	**0.9282 (±0.011)**	**0.9016 (±0.042)**	**0.7865 (±0.046)**
KCCA	0.9065 (±0.012)	0.8736 (±0.048)	0.7475 (±0.072)
CCA	0.8666 (±0.016)	0.8362 (±0.045)	0.7206 (±0.074)
AKCCA	0.9214 (±0.012)	0.8762 (±0.042)	0.7525 (±0.075)
LMKDCCA	0.9262 (±0.012)	0.8766 (±0.039)	0.7530 (±0.073)

5   Conclusion

In this paper, a method of sample factoring induced canonical correlation analysis (SF-KCCA) is proposed. Different from traditional KCCA methods, sample factors are introduced to impose penalties on the sample spaces to suppress the effect of corrupt data samples. By using cosine similarity as sample factoring strategy, the relationships among data samples and the principal projections are iteratively learned in order to obtain better correlation projections. By iteratively learning the relationships among data samples and the principal projections, the proposed SF-KCCA is able to discriminate samples if they are authentic or corrupt data samples thereby suppressing the impact of the latter. Comprehensive experiments were done on ORL and AR datasets to demonstrate the effectiveness of the proposed framework in classification, especially in corrupt datasets. This proves that SF-KCCA is able to suppress the impact of corrupt data samples much better than the comparative methods.

Acknowledgments. This work was supported in part by National Natural Science Foundation of China (Grant Nos. 61672268, and Grant Nos. 61170126) and Primary Research & Development Plan of Jiangsu Province of China (Grant No. BE2015137).

References

1. Chkifa, A., Cohen, A., Schwab, C.: Breaking the curse of dimensionality in sparse polynomial approximation of parametric PDEs. Journal de Mathématiques Pures et Appliquées **103**(2), 400–428 (2014)
2. der Maaten, L.J.P., Postma, E.O., den Herik, H.J.: Dimensionality reduction: a comparative review. Technical report TiCC TR 2009-005, pp. 1–35 (2009)
3. Belhumeur, P.N., Hespanha, J.P., Kriegman, D.J.: Eigenfaces vs. Fisherfaces: recognition using class specific linear projection. IEEE Trans. Pattern Anal. Mach. Intell. **19**(7), 711–720 (1997)
4. He, X., Niyogi, P.: Locality preserving projections. In: NIPS (2003)
5. He, X., Cai, D., Yan, S., Zhang, H.: Neighborhood preserving embedding. In: ICCV, pp. 1208–1213 (2005)
6. Dehon, C., Filzmoser, P., Croux, C.: Robust methods for canonical correlation analysis. In: Kiers, H.A.L., Rasson, J.P., Groenen, P.J.F., Schader, M. (eds.) Data Analysis, Classification, and Related Methods, pp. 321–326. Springer, Heidelberg (2000). https://doi.org/10.1007/978-3-642-59789-3_51
7. Liu, H., Sun, X.: Linear canonical correlation analysis based ranking approach for facial age estimation. In: 2016 IEEE International Conference on Image Processing (ICIP) (2016)
8. Zheng, W., Zhou, X., Zou, C., Zhao, L.: Facial expression recognition using kernel canonical correlation analysis (KCCA). IEEE Trans. Neural Netw. **17**(1), 233–238 (2006)
9. Alam, A., Nasser, M., Fukumizu, K.: Sensitivity analysis in robust and kernel canonical correlation analysis. In: Proceedings of 11th International Conference on Computer & Information Technology, pp. 399–404 (2008)
10. Van Vaerenbergh, S., Via, J., Santamaria, I.: Blind identification of SIMO Wiener systems based on kernel canonical correlation analysis. IEEE Trans. Signal Process. **61**(9), 2219–2230 (2013)
11. Tenenhaus, A., Tenenhaus, M.: Regularized generalized canonical correlation analysis. Psychometrika **76**(2), 257 (2011)
12. Lee, S.H., Choi, S.: Two-dimensional canonical correlation analysis. IEEE Signal Process. Lett. **14**(10), 735–738 (2007)
13. Foster, D.P., Johnson, R., Zhang, T.: Multi-view dimensionality reduction via canonical correlation analysis. Technical report TR-2009-5, TTI-Chicago (2008)

14. Wang, Y., Zhang, G., Jiang, B.: ApproxCCA: an approximate correlation analysis algorithm for multidimensional data streams. Knowl. Based Syst. **24**(7), 952–962 (2011)
15. Yuan, Y., Sun, Q., Ge, H.: Fractional-order embedding canonical correlation analysis and its applications to multi-view dimensionality reduction and recognition. Pattern Recogn. **47**(3), 1411–1424 (2014)
16. Sun, T., Chen, S.: Locality preserving CCA with applications to data visualization and pose estimation. Image Vis. Comput. **25**(5), 531–543 (2007)
17. Wang, F., Zhang, D.: A new locality-preserving canonical correlation analysis algorithm for multi-view dimensionality reduction. Neural Process. Lett. **37**(2), 135–146 (2013)
18. Chu, D., Liao, L., Ng, M.K., et al.: Sparse kernel canonical correlation analysis. Lecture Notes in Engineering and Computer Science, vol. 2202, pp. 322–327 (2013)
19. Li, Z., Jing, X., Wu, F., et al.: Cost-sensitive transfer kernel canonical correlation analysis for heterogeneous defect prediction. Autom. Softw. Eng. **25**(2), 201–245 (2018)
20. Wang, T., Zhang, Z., Jing, X., et al.: Multiple kernel ensemble learning for software defect prediction. Autom. Softw. Eng. **23**(4), 569–590 (2016)
21. Arora, R., Livescu, K.: Multi-view CCA-based acoustic features for phonetic recognition across speakers and domains. In: IEEE International Conference on Acoustics, Speech and Signal Processing, pp. 7135–7139 (2013)
22. Chen G., Lu J.: Localized multi-kernel discriminative canonical correlation analysis approach for video-based person re-identification. In: Proceedings of International Conference on Image Processing, pp. 111–115 (2017)

Biological Data Migration Method Based
on IPFS System

Changwen Zhang[1(✉)], Yi Man[1(✉)], Jin He[2(✉)], Jieming Gu[3(✉)],
and Xiao Xing[3(✉)]

[1] Beijing University of Posts and Telecommunications, Beijing, China
{zhangchangwen,manyi}@bupt.edu.cn
[2] Neusoft Corporation, Beijing, China
15575953840@163.com
[3] National Computer Network Emergency Response Technical
Team/Coordination Center of China, Beijing, China
1952902819@qq.com, 1521840143@qq.com

Abstract. IPFS (Interplanetary File System) is a peer to peer distributed file
system with fast downloading speed and high security. IPFS adopts the storage
method of equal-size fragmentation and DHT-based resource search. Through
investigation, it is found that the current IPFS system lacks the strategy of
resource migration. Therefore, in order to improve the efficiency of its search
and further improve the download speed of biological data, This paper adopts
the sharding and uploading method based on the characteristics of biological
data and propose the migration algorithm based on routing table and the
migration algorithm based on biological historical access table. The experimental and analytical results show that the proposed method is effective.

Keywords: IPFS · Routing table · P2P · Biological data history access table ·
Migration algorithm

1 Introduction

IPFS [1] is a distributed web, peer to peer hypermedia protocol, whose goal is to
replace the traditional Internet protocol HTTP [2]. In the aspect of file storage, it adopts
Merkle DAG [3] storage structure, and in the aspect of file transmission, it adopts P2P
protocol, which has the characteristics of fast transmission speed and high security.

Biological data [4] common categories include genes, chromosomes, nucleic acid,
protein, etc., with the continuous development of biological industry, biological data
shows a wide variety, high-dimensional complex internal structure, and relatively
scattered data, etc. Commonly biological data format has FASTQ [5], FASTA [6],
SAM [7], BAM [5], and so on.

IPFS system uses a structured P2P network [8], and based on DHT resource
lookup, structured P2P networks generally assumes that all resources evenly distributed
between nodes, but in actual operation, owing to the popularity of different without
resources will makes resource distribution is not uniform. So when the resource that the
node often accesses is far away from the node, the node will take a long time to find the

© Springer Nature Switzerland AG 2019
D. Milošević et al. (Eds.): HCC 2019, LNCS 11956, pp. 588–599, 2019.
https://doi.org/10.1007/978-3-030-37429-7_60

resource, which will affect the search efficiency of resources, In addition, the download speed of resources will be greatly reduced due to delay and other factors during the transmission of resources. Therefore, in order to improve the efficiency of resource lookup and then improve the transmission speed of resources, IPFS adopts caching mechanism, that is, IPFS nodes cache data locally when they download it, however, this method will greatly increase the load on the node and cause the waste of node space.

In order to solve this problem, some methods have been proposed successively. But some method is directly based on the uniform distribution, Some methods do not take into account the problem of nodes and the direct transmission delay of nodes, Some methods do not record the node's access to the resource, and even some method completely without considering the node load [9], and therefore on the basis of research above, this paper will be more comprehensive to consider how to deal with this problem.

In this paper, we propose a method of resource migration based on biological data based on IPFS. We upload biological data to the network by means of sharding and uploading, and encrypt each biological data block with MD5 [10] algorithm to ensure the safety of biological data. Moreover, we make real resource requests in the IPFS system. In addition, we record the query and download of biological data of each node in the form of table, that is, the historical access table of biological data. In order to improve its search efficiency and download speed, this paper proposes a migration algorithm based on routing table and a migration algorithm based on biological data history access table. The migration algorithm based on routing table, namely, through analyzing the load of routing table in the node, select the node with higher than average load as the migration node, The routing table information with the highest load on the node is then migrated to the neighbor node with the lowest load on the node, and update the routing table. Migration algorithm based on biological data access history table, that is, through the analysis of the historical biological data access table load of nodes, the nodes with higher than average load are selected as the moving nodes, select the most popular biological data block in the node, and then transfer the biological data to the neighbor node with less load in the target node through the method of resource migration [11], and update the routing table.

The rest of this article is laid out as follows. The second part describes the recent resource migration method. The third part explains the resource migration method based on biological data in detail. In the fourth part, the feasibility of the proposed method is verified by experiments. The fifth part summarizes the whole thesis.

2   Related Work

In order to solve the problem of reduced query efficiency in P2P networks under actual conditions, many related methods have been proposed.

In [12] Zhang et al. presents a routing algorithm based on multi-information integrated decision-making guidance. In the algorithm, indexing mechanism of neighbors' resource information is used to improve search probability and dynamic TTL adjustment mechanism is used to improve search efficiency.

In [13] Wang et al. suggest using the query answering heterogeneity to directly improve the search efficiency of P2P networks. The peers with high query answering capabilities will have higher priority to be queried.

In [14] Totekar et al. present an enhanced selective walk searching algorithm along with low cost replication schemes. Each node is used to selectively forward the query to a node having higher hit-ratio for the data of requested type, based on history of recently succeeded queries. Then explicates frequently accessed data objects on the nodes.

In [15] Busnel et al. propose PROXSEM, a refined proximity measure taking into account peer generosity and file popularity. Based on similar interests between peers in the context to improve search efficiency.

In [16] Sharifi et al. propose an adaptive sampling scheme to make a tradeoff between cost and accuracy. They apply a hybrid overlay that efficiently combines topology-aware and interest-based links instead of random or DHT invoked connections.

In [17] Yeferny et al. introduce a novel approach that aims to predict user profiles based on the shared documents and builds an initial knowledge base beforehand.

In [18] Youming et al. propose an improved range query algorithm, Upper Boundary Predictably Lookup (UBPL for short) to reduce the DHT-Lookup times during range query procedure in LIGHT. This paper also proved the effectiveness of UBPL by simulating experiments.

In [19] Napoli et al. propose to apply a mathematical model for the diffusion of fragments on a P2P in order to take into account both the effects of peer distances and the changing availability of peers while time goes on.

Through the above research, so in this paper, firstly, we will use the sharding method based on the characteristics of biological data to upload the biological data to the system, then we will fully consider the load of the nodes in the entire network, the delay between different nodes, and the historical resource access on each node under the real resource request condition of the IPFS system. And then uses the method of resource migration to improve the efficiency of resource searching. Further improve the transmission speed of resources.

3   Our Approach

The IPFS system is based on the P2P protocol. Each node maintains its own routing table (The PeerID, PeerIP, and biological data hash value are recorded in the routing table). Each node will have neighbor nodes (according the proximity metric), therefore, the basic principle of this algorithm is based on the sharding and uploading of biological data characteristics. By fully considering factors such as routing table information, biological data access table information, distance between required biological data and nodes, and delay between nodes, The migration algorithm based on routing table and the migration algorithm based on historical access table of biological data are used to improve the transmission speed of biological data.

3.1   Concepts and Definitions

Definition 3.1. Routing Table Load Metrics: The measurement of the load of each node's routing table is mainly composed of two parts, namely the routing forwarding quantity and the routing lookup quantity, and its definition is as follows:

$$\text{Routing Table Load} = \text{ForwardLoad} \times p + \text{QueryLoad} \times q \qquad (1)$$

ForwardLoad: That is, the number of forwarding biological data requests within the average time interval T of the routing table of the node.

QueryLoad: That is, the number of times a single node finds the required biological data in its routing table within the average time interval T.

Average load of routing table: which is the average load of all nodes in the whole network, can be summed by the load of routing table of each node in the network, and then divided by the number of nodes, which is defined as follows:

$$\text{Average load of routing table} = \frac{\sum_{i=1}^{n} peer_i.\text{Routing Table Load}}{n} \qquad (2)$$

Definition 3.2. biological data history access table load measurement: The biological data history access table records the biological data of other nodes accessing the node. The metrics of the load are as follows:

$$\text{History Table Load} = \text{sum of node biodata access times.} \qquad (3)$$

The average biological data historical access table load of nodes, that is, the average load of all nodes in the whole network, can be summed by the load of each node in the network, and then divided by the number of nodes, which is defined as follows:

$$\text{Average load of history table} = \frac{\sum_{i=1}^{n} peer_i.\text{History Table Load}}{n} \qquad (4)$$

Definition 3.3. Move in node and move out node: Move in node represents when the load on the node is higher than the average load on the network. Move out node indicates that the load on the node is lower than the average load on the network.

3.2   Sharding and Uploading Based on Biological Data Features

In order to improve the search efficiency of biological data and further improve the download speed of biological data, this paper proposes a sharding and uploading method based on the characteristics of biological data on the basis of investigation.

At present, the common representation formats of biological data include FASTQ, FASTA, SAM, BAM, etc., and each different biological format has obvious characteristics. Therefore, this paper proposes sharding and uploading method based on the characteristics of biological data based on the longest known FASTQ format.

FASTQ is a standard format for protecting biological sequences and their sequencing quality information. Each sequence usually has four lines, as shown in Fig. 1.

The first line is the sequence identifier and related descriptive information, beginning with '@'.

The second row is the sequence.

The third line begins with a ' + ' followed by a sequence identifier, description information, or nothing, but the ' + ' should not be missing.

The fourth line, quality information, refers to the logarithm value of the error probability of a base, corresponding to the sequence of the second line, each sequence has a quality score, and the meaning of each character varies according to the scoring system.

```
$ /opt/sratoolkit.2.5.7-centos_linux64/bin/fastq-dump SRR001666
@SRR001666.1 071112_SLXA-EAS1_s_7:5:1:817:345 length=72
GGGTGATGGCCGCTGCCGATGGCGTCAAATCCCACCAAGTTACCCTTAACAACT
TAAGGGTTTTCAAATAGA
+SRR001666.1 071112_SLXA-EAS1_s_7:5:1:817:345 length=72
IIIIIIIIIIIIIIIIIIII9IG9ICIIIIIIIIIIIIIIIDIIIIII>IIIII/
@SRR001666.2 071112_SLXA-EAS1_s_7:5:1:801:338 length=72
GTTCAGGGATACGACGTTTGTATTTTAAGAATCTGAAGCAGAAGTCGATGATAA
TACGCGTCGTTTTATCAT
+SRR001666.2 071112_SLXA-EAS1_s_7:5:1:801:338 length=72
IIIIIIIIIIIIIIIIIIIII6IBIIIIIIIIIIIIIIIIGII>IIII-I)8I
```

Fig. 1. Fastq

On the basis of the above characteristics of biological data, is proposed in this paper based on the biological characteristics of the data fragmentation, the way to upload that according to the biological data identifier "@" beginning to fragmentation of biological data, through the subdivision method will not only improve the search efficiency of biological data, and to further improve the biological data download speed, and as a result of segmentation is to separate biological sequences, which retained the biological significance of the data itself.

3.3 Routing Table Based Migration Algorithm (RTBMA)

The purpose of this method is to dynamically update the load of the node, and select the node whose node load is higher than the average load, that is, move in node. When a node (peera) uses this algorithm (RTBMA), peera detects whether there is a node with a higher average load in its routing table. If peerb is an overload node, that is, moves in node. Complete the migration of routing table information through the migration algorithm based on routing table. That is to say, a node that is not loaded but has a small delay with peera is selected as the move out node from the neighbor node of the move in node, and the routing table information is migrated to the move out node, and the biological data in the routing table information is migrated to this node, and then the routing table of the migration and migration node is updated. Through this algorithm, not only can the number of times the node queries the routing table can be reduced, but also the query efficiency of the biological data can be improved (Fig. 2).

Algorithm 1 Routing table based migration algorithm:

1:#algorithm at peera
2:#Find the moving in node
3:#Delay (): Calculate the delay between two nodes
4:#TargetRoutingTable : Select the routing table information
            With the largest load node
3:Overpeers = []
4:For peers[i] in peera's routing table:
5:    Peerload = peers[i].load
6:    If peerload > IPFS's averageLoad:
7:        Overpeers.append(peers[i])
8:#Migrate routing tables and biodata
9:For peers[j] in overpeers:
10:    GET TargetRoutingTable
11:    Delayinit = Delay (peer[j], peers [0])
12:    For peerneighbor[j] in peers[j]:
13:        If Delay(peerneighbor[j],peer[j]) < Delayint:
14:            Delayinit = Delay(peerneighbor[j],peer[j])
15:            TargetPeer = peerneighbor[j]
16:    Send the TargetRoutingTable to TargetPeer
17:    Send the biodata to targetPeer
18:    Update routing table

Fig. 2. Routing table based migration algorithm

When a node (peera) in the network runs the algorithm, firstly, it detects whether there is an overload node in the routing table, and uses it as the move in node to complete the migration of the routing table information through the routing table-based migration algorithm. Peera's neighbor node that is not overloaded and has a small delay with peera is selected as a move out node, and then the routing table information is migrated to the move out node, and the biological data in the routing table information is also migrated to the node, and then update the routing table.

3.4   Migration Algorithm Based on Historical Biodata Access Table

In the network, biological data are uploaded to the network based on the biological data feature sharding upload. When a node in the network becomes a move in node, through the analysis of its biodata history access table, the biodata records that need to be migrated are selected, and then the biodata is migrated to the target node's (that is, the PeerID corresponding to the most popular biodata in the move in node's access table) neighbor node with light load through the method of resource migration, and then update the routing table.

Historical Biological Data Access Table: That is, the case where other nodes access the biological data of the node is recorded.

Table 1. Historical biological data access table

Time	PeerIP	PeerID	Biodata hash	Count
1	192.168.125.128	Qmeuw1UULrZp8dQpC ms1dZGWa6z9h5gAfZH ctnryCd4U48	Qmdfv7PXLAFAHJEU4ZX kUyQYQMhgeQxym1q4x MWjXdHmJ9	8
1	192.168.125.133	QmU9bCpPYtWyufGfG3 uz8EcM4o4dV7WwpLQ yHZMWEvib2f	QmS6Mr3rhjQvKwQwz48 EFbekvaXQ21ofWPQpS7 Gn9xjJyX	7
2	192.168.125.134	QmZAacbg897VyFxPSY j41evFphWNP8yzxWBst nCh3ezPDh	QmSas6omskV4KPM5P3Xn FtaoLzySosiGyivEZaSaoyrBgV	4
3	192.168.125.135	Qmeu4KRm6kuTL9zbT9 QtUxFuwVNA6n4WgL3C wyLTCcW3LC	QmfYira7MEk4rzV1eXqv 1rQ2P1Wvm1jEV1EQXDi Rm89tbe	5

As shown in the Table 1, time represents the time interval, PeerIP represents the node IP of resource visitor, PeerID represents the IPFS node ID of resource visitor, biodata hash is the hash value of the accessed biological data, and count the number of times of biological data access.

A Measure of the Popularity of Biological Data: Based on the historical biological data access table, within a given time interval, the ranking of the number of visits to all biological data of each move in node is calculated. At this time, the biological data at the top of the list is the most popular biological data.

When a node is moved in node, based on the biological data history access table, we use the biological data popularity measure to select the most popular biological data, the target node (that is, the peerID of the most popular biodata in the move in node table) in our table is then selected as the move out node from the neighbor node with the lowest latency and light load, and then the move in node migrates its biological data to the move out node. And update the routing table, whose algorithm framework is as follows (Fig. 3):

Algorithm 2 MABOHBAT:

1: # Choose the most popular biological data
2: Biodata is a list;
3: If peer's load > peer's average load :
4:     TargetBiodata = biodata[0]
5:     For biodata[i] store in peerA:
6:         Biocount = biodata.count(biodata[i])
7:         TargetBiodatacount = biodata.count(biodata[0])
8:         If biocunt > TargetBiodatacount:
9:             TargetBiodat = biodata[i]
10:     Else:
11:         Continue
12: #Select the neighbor node of the target node
13: TargetPeer :Contains routing table information in TargetBiodata
14: Delayinit = Delay(TargetPeer[peerID],peera)
15: For peer[j] in TargetPeer's neighbors:
16:     If Delay(peer[j],TargetPeer) < Delayint:
17:         Delayinit = Delay(peer[j],TargetPeer)
18:         TargetPeerNeighbor= peer[j]
19: Send the TargetBiodat to TargetPeerNeighbor
20: Update routing table

Fig. 3. Migration algorithm based on historical biodata access table

When a move in node to run the algorithm, It will pick out one of the most popular biological data information and the move out node with minimum delay to the target node in the biological data information, and then transmit data to this move out node, and update the routing table information, with this algorithm, not only can reduce the number of node query routing table, reduce the load of move in nodes, but also can improve the query efficiency of biological data.

4 Performance Evaluation

In this experiment, a total of 5 IPFS nodes were set up for verification, including 1 node set up in the local computer and 4 node set up in virtual machine. The conditions of the nodes are shown in the following Table 2. 4 biological data were uploaded on each node, then random resource requests were made.

Table 2. The information of IPFS node

Name	PeerIP	PeerID	The name of the data	Data's hash value
Ipfs1	192.168.125.1	QmedpyL Wxe7krFo EKDRUk79 pNotC4Vp MGRkS1y3 PafP6fj	ESX000000115.fastq	Qmdfv7PXLAFAHJEU4ZXkUyQYQM hgeQxym1q4xMWjXdHmJ9
			ESX000000116.fastq	QmS6Mr3rhjQvKwQwz48EFbekva XQ21ofWPQpS7Gn9xjJyX
			ESX000006217.fastq	QmcHGxPTk5SKqXjQX9UtNu35wiB Rn3pcoRtJxxb56V6uK4
			ESX000000172.fastq	QmfYira7MEk4rzV1eXqv1rQ2P1Wv m1jEV1EQXDiRm89tbe
Ipfs2	192.168.125.128	Qmeuw1U ULrZp8dQ pCms1dZG Wa6z9h5g AfZHctnry Cd4U48	ESX000000173.fastq	QmVYpUJMQikTB4NV3SptvHWcn VFUKoWZSgk5HybNqpQobK
			ESX000000174.fastq	QmZQSc1QbmEqZeLyh3bnmxwEV N8pqeUhhVPJUptdBirzKN
			ESX000000175.fastq	QmUJf971jKdonUeLRHr44PJSUme xi8CeBW3Q6MzU5mTXuH
			ESX000000176.fastq	QmPZe5bNApiaqRv7j7NBy2DWx CswisbvJ1wTfUd5orumno
Ipfs3	192.168.125.133	QmU9bCp PYtWyufGf G3uz8EcM 4o4dV7Wwp LQyHZMWE vib2f	ESX000000185.fastq	Qmdu4dKKkurPhJHNy8AM3rLXtae rgSMMM3B2RutsDfe5Yf
			ESX000000186.fastq	QmW2YrC21L6i2sg8xasgEZwBTb UJ9ZaCJdzpawQyk75pp4
			ESX000000245.fastq	QmWyoPcmrTQHPxbNuCxjjrdktXE 5pDigQ6TKUYbnzUniBe
			ESX000000251.fastq	QmNhPUyUX8ipbqgtE8jYT9525bkk U7NHDkwksPJYwXsvLt
Ipfs4	192.168.125.134	QmZAacbg 897VyFxPSY j41evFphWN P8yzxWBstn Ch3ezPDh	ESX000000263.fastq	Qmak6LHfMfG6hzJTkeY8fTZD95ZM 9EDkcSW833R5Cjbc67
			ESX000000263.fastq	QmT1RehH22h2sUxgKYsWrSvdbEx 6EZLnoXe7GU8QxuBSSr
			ESX000000313.fastq	QmX7kYi4X2NcLBRJJPQHYXtmA EMZjxjNS7h5AeDBPLWHE7
			ESX000000316.fastq	QmawigMHndbrzuzVdLo9Ubz1KVT ZSbKeaHStASf5m7ot9u
Ipfs5	192.168.125.135	Qmeu4KRm 6kuTL9zbT9 QtUxFuwV NA6n4WgL 3CwyLTCc W3LC	ESX000000317.fastq	QmX4e2YyHkQMVrru3F2s9MmS8 BCwjQtAhaeVKY4KvKKJY6
			ESX000000317.fastq	QmdFhjRZPUJoDrejqgdCgY61CjNM LkDY5hkEF3STynVY7 s
			ESX000006251.fastq	QmfX4cBMfbeAbFD6V7nseZPihVh En9cV6TuuinR5AMY8qG
			ESX000001289.fastq	QmUsBhWMYevuwN5LuJCQkvZx RQvtTr6Co9w9W5ktHfiwPt

After the IPFS random resource access, we found that the overload of the routing table load occurred at node 192.168.125.134, and the overload of the biological data history resource access table occurred in node 192.168.125.1.

Figure 4 shows the initial download speed, the download speed using the IPFS caching mechanism, and the download speed running the routing table based migration algorithm of node 192.168.125.134.

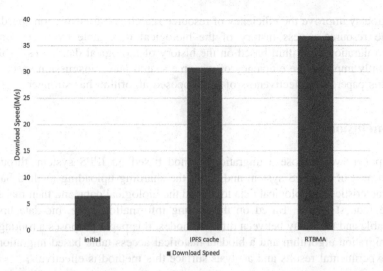

Fig. 4. Comparison of experimental results based on routing table migration algorithm

Figure 5 shows a comparison of the download speed in the initial case, the download speed using the IPFS caching mechanism, and the download speed of the migration algorithm running the historical biological data access table of the node 192.168.125.1.

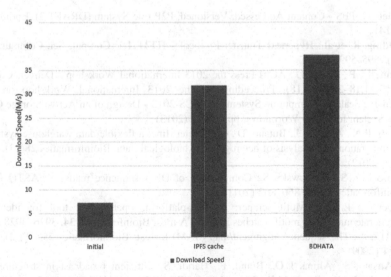

Fig. 5. Comparative experimental results based on biological data history access table algorithm

Through experimental verification shows that a node with a migration algorithm downloads faster than it would initially and with an IPFS cache, So when nodes in the routing table load overload situation, The migration algorithm based on routing table can obviously improve the efficiency of resource search and transmission speed, when the node resource access history of the biological data table overload condition, through migration algorithm based on the history of biological data access tables can significantly improve the efficiency of resource search and transmission speed, above all, in this paper, the effectiveness of the proposed algorithm has stronger.

5   Conclusion

In this paper, we propose a migration method based on IPFS system, through the establishment of an IPFS system and adopt the sharding uploading method based on the characteristics of biological data to upload the biological data, and then make a real resource request access. Based on the routing information table, bio-data historical access table and the delay between different nodes, this paper proposes a routing table-based migration algorithm and a biodata historical access table based migration algorithm, experimental results and analysis surface this method is effective.

Acknowledgments. The research of this article is supported by the national key research and development program "biological information security and efficient transmission" project, project letter no. 2017YFC1201204.

References

1. Benet, J.: IPFS - Content Addressed, Versioned, P2P File System (DRAFT 3). Eprint Arxiv (2014)
2. Fielding, R, et al.: Hypertext Transfer Protocol – HTTP/1.1. Comput. Sci. Commun. Dict. **7**(4), 595–599 (1999)
3. Donnelly, P., Thain, D.: ACM Press the 2013 International Workshop - Denver, Colorado (2013.11.18–2013.11.18). Proceedings of the 2013 International Workshop on Data-Intensive Scalable Computing Systems - DISCS-2013 - Design of an Active Storage Cluster File System for DAG Workflows, pp. 37–42 (2013)
4. Smith, R.N., Aleksic, J., Butano, D., et al.: InterMine: a flexible data warehouse system for the integration and analysis of heterogeneous biological data. Bioinformatics **28**(23), 3163–3165 (2012)
5. Deorowicz, S., Grabowski, S.: Compression of DNA sequence reads in FASTQ format. Bioinformatics **27**(6), 860–862 (2011)
6. Roberson, E.D.O.: Motif scraper: a cross-platform, open-source tool for identifying degenerate nucleotide motif matches in FASTA files. Bioinformatics **34**, 3926–3928 (2018)
7. Li, H.E.A.: The sequence alignment/map (SAM) format. Bioinformatics **25**(1 Pt 2), 1653–1654 (2009)
8. El-Ansary, S., Alima, L.O., Brand, P., Haridi, S.: Efficient broadcast in structured P2P networks. In: Kaashoek, M.F., Stoica, I. (eds.) IPTPS 2003. LNCS, vol. 2735, pp. 304–314. Springer, Heidelberg (2003). https://doi.org/10.1007/978-3-540-45172-3_28
9. Rao, A., Lakshminarayanan, K., Surana, S., Karp, R., Stoica, I.: Load balancing in structured P2P systems. In: Kaashoek, M.F., Stoica, I. (eds.) IPTPS 2003. LNCS, vol. 2735, pp. 68–79. Springer, Heidelberg (2003). https://doi.org/10.1007/978-3-540-45172-3_6

10. Prasad, K.K., Aithal, P.S.: A study on fingerprint hash code generation based on MD5 algorithm and freeman chain code. Int. J. Comput. Res. Dev. (IJCRD) **3**, 13–22 (2018)

11. Wu, D., Tian, Y., Ng, K.W.: Achieving resilient and efficient load balancing in DHT-based P2P systems. In: 2006 31st IEEE Conference on Local Computer Networks. IEEE (2007)

12. Zhang, Y., et al.: An efficient unstructured P2P network search algorithm based on multi-information. In: International Conference on Instrumentation. IEEE (2011)

13. Wang, C., Xiao, L.: An effective P2P search scheme to exploit file sharing heterogeneity. IEEE Trans. Parallel Distrib. Syst. **18**(2), 145–157 (2007)

14. Totekar, C.R., Thilagam, P.S.: An efficient search to improve neighbour selection mechanism in P2P network. In: Prasad, S.K., Routray, S., Khurana, R., Sahni, S. (eds.) ICISTM 2009. CCIS, vol. 31, pp. 119–127. Springer, Heidelberg (2009). https://doi.org/10.1007/978-3-642-00405-6_16

15. Busnel, Y., Kermarrec, A.M.: PROXSEM: interest-based proximity measure to improve search efficiency in P2P systems. In: European Conference on Universal Multiservice Networks. IEEE (2007)

16. Sharifi, L., Khorsandi, S.: A popularity-based query scheme in P2P networks using adaptive gossip sampling. Peer-To-Peer Netw. Appl. **6**(1), 75–85 (2013)

17. Yeferny, T., Arour, K.: Efficient routing method in P2P systems based upon training knowledge. In: International Conference on Advanced Information Networking & Applications Workshops. IEEE (2012)

18. Youming, F., et al.: An improved range query algorithm on over-DHT based P2P network. In: International Conference on Computer Science & Network Technology IEEE (2012)

19. Napoli, C., Pappalardo, G., Tramontana, E.: Improving files availability for bittorrent using a diffusion model. In: Wetice Conference, pp. 191–196. IEEE (2014)

Meter Detection of Substation Scene Based on Deep Learning

Guifeng Zhang[1], Zhiren Tian[1], Yongli Liao[1], Song Wang[1], and Jinchen Xu[2(✉)]

[1] Electric Power Research Institute, CSG, Guangdong 510663, China
[2] School of Electrical Automation and Information Engineering, Tianjin University, Tianjin 300072, China
1165969735@qq.com

Abstract. Automatic detection of the substation meters based on deep learning is of great importance. Meters are important equipment for monitoring substations safety. A large number of meters in the substation need to be detected and recognized, which is labor-intensive and time-consuming. Aiming at solving low detection accuracy problem in the existing methods, an efficient and accurate method based on YOLO for meter detection is proposed. First, we build a dataset of substation meters with 1432 images as the training set and 615 images as the test set. We use data augmentation strategy to solve the problem of data shortage and class-imbalance. Second, according to the characteristics of the meter image, feature pyramid is used to train the model to effectively improve the recognition performance. Finally, we use the training warm-up method to reduce the over-fitting of the training model. Experiments results show that, compared with other methods, our method can achieve better performance on several benchmarks. The detection recall improves from 88.03% to 96.08% and mAP improves from 89.64% to 93.84% on the premise of ensuring real-time performance.

Keywords: Meter detection · Deep learning · Data augmentation · Multi-scale training

1 Introduction

Substation is an important part of the power system, so it is necessary to find out the abnormal situation of each substation in time to effectively ensure the operation safety of the entire substation. At present, equipment inspections of substations are gradually completed by means of intelligent robot inspection. Compared with the traditional manual inspection, the use of intelligent robots for substation inspection has the advantages of saving manpower, quick and convenient. The meter is an important display device for status monitoring of sub-station equipment. By reading the real-time status or value displayed by each meter, it can effectively reflect the current operating status of the substation equipment.

Correct identification of the meter type is a necessary prerequisite for the real-time status and numerical reading of the substation meter. The images collected by the

© Springer Nature Switzerland AG 2019
D. Milošević et al. (Eds.): HCC 2019, LNCS 11956, pp. 600–607, 2019.
https://doi.org/10.1007/978-3-030-37429-7_61

intelligent robot inspection include various types of meters. in addition, the same type of meter has a variety of models. Therefore, designing a meter detection model capable of different types and multiple models is a huge challenge for the current implementation of robot intelligent inspection.

The use of intelligent robot inspection to complete the meter reading task can be divided into two steps in the implementation: First, through the identification of the inspection image, the detection and classification of the substation meter is completed; Second according to the classification results, different substation tables are realized. State judgment and numerical identification. For example, the dial pointer extraction for pointer-type substation meters and the readings for digital display substation instruments [1–4] have formed a more mature solution. Depending on this solution, the substation meter readings have been able to achieve higher accuracy. However, due to the lack of relevant detection schemes for different types of meters, this state judgment and reading method can only be realized by human recognition of different meters. Aiming at this problem, this paper proposes a substation instrument detection algorithm based on deep learning to realize the detection and classification of multi-meters, and perfects the intelligent robot inspection scheme to make up for the blanks of different instruments.

The detection and classification of the meter can be transformed into a target detection problem. Target detection is a fundamental problem in the field of image processing. Its main purpose is to determine the type and location of objects defined.

Deep learning is the main method of target detection algorithm. Deep learning mainly uses the powerful learning ability of neural networks instead of artificial feature extraction to achieve more abstract learning of data. At present, the target detection algorithms based on deep learning are mainly divided into two categories. One type is the target detection algorithm based on the pre-selection box idea [5–8], such as RCNN, Fast-RCNN, etc. These algorithms generally have strong target localization and high object classification ability, but the algorithm structure is relatively complex and runs. The rate is slow and not suitable for porting to embedded devices. The other is based on regression thinking to achieve target detection [9–12], such as YOLO, SSD and other algorithms. The YOLO and SSD algorithms are relatively simple in structure and highly portable. This kind of algorithm does not classify and return on the pre-selection box, but directly divides the image into a grid, and directly returns the object position and category information on the square at one time, so the running speed is faster.

Based on the network results proposed in other literatures, this paper proposes a network structure and training method suitable for substation metering detection based on actual needs. The second part will introduce the network structure and training process in detail. The third part compares the method of this paper with other classical algorithms and gives conclusions. The last part summarizes the whole text.

2 Meter Detection of Substation Scene

2.1 Construction of Meter Data Sets

The difficulty in meter detection lies in the fact that the different types of meters are similar in shape and difficult to distinguish. At the same time, there are some circular devices like meters in the substation, which will affect the classification effect of the meter. The data in this paper is taken from photos taken by intelligent inspection robots in China Southern Power Grid Corporation substation. The data set contains 2047 images of various types, all of which are 1920 × 1080. Label the data set using LabelImg. The experiment randomly divided the data set according to the ratio of 3:7, and used 1432 images as the training set and 615 images as the test set. The data set consists of 4 types of table and 10 different types of meters (5 kinds of pressure gauges, 3 kinds of thermometers, lightning rod detectors, 1 oil level meter) and a kind of equipment with similar appearance and meter. Such a data set design can effectively improve the effect of meter detection and avoid misidentification of non-meter devices as meter devices.

2.2 Data Augmentation

Before the data training, the mixup proposed by Zhang et al. [13] provided us with a data enhancement method that can enhance the detection effect. Mixup randomly selects two images at a time, with (x_i, y_i) and (x_j, y_j), respectively, and constructs a new image by linearly weighting the two images:

$$\widehat{x} = \lambda x_i + (1 - \lambda)x_j \tag{1}$$

$$\widehat{y} = \lambda y_i + (1 - \lambda)y_j \tag{2}$$

The values of λ are randomly selected in [0, 1], and the data is expanded in this way to complete the data enhancement process. This article also takes two basic data augmentation methods, random clipping and random rotation, which are listed later.

2.3 Network Structure

The network design structure of this paper imitates the YOLO algorithm and divides the entire network structure into four modules. In the training data input network structure, a downsampling is performed after the first convolutional layer. Each of the two convolutional layers adds a residual layer as a unit. The residual layer can be used to connect data hopping layers to achieve simultaneous output of multiple scales, improve network utilization efficiency, and more importantly, the residual layer can ensure network convergence and will not become difficult to converge due to deepening of the network. The deeper the network, the better the feature extraction effect and the stronger the detection capability. The first two modules are 1 and 2 units respectively, which is basically the same as the YOLO network framework. The latter two modules are composed of 8 units linearly superimposed, respectively outputting characteristic maps of 26 * 26 and 52 * 52. Achieve multi-scale training [14, 15]. The latter two modules consider training the default boxes of the two scales to make the network training better. Its network structure is shown in Table 1.

The network structure of this paper considers the default boxes of two scales to effectively improve the detection effect of the network on different size meters and improve the robustness of the network model.

Table 1. Network structure of this paper

	Network structure	Filter	Size	Output
	Convolution layer	32	3 × 3	256 × 256
	Convolution layer	64	3 × 3/2	128 × 128
1×	Convolution layer	32	1 × 1	
	Convolution layer	64	3 × 3	
	Residual layer			128 × 128
	Convolution layer	128	3 × 3/2	64 × 64
2×	Convolution layer	64	1 × 1	
	Convolution layer	128	3 × 3	
	Residual layer			64 × 64
	Convolution layer	256	3 × 3/2	32 × 32
8×	Convolution layer	128	1 × 1	
	Convolution layer	256	3 × 3	
	Residual layer			32 × 32
	Convolution layer	512	3 × 3/2	16 × 16
8×	Convolution layer	256	1 × 1	
	Convolution layer	512	3 × 3	
	Residual layer			16 × 16
	Average pooling layer		Global	
	Fully connected layer		1000	
	Softmax			

2.4 Implementation Process

This paper establishes a complete training method to realize the inspection task of the meter. The experiment uses DarkNet as the backbone network, and on this basis, it improves and proposes a network framework suitable for meter detection. During the training process, the experiment begins by following the steps below:

Step 1: Randomly sample the image and convert the image from its original pixel (0–255) to a 32-bit floating point number.

Step 2: Randomly crop the collected meter image, the aspect ratio is between 3/5–5/3, and the clipping area is randomly selected between 0.5–1.

Step 3: normalize the image of RGB channels.

In order to better complete the training process, this paper adopts the training strategy of learning rate warm-up. At the beginning of the training, all parameters in the network are random values, and the learning rate is set too high, resulting in unstable network parameters. In order to complete the training convergence quickly and accurately, Goyal et al. [15], assuming that the initial learning rate is set to, for the i-th training batch iteration $(0 \leq i \leq m)$ The learning rate should be:

$$\eta' = \frac{i\eta}{m} \tag{3}$$

The network frame parameters need to be fine-tuned before training. Due to the large size of the input image, this experiment will adjust the batch, subdivision is adjusted, the channel number is set to 3, and the initial learning rate is set to 0.001. The momentum is set to 0.9, the weight decay is set to 0.005, and the number of iterations is set to 50000.

3   Experiments

3.1   Evaluation Index

The technical measures of target detection are mainly average accuracy mean (mAP) and recall. The recall rate mainly reflects whether the test is comprehensive, and mAP mainly reflects the accuracy of the test target.

The formula for recall and precision is as follows:

$$Re = \frac{TP}{TP + FN} \tag{4}$$

$$Pr = \frac{TP}{TP + FP} \tag{5}$$

TP indicates the correct number of target classifications, FN indicates that the target error is divided into the number of other categories, and FP indicates the number of targets that misidentify the non-target object. The AP is the area under the Precision-recall curve. Generally speaking, the better the classifier, the higher the AP value. mAP represents the average AP value of multiple targets.

3.2   Meter Detection of Substation Scene Effect

The workstation operating system running this experiment is 64-bit Ubuntu 14.04, and the processor model is Intel Xeon CPU E5-2620 v3 @2.40 GHz×16, memory 32G. The graphics card model is GeForce GTX TITAN X/PCle/SSE2, and the memory is 32G.

Table 2 shows the comparison between the proposed method and other classical target detection algorithms. It can be seen that the algorithm in this paper has a significant effect on the meter detection. In addition, Fast-RCNN also has a good detection effect, but its network structure is complex. The running speed is slow and it is not suitable for porting to the intelligent robot.

Table 2. Comparison of the algorithm with other classical detection algorithms

Detection algorithm	Recall	mAP
SSD512	0.8842	0.8021
YOLOv2	0.8745	0.8320
YOLOv3	0.9323	0.9062
Fast-RCNN	0.9534	0.9202
Our Algorithm	**0.9608**	**0.9384**

Through the ablation experiment, Table 3 shows the improvement of the detection effect of the various training strategies mentioned above. The comparison benchmark algorithm is the test result obtained by directly training the first two modules of the network structure.

Table 3. Effect of various training strategies on detection results

Training strategy	Recall	mAP
Standard	0.9103	0.8964
+ Non-meter category	**0.9553**	**0.9218**
+mixup	0.9306	0.9012
+ Learning rate warming up	0.9142	0.9023
+ Multi-scale training	0.9522	0.9198
+ all method	**0.9608**	**0.9384**

It can be seen from Table 3 that the improvement of the detection effect of this paper mainly comes from the design of the data set and the multi-scale training method. In addition, the data enhancement method of the mixup also has a positive effect on the detection effect of the meter, data augmentation is generally in the case of insufficient data or small data sets, there is a significant effect, you can consider further expanding the data set in later practical applications. The learning rate warmup strategy also has a slight improvement in the detection effect. In theory, the learning rate warmup only makes the network accelerate convergence. However, the actual situation also has a positive impact on the detection results. This shows that the training strategy can also affect the final model to a certain extent.

Figure 1 shows the detection effect of the text algorithm on the substation meter. we propose a network structure and training strategy suitable for metering detection. Compared with the classical target detection algorithm, this method has obvious advantages in both recall and mAP values.

Fig. 1. Meter detection effect

4 Conclusion

This paper explores a substation metering method based on deep learning. The test results show the great potential of deep learning in safety monitoring. The experimental process verifies that the training strategy can also improve the detection effect. In addition, further expansion of the data set and improvement of target categories in the image, such as transformer boxes and other important equipment, can also improve the detection effect, and also enable the robot to undertake more substation monitoring tasks.

References

1. Zhang, W., Zheng, Y., Shan, H., et al.: Image recognition of pointer meter readings in substation based on labview. Equip. Maint. Technol. **3**, 66–70 (2015)
2. Anonymous: A method for meter reading identification, meter self-service meter reading system and meter reading method thereof, CN101209211A[P] (2013)
3. Li, X., Wei, S., Tian, F.: Pointer meter reading recognition system in patrol robot. Chin. J. Sci. Instrum. **38**(7), 1782–1790 (2017)
4. Anonymous: A digital meter recognition method based on machine learning:, CN104899587A[P] (2015)
5. Zhao, X., Wei, L., Zhang, Y., et al.: A faster RCNN-based pedestrian detection system. In: Vehicular Technology Conference (2017)
6. Sang, J., Guo, P., Xiang, Z., et al.: Vehicle detection based on faster-RCNN. J. Chongqing Univ. **40**(7), 32–36 (2017)

7. Zhu, C., Zheng, Y., Luu, K., et al.: CMS-RCNN: contextual multi-scale region-based CNN for unconstrained face detection. In: Bhanu, B., Kumar, A. (eds.) Deep Learning for Biometrics. Springer, Cham (2017). https://doi.org/10.1007/978-3-319-61657-5_3

8. Le, T.H.N., Zheng, Y., Zhu, C., et al.: Multiple scale faster-RCNN approach to driver's cellphone usage and hands on steering wheel detection. In: Computer Vision and Pattern Recognition Workshops (2016)

9. Liu, W., et al.: SSD: single shot multibox detector. In: Leibe, B., Matas, J., Sebe, N., Welling, M. (eds.) ECCV 2016. LNCS, vol. 9905, pp. 21–37. Springer, Cham (2016). https://doi.org/10.1007/978-3-319-46448-0_2

10. Redmon, J., Divvala, S., Girshick, R., et al.: You only look once: unified, real-time object detection (2015)

11. Shafiee, M.J., Chywl, B., Li, F., et al.: Fast YOLO: a fast you only look once system for real-time embedded object detection in video (2017)

12. Redmon, J., Farhadi, A.: YOLO9000: better, faster, stronger. In: IEEE 2017 IEEE Conference on Computer Vision and Pattern Recognition (CVPR), Honolulu, HI, 21–26 July 2017, pp. 6517–6525 (2017)

13. Zhang, H., Cisse, M., Dauphin, Y.N., et al.: mixup: beyond empirical risk minimization (2018)

14. Feiten, E.W., Martin, O., Otto, S.W., et al.: Multi-scale training of a large backpropagation net. Biol. Cybern. 62(6), 503–509 (1990)

15. Singh, B., Najibi, M., Davis, L.S.: SNIPER: efficient multi-scale training (2018)

16. Goyal, P., Dollár, P., Girshick, R., et al.: Accurate, large minibatch SGD: training ImageNet in 1 hour (2017)

Discretization of Laplace-Beltrami Operator Based on Cotangent Scheme and Its Applications

Qingqing Zhang and Chunmei Duan[✉]

Business School, Shandong Normal University, Jinan, China
qqzh710@163.com, cmduan@sdnu.edu.cn

Abstract. Laplace-Beltrami operator (LBO) is the basis of describing geometric partial differential equation. It also plays an important role in computational geometry, computer graphics and image processing, such as surface parameterization, shape analysis, matching and interpolation. Due to the different application fields of Laplace-Beltrami operator need to meet the mathematical properties of different, has produced many discretization methods but cannot replace each other discretization method, different discretization method can reduce the time complexity, at the same time can improve an inefficient algorithm. In this article, We are mainly aimed at discretization based on the Laplace-Beltrami cotangent operator. We improve the original discretization method and apply it to non-rigid 3D shape retrieval tasks. Experimental results shows the effectiveness of our discretization method.

Keywords: Laplace-Beltrami operator · Discretization · Cotangent scheme · Triangular mesh · Non-rigid 3D shape retrieval

1 Introduction

Laplace-Beltrami operator is the Laplace operator extension on Riemannian manifolds, and the discretization is more complex. Discrete representations are generally implemented in 3D model using triangular meshes, multi-deformed meshes, and point clouds. Therefore, the research on the discretization method of Laplace-Beltrami operator has always received great attention.

Since the curvature needs to be calculated, the Laplace-Beltrami operator begins to appear in geometry [1]. Later, LBO played a central role in many different fields, such as image processing [2, 3], signal processing, surface processing [5, 6] and PDE [7, 8]. Different properties need to be used in different fields, so there are many discretization methods. Different properties are required in different fields, so many discretization methods are produced, and there are certain differences between the methods and cannot be interchanged.

In recent years, the study and improvement of discretization methods have attracted extensive attention. Laplacian graphs are mainly used for polygon meshes, point clouds [9, 10]. The methods of cotangent weight and mean coordinate are mainly aimed at triangular mesh. Finite element method is mainly for polygon mesh; Belkin and Nyogi's

D. Milošević et al. (Eds.): HCC 2019, LNCS 11956, pp. 608–614, 2019.
https://doi.org/10.1007/978-3-030-37429-7_62

methods are mainly used for triangular mesh and point clouds; Discretization using cotangent scheme on triangular meshes has been applied more frequently, and the problem of convergence is also discussed more.

2 The Laplace-Beltrami Operator

In this section, we introduce the definition of LBO, we introduce the definition LBO on mesh surface and analyze their mathematical properties.

2.1 Definition

The LBO is a natural generalization of the Laplace operator for Riemannian manifolds. The Laplace-Beltrami operator Δ is defined as

$$\Delta_x f = div(\nabla_x f) \tag{1}$$

Where f is a smooth function, ∇ is gradient. For a detailed discussion on the main properties of the Laplace-Beltrami operator, we refer the reader to [11–13].

2.2 Discretization

Define a function f in the triangular mesh M whose number of vertices is n. The discrete LBO operator of this function at the vertex z_i is $Tf(z_i)$. Discrete Laplace-Beltrami operators is defined as [12, 14]:

$$Tf(z_i) = \sum_{j=1}^{n} w_{ij}(f(z_i) - f(z_j)) \tag{2}$$

In the above formula we can see that our focus is on how to choose the problem of w_{ij}, the mathematical properties it satisfies can affect the effect of the final discretization method. In the ideal state, the LBO should satisfy four properties, that is, the symmetry is $w_{ij} = w_{ji}$; the locality indicates that w_{ij} is not equal to 0 when z_j is adjacent to z_i. Linearity has $Tf(z_i) = 0$ for linear functions on the plane; $w_{ij} \geq 0$ when i and j are not equal, but in practical applications, this ideal state does not exist, we can only choose according to the actual situation.

3 Discretization of Laplace-Beltrami Operator Based on Cotangent Scheme

LBO discretization had experienced a long process, on the one hand, the theory constantly supplement and extension will appear new problems that it need to constantly research improvement and correction, on the other hand, it is used in different fields, so the focus of discretization of Laplace-Beltrami operator is not the same. Due to the difference of calculation theory and practical application, LBO discretization method has different details and frameworks for different problems. Therefore, the cotangent method and its improved method are summarized. Finally, we improve a discretization method.

3.1   The Theoretical Basis of the Cotangent Scheme

Discretization of Dirichlet's energy: f is defined on vertices, and it is mapped to the triangle in the case of linear mapping, $f : \Delta_1 \to \Delta_2$, which can obtain the Dirichlet energy [15] expression of f:

$$E_D(f) = \frac{1}{4} \sum_{i=1}^{3} \cot \beta_i \bullet \left| \overrightarrow{h_i} \right|^2 \tag{3}$$

Where β_i is the inner angle of Δ_1, $\left| \overrightarrow{h_i} \right|$ is the side length of Δ_2, the mapping relationship between the two triangles. If a vertex z is obtained for the surrounding area energy sum, and then the partial derivative is obtained, we obtain [15]:

$$\frac{\partial}{\partial f(z_i)} E_D(f_x) = \frac{1}{2} \sum_{i=1}^{S} (\cot \mu_i + \cot \theta_j)(f(z_i) - f(z_j)) \tag{4}$$

Where S is the set of z_i and adjacent vertices, and μ_i, θ_j are the opposite sides of the sides connecting z_i, z_j.

3.2   Calculation Method and Improvement

Using the cotangent scheme to improve, mainly to use:

$$w_{ij} = \frac{1}{2}(\cot \mu_{ij} + \cot \theta_{ij}) \tag{5}$$

However, there are many deficiencies in the cotangent scheme.

Delaunay Triangulation. A major problem with the cotangent scheme is that $\mu_{ij} + \theta_{ij} > \pi$, $\cot \mu_{ij} + \cot \theta_{ij}$ will have a negative value, and the result will affect the texture mapping. The solution for Delaunay triangulation is to preprocess the input model without changing the calculation method [16]. When $\mu_{ij} + \theta_{ij}$ is less than or equal to 1, $\cot \mu_{ij} + \cot \theta_{ij}$ is greater than or equal to 0, and it is extended to some extent on the mesh model, allowing the edge of the triangle to be a geodesic line along the surface of the mesh in the re-triangulated mesh. This method changes the input model and ultimately affects the output model, solving the problem of w_{ij} taking negative values.

Taubin's Discretization Method [17]. This is a class of discretization in the following form:

$$\Delta_x f(z_i) = \sum_{j \in s} w_{ij}[f(z_j) - f(z_i)] \tag{6}$$

where the weights w_{ij} are positive numbers. A simple way is to take $w_{ij} = 1/|S|$. A more general way is to define them by a positive function:

$$w_{ij} = \varphi(z_i, z_j) / \sum_{k \in s} \varphi(z_i, z_j) \tag{7}$$

and the function $\varphi(z_i, z_j)$ can be the surface area of the two faces that share the edge $[z_i, z_j]$, or some power of the length of the edge $\varphi(z_i, z_j) = \|z_i - z_j\|^\eta$. Fujiwara take $\eta = 1$. Polthier's discretization [16] takes $w_{ij} = \frac{1}{2}(\cot \mu_{ij} + \cot \theta_{ij})$, without imposing the normalization condition $\sum w_{ij} = 1$.

Mayer's Discretization [18]. On the triangular plane mesh X, he proposes the following formula:

$$\Delta_x f(z_i) = \frac{1}{M(z_i)} \sum_{j \in S(i)} \frac{\|z_{j-} - z_j\| + \|z_{j+} - z_j\|}{2\|z_j - z_i\|}[f(z_j) - f(z_i)] \tag{8}$$

Where $M(z_i)$ is the sum of areas of triangles around z_i.

Desbrun et al.'s Discretization [19]. Desbrun et al. proposed a direct choice of M_i, which is the sum of the areas of the triangle faces around this vertex. The expression can be written as:

$$M(z_i)M_i = \sum_{\Delta j \in F_N(z_i)} M(\Delta_j) \tag{9}$$

The problem with this approach is that there is overlap between the fields and may not be applicable to irregular grids. Desbrun et al. get the following discretization

$$\Delta_x f(z_i) = \frac{3}{M(p_{ij})} \sum_{j \in S(i)} \frac{\cot \mu_{ij} + \cot \theta_{ij}}{2}[f(z_i) - f(z_j)] \tag{10}$$

He is improving on the basis of $w_{ij} = \frac{1}{2}(\cot \mu_{ij} + \cot \theta_{ij})$, which using the normalization factor.

Meyer et al.'s Discretization [20]. Since the previous Voronoi finite volume region does not support obtuse angles, a new region is redefined for each vertex z, denoted as M_{mixed}. For any triangular mesh, the following formula can be used:

$$\Delta_x f(z_i) = \frac{1}{M(z_i)} \sum_{j \in S(i)} \frac{\cot \mu_{ij} + \cot \theta_{ij}}{2} [f(z_i) - f(z_j)] \tag{11}$$

His method is mainly for calculating the Voronoi diagram and calculating the area of the area. For the acute triangle, the calculation is relatively easy.

3.3   Improvement of Discretization Method

The cotangent scheme allows to computer the eigenvalues and eigenvectors as the solutions to the generalized eigenvalue problem $Af = \lambda Bf$, A is the stiffness matrix, B is the mass matrix, and a linear matrix is found to calculate the eigenvalues [21]. And eigenvectors, which can then produce different spectra. In 3D shape retrieval, we improve the performance of LBO and modify the stiffness matrix and mass matrix. We use piecewise affine, mass matrix A and stiffness matrix B are defined as:

$$A_{ij} = \prec \psi_i, \psi_j \succ B_{ij} = - \prec M(\nabla \psi_i), \nabla(\psi_j) \succ \tag{12}$$

where ψ_i are functions and it is affine on each triangle of the mesh. We assume that M is constant on each triangle.

4   Application

The LBO plays a role in many fields, especially in physics and mathematics. The implementation of these applications is closely related to the discretization of LBO. Different discretization methods will produce different properties, and the final application will also be different. In this chapter, we mainly introduce the application of cotangent scheme in non-rigid 3d shape retrieval. We give the experimental results, which show the effectiveness of our method and improve the retrieval precision.

In order to show the performance of our discretization method, we conduct several experiments in non-rigid 3D shape retrieval. We evaluate our discretization method of the LBO on datasets from SHREC'14. Of the real (scanned) 3D human models, datasets include 400 models, which are 40 classes. Each class means one kind of human model which has 10 different postures. The real 3D human model datasets are extremely challenging and even more challenging than the synthetic datasets in SHREC'14. Although we have tested our approach on all datasets in SHREC'14, we show the evaluations on the real (scanned) dataset (see Table 1). The shape is scaled as shown by the available code in [22]. We calculate 50-dimensional siHKS descriptors and 100-dimensional WKS descriptors, setting the variance to 6. We use the same training parameters as [22] which use 40% of the shape descriptors to train the transform matrix and test on the rest. The performance of our approach is evaluated by computing several accuracy metrics: nearest neighbor, first tier, second tier, discounted cumulative gain, e-measure, f-measure, precision.

Table 1. Cotangent scheme and Improved cotangent scheme evaluation on the SHREC'14 real dataset.

Metric	Cotangent scheme	Improved cotangent scheme
nn	0.9933	0.9953
ft	0.9735	0.9830
st	0.9974	0.9978
em	0.4389	0.4390
dcg	0.9938	0.9965
fm	0.9735	0.9830

We can get the effectiveness of our discretization method through Table 1. We compared our proposed LBO discretization method with the discretization method proposed in [23]. Our method gets a good result. The evaluation result of the discretization method of LBO proposed by us is only a little better than that of the original method, but the original method has given a relatively significant retrieval effect in 3D shape retrieval.

5 Conclusion

This paper mainly summarizes the LBO discretization based on the cotangent method, mainly the LBO discretization method existing at the present stage, and the theoretical knowledge, calculation method and final correlation of the LBO operator discretization based on the cotangent method. The application is extended according to the cotangent method. Some of the improvements are made on the basis of the cotangent method, and the improved application is briefly summarized and described. We improved the cotangent based discretization method of LBO, and applied it to 3d shape retrieval, proved the effectiveness of our method. The cotangent scheme is the basis of LBO discretization, and it is also a widely used method.

Acknowledgment. We acknowledge the National Natural Science Foundation of China for its financial support (Grant No. 61502284).

References

1. Hamann, B.: Curvature approximation for triangulated surfaces. Computing **8**(8), 139–153 (1993)
2. Wetzler, A., Aflalo, Y., Dubrovina, A., Kimmel, R.: The Laplace-Beltrami operator: a ubiquitous tool for image and shape processing. In: Hendriks, C.L.L., Borgefors, G., Strand, R. (eds.) ISMM 2013. LNCS, vol. 7883, pp. 302–316. Springer, Heidelberg (2013). https://doi.org/10.1007/978-3-642-38294-9_26

3. Afrose, Z., Shen, Y.: Mesh color sharpening. Adv. Eng. Softw. **91**(C), 36–43 (2016)
4. Tan, M., Qiu, A.: Spectral Laplace-Beltrami wavelets with applications in medical images. IEEE Trans. Med. Imaging **34**(5), 1005–1017 (2015)
5. Caissard, T., Coeurjolly, D., Lachaud, J.-O., Roussillon, T.: Heat Kernel Laplace-Beltrami operator on digital surfaces. In: Kropatsch, W.G., Artner, N.M., Janusch, I. (eds.) DGCI 2017. LNCS, vol. 10502, pp. 241–253. Springer, Cham (2017). https://doi.org/10.1007/978-3-319-66272-5_20
6. Boscain, U., Prandi, D.: Self-adjoint extensions and stochastic completeness of the Laplace-Beltrami operator on conic and anticonic surfaces. J. Differ. Equ. **260**(4), 3234–3269 (2015)
7. Burman, E., Hansbo, P., Larson, M.G.: A stabilized cut finite element method for partial differential equations on surfaces: the Laplace-Beltrami operator. Comput. Methods Appl. Mech. Eng. **285**, 188–207 (2015)
8. Bartezzaghi, A., Dedè, L., Quarteroni, A.: Isogeometric analysis of high order partial differential equations on surfaces. Comput. Methods Appl. Mech. Eng. **295**, 446–469 (2015)
9. Chou, P.A., Pavez, E., De Queiroz, R.L., et al.: Dynamic polygon clouds: representation and compression for VR/AR (2017)
10. Pavez, E., Chou, P.A.: Dynamic polygon cloud compression. In: 2017 IEEE International Conference on Acoustics, Speech and Signal Processing (ICASSP), pp. 2936–2940. IEEE (2017)
11. Reuter, M.: Laplace Spectra for Shape Recognition (2006). ISBN 3-8334-5071-1
12. Wardetzky, M., Mathur, S., Lberer, F., et al.: Discrete laplace operators: no free lunch. In: Eurographics Symposium on Geometry Processing (2007)
13. Trillos, N.G., Gerlach, M., Hein, M., et al.: Error estimates for spectral convergence of the graph Laplacian on random geometric graphs towards the Laplace–Beltrami operator (2018)
14. Wu, J.Y., Chi, M.H., Chen, S.G.: Convergent discrete Laplace-Beltrami operators over surfaces. Mathematics (2010)
15. Pinkall, U., Polthier, K.: Computing discrete minimal surfaces and their conjugates. Exp. Math. **2**(1), 22 (1993)
16. Bobenko, A.I., Springborn, B.A.: A discrete Laplace-Beltrami operator for simplicial surfaces (2005)
17. Taubin, G.: A signal processing approach to fair surface design. In: Proceedings of the 22nd Annual Conference on Computer Graphics and Interactive Techniques, SIGGRAPH 1995, pp. 351–358. ACM Press (1995)
18. Mayer, U.F.: Numerical solutions for the surface diffusion flow in three space dimensions. Comput. Appl. Math. **20**(3), 361–379 (2001)
19. Desbrun, M., Meyer, M., Schröder, P., Barr, A.: Implicit fairing of irregular meshes using diffusion and curvature flow. In: Proceedings of SIGGRAPH, pp. 317–324 (1999)
20. Meyer, M., Desbrun, M., Schröder, P., et al.: Discrete differential-geometry operators for triangulated 2-manifolds. Vis. Math. **3**(8–9), 35–57 (2002)
21. Wardetzky, M., Bergou, M., Harmon, D., et al.: Discrete quadratic curvature energies. Comput. Aided Geom. Des. **24**(8), 499–518 (2015)
22. Windheuser, T., Vestner, M., Rodolà, E., et al.: Optimal intrinsic descriptors for non-rigid shape analysis. In: BMVC (2014)
23. Chiotellis, I., Triebel, R., Windheuser, T., Cremers, D.: Non-rigid 3D shape retrieval via large margin nearest neighbor embedding. In: Leibe, B., Matas, J., Sebe, N., Welling, M. (eds.) ECCV 2016, Part II. LNCS, vol. 9906, pp. 327–342. Springer, Cham (2016). https://doi.org/10.1007/978-3-319-46475-6_21

Computer-Aided Diagnosis of Ophthalmic Diseases Using OCT Based on Deep Learning: A Review

Ruru Zhang[1,2], Jiawen He[1,2], Shenda Shi[1,2], Xiaoyang Kang[1,2],
Wenjun Chai[1,2], Meng Lu[1,2], Yu Liu[1,2], E. Haihong[1,2],
Zhonghong Ou[1,2], and Meina Song[1,2(✉)]

[1] School of Computer Science, Beijing University of Posts
and Telecommunications, Beijing, China
mnsong@bupt.edu.cn
[2] Education Department Information Network Engineering Research Center,
Beijing University of Posts and Telecommunications, Beijing, China

Abstract. Deep learning can effectively extract the hidden features of images and has developed rapidly in medical image recognition in recent years. Ophthalmic diseases are one of the critical factors affecting the healthy living. At the same time, optical coherence tomography (OCT) has the characteristics of non-invasive and high-resolution and has become the mainstream imaging technology in the clinical diagnosis of Ophthalmic diseases. Therefore, computer-aided diagnosis of ophthalmic diseases using OCT based on deep learning has caused a wide range of research craze. In this paper, we review the imaging methods and applications of OCT, the OCT public dataset. And we introduce in detail the computer-aided diagnosis system of multiple ophthalmic diseases using OCT in recent years, including age-related macular degeneration, glaucoma, diabetic macular edema and so on, and an overview of the main challenges faced by deep learning in OCT imaging.

Keywords: Deep learning · OCT · Ophthalmic disease

1 Introduction

The human eye is one of the most important sensory organs, the structure of the eye is elaborate, even if there is slight damage, it may cause changes in the structure of the tissue, resulting in vision decline or even injury, seriously affecting people's lives [1]. Early screening and treatment can minimize the harm caused by ophthalmic diseases and may cure ophthalmic diseases. Optical Coherence Tomography (OCT) is a non-contact, non-invasive imaging technique that visualizes the structure of the eye in a high-resolution, fast manner, providing precise pathological cross-sectional imaging [2], it can be repeated many times or used for early detection in patients undergoing surgery. OCT technology has become the gold standard for diagnostic imaging of ophthalmic diseases such as age-related macular degeneration (AMD) and diabetic macular edema (DME) [3].

© Springer Nature Switzerland AG 2019
D. Milošević et al. (Eds.): HCC 2019, LNCS 11956, pp. 615–625, 2019.
https://doi.org/10.1007/978-3-030-37429-7_63

However, OCT rapid imaging brings a large amount of high-resolution data, which dramatically increases the pressure on an ophthalmologist to read images. Deep learning technology can effectively analyze and quantify the pathological features in OCT images [4], quickly and accurately judge the pathological features and disease types, providing clinical assistance for doctors, and improving the accuracy and stability of diagnosis while reducing the diagnosis time. Computer-aided diagnosis system using OCT based on deep learning is of strategic significance for early diagnosis and treatment of ophthalmic diseases [5].

In this paper, we review the imaging methods and applications of OCT, OCT public dataset, the computer-aided diagnosis system of multiple ophthalmic diseases using OCT in recent years, and the substantial challenges faced by OCT.

2 Development Status of Application of Deep Learning in Computer-Aided Diagnosis of Ophthalmic Diseases Using OCT

2.1 General Procedures of Application of Deep Learning in Computer-Aided Diagnosis of Ophthalmic Diseases Using OCT

Computer-aided diagnosis using OCT based on deep learning is similar to the application of deep learning in other image fields, having some general procedures. These procedures chiefly include OCT pretreatment, OCT segmentation, and OCT classification, as shown in Fig. 1:

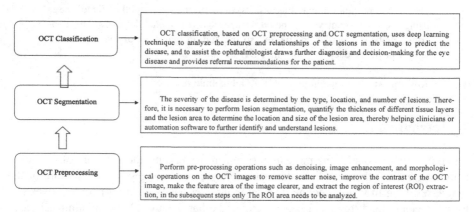

Fig. 1. General procedures based on deep learning for the diagnosis of ophthalmic diseases using OCT

2.2   Research Status of Application of Deep Learning in Computer-Aided Diagnosis of Ophthalmic Diseases Using OCT

Progress in Academia. Since the rapid development of deep learning and ophthalmic disease diagnosis technology, academic research has focused on the computer-aided diagnosis of ophthalmic diseases using OCT. So far, the research in this field has achieved good results.

In February 2018, Zhang Kang's team [6] from the University of Texas used the deep learning model to diagnose ophthalmic diseases using OCT and successfully classified four types of images, namely, vitreous wart, choroid neovasculogenesis, diabetic macular edema as well as the normal situation, with an accuracy rate of 96.6%.

In August 2018, Google DeepMind team [7] proposed a deep learning model based on the aided diagnosis of ophthalmic diseases using OCT, which can realize the segmentation of multiple focus areas and the classification of more than 50 diseases, and can accurately identify images acquired by various imaging devices with an accuracy rate up to 95%.

Progress in Industry. With the development of the computer-aided diagnosis of ophthalmic diseases using OCT, industry giants have begun to move into the field and translate their research results into clinical medical products.

BigVersion Medical Technology released MIAS 2000 [8], ophthalmic imaging storage, processing, and analysis system, in December 2018. The system can diagnose more than 20 high ophthalmic diseases with an overall accuracy of up to 96%, which has obtained the CFDA class II medical device registration certificate.

On May 15, 2019, Ping An Insurance Group Company of China [9] completed the first multi-center clinical trial of intelligent retinal disease screening system using OCT in the world. At present, the OCT image quality assessment, focus detection, referral emergency assessment, and other aspects of the system achieved 99.2%, 98.6%, 96.7% accuracy.

2.3   Public Datasets of OCT

Undoubtedly, the most direct way to obtain training samples is to use public dataset. Table 1 shows the current OCT available public dataset. Developers can choose a suitable data set according to different demand.

3   Related Research of Computer-Aided Diagnosis of Ophthalmic Diseases Using OCT

Deep learning technology has been widely used to aided diagnosis of ophthalmic diseases, by identifying, locating and quantifying the pathological features of almost all macular and retinal diseases, and classifying disease types and lesion grades. The following is an introduction to computer-aided diagnosis system of multiple ophthalmic diseases using OCT based on deep learning in recent years.

Table 1. OCT public dataset

Dataset	Source	Data format	Quantity	Link
DUKE dataset [10]	This database developed at DUKE University	MAT	There are 384 patients' OCT images, including 269 AMD and 115 normal	https://people.duke.edu/~sf59/RPEDCOphth2013dataset.htm
Kermany dataset [11]	This database was collected at the Shiley Eye Institute of the University of California San Diego, the California Retinal Research Foundation, Medical Center Ophthalmology Associates, the Shanghai First Peoples Hospital, and Beijing Tongren Eye Center	JPEG	This dataset consists of 109,312 OCT images, including 108,312 training images and 1,000 test images. Each set contains four types of labeled data: CNV, DME, DRUSEN and NORMAL	https://data.mendeley.com/datasets/rscbjbr9sj/2
Heidelberg OCT dataset [12]	This database is for corneal OCT images taken by Heidelberg imaging device	MAT	There are 15 patients, including 579 Mat files	https://sites.google.com/site/hosseinrabbanikhorasgani/datasets-1/corneal-oct
Noor Hospital OCT dataset [13]	This database was collected at Noor Hospital at Tehran	TIFF	This dataset includes 4388 OCT images from 50 DME patients, 50 normal patients, and 48 AMD patients	https://drive.google.com/file/d/1iSiFfD5LpLASrFUZu13uMFSRcFEjvbSq/view
OCTID dataset [14]	This database was collected at Sankara NethralyaEye Hospital, Chennai, India.	JPG, MAT	This dataset consists of 579 OCT images and 25 MAT, including normal, macular hole, AMD, central serous retinopathy, and DR. Ground truth labels are provided for 25 images	https://dataverse.scholarsportal.info/dataverse/OCTID
UK BioBank [15]	This database was collected by UK BioBank	PNG	256 OCT images	https://www.ukbiobank.ac.uk

3.1 Computer-Aided Diagnosis of AMD Using OCT

AMD is the leading cause of irreversible loss of vision in the elderly and a severe threat to the patients ability to live independently [16]. Therefore it is essential to assess the AMD accurately. With the development of deep learning technology, AMD intelligent diagnostic system technology has gradually matured, as shown below.

In 2017, Lee et al. [17] developed intelligent screening system of AMD. The system initializes the VGG16 model based on Xavier algorithm, and then trains and validates the modal on 48312 healthy images and 52690 AMD images. The AUROC, sensitivity, and specificity of this system reach 0.9746, 0.926, and 0.937, respectively.

Venhuizen et al. [18] proposed a deep learning model for automatic detection, segmentation, and quantification of intraretinal cystoid fluids (IRCs) in OCT in 2018. The IRCs volume was measured to assist doctors in diagnosing the severity of AMD. This model is composed of two FCNN cascades: the first FCNN aims to extract the region of interest (ROI). The second FCNN divides IRC in ROI. The model was validated on datasets of four different OCT equipment providers (Heidelberg, Zeiss, Topcon, Nidek) and achieved better performance than the observations of two experts.

In 2018, Lee et al. [19] developed computer-aided diagnostic system of AMD. This system can segment the IRF, SRF, PED, SHRM, and other lesions of neovascular AMD, providing the diagnostic basis for doctors to analyze the severity of AMD quantitatively. This system preprocesses the OCT images, and then segment them using

U-Net model. The accuracy of the system was verified on 2325 OCT datasets of the Konkuk University Medical Center.

In 2018, Treder et al. [20] used the Inception-V3 model for transfer learning on 1112 OCT images and applied the new model to classify health and exudative AMD images. The sensitivity, specificity, and accuracy of the model reached 100%, 92%, and 96%, respectively.

Schmidt-Erfurth et al. [21] used a combination of graph theory and a convolutional neural network segmentation algorithm to train 2,456 OCT images. Image segmentation was performed on the four regions of the retina, CNV, IRF, subretinal fluid (SRF), and pigment epithelial detachment (PED). Then the results of image segmentation and clinical visual acuity data are taken as inputs, and the random forest model is used for regression prediction. The system's highest predictive values for visual acuity at month 12, explaining as much as 70% of the variability.

Through the analysis of the above literature, it can be seen that the AMD screening system using OCT based on deep learning has achieved high accuracy and has practical effects in predicting the best-corrected vision for AMD patients. More and more studies tend to segment the relevant focus areas of AMD to help doctors better diagnose the severity of AMD, to make the best clinical decisions.

3.2   Computer-Aided Diagnosis of DME Using OCT

Diabetic macular edema (DME) is a chronic complication of diabetes and is one of the four major blind diseases. Moreover, DME symptoms have more obvious pathological features in OCT imaging. Therefore, the early diagnosis of DME based on deep learning technology is of great significance, as shown bellow.

In September 2017, Awais et al. [22] used convolutional neural networks to classify OCT images into DME and normal images. The model first performs image cropping and denoising by image preprocessing. Then, the author used the VGGNet-16 model to extract features from the preprocessed image. Finally, a 3-layer KNN classifier is connected for detection, and the OCT image is classified as DME or normal. The model achieved an accuracy of 87.5%, a sensitivity of 93.5%, and specificity of 81%.

In 2018, Perdomo et al. [23] proposed an end-to-end OCT-NET deep learning model to classify DME and normal OCT images. The OCT image is input into the OCT-NET model after being preprocessed by cropping, the region of interest (ROI) extraction, and the like. The model extracts and learns different OCT data characterizations by adjusting the number of filters. The model was validated at the Singapore Eye Institute (SERI) database with a calculation accuracy of 93.75%.

In 2018, Chan et al. [24] classified DME and normal OCT images through transfer learning techniques. The model uses AlexNet, VggNet, and GoogleNet three pre-trained convolutional neural networks to extract image features. The extracted features are subjected to spatial dimension reduction using a principal component analysis algorithm. Finally, these features are combined and classified by KNN, SVM, and random forest classifier. The model was validated on the OCT dataset obtained by the Singapore Eye Institute and the Chinese University of Hong Kong, achieving 90.63% and 93.75% accuracy and specificity, respectively.

In the 2018 ISBI Challenge, Vahadane et al. [25] proposed a two-stage framework for the intelligent diagnosis of DME by dividing hard exudates and fluid-filled regions (FFR). The first step is to detect candidate patches for hard exudates and FFR by image processing techniques. The second step is to predict the labels of these candidate patches using a deep convolutional neural network. In the final collation step, the system aggregates the confidences of the CNN model and uses a rule-based approach to predict the existence of DME, which has achieved better results than other participating methods.

In 2018, Kamble et al. [26] combined the Inception-v2 model with the ResNet model and proposed the Inception-ResNet-v2 model to perform DME detection and classification on OCT images. The Inception-ResNet-v2 framework is trained and tested on the Singapore Eye Research Institute's (SERI) public dataset, achieving 100% accuracy, sensitivity, and specificity. To verify the robustness of the model, the authors also tested on the Chinese University of Hong Kong (CUHK) dataset, and the accuracy of the test reached 100%.

It can be known from the analysis of the above-related documents that, because the lesion features of DME are relatively apparent in OCT imaging, most deep learning-based DME assisted diagnostic systems can achieve higher diagnostic accuracy only by migrating and combining existing deep learning models. Therefore, more and more studies are not satisfied with the intelligent diagnosis of DME diseases but tend to diagnose multiple ophthalmic diseases at the same time, such as AMD, DME, Drusen, and so on. The content of this section is summarized in Sect. 4.

3.3   Computer-Aided Diagnosis of Glaucoma Using OCT

Glaucoma is the first irreversible ophthalmic disease in the world. However, early diagnosis of glaucoma is challenging, and many factors should be considered comprehensively such as intraocular pressure, fundus morphology, visual field, and retinal nerve fiber layer (RNFL) changes [27]. Therefore, the development of Glaucoma computer-aided diagnosis system using OCT is more challenging and complicated than other ophthalmic diseases. In recent years, with the development of deep learning technology in the prevention and treatment of glaucoma, more and more promising results have appeared, as shown below.

In 2018, Wang et al. [28] established the S-D-net model for the diagnosis of glaucoma. S-net model enables retinal layer segmentation of OCT images (including vitreous, RNFL, ganglion cell complex (GCC), optic nerve epithelial cells, choroid, and sclera). D-net diagnoses glaucoma based on the thickness of the retinal layer. The model achieved 85.4% accuracy, 85.9% sensitivity, and 84.4% specificity and the diagnostic accuracy was better than that of human experts.

In 2018, Fu et al. [29] proposed a two-stage deep learning network model. The first stage was used for anterior segment OCT (AS-OCT) segmentation to extract anterior chamber Angle (ACA) region and measure its clinical parameters. The clinical parameters were then analyzed using a linear support vector machine (SVM) to predict the angular closure probability. In the second phase, the global image of the AS-OCT and the local video disc area image are obtained by the multi-context deep layer network (MCDN), which are then trained based on the pre-trained VGG-16 model to output a closed-angle glaucoma probability. Finally, the probabilistic results from the two-stage model are averaged as a final screening result.

In 2019, Asaoka et al. [30] proposed a deep learning method to diagnose early glaucoma by quantifying GCC thickness and RNFL defect. The method firstly pre-trains the model by obtaining a relatively large OCT dataset through the RS3000 OCT device, and then re-training the model by acquiring a relatively small OCT dataset through the OCT-1000/2000 device, and finally verified in a set of independent test data sets. The AROC value of the model reached 93.7%. Experiments have proved that this method can alleviate the problem of equipment differences.

In 2019, An et al. [31] developed an computer-aided diagnosis system of glaucoma patients by combining color fundus images of the optic disc region with OCT images in the macula region and optic disc region. The thickness maps of RNFL and GCC are obtained from the macula region and the optic disc region by OCT data, respectively. The color fundus image of the optic disc area and the RNFL and GCC thickness maps were respectively input into five VGG19-based pre-training models for feature extraction. The feature vectors of the five VGG19 model outputs are input into a random forest (RF) classifier for classification to obtain the probability of glaucoma.

According to the analysis of the above-related literature, since early diagnosis of glaucoma requires careful consideration of multiple factors, most models only consider the thickness of RNFL and GCC to diagnose glaucoma, and only a few models take into account clinical parameters and visual field data of glaucoma patients. The overall diagnostic accuracy is relatively low. Training high-precision models requires more extensive expert knowledge and more comprehensive data, including test data, optical disc images, OCT images, through multimodal analysis, comprehensive data characteristics, for the diagnosis and analysis of glaucoma patients.

3.4 Computer-Aided Diagnosis of Multiple Ophthalmic Diseases Using OCT

This section summarizes the research on deep learning models that enable OCT stratification or segmentation of various lesion features or classification of multiple disease types, as showed bellow.

In 2018, Google's DeepMind team [7] accurately detected more than 50 ophthalmic diseases that impair vision through a two-stage deep neural network. The first neural network is the 3D U-NET segmentation network, which analyzes OCT scans to enable segmentation of different types of ocular tissues and visible lesions. It finally outputs the tissue segmentation map of the OCT images, which improves the interpretability of the model. The second neural network is the CNN classification network, which provides diagnostics and referral recommendations for clinicians and patients by analyzing tissue segmentation map. The model is as accurate as of the world's top experts.

In 2018, Kermany et al. [6] used the Inception-V3 model for transfer learning for screening common treatable blinding ophthalmic diseases, including DME and AMD. They used 108,312 OCT images to train a deep learning model and used a separate 1000 OCT images to validate the model. The model proposes referrals based on the type of disease detected. The model has a sensitivity of 0.978, a specificity of 0.974, and an AUROC of 0.999 in disease classification tasks, with statistically similar performance compared to human experts.

On May 15, 2019, Ping An Insurance Group Company of China [9] completed the world's first multi-center clinical trial of intelligent retinal disease screening system using OCT. The clinical trials emphasize the integration of OCT retinal examination devices with artificial intelligence screening software. In the model training process, the generative adversarial nets (GAN) technology was used to analyze hundreds of thousands of OCT images of 784 patients with ophthalmic diseases. From the start of the OCT examination to the patient's acquisition of a smart screening report, the entire process can be completed in three minutes. In 2018, Lu et al. [32] used the 101-layer ResNet model to classify OCT images. The algorithm was validated on the OCT dataset of the Wuhan University Eye Center. The AUC of the intelligent system was 0.984, and the overall accuracy was 0.959. Among them, the accuracy of normal imaging, cystic macular edema, serous macular detachment, anterior retinal membrane, and macular hole were 0.973, 0.848, 0.947, 0.957, 0.978, and the system kappa value was 0.929. The authors compared the system performance with two experts. The kappa values of the two physicians were 0.882 and 0.889, respectively.

In 2019, Girish et al. [33] proposed a full convolutional network (FCN) model to segment IRCs, which can be used to detect DME, retinal vascular disease, AMD and inflammatory diseases by quantitative analysis of IRCs. The system uses the OPTIMA Cyst Segmentation Challenge Dataset to train to overcome the diversity of image noise. The system captures global and local features from the input image and obtains the exact position and relative size of the IRCs. Experiments show that the Dice correlation coefficient of this method is better than the published algorithm.

In 2019, Guan et al. [34] proposed a multi-ocular disease analysis and diagnos is system that can automatically detect and segment multiple retinal lesions. The system is divided into two phases. Firstly, the Mask-RCNN model is used to roughly segment the lesion area of the OCT images (including PED, SRF, drusen, CNV and macular hole (MH)), and then the accurate segmentation is performed by the improved distance regularized level set evolution (Improved DRLSE) model. The model was verified on the MSCOCO dataset and achieved excellent results.

In 2019, Masood et al. [35] proposed a deep learning method for intelligent diagnosis of common ophthalmic diseases. The vitreous membrane in the OCT images is first segmented by a series of conventional image processing methods and morphological operations. A deep convolutional neural network then segments the choroid, and the thickness map between the two films is quantified. Finally, according to the thickness map, AMD, DME, idiopathic polypoid choroidal vasculopathy and autoimmune diseases were diagnosed. The segmentation model was validated on the dataset of the Sixth People's Hospital affiliated to Shanghai Jiaotong University with an accuracy rate of approximately 97%.

Through the analysis of the above studies, deep learning technology has been widely used in the segmentation of multiple lesions and the classification of multiple disease types in OCT images. Academic and industrial circles have invested and achieved outstanding results. The ophthalmic disease computer- aided diagnostic system is developing in the direction of interpretability, multiclassification, generalization, and detection of uncommon eye diseases, which further promoted the industry of instrument and equipment. Therefore, research on the computer-aided diagnosis of multi ophthalmic disease using OCT based on deep learning is of great significance.

4   Challenges

Deep learning has a good application prospect in computer-aided diagnosis of ophthalmic diseases using OCT but still faces many challenges.

The quality of deep learning algorithms are limited by the variety of dataset, to obtain high-quality annotated data, it is necessary to hire several clinician experts to label OCT images, which can be time-consuming and expensive. Also, the OCT dataset collected should be diverse, including different races and devices, to improve the generalization ability of the model [36].

OCT images contain insufficient information: human doctors usually don't just use OCT images to make diagnoses, but also consider multi-modal data, including other data such as fundus images, patient electronic health records and so on.

Errors caused by transfer learning [37]: the feature extraction models currently used in transfer learning are often trained in the vast natural image datasets, such as the ImageNet dataset. Because of the substantial differences between medical images and natural images, there are some limitations in the application of model training.

The interpretability issue of the deep learning model: the internal mechanism of the deep learning model is not completely clear, and the interpretability is weak, which is one of the critical factors hindering the application of the deep learning based assisted diagnostic system in actual clinical practice.

Privacy issues: medical data often involves patient privacy, and privacy issues severely limit the sharing and use of medical data.

5   Conclusions

In summary, this article reviews recent research on deep learning based assisted diagnostic systems of multiple ophthalmic diseases using OCT, and shows the potential of deep learning applied in OCT, including lesion detection, disease screening, disease assisted diagnostic, helping ophthalmologists develop highquality treatment options. However, the clinical application of computer-aided diagnosis system based on deep learning is facing challenges due to the particularity of medical data and the interpretability of deep learning, which is causing widespread concern and research.

Acknowledgement. This work was supported by the National Key RD Program of China under Grant (2017YFB1400800).

References

1. Tortora, G.J., Derrickson, B.: Principles of Anatomy & Physiology. John Wiley, Incorporated (2017)
2. Schmidt-Erfurth, U., et al.: Artificial intelligence in retina. In: Progress in Retinal and Eye Research (2018)

3. Schmidt-Erfurth, U., Waldstein, S.M.: A paradigm shift in imaging biomarkers in neovascular age-related macular degeneration. In: Progress in Retinal and Eye Research, vol. 50, pp. 1–24 (2016)

4. Haloi, M.: Towards Ophthalmologist Level Accurate Deep Learning System for OCT Screening and Diagnosis. arXiv preprint (2018). arXiv:1812.07105

5. Rahimy, E.: Deep learning applications in ophthalmology. Curr. Opin. Ophthalmol. **29**(3), 254–260 (2018)

6. Kermany, D.S., et al.: Identifying medical diagnoses and treatable diseases by image-based deep learning. Cell **172**(5), 1122–1131 (2018)

7. de Fauw, J., et al.: Clinically applicable deep learning for diagnosis and referral in retinal disease. Nat. Med. **24**(9), 1342 (2018)

8. Big Vision: Big Vision ophthalmic image storage and analysis system (MIAS) was officially released. http://www.bigvisiontech.com/news_detail-1-31.html. Accessed 09 Jan 2019

9. Yang, X.: Ping an released the first intelligent OCT fundus disease screening system in the world. http://www.financialnews.com.cn/jigou/ssgs/201905/t20190530_160959.html. Accessed 30 May 2019

10. Farsiu, Sina, et al.: Quantitative classification of eyes with and without intermediate age-related macular degeneration using optical coherence tomography. Ophthalmology **121**(1), 162–172 (2014)

11. Goldbaum, M., Kermany, D., Zhang, K.: Labeled Optical Coherence Tomography (OCT) and Chest X-Ray Images for Classification. https://data.mendeley.com/datasets/rscbjbr9sj/2

12. Jahromi, M.K., et al.: An automatic algorithm for segmentation of the boundaries of corneal layers in optical coherence tomography images using gaussian mixture model. J. Med. Signals Sens. **4**(3), 171 (2014)

13. Rasti, R., et al.: Macular OCT classification using a multi-scale convolutional neural network ensemble. IEEE Trans. Med. Imag. **37**(4), 1024–1034 (2017)

14. Gholami, P., et al.: OCTID: Optical Coherence Tomography Image Database. In: arXiv preprint (2018). arXiv:1812.07056

15. Mitry, D., et al.: Crowdsourcing as a novel technique for retinal fundus photography classification: analysis of images in the EPIC norfolk cohort on behalf of the UKBiobank eye and vision consortium. In: PloS One, 8(8), e71154 (2013)

16. Bogunović, H., et al.: Machine learning of the progression of intermediate age-related macular degeneration based on OCT imaging. Invest. Ophthalmol. Vis. Sci. **58**(6), BIO141–BIO150 (2017)

17. Lee, C.S., Baughman, D.M., Lee, A.Y.: Deep learning is effective for classifying normal versus age-related macular degeneration OCT images. Ophthalmol. Retina **1**(4), 322–327 (2017)

18. Venhuizen, F.G., et al.: Deep learning approach for the detection and quantification of intraretinal cystoid fluid in multivendor optical coherence tomography. Biomed. Optics Express **9**(4), 1545–1569 (2018)

19. Lee, H., et al.: Automated segmentation of lesions including subretinal hyperreflective material in neovascular age-related macular degeneration. Am. J. Ophthalmol. **191**, 64–75 (2018)

20. Treder, M., Lauermann, J.L., Eter, N.: Automated detection of exudative age-related macular degeneration in spectral domain optical coherence tomography using deep learning. Graefe's Arch. Clin. Exp. Ophthalmol. **256**(2), 259–265 (2018)

21. Schmidt-Erfurth, U., et al.: Machine learning to analyze the prognostic value of current imaging biomarkers in neovascular age-related macular degeneration. Ophthalmol. Retina **2**(1), 24–30 (2018)

22. Awais, M., et al.: Classification of sd-oct images using a deep learning approach. In: 2017 IEEE International Conference on Signal and Image Processing Applications (ICSIPA). IEEE, 2017, pp. 489–492 (2017)

23. Perdomo, O., et al.: Oct-net: a convolutional network for automatic classification of normal and diabetic macular edema using sd-oct volumes. In: 2018 IEEE 15th International Symposium on Biomedical Imaging (ISBI 2018), IEEE 2018, pp. 1423–1426 (2018)

24. Chan, G.C.Y., et al.: Fusing results of several deep learning architectures for automatic classification of normal and diabetic macular edema in optical coherence tomography. In: 2018 40th Annual International Conference of the IEEE Engineering in Medicine and Biology Society (EMBC). IEEE, pp. 670–673 (2018)

25. Vahadane, A., et al.: Detection of diabetic macular edema in optical coherence tomography scans using patch based deep learning. In: 2018 IEEE 15th International Symposium on Biomedical Imaging (ISBI 2018), IEEE. 2018, pp. 1427–1430 (2018)

26. Kamble, R.M., et al.: Automated diabetic macular edema (DME) analysis using fine tuning with inception-Resnet-v2 on OCT Images. In: 2018 IEEE-EMBS Conference on Biomedical Engineering and Sciences (IECBES). IEEE, pp. 442–446 (2018)

27. Sengupta, S. et al.: Ophthalmic Diagnosis and Deep Learning–A Survey. In: arXiv preprint (2018). arXiv:1812.07101

28. Wang, J., et al.: SD Net: Joint Segmentation and Diagnosis Revealing the Diagnostic Significance of Using Entire RNFL Thickness in Glaucoma (2018)

29. Fu, H., et al.: Multi-context deep network for angle-closure glaucoma screening in anterior segment OCT. In: International Conference on Medical Image Computing and Computer-Assisted Intervention. Springer, pp. 356–363 (2018)

30. Asaoka, R., et al.: Using deep learning and transfer learning to accurately diagnose early-onset glaucoma from macular optical coherence tomography images. Am. J. Ophthalmol. **198**, 136 145 (2019)

31. An, G., et al.: Glaucoma diagnosis with machine learning based on optical coherence tomography and color fundus images. J. Healthcare Eng. **2019**, 4061313 (2019)

32. Wei, L., et al.: Deep learning-based automated classification of multi- categorical abnormalities from optical coherence tomography images. Transl. Vis. Sci. Technol. **7**(6), 41–41 (2018)

33. Girish, G.N., et al.: Segmentation of intra-retinal cysts from optical coherence tomography images using a fully convolutional neural network model. IEEE J. Biomed. Health Inform. **23**(1), 296–304 (2018)

34. Guan, L., Yu, K., Chen, X.: Fully automated detection and quantification of multiple retinal lesions in OCT volumes based on deep learning and improved DRLSE. In: Medical Imaging 2019: Image Processing, vol. 10949. International Society for Optics and Photonics, p. 1094933 (2019)

35. Masood, S., et al.: Automatic choroid layer segmentation from optical coherence tomography images using deep learning. Sci. Rep. **9**(1), 3058 (2019)

36. Goodman, B., Flaxman, S.: European union regulations on algorithmic decision-making and a right to explanation. AI Magazine **38**(3), 50–57 (2017)

37. Dou, Q., et al.: Unsupervised cross-modality domain adaptation of convenets for biomedical image segmentations with adversarial loss. In: arXiv preprint (2018). arXiv:1804.10916

An Improved Spectral Clustering Algorithm Based on Cell-Like P System

Zhe Zhang and Xiyu Liu[✉]

Business School, Shandong Normal University, Jinan, China
zaq1230123@163.com, sdxyliu@163.com

Abstract. When using spectral clustering algorithm to perform clustering, there are some shortcomings, such as slow convergence rate, and the clustering result is easily affected by the initial center. In order to improve this problem, this paper proposes an improved spectral clustering algorithm based on cell-like P system, called SCBK-CP algorithm. Its main idea is to use the bisecting k-means algorithm instead of k-means algorithm and construct a cell-like P system as the framework of the bisecting k-means algorithm to improve the spectral clustering algorithm. The maximum parallelism of the P system improves the efficiency of the bisecting k-means algorithm. The algorithm proposed in this paper improves the clustering effect of spectral clustering, and also provides a new idea for the application of membrane computing. The SCBK-CP algorithm uses three UCI datasets and an artificial dataset for experiments and further comparison with traditional spectral clustering algorithms. Experimental results verify the advantages of the SCBK-CP algorithm.

Keywords: Spectral clustering · P systems · Bisecting K-means · SCBK-CP algorithm

1 Introduction

Membrane computing (P system) is a computational model abstracted by the Păun [1] based on the cellular structure and function of the organism. It has the characteristics of maximum parallelism and uncertainty. The P system can be broadly classified into three types, cell-like P systems, tissue-like P systems, and neural-like P systems. These computational models have proven to have powerful computing power and high efficiency. Since the introduction of membrane computing, it has received extensive attention from scholars in related fields, and many membrane computing variants have been proposed to solve different problems. Păun [2] gave a review of the current progress in membrane computing. Rudolf et al. [3] investigated the computing power of tissue-like P systems with states associated with the links (synapses) between cells. Pan [4] proposed spiking neural P systems with astrocytes. In recent years, membrane computing has begun to apply to clustering problems. Zhao et al. [5] proposed a special P system whose rules are used to represent improved k-medoids algorithms. Peng [6] applied evolutionary communication P system to solve fuzzy clustering problem. In [7] and [8], two computational models based on P system have been proposed to solve the problem of automatic clustering. Using P system as a framework, Liu et al. [9]

© Springer Nature Switzerland AG 2019
D. Milošević et al. (Eds.): HCC 2019, LNCS 11956, pp. 626–636, 2019.
https://doi.org/10.1007/978-3-030-37429-7_64

proposed a consensus clustering algorithm based on K-medoids. Gong [10] combines the MST clustering algorithm with the membrane calculation to effectively improve the quality of the cluster. Based on the idea of multi-layer framework of cells, Liu [11] proposed the chain structure of membrane system and improved the performance of clustering algorithm. Liu [12] used the combination of organizational P system and Hopfield networks to cluster. In [13], a membrane clustering algorithm was proposed to deal with automatic clustering problem.

Spectral clustering is a clustering algorithm evolved from graph theory. By dividing the undirected graph into different subgraphs, the similarity between the subgraphs is as small as possible. Between the subgraphs and the subgraphs, the similarity is as large as possible to complete the clustering. Unlike the most commonly used k-means algorithm, spectral clustering is able to divide datasets of any shape and converge to global optimality. Its basic idea is to use the Laplacian matrix corresponding to the data to perform feature decomposition to obtain feature vectors to achieve the purpose of clustering. In 2000, according to the theory of spectral partitioning, Shi [14] proposed a standard cut set criterion based on 2-way partitioning (Normalized Cut). Hagen and Kahng [15] proposed the proportional secant objective function (Ratio Cut). Andrew Y. Ng et al. [16] presented a simple spectral clustering algorithm that can be implemented using a few lines of Matlab. Francis [17] derived a new cost function for spectral clustering based on a measure of error between a given partition and a solution of the spectral relaxation of a minimum normalized cut problem. The paper [18] deals with a new spectral clustering algorithm based on a similarity and dissimilarity criterion by incorporating a dissimilarity criterion into the normalized cut criterion. The traditional spectral clustering algorithm usually uses the k-means algorithm to cluster the data sets corresponding to the constructed feature matrix after dimension reduction. However, the k-means algorithm is sensitive to the choice of the initial center point, and the poor selection of the initial center may cause the result to fall into local optimum.

In order to improve this problem, this paper proposes an improved spectral clustering algorithm based on cell-like P system, called SCBK-CP algorithm. Its main idea is to use the bisecting k-means algorithm instead of k-means algorithm and construct a cell-like P system as the framework of the bisecting k-means algorithm to improve the spectral clustering algorithm. The maximum parallelism of the P system improves the efficiency of the bisecting k-means algorithm.

The content of this paper is arranged as follows. The second part introduces the basic knowledge and steps of P system and spectral clustering. The third part proposes a hybrid algorithm that combines spectral clustering and bisecting k-means effectively, and the second half of the algorithm is placed in the P system to improve computational efficiency. The fourth part is used to show the experimental results and compare with other algorithms. The conclusion will be presented in the fifth part.

2 Cell-Like P Systems and Spectral Clustering

2.1 Cell-Like P Systems

In the cell-like P systems, the outermost membrane is called the skin. A membrane that does not contain other membranes is called the elementary. The space delimited by the membrane is called the region. Each region contains a multiset of objects and a set of rules.

A cell-like P system with an output set of objects based on evolution and communication rules is formally defined as follows:

$$\prod = (O, \mu, w_1, \cdots, w_n, R_1, \cdots, R_n, i_0)$$

Where

1. O is the alphabet of objects;
2. μ is a membrane structure consisting of n membranes with membranes injectively labeled with $1, 2, \cdots, n$; $n \geq 1$ is denoted as the degree of system $\prod$;
3. $w_i, 1 \leq i \leq m$, represents the strings that exist in the region i in the initial state, and these strings represent the multisets on the alphabet O;
4. $R_i, 1 \leq i \leq m$, is a finite set of evolution rules on the alphabet O; R corresponds to the set of evolution rules of the region i in μ;

We convert the calculation state of the membrane system into the transition between the patterns, and a series of patterns constitute a calculation. The calculation process of the membrane calculation starts from the initial pattern, and the calculation process is a process from one pattern to another. If there is a pattern without any rules to use, the pattern is called the termination pattern, that is, the system is halting, and the final result of the output in the shutdown system is the calculation result of the membrane calculation model (Fig. 1).

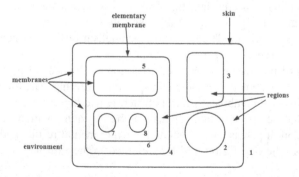

Fig. 1. The membrane structure of a cell-like P systems

2.2 Spectral Clustering

Spectral clustering is an algorithm that evolved according to graph theory and has been widely used in clustering. Its main idea is to regard the data in the dataset $X = \{x_1, x_2, \cdots, x_n\}$ as the points in the space. These points are connected by edges to form an undirected graph with weights $G = (V, E)$. $V = \{x_i | 1 < i < n\}$ is a vertex corresponding to n data. $E = \{w_{ij} | 1 < i, j < n\}$ is an edge with weights, and the weight represents the similarity between x_i and x_j. By cutting the undirected graph, the sum of the weights between the cut subgraphs is as small as possible, and the sum of the weights of the edges included in the subgraph is as large as possible, thereby achieving the purpose of clustering.

Typically, a Gaussian kernel function is used to construct the adjacency matrix W,

$$w_{ij} = \exp\left(-dist\left(x_i, x_j\right)/2\sigma^2\right) \tag{1}$$

The degree matrix D is a diagonal matrix $(d_{ii} = \sum_{j=1}^{n} w_{ij})$, $D = diag(d_i)$. The Laplacian matrix L can be obtained from the NCUT. The first k largest eigenvectors of the Laplacian matrix L are calculated to form a new feature matrix U. And normalize U to get a new matrix Y. Each row of the Y matrix is treated as a single data, and clustering is performed using k-means or other clustering methods to obtain k clusters.

3 Improved Spectral Clustering Algorithm Based on Cell-Like P System

In the ordinary spectral clustering algorithm, the last step is usually clustered using the k-means algorithm. However, the k-means algorithm is sensitive to the choice of the initial center, and the poor selection of the initial center may cause the result to fall into local optimum. In this paper, we choose the bisecting k-means algorithm to replace the k-means algorithm in spectral clustering, in order to weaken the influence of the initial center selection on the clustering results. The basic idea of the bisecting k-means algorithm is to split all sets of points into two clusters, and then select one of the clusters to continue splitting according to the sum of squared error (*SSE*) values, and repeat the process until the specified k clusters are generated. Among them, the error square sum *SSE* is defined as:

$$SSE = \sum_{i=1}^{n} \|x_i - u_i\|^2 \tag{2}$$

Where x_i is the data point within the cluster and u_i is the centroid point of the cluster. The *SSE* is used to calculate the squared difference from each point in the cluster to the centroid point of the cluster.

Since the bisecting k-means algorithm process needs to continuously perform k-means and calculate the *SSE* values associated with each cluster, parallel computing can save a lot of time. P system has the characteristics of maximum parallelism, so the entire bisecting k-means algorithm is placed in a cell-like P system.

3.1   Object Initialization

Membrane Structure. The membrane structure is as follows:

$$\prod = (\sigma_1, \sigma_2, \cdots, \sigma_i, r_0, r_1, \cdots r_i, i_0)$$

Where σ_i is the object in the ith cell, r_i is the rule in the ith cell, i_0 indicates the environment, when $i_0 = 0$, it indicates the output environment of the system. The P system contains K elementary, K is the number of classifications. This P system has a two-layer membrane structure, the basic membrane label is from 1 to k. The skin labeled 0 is used as the output environment (Fig. 2).

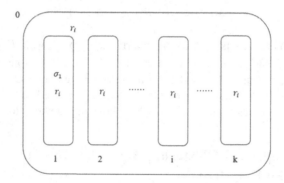

Fig. 2. Initial state of the membrane structure

Initial Objects. Initially only the first cell has object σ_1, and object σ_1 refers to a new data set consisting of the first K largest eigenvectors of the normalized Laplacian matrix. Initially, σ_1 is treated as a cluster.

3.2   Description of Rules

For each cell, when an object is present in the cell, the rule r_1 will be executed to generate an object σ_0 to be sent to the membrane 0, and the number of objects σ_0 in the membrane 0 represents the number of current clusters. When the number of objects σ_0 in the membrane 0 reaches the required number of classifications, the system is stopped. Otherwise, rule r_2 is executed. For objects (that is, a cluster) in each cell, rule r_2 is used to calculate the SSE value of this cluster that is divided into two clusters by using k-means, denoted as SPSSE, and calculate the SSE of other undivided clusters. The value denoted as NOSPSSE, the total SSE value corresponding to the cluster is SPSSE + NOSPSSE. All the clusters are selected in turn, and the total SSE value corresponding to each cluster is calculated and sent to the cell membrane 0. Rule r_3 in membrane 0 is used to select a cluster that specifies a minimum total SSE value. Rule r_4

then uses the k-means algorithm to divide the selected clusters, and one of the two clusters produced after the partitioning is retained in the original cells and one is sent to the other cells. When an object already exists in the cell, rule r_5 is used to prohibit accepting objects sent by other cells (Fig. 3).

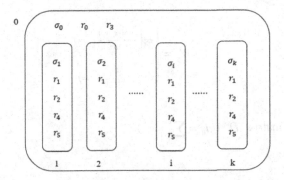

Fig. 3. Termination state of P system

3.3 Halting

In the whole system, the rule r_0 is continuously used to judge the number of cells to determine whether the number of classifications is reached. When the required number of classifications is reached, the system stops. At this time, there is one cluster in each cell, and there is a total of k clusters in the whole system. These clusters are the clustering results obtained by the whole algorithm.

3.4 Description of SCBK-CP Algorithm

As mentioned above, the bisecting k-means algorithm is used as the second part of the SCBK-CP algorithm, and a P system is constructed as a framework, which effectively improves the efficiency of the algorithm. The complete SCBK-CP algorithm process is as follows:

SCBK-CP algorithm

1.Use the Gaussian kernel function to construct the similar matrix W .

$$
W_{ij} = \begin{cases} \exp\left(-\dfrac{dist\left(x_i, x_j\right)}{2\sigma^2}\right), & i \neq j \\ 0, & i = j \end{cases}
$$

2. Degree matrix D ($d_{ii} = \sum_{j=1}^{n} w_{ij}$).

3. Construct a normalized symmetric Laplacian matrix L , $L = D^{-\frac{1}{2}} W D^{-\frac{1}{2}}$.

4. Calculate the feature vector v corresponding to the first k largest eigenvalues of L , and construct the feature matrix U .

5. Normalize the feature matrix U to obtain a normalized matrix Y , which contains n points in space reduced to k dimensions.

6. Clustering the n points corresponding to the Y matrix with the improved bisecting k-means algorithm based on P system.

7. Consider all data points as a cluster.

8. For each cluster, calculate the sum of the corresponding SSE value after dividing the cluster and the SSE value of other undivided clusters at this time, that is, the total error.

9. Use the k-means algorithm for clustering the cluster with the smallest total error.

10. Repeat steps 8 and 9 until k clusters are divided and the algorithm ends.

4   Experimental Results and Analysis

4.1   Data Set

In order to evaluate the performance of the SCBK-CP algorithm proposed in the paper, three UCI datasets and an artificial dataset were used for the experiment. The specific information of the data set is shown in the following table (Table 1).

Table 1. Data sets information

Data Sets	Objects	Attributes	Classes	Source
Iris	150	4	3	UCI
Wine	178	13	3	UCI
Ecoli	327	7	5	UCI
Twocircle	314	2	2	Artificial

As a contrast, this paper uses k-means algorithm and ordinary spectral clustering algorithm to experiment on the same data sets. The experimental results will be compared together. In order to ensure the validity of the results, all experiments are carried out in the same system.

4.2   Evaluation Index

In this paper, Purity Index (*PUR*) is used as the evaluation index, and its calculation formula is as follows:

$$\rho_{PUR} = \frac{1}{N} \sum_{i=1}^{k} \max_{j} |v_i \cap u_j| \tag{3}$$

4.3   Experimental Result

In this paper, the k-means algorithm, the spectral clustering algorithm and the SCBK-CP algorithm proposed in this paper are used on three UCI datasets and an artificial dataset respectively. In order to obtain a result that can reflect the average level of each algorithm, each algorithm runs 100 times for each data set. In this paper, Purity Index (*PUR*) is used as an evaluation index to test the accuracy of clustering of different algorithms. The experimental results of UCI dataset are as follows (Table 2):

Table 2. Experimental result

Purity	k-means	Spectral clustering	SCBK-CP
Iris	0.893	0.893	0.927
Wine	0.653	0.674	0.692
Ecoli	0.728	0.741	0.823

The experimental results of the artificial data set are shown in the figure (Figs. 4, 5, 6, 7):

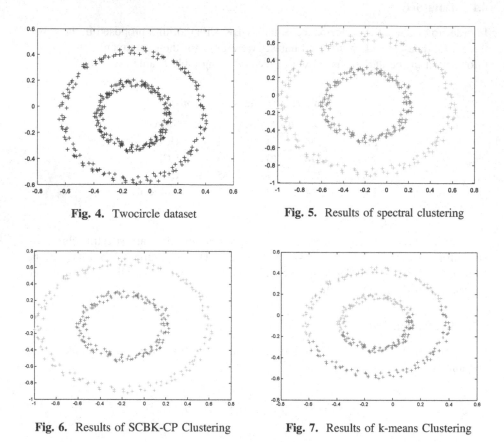

Fig. 4. Twocircle dataset   **Fig. 5.** Results of spectral clustering

Fig. 6. Results of SCBK-CP Clustering   **Fig. 7.** Results of k-means Clustering

It can be seen that in the data set under test, the SCBK-CP algorithm clustering accuracy is slightly higher than the k-means algorithm and the spectral clustering algorithm for the Iris dataset and Wine dataset. For the Ecoli dataset, the clustering accuracy of the SCBK-CP algorithm is significantly higher than that of the k-means algorithm and the spectral clustering algorithm. From the experimental results of the artificial dataset, it can be seen that the spectral clustering algorithm and the SCBK-CP algorithm are better than the k-means algorithm in clustering the circular dataset.

5   Conclusion

In this paper, we propose an improved spectral clustering algorithm based on P system, called SCBK-CP algorithm. It utilizes the maximum parallel features of the P system to cluster the new datasets after spectral clustering and dimension reduction by using bisecting k-means. On the one hand, it weakens the influence of the initial centroid selection on the clustering results. On the other hand, the characteristics of maximum parallel computation of bisecting k-means algorithm coincide with the characteristics of maximum parallel of P system., which can greatly improve the computational efficiency. The experimental results show that the SCBK-CP algorithm proposed in this paper is superior to the k-means algorithm and the ordinary spectral clustering algorithm in both the UCI dataset and the artificial dataset. Since there are more and more researches on P systems, many P system models with different functions have been proposed. In the future, we will further discuss the research on the P system model and the method of applying it to the other algorithm.

Acknowledgment. Project is supported by National Natural Science Foundation of China (61472231, 61502283, 61876101, 61802234, 61806114), Social Science Fund Project of Shandong Province, China (16BGLJ06, 11CGLJ22), Postdoctoral Project, China(2017M612339).

References

1. Păun, G.: Computing with membranes. J. Comput. Syst. Sci. **61**(1), 108–143 (2000)
2. Păun, G.: Introduction to membrane computing. J. Logic Algebraic Program. **79**(6), 1–42 (2006)
3. Freund, R., Păun, G., Pérez Jiménez, M.J.: Tissue-like P systems with channel-states. Theoret. Comput. Sci. **330**(1), 101–116 (2005)
4. Pan, L., Wang, J., Hoogeboom, H.J.: Spiking neural P systems with astrocytes. Neural Comput. **24**(3), 805–825 (2014)
5. Zhao, Y., Liu, X., Qu, J.: The K-medoids clustering algorithm by a class of P system. J. Inf. Comput. Sci. **9**(18), 5777–5790 (2012)
6. Peng, H., Wang, J., Pérez-Jiménez, M.J., et al.: An unsupervised learning algorithm for membrane computing. Inf. Sci. **304**, 80–91 (2015)
7. Peng, H., Wang, J., Shi, P., Pérez-Jiménez, M.J., Riscos-Núñez, A.: An extended membrane system with active membrane to solve automatic fuzzy clustering problems. Int. J. Neural Syst. **26**(3), 1–17 (2016)
8. Peng, H., Wang, J., Shi, P., et al.: An automatic clustering algorithm inspired by membrane computing. Pattern Recogn. Lett. **68**, 34–40 (2015)
9. Liu, X., Zhao, Y., Sun, W.: K-medoids-based consensus clustering based on cell-like p systems with promoters and inhibitors. Commun. Comput. Inf. Sci. **681**, 95–108 (2016)
10. Gong, P., Liu, X.: An improved MST clustering algorithm based on membrane computing. In: Zu, Q., Hu, B. (eds.) HCC 2017. LNCS, vol. 10745, pp. 1–12. Springer, Cham (2018). https://doi.org/10.1007/978-3-319-74521-3_1
11. Xiyu, L., Zhenni, J., Yuzhen, Z.: Chain membrane system and the research progress of direct (indirect) membrane algorithm in clustering analysis. J. Anhui Univ. (Nat. Sci. Ed.) **3**, 3 (2018)
12. Liu, X., Xue, J.: A cluster splitting technique by hopfield networks and tissue-like P systems on simplices. Neural Process. Lett. **46**(1), 171–194 (2017)
13. Graciani, C.: An automatic clustering algorithm inspired by membrane computing. Pattern Recogn. Lett. **68**(P1), 34–40 (2015)
14. Shi, J., Malik, J.: Normalized cuts and image segmentation. IEEE Trans. Pattern Anal. Mach. Intell. **22**(8), 888–905 (2000)

15. Hagen, L., Kahng, A.B.: New spectral methods for ratio cut partitioning and clustering. IEEE Trans. Comput. Aided Des. Integr. Circuits Syst. **11**(9), 1074–1085 (1992)
16. Weiss, Y.: On spectral clustering: analysis and an algorithm. Proc. Nips **14**, 849–856 (2001)
17. Bach, F.R., Jordan, M.I.: Learning spectral clustering. Neural Inf. Process. Syst. **16**(2), 305–312 (2003)
18. Wang, B., Zhang, L., Wu, C., et al.: Spectral clustering based on similarity and dissimilarity criterion. Pattern Anal. Appl. **20**(2), 495–506 (2017)

Diagnosis of Depression Based on Short Text

Jinghua Zheng[1], Jianli Bian[1($\boxtimes$)], and Jincheng Jia[2]

[1] National University of Defense Technology, Hefei 230037, China
zhengjhl001@163.com
[2] Institute of Remote Sensing Information, Beijing, China

Abstract. Depression has become a serious global disease which can harm to mental health. Diagnosis of depression is a complicated process, and needs a great deal of specialist knowledge. A person's social network behavior can reflect his/her real psychological. The traditional process of diagnosing depression requires not only close coordination of depressed people but also too much professional knowledge. And depressed people maybe have fewer and fewer social activities. In this paper, machine learning is used to diagnosis whether a person has depression. We use a SVM algorithm to create the depression diagnosis model based on short text. And we validate this model using a data set of Sina Micro-blog users' Micro-blog content and depression label data gotten by certificate of diagnosis. Our work makes three important contributions. Firstly, we show how to diagnosis depression based on short text published on the social network. Secondly, we build diagnosis model of depression based on SVM. And last and thirdly, we give the experimental results that validate our method, the accuracy of whether or not depression can reach 93.47%.

Keywords: Depression prediction · Sina Micro-blog · Machine learning · Prediction accuracy

1 Introduction

Depression has become the serious global public-health issue which can lead disease burden. World health Organization (WHO) predicted that depression would be the most serious mental health disease in our world. It declared that the number of depression has reached 322 million. And the number of patients increased by 18.4% between 2005 and 2015. Depression is a state of low mood to activity that can affect a person's thoughts, behavior, ideas, feelings and healthy. It may include feelings of sadness, anxiety, emptiness, hopelessness, worthlessness, guilt, irritability, or restlessness. Depression is characterized by low mood, slow thinking, and few words. But the medical understanding of depression is not so profound. Researchers have concluded that brain electrical signals of the depressed people is different from healthy people in the indication of rhythm, waveform amplitude and power spectrum [1]. So it is necessary to rely on drugs to recover from his/her depression. Depression can be easily cured with appropriate psychological guidance and drugs in the early stages. However, it is so annoying that depression is very hard to be found in the early depression stages so that we do not pay more attention to it.

© Springer Nature Switzerland AG 2019
D. Milošević et al. (Eds.): HCC 2019, LNCS 11956, pp. 637–646, 2019.
https://doi.org/10.1007/978-3-030-37429-7_65

Depression people have the highest suicide rate, with more than 200,000 people committing suicide every year in China, more than half of which are related to depression [2].

With the development of Internet, much research has been done to identify the relationship between social network and psychological activity, for example depression prediction [3–13]. Rosenquist et al. [9] showed that depression could be spread between friends and neighbors and depressed patients were often in a small closed social network. Wang et al. [6] proposed their improved depression prediction model by the connection relationship based on the idea that the law of depression spread could improve the performance of prediction. Qiu et al. [8] created a depression emotional inclination recognition model based on the college students depression problem hidden in micro-blog text. The experiment on Twitter data shows the recognition rate of their model could get 83.82%. Wang et al. [10] proposed a sentiment analysis method based on Micro-blog content, interactions with others and users behaviors. And a depression detection model was constructed using this method, whose precision are all around 80%. Cohn et al. [11] compared clinical diagnosis of depression with facial actions and vocal prosody in depression patients, they created the classifier model using SVM and Logistical Regression. Reece et al. [13] extracted some statistical features from 43950 Instagram photos of 166 individuals to identify markers of their depression with machine learning. They showed that it is helpful to diagnose depression for general practitioners.

Yang et al. [7] designed a hybrid depression classification and depression estimation framework from audio, video and text. This reference first used deep learning to predict depression. In this study, the Deep Convolutional Neural Network (DCNN) was created to predict depression based on 6902 dimensional feature of video and audio, and the result of their model was the score of Patient Health Questionnaire (PHQ-8). Their experiments were carried on public dataset, showing their depression recognition model obtained promising result.

The traditional process of diagnosing depression requires not only close coordination of depressed people but also too much professional knowledge. It is impossible to complete the diagnosis process without a point. We can see the research about depression prediction based on social network is getting more and more attention. Some studies are using simple machine learning or statistical methods to predict directly individual depression using their social network data, such as text, photo or audio and video. Some studies use so complex models, such as deep learning. For simple machine learning model, feature selection is the most critical process. And for deep learning, the result of predictions is not the best. There is a problem: most depressed people maybe have fewer and fewer social activities. So it may be very hard to collect the training data. But we find some depressed people like to express his/her depressed emotion in short text, such as "If I jump off the building now, is it free?". Our goal is to create a diagnosis model of depression using the short text from Sina Micro-blog.

In this paper, we extract depression feature of Sina Micro-blog users with Micro-blog data. Sina Micro-blog is currently the largest social platform in China. By June 2018, the number of monthly active users has increased to 431 million and the number of active users has increased to 190 million. At present, the content of Sina Micro-blog has exceeded 100 billion. So Sina Micro-blog has become the object of academic research on which people can enjoy expressing feelings and releasing information through texts, imagines, voice and even videos. The information contains a mass of emotional. These emotional information from depressed user is different from healthy user. And the text content from depressed user is found to be largely different from healthy user. So it is feasible to infer depression based on Micro-blog short text.

Hence, We propose the depression diagnose model based on short text. And the paper makes three important contributions:

(1) We get the labeled data of depression manually. The label is based on the fact that some depressed patients provide hospital diagnostic certificates.
(2) We extract the 102 psychology features using the TextMind system researched by Institute of psychology, CAS, and created a depression diagnosis model based on user's short text.
(3) The accuracy of depression inferring reaches 93.47%. The rest of this paper is organized as follows: In Sect. 2, we will introduce the data and method. In Sect. 3, we will explicate our dataset and experiment, and give the experiment results followed by a discussion about the results. Finally, we will make conclusion in Sect. 4.

2 Data and Method

2.1 Data Processing and Feature Extraction

Our experiment will use the dataset of Sina-blog users related to depression label, and their Micro-blog nickname includes the word depression. We crawled 449932 posts from 316 sample data of Sina Micro-blog users by Internet worm and API which Sina Micro-blog provided. And we labeled manually 16 depression users who provide hospital diagnostic certificates.

First, we filter valid data and the screening method is to determine the active user whose status is greater than 10 within one year. We filter out users who have not post a Micro-blog in the first three months of our collection.

Then, we preprocess the text of Micro-blog. Filtering all Micro-blog text which do not meet the requirement, i.e., whose content is empty, just only hyperlinks, or pictures, or video contents.

Then, Micro-blog text features are extracted. We extract 102 dimension text features using the TextMind system (http://ccpl.psych.ac.cn/textmind/), including first/second/third person, positive/negative emotional words, word expressing anxiety, words indicating physiological history, psychological words, words showing love, @ numbers, and so on. The TextMind system is a Chinese language psychological analysis system developed by Computational Cyber Psychology Lab, Institute of Psychology, Chinese Academy of Sciences. The dictionary, text and punctuation used in TextMind are optimized to Simplified Chinese, and the categories are compatible to LIWC.

And finally, we preprocess the extracted feature data including data cleaning, missing value processing, and data combinations. Random forest algorithm is used to fit missing values. But the dataset feature vectors are sparse. We combine the dataset according to different time.

We got the Pearson correlation coefficient between feature and depression using SPSS. Table 1 gives us the Pearson correlation coefficient, where * shows the correlation on 0.05 and ** on 0.01.

Table 1. The Pearson correlation coefficient between feature and gender

Feature	Gender	Feature	Gender	Feature	Gender
Function person	$-.072^{**}$	Sentiment	$-.037^{**}$	Time	$-.022^{**}$
Pronoun	$-.042^{**}$	Positive emotion	$-.045^{**}$	Leisure	$.009^{*}$
Specific personal pronouns	$-.063^{**}$	Negative emotion	$.024^{**}$	Home	$-.024^{**}$
First person single	$-.067^{**}$	Anxiety	$.010^{**}$	Money	$.026^{**}$
First person plural	$.016^{**}$	Psychology	$.010^{**}$	Death	$-.032^{**}$
Second person	$-.039^{**}$	Angry	$-.014^{**}$	Assent	$-.115^{**}$
Third person single	$-.014^{**}$	Sadness	$.032^{**}$	Pause	$-.058^{**}$
Not specific personal pronouns	$.037^{**}$	Cognition	$-.044^{**}$	Filler	$-.014^{**}$
Verb	$-.051^{**}$	Disparity	$-.019^{**}$	@ count	$-.145^{**}$
Aux verb	$-.033^{*}$	Inhibition	$-.020^{**}$	Topic count	$-.034^{**}$
Adverb	$-.043^{**}$	Inclusive	$-.023^{**}$	Love	$-.008^{*}$
Quantifier	$-.025^{**}$	Exclusive	$-.014^{*}$	Comma	$.065^{**}$
Number	$-.033^{**}$	Perceptive	$-.025^{**}$	Colon	$-.245^{**}$
Special words	$-.009^{*}$	Sight	$-.010^{**}$	Semicolon	$-.036^{**}$
Quanty unit	$-.029^{**}$	Hearing	$-.030^{**}$	Question mark	$.102^{**}$
Token word	$-.098^{**}$	Feeling	$-.021^{**}$	Exclamation	$-.089^{**}$
Multi-function	$-.058^{**}$	Physiology	$-.009^{*}$	Dash	$-.080^{**}$
Tense	$-.071^{**}$	Body	$-.040^{**}$	Quote	$.013^{**}$
Present tense	$-.017^{**}$	Health	$.016^{**}$	Abbreviation	$.030^{**}$
Future tense	$-.020^{**}$	Feeding	$.010^{**}$	Brackets	$-.034^{**}$
Continue	$-.075^{**}$	Relative	$-.034^{**}$	Other punctuation	$-.111^{**}$
Family	$-.032^{**}$	Motion	$-.015^{**}$	Words count	$-.127^{**}$
Humans	$.038^{**}$	Space	$-.025^{**}$		

**the correlation is marked on 0.01, *the correlation is marked on 0.05.

As can be seen from Table 1 that depression is positively related to words expressing negative emotional, anxiety, sad and healthy, psychology, money, and non-specific personal pronoun, comma, question mark etc. And depression is negatively correlated with words expressing tense, continue, family, angry, positive emotion and so on. Depression people are often in a bad emotional state, preferring to express his depressed mood with negative energy words, and have suicidal tendencies.

2.2 Machine Learning Framework

SVM (Support vector machine) is a supervised learning model commonly used for pattern recognition, classification and regression analysis. SVM is a good method for large-scale training samples. And SVM can map the sample space into a feature space of high-dimensional or even infinite dimension (Hilbert space) through a kernel function, so that the problem of nonlinear separable problem in the original sample space is transformed into linear separable problem in the new high dimensional feature space.

There are four commonly used kernel functions.

① Linear kernel function: $K(x, y) = x * y$

② Polynomial kernel function: $K(x, y) = [(x * y) + 1]^d$

③ RBF kernel function: $K(x, y) = \exp(\frac{-|x-y|^2}{d^2})$

④ Sigmoid kernel function: $K(x, y) = \tanh(a(x * y) + b)$

The Linear kernel function is mainly used in the linear separable training samples. The parameters are small and the speed is fast. The RBF kernel function is mainly used in the linear inseparability training samples. There are many parameters, and the classification result is dependent on the parameters. The polynomial kernel function can map the low-dimensional space to the high-dimensional feature space. But there are so many parameters. With the increase of the polynomial order, the computational complexity is too large to calculate. For Sigmoid kernel function, SVM is equivalent to multi-layer neural network.

3 Experiments and Comparison

3.1 Data Acquisition

Our experiment will use the dataset of Sina Micro-Blog Users including 102 psychology features, such as positive/negative emotional, psychological words etc. (extracted by the TextMind system) and depression feature (1 represents depression and 0 represents no) by hand-labeled.

After data processing, we get 39200 posts from 267 Sina Micro-blog uses in which 140 people over 25 years old, accounting for 52.4%. There are 130 men, accounting for 48.7%. And the average number of posts is 1685. 170 people have received university education or above, account for 63.7%. Therefore the practical dataset is infective. The general info about these subjects is described in Table 2.

Table 2. Descriptive statistics of subjects (n = 267)

Variables		Samples	Percentage
Gender	Male	130	48.7%
	Female	137	51.3%
Age	<25	127	47.6%
	>=25	140	52.4%
Education	University or above	170	63.7%
	High school	34	12.7%
	Junior high school	11	4.1%
	No	52	19.5%
Number of posts	<=1685	196	73.4%
	>1685	71	26.6%

3.2   Diagnosis Model Based on SVM

Depression have some reflections in his language both in the real world and on the Internet. We try to learn this mapping of depression to language in social network. So we train the diagnosis model based on LIBSVM (an open source machine learning library). LIBSVM implements the SMO algorithm for kernelized SVMs, supporting classification and regression. The training steps are as follows:

The depression diagnosis model based on SVM
Input: X: training data matrix
y: label vector on depression
1. Process input data including formatting and scale
2. Select the RBF function as a kernel function
3. Select the best parameters pair <c, g> by cross validation
4. Train the model using the best parameters

Select the optimal parameter pair <c, g> by traversing. The shorter the step size, the more time it takes. Figures 1, 2 and 3 shows the parameter selection process for different step size.

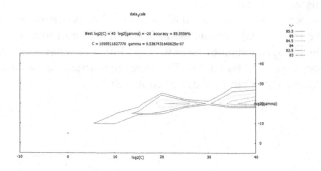

Fig. 1. The accuracy rate result of step size = 10

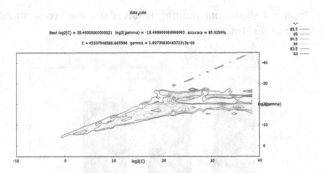

Fig. 2. The accuracy rate result of step size = 2

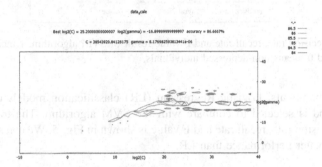

Fig. 3. The accuracy rate result of step size = 1

3.3 Result

Diagnose whether a person is depression using his/her Sina Micro-blog content is the goal of our experiment. We chose SVM algorithm as the classification inferring model. We use RBF kernel function and 10-fold cross validation to test the performance of the prediction mode.

Table 3. The training result of SVM algorithm

Step size	C	Gamma	Kernel function	Training time (s)	Accuracy
10	1099511627776	9.5e−7	Linear	2155.95	49.1875%
10	1099511627776	9.5e−7	RBF	4815.97	80.8949%
2	45337946588.7	2.7e−6	RBF	3245.21	93.0228%
1	38543920.8	8.1e−6	RBF	3210.58	93.4718%

Our experimental results are showed in Table 3 from different parameters and kernel functions. As seen from the Table 3, when we chose linear kernel function, a bad model is got. And there is a nice performance to depression diagnose when c = 38543920.8, g = 8.1e−6 and RBF kernel function is selected, and model training time is shorter. Figure 4 shows our training result about the precision rate, the recall rate and the F value at this time.

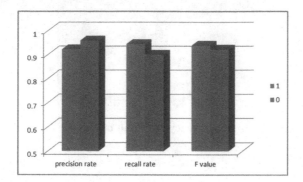

Fig. 4. The precision rate, recall rate and F value result of SVM algorithm. 1 means depressed individuals, and 0 means non-depressed individuals.

In our experiments, Logistic Regression (LR) classification models on the same dataset method is selected to compare with the SVM algorithm. The training result about the precision rate, recall rate and F value is shown in Fig. 5. We can see the SVM algorithm is better performance than LR.

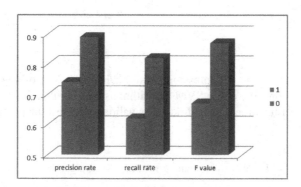

Fig. 5. The precision rate, recall rate and F value result of LR algorithm. 1 means depressed individuals, and 0 means non-depressed individuals.

The LR algorithm is simple and easy. We try to make a fuss about feature selection that is to remove some features that are strongly uncorrelated with depression. But the training result is still not good because of the class imbalance in our training samples.

3.4   Discussion

With the development of the Internet technology, people have transferred the real life of the physical world to virtual space. The expansion of diagnosis of depression is of great technical and social significance. Many researches used classic simple or deep machine learning, based on different social network characteristics, such as photo, audio, text, video or users social network behaviors. The precision in these studies is up to around 80% [8, 10]. The accuracy of whether or not depression can reach 93.47%. It is proved that the proposed method in this paper could effectively help for detecting depressed ones and preventing suicide in online social networks.

4   Conclusion

In this paper, we analyzed depression characteristics on Micro-blog users and created the depression diagnosis model based on short text. Experiment was conducted to demonstrate that behavior in the virtual world can also represent individual characteristics in reality. The Micro-blog text can also predict users' mental health, such as depression. Experiment was conducted to demonstrate the good performance of the traditional classification/regression machine learning by real Sina Micro-blog data.

There is a great application value to infer Micro-blog users' depression based on their public information. In clinical, the diagnosis of depression is also a complex undertaking. Our method may inspire the study about inferring depression based on social network. But there maybe two problems: (1) Depressed people maybe have fewer and fewer social activities with the development of depression, so it is very hard to collect the training data. (2) Depression is the accumulative effects of long time. So we think that the best way to cure the depression is through psychological intervention or drug treatment in the early stage or stages of depression. In the future, so we will do further work on depression pre-diagnosis. And we will look for better models on a larger scale to improve the accuracy of diagnosis models. We try to use deep learning method to create the prediction model of depression [14], and apply our method to a multilingual platform, such as Twitter, Facebook and so on.

Acknowledgement.   The author would like to thank the anonymous reviewers for their valuable comments. This work is supported by the National Natural Science Foundation of China (Grant Number: 61602491).

References

1. Lou, E.: Study on Feature Extraction and Classification of Melancholia EEG. Zhejiang Normal University, pp. 7–9 (2009)
2. Depression and Other Common Mental Disorders: Global Health Estimates (2017). http://apps.who.int/iris/bitstream/10665/254610/1/WHO-MSD. Accessed Feb 2017
3. Balani, S., DeChoudhury, M.: Detecting and characterizing mental health related self-disclosure in social media. In: Proceedings of the 33rd Annual ACM Conference Extended Abstracts on Human Factors in Computing Systems, pp. 1373–1378. ACM, Seoul (2015)

4. De Choudhury, M., Kiciman, E., Dredze, M.: Discovering shifts to suicidal ideation from mental health content in social media. In: Proceedings of the 2016 CHI Conference on Human Factors in Computing Systems, pp. 2098–2110. ACM, California (2016)
5. Hu, Q., Li, A., Heng, F.: Predicting depression of social media user on different observation windows. In: International Conference on Web Intelligence and Intelligence Agent Technology (WI-IAT), pp. 1361–1364. IEEE, Singapore (2016)
6. Wang, X., Zhang, C., Sun, L.: An improved model for depression detection in micro-blog social network. In: 13th International Conference on Data Mining Workshops, pp. 80–87. IEEE, Dallas (2013)
7. Yang, L., Sahli, H., Xia, X., Pei, E., Oveneke, M.C., Jiang, D.: Hybrid depression classification and estimation from audio video and text information. In: Proceedings of the 7th Annual Workshop on Audio/Visual Emotion Challenge, AVEC 2017, pp. 45–51. ACM, New York (2017)
8. Qiu, J., Gao, J.: Depression tendency recognition model based on college student's microblog text. In: Shi, Z., Goertzel, B., Feng, J. (eds.) ICIS 2017. IAICT, vol. 510, pp. 351–359. Springer, Cham (2017). https://doi.org/10.1007/978-3-319-68121-4_38
9. Rosenquist, J.N., Fowler, J.H., Christakis, N.A.: Social network determinants of depression. Mol. Psychiatry **16**(3), 273–281 (2010)
10. Wang, X., Zhang, C., Ji, Y., Sun, L., Wu, L., Bao, Z.: A depression detection model based on sentiment analysis in micro-blog social network. In: Li, J., et al. (eds.) PAKDD 2013. LNCS (LNAI), vol. 7867, pp. 201–213. Springer, Heidelberg (2013). https://doi.org/10.1007/978-3-642-40319-4_18
11. Cohn, J., Kruez, T.S., Matthews, I., et al.: Detecting depression from facial actions and vocal prosody. In: The 3rd International Conference on Affective Computing and Intelligent Interaction and Workshops, pp. 10–12. IEEE, Amsterdam (2009)
12. Yang, L., Jiang, D., He, L., Pei, E., Oveneke, M.C., Sahli, H.: Decision tree based depression classification from audio video and language information. In: Proceedings of the 6th International Workshop on Audio/Visual Emotion Challenge, AVEC 2016, pp. 89–96. ACM, Amsterdam (2016)
13. Reece, A.G., Danforth, C.M.: Instagram photos reveal predictive markers of depression. EPJ Data Sci. **16**(1), 1–15 (2017)
14. Hochreiter, S., Schmidhuber, J.: Long short-term memory. Neural Comput. **9**(8), 11735–11780 (1997)

An Improved CNN-Based Pneumoconiosis Diagnosis Method on X-ray Chest Film

Ran Zheng$^{(\boxtimes)}$, Kui Deng, Hai Jin, Haikun Liu, and Lanlan Zhang

National Engineering Research Center for Big Data Technology and System Services,
Computing Technology and System Lab Cluster and Grid Computing Lab,
School of Computer Science and Technology,
Huazhong University of Science and Technology, Wuhan 430074, China
zhraner@hust.edu.cn

Abstract. Pneumoconiosis is one of the most serious occupational diseases in China, which seriously endangers the health of most workers in dust environments. The diagnosis of pneumoconiosis is very complex and cumbersome, which relies mostly on doctor's medical knowledge and clinical reading experiences of X-ray chest film. Traditional image processing approach has helped doctors to reduce the misdiagnosis but with lower accuracy. An improved CNN-based pneumoconiosis diagnosis method on X-ray chest films is proposed to predict pneumoconiosis disease. The CNN structure is decomposed from 5×5 convolution kernel into two 3×3 convolution kernels to optimize the execution. Compared with GoogLeNet, the proposed GoogLeNet-CF achieves higher accuracy and gives a good result in the diagnosis of pneumoconiosis disease.

Keywords: Convolutional neural network · Deep learning · Pneumoconiosis · Auxiliary diagnosis

1 Introduction

Pneumoconiosis is the most serious occupational lung disease caused by the inhalation of dust, often in coal mines, metal and non-metal mines, and from agriculture [1]. It can not be terminated by medical treatment at present, and it is a disability disease. According to the report, there are about 20,000 new pneumoconiosis cases every year, which seriously endanger the health of most workers. The mortality rate of pneumoconiosis is as high as 22.04%, and the death rate of pneumoconiosis in coal mines is much higher than that of safety accidents in coal mines.

The diagnosis of pneumoconiosis is very complex and cumbersome. Doctors read X-ray chest films, referring to the history of dust exposure, epidemiological

Supported by the project of medical and health big data center from Hubei Provincial Development and Reform Commission.

investigation of pneumoconiosis, clinical manifestations, and laboratory examinations to give out the results, which mainly rely on doctors' medical knowledge and clinical experiences. However in China, medical resources are obviously not balanced, and there is a certain rate of misdiagnosis in many grass-roots medical institutions because of doctors' limited medical abilities, which causes a higher death rate of pneumoconiosis.

Traditional *computer-aided diagnosis* (CAD) method of pneumoconiosis is based on image processing technology. X-ray chest films are processed with various transformations to obtain features for the diagnosis of pneumoconiosis disease. It helps doctors to diagnosis the disease rapidly and reduce the rate of misdiagnosis, but long processing time and limited diagnosis accuracy are problems for this method. In the era of artificial intelligence, deep learning has been used in many fields, especially in the fields of medical and health. *Convolution Neural Network* (CNN) is a popular method of deep learning, which is mainly used in various medical image processing, such as colonic polyps detection [4], interstitial lung pattern classification [5], and pulmonary embolism detection [6], to assist doctors diagnosis some diseases. Therefore, CNN can be suitable for pneumoconiosis diagnosis. However, deep network suffers from huge parameters to cause more computation complexity. In this paper, we use CNN for efficient diagnosis of pneumoconiosis. We reconstruct our model with the idea of convolutional kernel decomposition to reduce computational complexity and improve diagnosis accuracy.

The rest of the paper is organized as follows. Section 2 describes related works about deep learning in medical imaging. Section 3 gives a description about motivation and consideration. Section 4 describes the design of image preprocessing and improved CNN approach. Section 5 exhibits the experimental results. Conclusions and future work are shown in Sect. 6.

2   Related Work

Traditional computer-aided diagnosis refers to the use of imaging and medical image processing technologies, combined with computer analysis and calculation, to assist the detection of lesions. In 1976, the first CAD system, MYCIN [2], was developed for the diagnosis of blood infection patients and the development of related antibiotic treatment programs. Raithel and Kraus [3] developed extraction, labeling, and classification of pneumoconiosis lesion feature points in chest films in 1997.

In recent years, deep learning is widely used in the field of computer vision, especially in medical imagings. Roth et al. [4] used CNN and random view aggregation to detect colonic polyps in colonoscopic CT images. Anthimopoulos et al. [5] used CNN on 2D chest CT images to classify lung pattern for interstitial lung disease. Tajbakhsh et al. [6] did more work on computer-aided pulmonary embolism detection in CT images with a novel vessel-aligned multi-plannar image representation and CNN method. DeepMind [7] cut into the field of early diagnosis of eye diseases. Liu et al. [8] proposed an artificial neural network to detect lung nodules in thoracic CT images. However, there are few deep learning methods used in the diagnosis of pneumoconiosis.

At present, there are a variety of deep learning frameworks, such as Caffe [9], Convnet [10], MXNet [11], and TensorFlow [12]. Currently, TensorFlow framework is most popular between researchers and organizations for deep learning implementations. The classical CNNs are LeNet, AlexNet, GoogLeNet, and ResNet.

LeNet [13] serves as a milestone in the beginning of CNN, which uses 32×32 size for digital classification. However, its performance cannot be further improved because of its inherent bottleneck. In 2012, the outstanding performance of AlexNet [14] in ImageNet competition opened the door of CNN in other application. AlexNet uses nonlinear activation function, ReLU, to prevent overfitting. In 2014, GoogLeNet network structure [15] was proposed to convert general convolution into sparse connection. That means, any $N \times N$ convolution can be used to replace $1 \times N$ convolution followed by $N \times 1$ convolution to build a larger network by simply amplifying inception structure with lower computational cost [17]. In 2015, ResNet model [16] was proposed which is a leap in performance in terms of the network depth and convergence acceleration. Our work is to find a suitable deep learning method for pneumoconiosis diagnosis.

3 Motivation and Consideration

Traditional image processing method of pneumoconiosis diagnosis is designed for the comparison with deep learning method. After image pre-processing of redundant noise removal and image segmentation, feature extraction is done to detect and quantify lesion area, and then classification operation is done to classify feature data into different classifications.

Then deep learning method is used to diagnosis pneumoconiosis. CNN and GoogLeNet are used as the learning algorithm and model. The compared experiments are done on 50 X-ray chest films to distinguish the differences between the two methods, shown in Table 1. We can find that the deep learning method is obviously more superior than traditional image processing method in terms of time and accuracy.

Table 1. The comparison of two different methods for pneumoconiosis diagnosis

	Diagnostic time	Diagnostic accuracy
Image processing method	4.6 min	83%
Deep learning method	1 min	91.3%

The reason is that convolutional layer handles the feature learning automatically, so that the accuracy is higher than traditional image processing method, and the diagnosis time is also shorter. We can conclude that deep learning method is better than traditional image processing method for pneumoconiosis diagnosis.

However, in so many classic neural network models, which one is suitable for pneumoconiosis diagnosis? In order to choose a best one, the experiments are done to test their effectiveness. The experimental data is shown in Table 2, in which the numbers after the word "Abnormal_" mean different disease stages of pneumoconiosis. The experimental images come from health examination data of workers in a coal mine. The volume of early pneumoconiosis (Abnormal_01) is much larger than healthy (Normal) and middle and late (Abnormal_02 and Abnormal_03) workers. This is because it is easy to inhale a large amount of dust and get pneumoconiosis in dust environment. The experimental results are shown in Table 3.

Table 2. Experimental data volumes for deep learning

	Normal	Abnormal_01	Abnormal_02	Abnormal_03
Quantity	346	1100	137	42

Through the experiments, we can find that different network models have different network layers. With the continuous increase of network layers, diagnostic performance can be effectively improved. However, the accuracy of classification and recognition is decreased unsteadily from 22-layered GoogLeNet to 46-layered GoogLeNet-V2. Serious over-fitting will even occur when ResNet are used, which makes the performance not rise but fall. GoogLeNet is the most effective model for pneumoconiosis diagnosis among these models.

Table 3. Experimental results on various classical networks

	Layers	Efficiency	Remark
LeNet	7	71.6%	–
AlexNet	8	87.9%	–
GoogleNet	22	91.6%	–
GoogLeNet-V2	46	90.7%	Performance decline
ResNet	100+	–	Overfitting severe

Normally the number of deep neural network layers and the number of units at each layer can be increased to improve the performance of training models with massive labeled training data. But the larger number of parameters will prone to the overfitting problem. It will be even more serious for limited labeled training data. At the same time, more computational resources are also network training concern when network size is increased. The problems can be solved by ultimately moving from fully-connected convolution to sparsely connected architectures. That means we can construct an improved deep learning model to get higher performance.

4   Design of Model Architecture

4.1   Analysis of Medical Data and Method

There are three steps in pneumoconiosis diagnosis with deep learning method:

- The labeled original X-ray chest films are preprocessed firstly. The pre-processing operations are noise filtering, centering, and image enhancement.
- Based on Caffe framework, deep neural network and suitable training algorithms are used to train preprocessed images to generate a trained model.
- With the trained model, target X-ray chest films are identified to help doctors for disease screening and studying by doctors.

4.2   Image Preprocessing

X-ray chest film is filmed with different gray levels based on the density of human tissues, in which the information of tissue structure may not show clearly or have little difference in gray level with adjacent areas. So gray-scale enhancement is adopted to stretch and enhance the difference between target area and adjacent areas. Gray histogram processing is realized to achieve sparse X-ray chest films with denser gray levels. The corresponding image effect is to enhance the target area and improve the recognition of lung region.

Formula (1) is used to obtain image histogram, in which i represents a gray value in the range of 0 to $K - 1$ in the image; n represents the total number of image pixels and n_i represents the number of gray value of i pixels; $p(i)$ represents the probability of gray value of i pixels in the image.

$$P(i) = \frac{n_i}{n} \quad (i = 0, 1, \ldots, K - 1) \tag{1}$$

Image enhancement in spatial domain is done with Formula (2) and Formula (3), in which i and j represent gray values in the range of 0 to $K - 1$; n_j represents the number of gray value of j pixels; $p'(i)$ represents the probability of i level pixel gray in the transform; N' represents the sum of the first i gray level pixels.

$$P'(i) = \frac{\sum_{j=0}^{i} n_j}{N'} \quad (i, j = 0, 1, \ldots, K - 1) \tag{2}$$

$$N' = \sum_{i=0}^{K-1} \sum_{j=0}^{i} n_j \quad (i, j = 0, 1, \ldots, K - 1) \tag{3}$$

The gray level distribution of the image is transformed into an approximate uniform distribution by formula (4), in which $I(i)$ is the pixel value with desired gray level i in original image, $P'(i)$ represents gray probability of the transformed pixels, and $W(I)$ is image gray scale. Regardless of the distribution of original image, the histogram will be transformed into an approximately uniform distribution.

$$I(i) = P'(i) \times W(I) \tag{4}$$

The above steps are applied to each gray level pixel to complete the histogram equalization of original image. The results are shown in Fig. 1.

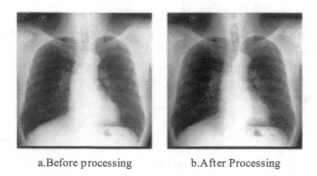

a.Before processing   b.After Processing

Fig. 1. The effect of image pre-processing on X-ray film

4.3   CNN Reconfiguration and Optimization

From Sect. 3, we can conclude that the performance of deep neural network can be improved by enlarging the scale of CNN including the depth (network layers) and the width (the units at each layer). But overfitting and expansive resource requirements become more and more serious with the increasing of network layers. The basic solving method is to transfer fully-connected convolution to sparsely-connected architecture. The kernel structure of GoogLeNet, Inception [17], can release the problem to an extent, but overfitting is still existed when enlarging Inception simply.

The kernel structure of GoogLeNet, Inception, is shown in Fig. 2. After analysis, we can get the conclusions as follows:

– The sizes of receptive fields are different with different scales of convolution kernels. Different scale features must be fuzed when splicing multi convolution kernels with different sizes.
– There are three sizes for convolution kernel: 1×1, 3×3, and 5×5. For example, suppose the convolution step stride $= 1$, and the pad $= 0$, 1, and 2 correspondingly, the convolutions can be processed to obtain features with same dimension, and then these features can be directly spliced together.

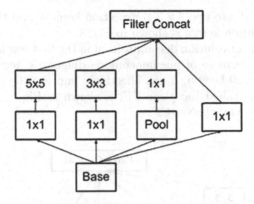

Fig. 2. Original Inception structure of GoogLeNet

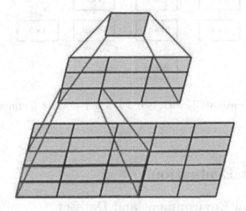

Fig. 3. The degraded representation of 5×5 convolution kernel

- Deeper network gives more abstract features with higher receptive field features. Therefore, with the increase of the layers, the ratio of 3×3 and 5×5 convolution will be increased.

Because of the limited X-ray chest films, the Inception structure is reconstructed according to the characteristics of pneumoconiosis diagnosis to improve its performance. However, the 5×5 convolution kernel will still bring more and more computation, so we take into account that the 1×1 volumes are connected to the 3×3 convolution kernel and the 5×5 convolution kernel in the same layer, and the large size convolution kernel provides larger receptive fields and more parameters. In particular, the kernel parameter of 5×5 convolution is 2.78 times of the convolution kernel of 3×3, and high computing resources are needed to process more computing parameters.

In the visual network, the structure of adjacent layers is highly correlated. We can maintain the receptive field with small amount of parameters with the method of convolution kernel substitution. In this paper, the 5×5 convolution kernel is decomposed into two 3×3 convolution kernels, and the representation of the 5×5 convolution kernel is shown in Fig. 3.

But the effect of convolution decomposition in the first few layers is not good, which leads to the decrease of pneumoconiosis diagnostic accuracy. But in the middle layers (12 to 20 layers), it can effectively improve the performance. The optimized structure, called Inception-CF, is shown in Fig. 4. Correspondingly, we call the model as GoogLeNet-CF.

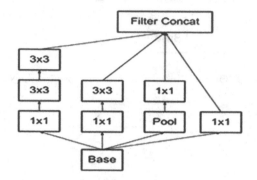

Fig. 4. Inception-CF with two 3×3 but not 5×5 convolutions

5   Experiment Evaluation

5.1   Experimental Environment and Dataset

Improved CNN model is implemented to replace the normal CNN model in the developed pneumoconiosis diagnosis system. In order to test the model performance, the experiments are run on a server configured by Intel CPU i7-950 with 4 cores and 8 threads, 8 GB memory, and two NVIDIA GeForce GTX TITAN Z cards (5760 CUDA cores, 12 GB GDDR5 memory with 768-bit memory interface width). The software environment is Ubuntu 14.04 Linux with GCC 4.8.2 and CUDA 7.5.

The experimented X-ray chest films are from health examination workers in dust environment, including abnormal X-ray chest films of pneumoconiosis patients and normal data from healthy workers. For abnormal X-ray chest films, they are staged into 3 stages: initial (stage 1), middle (stage 2), and late (stage 3) pneumoconiosis. The attributes of these images are shown in Table 4.

5.2   Experiments on the Accuracy of Pneumoconiosis Diagnosis

GoogLeNet-CF model is used to train with labeled images for pneumoconiosis diagnosis. GoogLeNet model is used to compare the object for the verification of GoogLeNet-CF.

Table 4. The attributes of X-ray films for testing

Attribute of image	Value
Quantity	1925
Width and Height	3372 × 3580
Color	256 gray level
Format	.bmp and .jpg

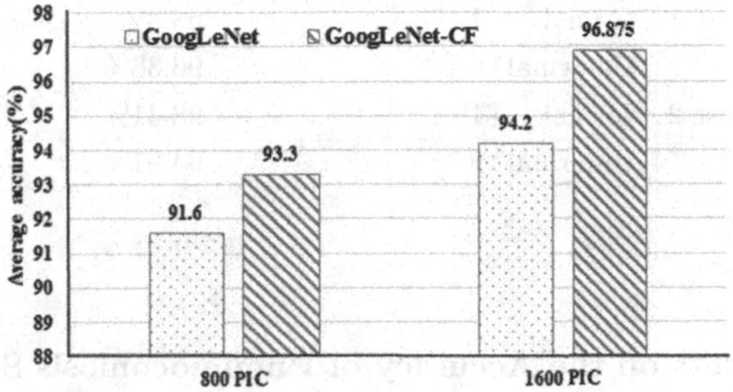

Fig. 5. Performance comparison between GoogLeNet and GoogLeNet-CF model in different data sizes

The experiments are done based on two different data scales: small scale with 800 images and big scale with 1600 images. The final results are shown in Fig. 5. After training, the accuracies of the two methods with 800 training images are 91.6% (GoogLeNet) and 93.3% (GoogLeNet-CF) respectively, while 94.2% (GoogLeNet) and 96.875% (GoogLeNet-CF) can be obtained when the training images are increased to 1600 images. The experimental results show that the accurate diagnosis depends on more images. At the same time, GoogLeNet-CF neural network model achieves higher improvement than GoogLeNet.

For pneumoconiosis diagnosis, that is to determine whether the patient suffers from pneumoconiosis, 1600 images are used for the experiments. In order to ensure the validity of the experiments, we divide the testing images into two parts for two experiments. The detailed information of the testing images is shown in Table 5.

Table 5. The composition of experimental images

Type	Training set	Validation set	Test 1	Test 2
Normal	236	19	60	63
Abnormal	1382	135	166	177

123 normal and 343 abnormal X-ray chest films are provided for testing, which are divided into two testing data sets randomly. In test 1, there are 60

normal and 166 abnormal X-ray chest films as testing data. During the experiment, one normal image is error, so that the 'normal' accuracy is 98.3%; there are 6 errors in 166 abnormal images with the 'abnormal' accuracy of 96.38%. The two experimental results are shown in Table 6.

Table 6. Two experiment results of pneumoconiosis diagnosis

	Type	Recognition errors	Accuracy rate
Test 1	Normal	1	98.3%
	Abnormal	6	96.38%
Test 2	Normal	1	98.41%
	Abnormal	9	94.91%

5.3   Experiments on the Accuracy of Pneumoconiosis Staging

We also do the experiments on pneumoconiosis staging, which means the abnormal images must be further analyzed to determine the stage of pneumoconiosis. Although the pneumoconiosis is divided into three stages: early (Abnormal_1), middle (Abnormal_2), and late stages (Abnormal_3), because there are not enough labeled images in middle and late stages, which can not meet the needs of individual learning and training, we have to merge the two stages into one called Abnormal_23, which means the operation of pneumoconiosis staging is to classify images into three categories: Normal, Abnormal_1, and Abnormal_23. The test data set is made up of 306 normal X-ray films, 986 abnormal_01 X-ray films and 179 abnormal_23 X-ray films, shown in Table 7.

Table 7. The composition of experimental images for pneumoconiosis staging

Type	Image number of training set	Image number of validation set	Image number of test set
Normal	306	40	50
Abnormal_1	986	114	76
Abnormal_23	179	23	54

The results of the experiment show, there are 2 errors in 50 normal X-ray chest films with 96% normal accuracy, 5 errors in 76 early staging X-ray chest films with the accuracy of 93.4%, and 4 errors in 54 sick X-ray chest films whose accuracy is 92.59%. The overall accuracy rate of the diagnosis is as high as 93.88%, shown in Table 8.

Table 8. Experimental results of pneumoconiosis staging

Type	Test number	Recognition errors	Accuracy rate
Normal	50	2	96%
Abnormal_01	76	5	93.4%
Abnormal_23	54	4	92.59%

6   Conclusion

In this paper, an improved CNN-based pneumoconiosis diagnosis method is introduced. X-ray chest films can not only be processed with traditional image processing technology to obtain the image features for medical analysis, but also be diagnosed with deep learning method. According to the characteristics of pneumoconiosis diagnosis, an improved CNN-based method, named GoogLeNet-CF, is proposed. The depth of convolutional neural network model is reconstructed to improve the effectiveness of diagnosis. Before the training of GoogLeNet-CF, labeled X-ray chest films are pre-processed to enhance the interesting areas. With the preprocessing and optimized neural network structure, the diagnostic accuracy of pneumoconiosis is increased by 2%. Therefore, it can help doctors to diagnose pneumoconiosis rapidly and improve the diagnostic accuracy.

Misdiagnosis will have a fatal impact on the patient if the method is applied into reality. We still need to do more work to improve the accuracies of pneumoconiosis diagnosis and staging.

References

1. Schenker, M., Pinkerton, K., Mitchell, D., Vallyathan, V., Elvine, B., Green, B.: Pneumoconiosis from agricultural dust exposure among young California farmworkers. Environ. Health Perspect. **117**(6), 988 (2009)
2. Melle, W.V.: MYCIN: a knowledge-based consultation program for infectious disease diagnosis. Int. J. Man Mach. Stud. **10**(3), 313–322 (1978)
3. Kraus, T., Raithel, H., Lehnert, G.: Computer-assisted classification system for chest X-ray and computed tomography findings in occupational lung disease. Int. Arch. Occup. Environ. Health **69**(6), 182–486 (1997)
4. Roth, H., et al.: Improving computer-aided detection using convolutional neural networks and random view aggregation. IEEE Trans. Med. Imaging **35**(5), 1170–1181 (2016)
5. Anthimopoulos, M., Christodoulidis, S., Ebner, L., Christe, A., Mougiakakou, S.: Lung pattern classification for interstitial lung diseases using a deep convolutional neural network. IEEE Trans. Med. Imaging **35**(5), 1207–1216 (2016)
6. Tajbakhsh, N., Gotway, M.B., Liang, J.: Computer-aided pulmonary embolism detection using a novel vessel-aligned multi-planar image representation and convolutional neural networks. In: Navab, N., Hornegger, J., Wells, W.M., Frangi, A.F. (eds.) MICCAI 2015. LNCS, vol. 9350, pp. 62–69. Springer, Cham (2015). https://doi.org/10.1007/978-3-319-24571-3_8

7. Powles, J., Hodson, H.: Google DeepMind and healthcare in an age of algorithms. Health Technol. **7**(4), 351–367 (2017)
8. Liu, X., Hou, F., Qin, H., Hao, A.: A CADe system for nodule detection in thoracic CT images based on artificial neural network. Sci. China (Inf. Sci.) **60**(7), 072106 (2017)
9. Jia, Y., et al.: Caffe: convolutional architecture for fast feature embedding. In: Proceedings of the 22nd ACM International Conference on Multimedia (MM), pp. 675–678. ACM, USA (2014)
10. Long, J., Shelhamer, E., Darrell, T.: Fully convolutional networks for semantic segmentation. In: Proceedings of of 2015 IEEE Conference on Computer Vision and Pattern Recognition (CVPR), pp. 3431–3440. IEEE, USA (2015)
11. Chen, T., et al.: MXNet: a flexible and efficient machine learning library for heterogeneous distributed systems. Comput. Sci. **6**(2), 2433–2445 (2015)
12. Ku, C., Lee, K.: Diagonal recurrent neural networks for dynamic systems control. IEEE Trans. Neural Netw. **6**(1), 144–156 (1995)
13. Lecun, Y., Bottou, L., Bengio, Y., Haffner, P.: Gradient-based learning applied to document recognition. Proc. IEEE **86**(11), 2278–2324 (1998)
14. Krizhevsky, A., Sutskever, I., Hinton, G.: ImageNet classification with deep convolutional neural networks. In: Proceedings of the 25th International Conference on Neural Information Processing Systems (NIPS), pp. 1097–1105 (2012)
15. Szegedy, C., et al.: Going deeper with convolutions. In: Proceedings of of 2015 IEEE Conference on Computer Vision and Pattern Recognition (CVPR), pp. 1–9. IEEE, USA (2015)
16. Hei, K., Zhang, X., Ren, S., Sun, J.: Deep residual learning for image recognition. In: Proceedings of of 2016 IEEE Conference on Computer Vision and Pattern Recognition (CVPR), pp. 770–778. IEEE, USA (2016)
17. Szegedy, C., Vanhoucke, V., Ioffe, S., Shlens, J., Wojna, Z.: Rethinking the inception architecture for computer vision. In: Proceedings of 2016 IEEE Conference on Computer Vision and Pattern Recognition (CVPR), pp. 2818–2826. IEEE, USA (2016)

Modelling Mental States via Computational Psychophysiology: Benefits and Challenges

Weihao Zheng[1,2] ![ORCID], Hanshu Cai[1], Zhijun Yao[1], Xiaowei Zhang[1],
Xiaowei Li[1], and Bin Hu[1(✉)]

[1] School of Information Science and Engineering,
Lanzhou University, Lanzhou 730000, People's Republic of China
bh@lzu.edu.cn
[2] College of Biomedical Engineering and Instrument Science,
Zhejiang University, Hangzhou 310027, People's Republic of China

Abstract. The human psychophysiological processes are complex phenomenon built upon the physical scaffolding of the body. Machine learning approaches facilitate the understanding of numerous physiological processes underlying complex human mental states and behavior, leading to a new research direction named Computational Psychophysiology. Computational Psychophysiology aims to reveal the psychophysiological processes underlying complex human emotion and mental states from a computational perspective, and can be used to predict affective and psychological outcomes based on different physiological features or experimental manipulations. In this paper, we discuss the benefits and challenges in the future of bringing computing technologies into decoding human mental states.

Keywords: Computational Psychophysiology · Psychophysiological processes · Machine learning · Benefits · Challenges

1 Background

The quantification of relationship between psychological processes and physiological changes is a perennial question that plagues human beings. Early explorations focused on observing animal behavior and emotional changes by altering the physiological basis, such as by cutting a part of the brain tissue to observe its behavioral changes. By the 1960s, research turned to study the physiological changes that induced by man-made modulation of psychological or emotional response [1], that is, to build a quantitative model with psychological response as the independent variable and physiological changes as the dependent variable, respectively [2]. For instance, when perceiving intense mental stimuli (e.g., threats of pain), the hormones are released and the sympathetic nervous system is activated, which lead to the release of catecholamines (including adrenaline and norepinephrine) from adrenal gland, and as a result, cause increases in heart rate, galvanic skin response, blood pressure, and respiratory rate. Such evaluation mode emphasizes the use of physiological information to

© Springer Nature Switzerland AG 2019
D. Milošević et al. (Eds.): HCC 2019, LNCS 11956, pp. 659–670, 2019.
https://doi.org/10.1007/978-3-030-37429-7_67

reflect psychological activities and facilitates the inference of mental states, therefore is called Psychophysiology [1]. Psychophysiology brings a new perspective into the assessment of mental states, which facilitates the shift of mental assessment from subjective judgment to objective reasoning model based on physiological information.

Both the terms psychophysiology and computation have long histories within the study of human behavior. A number of peripheral recording measures such as heart rate, skin conductance and respiration have been used to study topics as diverse as truthfulness, fear and attention [3–5]. However, traditional psychophysiology studies are increasingly suspected to misrepresent the relationships between psychological and physiological factors, because of the complex structure and high dimension of the physiological data were not completely taken into account. The rapid development of machine learning (ML) technology (e.g., deep learning and manifold learning) and high-performance computing provide more effective solutions to these issues, and allow us to analyze psychophysiology from macroscopic (emotional, cognitive level, etc.), mesoscopic (electrophysiological signals, neuroimaging, etc.) and microscopic (neural system, molecular level, etc.) levels, thus benefiting the investigation of the complex mechanisms in psychophysiological processes [6].

The concept of Computational Psychophysiology was first introduced in [7], in which it was defined as "an interdisciplinary research field that employs methods from the disciplines of psychology, physiology, neuroscience, computer science, mathematics, physics, and others to model physiological activity in relation to the psychological components of human behavior". Since applying computational data analysis allows to mine structured knowledge from extensive data in a more objective manner [8], it may help to identify new objective phenotypes and build effective models for emotion, mental states and behavior.

2 Fundamental Assumptions of Computational Psychophysiology

Since Computational Psychophysiology aims to modelling the complex mapping relations between mental, body and behavior, this concept should be built upon two fundamental assumptions, that are, (1) the existence of physiology-to-psychology path, and (2) the computability of psycho-physiology [6].

2.1 Existence of Physiology-to-Psychology Path

The assumption of existence of physiology-to-psychology path is the foundation of Computational Psychophysiology, which assumes a particular mental state can be estimated by multiple physiological measurements. This is a backward inference model that infers the psychological projection mechanism through the evoked-physiological activity. Actually, the ideal solution is to find an 'one-to-one' mapping strategy, in which each mental state has a corresponding pattern of physiological changes.

Suppose a specific psychological state is denoted as independent variable (A) and the physiological feature is denoted as dependent variable (B), we can defined the probability of physiological changes caused by psychological activities as a conditional probability $P(B|A)$; then the psychological states could be inferred if the conditional probability $P(A|B)$ can be calculated [9]. However, it is difficult to find an exact 'one-to-one' correspondence for each mental state in the real world. For example, although functional co-activation of thalamus and cortical brain regions were significantly associated with perception and consciousness [10], the co-activation pattern cannot directly reflect mind of consciousness, other factors, such as hormone secretion and blood pressure, still need to be considered as the covariates. Therefore, a more reasonable choice is to identify a series of physiological signals that are highly associated with the target psychological state, in other words, it is necessary to expand the 'one-to-one' correspondence to an 'many-to-one' mapping strategy.

2.2 Computability of Psychophysiology

The computability of psychophysiology assumes that psychological state can be represented as a function of a set of high dimensional heterogeneous physiological data. This requires proper representation of physiological signals and effective quantification of psychological states. Actually, bring computational thinking into simulating and calculating human mind and behavior is not new, other perspectives, such as Cognitive Computing [11, 12] and Computational Psychiatry [13], also rely on the assumption of the computability of the brain. Compare to Cognitive Computing that aims to make the computer "think" like a man [14, 15], and Computational Psychiatry that aims to improve prediction and treatment of mental illness [13, 16], Computational Psychophysiology, however, focuses on modeling the internal connections between various physiological indicators (e.g., brain activity and metabolism, bioelectrical signals, heart rate, respiration, and behavior, etc.) and the physiological effect of mentation under different mental states [7]. In view of this, we will argue that Computational Psychophysiology may bring unprecedentedly benefits for understanding the underlying mechanisms of psychophysiological processes. A basic research process of Computational Psychophysiology is given in Fig. 1.

3   Benefits

ML technologies primarily benefits Psychophysiological research in three domains, including data re-expression, dimensionality reduction, and the identification of mental states. These applications of ML algorithms have led to several impressive feats of mind reading, therefore, improve our understanding of the relationship between psychology and physiology. Some typical ML algorithms for facilitating the three research domains are summarized in Table 1.

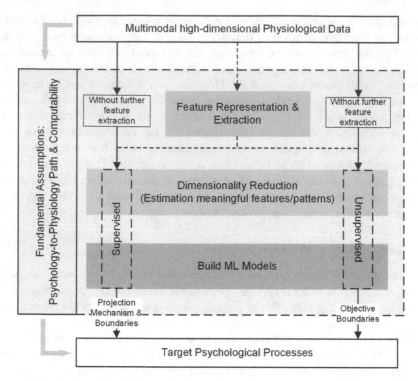

Fig. 1. Basic pipeline of applying ML technologies to explore psychophysiological processes. All these process are based on two foundational assumptions that are the existence of psychology-to-physiology path and the computability of Psychophysiology. Both supervised and unsupervised methods can be used to reduce the dimensionality of multimodal physiological data. The feature subset after dimensionality reduction is used to train ML models, where applying supervised algorithms and unsupervised algorithms may have different purposes. The supervised algorithms focus on find optimal boundaries between mental states by using only physiological data, whereas, supervised algorithms are also able to discover the projection mechanism between physiological data and psychological process in addition to classification. Notably, this figure only gives a basic pipeline of Computational Psychophysiological, which can be flexibly changed according to the research demands.

3.1   Advanced Feature Extraction and Representation

ML technologies have been widely used to the investigation of how emotions and mental disorders influence the physiological properties. Compared to theory-driven models that specify mechanistically interpretable relations between physiological and psychological variables [13], data-driven approaches allows us to explore more complex physiological patterns result from psychological activity and identify mental state of an individual via unseen physiological signals (e.g., brain activity). A good example is independent component analysis (ICA), which is an efficient tool to identify the essential data structure by decomposing data into statistically independent components [17], and has been broadly applied to reveal the intrinsic functional architecture in the brain and its abnormal changes in mental disorders (e.g., autism spectrum disorders) [18, 19]. ML algorithms have also been applied to extract biomarkers (e.g., adjacent matrix and its topological structure), which exhibited superior performance relative to traditional physiological features (e.g., electrodermal activity, morphology and activity of the brain) in representing alterations induced by emotion [5, 20–22], pain perception [23, 24], and mental disorders [25–34]. Other studies utilized ML to build multivariate statistical model, such as multi-voxel pattern analysis (MVPA) that is able to reveal the significance of overlapping functional activations in the brain [35, 36]. These ML algorithms enhance the representation of characteristics of observations under different psychological conditions, and may help to improve the interpretation of mental disorders, emotion and behavior.

Table 1. Typical ML technologies and their purposes

Notion	Purpose	Training strategy	Common methods
Feature extraction	Extract new features from the initial set to facilitate the subsequent learning and generalization steps	–	FFT, Graph theoretical analysis, deep learning, latent semantic analysis, data compression, etc.
Dimensionality reduction	To represent subjects more clearly in low dimensional feature space by removing unimportant features	Supervised	Mutual information, correlation, SVM-RFE, LASSO, Bayesian models, etc.
		Unsupervised	PCA, locality preserving projection, Laplacian eigenmaps, etc.
Classification	Identifying which categories a new subject belongs to	Supervised	SVM, random forest, logistic regression, neural network, naïve Bayes classifier, linear discriminant analysis, etc.
		Unsupervised	Clustering, self-organizing mapping, etc.
Regression	Represent one (possibly multiple) dependent variable (e.g., mental state) via a series of independent variables (physiological features)	Supervised	GLM, LASSO, elastic net, support vector regression, multi-task regression, etc.
Cross validation	Assess the robustness and generalizability of an ML model	–	Leave-one-out, ten-fold cross validation (repeat for at least 10 times)

Note: *FFT = fast Fourier transform; SVM-RFE = SVM-based recursive feature elimination; LASSO = least absolute shrinkage and selection operator; PCA = principal component analysis; GLM = generalized linear model.*

3.2 Handle High Dimensional Data

The inference of psychological states requires diversified physiological features. However, the risk of overfitting dramatically increases with the addition of features when the number of subjects is far less than the number of features [37]. Although p values from statistics can be used as criteria to filter insignificant features (e.g., exclude features with p > 0.05) [38], statistical methods always assume that each feature is independent from the others, and the relationships among features (e.g., complementarity) are usually ignored because of the methodological limitations. Compared to the statistical methods, the advances of ML approaches in dimensionality reduction are mainly reflected in the consideration of the feature set as a whole, by which optimal feature subset would be extracted based on both discriminating power and complementarity of features. Since ML and statistical analysis do not assess data in the same way, leading to distinct assessment of the effects of physiological data, for instance, a features that contribute highly to discrimination may not show statistically significant correlation with psychological state, and vice versa [8, 39].

The dimensionality reduction algorithms can be divided into supervised and unsupervised methods by judging whether prior knowledge are incorporated in the process. Unsupervised methods, such as principal component analysis (PCA) [40], are usually used to explore unknown configuration of the data meanwhile reduce the dimensionality by excluding features showing low variance between groups. Supervised methods are specialized for finding optimal feature subset that highly contribute to the outcome of classification or regression process. For example, support vector machine (SVM) can be used to quantify the importance of features and their complementarity [41], and reduce the dimensionality by removing features that have negative or possibly null effect to classification accuracy; regularization technology is usually embedded into regression models to remove features that are not relevant to regression process [42, 43]. These approaches have been broadly utilized to reveal valuable information that are hidden in the high-dimensional heterogeneous physiological data, and are improved to adapt to the rapid update of the requirements in feature selection [44, 45]. In summary, ML algorithms can substantially improve the analytical ability of high-dimension data beyond traditional statistical methods, and are effective tool in reducing the danger of overfitting.

3.3 Psychophysiological Modelling

Another big advance of bringing computation to psychophysiological research is that it brings unprecedented opportunity to 'calculate' the psychological processes. By training ML models (e.g., SVM and neural network), we may able to estimate the psychological processes and mental states of an unknown individual by using physiological features (e.g., facial expression [46, 47]). In contrast, statistical analysis (e.g., student's t-test and ANOVA test) that grounds in the notion of testing statistical significance of group effects and cannot be used for the estimation of a single subject [8]. State-of-the-art research of utilizing ML to recognize emotion and mental disorders have achieved remarkable performance [48–52], however, most of them only focused on improving the final accuracy, omitting to discover the reasons that lead to the high recognition accuracy. Actually, a high accuracy always results from an optimal model, which depends on how much we have known of the mechanism of mapping psychology-to-physiology. Therefore, it is necessary to investigate the mechanism behind high recognition accuracy, which is also what the Computational Psychophysiology mainly focuses on. This goal can be achieved through the combination of more physiological modalities and algorithmic advances (such as the interpretation of deep learning models [53]).

4 Challenges

4.1 Categorization Criteria for Mental States

Emotions (e.g., happy, fear, and anger) can be evoked by experimental designs using different external stimulations (such as award, music, pictures, and videos [54–56]). Although individual differences evidently reduce during tasks relative to resting-state studies [57, 58], very few experimental designs yet are sufficiently optimal to evoke specific emotion without simultaneously evoking other psychological processes. In other words, the current criteria cannot perfectly define the boundaries between different mental states. Similar problem also exists in the identification of mental disorders (e.g., autism, generalized anxiety, and depression), because of categorization defined in the Diagnostic and Statistical Manual (DSM) [59] and International Classification of Diseases (ICD) [60] may not align well with some newly identified physiological and behavioral features [8], as well as the inevitable influences from comorbidity. The identification of novel symptoms/biomarkers that can defines clear demarcation of mental states is of great importance, which requires the combination of multimodal physiological information as well as additional features behind DSM and ICD (e.g., suicidal thoughts, social ability, and anhedonia).

4.2 Physiological Data Acquisition

Computational Psychophysiology largely relies on the acquisition of physiological data. However, the technical limitations of current devices, to some extent, limited the data collection. For example, electroencephalo-graph (EEG) cannot reach the signals in deep brain, and its spatial resolution is relative low; magnetic resonance imaging (MRI) provides both structural and functional information with high spatial resolution but has high requirement for collection environment, and its temporal resolution remains low; other highly accessible electrical signals (e.g., electrodermal activity) can be easily influenced by breathing, motion, and environment. The requirement of large number of diversified physiological data inevitably magnifies the influence of such technical limitations, which thus become an unnegligible obstacle in the development process of Computational Psychophysiology.

4.3 Multimodal Data Analysis

The integration of features across multiple modalities would improve the accuracy of the inference of psychological states. Although amount of approaches for multimodal data analysis have been introduced in literatures [61–64], such as multi-task feature learning [65], multiple kernel learning classifier [66, 67], and joint ICA [68], these methods typically omit the characteristics of each modality. Since multimodal features are typically complex and heterogeneous, it is important to investigate suitable algorithms to diminish the differences in data structure between different modalities, meanwhile, preserve the unique information of each modality. In addition, in the investigation of the physiological substrates of psychological states, some features are not meaningful in isolation but may contain important complementary information when combine with other features, therefore, multimodal fusion algorithms that can effectively identify such complementary relationship must be developed in order to compute reliable physiology-to-psychology mapping strategy.

5 Conclusion

ML approaches have promoted the development of various research fields. Computational Psychophysiology that brings data-driven algorithms to modelling psychological processes through multimodal multidimensional physiological data will broaden the field of psychophysiology. In the present article, we indicated two fundamental assumptions of Computational Psychophysiology, and highlighted the primary difference between Computational Psychophysiology and other related research directions (the main focus of Computational Psychophysiology is placed on understanding of the psychophysiological processes). Though there are a number of challenges from various research field, such as psychology (e.g., task design), equipment technology (e.g., data acquisition), and computation (e.g., algorithm advances) in the current stage, the increasing cooperation of multidisciplinary research worldwide will have great contribution to address these issues. Overall, Computational Psychophysiology will lead to a better understanding of the complex psychophysiological processes, and will facilitate the development of new computational models that may provide the basis for a multidimensional diagnostic system of mental disorders.

Acknowledgment. This work was supported by the National Natural Science Foundation of China (grant number: 61210010 and 61632014, to B.H.), the National key research and development program of China (grant number: 2016YFC1307203), and the Program of Beijing Municipal Science & Technology Commission (grant number: Z171100000117005, to B.H.). The authors declare there is no conflict of interest in relation to this work.

References

1. Fowles, D.C.: Psychophysiology and psychopathology: a motivational approach. Psychophysiology **25**, 373–391 (1988)
2. Stern, R.M., Ray, W.J., Quigley, K.S.: Psychophysiological Recording. Oxford University Press, Oxford (2001)
3. Allen, J.J., Kline, J.P.: Frontal EEG asymmetry, emotion, and psychopathology: the first, and the next 25 years. Biol. Psychol. **67**, 1–5 (2004)
4. Dolan, R.J.: Emotion, cognition, and behavior. Science **298**, 1191–1194 (2002)
5. Picard, R.W., Vyzas, E., Healey, J.: Toward machine emotional intelligence: analysis of affective physiological state. IEEE Trans. Pattern Anal. Mach. Intell. **23**, 1175–1191 (2001)
6. Hu, B., Zheng, W.: A review of computational psychophysiology: the progress and trends. Commun. CCF **14**, 31–34 (2018)
7. American Association for the Advancement of Science: Advances in computational psychophysiology. Science **350**, 114 (2015)
8. Bzdok, D., Meyer-Lindenberg, A.: Machine learning for precision psychiatry: opportunities and challenges. Biol. Psychiatry Cogn. Neurosci. Neuroimaging **3**, 223–230 (2018)
9. Cacioppo, J.T., Tassinary, L.G., Berntson, G.: Handbook of Psychophysiology. Cambridge University Press, Cambridge (2007)
10. Boveroux, P., et al.: Breakdown of within- and between-network resting state functional magnetic resonance imaging connectivity during propofol-induced loss of consciousness. Anesthesiol. J. Am. Soc. Anesthesiol. **113**, 1038–1053 (2010)
11. Kelly, J.E.: Computing, cognition and the future of knowing. Whitepaper, IBM Reseach 2 (2015)
12. Modha, D.S., Ananthanarayanan, R., Esser, S.K., Ndirango, A., Sherbondy, A.J., Singh, R.: Cognitive computing. Commun. ACM **54**, 62–71 (2011)
13. Huys, Q.J., Maia, T.V., Frank, M.J.: Computational psychiatry as a bridge from neuroscience to clinical applications. Nat. Neurosci. **19**, 404 (2016)
14. Solo, A.M., Gupta, M.M.: Uncertainty in computational perception and cognition. In: Nikravesh, M., Kacprzyk, J., Zadeh, L.A. (eds.) Forging New Frontiers: Fuzzy Pioneers I, pp. 251–266. Springer, Heidelberg (2007)
15. Chen, Y., Argentinis, J.E., Weber, G.: IBM Watson: how cognitive computing can be applied to big data challenges in life sciences research. Clin. Ther. **38**, 688–701 (2016)
16. Adams, R.A., Huys, Q.J., Roiser, J.P.: Computational psychiatry: towards a mathematically informed understanding of mental illness. J. Neurol. Neurosurg. Psychiatry **87**, 53–63 (2016)
17. Hyvärinen, A., Oja, E.: Independent component analysis: algorithms and applications. Neural Netw. **13**, 411–430 (2000)
18. Calhoun, V.D., Liu, J., Adali, T.: A review of group ICA for fMRI data and ICA for joint inference of imaging, genetic, and ERP data. Neuroimage **45**, S163–S172 (2009)
19. Yao, Z., et al.: Resting-state time-varying analysis reveals aberrant variations of functional connectivity in autism. Front. Hum. Neurosci. **10**, 463 (2016)
20. Dhall, A., Asthana, A., Goecke, R., Gedeon, T.: Emotion recognition using PHOG and LPQ features. In: Face and Gesture 2011, pp. 878–883. IEEE (2011)
21. Jenke, R., Peer, A., Buss, M.: Feature extraction and selection for emotion recognition from EEG. IEEE Trans. Affect. Comput. **5**, 327–339 (2014)
22. Kim, K.H., Bang, S.W., Kim, S.R.: Emotion recognition system using short-term monitoring of physiological signals. Med. Biol. Eng. Comput. **42**, 419–427 (2004)

23. Wager, T.D., Atlas, L.Y., Lindquist, M.A., Roy, M., Woo, C.-W., Kross, E.: An fMRI-based neurologic signature of physical pain. N. Engl. J. Med. **368**, 1388–1397 (2013)
24. Wager, T.D., Atlas, L.Y., Leotti, L.A., Rilling, J.K.: Predicting individual differences in placebo analgesia: contributions of brain activity during anticipation and pain experience. J. Neurosci. **31**, 439–452 (2011)
25. Querbes, O., et al.: Early diagnosis of Alzheimer's disease using cortical thickness: impact of cognitive reserve. Brain **132**, 2036–2047 (2009)
26. Tong, T., et al.: A novel grading biomarker for the prediction of conversion from mild cognitive impairment to Alzheimer's disease. IEEE Trans. Biomed. Eng. **64**, 155–165 (2017)
27. Wee, C.Y., Wang, L., Shi, F., Yap, P.T., Shen, D.: Diagnosis of autism spectrum disorders using regional and interregional morphological features. Hum. Brain Mapp. **35**, 3414–3430 (2014)
28. Yao, Z., Hu, B., Nan, H., Zheng, W., Xie, Y.: Individual metabolic network for the accurate detection of Alzheimer's disease based on FDGPET imaging. In: 2016 IEEE International Conference on Bioinformatics and Biomedicine (BIBM), pp. 1328–1335. IEEE (2016)
29. Zhang, Z., et al.: Frequency-specific functional connectivity density as an effective biomarker for adolescent generalized anxiety disorder. Front. Hum. Neurosci. **11**, 549 (2017)
30. Zhao, Y., et al.: Predicting MCI progression with individual metabolic network based on longitudinal FDG-PET. In: 2017 IEEE International Conference on Bioinformatics and Biomedicine (BIBM), pp. 1894–1899. IEEE (2017)
31. Zheng, W., et al.: Multi-feature based network revealing the structural abnormalities in autism spectrum disorder. IEEE Trans. Affect. Comput. 1 (2019)
32. Zheng, W., Yao, Z., Hu, B., Gao, X., Cai, H., Moore, P.: Novel cortical thickness pattern for accurate detection of Alzheimer's disease. J. Alzheimers Dis. **48**, 995–1008 (2015)
33. Zheng, W., Yao, Z., Li, Y., Wu, D., Hu, B.: Prediction of Alzheimer's disease in patients with mild cognitive impairment using connectivity extracted from multi-modal brain imaging (under review)
34. Zheng, W., Yao, Z., Xie, Y., Fan, J., Hu, B.: Identification of Alzheimer's disease and mild cognitive impairment using networks constructed based on multiple morphological brain features. Biol. Psychiatry Cogn. Neurosci. Neuroimaging **3**, 887–897 (2018)
35. Norman, K.A., Polyn, S.M., Detre, G.J., Haxby, J.V.: Beyond mind-reading: multi-voxel pattern analysis of fMRI data. Trends Cogn. Sci. **10**, 424–430 (2006)
36. Peelen, M.V., Downing, P.E.: Using multi-voxel pattern analysis of fMRI data to interpret overlapping functional activations. Trends Cogn. Sci. **11**, 4 (2007)
37. Hawkins, D.M.: The problem of overfitting. J. Chem. Inf. Comput. Sci. **44**, 1–12 (2004)
38. Emerson, R.W., et al.: Functional neuroimaging of high-risk 6-month-old infants predicts a diagnosis of autism at 24 months of age. Sci. Transl. Med. **9**, eaag2882 (2017)
39. Maia, T.V.: Introduction to the series on computational psychiatry. J. Clin. Psychol. Sci. **3**, 374–377 (2015)
40. Jolliffe, I.: Principal Component Analysis. Springer, New York (2011). https://doi.org/10.1007/b98835
41. Guyon, I., Elisseeff, A.: An introduction to variable and feature selection. J. Mach. Learn. Res. **3**, 1157–1182 (2003)
42. Zou, H., Hastie, T.: Regularization and variable selection via the elastic net. J. R. Stat. Soc. Ser. B (Stat. Methodol.) **67**, 301–320 (2005)
43. Tibshirani, R.: Regression shrinkage and selection via the lasso. J. R. Stat. Soc. Ser. B (Methodol.) **58**, 267–288 (1996)

44. Liu, H., Motoda, H., Setiono, R., Zhao, Z.: Feature selection: an ever evolving Frontier in data mining. In: Feature Selection in Data Mining, pp. 4–13 (2010)
45. Saeys, Y., Inza, I., Larrañaga, P.: A review of feature selection techniques in bioinformatics. Bioinformatics **23**, 2507–2517 (2007)
46. Wang, W., Zheng, W., Ma, Y.: 3D facial expression recognition based on combination of local features and globe information. In: 2014 Sixth International Conference on Intelligent Human-Machine Systems and Cybernetics, pp. 20–25. IEEE (2014)
47. Zheng, W.H., Wang, W., Ma, Y.D.: Facial expression recognition based on the texture features of global principal component and local boundary. Appl. Mech. Mater. **548–549**, 1110–1117 (2014)
48. Ali, M., Mosa, A.H., Machot, F.A., Kyamakya, K.: Emotion recognition involving physiological and speech signals: a comprehensive review. In: Kyamakya, K., Mathis, W., Stoop, R., Chedjou, J.C., Li, Z. (eds.) Recent Advances in Nonlinear Dynamics and Synchronization. SSDC, vol. 109, pp. 287–302. Springer, Cham (2018). https://doi.org/10.1007/978-3-319-58996-1_13
49. Han, J., Zhang, Z., Cummins, N., Schuller, B.: Adversarial training in affective computing and sentiment analysis: recent advances and perspectives. arXiv preprint: arXiv:1809.08927 (2018)
50. Krueger, R.F., et al.: Progress in achieving quantitative classification of psychopathology. World Psychiatry **17**, 282–293 (2018)
51. Swain, M., Routray, A., Kabisatpathy, P.: Databases, features and classifiers for speech emotion recognition: a review. Int. J. Speech Technol. **21**, 93–120 (2018)
52. Wiecki, T.V., Poland, J., Frank, M.J.: Model-based cognitive neuroscience approaches to computational psychiatry: clustering and classification. Clin. Psychol. Sci. **3**, 378–399 (2015)
53. Dong, Y., Su, H., Zhu, J., Zhang, B.: Improving interpretability of deep neural networks with semantic information. In: Proceedings of the IEEE Conference on Computer Vision and Pattern Recognition, pp. 4306–4314 (2017)
54. Cui, X., Bray, S., Bryant, D.M., Glover, G.H., Reiss, A.L.: A quantitative comparison of NIRS and fMRI across multiple cognitive tasks. Neuroimage **54**, 2808–2821 (2011)
55. Fan, T., Wu, X., Yao, L., Dong, J.: Abnormal baseline brain activity in suicidal and non-suicidal patients with major depressive disorder. Neurosci. Lett. **534**, 35–40 (2013)
56. Schaefer, R.S., Vlek, R.J., Desain, P.: Music perception and imagery in EEG: alpha band effects of task and stimulus. Int. J. Psychophysiol. **82**, 254–259 (2011)
57. Finn, E.S., Scheinost, D., Finn, D.M., Shen, X., Papademetris, X., Constable, R.T.: Can brain state be manipulated to emphasize individual differences in functional connectivity? Neuroimage **160**, 140–151 (2017)
58. Geerligs, L., Rubinov, M., Henson, R.N.: State and trait components of functional connectivity: individual differences vary with mental state. J. Neurosci. **35**, 13949–13961 (2015)
59. American Psychiatric Association: Diagnostic and Statistical Manual of Mental Disorders (DSM-5®). American Psychiatric Publishing (2013)
60. World Health Organization: International Classification of Diseases. World Health Organization Press (1990)
61. Acar, E., Levin-Schwartz, Y., Calhoun, V.D., Adali, T.: ACMTF for fusion of multi-modal neuroimaging data and identification of biomarkers. In: 2017 25th European Signal Processing Conference (EUSIPCO), pp. 643–647. IEEE (2017)

62. Bänziger, T., Grandjean, D., Scherer, K.R.: Emotion recognition from expressions in face, voice, and body: the Multimodal Emotion Recognition Test (MERT). Emotion **9**, 691 (2009)
63. Sebe, N., Cohen, I., Gevers, T., Huang, T.S.: Multimodal approaches for emotion recognition: a survey. In: Internet Imaging VI, pp. 56–68. International Society for Optics and Photonics (2005)
64. Tzirakis, P., Trigeorgis, G., Nicolaou, M.A., Schuller, B.W., Zafeiriou, S.: End-to-end multimodal emotion recognition using deep neural networks. IEEE J. Sel. Top. Sig. Process. **11**, 1301–1309 (2017)
65. Argyriou, A., Evgeniou, T., Pontil, M.: Multi-task feature learning. In: Advances in Neural Information Processing Systems, pp. 41–48 (2007)
66. Poria, S., Chaturvedi, I., Cambria, E., Hussain, A.: Convolutional MKL based multimodal emotion recognition and sentiment analysis. In: 2016 IEEE 16th International Conference on Data Mining (ICDM), pp. 439–448. IEEE (2016)
67. Zhang, D., Wang, Y., Zhou, L., Yuan, H., Shen, D., Alzheimer's Disease Neuroimaging Initiative: Multimodal classification of Alzheimer's disease and mild cognitive impairment. Neuroimage **55**, 856–867 (2011)
68. Sui, J., et al.: Multimodal neuromarkers in schizophrenia via cognition-guided MRI fusion. Nat. Commun. **9**, 3028 (2018)

Resource Allocation in HetNets with Green Energy Supply Based on Deep Reinforcement Learning

Weijun Zheng[1], Jinghui Fang[1], Siyu Yuan[2(✉)], Da Guo[2], and Yong Zhang[2]

[1] Jiaxing Power Supply Company, State Grid Zhejiang Electric Power Co. Ltd., Jiaxing, Zhejiang Province, People's Republic of China
{zhengweijun, fangjinghui}@Zj.SGCC.COM.CN
[2] School of Electronic Engineering, Beijing University of Posts and Telecommunication, Beijing, People's Republic of China
{yuanisyu, guoda, Yongzhang}@bupt.edu.cn

Abstract. Heterogeneous network (HetNet) is the main networking form of the fifth-generation mobile communication system. In this paper, we propose a heterogeneous network resource management algorithm based on deep reinforcement learning (DRL). This algorithm uses deep Q-network (DQN) to solve resource allocation problem in heterogeneous network. This algorithm encourages the use of green energy to power the base station as much as possible, minimizing the use of the power grid to power the base station and achieve maximum energy efficiency. The simulation results show that this algorithm has efficient learning ability, can effectively improve the energy efficiency of the network, thereby realize great resource management.

Keywords: Heterogeneous networks · Resource allocation · Deep reinforcement learning · Green energy · Energy efficiency

1 Introduction

With the rapid development of wireless communication, the realization of wireless heterogeneous networks is considered as a promising technique to solve this problem. The realization of wireless heterogeneous networks is conducive to the improvement of resource utilization. However, when macro-cell users and femto-cell users are present in a wireless heterogeneous network, this will introduce serious intra-interference and inter-interference, and thus, spectrum resource allocation becomes difficult to implement.

The future wireless communication environment is a network in which multiple heterogeneous wireless access technologies coexist. Overlapping network coverage, diverse business needs and complementary technical features make it especially important to coordinate heterogeneous wireless network resources. Therefore, I propose a heterogeneous wireless network resource allocation algorithm based on reinforcement learning to maximize the energy efficiency of heterogeneous wireless network.

A heterogeneous network is a network connecting computers and other devices with different operation systems and/or protocols. In this paper, the heterogeneous network is a network composed of the communication network of MBSs and SBSs.

Reinforcement learning (RL) is an algorithm for learners to learn by interacting with the environment. The goal of RL is to understand what actions are taken in each state to maximize a specific metrics. Learners evaluate their performance by interacting repeatedly with the control environment to obtain the best decision. RL is widely used in robot and automatic control [1], and its flexibility and adaptability are also introduced into resource management of wireless communication system [2]. The essence of RL is to solve the decision-making problem. In other words, with RL, decisions can be made automatically and continuously.

Through the investigation of existing researches, the methods of the researches are still the traditional optimization methods, including graph theory [3], dual Lagrangian method [4] and optimization decomposition [5].

In [6], They propose a small cell network interference management framework based on RL. In particular, the application of small cell networks is proposed in order to reduce the disturbance to the macro cell network while maximizing its own spectral efficiency. The proposed self-organizing algorithm is completely decentralized, in which femtocell only relies on local information (in the form of feedback), jointly estimates its long-term utility, optimizes the probability distribution of its policy choice, and obeys the cross-interference constraint set by the macro cell.

In [7], they formulate joint multi-cell scheduling, UE association and power allocation for 5G network, aiming at maximizing system throughput while ensuring fairness among UEs. Power control is solved in a centralized manner, and UE correlation is handled in centralized and attractive convex optimization. Through non-cooperative game theory in a decentralized way.

In [8], they explore the reinforcement learning approach in the multi-agent field. Since the environment state of multiple agents is determined by the behavior of multiple agents, it has non-stationarity, and the Q-learning algorithm is difficult to train. The variance of the policy gradient algorithm will increase with the number of agents. Bigger. The author proposes a variant MADDPG of the actor-critic method. The reinforcement learning of each agent considers the action strategies of other agents, and carries out centralized training and decentralized execution, and has achieved remarkable results. In addition, based on this, a training method of strategy integration is proposed, which can achieve more stable effects.

The rest of this paper is organized as follows. Section 2 introduces the system model and problem formulation for the resource allocation problem in HetNets. The HETNETDQN is proposed in Sect. 3. The performance evaluations are presented in Sect. 4. The conclusions drawn from this study are presented in Sect. 5.

2 System Model and Problem Formulation

2.1 System Model

In this paper, we mainly consider a downlink heterogeneous network model. The network model which we use is shown in Fig. 1.

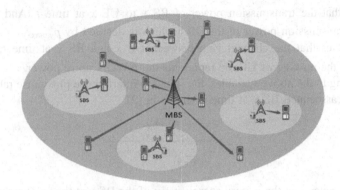

Fig. 1. HetNet system model

This HetNet system model is composed of several user equipments (UEs) and base stations (BSs).

For UEs, they are randomly deployed within a given range and obey uniform distribution. Each UE has its own moving direction and moving speed. With the moving of UE, UE will select the BS which can provide an optimal service to the UE and make a switching.

For BSs, in this model, there is one MBS at the center of the given area and several SBSs which can be powered by green energy (e.g. solar, wind). We assume that the SBSs are deployed by the same network operator. What's more, the SBSs are covered by the central MBS and logically connect with the MBS. All BSs share all spectrum resources by OFDM.

2.2 Problem Formulation

We assume that there are X UEs and $Y + 1$ BSs (one MBS and Y SBSs) in the HetNet system model. The set $\mathbf{X} = \{1, 2, \ldots, X\}$ represents the set of the whole UEs in the given area. The set $\mathbf{Y} = \{0, 1, 2, \ldots, Y\}$ represents the set of the whole BSs in the given area and the element 0 represents the central MBS. What's more, we assume that the state of the BS and UE are constant in a small time slot, and the HetNet system only changes between time slots.

Each UE has a moving speed v_x^t and a moving direction which can be represented by the moving angle of the UE $\theta_x^t \in [0, 2\pi)$. And there is a maximum moving speed $v_{\max}$ for the UEs.

n_y^t denotes that the number of the UEs which connects with the BS y at time t. We assume that the UE always remains connection to the BS and is associated with only one BS at the same time. Hence,

$$\sum_{y \in Y} n_y^t = X \tag{1}$$

p_y^t denotes that the transmission power of BS y to UE x at time t. And there is a maximum transmission power of the BS y which is denoted by $p_{y,max}$.

$g_{x,y}^t$ denotes that the channel gain between UE x and BS y at time t, which is changing as the moving of UE. σ^2 represents the AWGN noise power.

According to the above variables, the signal to interference plus noise ratio (SINR) of the UE x associated BS y at time is shown in (2).

$$\gamma_{x,y}^t = \frac{g_{x,y}^t p_y^t}{\sum_{y' \in Y, y' \neq y} g_{x,y'}^t p_{y'}^t + \sigma^2} \tag{2}$$

$b_y^t \in [0, 1]$ denotes that the battery energy status of the BS y at time t. 0 means that the battery energy status of the BS y is empty and 1 means for full battery energy status. We assume that the central MBS is always powered by the power grid, so $b_0^t = 0$ all the time.

For the SBSs, they can be powered by the power grid or the green energy. We assume that the static baseline power $p_{y,stat}^t$ of each SBS is powered by the power grid.

At the time slot switching, if the remaining amount of green energy of the SBS cannot meet the energy demand of the next time slot, the SBS will use the power grid to supply power in the next time slot.

In addition, in order to maximize the utilization of green energy, the transmission power obtained from the power grid is set to positive, and the power obtained from green energy is set to negative, which is shown in (3).

$$p_{y,signed}^t = \begin{cases} p_y^t, & \text{if } p_y^t \text{ is obtain from power grid} \\ -p_y^t, & \text{if } p_y^t \text{ is obtain from green energy} \end{cases} \tag{3}$$

Hence, the total power of the BS y at time t is denoted by $p_{y,total}^t = p_{y,stat}^t + \eta p_{y,signed}^t$, where η is a coefficient factor which is related with wireless power amplifier efficiency.

The power of the BS y which is obtained from the power grid at time t is shown in (4).

$$p_{y,grid}^t = p_{y,stat}^t + \eta \left(\frac{p_{y,signed}^t + \left| p_{y,signed}^t \right|}{2} \right) \tag{4}$$

We assume that the total bandwidth of the HetNet is W. And r_x^t represents the ratio of the bandwidth which is assigned to the UE x to the total bandwidth W.

According to Shannon capacity formula, the information rate of UE x at time t is shown in (5).

$$R_x^t = r_x^t W \log_2(1 + \gamma_{x,y}^t) \tag{5}$$

And the total information rate of the overall HetNet system model at time t is shown in (6).

$$R^t = \sum_{x=1}^{X} R_x^t = \sum_{x=1}^{X} r_x^t W \log_2(1 + \gamma_{x,y}^t) \tag{6}$$

To ensure communication quality, we set a minimum value for the information rate $R_{x,\min}$. The energy efficiency is defined as the ratio of the information rate of the overall HetNet system and the power which is powered by the power grid, and shown in (7).

$$\delta = \frac{\sum\limits_{x=1}^{X} r_x^t W \log_2(1 + \gamma_{x,y}^t)}{\sum\limits_{y \in \mathbf{Y}} \left(p_{y,stat}^t + \eta \left(\frac{p_{y,signed}^t + \left| p_{y,signed}^t \right|}{2} \right) \right)} \tag{7}$$

$$s.t.\ R_x^t = r_x^t W \log_2(1 + \gamma_{x,y}^t) \geq R_{x,\min} \tag{7a}$$

$$0 \leq \left| p_{y,signed}^t \right| \leq p_{y,\max} \tag{7b}$$

$$\gamma_{x,y}^t = \frac{g_{x,y}^t p_y^t}{\sum\limits_{y' \in \mathbf{Y}, y' \neq y} g_{x,y'}^t p_{y'}^t + \sigma^2} \tag{7c}$$

The above problem is a non-convex optimization problem. In order to solve this problem, in the next chapter, a deep reinforcement learning algorithm is proposed to better solve the existing problems.

3 Algorithm Description

In this chapter, the traditional principles of reinforcement learning will be introduced firstly. Secondly, deep reinforcement learning will be introduced. Finally, HETNETDQN algorithm will be proposed to solve resource allocation problem.

3.1 Reinforcement Learning

Reinforcement learning (RL), as an important branch of machine learning, is a product of multidisciplinary and multi-domain intersection. The essence of RL is to solve the decision-making problem. In other words, with RL, decisions can be made automatically and continuously. There are four key elements of RL that are agent, action, status, reward, and the goal of RL is to get the most cumulative rewards. Reinforcement learning learns a set of strategies based on the environment and maximizes the rewards of expectations. It is studied in many fields because of its universality, such as autonomous driving, game theory, cybernetics, genetic algorithms, statistics, swarm intelligence, multi-agent systems learning, simulation optimization, information theory and operations research.

The basic principle of RL is shown in Fig. 2.

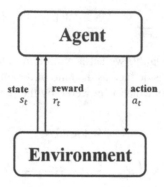

Fig. 2. Basic principle of RL

As shown in Fig. 2, when agents perform an action a_t, it first interacts with the environment. Then the environment will generate the next state S_{t+1} and give a corresponding reward r_t. As this cycle continues, the agent and the environment interact to generate more new data. The reinforcement learning algorithm interacts with the environment through a series of action strategies to generate new data, and then uses the new data to modify its own action strategy. After several iterations, the agent learns the action strategy needed to complete the task.

3.2   Deep Reinforcement Learning – DQN

Deep reinforcement learning (DRL) combines reinforcement learning (RL) with deep learning (DL) to learn control strategies from high-dimensional raw data. DRL is formed by combining the decision-making ability of RL with the perception of DL. The basic idea of DRL is to use depth learning to automatically learn abstract representations of large-scale input data, and then to use reinforcement learning based on depth learning representations to learn and optimize problem solving strategies.

DeepMind team first proposed the Deep Q Network (DQN) [9] in 2013, which was applied to the Playing Atari 2600 video game and scored higher scores than human players in some games.

As the representative algorithm of DRL, DQN combines CNN and Q-learning. The input of CNN is the original image data (as the state), and the output is the value function (Q value) corresponding to each action.

In order to solve these problems, DQN adopts the following methods. The reward obtained by Q-Learning is used as the label for deep learning. To solve the stability of the algorithm and make the algorithm converge quickly, DQN uses experience replay and Fixed-Q target network.

The role of CNN in DQN is to perform function fitting on Q-Table in high-dimensional and continuous state. For function optimization problems, the general method of supervised learning is to first determine the loss function, then the gradient, and update the parameters using methods such as random gradient descent. DQN determines the loss function based on Q-Learning. The update formula for Q-Learning is shown in (8).

$$Q^*(s,a) = Q(s,a) + \alpha(r + \gamma \max_{a'} Q(s',a') - Q(s,a)) \tag{8}$$

The loss function of DQN is shown in (9),

$$L(\theta) = E[(\text{Target}Q - Q(s,a;\theta))^2] \tag{9}$$

where θ is the network parameter and the Target Q is shown in (10).

$$\text{Target}Q = r + \gamma \max_{a'} Q(s',a';\theta) \tag{10}$$

The function of the experience replay is mainly to solve the problem of correlation and non-static distribution. In the DQN learning process, the agent stores the data in a database, and then extracts the data from the database by means of uniform random sampling, and then uses the extracted data to train the neural network. The format of the sample in the database is $< s_t, a_t, r_t, s_{t+1} >$.

3.3   HETNETDQN

This paper proposes a DQN algorithm that can solve the resource allocation problem. The purpose of the HETNETDQN is to maximize the energy efficiency of the power grid.

HETNETDQN algorithm is running in MBS and control the operating power of the all base stations. In this way, the action of the agent in DQN algorithm is the operating power of BSs. In order to simplify the power adjustment of the BSs, we assume that there is a minimum value of adjustment in operating power which is represented by Δ. In this way, the adjustment in operating power should be the integer multiple of Δ. The action space of BS y can be represented as (11).

$$actions = \{-p_{y,\max}/\Delta, \ldots, -\Delta, 0, \Delta, \ldots, p_{y.\max}/\Delta\} \tag{11}$$

The purpose of this algorithm is to maximize the energy efficiency of the HetNet model. Therefore, we directly use the energy efficiency of the HetNet model as a reward function for HETNETDQN algorithm. In addition, the static baseline power of the BS is provided by the power grid, and the purpose of the resource allocation problem is to maximize the utilization of green energy. The reward function can be represented in (12).

$$rewards = \delta = \frac{\sum\limits_{x=1}^{X} r_x^t W \log_2(1 + \gamma_{x,y}^t)}{\sum\limits_{y \in Y} \left(p_{y,stat}^t + \eta \left(\frac{p_{y,signed}^t + |p_{y,signed}^t|}{2} \right) \right)} \tag{12}$$

The input of CNN needs to represent the state of the HetNet model. To represent the state of HetNet model, the channel gain matrix is a significant information matrix. What's more, using green energy is encouraged and will courses a high reward in DQN algorithm. So, the battery energy status matrix of BSs is also an important information matrix for the HetNet model. As a result, the input of CNN consists of channel gain matrix and battery energy status matrix.

According to the above analyzation, we use the VGG network structure [10]. We put three convolutional layers, three max pooling layers, two fully connected layers and one output layer. The construction of CNN is shown in Fig. 3. The convolution kernel size is k = [3, 3], the stride is equal to 1, and the padding uses the VALID model. The size of the pool layer is [2, 2] and the stride is s = 2.

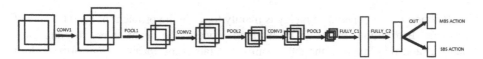

Fig. 3. Construction of CNN

4   Simulation Results

There is one MBS and eight SBSs in the HetNet model. The size of HetNet model is set to 200 m * 200 m. There are 400 UEs in the HetNet model area. The total bandwidth of the HetNet system is 10 MHz. The SBS is randomly placed in the area.

For MBS, the maximum transmission power is $p_{0,max} = 50\,dBm$, the cover radius is 65 m. For SBS, the maximum transmission power is $p_{y,max} = 38\,dBm, \forall y \in \{1, 2, 3, \ldots, 8\}$, the cover radius is 35 m. The minimum value of adjustment in operating power is set to $\Delta = 2\,dBm$. For UE, the maximum moving speed is 5 m/s. For AWGN channel, the noise power is $\sigma^2 = 1e - 7$. For channel loss, the path loss can be calculated by $path\,loss = 37 + 30 * \log_2(l)$ [11], where l is the distance between the UE and BS. The standard deviation of shadow fading is 8 dB.

Firstly, we observe the function image of the energy efficiency at different learning rates with respect to the number of iterations which is shown in Fig. 4. We can find that in the case of low learning rate, HETNETDQN obviously converges slowly. In the case of moderate learning rate, the convergence speed of HETNETDQN obviously improves. In the case of high learning rate, the convergence speed of HETNETDQN improves continuously, but the convergence value becomes low. As a result, a suitable learning

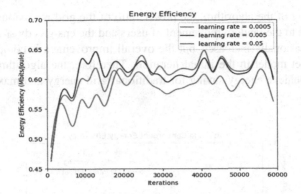

Fig. 4. Energy efficiency

rate can achieve a faster convergence rate without affecting the final reward convergence value. What's more, because reinforcement learning is an algorithm that learns strategies through continuous exploration. The reward (energy efficiency) at the beginning of the iteration is low and unstable. As the iteration progresses, the reward continues to grow. And after a certain number of iterations, the reward function value converges near a maximum. This means that the DQN algorithm has learned a certain action policy. The jitter after reward converges is due to the eps-greedy strategy which means there is a small probability that the action is not selected according to the learned strategy.

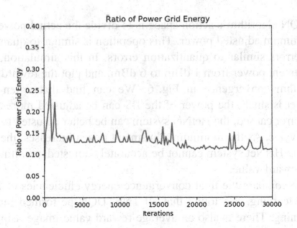

Fig. 5. Ratio of power grid energy

The purpose of HETNETDQN is to achieve energy savings for traditional grid energy. Figure 5 shows the ratio of power grid supply power to total power consumption. At the beginning of the algorithm's iteration, the algorithm randomly generates a battery state value. This makes SBS have enough green energy storage in the first few iterations. With the initial storage of green energy exhausted, the power of the grid power of the untrained HetNet increases rapidly, with a maximum of around 27%.

After the training of the algorithm, the power ratio of the grid is obviously reduced to about 13%. Due to the random movement of users and the eps-greedy strategy, the grid power ratio occasionally increases, but the overall improvement is compared with the untrained HetNet model in the initial iteration. Therefore, the algorithm proposed in this paper can achieve energy saving for traditional grid energy and maximize the use of green energy.

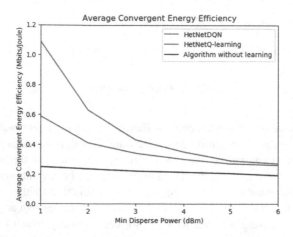

Fig. 6. Average convergent energy efficiency

HETNETDQN algorithm's action space is to divide the actual power generated by the BS by a minimum adjusted power. This operation is similar to quantization, which will introduce errors similar to quantization errors. In this simulation, we adjust the minimum adjustment power from 1 dBm to 6 dBm, and plot the reward function value for each algorithm convergence in Fig. 6. We can find that when the minimum adjustment power is small, the power of the BS can be adjusted more accurately, and the smaller the error caused, the HetNet system can be better adjusted to obtain a larger average reward value. As the minimum adjustment power increases, the resulting error increases, and the HetNet system cannot be accurately adjusted, resulting in a decrease in the average reward value.

In Fig. 6, we compare the final convergence energy efficiencies of HETNETDQN and HETNETQ-learning. We found that HETNETDQN performed much better than HETNETQ-learning. There is also an average reward value image using an algorithm that does not use reinforcement learning. The scheme without learning process uses the same minimum adjustment power values as HETNETDQN, but it selects action level with the best system reward and simply depends the current observed state level. By comparing with the curve of the HETNETDQN algorithm, the HetNet system trained by the DQN algorithm has better traditional grid energy saving rate and green energy utilization rate.

5 Conclusion

This paper is to propose a solution to resource allocation in heterogeneous network. To realize the resource allocation, we need to formulate a resource allocation policy to help the BSs to manage the association with the UEs and the transmission power to the UEs. Due to the randomness of the UEs, this resource allocation policy is formulated without a data set and without a deterministic model. Therefore, we decided to use a reinforcement learning algorithm which has been tried in the environment to develop the policy. However, this problem is a nonconvex problem. Traditional machine learning algorithms may be limited to a local optimum and cannot reach the global optimum. Hence, we decide to use a reinforcement learning algorithm which combines with deep learning to solve the problem.

In this paper, we propose a resource allocation based on deep reinforcement learning algorithm. We use the classic DQN algorithm to solve this problem in deep reinforcement learning. We propose a HetNetDQN algorithm to solve the problem of resource allocation so as to maximize energy efficiency. After completing the algorithm, we simulate the proposed algorithm. The simulation results show that the algorithm can make HetNet system achieve optimal energy efficiency. By modifying the environmental parameters, we can also use the proposed algorithm to predict the optimal capacity of HetNet system and other data.

References

1. Kaelbling, L.P., Littman, M.L., Moore, A.W.: Reinforcement learning: a survey. J. Artif. Intell. Res. **4**, 237–285 (1996)
2. Nie, J., Haykin, S.: A Q-learning-based dynamic channel assignment technique for mobile communication systems. IEEE Trans. Veh. Technol. **48**(5), 1676–1687 (1999)
3. Song, Q., Wang, X., Qiu, T., Ning, Z.: An interference coordination-based distributed resource allocation scheme in heterogeneous cellular networks. IEEE Access **5**, 2152–2162 (2017)
4. Saeed, A., Katranaras, E., Dianati, M., Imran, M.A.: Dynamic femtocell resource allocation for managing inter-tier interference in downlink of heterogeneous networks. IET Commun. **10**(6), 641–650 (2016)
5. Li, Y., Luo, J., Xu, W., Vucic, N., Pateromichelakis, E., Caire, G.: A joint scheduling and resource allocation scheme for millimeter wave heterogeneous networks. In 2017 IEEE Wireless Communications and Networking Conference (WCNC), pp. 1–6. IEEE, March 2017
6. Bennis, M., Perlaza, S.M., Blasco, P., Han, Z., Poor, H.V.: Self-organization in small cell networks: a reinforcement learning approach. IEEE Trans. Wirel. Commun. **12**(7), 3202–3212 (2013)
7. Maaz, B., Khawam, K., Tohmé, S., Lahoud, S., Nasreddine, J.: Joint user association, power control and scheduling in multi-cell 5G networks. In: 2017 IEEE Wireless Communications and Networking Conference (WCNC), pp. 1–6. IEEE, March 2017

8. Lowe, R., Wu, Y., Tamar, A., Harb, J., Abbeel, O.P., Mordatch, I.: Multi-agent actor-critic for mixed cooperative-competitive environments. In: Advances in Neural Information Processing Systems, pp. 6379–6390 (2017)
9. Mnih, V., et al.: Playing atari with deep reinforcement learning. arXiv preprint: arXiv:1312.5602 (2013)
10. Simonyan, K., Zisserman, A.: Very deep convolutional networks for large-scale image recognition. arXiv preprint: arXiv:1409.1556 (2014)
11. Yang, Z., Xu, W., Xu, H., Shi, J., Chen, M.: User association, resource allocation and power control in load-coupled heterogeneous networks. In: 2016 IEEE Globecom Workshops (GC Wkshps), pp. 1–7. IEEE, December 2016

Deep Reinforcement Learning for Joint Channel Selection and Power Allocation in Cognitive Internet of Things

Weijun Zheng[1](✉), Guoqing Wu[1](✉), Wenbo Qie[2](✉), and Yong Zhang[2](✉)

[1] State Grid Zhejiang Electric Power Company,
Jiaxing Power Supply Company, Jiaxing, People's Republic of China
{zhengweijun, wuguoqing}@zj.sgcc.com.cn
[2] School of Electronic Engineering, Beijing University of Posts
and Telecommunication, Beijing, People's Republic of China
{qwb, Yongzhang}@bupt.edu.cn

Abstract. With the development of wireless communication technology and the lack of spectrum resources, it is very meaningful to study the dynamic spectrum allocation in the cognitive Internet of Things. In this paper, the system model is firstly established. In an underlay mode, considering the interference between primary and secondary users, jointing channel selection and power allocation, aiming to maximize the spectrum efficiency of all secondary users. Different from the traditional heuristic algorithm, the underlay-cognitive-radio-deep-Q-network frame-work (UCRDQN) based on deep reinforcement learning, is proposed to find the optimal solution efficiently. The simulation results show that the UCRDQN algorithm can achieve higher spectrum efficiency and is more stable and efficient than other algorithms.

Keywords: Cognitive Internet of Things · Dynamic spectrum allocation · Underlay mode · Deep reinforcement learning

1 Introduction

With the rapid development of the fifth-generation (5G) wireless communication, the data transmission rate is greatly improved, which makes it possible for more devices to be connected to the Internet. In [1], this greatly promotes the development of the Internet of Things and requires a large number of spectrum resources at the same time. The traditional approaches of the fixed spectrum allocation to licensed networks (primary networks) lead to spectrum underutilization. Since spectrum usage is regulated by the government due to security, safety and stability considerations [2]. In the licensed spectrum, there are some idle spectrums which result in very low spectrum utilization. Therefore, how to improve the utilization of the existing licensed spectrum becomes a problem that we need to solve.

Cognitive radio has emerged as a promising technology to alleviate the imbalance between spectrum allocation and its use created by the increasing demand for high data rate wireless services [3, 4]. Its core idea is to allow unlicensed users to access licensed

© Springer Nature Switzerland AG 2019
D. Milošević et al. (Eds.): HCC 2019, LNCS 11956, pp. 683–692, 2019.
https://doi.org/10.1007/978-3-030-37429-7_69

frequency bands without causing interference to authorized users, thereby improving spectrum utilization. In cognitive radio, cognitive radio modes can be divided into two categories: overlay and underlay. Overlay mode is the simplest cognitive radio mode, and a secondary user can randomly access channels which are not occupied by the primary user. The secondary user needs to have sensitive spectrum sensing capability, which can accurately and efficiently perceive the spectrum that is not occupied by the primary user, and the secondary user occupies all the bandwidth to transmit data with maximum transmitted power. Underlay mode is an interference control technology based on power control. As long as the interference of the secondary user to the primary user is within a certain threshold, the secondary user can occupy the spectrum resources together with the primary user. This method does not need the secondary user to the spectrum that will be occupied, which is beneficial to the secondary user.

Resource allocation in cognitive radio networks (CRNs) has been widely studied [5, 6]. In [7], the authors consider interference from the secondary user to the primary user and the Quality of Service (QoS) of the secondary user and formulate the joint power and rate allocation with proportional and max-min fairness criteria as optimization problems. In [7], a heuristic-based channel and power allocation algorithm are proposed which considers the interference constraints for the primary user. In [8], a user removal algorithm based on the tree-pruning algorithm is proposed to satisfy the QoS constraints for the secondary user and interference temperature constraints for the primary user. However, the proposed removal algorithm is computationally extensive. In [9], the authors consider maximizing the sum rate of the secondary users under the interference power constraint enforced by the primary user and the transmit power constraints and the optimal joint power control for the Wireless Power Transmission (WPT) phase and time allocation among the secondary users are obtained at the same time. In [10], the authors study the relationship between energy efficiency and spectral efficiency in cognitive network systems and the impact of interference thresholds on them. In [11], the authors propose a learning-based approach to share spectrum resources with the primary user, in the case where the secondary user does not understand the transmit power and power control strategy of the primary user. However, the paper doesn't consider the secondary user's choice of channel.

Inspired by deep reinforcement learning [12], we propose a framework called UCRDQN for joint optimization of channel and power in underlay awareness mode to maximize the spectral efficiency of cognitive networks. The algorithm adopts a learning method to establish a mapping between the primary user channel power control strategy and the secondary user channel power allocation strategy, which can efficiently make decisions and obtain the best solution. Firstly, the constraints and optimization problems existing in cognitive wireless network scenarios are analyzed. Secondly, the reinforcement learning algorithm is introduced, and Q-learning and deep Q-learning network (DQN) algorithms are introduced. Then the detailed DQN algorithm design based on the cognitive wireless network is given. Finally, a simulation experiment is carried out to verify the stability and effectiveness of the framework.

The rest of this paper is organized as follows. The next section presents the system model and problem formulation for the optimization problem that joint channel and power in CRNs. The UCRDQN algorithm is proposed in Sect. 3. Simulation and numerical results are provided in Sect. 4. The conclusions drawn from this study are presented in Sect. 5.

2 System Model and Problem Planning

2.1 System Model

This paper considers that there are M primary users $PU_m = PU_1, PU_2, \ldots, PU_M$ and N secondary users $SU_n = SU_1, SU_2, \ldots, SU_N$ in one area, as shown in Fig. 1.

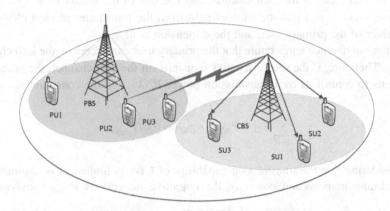

Fig. 1. System model (Color figure online)

Assume that in this cellular network, a base station is used as a transmitter by all primary users, called PBS (Primary Base Station), and each primary user has a receiver. CBS (Cognitive base station) is used by all secondary users, and each secondary user has a receiver. In this scenario, the underlay mode transmission is used, allowing the secondary user to use the licensed band resources allocated to the primary user. When the primary user occupies the bandwidth resource for information transmission, the secondary user can use the same bandwidth resource as the primary user to transmit when the interference to the primary user is within a certain range. At this time, the primary user and the secondary user perform spectrum sharing, that is, the cognitive user adopts the underlay mode. The red connection part of Fig. 1 is the interference caused by the secondary user to the receiving end of the primary user at the same frequency.

2.2 Problem Description

Assume that the total bandwidth of the licensed spectrum for the primary user is B, the OFDM method is adopted, and the total number of channels is K, where n represents the n-th SU and k represents k-th channel is occupied. Assume that the primary user can occupy up to several channel resources, while the secondary user can occupy only one channel [13].

Define a $M \times K$ matrix $PA = \{pa_{m,k}\}$ to represent the channels occupancy by all primary users. If the m-th PU occupies the k-th channel, $pa_{p,k} = 1$, otherwise $pa_{p,k} = 0$.

Define a $N \times K$ matrix $CA = \{ca_{n,k}\}$ to represent the availability of all secondary users on all channels. If the n-th SU can use the k-th channel, $ca_{n,k} = 1$, otherwise $ca_{n,k} = 0$.

Define a matrix $A = \{a_{n,k}\}$ of size $N \times K$ to indicate channel allocation. If the n-th secondary user is assigned to the k-th channel for communication, $a_{n,k} = 1$, otherwise $a_{n,k} = 0$.

Define $G_{SS} = \{g_{n,k}\}$ as the channel gain from the secondary user transmitter to the secondary user receiver for each channel, and the size of the matrix is $N \times K$.

Define $G_{SP} = \{g_{m,k}\}$ as the channel gain from the transmitter of each channel to the receiver of the primary user, and the dimension is $M \times K$.

Set the interference temperature that the primary user can accept on the k-th channel to IT_{max}^k. Therefore, if the secondary user transmits on the k-th channel, the secondary user needs to control its own transmission power $P_n(k)$, and the constraint is:

$$\sum_{n=1}^{N} a_{n,k} * P_n(k) * g_{SP}(m) \leq IT_{pmax}^k \qquad (1)$$

Considering that the transmission capability of CBS is limited, it is assumed that the maximum transmission power of the cognitive network is P_{CBS}^{max}, satisfying the constraint $0 \leq \sum_{n=1}^{N} P_n(k) \leq P_{CBS}^{max}$. $\eta^{n,k}$ denotes the signal to interference plus noise ratio (SINR) of the secondary user n on the k-th channel, expressed as follows

$$\eta^{n,k} = \frac{g_{n,k} P_n(k)}{\sum\limits_{m=1}^{M} g_{m,k} P_m(k) + \sum\limits_{n=1}^{N} g_{n,k} P_n(k) + \sigma^2} \qquad (2)$$

According to the Shannon formula, the spectrum efficiency of the n-th secondary user can be obtained as

$$Se_{n,k} = \log(1 + \eta^{n,k}) \qquad (3)$$

Therefore, the spectrum efficiency of the entire cognitive network is

$$SE = \sum_{n=1}^{N} \sum_{k=1}^{K} a_{n,k} * Se_{n,k} \qquad (4)$$

Finally, our goal is to achieve how to dynamically allocate effective spectrum resources for secondary users in underlay mode to maximize the spectrum efficiency of the entire cognitive network. The whole problem is planned as follows

$$\begin{cases} \max SE = \sum_{n=1}^{N} \sum_{k=1}^{K} a_{n,k} * Se_{n,k} \\ St : \\ \sum_{n=1}^{N} a_{n,k} * P_n(k) * g_{SP}(m) \leq IT_{\max}^k \\ 0 \leq P_n(k) \leq P_{CBS}^{\max} \end{cases} \quad (5)$$

The above problem is a multi-objective non-convex optimization problem, and most of the algorithms were proposed and improved based on the heuristic search algorithms. But most of this algorithm is very inefficient, and it takes a long time to find the optimal solution. To solve this problem, in the next chapter, a deep reinforcement learning algorithm is proposed to better solve the existing problems.

3 Algorithm Description

This chapter is divided into three parts. Firstly, the general reinforcement learning algorithm is introduced. Secondly, deep reinforcement learning is deeply explored. Finally, the HCRDQN algorithm is proposed for the spectrum allocation problem of underlay cognitive radio networks.

3.1 Reinforcement Learning – Q Learning

The definition of reinforcement learning is the reward-guided behavior that the agent learns in a "trial and error" manner, and the goal is to maximize the reward of the agent through the reward-guided behavior obtained by interacting with the environment.

In reinforcement learning, there are two objects that can interact: the agent and the environment. Agents can learn and make decisions by sensing the state of the external environment and the rewards of feedback. The environment is all things outside the agent, and is changed by the action of the agent to change its state, and feedback to the corresponding reward of the agent.

The five basic elements in reinforcement learning include:

$N = \{1, 2, \ldots, n\}$ denotes a set of agents, such as secondary users in a cognitive radio network, where n represents the number of agents.

$S = \{s_1, s_2, \ldots, s_z\}$ represents a collection of all states that the entire system may produce, where z represents the number of states.

$A = \{a_1, a_2, \ldots, a_q\}$ represents a collection of all actions of the system, where q represents the number of actions.

$P(s'|s, a)$ is a probability transfer matrix, which represents the probability value that the system moves from one state to another.

$r(s, a, s')$ indicates instant reward, which is a scalar function. After the agent makes an action a according to the current state s, the environment will feed back a reward to the agent. This agent is also often related to the state s' of the next moment, where $s, s' \in S, a \in A$.

In the application scenario close to the real world, such in cognitive radio network, it is difficult to model this, so the model-free based method is adopted, and the most classic reinforcement algorithm is Q-Learning [13].

Q-Learning is a time-series differential learning algorithm based on table-based heterogeneous strategy. The value function is represented by a Q-value table of state-action coordinates. The estimation method of the value function is

$$Q(s,a) \leftarrow Q(s,a) + \alpha(r + \gamma \max_{a'} Q(s',a') - Q(s,a)) \tag{6}$$

where α denotes the learning rate and γ denotes the discount factor.

As the Q value table is continuously updated and iterated, it will eventually converge to an optimal value. At this point, the agent can get the optimal strategy based on the converged Q value table.

3.2 Deep Reinforcement Learning - DQN

In the Q-Learning algorithm, states and actions are discrete and finite, and the Q values are recorded in a table. However, in most practical problems, the number of states and actions of the agent is large, and continuous.

To effectively solve these problems, DeepMind proposes that deep neural network functions which can be used to make agents aware of more complex environments and build more complex strategies [12]. It can be understood that the update problem of the Q table becomes a function fitting problem, the state and the action are function-mapped, and the similar state obtains a similar output action, as shown in the formula.

$$Q*(s,a) = \max_{\pi} E[r_t + \gamma r_t + \gamma^2 r_t + \ldots | s_t = s, a_t = a.\pi] \tag{7}$$

where r_t denotes the reward, γ denotes the discount factor and $\pi = P(a|s)$ denotes the action policy function.

Using a stochastic gradient descent, the neural network objective function is

$$L(s,a,s'|\phi) = (r + \gamma \max_{a'} Q_\phi(s',a') - Q_\phi(s,a))^2 \tag{8}$$

where s' denotes the state at next moment and a' denotes the action at next moment.

For this function, there is a problem that the target is not problematic and there is a strong correlation between the samples. DQN mainly has two key steps [14]. One is freezing target networks, which keeps the parameters in the target network from being updated within certain iteration steps, which can stabilize the learning objectives and remove the correlation of data. The second is to use experience replay, collect some interactive data of the agent and the environment in the experience pool, and then randomly extract to break the correlation between the data.

The DQN algorithm has shown good performance advantages in many practical applications. With its continuous development and wider application fields, it has been recognized in many industries and fields. Therefore, this paper designs a deep reinforcement learning algorithm based on underlay cognitive radio network.

3.3   Underlay-Cognitive-Radio-Deep-Q-Network (UCRDQN)

This paper proposes a DQN algorithm based on underlay cognitive radio network, which aims to learn a better strategy through the continuous "trial and error" and "learning" of the agent in the environment, joint channel selection and power allocation. In this algorithm, the basic elements and actual meanings of the algorithm are defined as follows:

Agent: SU. $1 \leq n \leq N$. Where every secondary user sees it as an agent.

States: $s_t^{n,k} = \{PA_t^{m,k}, P_t^m(k)\}$. The state consists of the situation in which the primary user occupies the channel and he transmitted power.

Action: $a_t = \{A_t^{n,k}, \vec{p}_n(k)\}$. The action consists of the secondary user occupied channel condition and the transmitted power size.

Reward: $r_t = SE + \lambda(IT_{\max}^k - \sum_{n=1}^{N} a_{n,k} * P_n(k) * g_{sp}(m))$. The purpose of the interference temperature constraint in reward is to maximize the spectrum efficiency while making the interference of the secondary user to the primary user not exceed the threshold, and the smaller the better, the better.

In this algorithm, there are two neural networks, including the training network and the target network. The two networks have the same structure, but the parameter update times are different. The design of the deep neural network is shown in Fig. 2. The three-layer convolution structure is used, for the first layer, 32 filters of size 1×5 are used to generate 32 feature maps with the step of size 2, and rectified linear units (ReLU) are then utilized for nonlinearity. The second layer, 64 filters of size 1×3 are used to generate 64 feature maps with the step of size 1, which is filled by the VALID method. The first two layers extract the deep feature of the input layer, and the latter layer uses the 1×1 convolution kernel to reduce the size of data and perform nonlinear processing to improve expressive ability of the network [15].

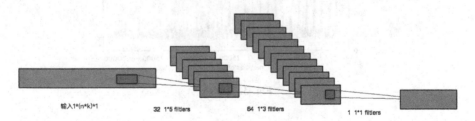

输入1*(n*k)*1   32 1*5 filtiers   64 1*3 filtiers   1 1*1 filtiers

Fig. 2. Deep neural network model

4   Simulation Results

Considering an underlay CRN deployed in a square area with size of $100 * 100$ m^2, where all primary and secondary users are located within this area. The PBS is located at (40, 50), the CBS is located at (60, 50), and it is assumed that the five primary users and ten secondary users are uniformly distributed. In this square area, 10 channels are available for selection, and the simulation parameters are shown in Table 1. The

Table 1. Parameters of the simulated cognitive radio network

Parameter name	Parameter value
Number of primary users, M	5
Number of secondary users, N	10
Number of channels, K	10
Power of primary base station, PBS	46 dbm
Power of cognitive base station, CBS	46 dbm
Interference temperature threshold, IT_{max}^{k}	5 dB

channel adopts a flat fading model. The primary user's selection strategy of the channel and the transmit power are uncertain and the secondary user makes decisions by sensing channels and transmission power occupied by the primary users, aiming to maximize the spectrum efficiency of secondary users.

After 30,000 iterations by training the neural network, the loss value of the neural network is shown in Fig. 3, and it performs converging so well. The result shows that the design of neural network and parameters of this algorithm are reasonable.

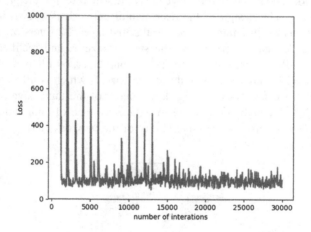

Fig. 3. Loss

Later CRN deployments were independently simulated 18 times in accordance with the parameters in Table 1. Each time, the channel occupancy and emission power matrix of different primary users is randomly input. As shown in Fig. 4, the average spectrum efficiency of cognitive system is obtained. Otherwise the experimental results are quickly obtained and stable, the performance of this algorithm is verified at the same time.

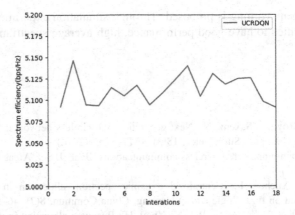

Fig. 4. Average spectral efficiency

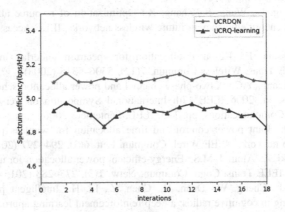

Fig. 5. Comparison of spectrum efficiency

Figure 5 shows that comparing Q-learning algorithm to our UCRDQN algorithm based on deep reinforcement learning, which can see obviously that the UCRDQN algorithm was stable in this simulation scenario. And the average spectrum efficiency of CRN performed much better than UCRQ-learning.

5 Conclusion

In the context of the Internet of things, how to allocate resources reasonably to the cognitive system is one of the key issues to improve spectrum utilization. Resource allocation includes spectrum and power. Joint channel selection and power allocation are non-convex problems, and it is difficult to find the optimal solution quickly with traditional optimization methods. In this paper, under the underlay spectrum sharing mode, aiming at the average spectrum efficiency of the cognitive system and jointly optimizing channel selection and power distribution, a UCRDQN algorithm based on deep reinforcement learning is proposed. Through simulation experiments, UCRDQN algorithm is verified to have good performance, high average spectrum efficiency and stable system.

References

1. Agiwal, M., Roy, A., Saxena, N.: Next generation 5G wireless networks: a comprehensive survey. IEEE Commun. Surv. Tutor. **18**(3), 1617–1655 (2016)
2. Radio, C.: Brain-empowered wireless communications. IEEE J. Sel. Areas Commun. **23**(2), 201–220 (2005)
3. Wenzhu, Z., Xuchen, L.: Centralized dynamic spectrum allocation in cognitive radio networks based on fuzzy logic and Q-learning. China Commun. **8**(7), 46–54 (2011)
4. Wang, Y., Wang, Y., Zhou, F., Wu, Y., Zhou, H.: Resource allocation in wireless powered cognitive radio networks based on a practical non-linear energy harvesting model. IEEE Access **5**, 17618–17626 (2017)
5. Yang, K., Wang, L., Wang, S., Zhang, X.: Optimization of resource allocation and user association for energy efficiency in future wireless networks. IEEE Access **5**, 16469–16477 (2017)
6. Le, L.B., Hossain, E.: Resource allocation for spectrum underlay in cognitive radio networks. IEEE Trans. Wireless Commun. **7**(12), 5306–5315 (2008)
7. Hoang, A.T., Liang, Y.C.: A two-phase channel and power allocation scheme for cognitive radio networks. In: 2006 IEEE 17th International Symposium on Personal, Indoor and Mobile Radio Communications, pp. 1–5. IEEE, September 2006
8. Xu, D., Li, Q.: Joint power control and time allocation for wireless powered underlay cognitive radio networks. IEEE Wirel. Commun. Lett. **6**(3), 294–297 (2017)
9. Sboui, L., Rezki, Z., Alouini, M.S.: Energy-efficient power allocation for underlay cognitive radio systems. IEEE Trans. Cogn. Commun. Netw. **1**(3), 273–283 (2015)
10. Li, X., Fang, J., Cheng, W., Duan, H., Chen, Z., Li, H.: Intelligent power control for spectrum sharing in cognitive radios: a deep reinforcement learning approach. IEEE Access **6**, 25463–25473 (2018)
11. Mnih, V., et al.: Human-level control through deep reinforcement learning. Nature **518**(7540), 529 (2015)
12. Szegedy, C., et al.: Going deeper with convolutions. In: Proceedings of the IEEE Conference on Computer Vision and Pattern Recognition, pp. 1–9 (2015)
13. Zhang, X., Zhang, X., Han, L., Xing, R.: Utilization-oriented spectrum allocation in an underlay cognitive radio network. IEEE Access **6**, 12905–12912 (2018)
14. Wang, S., Liu, H., Gomes, P.H., Krishnamachari, B.: Deep reinforcement learning for dynamic multichannel access in wireless networks. IEEE Trans. Cogn. Commun. Netw. **4**(2), 257–265 (2018)
15. Zhang, Y., Kang, C., Ma, T., Teng, Y., Guo, D.: Power allocation in multi-cell networks using deep reinforcement learning. In: 2018 IEEE 88th Vehicular Technology Conference (VTC-Fall), pp. 1–6. IEEE, April 2019

A Random-Based Approach to Social Influence Maximization

Huie Zou and Mingchun Zheng[✉]

Business School, Shandong Normal University, Jinan, China
sdnuzhe@163.com, zhmc163@163.com

Abstract. The problem of influence maximization is to find a small subset of nodes (seed nodes) as influential as the global optimum in a social network. Despite researchers have achieved fruitful research result which can be applied widely, a key limitation still remains, and that is this work is generally too time consuming to find seeds and difficult to be apply into the large-scale social network. In this paper, we propose a new random-based algorithm which combines "random selection" and "the optimal neighbor", which idea both greatly reduces the computational complexity and achieves the desired effect. Our algorithm is able to avoid overlapped information and thus determine high-quality seed set for the influence maximization problem. Our empirical study with real-world social networks under independent cascade model (ICM) demonstrates that our algorithm significantly outperforms the common used algorithms in terms of efficiency and scalability, with almost no compromise of effectiveness.

Keywords: Social networks · Influence maximization · Information spread

1 Introduction

Social networks play a fundamental role in propagating information, ideas and innovation. Influence maximization problem is one of the key research question in the field of social network information transmission, its purpose is to find the most influential node set in social networks to spread the influence, which has been widely used in many important scenes, such as marketing, advertising, early warning, epidemic monitoring, etc., so it has the very high research value and application value. the influence maximization problem was firstly regarded as an algorithm problem by Domingos and Richardson [1]. Kempe et al. [2] refined it into a discrete optimization problem, which is to find k influential nodes under the given propagation model to maximize the final impact range, and it has been proved to be NP-hard. At present, the research on the algorithm of the optimization issue attempts to find the approximate solution instead of the exact solution. Heuristic measures are widely used to find the seed set. The most straightforward idea is to rank all nodes according to specific centrality measure and directly extract the top-k nodes, such as degree, PageRank [3] and Eigenvector method [4]. However, as mentioned above, this method could be impractical. There are two reasons, one is that all measure based on the knowledge of global network structure what is unavailable in gain networks, and the other is because

© Springer Nature Switzerland AG 2019
D. Milošević et al. (Eds.): HCC 2019, LNCS 11956, pp. 693–699, 2019.
https://doi.org/10.1007/978-3-030-37429-7_70

the nodes with highest degrees or other centrality measures may be highly clustered. However, the effect of the algorithm is unstable and easy to fall into the "richer club". Based on Degree Heuristic, Chen et al. [5] proposed DegreeDiscount algorithm for independent cascading model, whose precision is greatly improved compared with Degree Heuristic. In recent years, the scalability of influence maximization algorithm in large-scale networks and the problem of overlapping influence have attracted the attention of scholars. Another extensively studied direction is the greedy algorithms, which add only one node at a time, and the node provides the largest increase of influence to the previous set. Leskovec et al. [6] proposed CELF using the submodularity property of influence maximization which improves the efficiency by 700 times. Chen et al. [5] improve the Greedy algorithm to form two new algorithms called NewGreedy and MixGreedy. Although the advantage of greedy algorithm is high precision, it is difficult to apply to large-scale social networks due to the high complexity.

For the above questions, this paper focuses on resolving the scalability, accuracy, and memory cost dilemma of influence maximization and put forward a new efficient algorithm for the influence maximization problem.

The most novel of the paper is that combines "random selection" and "the optimal neighbor", which idea both greatly reduces the computational complexity and achieves the desired effect. The contribution of this paper can be summarized as follows. In this article, we develop a new random-based approach, to tackle the problem of the influence maximization problem, with an emphasis on the time efficiency issue. our algorithm that only require to traverse finite nodes other than whole nodes in the entire network, so it can bring down algorithm complexity. The experimental results demonstrate that the proposed influence estimation method can generate superior quality seed set compared to other famous approach.

The rest of the paper is organized as follow. The thought, specific operation steps and pseudo-code of the new random-based algorithm is shown in Sect. 2. In Sect. 3, we propose the experimental results. Sect. 4 concludes.

2 Random Node Recommend

2.1 The Thought of Random Node Recommend

We now introduce our main contribution. In the new random-based algorithm, we use the popular "Jump-Crawl" model which was introduced by Brautbar and Kearns in 2010, to select nodes randomly and find the nodes' neighbors, the second-order neighbors and the third-order neighbors [7, 8]. The algorithm is permitted only two graph operations. In the Jump operation, we visit a uniformly random node from the N nodes. In the Crawl operation, we can crawl along an edge from a currently known vertex to any or all of its neighbors. When visited, a node reveals all of its edges. For example, in Facebook, once you get one user's profile, you can visit all of his/her friends. The basic idea behind our algorithm is that we divide the network into X communities, then to select a set of vertexes chosen uniformly at random from all graph nodes of each community as target nodes(Jump queries), explore the most influential neighbor node around each target nodes as seeds (Crawl queries). Popular speaking it: in real social network, firstly, we need to divide people into groups according to certain rules, and then select a certain number of people randomly from each group and ask them to recommend one of the most influential people in his circle of friends as the seeds to spread information.

In a social network, a community is a subset of individuals who within the community are more connected than outsiders. Johan Ugander et al. discussed that the invitations sent by friends of the same community have less influence than those of friends of different societies [9]. Through community detection, it can disperse seed nodes across different communities, thus avoid information overlapping and to find the influential nodes according to the community structure. Blondel et al. [10] studied a heuristic method that is based on modularity optimization used to partition large networks in short time. Further, the scale of the social network is very large, and it is unknown how many communities can be divided before community detection. If the number of communities divided is greater than X, the larger X communities will be retained, and other smaller communities will be merged into larger ones.

2.2 The Step of the New Random-Based Algorithm

In practice, our method is designed as follows:

Step 1: Community detection. We divided the network into X communities and save the set of target node which randomly selected the u nodes from each community as C_x, where $x = 1, 2, \ldots, X$.

Step 2: Select candidates. We search the neighbor nodes of each target nodes from C_x, and then calculate influence (*Influ*) of their neighbor based on Eq. 1.

$$Influ_{(i)} = k_i \times d_i \tag{1}$$

Where k_i is the degree of node i, and d_i is the number of the third-order neighbors. We refer to the neighbor node with the highest *Influ* of each target nodes as candidate notes, and save them as U_x. where $x = 1, 2, \ldots, X$.

Step 3: Select seeds. Sort the nodes in the set U_x by *Influ*. According to the number of seeds required, we select the first v nodes with the highest *Influ* from U_x as seeds set S_x. Finally, $S = \cup_{X=1}^{X} S_X$.

Notice that, the relationship between variables X, u and v is as follows:

$$k = \lambda \cdot X, (\lambda \in N*)$$

$$v = \left\lceil \frac{k}{K} \right\rceil \tag{2}$$

$$u = \lceil \gamma \cdot v \rceil, (\gamma > 1)$$

Where λ, γ are the adjustment factors, depending on the precision.

The steps of the algorithm are described above, and the specific pseudo-code of the algorithm is given below.

Algorithm:

Input:

Network G(V,E), The number of nodes randomly selected for each community:T, the number of seeds for each community:n

Output:Seed Set S

Initialize: Let $S \leftarrow \phi$

$C = \{C1, C2, \cdots, Cx\}(x = 1,2,\ldots X)$ // community detection (G)

for i=1... T do

    sample vi uniformly random from C_x // a Jump query

    for each vi do

        Hi=neighbors(G,vi) // Crawl queries

        Compute $Influ$, $\forall u \in H_i$ with Eq.1

        $u = \arg\max_{u \in H_i}\{Influ\}$

        $Ux \leftarrow Ux \cup \{u\}$ // candidate set for community for C_x

    end for

end for

Select the top-n nodes with highest $Influ$ in U_x as the seed sets S_x

$S \leftarrow S \cup S_x$

end for

return S

---

3   Experiment Results

We now show experiments comparing our random-based algorithm with several baselines on four real networks.

At first, Facebook: the data was collected from survey participants using this Facebook app. It is obtained from Stanford Large Network Dataset Collection. The second network is CA-GrQc: a collaboration network gathered from arXiv General Relativity and Quantum Cosmology. We download from Stanford Large Network Dataset Collection. Third, US power grid: This undirected network contains information about the power grid of the Western States of the United States of America. It is downloaded from the Koblenz Network Collection. Fourthly, Caida: This is the undirected network of autonomous systems of the Internet connected with each other from the CAIDA project.

To investigate the performance of our random-based algorithm, we perform some experiments with different algorithm and datasets under IC model. In our experiment, we mainly run the following algorithms.

NewGreedyIC: The greedy algorithm proposed for the IC model by Chen et al. We set R = 200.

DegreeDiscountIC: This is a heuristic algorithm for the IC model proposed by Chen et al.

SnowBall [11]: Randomly selects a node and seeds that node's highest degree neighbor and then seeds the highest degree neighbor of the first seed, and so on.

Random Recommendation: Randomly selects several nodes at random, then seeds top-k highest *Influ* (Table 1).

Table 1. The basic topology characteristics of real data sets

Network	n	m	<k>	kmax	C	<d>
Facebook	3959	84243	42.558	293	0.557	5.526
CA-GrQc	5242	14496	5.531	81	0.687	6.049
US power gird	4941	6594	2.669	19	0.107	18.989
Caida	26475	53381	4.033	2628	0.333	3.876

As the initial active node, the seeds selected by the above 8 node selection methods were respectively spread according to independent cascade model (ICM) [12, 13], and the final transmission effect was compared.

The experiment results are as follows:

In Fig. 1, we show the propagation effect of seed nodes of different algorithms with k value change in the independent cascade model in four real networks. It can be clearly seen from the figure that different algorithms have different effects when k values are the same in these four networks. It can be seen from the figure that the our algorithm is generally better than other algorithms. From the perspective of algorithm accuracy, the effect of our algorithm is only a little bit better than that of NewGreedyIC and DegreeDiscountIC. But the algorithm complexity of NewGreedyIC is difficult to be used to the large-scale network. However, the proposed our algorithm does not need to calculate the *Influ* of each node, but only needs to calculate the *Influ* of the neighbor of the randomly selected node, so this greatly reduces the complexity of the algorithm and saves a lot of computing time and space storage. However, because our algorithm combines "random selection" and "nearest neighbor recommendation", the algorithm greatly avoids the phenomenon of influence overlap. Although SnowBall algorithm use fewer queries, the accuracy is not very good. As Random Recommendation algorithm randomly selects nodes in the whole network, it is inevitable that the influence between seed nodes overlaps, so its performance is unstable.

The above experimental results reflect that the influence effect of Random algorithm is sometimes good and sometimes bad, because the randomly selected nodes are likely to gather together, which results in overlapping influence, so the influence effect is not ideal. This is also why the first step of our experiment is to divide the community. Overall, the our algorithm has performed well on four networks, particularly on the Facebook network. Whether under the IC model, the node set obtained by our algorithm has excellent propagation effect, and the comparison of propagation effect reflected in the two simulation models is consistent.

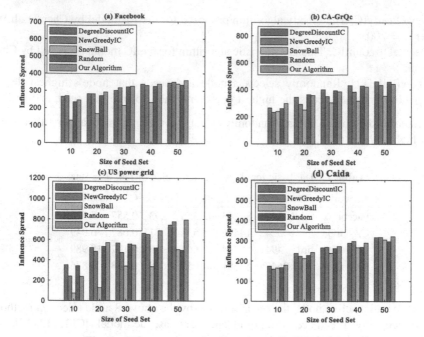

Fig. 1. Influence spread as k varies under IC model.

4   Conclusion

In this paper, we have further studied the influence maximization problem where k seed nodes can be selected, aiming to maximize the expected size of the resulting influence cascade. We proposed a new random-based algorithm to select the seed nodes to maximize their influence spread. Our algorithm greatly reduces the time complexity. We also have conducted extensive experiments under IC model using four real networks to test the influence performance of the random-based algorithm, and the results strongly confirm the superiorities and effectiveness of our measure.

Acknowledgment. We acknowledge the National Natural Science Foundation of China for its financial support (Grant No. 61402266). We thank Mingchun Zheng for discussion.

References

1. Domingos, P., Richardson, M.: Mining the network value of customers. In: Proceedings of the seventh ACM SIGKDD International Conference on Knowledge Discovery and Data Mining, pp. 57–66. ACM (2001)
2. Kempe, D., Kleinberg, J., Tardos, É.: Maximizing the spread of influence through a social network. In: Proceedings of the Ninth ACM SIGKDD International Conference on Knowledge Discovery and Data Mining, pp. 137–146. ACM (2003)

3. Brin, S., Page, L.: The anatomy of a large-scale hypertextual web search engine. Comput. Netw. ISDN Syst. **30**(1–7), 107–117 (1998)
4. Bonacich, P.: Power and centrality: a family of measures. Am. J. Sociol. **92**(5), 1170–1182 (1987)
5. Chen, W., Wang, Y., Yang, S.: Efficient influence maximization in social networks. In: Proceedings of the 15th ACM SIGKDD International Conference on Knowledge Discovery and Data Mining, pp. 199–208. ACM (2009)
6. Leskovec, J., Krause, A., Guestrin, C., Faloutsos, C., VanBriesen, J., Glance, N.: Cost-effective outbreak detection in networks. In: Proceedings of the 13th ACM SIGKDD International Conference on Knowledge Discovery and Data Mining, pp. 420–429. ACM (2007)
7. Brautbar, M., Kearns, M.J.: Local algorithms for finding interesting individuals in large networks (2010)
8. Wilder, B., Immorlica, N., Rice, E., Tambe, M.: Maximizing influence in an unknown social network. In: Thirty-Second AAAI Conference on Artificial Intelligence (2018)
9. Singh, S.S., Kumar, A., Singh, K., Biswas, B.: C2IM: community based context-aware influence maximization in social networks. Phys. A **514**, 796–818 (2019)
10. Blondel, V.D., Guillaume, J.L., Lambiotte, R., Lefebvre, E.: Fast unfolding of communities in large networks. J. Stat. Mech: Theory Exp. **2008**(10), P10008 (2008)
11. Singh, S.K.: An evaluation of the sampling methods in social research. Bus. Sci. Int. Res. J. **4**(1), 193–196 (2016)
12. Chen, W., Wang, C., Wang, Y.: Scalable influence maximization for prevalent viral marketing in large-scale social networks. In: Proceedings of the 16th ACM SIGKDD International Conference on Knowledge Discovery and Data Mining, (pp. 1029–1038). ACM (2010)
13. Wu, H., Shang, J., Zhou, S., Feng, Y., Qiang, B., Xie, W.: LAIM: a linear time iterative approach for efficient influence maximization in large-scale networks. IEEE Access **6**, 44221–44234 (2018)

Order Batch Optimization Based on Improved K-Means Algorithm

Qiaohong Zu[(⊠)] and Rui Feng

School of Logistics Engineering, Wuhan University of Technology,
Wuhan 430063, People's Republic of China
1217438053@qq.com

Abstract. Aiming at the shortcomings of the K-Means algorithm in the traditional K-Means algorithm, the DBSCAN algorithm is used to divide the order set according to the density, and obtain the batch number K value and the initial cluster center point. Based on this, the improved K-Means algorithm is used for optimization. Based on the real environment and instance data, the established batch assignment batch model is simulated. The experimental results show that the density-based K-Means clustering algorithm can effectively shorten the picking time and improve the warehouse logistics operation.

Keywords: Clustering algorithm · Order batch

1 Introduction

The order sorting process is a bottleneck that restricts the distribution center from meeting customer orders. Based on the order characteristics of the e-commerce environment, this paper shortens the demand for picking time for the parallel partition sorting system, and completes the order picking process with the batch strategy as the guide [1]. According to the actual problem that the order processing time is too long caused by the huge difference in the length of picking of different partitions, an order batch mathematical model with processing time as the objective function and order segmentation as the constraint condition is constructed [2]. The DBSCAN algorithm and the K-Means algorithm are combined to solve the batch model, and the best order batch result is obtained [3]. The result is substituted into the established order batch optimization mathematical model to solve the shortest time-consuming order set.

2 Problem Description and Model Building

2.1 Order Batch Problem Description

The warehouse adopts parallel sorting mode based on the partition layout. The order batching problem based on parallel partitioning is to divide all the orders to be sorted into different combinations for batch picking operations. Before the picking task starts, the order is divided into sub-order sets according to the picking area, a list of picking list items of each area is generated, and then the work is performed simultaneously until

© Springer Nature Switzerland AG 2019
D. Milošević et al. (Eds.): HCC 2019, LNCS 11956, pp. 700–705, 2019.
https://doi.org/10.1007/978-3-030-37429-7_71

all the partitioning operations are completed [4]. The longest picking time in the partition is the total picking time of the batch, so based on maintaining the consistency of the picking operation and maintaining the balance of the number of picking of each partition, a batch model is established to obtain the picking batch. The model can make the order set processing time the shortest.

2.2   Order Batch Model Construction

Model Hypothesis

The construction of the order batch model combines various factors such as the layout of the warehouse and the capacity of the sorting equipment to make the following assumptions.

(1) The picking area is divided into partitions of the same size and the channel entrance is at the leftmost end of each partition, and each partition has one and only one picking person is responsible. The staff has no channel obstruction when picking the goods;

(2) Each item corresponds to a storage compartment in the warehouse, and the picking personnel walk in the center of the passage while picking the goods on the shelves on both sides of the current lane. The picking device is able to pass through the channel in both directions and is able to change direction within the channel.

Order Processing Time

(1) Pick up time

$$t_{zj}^{pick} = Q_{zj}/v_z^{pick} \tag{1}$$

(2) Walking time

$$t_{zj}^{travel} = D_{zj}^{travel}/v_z^{travel} \tag{2}$$

(3) Total picking time

$$T = \sum_{j=1}^{n} T_j \tag{3}$$

$$T_j = \max_{1 \le z \le Z} t_{zj}, j \in n \tag{4}$$

$$t_{zj} = t_{zj}^{travel} + t_{zj}^{pick} \tag{5}$$

Order Batch Model

$$\min T \tag{6}$$

$$\min R \tag{7}$$

$$\sum_{j=1}^{n} x_{ij} = 1 \qquad i = 1, 2, \cdots, m \tag{8}$$

$$\sum_{i=1}^{m} v_i x_{ij} \leq V \qquad j = 1, 2, \cdots, n \tag{9}$$

$$\sum_{i=1}^{m} w_i x_{ij} \leq W \qquad j = 1, 2, \cdots, n \tag{10}$$

$$x_{ij} \in \{0, 1\} \tag{11}$$

$$x_{ij} = \begin{cases} 1 & \text{The order is assigned to the batch} \\ 0 & \text{Otherwise} \end{cases} \tag{12}$$

The parameter symbols and variable descriptions involved in the model are shown in Table 1.

Table 1. Parameter symbol description table

Model parameter	Meaning
t_{zj}	The time required to complete the picking after the batch j is assigned to the partition z
v_i	Total capacity of the order i
w_i	Total weight of the order i
q_{ia}	The number of items a in the order i
t_{zj}^{travel}	Complete the walking time of batch j in the partition z picking task
D_{zj}^{travel}	Complete batch j in partition z pick task
x_{ij}	Indicates which batch the order was assigned to complete the picking

3   Density-Based K-Means Clustering Algorithm Design

Firstly, the DBSCAN algorithm is used to divide the order set according to the density, and the batch number value and the initial cluster center point to be divided are obtained [5]. Based on this, the K-Means algorithm is used for optimization, and finally the clustering results are sorted [6]. Pick operations by batch. The density-based improved K-Means clustering algorithm used in this paper is abbreviated as DBKM. Improved K-Means algorithm clustering.

3.1   Improved K-Means Algorithm Clustering

In order to meet the constraints that need to be met in the actual picking process, the K-Means algorithm is improved. The improved clustering process is divided into two parts: the first part is the traditional K-Means clustering process, after several iterations. The order object is divided into clusters; the second part is to adjust the divided cluster classes to meet the constraints, and to ensure that the workload of each partition is roughly the same. The specific process is as follows.

(1) *Convergence test:*
   (a) input K clusters and cluster center points;
   (b) separately calculate the distance of the centroid of each order object, and re-divide the cluster according to the distance;
   (c) If all the orders have been divided, go to step (d), otherwise continue to step (b);
   (d) output clustering results.

(2) *Balance indicator calculation:*
   (a) for each cluster obtained, calculate the number of orders included and the picking time of each partition;
   (b) If the item volume included in the order is less than the constraint, and the picking time difference is satisfied for each batch, the algorithm ends and the clustering result is output, otherwise step (c);
   (c) For clusters that do not satisfy the condition, pop up the farthest point of the cluster and mark it, add it to another nearest cluster class, and repeat steps (a)–(b).

4   Simulation Experiment

The order data for this case is from a B2C e-commerce website. The data was taken from the 10,000 daily consumer goods orders of the e-commerce website in 2018, and the most recent 1000 orders were used for the verification of the order batch algorithm in this paper.

The warehouse information of the case design includes: the warehouse contains three divisions, each division assigns one picking employee, and is responsible for three picking lanes respectively. The picking speed and walking speed of the employees are different, and the picking speed is 5 s/piece, 4 s/piece, 6 s/piece, the walking speed is 1 m/s, 1.2 m/s, 1.5 m/s, and the maximum product capacity of each picking car is 10. Shelves are placed on each side of each lane, each shelf containing 30 storage spaces. The width of the roadway and the shelf are 1 m. The order picker follows the return strategy path to pick up all items on the list. In this experiment, three different order sets of small batch, medium batch and large batch are selected, and the order rows satisfy the random distribution of [1, 2], [1, 3], [1, 7] respectively.

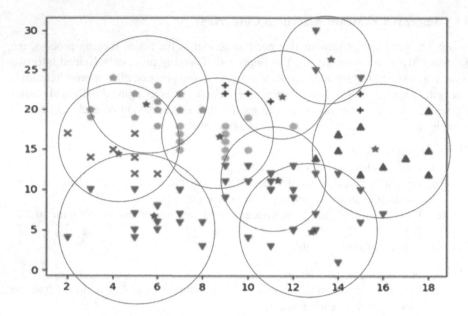

Fig. 1. Clustering diagram when the order quantity is 100

Extract 100 orders from the data containing 1000 orders for calculation. The initial parameter is calculated $\varepsilon = 8.63$, *MinPts* = 4.64. The order point is the number of points included in the circle with the truncated radius. The initial clustering results in 9 clusters, that is $K = 9$, using the improved density-based K-Means clustering algorithm to obtain the final optimal solution after 14 iterations. As shown in Fig. 1, the different shaped points in the graph represent different clusters. Points of the same shape represent the same class, that is, the corresponding orders are sorted in the same batch, wherein the five-pointed star represents the cluster center after the iteration.

The batch result with the order quantity of 100 fully proves the effectiveness of the algorithm. In addition, 200, 500, and 1000 orders are selected respectively, and FCFS and K-Means are compared with the algorithm, mainly from the selection of the length of time and the number of batches. consider. When the K-Means algorithm is used, the K value is set to the same K value as DBKM, and the initial centroid is randomly specified.

For different size order sets, Table 2 compares the batch results of FCFS, K-Means, and DBKM algorithms. It can be seen intuitively that the more orders, the longer the sorting takes, and the same size orders. Different algorithms can present better results. The FCFS algorithm is applicable to the order concentration and the order quantity is small. The larger the order set, the better the DBKM algorithm can obtain the optimal batch result.

Table 2. Experimental results and analysis

Order quantity	Batch method	Batch number	Picking time
100	FCFS	13	2367
	K-Means	9	1033
	DBKM	9	1178
200	FCFS	19	3042
	K-Means	12	1865
	DBKM	12	2001
500	FCFS	40	5663
	K-Means	30	6556
	DBKM	30	5997
1000	FCFS	86	13350
	K-Means	79	12493
	DBKM	79	11797

5 Conclusion

For the manual picking system, order batch optimization is to sort and divide the quantitative orders into multiple batches for sorting operations, shortening the sorting operation cycle required for a single batch. Based on the requirements of multi-level modern logistics system, the order batch mathematical model and the targeted solution algorithm with processing time as the objective function and order segmentation as constraints are used to compare and analyze the running results of the algorithm. The simulation results show that the batching effect is better, the picking time can be shortened significantly, the working intensity of the picking operation can be reduced, and the effectiveness of the improved algorithm is verified.

References

1. Gils, T.V., Ramaekers, K., Braekers, K., et al.: Increasing order picking efficiency by integrating storage, batching, zone picking, and routing policy decisions. Int. J. Prod. Econ. **197**, 243–261 (2018)
2. Quader, S., Castillo-Villar, K.: Design of an enhanced multi-aisle order-picking system considering storage assignments and routing heuristics. Robot. Comput.-Integr. Manuf. **50**, 13–29 (2017)
3. Zhang, J., Wang, X., Chan, F.T.S., et al.: On-line order batching and sequencing problem with multiple pickers: a hybrid rule-based algorithm. Appl. Math. Model. **45**, 271–284 (2017)
4. Koch, S., Wäscher, G.: A grouping genetic algorithm for the order batching problem in distribution warehouses. J. Bus. Econ. **86**(1/2), 131–153 (2016)
5. Lam, C.H.Y., Choy, K.L., Ho, G.T.S., et al.: An order-picking operations system for managing the batching activities in a warehouse. Int. J. Syst. Sci. **45**(6), 1283–1295 (2014)
6. Pérez-Rodríguez, R., Hernández-Aguirre, A., Jöns, S.: A continuous estimation of distribution algorithm for the online order-batching problem. Int. J. Adv. Manuf. Technol. **79**(1/4), 569–588 (2015)

Author Index

Printed in the United States
By Bookmasters